PRINCIPLES OF SUSTAINABLE ENERGY SYSTEMS

Second Edition

MECHANICAL and AEROSPACE ENGINEERING

Frank Kreith & Darrell W. Pepper
Series Editors

RECENTLY PUBLISHED TITLES

PRINCIPLES OF SUSTAINABLE ENERGY SYSTEMS

Second Edition

FRANK KREITH, EDITOR
University of Colorado (retired)

SUSAN KRUMDIECK, CO-EDITOR
University of Canterbury

CRC Press
Taylor & Francis Group
Boca Raton London New York

CRC Press is an imprint of the
Taylor & Francis Group, an **informa** business

CRC Press
Taylor & Francis Group
6000 Broken Sound Parkway NW, Suite 300
Boca Raton, FL 33487-2742

Printed on acid-free paper
Version Date: 20130701

International Standard Book Number-13: 978-1-4665-5696-6 (Hardback)

Library of Congress Cataloging-in-Publication Data

Kreith, Frank.
    Principles of sustainable energy systems / edited by Frank Kreith and co-edited by Susan Krumdieck. -- Second edition.
        pages cm
    Revised edition of: Principles of sustainable energy / Frank Kreith, Jan F. Kreider. 2011.
    Includes bibliographical references and index.
    ISBN 978-1-4665-5696-6
    1. Renewable energy sources. I. Krumdieck, Susan. II. Title.

TJ808.K74 2013
621.042--dc23   2013003651

Visit the Taylor & Francis Web site at
http://www.taylorandfrancis.com

and the CRC Press Web site at
http://www.crcpress.com

We, the people, still believe that our obligations as Americans are not just to ourselves, but to all posterity. We will respond to the threat of climate change, knowing that the failure to do so would betray our children and future generations. Some may still deny the overwhelming judgment of science, but none can avoid the devastating impact of raging fires, and crippling drought, and more powerful storms. The path towards sustainable energy sources will be long and sometimes difficult. But America cannot resist this transition; we must lead it.

President Barack Obama
Second inaugural address, 2013.

Contents

4. Capturing Solar Energy through Biomass 199
Contributing Authors: Robert C. Brown and Mark M. Wright

Foreword

The book you are now holding, *Principles of Sustainable Energy Systems*, arrives at an opportune moment. The growing worldwide realization of the need to leave behind our dependence on fossil fuels, and set course for a sustainable energy future, gives this volume abundant timeliness and relevancy. While support for clean energy alternatives has never been higher, the obstacles to achieving a true energy transformation are equally formidable. For those of us who have long known the value of clean energy technology, these are both exciting and challenging times.

This book examines the practical issues surrounding energy efficiency and renewable energy concepts and systems. And a more appropriate author for such a work could not be found. Professor Frank Kreith has shown a lifelong devotion to exploring renewable technologies and appropriately pushing them into mainstream use.

I first came to appreciate Prof. Kreith's depth of knowledge as an engineering student at Stanford. I benefited greatly from his authoritative text on heat transfer systems—it served as a foundation for much of my own early professional research. And later, Prof. Kreith's work as a pioneering solar energy engineer became even more pertinent when I was named director of the same Colorado laboratory where he had served as a senior research fellow.

To those who choose renewable energy as an academic field or career path, this text and the insights you gain from it will likely become fundamental to the achievements that await you. But this text is not merely academic; it serves a valuable purpose in its clear-headed look at the energy picture from the ground up, and the environmental, economic, and sustainability benefits that renewable energy systems can provide.

Principles of Sustainable Energy Systems combines the expertise earned from decades of practical and scholarly research with the most up-to-date analysis of the energy scene. For today, sustainable energy is no longer just a dream for the future; it is a growing field with incredible opportunity for anyone with the foresight to make its vision their own.

Dan E. Arvizu
President, Alliance for Sustainable Energy
Manager and operator of the National Renewable Energy Laboratory

Preface

The field of sustainable energy is changing fast, and new information about technologies and analytic processes are constantly appearing in the literature. The rapid pace of change has made it necessary to prepare a second edition to the original text, which was based on information available until about six years ago.

The second edition responds to comments and suggestions made by users of the original textbook and introduces several new ideas. I have made a serious effort to correct all of the printing errors in the first edition and also update, as much as possible, the data on energy use worldwide.

Furthermore, to adapt the book, which is over 700 pages long, as a text for a one-semester course in engineering and/or environmental sciences, I have marked sections that can be omitted without breaking the continuity of presentation with an asterisk, and thereby reduced the important technical contents to less than 500 pages. More importantly, I have introduced several new issues:

1. Coverage of the availability of renewable and fossil energy sources, including coal and natural gas, and the effect of fracking has been expanded.

2. A section on global carrying capacity based on the interaction between population, water, food, energy, and the environment has been added.

3. A system analysis approach to sustainability has been taken throughout the book. This approach includes an energy return on energy invested (EROI) strategy for sustainability analysis and design.

4. Coverage of the nuclear power option has been expanded.

5. A quantitative analysis and critique of a hydrogen-based energy system has been added.

6. Calculations of potential emission of air pollutants that can lead to global warming and methods for their reduction have been interspersed throughout the book.

7. The chapter on energy storage has been expanded to include virtual storage potential in the electric grid. This combined water–wind–solar (WWS) approach can attain, with some pumped storage capacity, up to 83% of the U.S. electric load in a robust distribution network.

8. A new chapter presenting approaches for a transition from a fossil-based to a renewable and sustainable energy future has been added. This chapter, authored by Professor Susan Krumdieck of the University of Canterbury in New Zealand, brings her insight, experience, and broad perspective to the new edition.

9. During the past five years, the National Renewable Energy Laboratory, in conjunction with Sandia National Laboratories, has developed the system advisory model (SAM), a web-based tool that simplifies the economic and technical analysis of sustainable and renewable energy systems. The solar radiation impinging on the Earth and the resulting wind and water motions vary substantially both during the year and during the day. Consequently, predictions about the performance of such systems must be based upon a collection of past weather data that need to be encapsulated and then fed into a performance prediction model. This approach

results in rather tedious calculations using spreadsheets and similar procedures. SAM has short-circuited the need for these calculations and can be used by engineers evaluating and designing renewable energy systems to predict their performance and the levelized cost of the energy generated. To utilize these numerical methods, however, it is necessary to gain a basic understanding of the technology, as well as the assumptions and nomenclature used to exercise the model. The model demonstrates the steps in the solution for a given technology by means of default values that represent typical values for given locations and conditions. In order to use the model properly, however, it is necessary to input the correct values for the system, location, and time that are being analyzed. Thus, by combining the reading of the chapter with the illustrative problems, it is possible to reduce the time necessary to cover the performance and cost estimates for various technologies. This should prove of benefit both to the students and the instructors.

In view of the many changes taking place continuously in the field of sustainable energy and the teaching tools constantly being made available by various sources, a website has been established for the book:

http://www.crcpress.com/product/isbn/9781466556966

This website will contain lengthy tables that cluttered many pages of the first edition and also provide the instructor with teaching tools such as a PowerPoint presentation of the illustrations, videos, and additional reading material that could be useful for a course. With the availability of background material on the website, it is then possible to present the course at various levels of technology so that it can be used in different departments and colleges of a university interested in offering a course in sustainable energy.

Acknowledgments

Frank Kreith would like to express his sincere appreciation to many people, without whose assistance the new edition could not have been completed. First of all, I want to thank the following contributing authors: Dr. Gary Pawlas and Professors Mark Wright, Robert Brown, Jeffrey Morehouse, and Susan Krumdieck. In addition to writing Chapter 13 of the new edition, Professor Krumdieck also provided important input to Chapters 1 and 2, which corrected some errors and expanded the presentation of some of the topics. Professor Ron West, my old friend and colleague, has reviewed the chapters on biomass and economics and made valuable suggestions. Dr. Rita Klees has given guest lectures on the interaction between water and energy, and her ideas have found a place in the book. Professor Michael Hannigan of the University of Colorado, Paul Norton of the National Renewable Energy Laboratory in Golden, Colorado, and Dr. Isaak Garaway of the Technion, Israel, assisted by coteaching versions of the sustainable energy course at the University of Colorado. Paul Gilman, consultant to NREL on the SAM, reviewed the sample problem using SAM. Hayley Schneider of Tesla Motors reviewed the transportation chapter, and Robert Kennedy of the ASME Energy Committee made helpful comments on material dealing with nuclear energy. Charles Tse was my editorial assistant and helped by integrating the material on SAM into the book, by providing checks on the references and figures, and assisted in the preparation of the solutions manual. Professor Kevin Scoles of Drexel University provided corrections for the first edition of the book. Bev Weiler has given invaluable assistance by taking my dictations and drafting them into a coherent story that I could edit with minimum effort, as well as preparing the index for the text. As always, my wife Marion has been of enormous help in providing a setting that gave me the opportunity to work without having to attend to any of the daily chores. Finally, I want to thank all of the students who have worked with me in various capacities and endeavors related to my experimenting with new ideas in the sustainable energy courses I have taught for the past six years at the University of Colorado. I would also like to express my gratitude to the Mechanical Engineering Department at the University of Colorado for having given me the opportunity to continue my work during the twilight of my professional career.

Contributors

Robert C. Brown
Department of Mechanical Engineering
Iowa State University
Ames, Iowa

Susan Krumdieck
Advanced Energy and Material
  Systems Lab
and
Department of Mechanical Engineering
University of Canterbury
Christchurch, New Zealand

Jeffrey H. Morehouse
Department of Mechanical Engineering
University of South Carolina
Columbia, South Carolina

Gary E. Pawlas
Department of Mechanical Engineering
University of Colorado
Boulder, Colorado

Mark M. Wright
Department of Mechanical Engineering
Iowa State University
Ames, Iowa

1

Introduction to Sustainable Energy

It is evident that the fortunes of the world's human population, for better or for worse, are inextricably interrelated with the use that is made of energy resources.

M. King Hubbert (1969)

The purpose of this chapter is to give an overview of the principles of sustainability in the context of energy engineering. This chapter can be used as a stand-alone supplement to any engineering course or the introduction to an energy engineering course. Section 1.1 gives a historical review of sustainability principles. Section 1.2 presents the critical issues for sustainability: population, water, food and growth, and the relationships between these systems and energy use. Section 1.3 presents some context for the complex nature of sustainable energy by looking at the world energy system of the past, present, and future. Section 1.4 explains some of the most important considerations you must understand in order to effectively deal with sustainability in engineering, including energy return on energy invested (EROI), the levelized cost of energy (LCOE), efficiency, environmental impacts, social values, depletion, measures of well-being, and the orders of magnitude that separate different end uses and sources of energy. Section 1.5 explains the most promising of all sustainable energy strategies—reducing energy demand by improving efficiency, changing behavior, and adaptive design. Section 1.6 deals with the sustainability of fossil fuels, carbon dioxide and other emissions, and the prospects for carbon capture and sequestration (CCS). Section 1.7 gives an overview of the technology and prospects for the sustainability of power generation from uranium. Section 1.8 reviews the technology, history, current status, and future prospects for renewable resources including geothermal, hydro, wind, solar, biomass, and the ocean. Section 1.9 discusses the prospects for hydrogen, and Section 1.10 gives an introduction to energy system modelling using readily available software.

1.1 Sustainability Principles

This book deals mostly with the engineering challenges confronting our efforts to achieve sustainability, but it is important to keep in mind that the development of energy sustainability is taking place in an era when at least four other relevant forces impinge on one another. The energy challenge also involves societal issues such as population explosion, resource depletion, and environmental degradation. As shown in Figure 1.1, all of these issues impinge upon the engineering design and economics of a sustainable energy future.

Although this book does not directly deal with these other issues, it is important to keep them in mind because, unless the political and social systems support the road map toward a sustainable energy future, social obstacles can impede the implementation of long-term technical solutions that are technically reasonable as well as economically acceptable.

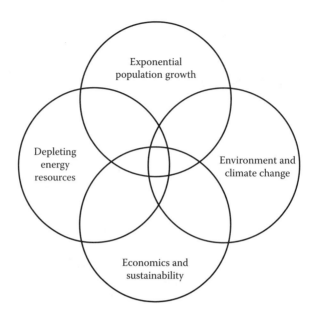

FIGURE 1.1
Complex and interrelated nature of engineering, social, and environmental issues. (From Alliance for Water Efficiency and American Council for an Energy-Efficient Economy, *Addressing the Energy–Water Nexus: A Blueprint for Action and Policy Agenda*, May 2011; ASME, *ETP: Energy-Water Nexus—Cross-Cutting Impacts.*)

For the reader interested in these social issues, several books in the supplemental reading list at the end of the chapter will provide background information.

The energy choices made in the near future are among the most important of any choices in human history. While energy development questions may appear to be mostly technical and economic, there are broader considerations. Sustainability considerations reflect priorities in our society as well as our attitude toward future generations. Development of more renewable energy is not a panacea for our energy problems. A sustainable energy future appears feasible only if we develop an overall social and political strategy that includes renewable energy development, combined with energy conservation and adaptation of our affluent lifestyles to greatly reduce energy consumption.

1.1.1 Energy Crisis: Security Issues

The world oil production capacity increased apace with demand for liquid fuels most of the last 100 years. Consider the perspective of people like your grandparents in 1950. At that time, the potential future oil supply would have been much larger than any conceivable demand, oil was abundant and cheap, and few were thinking about global, irreversible environmental impacts. The production of oil in the United States has been controlled at affordable prices by various organizations since 1910. In this context, the decisions made about building roads, dismantling urban trams, underinvestment in rail, and promotion of low-density land use reflect the perception that oil supply would last forever.

The average price of oil remained below $3.00 per barrel until 1970. On October 5, 1973, the Yom Kippur War started when Syria and Egypt attacked Israel. The United States and most Western countries showed support for Israel. Several Arab nations, members of the Organization of Petroleum Exporting Countries (OPEC), plus Iran imposed an oil embargo on Western nations and curtailed their production by 5 million barrels per day (mbpd). There had been no spare production in the United States, the largest oil-producing nation, since 1971, and other non-OPEC producers were able to increase production by only 1 mbpd, resulting in an overall 7% decline in world oil supply. By the end of 1974, the price of oil was over $12 per barrel. OPEC did not increase production again after the crisis, but there was a rapid increase in world supply from other areas like the North Sea, Mexico, and Alaska. Despite the increased supply, the price remained in the new price range of $12.50–$14.50 per barrel. The Iranian revolution followed shortly after the invasion by Iraq in September 1980 and led to a decrease in production of 6.5 mbpd from the two warring nations, representing a further 10% decline in world supply from 1979. This led to a second oil supply shortage with long lines at gas stations and shortages around the world. By 1981, crude oil prices had more than doubled to $35 per barrel, and the global economic recession of the early 1980s was the most severe economic crisis since the Great Depression.

In the United States, this series of events caused a serious energy crisis in the 1980s. Electricity demand had grown rapidly since World War II with a building boom and an explosion of new energy uses from conveniences like air-conditioning, refrigeration, washing machines, and other appliances. Significant generation capacity had been added using diesel and nuclear fired power plants. With no spare oil production capacity in the United States, the OPEC Oil Embargo also caused electricity price increases and shortages in some places. The energy crisis generated great interest in alternative energy sources, spurred research and development, and resulted in new standards and regulations on energy efficiency for buildings, appliances, and vehicles. The energy crisis also started a political obsession with energy independence as a national security objective that persists to this day. The realization that all critical activities relied on finite fossil fuels like oil and gas influenced thinking about sustainability to extend beyond environmental impacts and to look at the future implications of depletion. Energy engineering emerged as a new field, primarily within mechanical engineering with courses on energy conversion, air pollution, and energy management taught at many major universities. However, as new oil fields in the North Sea and other non-OPEC nations were developed, world oil production capacity increased again, the oil price declined to a new price range below $30 per barrel, and it became politically unpopular to worry about energy shortages. In fact, the discussion around fossil fuel use and the development of the resources in remote and higher-risk environments became polarized and framed in terms of the environment vs. the economy.

Sustainability in terms of fossil fuel energy supply is often seen as a short-term economics and political issue. There have been 11 recessions in the U.S. economy since World War II. One of them followed the bursting of the dot-com investment bubble, and the rest were preceded by an oil price spike. All of the oil price spikes except the 2007–2008 price run-up have resulted from political conflict causing reduced spare production capacity. The best economic mitigation for the short-term risks of oil supply insecurity is the ability to reduce demand as demonstrated in the previous energy crises. In the long term, there is a convergence of downward pressure on energy supplies including the retirement of aging nuclear power plants, decreased coal use, and high cost and environmental risks limiting the development of deep sea oil, tar sands, and shale gas. The ability to adapt to lower the energy usage will be the most cost-effective and timely mitigation for risks to our economy. Class discussion using the questions at the end of this chapter will help to keep you informed about current issues and future trends in energy fields.

1.1.2 Sustainable Development

In 1983, the United Nations set up an independent body headed by Gro Harlem Brundtland, the former prime minister of Norway, entitled the World Commission on Environment and Development. The commission was asked to examine the critical environmental and developmental problems facing our planet and to formulate realistic proposals to solve them. The proposed solution was to make sure that human progress can be sustained, but without bankrupting the resources of future generations. The outcome of this study was an important book entitled *Our Common Future* [1]. The Brundtland Commission concluded that worldwide efforts to guard and maintain human progress to meet human needs and to realize human ambition are unsustainable, both in developed and developing nations, because they rely on ever-increasing use of already overdrawn environmental resources. The commission provided a widely quoted definition of sustainability.

> Sustainable development should meet the needs of the present without compromising the ability of future generations to meet their own needs.

The commission stated that sustainable global development requires that those who are more affluent adopt lifestyles that can be accommodated by the planet's ecological means, particularly their use of energy. Further, rapidly growing populations increase the pressure on resources. Sustainable development can, therefore, only be achieved if population size and population growth rate are reduced to a point where they no longer exceed the productive potential and maintenance level of the earth's ecosystems. The commission realized that future economic growth must be less energy intensive than growth in the past. Energy efficiency policy must be the primary vehicle of national energy strategies for sustainable development, but the commission also realized that energy efficiency can only buy time for the world to develop energy paths based on renewable sources and that this goal must be the foundation of any global energy strategy for the twenty-first century.

1.1.3 Sustainability Principles in Practice

Allocation of the use of common resources is an ancient practice that is essential for sustainability. There are villages in the Old World where the same families have farmed the same land and lived in the same homes for over 600 years. All of the people in an area shared grazing areas, forest resources, and water. Management of the *commons* was part of the operation of the village to ensure that users did not overtax resources. In American history, abundant resources were always available for development, and the traditional methods of allocation and management of individual use of common resources were not developed to the same degree as in Europe. The political climate during the westward expansion was not generally willing to limit a corporation's ability to extract resources in order to preserve the environment or to protect the resources of others. However, where resources have become scarce, protection and allocation systems have often been developed and implemented. Water is an essential resource that is in short supply in the Western United States. Systems have been developed to allocate water to end users and to regulate and limit taking more water than the allocation. Water allocation systems are engineered systems that include monitoring, modeling, and communication between users and the supplier that involves more than just price signals. Energy supply shortfall causes economically destructive price spikes, partly because there are no engineered allocation systems.

Development of allocation systems for electricity and fuels may become a future innovation in order to bring demand into balance with supply without economically destructive price shocks.

Regulations to limit environmental impacts and protect public health are key sustainability principles. Punishment of people who pollute rivers and lakes is an ancient practice around the world. In 1899, the United States passed the River and Harbor Act banning pollution of navigable waterways. As industries grew up around the country, states, local counties, and cities have worked with health boards to monitor and regulate pollution discharged into rivers. The 1970s saw a raft of federal environmental legislation including the Clean Air Act (1972) and the Safe Drinking Water Act (1974), which were followed by great improvements in air and water quality. In 1987, the Montreal Protocol on Substances that Deplete the Ozone Layer was ratified by most of the nations of the world and required corporations to stop producing certain widely used refrigerants and propellants that were extremely useful and profitable, but which had been shown by scientists to be causing a hole in the earth's protective ozone layer at the South Pole. The ozone hole has stopped growing, but people who live in South America and New Zealand now live with dangerous UV exposure and increased skin cancer rates. The most important lesson to learn from the history of environmental regulation is that it comes after the problem arises and technical remedial processes are developed. Scientists observe and measure the harm being caused by a given practice or pollutant, engineers undertake research and development of alternative processes, and finally, when the environmental and health issues get sufficiently alarming and the alternative engineering solutions are available at a reasonable cost, the legislation is enacted that requires alteration of an existing industrial process. This pattern has many examples: banning of the pesticide DDT, lead additive being removed from gasoline, particulate and sulfur dioxide scrubbing being required for coal-fired power plants, and catalytic converters on automobiles. Thus one of the guiding principles of sustainable energy engineering is to seek out scientific evidence and work on changes to existing processes to reduce the environmental risks regardless of the current environmental regulation requirements.

Triple bottom line (TBL) accounting is an idea that was developed in the late 1980s and 1990s as an economic approach to sustainability. The TBL practice accounts for the financial, social, and environmental benefits or costs of corporate operations for all stakeholders. Stakeholders are workers, customers, and people who live near factories or mines. The financial performance of the corporation is obviously one of the bottom lines as all corporations are operated to return profit to shareholders. Social and environmental bottom lines are not as easy to quantify, so corporations normally set out principles for their practices so that they do not exploit resources or people in a way that negatively impacts well-being. Some companies monitor the environmental and labor practices of their suppliers to make sure that fair wages are paid, child labor is not exploited, and environmental regulations are not violated. Fair Trade certification has been developed to ensure that suppliers in less developed countries receive a fair price for their goods and labor. Ecological footprint, life cycle assessment, and cradle-to-grave analysis of goods are techniques that have been developed to quantify the energy and material use and emissions associated with products. These types of accounting approaches have yielded real improvements in sustainability in instances where the company actually directs, allows, or rewards its engineering staff to carry out audits, propose action plans, and make changes to supply chains, materials, designs, processes, and operations to meet quantitative sustainability goals.

Companies that actively follow sustainability practices are referred to as *social businesses*. People in organizations who advocate for TBL, recycling, bike parking, etc., are called *sustainability champions.* At this time, there are no regulations requiring corporations to consider sustainability or to ensure the well-being of stakeholders. Consideration of sustainability is currently treated more as an option or a corporate marketing approach than a necessary social requirement.

1.1.4 Challenges for Sustainability Engineering

The challenge for the engineering professions is to institute sustainability considerations in all projects and operations in much the same way that safety issues are included. Safety engineering requires honesty with employers and the public about the previous safety failures, the potential hazards, and effective risk management measures. Energy engineers will need to similarly be honest with business leaders, policy makers, and the public about the issues of supply security, the potential for alternative technologies, and the environmental risks. The sustainable energy challenge involves aspects of human behavior and perception that most engineers are not familiar with. However, a major part of the work of safety engineering is achieving safe behavior by design, and by safety signals and equipment that manage operating errors. Another challenge is the perceived conflict between economic benefit and sustainability. Again, in safety engineering, while cost-effective solutions are sought, protection of health and well-being take precedence over profit except in cases of neglect. Safety engineering can provide many examples of the ways that engineering practice can and does take on projects to manage risks to well-being as part of the fundamental ethical responsibility of the profession. The engineering professions have the responsibility to consider the sustainability context and the carrying capacity implications of all developments, even if the regulatory, social, and economic systems are not currently structured for sustainable development. The lessons of history show that where engineering leads, the rest of society follows.

1.2 Carrying Capacity and Exponential Growth

1.2.1 Population Issue

World population started to increase rapidly at the end of the eighteenth century. According to data from the United Nations [2], the world's population increased in the 118 years from 1804 to 1922 from 1 to 2 billion and doubled once again in the following 52 years reaching 4 billion in 1974. Although not thought of as an engineering topic, the challenge of sustainable development is intimately linked to global population growth. Figure 1.2 shows historical data and growth projections for world population from 1800 to 2050. The world population reached 7 billion in 2011. The United Nations predicts that the world population may reach 8 billion by 2028.

Determining or predicting population is not an exact science. Population figures are derived from census and other government data, but also involve statistical analysis of these data. Consider that the margin of uncertainty in the population estimate for China is of the same magnitude as the total population of the United States. Current population is a function of the number of people alive last year, minus the number who died, plus the number who were born. If the average life expectancy increases, then the population grows. But the population growth rate is also lowest for countries with

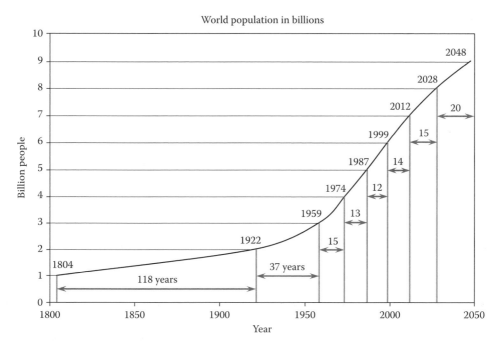

FIGURE 1.2
Annual additions and the annual growth rate of global population. (From Population Division of the Department of Economic and Social Affairs of the United Nations Secretariat, World population prospects: The 2008 revision, March 2009, http://esa.un.org/unpp)

high life expectancy. Life expectancy includes the deaths at all ages, including people who die in childhood. Thus, a higher life expectancy means more elderly people, but it also means more females are living to childbearing age. Life expectancy is currently around 80 years in Europe, 75 in China, 67 in India, and 50 years in Africa, largely reflecting vastly different childhood mortality. In the United States, the current life expectancy is 75–80 depending on whether you are a man or a woman. Consider that in 1900, the life expectancy in America was 46–48.

The fertility rate, or the number of children born in a year to a given population of women, depends on the number of women, the age at which they start having children, and the number of children they have during their lifetime. The dramatic increase in world population in the last century is largely due to the reduced infant and childhood mortality, an increasing number of fertile women, as well as increased longevity. The rate of population growth has slowed appreciably in many developed countries, but not in Africa, the Middle East, and South America. In the United States, the native born population decreased for the first time in its history during 2009. In Western countries including Italy, Germany, and Canada, the percentage of women having children in a given year has dropped to a point where the population is no longer growing, even though the longevity is very high and the infant mortality is very low.

A large body of research on the sociology of population and fertility has unequivocally concluded that if women have any socially acceptable alternatives to childbearing, they will voluntarily limit the number of children they have, regardless of the access to modern birth control services. Education has been shown to be the most effective way to reduce population growth. Where girls are given the chance to go to school, the average age of first pregnancy shifts upward. Where women have the opportunity for employment, the number of children per woman declines. Adding availability of contraception and social acceptability of small families brings the population growth rate down to replacement levels. In places that Westerners would consider "un-developed" like the Pacific Islands, voluntary control of fertility has been an ancient practice. The land and coastal resources are obviously limited on a small island. Anthropologists have observed that family groups will manage their new births to coincide with the death of an elder, thus maintaining a stable population on the island.

1.2.2 Water Issue

Less than 1% of the water on the planet is suitable for drinking and agriculture; the rest is salt-water, brackish, or frozen. Freshwater refers to rivers and lakes fed by seasonal precipitation. Aquifers are underground freshwater reservoirs in permeable gravels or sand. Some aquifers, called unconfined, are replenished by surface precipitation, but other confined aquifers like the massive Ogallala aquifer under the Great Plains in the United States are actually fossil water, deposited over one million years ago. The sustainable water consumption in any given location depends on the precipitation rate, storage systems, and the requirements to sustain local eco-systems. Confined aquifers have a finite lifetime for use rather than a sustainable usage rate. In Texas, artesian wells were plentiful in the early twentieth century, but wells must now be pumped and in some places have run dry. A recent study found that 30% of the Southern Great Plains will have exhausted the groundwater resource within the next 30 years [3].

The World Health Organization stipulates that the basic requirement for water is that 20 L/day per person be accessible within 1 km of the home. In an industrial society such as the United States, personal water consumption is 10 times as large, somewhere between 200 and 300 L/day for household uses. But if the industrial and energy production usage are added, freshwater usage can exceed 5000 L/day on a per capita basis. The World Bank [4] suggests that an indicator of water scarcity is an annual availability of less than 1000 m^3/person, while the Water Footprint Organization, in a 2008 study by Hoekstra and Chapagain [5], suggests that 1200 m^3/person is the minimum for a modest consumption footprint. We can easily calculate the carrying capacity at minimum needs for available freshwater, as well as for a modest consumption. Figure 1.3 shows the population carrying capacities in millions based upon data for 1990 for a range of African and Mediterranean countries. Jordan, Kenya, Algeria, and Saudi Arabia and Israel already had populations exceeding the water carrying capacity in 1990. There is a stark difference in the economic ability of these countries to deal with this shortfall to support different levels of lifestyle. Under projections of expected population growth predicted by the United Nations [6], all of these countries will have substantially exceeded the water carrying capacity for even minimum survival needs by 2025.

There are more than 7000 desalination plants in operation worldwide, with 60% located in the Middle East. The levelized cost of water produced by various desalination plants is between $0.60/m^3 for large plants and $3.00/m^3 for small remote plants. The energy requirement varies greatly with the type of process and is much lower for brackish water than for seawater. Thermal processes like multistage flash desalination use 10–15 kWh/m^3 of water while reverse osmosis uses 4–13 kWh/m^3 of electricity for seawater and 0.5–2.5 kWh/m^3 for brackish water [7]. Desalination is currently an option in those parts of the world where energy supplies are plentiful and the value of water is extremely high. For example, in Saudi Arabia, the world's largest desalination plant produces 128 million gallons per day (0.5 million m^3) of potable water.

It is estimated that around 3% of the electricity demand in the United States (56 billion kWh/year) is for treating and pumping freshwater. California has a large agricultural

Country	Available Freshwater (m³/Capita)	1990 Population	Population Carrying Capacity for Minimum and Modest Consumption	
			1000 m³/person	1200 m³/person
Egypt	1123	52.0	57.0	49.0
Ethiopia	2207	50.0	110.0	92.0
South Africa	1317	38.0	50.0	42.0
Israel	461	4.6	2.1	1.8
Jordan	327	4.0	1.3	1.1
Algeria	689	25.0	17.2	14.3
Kenya	636	24.0	15.2	12.5
Morocco	1117	25.0	27.9	23.3
Saudi Arabia	306	15.0	4.6	3.8

Note: Dark shading denotes the population is already beyond the carrying capacity.

FIGURE 1.3
Population carrying capacity (in millions of people) for nations based on water availability

sector, and a large population living far from freshwater sources resulting in 19% of electricity demand associated with provision of water. The biggest user of freshwater from lakes and rivers is for cooling thermoelectric power plants. Figure 1.4 shows a breakdown of the water withdrawals according to the U.S. Department of Interior [8]. Much of the water used by a power plant is returned to the river or lake, albeit warmed by 4°C–10°C as allowed by permits to protect the aquatic ecosystem. However, 3% of all water consumed in the United States is evaporated in the cooling towers in power plants and lost as water vapor in the atmosphere. Power plants constructed near the coast can use seawater for cooling. A Department of Energy/National Energy Technology Laboratory (DOE/NETL) analysis for 2005 indicates that the thermoelectric power generation sector withdrew 147 billion

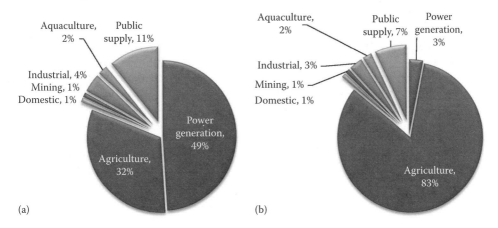

FIGURE 1.4
U.S. freshwater withdrawal in 2005 was 410 billion gallons per day. (a) Freshwater withdrawals and (b) freshwater consumption distribution. (From Alliance for Water Efficiency and American Council for an Energy-Efficient Economy, *Addressing the Energy–Water Nexus: A Blueprint for Action and Policy Agenda*, May 2011; ASME, *ETP: Energy-Water Nexus—Cross-Cutting Impacts*.)

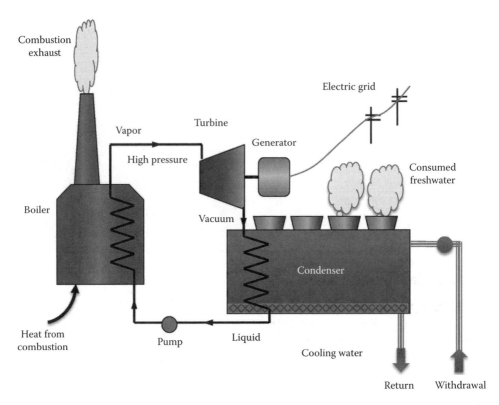

FIGURE 1.5
Basic schematic figure of a Rankine cycle power plant using freshwater for cooling.

gallons (556 million m³) and consumed 3.7 billion gallons (14 million m³) of freshwater per day. The connection between electrical power generation and water use is particularly important for long-term planning of sustainable energy systems [9].

Thermoelectric power plants need cooling water to operate the Rankine cycle as illustrated in Figure 1.5. In the Rankine cycle, heat is used to boil a working fluid to produce a high-pressure vapor, which is expanded through a turbine to produce electricity. The generation capacity of the turbine depends on the mass flow rate of the vapor and on the pressure drop from the inlet to the outlet. A vacuum is created at the turbine outlet by condensing the exiting vapor to a liquid. The efficiency of the cycle is defined as the power generated divided by the heat input and is directly related to the ratio of the vapor boiler temperature and the condenser liquid temperature. The power cycle efficiency increases, the higher the boiler temperature and the lower the condenser temperature.

The heat for the boiler can be supplied by combustion of coal or other fuel, radioactive decay of uranium, geothermal fluid, waste industrial heat, or even concentrated solar energy. In the United States, in 2011 around 39% of all electricity (132 TWh) was generated by pulverized coal combustion (PCC) where the coal is crushed and mixed with water to form a slurry that is sprayed into the combustor. Natural gas is used to generate electricity, but it is normally used to run a gas turbine. The hot exhaust gas provides the heat for the boiler of a Rankine cycle. This is called natural gas combined cycle (NGCC) power generation and can use the fuel more efficiently, with total plant efficiency of 50%–65% compared to 35%–45% for pulverized coal power plants. PCC and NGCC power plants are called steam power plants as they use treated water as the working fluid. Geothermal power plants use

TABLE 1.1

Water Consumption for Different Types of Power Plants

Plant Type	Water Consumption per Unit Electricity (gal/MWh$_e$)
Solar thermal	700–1000
Nuclear	700–800
Subcritical pulverized coal	520
Supercritical pulverized coal	450
Integrated gasification combined cycle	310
Natural gas combined cycle	190
Geothermal power plant	0 (if air cooled and geothermal fluid is re-injected)
Waste heat power plant	0 (if air cooled condenser used)

hydrothermal resources that are seldom above 300°C. Depending on the pressure of the resource, some of the geothermal brine is flashed and expanded through a steam turbine. The remaining brine and condensate from the flash steam turbine provide the boiler heat for an organic Rankine cycle (ORC), so-called because the working fluid is a hydrocarbon like pentane. Organic fluids, ammonia, and refrigerants boil at much lower temperatures than water and also condense at higher temperatures. ORCs are also used for solar thermal power plants and waste heat power conversion.

The amount of water consumed or withdrawn depends on the kind of cooling system used for the condenser. In coastal locations or where a large river or lake is available and where the water temperature is cool year-round, the withdrawn water is returned to the freshwater body. In hot weather, the freshwater is sprayed over evaporation pads, air is blown through the cooling tower, and the evaporation cools the water stream. Thus some of the freshwater is consumed and exhausted to the atmosphere. Table 1.1 gives the amount of water withdrawn and consumed by various types of thermal electric power plant. The U.S. average water consumption for thermoelectric power generation was about 23 gal per kWh in 2005 for all types of cooling [10]. Water use in electricity generation is an issue of long-term sustainability. The U.S. DOE in an extensive report, "Energy demands on water resources," prepared by DOE [11] for the U.S. Congress in December 2006 stated that:

> Populations continue to grow and move to areas with already limited water supplies. The growing competition of water availability for energy production and electric-power generation has already been documented in many river basins. … Available surface water supplies have not increased in 20 years, and groundwater tables and supplies are dropping at an alarming rate. New ecological water demands and changing climate could reduce available freshwater supplies even more.

Example 1.1: Water Use in Electricity Generation

Water at 60°F (15.6°C) is used to condense steam in the Rankine cycle of a nuclear power plant. Given a condensation pressure of 4 in Hg (abs.) (135.5 mbar), the condensate temperature of the condensing steam is 125.4°F (51.9°C).* This medium-sized plant is designed to produce 1000 MW of electric power at an efficiency of 31.5%. To eliminate

---

* This value can be found in the steam property tables of any standard thermodynamics textbooks or online at many engineering resource sites.

the high cost of building long transmission lines, and the additional associated losses, a central location is generally desired for any power plant. If the temperature of the cooling water returned to a river is not to exceed 110°F (43.3°C),

    (a) Estimate the rate of heat transfer in the condenser (MW).
    (b) Estimate the rate of flow of cooling water (lbm/h).
    (c) Discuss the implications of putting such power plants in the interior of the country near a population center. Consider environmental and other concerns.

Solution

    (a) The definition of the cycle efficiency gives

$$Q_{boiler} = \frac{P_{electric}}{\eta} = \frac{1000 \text{ MW}}{0.315} = 3175 \text{ MW}$$

The energy balance on the cycle gives

$$Q_{cond} = Q_{boiler} - W_{gen} = 3175 \text{ MW} - 1000 \text{ MW} = 2175 \text{ MW} \ (7.43 \times 10^9 \text{ Btu/h})$$

    (b) The temperature increase in the cooling water in the condenser is obtained from an energy balance on the condenser freshwater flow:

$$Q_{cond} = \dot{m}c_p(T_{out} - T_{in})$$

With the specific heat of water $c_p = 1.0$ Btu/lbm°F (4.186 kJ/kg °C), the mass flow rate of the cooling water is

$$\dot{m} = \frac{Q_{cond}}{c_p(T_{out} - T_{in})} = \frac{7.43 \times 10^9 \text{ Btu/h}}{1.0 \text{ Btu/1bm }°F \times (110°F - 60°F)} = 149 \times 10^6 \text{ lb/h} \ (67.6 \times 10^6 \text{ kg/h})$$

    (c) The water issues with the cooling water for this plant are the heating of the freshwater resource and the consumption of water if wet cooling towers were used. We would need a specific location with known water resources and aquatic ecosystems in order to proceed to quantify the issues posed by this water use for power generation. In many places, power plants on rivers must reduce generation capacity in the summer months in order to stay within their permitted heat addition to rivers.*

Hydropower generation uses the flow of freshwater to produce electricity without consuming or withdrawing water, although the storage lakes do increase the evaporation losses from the watershed. Hydroelectricity supplies about 7%–10% of the electricity in the United States or about 270 billion kWh at the lowest average generation cost of all sources at $0.02 per kWh [12]. Dams have been built on rivers where a suitable site provides for a storage lake and a sufficient height for driving the flow through the turbines. It is estimated that it takes about 14,000 L of water, with lake storage area of 185 ac to produce 1 MWh of electricity.

---

* http://www.guardian.co.uk/world/2003/aug/12/france.nuclear

Production of fuels also places pressure on water resources. Ethanol processing from corn requires 3 gal of water per gallon of fuel and anywhere from 10 to 330 gal water/gal fuel for agriculture depending on irrigation requirements [13]. Between 40% and 70% of the U.S. corn crop, depending on weather and market conditions, can be processed into ethanol, which in itself is a food supply issue. There are no commercial processes for producing cellulosic ethanol from wood, but the biochemical processing currently requires 10 gal (w)/gal (f). Biodiesel processing does not require water, but the soybean crops are more heavily irrigated resulting in an interesting figure of 12 gal of water per mile driven in a car fuelled with biodiesel. American petroleum production requires 2–6 gal (w)/gal crude oil depending on the amount of recycling of produced water as injection fluid. Saudi Arabia consumes 1.4–4.6 gal of desalinated water per gal of crude. Refining crude oil into gasoline consumes 1.5 gal (w)/gal product. Some of the water used to produce gasoline from tar sands is heated and used as steam, with overall water use of 2.6–6.2 gal (w)/gal gasoline. Water pollution is a major sustainability issue with tar sands extraction operations.

1.2.3 Food Supply Issues

In order to have sustainable development, the global food production must be adequate to feed the world's population. Food comes from the sea and the land. It has been estimated that the earth has 8.2 billion acres (ac) of potentially arable land. According to a study by the Millennium Institute [14], this acreage is divided according to the levels of productivity shown below:

World Agricultural Land Endowment:

Highly productive—1.1 billion acre

Total somewhat productive—2.2 billion acres

Total slightly productive—4.9 billion acres

Currently productive—3.6 billion acres

Potentially productive—8.2 billion acres

In the United States, about 1.5 ac of land is used to produce the food for one person per year, but a modest per capita global footprint based upon diets with less meat and more grain is estimated to be about 1 ac per capita. At this modest per capita global footprint, the global carrying capacity would be somewhere between 3.6 and 8.2 billion people. In 2010, the United Nations Food and Agriculture Organization estimated that 13%, or one in seven people, most of whom live in Asia and Africa, had chronic malnourishment in either total calories or protein [15]. Global limits on food production and water availability indicate that more than 9 or 10 billion people are an unsustainable global population. This is because the mortality rate for a large proportion of people would be extremely high, particularly given the increasing risks of devastating effects of extreme weather events in all countries.

We are clearly approaching the population capacity limit based on food production. The food distribution and market systems are already failing for one in seven people. Soil depth and quality are other issues that pose risks to land productivity, and agricultural land loss to urbanization is on the order of 1.5 million acres per year in the United States. It is therefore apparent that, even if the world diet would shift toward more grain and less meat, the global population predicted for 2025 in Figure 1.2 would be in excess of the maximum available land and global food production capacity.

There are enormous differences in agriculture, diets, and food market systems in different parts of the world. Clearly, it is possible to have food production systems that do not rely on fossil fuel, and much of Asia, Africa, and South America still rely heavily on human and animal inputs. The food supply system in the United States and other industrial countries requires fossil fuel. The current food supply system for the typical American diet requires approximately 2000 L oil equivalent per year per person and accounts for 19% of the total national energy use, with 14% for production, processing, packaging and storage, and 5% for transportation [16].

Corn, soybeans, wheat, and sorghum are the basis of the American food production system. In 2000, the United States produced about half of the world corn and soybean supply, one-fourth of the wheat supply, and 80% of the sorghum supply. In the United States, about 12% of the domestic corn market is for food: the majority of corn is used for animal feed and ethanol. Biofuel production grew rapidly from 2005 in response to U.S. government mandates and incentives. Ethanol production was 175 million gallons in 1980 and 14 billion gallons (bgal) in 2011. The consequences of the bioethanol boom from 2000 to 2011 have been that the percentage of corn crop used for food in the United States dropped from 13% to 11%, exports dropped from 20% to 13%, and feed for U.S. cattle and chickens plummeted from 60% to 36% of the annual crop. In 2011, ethanol processing consumed 40% of the total U.S. corn crop, and the monthly farm price for corn hit an all-time record, three times higher than that in 2006. Although the U.S. government has mandated continued growth in biodiesel consumption to over 35 bgal by 2022, it is not anticipated that any further growth in corn ethanol is possible. The government policy was based on anticipated commercial cellulose or algae ethanol plants beginning production by 2012, which has not eventuated. Biodiesel also expanded from 0.5 million gal in 1999 to 800 million gal in 2011. Despite this rapid growth, biofuel in total accounted for 6% of U.S. fuel supply on a gasoline-equivalent basis in 2011 [17].

Competition between agriculture for food and for biofuels has become an international issue. One gallon of ethanol requires 12 kg of corn. A Ford Explorer has a 22.5 gal capacity fuel tank. Thus, to fill up a flex-fuel explorer with E100 (100% bioethanol) would require 270 kg of corn. The average diet for Western people includes about 23 g per day of grains, like cornflakes. We can see that one fill-up of our favorite vehicle could meet the grain dietary needs of 32 people for a whole year. People in Asia, India, and South America eat more grain, 70 g per day, and less protein and fats. Thus, the 270 kg of corn could provide the tortillas for 10 people in Mexico for 1 year. If the 270 kg of corn were used as seed, it could plant about 40 ac and, with the yield of 3225 kg of corn per acre, could grow enough grain for 5000 people in Belize for one year.

1.2.4 Energy Issue

Unlike food and water, the basic level of energy to sustain a given human population is more difficult to specify. The daily energy requirement for a family living a traditional lifestyle on a remote island in Fiji is two coconut husks dried for two days in the sun that are used to cook the day's taro in a pit fire and fry the day's fish catch on a simple grill. On the same island, there are homes connected to village electric grids and the island water supply. Their daily energy demand also includes 72 Wh for two small lights for 6 h per day when the village generator functions and a small share of the 20 L of diesel fuel used to run the island water pump for 3 h each day. Also on the same island, a home built by an Australian has a much different lifestyle with continuous electricity, a refrigerator, TV, lights in every room, and air-conditioning. This home supplies its own 11 kWh per day by solar panels, a small wind generator, a large battery bank, and several diesel generators, and compressed natural gas is used for cooking. While the needs of this house are only about half the average in Australia, they are still vastly different from their neighbors on the island [18]. The standard of living for all of the island residents is roughly equal in that they all have adequate nutrition, access to the same quality of air and water, access to medical treatment and education, and the same life expectancy. Clearly, the energy use of the traditional lifestyle is sustainable while the Australian lifestyle depends heavily on finite resources, and even though about half of the energy is provided by renewable sources, the technology was manufactured using fossil fuels.

A study conducted some time ago by the United Nations proposed a measure of standard of living, termed the Human Development Index (HDI), based upon various criteria [6]. Figure 1.6 is a plot of energy consumption per capita for various countries as a function of a composite indicator, developed by UNDP, to show countries' relative well-being using the HDI. It is apparent that the United States and Iceland have the highest energy consumption per capita, but their human satisfaction level is not much different from other developed countries such as France or Germany, which provide their populations with a good standard of living at much lower energy consumption per capita.

There are large variations in the amount and type of food and the amount of water used according to lifestyle, just as there are wide variations with energy. However, we can set physiological requirements for food and water where it is hard to set such a requirement for energy. In considering the data in Figure 1.6, we could set 1000 kg of

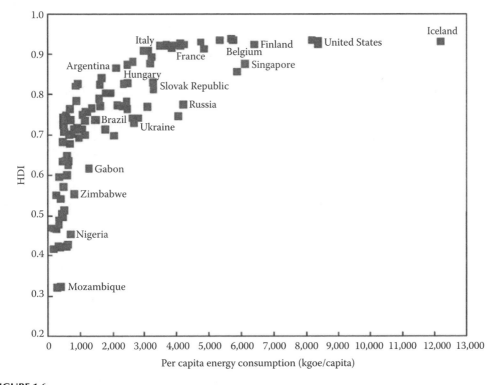

FIGURE 1.6
Relationship between HDI and per capita energy use, 1999–2000. (From Goldemberg, J. and Johansson, T.B., eds., *World Energy Assessment: Energy and the Challenge of Sustainability*, United Nations Development Programme, New York, 2004.)

oil equivalent (kgoe) as a range for determining the modest energy requirement, or the fossil energy budget plus an allowance for shelter in more extreme climates. This level of energy use seems draconian in relation to the current U.S. average per capita consumption of over 9000 kgoe. However, in quantifying carrying capacity, we are not considering lifestyle, rather we are considering necessity.

1.2.5 Exponential Growth Issue

Energy use has grown during the past 150 years in lock step with growth in population, and with water, food, and land use. The relevant history and experience of everyone today is that growth is inevitable; sometimes there is a correction, but we always get growing again. The history of growth in land and resource use in the United States has greatly affected the economic perspective. This perspective, however, discounts the value of both benefits and expenses in the future, seldom assigns costs to externalities, and ignores the physical limitations of growth.

Cognitive dissonance is a situation where belief and observation do not agree. We rationally know that continuous growth in a finite system is not possible, and yet it is our historical experience as a society that somehow growth always continues. In fact, the cultural consensus of economists and politicians is that growth is necessary. Perhaps we do not distinguish clearly enough between surplus and growth. Surplus is necessary for survival. Every organism must achieve a positive return on their efforts to forage for food and to reproduce in order to have enough surplus to invest in the next foraging excursion and in nurturing the next generation. Farmers must raise enough crops to feed themselves, plus surplus to trade for other goods, maintain equipment, and invest in the next planting. Businesses must operate at a profit at least equal to their depreciation in order to maintain and replace the means of production. Growth can only occur if there is surplus after all the current needs are met. However, growth, like increasing population, expanding a business, building more buildings, putting more land under cultivation, etc., locks in a higher base level of consumption needed for basic operations and maintenance.

For example, a particularly good season for grass growth results in a boom in the number of rabbits. Then the boom in rabbits causes a boom in coyotes. But when the subsequent year returns to average grass-growing conditions, the high rabbit population stresses the food supply, and the higher predation from the increased coyote population brings an end to the boom in rabbits. Growth is not necessary for survival, growth is necessary for more growth, and obviously, continuous growth is not sustainable. Growth is always the front half of a boom and bust cycle.

Perhaps we mistake prosperity for growth. Prosperity also requires surplus energy, food, or materials in excess of meeting minimum standard of living. However, prosperity has many other aspects than simply consuming more and building more consumption capacity. Prosperity can also mean higher quality of life, more leisure time, higher education level, more specialization of labor, and more arts and social activities. We will put forward an important distinction between growth and prosperity; growth is a cumulative increase in the energy and material intensity of lifestyle, while prosperity is a higher quality of life than the minimum living standard.

Perhaps we mistake inflation for growth. Inflation is a human artifact associated with the way money is used. Inflation does not occur in the natural environment. It does not take some percentage more solar energy every year to produce an apple. Inflation is also a very recent phenomenon in human economics. In the 1950s, middle-class salaries in America were in the range of $3000–$5000 per year. We think of our salaries as growing, we think of our home values as appreciating. Inflation is calculated from the growth in consumer prices. The actual ability of people to meet their needs and to prosper depends on the ratio of wages and consumer prices. The engineering approach to sustainability must be based on physical energy and resource systems rather than focusing on the human economic system. We can understand that the economic system will have to respond to the sustainability of resources, but we cannot analyze or predict economic behavior the same way we can analyze or predict physical systems.

The simple and clear arithmetic of the exponential function is the most important tool that engineers can use to analyze, understand, and communicate the issues associated with past and future growth. Suppose the quantity, N, periodically increases from an initial value, N_0, at a continuous growth rate, r, per unit time, t. The future value of the quantity, $N(t)$, is given by

$$N(t) = N_0(1+r)^t \tag{1.1}$$

After one period, the quantity will be equal to $N(1) = N_0(1 + r)$, after two periods $N(2) = N_0(1 + r)^2$, and so on. The time it would take for the original quantity to double, or the doubling time, D, is of interest in the studies of growth. Setting $N(t) = 2N_0$ in Equation 1.1 and solving for t gives

$$D = t = \frac{\ln(2)}{\ln(1+r)} \tag{1.2}$$

For most practical situations, r is small, say less than 10%, and the quantity $\ln(1 + r) \cong r$. Noting that $\ln(2)$ is equal to 0.693, Equation 1.2 simplifies to the rule-of-thumb approximation for doubling time:

$$D \cong \frac{0.7}{r} \tag{1.3}$$

Thus, for a growth rate of 7%, the doubling time would be about 10 years.

Now suppose that a finite resource would have a total useable quantity, N_T. The amount that has been consumed at the present is N_0, and the present consumption rate per year is $N(0)$. If the consumption rate does not change, then the lifetime of the resource, t_L, would be found by $t_L = (N_T - N_0)/N(0)$. However, when the consumption rate of the resource is increasing every year with an annual growth rate, r, the lifetime of the resource is given by

$$t_L = \frac{\ln\left(\dfrac{N_T}{N_0}\right)}{\ln(1+r)} \tag{1.4}$$

Growth is a natural part of all living systems. However, growth is only the first part of the inevitable cycle of growth and decay. In human societies, and in particular with regard to economics, we need to acknowledge how powerful belief is in shaping perceptions and consensus. Fiat money is a human construction, the value of which is determined by consensus. Money is not a natural resource, so the supply of money can indeed grow as long as all the participants in the economy continue to agree on its value as a proxy for their goods and labor. However, when we are considering the carrying capacity of a region or the lifetime of a finite resource, we need to be able to perform quantitative analysis in order to assess the depletion risks and the time frames associated with these issues.

Example 1.2: Continuous Growth, Finite Resources, and Perception

In 1950, the city of Fort Collins, Colorado, had a population of 15,000 and occupied 3,750 ac. The city was surrounded by 50,000 ac of prime farmland within a 20 min drive, which grew most of the food sold in the city markets plus grains, fruits, and vegetables and dairy that were transported to the capital city of Denver. Like the rest of the country, Fort Collins was about to experience a prolonged period of population growth at an annual rate of 4.2%. New housing for the population was typical suburban single-family homes, and 158 ac was converted from agriculture to urban use in 1950.

In 1950, people started to wonder: if this current rate of land conversion continued, how long until the farmland was consumed? The answer (50,000/158) is 317 years. No worries! But the next year 164 ac was developed, and the next year 171 ac. A professor at CSU used Equation 1.2 and determined that the city seemed to be increasing its land use by 4.2% each year, the same rate as the population growth. If the 3750 ac of urban land in 1950 were to increase apace with population, then Equation 1.4 gives the year that the city would have grown to 53,750 ac and consumed all of the farmland as 64 years later in 2014 as shown in Figure 1.7. The professor in 1953 would surely have looked at the map and the farmland around his city and not really been able to picture 118,000 people living there and all of the farmland gone. The mathematical model is clear, but the professor's perception of the issues associated with future growth would likely be limited by his lack of experience with this kind of situation. He might think that something would change in the future to slow the growth or manage the preservation of some of the agricultural productivity.

The professor would have been called an alarmist if he suggested that population and land use were an issue. The mayor would not appreciate the suggestion that population

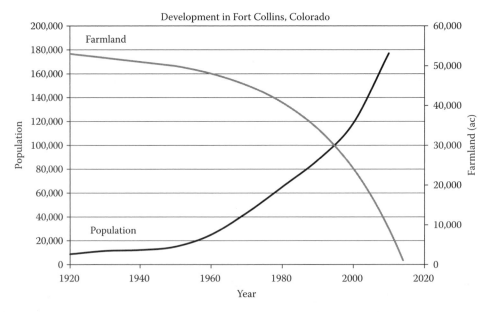

FIGURE 1.7
Annual population growth of a city is 4.2% resulting in consumption of agricultural land for urban development.

growth could be a problem when it is actually necessary to pay for the new schools, sewer system, and library built to meet the needs of previous growth. In 1997, there was still half of the farmland remaining around the city. But in fact the steady, "healthy" growth rate of the city continues to the present day, and all of the original farmland is now within urban city limits.

Let us say that the professor had argued for more "efficient" urban development and championed regulations to increase the housing density so that the land use per person could be cut in half? If the council had taken this advice, then today a bit less than half of the farmland would still remain. However, if the population growth rate continues unabated into the future, the farmland would still be consumed by 2035. The improved efficiency would have bought 23 years before total consumption of the farmland resource, but it would not have changed the inevitable outcome.

1.3 Context for Sustainable Energy

1.3.1 Historical Energy Development in the United States

America has made two major energy transitions in its relatively short history as shown in Figure 1.8 [19]. Before the Civil War, American industry depended on wood, wind, water, and animals for its heat, energy, and power. Charcoal from hardwoods had been used for processing metals and ceramics since ancient times. The industrial revolution outstripped the wood supply in Europe and contributed to the clearance of forests in North America. In the first energy transition, coal became the major supply for U.S. energy needs between 1885 and the start of World War II in 1941. New steam engines to de-water and ventilate

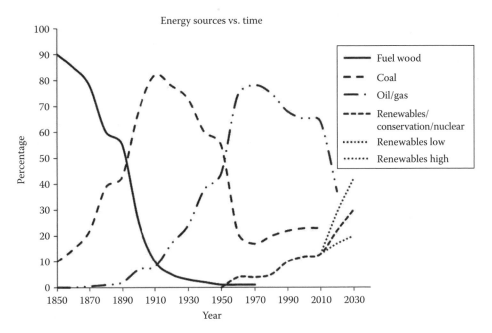

FIGURE 1.8
Diagram illustrating energy transitions in share of total energy supplied by wood, coal, oil and gas, and renewables. (From U.S. Energy Information Administration, Primary energy production by source, 1949–2010.)

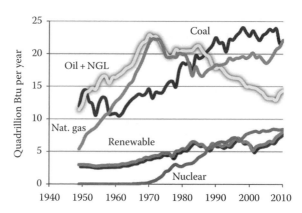

FIGURE 1.9
United States primary energy production by source, 1949–2010. (From Annual Energy Outlook 2012 with Projections to 2035, www.eia.gov/forecasts/aeo/er/pdf/0383(2012).pdf)

coal mines stimulated a transition to coal followed by a dramatic increase in total energy consumption. Coal was found to be a vastly superior fuel to wood in both running the new steam engines and high materials processing. However, the air pollution from coal burning in factories and homes, and gasification of coal for lighting was a serious public health issue in most cities around the turn of the century.

The second transition from coal was not due to a shortage of coal; rather, there were great advantages to using oil and gas. The development of oil and natural gas, and coal-fired power generation that could supply electricity via the grid provided a vast improvement in air quality in cities. Beginning in the 1970s, the United States transitioned from being a net exporter of oil to being an importer. Total U.S. energy production has leveled off, conversion efficiency has improved, but consumption continued to grow. The enormous increase in energy utilization since the end of World War II in 1945 is the result of rebuilding parts of the world devastated by the war, improvement in the living standards of the middle class, and an exponential growth in population. In the 1960s, it was thought that nuclear energy would be the next major energy source of the future. However, the costs of nuclear energy turned out to be much higher than expected, and the anticipated safe long-term waste storage solution did not emerge. In the 1980s, concerns about nuclear plant safety caused public opposition, and excessive cost overruns derailed new nuclear power developments.

Figure 1.9 shows the primary production history for the United States since 1949 [20]. Until the 1970s, energy prices remained low, thus putting no damper on ever-increasing demand. The oil crisis caused higher oil prices, but it is important to note that even though market conditions were favorable and policy incentivized oil exploration, the U.S. oil production peaked and has since declined continuously. While the share of consumer energy provided by coal declined after the surge in energy use in the 1950s, the total amount of coal production increased, until recently, when cheap natural gas from fracking operations has become available.

1.3.2 Current Energy Use

The current energy crisis for America and the rest of the world is not related to the total quantity of fuel, but rather to the end of growth and surplus in the lowest cost fuels, conventional oil, and natural gas. To address this crisis, the world will make a new kind of energy transition that is quite different from any past experience. It will be a transition from growth to decline in surplus energy, primarily for transportation. All energy end-use infrastructure and activities will adapt to using less energy. Those businesses and cities that adapt the fastest will be the most prosperous during this transition.

In order to get an idea of the scale of the transition from current energy systems to renewable systems, one needs to examine the detailed pattern of energy supply and consumption. It is useful to look at end-use consumption by dividing it into four major sectors: transportation, residential, commercial, and industrial. These sectors consume, respectively, 27%, 22%, 18%, and 30% of total U.S. energy, as shown in Figure 1.10. Total U.S. consumption in 2009 was about 102 quads, which is about one-fifth of the world's energy use, with fossil fuel accounting for about 85% of the total. The total contribution of renewable energy is only about 7%, with the majority derived from hydroelectric power generation that contributes 31% of the renewable supply, wood 25%, and biofuels 23%, and wind and solar energy together is less than 12%. Thus, a transition to renewable energy systems today would mean 93% reduction in energy demand plus very high demand tolerance for variability in supply.

The small contribution from solar and wind energy sources illustrates the dramatic change in the energy use pattern that will be necessary for a shift from fossil to renewable sources. An ancillary to that observation is the fact that the hardware necessary to begin producing energy from solar, biomass, and wind sources will have to be built with energy input from fossil fuels and nuclear power due to the energy intensity of the materials manufacturing. Planning for the renewable energy transition will therefore have to include the energy required to build the renewable and possibly also nuclear energy production facilities. This energy would be in addition to the normal needs of the world economy.

1.3.3 Future Energy Scenarios for the United States

Future scenarios are typically developed using historical information, interpreting trends, and modeling future behavior under different assumptions. Nearly every nation has a governmental energy-modeling unit dedicated to developing the national energy outlook. Figure 1.11 is a summary of the projection for U.S. energy use by fuel type produced by the Energy Information Administration (EIA) [20]. The outlook is for oil consumption to reduce by 14% by 2035 largely through much better vehicle efficiency. Coal, nuclear, and natural gas end uses are expected to remain virtually unchanged, while liquid biofuel and other renewables are seen to provide the expected energy demand growth. It is interesting to consider the EIA Energy Outlook projections from previous years. This 2012 outlook has the lowest growth rate in future energy use of any scenario the EIA has ever produced. For example, the 2006 Energy Outlook projection had total energy demand toping 125 quads in 2025. If major infrastructure had been built in anticipation of this high demand, such as upgrading the national grid or building new freeways, then those assets would now be at risk of not returning the anticipated economic benefits.

The current projection for continued nuclear production over the next 25 years would require either massive investment in replacement power plants or further extensions of operating permits beyond the original design life for existing plants. Both of these options present economic and safety issues.

The projected consumption of coal demonstrates no response to the issue of atmospheric CO_2 accumulation and the associated risks, or it assumes the timely research and development of CCS technology. The probability is very low that CCS technology will be installed

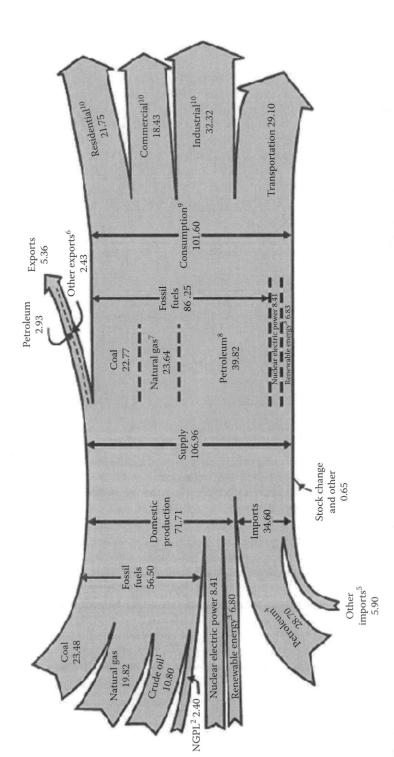

FIGURE 1.10

Energy flow 2007 (quadrillion Btu). (1) Includes lease condensate. (2) Natural gas plant liquids. (3) Conventional hydroelectric power, biomass, geothermal, solar/photovoltaic, and wind. (4) Crude oil and petroleum products. Includes imports into the Strategic Petroleum Reserve. (5) Natural gas, coal, coal coke, biofuels, and electricity. (6) Coal, natural gas, coal coke, electricity, and biofuels. (7) Natural gas only; excludes supplemental gaseous fuels. (8) Petroleum products, including natural gas plant liquids, and crude oil burned as fuel. (9) Includes 0.09 quadrillion Btu of electricity net imports. (10) Total energy consumption, which is the sum of primary energy consumption, electricity retail sales, and electrical system energy losses. Losses are allocated to the end-use sectors in proportion to each sector's share of total electricity retail sales. *Notes:* Data are preliminary. Values are derived from source data prior to rounding for publication. Totals may not equal sum of components due to independent rounding. (From Energy Information Agency, U.S. DOE annual energy review 2010, Washington, DC, October 2011.)

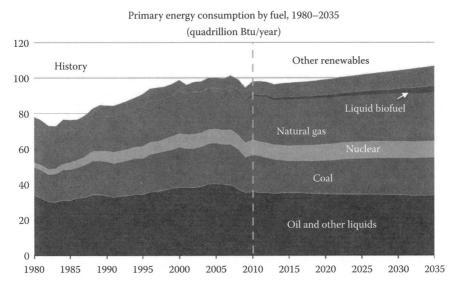

FIGURE 1.11
U.S. Energy end-use projection (quadrillion Btu/year) by fuel type reflects scenario for slower economic growth than historical trends and improving efficiency. (From Annual Energy Outlook 2012 with Projections to 2035, www.eia.gov/forecasts/aeo/er/pdf/0383(2012).pdf)

on any coal power plant in the United States by 2025. According to the Inter governmental Panel on Climate Change (IPCC) Working Group 3, the decline rate of fossil energy use must be dramatic and sustained in order to manage the risks of catastrophic global climate change [21]. According to the report, the most cost-effective approach is improved energy efficiency in buildings and reduction of coal combustion for power generation. However, the U.S. EIA scenario projects that CO_2 emissions will grow continuously at 1.2% annually.

The EIA projects further expanded oil imports into the future. The expectation for continuation of current levels of oil consumption over the next 20 years could put the U.S. transportation systems at risk. The International Energy Agency's (IEA) *World Energy Outlook 2011* projects world petroleum production to drop by more than 70% due to rundown in existing fields unless massive investment is made in discovering and developing new resources. Either way, oil price will be a drain on the U.S. economy if the transport system is not prepared to adapt to reduced fuel use.

The rapid growth in biofuel experienced to date has been in response to government mandates and incentives that are due to end in 2014. Given that the projection includes three to four times more biofuel production than today, the scenario for supply growth due to biofuel expansion may be unlikely. Future planning will also have to deal with resistance to industrialization of the few remaining wild rivers and landscapes for new large hydropower dams or wind energy farms. High-value wind resources are often located in sparsely inhabited regions requiring investment in upgraded transmission systems to transport the power from the source to where it is needed in industrial parts of the country. The same applies, although to a lesser extent, to good solar radiation sources, many of which are located in parts of the country with limited water resources and some distance from industrial centers and large cities. Thus, we conclude that the historical development, the current status, and the future development scenarios for the United States are not sustainable.

Lower carbon–emitting renewable energy resources can and will be developed, but this will not change the fact that the most likely future energy transition will actually be to reduced energy use. Aggressive development of alternative fuels such as biofuels in the United States has not mitigated the risk that volatile world oil prices pose to the economy. Development of wind and solar resources has not yet led to a decline in coal or gas use for power generation. The risks to the economy and society are not from actually *using* less fossil energy; rather, the risks are from not *adapting* to a lifestyle that uses less energy.

1.4 Key Sustainability Considerations

1.4.1 Energy Economic Efficiency

Energy efficiency is a widely used term in the public realm. In engineering, there are technical meanings: generation efficiency, conversion efficiency, or appliance efficiency. There is also a concept of economic efficiency. Economic efficiency refers to the ratio of resources and energy used to produce goods and services. Economic efficiency always improves profitability and is a necessary part of the strategy for adapting to resource constraints in a prosperous manner.

There is currently a direct coupling between the wealth of a country and its energy use. The data in Figure 1.12 show the gross national product (GNP) and energy consumption per capita for several countries. While this might seem to indicate that more energy is necessary to generate wealth, the United Kingdom and France have twice the energy economic efficiency with the same standard of living as the United States.

The low energy economic efficiency of the United States has resulted in an enormous discrepancy in relative energy use with the rest of the world. Table 1.2 shows the percentage shares of world population, world gross domestic product (GDP), and energy consumption for some of the major countries of the world. The United States, with less than

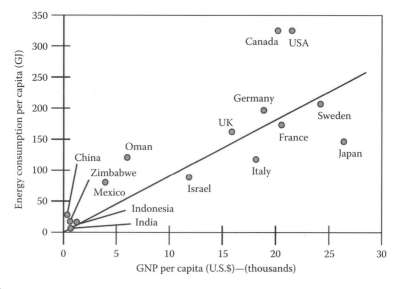

FIGURE 1.12

Comparison of energy use per capita per year versus GNP per capita for various countries. (Lines are provided only to show a general trend. 1 GJ = 10^9 J.) (From World Bank, *Development Data and Statistics*, 2003, www.world-bank.org/data, accessed May, 2007.)

TABLE 1.2

Percentage Shares of World Population, World GDP, and World Commercial Energy Consumption for Selected Countries

Country	% of World Population 2011[a]	% of World GDP 2011[b]	% of World Energy Consumption 2011[c]
United States	4.5	19	19
Japan	1.8	5.6	4.2
France	0.9	2.8	2.1
Germany	1.2	3.9	2.7
United Kingdom	0.9	2.8	1.7
China	19	14	20
India	17	5.6	4.4

[a] World population 2011 was 7.0 billion. Country data from The World Factbook [70].
[b] World GDP 2011 was $79 trillion. Country data from The World Factbook [70].
[c] World primary energy consumption 2011 was 12 billion tonnes of oil equivalent. Country data from BP [71].

5% of the world's population, consumes almost a quarter of the total world energy and produces almost one-third of the world's GDP. In contrast, China with 19% of the world's population, contributes 14% of the GDP, while using 11% of the world's energy.

1.4.2 Energy Return on Energy Invested

Energy transformation systems take natural resources from the environment and produce fuels and electricity for sale in the economy. They initially require an investment in money and materials as well as in energy for manufacture and installation. Once installed, the energy production system will use or consume natural resources and fuel to generate electricity, or it will extract and process resources into fuel. During the lifetime of the energy transformation plant, there will also be maintenance and replacement requirements. If we put all of these inputs and outputs in terms of energy, we can compare the consumer energy produced for the market to the energy required either from the economy or parasitic internal use. The main point of examining the EROI is to make sure that energy investments that have a very low return, usually under EROI = 8, are carefully scrutinized and decision-makers informed of the risk to long-term sustainability of the energy flow to the economy from such investments.

One effective way to look at this EROI for renewable electricity generation is to construct a timeline of the energy inputs and outputs using a spreadsheet and then divide the cumulative production by the cumulative inputs. This concept is illustrated in Figure 1.13, where the solid line represents the energy input and the dotted line the energy output for a renewable system like solar PV that does not require any fuel for its operation. The construction begins at time 0, energy generation at time 1, energy input equals energy produced at time 2, and lifetime of the system ceases at time 3. After the construction of the system is completed and energy generation begins, it takes some time before the energy input is repaid and net energy delivery begins. For this hypothetical system, the delivered energy is 5 units and the energy input is 1 unit giving an EROI is 5/1 = 5.

Consider a thin-film PV system described by the National Renewable Energy Laboratory (NREL) [22]. Thin-film PV modules use very little semiconductor material. The major energy

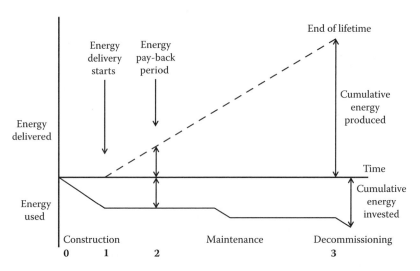

FIGURE 1.13
Schematic diagram of EROI for renewable technology's lifespan.

costs for manufacturing are the substrate on which the thin films are deposited, the film-deposition process, and facility operation although there can be substantial energy used in mining and processing rare earth minerals. NREL reported that it takes 120 kWh/m² to make frameless, amorphous-silicon PV modules. Adding another 120 kWh/m² for a frame and support structure for a rooftop-mounted, grid-connected system, assuming 6% conversion efficiency (standard conditions) and 1700 kWh/m² per year of available sunlight energy, the energy payback is about 3 years and the EROI = 11.75 assuming a 30 year life, but EROI is only 7.8 if the lifetime is 20 years.

For conventional electricity generation such as coal or nuclear power plants, the EROI is vastly different depending on whether the energy input from the fuel consumed is considered as a free resource taken from the environment like solar or wind energy rather than a consumer fuel taken from the economy. The economics of a coal-fired power plant that has its own dedicated mining operation are much different from a plant that must purchase coal in the market. Likewise, if electricity is used for heat when coal could be used directly, then the EROI on *thermal* energy would be highly negative for the electricity system compared to the coal fuel. This is because two-thirds of the thermal energy in the coal is rejected heat in the power plant that produces electricity. There is currently no standardized approach for calculating EROI, so all reported values should be examined closely to ascertain the underlying assumptions [23].

The diesel fuel for mining and electricity for processing and handling the fuel are usually a small energy input for a coal power plant. The embedded energy in the plant and maintenance are energy inputs. The biggest difference between fossil fuel generation and renewables is that the fossil EROI is often calculated on an annual basis using annualized values of embedded energy in equipment, whereas the EROI for wind and solar is determined over the generating lifetime. The EROI for coal power generation is in the range of 10. For nuclear power plants, the EROI would depend greatly on whether the energy needed for the storage of spent fuel and decommissioning is counted. Nuclear power plants are vastly energy intensive to build, but the EROI not counting the uranium processing and waste storage is in the range of 7–10. However, even a small energy requirement for long-term safe storage of spent fuel, given the very long storage period, can make

EROI of nuclear power highly negative to future generations that will have to provide the decommissioning and the electricity and cost for storage without receiving any electricity from the power plant.

The EROI for bringing fuels to the market would be based on the investment of exploration and refining, and the return of diesel and gasoline. The EROI for coal fuel is by far the highest of any energy resource. The EROI of coal fuel is estimated to be dropping due to the best coal being mined first, but it is still at least 50. The EROI for finding and producing oil in the United States was vastly positive until the 1970s when mechanical pumping, enhanced recovery including steam injection, and offshore production began to represent a larger share of the production. The EROI for oil has declined from a world aggregate value across the industry from over 80 early in the last century to 15–25 from conventional oil to fuel products. This drop is a result of oil reservoirs being depleted and energy investment costs increasing as exploration and development are shifted to lower-grade crude or offshore reservoirs [10].

Estimates of the range of the EROI for various energy transformation technologies, including those for providing liquid fuels, are shown in Figure 1.14. There have been a limited number of studies reporting EROI of existing systems, but the solar power tower, cellulosic ethanol, and coal with CO_2 sequestration are based merely on analysis because no commercial systems exist as yet.

EROI is important for achieving sustainability because it is a measure of low-hanging fruit. If a farmer based his income projections only on the rate of production during the time when workers were picking the low-hanging fruit, the farmer might make some decisions that could put his farm's financial position at risk. The rate of return is very good at first, but as the workers have to climb up in the trees and use ladders, the investment in the harvest goes up and the harvest rate goes down. Luckily, farmers are familiar with the boom and bust nature of the farm production cycle and use a long-term perspective in making financial decisions. EROI is a quantitative means to understand the past, present, and future net flows of energy to the economy for different energy transformation technologies.

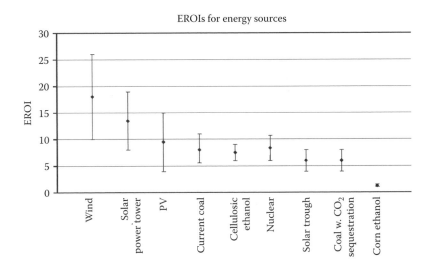

FIGURE 1.14
A comparison of the ratios of electric power or transportation fuel output to fossil fuel input for various technologies. (These estimates do not include decommissioning, social costs, or the potential of recycling of the material in the generating facilities.)

The calculation is straightforward, and all energy engineers must be familiar with the concept and understand the ramifications of different policies and investments on risks to future energy prosperity.

1.4.3 Cost of Energy Production

The capital cost per kW capacity is a way to compare the cost of building an energy production system to the market value of the energy produced. The investment costs include the capital cost, which is the cost of engineering and building the plant, plus maintenance and cost of fuel. The amount of energy produced for sale depends on the plant capacity and the number of hours per year the plant is generating at capacity, called the utility factor, and the ability of the plant to generate on demand during peak times. A broad perspective of the capital and levelized energy costs of renewables, as well as more traditional energy technologies, has been prepared by Lazard, a well-known financial advisory and asset management firm. The firm provides advice on mergers and acquisitions, as well as asset management services for large institutions and governments. The technology assessments shown here were prepared to advise its customers about investments in energy technologies. The details of the approach taken by Lazard are available in Ref. [24], but in subsequent chapters of the book, you will learn how to make such estimates on your own and how these estimates depend on assumptions such as the cost of money or the assumed life of the system. The cost figures provided by an impartial investment company will be useful for preliminary estimates, and particularly for making what are called "back-of-the-envelope" calculations. In Figure 1.15, the capital cost estimates (i.e., the cost of installing a given technology) are presented for conventional and alternative energy systems. Using the assumed

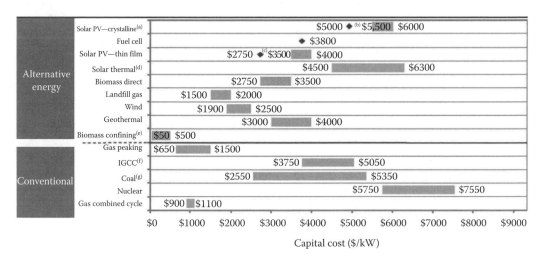

FIGURE 1.15
Capital costs for various electric power generation technologies in 2008. *Notes:* (a) Low end represents single-axis tracking crystalline. High end represents fixed installation. (b) Based on a leading solar-crystalline company's guidance of 2010 total system cost of $5.00 per watt, company guidance for 2012 total system cost is $4.00 per watt. (c) Based on the leading thin-film company's guidance of 2010 total system cost of $2.75 per watt, company guidance for 2012 total system cost is $2.00 per watt. (d) Low end represents solar trough. High end represents solar tower. (e) Represents retrofit cost of coal plant. (f) High end incorporates 90% carbon capture and compression. (g) Based on advanced supercritical pulverized coal. High end incorporates 90% carbon capture and compression. (From Lazard, Ltd., *Energy Technology Assessment*, Lazard, Ltd., New York, 2009, www.lazard.com; Palz, W. and Zibetta, H., *Int. J. Solar Energy*, 10(3–4), 211, 1991. With permission.)

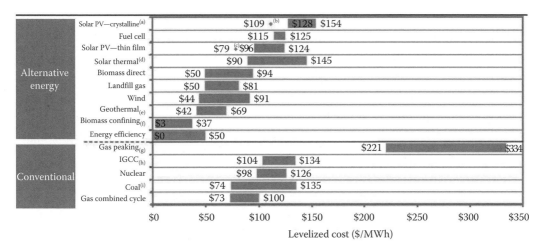

FIGURE 1.16
Levelized cost of energy comparison. *Notes:* Reflects production tax credit, investment tax credit, and accelerated asset depreciation as applicable. Assumes 2008 dollar, 60% debit as 7% import rate, 40% equity at 12% cost, 20-year economic life, 40% tax, and 5–20 year tax life. Assumes coal price of $2.50 per MMBtu and natural gas price of $8.00 per MMBtu. (a) Low end represents single-axis tracking crystalline. High end represents fixed installation. (b) and (c) Represents a leading solar crystalline company's targeted implied levelized cost of energy in 2010 assuming a total system cost of $5.00/watt. Company guidance for 2012 total system cost of $4.00/watt would imply a levelized cost of energy of $90/MWh. (d) Low end represents solar tower. High end represents solar trough. (e) Represents retrofit cost of coal plant. (f) Estimates per National Action Plan for Energy Efficiency; actual costs for various initiatives vary widely. (g) High end incorporates 90% carbon capture and compression. (h) Does not reflect potential economic impact of federal loan guarantees or other capture subsides. (i) Based on advanced supercritical pulverized coal. High end incorporates 90% carbon capture and compression. (From Lazard, Ltd., *Energy Technology Assessment*, Lazard, Ltd., New York, 2009, www.lazard.com; Palz, W. and Zibetta, H., *Int. J Solar Energy*, 10(3–4), 211, 1991.)

capital costs in Figure 1.15, Lazard estimated the levelized cost of energy (LCOE) from conventional and alternative energy technologies, shown in Figure 1.16. These estimates do not include social and environmental costs, and as a result, there are large uncertainties in what the real costs of energy in the future will be.

It is important to note that these estimates are based on current generation technologies for the conventional systems, which are considered mature. It should also be noted that the conventional systems, such as coal, nuclear, and gas, can operate continuously and can be located in many different localities, while most alternative energy systems operate only intermittently, may require extensive storage, and some do not have a proven track record. Moreover, their performance is limited to certain geographic locations, such as those with good wind or geothermal resources.

1.4.4 Cost of Time and Uncertainty

Integrated gas combined cycle (IGCC) is an emerging technology using coal to generate a syngas, which is then used to power a conventional gas turbine. In Figure 1.16, the high-cost figures are for IGCC and for coal with 90% carbon capture and disposal. However, these numbers are estimates. The technologies for carbon capture systems (CCS) have not been commercially demonstrated, and the general consensus is that CCS will not be available until 2025. Thomas D. Shope, the acting assistant secretary for fossil energy of the U.S. DOE, testified on April 16, 2007, before the Senate Energy and Natural Resources Committee, stating that CCS would be available in 2020–2025, but that "common, everyday employ-ment" will not be seen until 2045 [25]. These projections assume success in several demon-stration projects that have been proposed for 2010 and beyond.

The high-end capital cost for solar thermal is for a power tower, which has not been fully commercialized and may be too high. The cost estimates for nuclear power do not include a facility for disposal of radioactive waste and the safe disposal of the plant after decommis-sioning. It should be noted that the levelized cost of electric power from a fuel cell assumes that hydrogen is generated from natural gas, which is not a renewable resource. The level-ized cost is approximately three times higher if the hydrogen was produced by electrolysis from water and the required electricity would come from solar, fossil, or nuclear sources.

There are other costs associated with power plants, such as insurance, potential liabil-ity, safety, and security. Figure 1.17 illustrates a factor that incorporates these issues and is important to investors, the development time. The time to design and install solar PV systems is very short, less than 6 months, if the system is for summer demand peak reduc-tion and placement is on the roof of a commercial building that can use the power directly, meaning no feed-in to the grid and no battery storage system. If the resource is already characterized and available, then installation of landfill gas generators is also short with relatively easy consent processes because of the previous industrial use of the site and proximity to grid tie-in. Wind farms can be constructed quickly, but often have issues with acceptance of the site and intermittency, and location affects the market price of the power and can increase costs and time for grid tie-in.

It should be noted that one of the biggest problems with ocean energy conversion and nuclear power is insurance and liability. While the time-to-build estimates for IGCC and

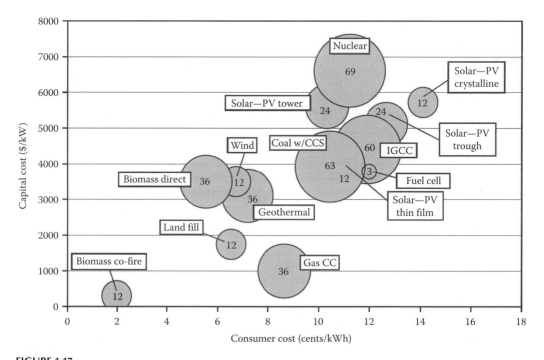

FIGURE 1.17
Consumer cost, capital cost, and construction times for various supply-side alternatives (circle size indicates construction time in months). (From Lazard, Ltd., *Energy Technology Assessment*, Lazard, Ltd., New York, pp. 11–13, 2009, www.lazard.com. With permission.)

coal with CCS are large according to Lazard, they also caution that no commercial facilities have been constructed, and there are no suppliers for the equipment, so the project development time may actually be prohibitively large for many years to come. Finally, the energy engineer will note that fuel cells are not energy conversion systems per se if they are conceived as converting hydrogen to electricity when the energy used to make the hydrogen was already produced and could have been sold in the market.

1.4.5 Other Costs of Energy Development

Land has a value and a potential productivity for producing food, fibers, and wood. Fossil fuel sources have a great energy density, while all of the renewable sources are very diffuse and therefore require large areas to deliver power needed in an industrial society. Table 1.3 shows the power generation per unit land or water area for various renewable energy technologies.

Example 1.3: Back-of-the-Envelope Calculations

Recently, the Arizona legislature passed a law requiring that by the year 2025, 15% of all the electric energy generation be from renewable sources. Calculate the collector area required and estimate land area for this target to be met by solar PV.

Solution
From the Arizona Energy Commission website, we find that in 2009, electricity demand was 30,800 GWh/year and growth to 50,000 GWh/year in 2025 is expected. From the NREL U.S. Solar Radiation Resource Map, annual average the solar radiation impinging on a flat plate facing south and inclined at the latitude is 5.5 kWh/m²/day, which is equivalent to about 2×10^3 kWh/m²/year. Assuming a PV efficiency of 12%, the amount of electric power delivered per year is 240 kWh/m². Hence, the collector area is

$$A = \frac{50{,}000\,\dfrac{\text{GWh}}{\text{year}} \times 0.15 \times 10^6\,\dfrac{\text{kWh}}{\text{GWh}}}{240\,\dfrac{\text{kWh}}{\text{year} \times \text{m}^2}} = \frac{7.5 \times 10^9\,\text{kWh/year}}{240\,\dfrac{\text{kWh}}{\text{year} \times \text{m}^2}} = 3.1 \times 10^7\,\text{m}^2$$

TABLE 1.3

Land or Water Area and Efficiency (η) for Different Renewable Electric Power Generation Systems

Power per Unit Land or Water Area	W/m²	Availability	η (%)
Wind	2	Intermittent	24–32
Offshore wind	3	Intermittent	24–32
Tidal barge	3	Periodic	20–40
Solar PV panels	5–20	Intermittent, periodic	10–20
Bioplantation	0.5	Seasonal	1 harvest/year
Hydroelectric	11	On demand within limits	80–90
Ocean thermal (tropics)	5	Continuous	85
Concentrating solar power (desert)	15	Periodic, intermittent	20–30

The rule of thumb for PV installation to avoid shading is spacing so the total area is twice the collector area, giving a land requirement of 6.2×10^7 m². The land area of Arizona is 29.5×10^{10} m², hence the required PV power plant area would be 0.021% of the state of Arizona. Using Figure 1.16, we can estimate that the cost of such a system would be about $40 billion today.

Discussion: Are there any sustainability issues with the proposed solution? One of Arizona's biggest electricity loads is summer air-conditioning. Is this resource a good match with end-user demand? Studies have shown that shading of windows and walls in Arizona can reduce cooling loads by 15%–30%. If the cost of shading louvers, awnings, and screens is about $2/kWh saved over the year, then discuss the cost of reducing energy demand.

Example 1.4: Back-of-the-Envelope Policy Evaluation

On July 17, 2008, former Vice President Al Gore proposed that by 2018 all U.S. electric power generation be carbon-free (transcript at http://www.npr.org/templates/story/story.php?storyId=92638501). Senator John McCain has suggested that nuclear power plants be installed in order to reduce the effects of global warming. In the summer of 1979 at the height of the energy crisis, President Jimmy Carter suggested that Americans should conserve energy. Assume that conservation could reduce fossil generation demand by 25%. Estimate how many 2.5 MW wind turbines (average yearly capacity factor of 30%) would meet Mr. Gore's suggestion. Then estimate the total costs and land area of these wind turbines assuming that wind turbines cost about $1100/kW$_e$ installed. Then estimate how many 1000 MW nuclear power plants with a 90% capacity factor would be required and what would be the capital cost if nuclear power installation cost is $7000/kW$_e$.

Solution

U.S. Projected Electricity Generation 2008 from the EIA—*Annual Energy Outlook* gives the following:

Fossil-based electricity generation in year 2020: 31.53 quads (31.53×10^{15} Btu)
Total electricity generation in year 2020: 45.18 quads (45.18×10^{15} Btu)
Assuming conservation can reduce the 2020 electricity needs by 8 quads, we are left with 23.53 quads of electric energy that must be produced by renewable sources.

Wind power: Given a 30% annual capacity factor, the 2.5 MW wind turbine would be expected to generate

$$(0.30) \times (2.5 \text{ MW}) \times \left(8760 \frac{\text{h}}{\text{year}}\right) \times \left(3.4144 \times 10^{-9} \frac{\text{quads}}{\text{MWh}}\right) = 2.24 \times 10^{-5} \frac{\text{quads}}{\text{year}}$$

requiring over 1 billion wind turbines! We can now determine the total land area required for these wind farms with about 50 ac of land area needed per turbine:

$$\left(50 \frac{\text{ac}}{\text{turbine}}\right) \times 1 \times 10^6 \text{ turbines} = 50 \text{ million acres}$$

At $1100 per kW for wind turbines, the capital needed for the installation of these farms is

$$\left(\frac{\$1100}{\text{kW}}\right) \times \left(\frac{1000 \text{ kW}}{\text{MW}}\right) \times \left(\frac{2.5 \text{ MW}}{\text{turbine}}\right) \times (1 \times 10^6 \text{ turbines}) = \$2.75 \text{ trillion}$$

Nuclear power: The yearly electric energy produced by one 1000 MW nuclear power plant operating at 90% capacity is

$$(0.90) \times (1000 \text{ MW}) \times \left(8760 \frac{\text{h}}{\text{year}}\right) \times \left(3.4144 \times 10^{-9} \frac{\text{quads}}{\text{MWh}}\right) = 0.0269 \frac{\text{quads}}{\text{year}}$$

This would require 875 power plants, and at $7000 per kW for nuclear plants, we determine the capital investment for the construction of these plants:

$$\left(\frac{\$7000}{\text{kW}}\right) \times \left(\frac{1000 \text{ kW}}{\text{MW}}\right) \times \left(\frac{1000 \text{ MW}}{\text{plant}}\right) \times (875 \text{ plants}) = \$6.13 \text{ trillion}$$

1.5 Energy Efficiency and Conservation

This section describes how reducing energy consumption through efficiency and conservation provides the most important and most economically strategic projects for transition to a sustainable energy system. The present per capita energy consumption in the United States is 281 GJ, which is equivalent to about 9 kW per person. The average consumption in Europe is 4.2 kW and the average for the whole world is only 2 kW. The Board of the Swiss Federal Institutes of Technology has developed a vision of a 2 kW per capita European society by the middle of the twenty-first century [6]. The vision is technically feasible, will require a combination of increased R&D on energy efficiency and policies that encourage conservation and use of high-efficiency systems. It will also require structural and behavior changes in transportation systems.

According to the 2004 World Energy Assessment by UNDP, a reduction of 25%–35% in primary energy in the industrialized countries is achievable cost-effectively in the next 20 years, without sacrificing the level of energy services. The report also concluded that similar reductions of up to 40% are cost-effectively achievable in the transitional economies and more than 45% in developing economies. As a combined result of efficiency improvements and structural changes such as increased recycling, substitution of energy-intensive materials, etc., energy intensity could decline at a rate of 2.5% per year over the next 20 years [6].

The estimated economic energy efficiency potentials in North America up to the year 2010 are shown in Table 1.4. The greatest energy savings potential is in the transportation industry, followed by residential heating and cooling. A similar estimate for the economic energy efficiency potential for Western Europe for the years 2010 and 2020 is presented in Table 1.5, where the resource references refer to the bibliography in [11]. Similar estimates for the energy-saving potential in Japan, Asia, and Latin America are presented in [11].

1.5.1 Energy End-Use Demand Reduction in Buildings

Buildings account for 35%–40% of all end-use energy demand in the United States. Buildings use 65% of the electric power and are responsible for 30%–40% of the nation's greenhouse gas emissions. Residential buildings account for about half of the building energy use. Over half of all energy used in residences is for heating and cooling and a further 20% for heating water. Residential air-conditioning is a main contributor to the peak

TABLE 1.4

Economic Energy Efficiency Potentials in North America
Projected from 2000 to 2010

Sector and Area	Economic Potential Demand Reduction (%)	
	United States	Canada
Industry		
Iron and steel	4–8	29
Aluminum (primary)	2–4	
Cement	4–8	
Glass production	4–8	
Refineries	4–8	23
Bulk chemicals	4–8	18
Pulp and paper	10–18	9
Lighting manufacturing		
Mining		7
Industrial minerals		9
Residential		
Lighting	53	
Space heating	11–25	
Space cooling	16	
Water heating	28–29	
Appliances	10–33	13
Overall		
Commercial and public buildings		
Space heating	48	
Space cooling	48	
Lighting	25	
Water heating	10–20	
Refrigeration	31	
Miscellaneous	10–33	9
Overall		
Transportation		
Passenger cars	11–17	
Freight trucks	8–9	
Railways	16–25	
Aeroplanes	6–11	
Overall	10–14	3

Source: Jochem, E., Energy end-use efficiency, in *World Energy Assessment, Energy and the challenge of sustainability*, UNDP, New York, 2000, pp. 173–218, Chapter 6. http://www.undp.org/content/dam/undp/library/Environment%20and%20Energy/Sustainable%20Energy/wea%202000/chapter6.pdf

demand on the power grid. Peak power is the most expensive to provide as the entire grid must be able to meet the peak load, but the peaking power plants only run, and thus only earn revenues, during the peak periods resulting in very low utilization factors for investors. Thus, economic residential energy efficiency and conservation programs focus on reducing space conditioning and water usage. Figure 1.18 shows the relative effectiveness

TABLE 1.5

Economic Energy Efficiency Potentials in Western Europe
Projected from 2000 to 2010 and 2020

Sector and Area	Economic Potential Demand Reduction (%)	
	2010	2020
Industry		
Iron and steel	9–15	13–20
Construction materials	5–10	8–15
Glass production	10–15	15–25
Refineries	5–8	7–10
Basic organic chemicals	5–10	
Pulp and paper		50
Investment and consumer goods	10–20	15–25
Food	10–15	
Cogeneration in industry		10–20
Residential		
Existing buildings		
Boilers and burners	15–20	20–25
Building envelopes	8–12	10–20
New buildings		20–30
Electric appliances	20–30	35–45
Commercial, public, agriculture		
Commercial buildings	10–20	30
Electricity	10–25	20–37
Heat		15–25
Public buildings		30–40
Agriculture and forestry		15–20
Horticulture		20–30
Decentralized cogeneration		20–30
Office equipment		40–50
Transportation		
Passenger cars	25	
Railways		20
Aircraft, logistics	15–20	25–30

Source: Jochem, E., Energy end-use efficiency, in *World Energy Assessment, Energy and the challenge of sustainability*, UNDP, New York, 2000, pp. 173–218, Chapter 6. http://www.undp.org/content/dam/undp/library/Environment%20and%20Energy/Sustainable%20Energy/wea%202000/chapter6.pdf

of a range of residential energy retrofit options. Energy design and construction standards were greatly increased after the energy crisis in the late 1970s. Homes built before this time can benefit the most from major remodeling work to bring windows, doors, walls and the ceiling insulation, and weather tightness up to standard. Many older homes are remodeled for stylistic reasons. Incentives to include the energy audit and energy retrofit plan in remodeling projects would be cost-effective policy for both the homeowners and the utilities. With over 130 million residential units in the United States, a nationally coordinated

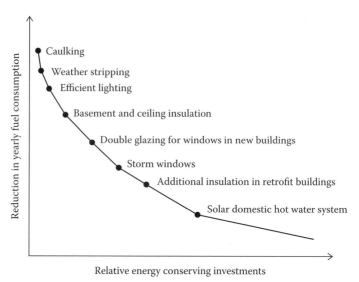

FIGURE 1.18
Relative energy savings of various conservation measures.

effort, together with the growth of the energy retrofit industry, could deliver demand reduction in the range of 2 kW per household—260,000 MW, or more than twice the total current U.S. nuclear generation capacity. Each year, 1.5–2 million new housing units are built in the United States. It is vital that every new home built use passive solar design and the highest energy efficiency standards. Water-efficient fixtures and faucets would also reduce demand for natural gas. The installation of a solar heating, and particularly a solar PV system, would not be economically appropriate until other less expensive and higher return improvements have been completed.

Regulations requiring energy-efficient performance for all appliances are economically beneficial policy. In the United States, national efficiency standards were imposed by Congressional legislation for a variety of residential and commercial appliances since 2004. Updated standards will take effect in the next few years for several more products. Outside the United States, over 30 countries have also adopted minimum energy performance standards for buildings and appliances. These measures have been economically attractive in the long term and can not only provide an appreciable reduction in energy consumption, but also reduce adverse environmental impacts. Consider a refrigerator, which is normally the largest single user of electricity in a home. Better insulation and other design changes could cut the average power consumption of a refrigerator considerably. The replacement of an old refrigerator with a faulty thermostat (a very common occurrence) that uses 250 W that runs 5256 h/year, with a new one that uses 100 W and runs 1752 h/year results in a demand reduction of 525.6 kWhe/year, much of this at peak demand time during hot summer days when power can cost four or five times more to produce.

Commercial buildings are demolished and new ones built on a shorter cycle than residential buildings, with around 1% of buildings turned over each year. Heating, ventilating and air-conditioning, and lighting are major loads in commercial buildings. Professional energy auditing and energy management are required for commercial buildings. Commercial sites like shopping centers, offices, medical centers, and schools also have a transport impact because they become attractors for travel demand. Thus, the location and the context of commercial buildings are part of the overall U.S. Green Building Council's

Leadership in Energy and Environmental Design (LEED) framework. LEED is a building certification that identifies and implements green design, construction, operation, and maintenance measures yet is flexible for each existing or new project. The approach is to consider the building life cycle as well as the surrounding community and environment.

Improving energy efficiency across all sectors of the economy should become a worldwide objective [6]. Engineering and architecture for low-energy buildings are mature fields. However, free market price signals may not always be sufficient to increase energy efficiency. The potential of conservation was recently demonstrated in Japan. A year after the Fukushima disaster, all of Japan's nuclear reactors were temporarily closed down for safety inspection, resulting in a loss of one-third of its electricity generation capacity. The response of Japan to this major loss in electric power has been truly amazing. Japan succeeded in avoiding a drastic economic crisis by means of a radical conservation program that includes turning off air conditioners in the summer and office lighting during the day and many other participatory measures [26].

1.5.2 Energy End-Use Demand Reduction in Transportation

Of every 5 dollars spent in the United States, 1 dollar is estimated to be related to transport: fuel, vehicles, road, and parking infrastructure or related services. The U.S. personal transportation system has many features that make it the ideal target for energy demand reduction. Transportation accounts for nearly 30% of total U.S. energy use, with 93% of that energy from petroleum. Cars and light trucks use 60%, heavy trucks 18%, and medium trucks 4% of the energy used in transport. There are 134.9 million cars and 100.2 million light trucks in the United States. The average fuel economy for the car fleet is 22.5 mpg, and for the light trucks in service is 18.0 mpg. In 2010, 5.6 million cars were sold, which was 49% of sales with light truck sales at 5.9 million. The average annual vehicle miles traveled (VMT) was 11,300 miles for each vehicle [27].

These and other statistics can be used to understand the relative merits and costs of different transitional change projects. For example, the U.S. ethanol fuel program costs more than $4.00/gal and replaces 8% of personal transport fuel. This amount of fuel could be saved by VMT reduction of 2.5 mi per car per week at an average fuel *savings* of $150 for the year. A disincentive tax on light trucks for city driving could drop the sales of sport utility vehicles (SUVs) and trucks for personal use to 20% of sales. This policy would not require investment in any alternative vehicle technology, would collect new revenue from a gas guzzler tax, and in the first year would reduce fuel use by 9040 million gallons, saving $33.9 billion (at $3.75/gal). If lowering the speed limit to 55 mph resulted in a 7% fuel savings, then the savings to the U.S. economy would be in the range of $34 billion, not to mention the savings in accidents and loss of life. The potential carbon emission reduction, fuel demand reduction, and economic savings from modest, low-cost changes are truly astounding. Use of gasoline and diesel in personal vehicles for travel is not a productive use of this energy in the economy; it provides a few health and safety benefits and produces many social costs. The high economic cost is partly because nearly 45% of U.S. petroleum consumption is imported (as at 2011). The inelastic economic dependence on oil caused a wealth transfer out of the United States and GDP losses combined amounting to at least half a trillion dollars in 2008 due to the oil price rise. When the price fell to an average of $60 per barrel, the oil dependence costs fell to the range of $300 billion in 2009 and 2010 [28]. Thus, the rate of return on projects that reduce oil dependence is potentially very large.

The standard approach and techniques of energy auditing and energy management normally used for commercial buildings can be adapted to help companies manage fuel consumption and plan for adaptation to reduce fuel use when fuel price spikes occur. Efficiency engineering can be useful for companies to explore long-term plans and changes toward lower energy operations. However, the amount of oil used for production is very small compared to personal travel and recreational use. In 2010, all U.S. agriculture used 563 trillion Btus of oil, construction used 902, logging 25, road maintenance 3.6, and airport ground equipment used 14.6 trillion Btus. These seem like large numbers until they are compared to personal travel and recreation that used 16,383 trillion Btus. Air travel used 2,148 trillion Btus in 2010 while the nation's railways used 581, and barges and other water transport used 1374. Another form of transportation is pipelines, which used 933 trillion Btus in 2010.

1.5.3 Energy Management in Industry and Manufacturing

Industry and the primary production sector are good candidates for energy management projects. The best way for energy engineers to tap into the energy savings in this sector is to study the best examples of energy management projects that have already been achieved and develop new projects for the vast majority of companies that currently do not have energy management practices. As a flagship example, you can study the example of 3M [29]. The Corporate Energy Management Team sets annual strategic plans and works to provide oversight and guidance at the plant level. In 2006, 3M completed more than 185 projects that delivered $18.2 million in savings. The energy management goals are to control energy costs, improve operational efficiency, and reduce environmental impacts. Engineering teams in the company are rewarded based on meeting targets and goals. The U.S. DOE has a program that can partner with companies to help with energy assessments: the Energy Industrial Technologies Program—Save Energy Now LEADER initiative.

1.6 Energy from Fossil Fuels

This section will examine the current and future outlook for the three primary fossil fuel resources: natural gas, oil, and coal. The production rate of these fuels is, from a historical perspective, truly enormous. The near-term supply security issues with fossil fuels are not so much depletion, for example, running out, as depletion effects, for example, no further growth in supply and tight spare capacity. The long-term sustainability issues with fossil fuels arise more from EROI decline and "gold rush" type decision-making resulting in irreversible environmental degradation. Of course, fossil fuels are finite resources; clearly, there are enormous reserves remaining, and just as obviously, the most profitable resources have been developed first. The conditions facing the energy engineering field in this century are not the same as last century. In the twentieth century, the energy supply could grow, but in the twenty-first century, the net energy supply to the economy has peaked and will continue to decline. Figure 1.19 shows the result of a biophysical system analysis that calculated the total energy produced and the net energy delivered to end users using the current EROI of the different energy resources and assuming that all resources would be developed at the earliest possible time. Because of lower EROI resources beginning to be developed, the net energy yield has begun to diverge from the total energy production. The rush to development of remaining fossil resources in the first half of the century will lead to a rapid decline of net energy to the economy over the second half of the century [30]. The net energy represents the consumer fuels and electricity that are made available in the market. Although the total energy production could increase, a larger amount of

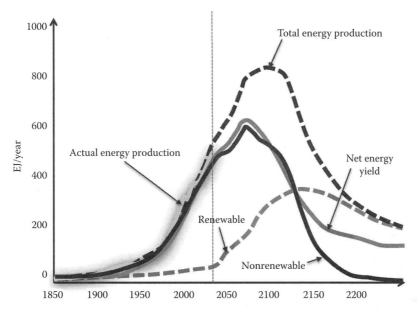

FIGURE 1.19
Net energy analysis assuming aggressive development of all energy resources and using the EROI of the resources shows the transition from fossil fuels to renewable energy results in a decline of net energy to end users over the current century. (From Dale, M. et al., *Ecol. Econ.*, 73, 158, 2012. With permission.)

that energy is used in the energy transformation industries for lower EROI resources. In the market, the net energy yield is what is available to meet demand.

Fossil fuels will always be the highest energy density resource and will always have high utility for industry, transport, and domestic services even at extremely high prices. The energy density in the fuel products are

- Natural gas   1 MBtu/ft³   36.4 MJ/m³   50 MJ/kg
- Petroleum   4.9 MBtu/barrel   5.2 GJ/barrel   50 MJ/kg (*gasoline*)
- Coal   20 MBtu/short ton   1365–2727 GJ/rail car   35 MJ/kg

1.6.1 Emissions from Fossil Fuels

The main focus of this section will be carbon dioxide emissions, but it is important to know that burning fossil fuels has always presented air pollution issues. Soot from coal and diesel and NO_x from petroleum and gas directly cause negative health effects. Combustion science and air pollution engineering have provided great improvements in air quality since the 1970s, but it is important to be aware that acid precipitation from the sulfur dioxide and the deposition of mercury, arsenic, and radioactive minerals from coal combustion continue to be major environmental issues. Mercury is a potent neurotoxin that has been accumulating in biological systems. Today it is estimated that one in six American women of childbearing age have blood mercury levels that exceed the safe exposure limit for fetal development. This means that as many as 600,000 babies are born each year at serious risk of prenatal methylmercury exposure, which has been determined to cause permanent and irreversible impairment with language, memory, attention, visual skills, learning, and reasoning. Every state in the United States has issued health advisories banning fishing and consumption of fish due to toxic mercury exposure. By far, the largest source of mercury pollution in America is coal-fired power plants, but combustion of coal in industry has also caused some very high local mercury contamination problems. Mercury accumulates in the fatty tissues and oils in all animals and is never broken down or dispersed. Thus, release of mercury into the atmosphere from coal combustion is a serious sustainability issue for humans, agriculture, and wild animals, particularly aquatic predator species like tuna and dolphins. Coal-fired power plants in the United States are currently not required to limit their mercury emissions, even though the abatement technology is available. The Environmental Protection Agency (EPA) is working to enforce restrictions on mercury emissions under the Clean Air Act.

The hydrocarbons that are the basis for our energy system were deposited in sediments between 240 and 65 million years ago during a time when the Earth's atmosphere was much higher in CO_2 and the climate was warm enough that there were no ice caps or glaciers. This period of biological sequestration gradually depleted the CO_2 in the atmosphere and contributed to the cooling climate of the planet. Even though this climate change happened gradually over tens of millions of years, the end of the Mesozoic era had already seen the extinction of most of the land and marine dinosaur species by the time of the catastrophic event thought to be caused by a meteorite impact.

The issue of climate disturbance from fossil fuel combustion has taken on highly political and social aspects even though the effects of man-made emissions and land-use changes have long been well understood as demonstrated in Table 1.6. Sustainable energy is normally thought of as increasing the renewable energy production, but the most serious risk that must be managed at the current time is reduction of fossil fuel use and the attendant emissions so that climate systems do not become catastrophically destabilized. The most straightforward way to approach this important work is to focus on the targets set by scientists for climate stability.

The energy engineer should become well acquainted with the sources of greenhouse gases, and there are numerous reports explaining the issue and the risks involved. Table 1.6 gives an abbreviated history of some of the scientific developments alongside the atmospheric CO_2 concentration over the past century. It is clear that the issue of liberating CO_2 sequestered for over 60 million years back into the planetary system within a short space in time has been a subject of concern and study for over a century.

The main human CO_2 additions to the atmosphere are from burning fossil fuels and from manufacture of cement, plus forest clearance and soil disturbance. The ocean is estimated to be adsorbing 2.0 ± 0.8 Gt/year, which is causing acidification effects that are harming coral reefs and other systems. Some forest areas that were previously cleared are being allowed to regenerate and are estimated to be absorbing 0.5 ± 0.5 Gt/year. The imbalance has led to the current CO_2 atmospheric concentration that has not existed for over one million years. The increased energy retention in the earth system has to date resulted in a global average 0.8°C temperature increase. This seemingly small change in average conditions has already produced an increased risk of occurrence of extreme weather events like floods and droughts, heat waves, and severe storms. Climate modeling indicates that the risk of catastrophic climate change, in particular melting of the Greenland ice sheet and the resulting sea level rise, is very high if the global average temperature increases by 2°C. The probability of initiating catastrophic climate change will be at least one in five if 565 more gigatons of CO_2 are added to the atmosphere. Fossil fuel combustion must be reduced by 90% by 2050 in order to have a chance of avoiding run-away climate change. The IEA estimates that 2011 CO_2 emissions were the highest on record at 31.6 Gt, at least a 3% increase from the previous year. Recent estimates for the known economically recoverable reserves of coal, oil, and natural gas would produce 2795 Gt of CO_2. This represents

TABLE 1.6

Abbreviated History of Climate Science and the Atmospheric Concentration of the Greenhouse Gas Most Directly Associated with Energy Use from Fossil Fuels

Date	Parties	Statement	Atmospheric CO_2 (ppm)
1896	Svante Arrhenius, Swedish Scientist	Global warming is possible due to CO_2 released from burning coal in industry	280
1924	A Lotka, Physicist	Calculates that burning coal at current rates will double the atmospheric CO_2 within 500 years	286
1948	Scientific Debate	Earth and oceans are "self-regulating," and natural balance may absorb anthropogenic CO_2 emissions	300
1958	C.D. Keeling	First continuous measurement of CO_2 at Mana Loa, Hawaii	315
1975	Manabe and Wetherald, NOAA	First three-dimensional model of Earth climate including effects of greenhouse gases	330
1985	UN Conference ICSU	Global warming is inevitable due to existing CO_2 accumulation	345
1990	49 Nobel Prize Winners	"There is broad agreement within the scientific community that amplification of the Earth's natural greenhouse effect by the build-up of various gases introduced by human activity has the potential to produce dramatic changes in climate…"	352
1990	IPCC: 170 scientists from 25 countries	60% reduction in CO_2 emissions needed to stop temperature increase	352
1995	IPCC consensus of climate scientists	There is a discernable human influence on global climate (*hottest year on record*)	358
1997	Kyoto Protocol	International agreement to limit emissions to 1990 levels by 2012 (*1998 hottest year on record*)	362
2001	IPCC 3rd assessment report	Most of the warming over the last 50 years is attributable to human activities	370
2001	U.S. National Academy of Sciences	Temperatures are in fact rising due to human activities	371
2001	U.S. Global Change Research Program	Higher probability of extreme weather, drought, flood, storm, and sea-level rise is now occurring	372
2002	Europe, Canada, Japan, New Zealand	Ratify Kyoto agreement, Europe sets up carbon trading, Japan reduces CO_2 emissions	373
2004	Numerous Scientists	Arctic sea ice has thinned by 3.1–1.8 m, and extent of summer ice shelf has shrunk by 5%	378
2012	NOAA	Hottest summer on record and most severe drought over most of the United States	401
2012	Dr. James Hansen, Chief Climate Scientist, NASA	"The target that has been talked about in international negotiations for two degrees of warming is actually a prescription for long-term disaster."	402
…			

the fossil energy supply that producers are planning to bring to the market. This level of CO_2 evolution would leave no uncertainty of catastrophic climate disruption.

The carbon emissions per unit of energy for the three fossil fuel products are

- Natural gas   18.3 MJ/kg CO_2
- Gasoline   16.1 MJ/kg CO_2
- Coal (average)   8.2 MJ/kg CO_2

1.6.2 Natural Gas

Natural gas is one of the most important fossil fuels for industry, heating buildings, and electricity production, particularly during peak hours. Natural gas combustion technology is mature. Natural gas can be burned at stoichiometric air to fuel ratio, which means a high flame combustion gas temperature is possible. Diffusion flames are common as the diffusion rates of O_2 and CH_4 into the flame zone are sufficiently high. High combustion temperature produces more NO_x, but gas combustion normally has no soot and low unburned hydrocarbons. As with all fuels, the heat extracted from the combustion gas is limited by the acid condensation point where condensed water vapor in the presence of high CO_2 gas concentration forms carbonic acid, which presents corrosion issues.

Over the past half-century, natural gas has increased its market share continuously, growing from about 15.6% global energy consumption in 1965, to around 24% in 2012. The natural gas energy flow in 2009 in the United States is shown in Figure 1.20. A total of about 23 trillion ft³ was consumed, with about 4 trillion ft³ coming from imports. To date, only an insignificant amount of natural gas is used for transportation, only about 3% of the total consumption.

Natural gas was flared off as an unwanted by-product of oil production for many years. Like oil, natural gas prices were regulated for many years to keep the price low. However, after the energy crisis of the 1970s, distribution pipelines were extended to Canada and power generation using the clean and plentiful resource became increasingly popular. However, natural gas production and proven reserves fell dramatically from a peak in 1973

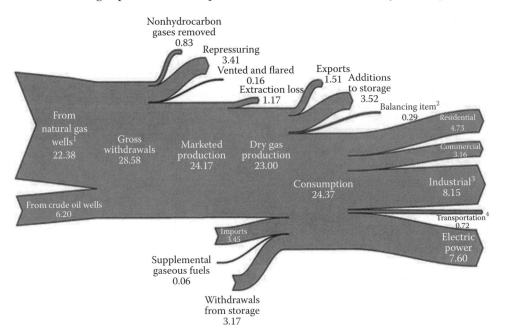

FIGURE 1.20
U.S. natural gas energy flow in trillion cubic feet (tcf) in 2011. (From U.S. EIA Annual Energy Review, 2011, http://www.eia.gov/totalenergy/data/annual/pdf/sec6_3.pdf (accessed April 2013)). (1) Includes natural gas gross withdrawals from coalbed wells and shale gas wells. (2) Quantities lost and imbalances in data due to differences among data sources. (3) Lease and plant fuel, and other industrial. (4) Natural gas consumed in the operation of pipelines (primarily in compressors), and as fuel in the delivery of natural gas to consumers; plus a small quantity used as vehicle fuel. *Notes:* Data are preliminary. Values are derived from source data prior to rounding for publication. Totals may not equal som of components due to independent rounding. *Sources:* (Tables 6.1, and 6.2 and 6.5.)

and remained flat until about 2003 when shortages caused high prices that in turn spurred building of liquefied natural gas (LNG) import terminals, deregulation, and a rush to develop shale gas. The United States has an extensive pipeline distribution network, but international trade in natural gas is much more expensive if it must be compressed to LNG and then transported by ship. The resource estimate in 2006 was 1530 tcf, inclusive of about 140 tcf of shale gas, giving almost 100 years' worth of production at the 2006 consumption rate. With extended exploration and aggressive development, EIA's preliminary estimate of natural gas resources in 2010 was 2550 tcf, including 827 tcf of natural gas from shale.

Natural gas production from deep and tight shale formations is carried out by hydraulic fracturing with horizontal drilling, called fracking by the industry. Experts have known for a long time that natural gas and oil deposits exist in deep shale formations, but until recently, these resources were not thought to be economically recoverable. Fracking is the process of creating fissures in underground formations to allow natural gas to flow more easily. A fluid is pumped under high pressure into the shale formation. The fluid used is approximately 99% water and sand, along with a small amount of proprietary chemicals. Concerns have been raised about the potential deleterious environmental effects of using these chemicals, disposal of the contaminated water, and leaking of methane and chemicals into groundwater. MIT has conducted, in 2011, an interdisciplinary study of the future of natural gas and claims that environmental impacts of shale gas development by fracking "are challenging but manageable" [31].

One of the most extensive studies of the EROI and net energy output of natural gas was conducted in 2010 by the Canadian National Energy Board for the Western Canadian natural gas wells drilled in the period between 1993 and 2009. The EROI was based on the ratio of the energy produced during the entire lifetime of the gas wells to the energy expended in exploration, drilling, and operation of the wells. The results are quite remarkable. The EROI in the early years of exploration was on the order of 42, but decreased continuously until it reached 13 in 2009. The EROI of the combined oil and gas industry fell during that same period from a high of 79 to a low of 15. There are no reliable data of the EROI for natural gas production by fracking. From preliminary estimates, it is clear, however, that the EROI of fracking will be considerably below that of natural gas production for three reasons:

1. The amount of effort involved in drilling the well and encasing the pipes leading to the fractured formation, as well as any efforts to separate the polluted oil shale from groundwater, will involve considerably more energy than simply drilling an ordinary gas well.

2. According to experts' estimates, the time during which a fracked well will operate and the total amount of natural gas extracted during its lifetime are likely to be less than from an ordinary natural gas well and the water for injection may have to be recycled.

3. More complex wells must be drilled than for conventional gas reservoirs to produce the same quantity of gas, and water, waste chemicals, as well as some of the produced gas must be trucked to and from the site rather than transported by consolidated pipeline.

For these reasons, it would not be surprising if the EROI of fracking technology would be considerably less. Consequently, caution should be exercised in expecting the amount of gas accessible by fracking to be of the same extraction EROI as that for natural gas from ordinary drilling.

Although the estimates of total available resources are enormous and could exceed a 100 year supply at 2007 levels of consumption, the picture changes drastically if a 2% of yearly growth in demand occurs. Because natural gas has the lowest carbon content of all fossil fuels, there is a push to increase power generation from gas and to look at

TABLE 1.7

Projected Lifetime of Natural Gas with and without Its Use for Transportation
as Well as with and without Nontraditional Sources

			Projected U.S. Natural Gas Resource Lifetime			
Source	Estimated Reserves	Trillion ft^3	Scenario A Years	Scenario B Years	Scenario C Years	Scenario D Years
EIA[a]	With fracking	2119	88	45	51	32
PGC[b]	With fracking	2170	90	46	52	33
NPC[c]	Without fracking	1451	60	31	39	24

Notes:

Nontraditional source refers to natural gas obtained with hydraulic fracturing of oil shale or other
   tight formations (fracking).

A—At current rate of natural gas consumption, no future growth.

B—At current rates of natural gas plus light vehicle energy consumption, no future growth.

C—With 2% growth in annual natural gas consumption rate.

D—With 2% growth in annual natural gas plus 1.9% growth in light vehicle energy consumption rates.

[a] U.S. Energy Information Administration [73].

[b] Potential Gas Committee [74].

[c] National Petroleum Council [75].

transportation uses. The life time of the natural gas reserve under those two scenarios for
estimates that include the gas from oil shale is shown in Table 1.7. These results indicate
that the natural gas availability will decrease substantially below the optimistic estimates
of 90 years, and its cost is bound to increase in the long term.

1.6.3 Petroleum

Petroleum refers to crude oil and natural gas liquids. Petroleum is refined by high-temperature catalytic reactions and separations into a wide range of liquid fuels and tar or
bitumen. Combustion of liquid fuels requires atomization into a hot combustion chamber, mixing with air, and sufficient residence time to vaporize the liquid and completely
burn all of the different chemical components in the fuel. Light fuels like kerosene are
used in jet engines and cook stoves. Petroleum and diesel fuel can produce unburned
volatile organic compounds, soot, and CO if engines are slightly out of tune. Heavy oil,
used engine oil, and diesel are used in boilers. The acid dew point and natural draught of
combustion products out of flue limit the heat extraction.

Petroleum reserves have been formed by accumulation of biological sediments in costal
waters where low oxygen content slowed decay. These areas with hydrocarbon sediments were
then buried under further sedimentation and heated by geothermal energy causing chemical
breakdown of the organic matter. If the matrix rock was porous enough, the lighter organic
material could become mobile and, due to its lower density, could migrate upward. Where
nonporous sediment layers were folded to form anticlines or where faults caused caprock
traps, the organic material could accumulate into an oil field. If there was no caprock structure,
then the oil, gas, or CO_2 would percolate up through the soils and evolve on the surface.

The productivity of the oil field depends on how porous the resource rock is and how
light the hydrocarbons are. Hydrocarbons that have been subjected to the highest temperature for the longest time may be decomposed to CO_2, and the lightest energy deposits of natural gas and oil normally have some fraction of CO_2 in the resource that must
be separated in production. A "sweet" gas or petroleum resource refers to a low CO_2

content. The lowest grade of petroleum resource is sediment hydrocarbons in nonporous source rocks such as shale, or bituminous hydrocarbons such as tar sands, which are called "heavy" because they have not been thermally broken down. The "low-hanging fruit" of petroleum resources is the light, sweet crude oil found in highly porous source rock under massive anticline features near the land surface in the Middle East. When the caprock of a high-quality oil field is first breached by drilling, the oil and gas are at very high pressure and can be produced without pumping or injection of natural gas or CO_2 to enhance recovery. Later in the production cycle, all oil wells lose pressure, start to produce water mixed with the oil, and require enhanced recovery to continue production.

Proven reserves are classified as a resource that has produced hydrocarbons from test wells with geological mapping of source rock indicating the recoverable fraction of hydrocarbons in the rock matrix. Approximately 40% of the proven petroleum resource has been extracted and used. The remaining conventional petroleum resource is estimated at 1–1.6 trillion barrels, with more than 70% in OPEC member countries. Burning 1 barrel of oil produces 0.43 ton of CO_2, not counting the emissions from production.

Oil spills from wells, pipelines, and tankers are quite common. Sustainability energy engineers should be well aware of the risks of oil handling and insist on the highest standard of operations and safety. While hundreds of millions of gallons of oil have been spilt in innumerable locations, there are two catastrophic oil spills to be familiar with. On March 24, 1989, the oil tanker *Exxon Valdez* struck a reef in Prince William Sound in Alaska. In all, more than 11 million gallons of oil were dumped into the pristine environment impacting over 1300 mi of coastline in Alaska. Scientists are still studying the long-term impacts of the disaster on fish, marine mammals, and birds. There are still pockets of crude oil in some locations and damage to the ecosystem continues to be observed. The Exxon disaster was caused by human error of the tanker captain, but a push for double-hulled tankers was instigated after the disaster to limit the risk of future catastrophic spills. The second incident was the *Deepwater Horizon* oil spill from a BP oil platform that flooded one of the most productive fishing regions with at least 170 million gallons of crude oil over the course of nearly 5 months. On April 20, 2010, the $560 M oil rig exploded and sank, killing 11 workers. The cause of the BP disaster was the result not of one human error, but rather of systemic negligence of engineering practice and operations in the oil industry. The scale of the disaster is still unknown.

1.6.4 Coal

Coal is by far the most abundant fossil fuel. Coal is crushed before use as a fuel, the most common fuel being "pea coal" with average particle size of ¼ in. Coal must be burned with excess combustion air in a high-temperature combustion chamber with a long fuel residence time. All coal has a mineral content that will remain in the combustion chamber as slag and also be carried out in the combustion gas stream in small particles as fly ash. Fly ash is spheres of silica and other minerals, but also has deposition on the particles of other compounds liberated during the combustion. The fly ash can be collected in fabric filters called a bag house and by electrostatic collection onto charged plates called electrostatic precipitation. Both slag and fly ash pose storage and disposal issues. Some fly ash can be used in concrete. Incomplete combustion of coal causes serious air pollution.

Coal was formed largely on land in areas of swamps and bogs where vegetation decay was slower than the sedimentation rate. Coal is found in nearly every country. High energy content coal was lithified by being buried by other sediments and exposed to elevated geothermal temperature and pressure. The conditions for coal formation changed over tens of millions of years so coal is bedded in other geologic sediment layers. Thick coal seams less than 200 ft below the surface are normally extracted in open cast mines. The coal is blasted and fractured in strips, thus the term strip-mining. In the United States, more than 65% of coal is surface mined. Underground coal mines have been developed up to 1500 ft deep, but some mines in China have developed to depths of 3900 ft. Spoil refers to the overburden, top soil, and sedimentary layers between coal seams that must be placed and stored. The overburden rate refers to the depth of overburden compared to the coal seam depth. Typical overburden rates exceed 10:1 (10 ft of soil for 1 ft of coal). The spoil materials are acidic, and the spoil is placed in engineered land formations in order to manage acid mine drainage. The blasting and moving of coal produce dust that is blown and deposited in surrounding areas. Sediment run-off of coal fines is a serious water pollution issue.

Coal has different grades depending on the ash content and the carbon content. The highest grade of coal is coking coal that is not used for power generation but rather is used in making steel. Hard coal, also called anthracite, has the highest energy content (35 MJ/kg), brown coal, also called lignite, can have an energy content not much better than wood (28–18 MJ/kg). The depth of the seam, the amount of overburden, and the quality of the coal limit marketable resources of coal. The global coal resource has about four times the energy content as the world oil resource. The United States has the largest resource of coal, but China is currently producing three times as much coal as the United States. While the amount of coal is vast, coal is a resource that can be depleted locally as already seen in areas of Europe. Coal is also subject to the low-hanging fruit effect where the most economical resources have been developed first.

1.6.5 Carbon Capture and Sequestration

The first point to make in considering CCS is that the hydrocarbons used as fuels have been sequestered for tens of millions of years, posing little risk of climate disruption. People concerned about climate change have been pointing out that leaving fossil fuels in the ground is the most fail-safe approach for carbon emission reduction [32]. It is possible that at least half of the "low-hanging fruit" of relatively easy to access and economical to produce fossil fuels has been consumed over the past 70 years. Considering that all industry, construction, primary production, and manufacturing, depends on fossil fuels, a new dialogue in sustainability is emerging around planning to reduce current energy demand in order to ensure economical energy supplies for the future.

The second point is that forests sequester carbon. Less than 5% of America's native forests remain. About half of the original forest area has regrown or been replanted as plantation; the remainder has been converted to agriculture or has become desert or scrub. Thus, it is not correct to say that we can "offset" fossil carbon emissions by planting trees. Trees planted and forests allowed to regenerate are actually restoring the carbon released when the forests were originally disturbed. Currently, the world continues to lose primary forest at a rate of 0.2%–0.14% to logging and development with the highest rates in Africa and Latin America. Demand for wood in OECD countries is one of the biggest driving forces behind the rush to log remaining forests in Tasmania, the Congo, the Amazon, and other forests critical to climate and hydrology systems.

The IPCC has identified geologic sequestration of CO_2 from stationary power plants as necessary to reduce carbon emissions enough to avoid catastrophic climate change. Geologic sequestration essentially means putting CO_2 gas into geologic structures with porous reservoir rocks and nonporous caprock structures that will trap and permanently hold the gas. The types of formations with these characteristics that we know about are depleted oil and gas reservoirs. In many enhanced oil recovery processes, CO_2 and methane produced with the oil are compressed and re-injected into the oil field to keep the reservoir pressure up and keep the oil production going. However, there have been a limited number of studies of the long-term storage in such formations. Another problem with geologic sequestration is that the locations of the possible depleted oil and gas fields are not often near coal-fired power plants. Saline aquifers have been proposed for geologic storage as these formations are more widespread and not considered to be of use as ground water. The technology to compress CO_2 and transport it via pipelines is known, but the economics are challenging. Other large pipeline projects have been constructed to transport valuable energy products, but not waste products. Also the pumps that keep natural gas pipelines pressurized normally use some gas from the pipeline as part of the production investment.

The one project most often cited as a promising demonstration for CCS is the Sleipner gas platform in the North Sea. The natural gas produced from this field has a high CO_2 fraction. In order to be compressed for shipment to market, the CO_2 must be removed using the monoethanolamine (MEA) temperature swing absorption–desorption separation process widely used in industry for both natural gas sweetening and production of CO_2 from diesel combustion for the soft drink industry. The Norwegian government has instituted a relatively high carbon emission tax. Thus it was economically reasonable for the CO_2 to be compressed and injected into an exhausted gas well. The lesson learned from Sleipner is that if the energy product cannot be brought to the market without dealing with the CO_2, then the CO_2 will be dealt with. Some groups are proposing an international treaty that bans any new power plant that does not capture and sequester the CO_2 emissions.

Other ideas about where to put CO_2 include deep on the seabed where it might freeze, and reacted with lime minerals. Calcite and magnesium oxide, for example, will react with CO_2 to form calcium carbonate ($CaCO_3$) and magnesium carbonate ($MgCO_3$). It would require about 1 ton of calcite to be mined, crushed, and heated to a high temperature (above 900°C) in order to react with 1 ton of CO_2. The reaction is exothermic, but in order to be efficient would need to be accomplished with separated CO_2 and not the whole combustion waste stream. Separation of the CO_2 from a coal combustion product stream has never been attempted on the scale of a typical power plant.

The probability that CCS will reduce emissions from any existing coal power plants is vanishingly low. Given that existing coal power plants currently generate more than 8 Gt-CO_2 per year, even if CCS technology were to be developed over the next 5–10 years, it would not make any contribution to reducing emissions from coal, only from new plants built after that time. Capturing the CO_2 from the combustion product gas stream of existing coal-fired power plants has been recognized as uneconomical and technically unfeasible [33]. CO_2 separation technologies are simply not viable due to the excess air required for coal, the ash content, and the sheer volume of the gas streams involved. No commercial initiative to develop CCS for existing coal power plants is currently underway. The main emphasis for research in CCS is modeling of precombustion CO_2 separation from the integrated gasification combined cycle (IGCC) power plant design. One issue is that there are only a few IGCC power plants, and none of the plants planned in China are IGCC. There have been only a few CCS trials involving a small portion of the gas stream of an IGCC plant. The basic thermodynamics of CCS carry a high energy penalty. First the stack gas stream must be cooled, and then pumped through a chemical reaction column where a chemical solution, typically MEA, would absorb CO_2. Then heat energy would be required to thermally desorb the CO_2, and more cooling and compression would be needed to at least 2 MPa for transport. The energy penalty is defined as the difference between the power output of the plant with and without CCS compared to the power output without CCS. Basic thermodynamics for the processes involved in CCS assuming ideal conditions in an IGCC plant indicate that the energy penalty is between 27% and 43% depending on assumptions about plant efficiency.

Example 1.5: CCS Energy Penalty

A new IGCC coal-fired power plant is being explored. Without CCS, the power plant design would burn lignite coal with an energy content of 10.5 MJ/kg-CO_2 with a plant efficiency of 40%. The plant would generate 1000 MW using brown coal (28.5 MJ/kg). The CCS system would divert steam from the power plant boiler to regenerate the MEA solution (0.5 MJ/kg-CO_2), and electricity to run pumps and fans (0.47 MJ/kg-CO_2), a compressor to 2 MPa (0.3 MJ/kg-CO_2) and cooling (0.33 MJ/kg-CO_2). What is the energy penalty for the CCS system? How much more coal would have to be mined and burned for the plant with CCS to provide the same power as the plant without CCS?

Solution
The power plant produces 0.4 × 10.5 MJ/kg-CO_2 = 4.2 MJ/kg-CO_2 without CCS.
   With CCS, the power plant produces 4.2 − (0.5 + 0.47 + 0.3 + 0.33) = 2.6 MJ/kg-CO_2. Therefore, the energy penalty is

$$EP = 100\left(\frac{4.2 - 2.6}{4.2}\right) = 38\%$$

The plant without CCS is burning coal at a rate of

$$(1{,}000 \text{ MW}/0.4) \times 3{,}600 \text{ MJ/MWh}/28.5 \text{ MJ/kg}$$

$$= 315{,}789 \text{ kg/h or } 316 \text{ kg coal/MWh electricity}$$

With an energy penalty of 38%, the plant with CCS would need to have a generation capacity of 1380 MW in order to deliver 1000 MW to the grid. Thus the coal consumption rate with CCS would be

$$(1380 \text{ MW}/0.4) \times 3600 \text{ MJ/MWh}/28.5 \text{ MJ/kg}$$

$$= 435{,}789 \text{ kg/h or } 436 \text{ kg coal/MWh electricity}$$

Thus, the plant with CCS burns 120 more kg coal/MWh electricity.

1.6.6 Industrial Waste Heat

Coal and other fossil fuels are essential fuels for manufacturing and processing plants. Much of the fuel is burned to produce high temperatures for material processing. Nearly all of our materials require thermal processing, particularly metals, glass, petroleum, cement, fertilizer, methanol, and most processed foods. Once the process is completed, the product needs to be cooled. The industrial process to remove this heat and exhaust it to the environment produces a thermal resource larger than all renewable energy resources combined. The U.S. EPA has estimated the potential resource at 420 trillion Btus. Rather than exhausting industrial waste heat to cooling water or the atmosphere, the heat is put through the vaporizer heat exchanger of an ORC power generation plant. This technology is already used in geothermal power generation. While the thermal science and the plant technology are mature, prospecting and developing these waste heat resources will require further research and development and more experienced thermal systems engineers. This is because each waste heat stream is part of an existing production plant, and the particular details of the extraction and power plant design and construction must be determined for those site conditions. Development of industrial waste heat does not require new land or new water resources because these are already being used by the industry. Most of these industries have sufficient on-site power demand that the power generated can be directly used and does not need to be tied into the grid. Waste heat power plants would not need new environmental assessments and would probably have no public opposition, as they would be built on existing industrial sites. Due to the relatively low temperatures involved, the plant efficiencies are not as high as combined cycle gas-fired power plants. However, the power is produced at the site where it can be consumed, and it does not need new transport of fuels or wastes. Developing the nation's vast resource of industrial waste heat will require more advanced engineering, but it will not require any new materials or unknown technologies. Hence, it could become a major power resource in the very near future.

1.7 Nuclear Energy

The majority of nuclear power plants are located in the United States (104), France (58), Japan (50), and Russia (33). Table 1.8 shows the number of plants, their net electric output in MW, and the number of plants under construction at this time in all the countries worldwide. The EPR™ plant, built on the Finnish island of Olkiluoto, was intended to become the showpiece of a nuclear renaissance. This plant was built by a consortium formed by the French company of AREVA MP and the German company Siemens AG (investments of 73% and 27%, respectively), led by the French company that had successfully designed and built many of the French nuclear power units [34,35]. It has the most powerful reactor, with a net electric output of 1600 MWe and a reactor thermal output of 4300 MW. Its modular design was intended to make it faster and cheaper to build and safer to run. But during 4 years of construction, hundreds of defects and deficiencies delayed the completion of the power plant and the cost estimates climbed at least 50% [36].

The other European plant under construction is located in Flamanville, France. It is a third-generation nuclear reactor, very similar to the one in Finland, and was started in 2006 with a projected completion date of 2012. However, the company, Electricity of France (EDF), announced in March 2012 that construction completion would be delayed until 2014, with electricity production to begin in 2016. Thus, the total construction time of the plant has increased from 6 to 8 years, and even that time may be optimistic. Furthermore, the initial cost estimate of $4.7 billion, increased by 2011 to $5.7 billion and further increased by 2012 to $8.55 billion [37]. EDF attributed some of the delays, and the increase in cost to additional safety requirements instituted as a result of the Fukushima nuclear accident.

In the United States, two nuclear reactors were approved for construction in 2012. These two reactors are Westinghouse AP1000 (Advanced Pressurized nuclear units) with a maximum generating capacity of approximately 1100 MW each. The Westinghouse AP1000 design was developed in anticipation of more nuclear power plants being ordered worldwide, and several of those units are under construction in China. The units were designed and will largely be built by the Japanese nuclear company Toshiba, which acquired the Westinghouse Electric Company in 2006. The two new U.S. units, VOGTLE 3 and 4, will be owned by Southern Company on behalf of Georgia Power, which has operated two nuclear power plants called VOGTLE 1 and 2, for many years on the same site [38]. Despite the widespread interest in nuclear power, private investors considered the risks of building new nuclear power plants too high for investment on their own. However, the U.S. government authorized loan guarantees to utilities willing to build nuclear power plants in accordance

TABLE 1.8

Nuclear Power Plants, Worldwide

Country	In Operation		Under Construction	
	Number	Electric Net Output (MW)	Number	Electric Net Output (MW)
Argentina	2	935	1	692
Armenia	1	375	—	—
Belgium	7	5,927	—	—
Brazil	2	1,884	1	1,245
Bulgaria	2	1,906	2	1,906
Canada	18	12,604	—	—
China				
Mainland	16	11,688	26	26,620
Taiwan	6	4,981	2	2,600
Czech Republic	6	3,678	—	—
Finland	4	2,736	1	1,600
France	58	63,130	1	1,600
Germany	9	12,068	—	—
Hungary	4	1,889	—	—
India	20	4,391	6	4,194
Iran	1	915	—	—
Japan	50	44,215	2	2,650
Korea, Republic	21	18,751	5	5,560
Mexico	2	1,300	—	—
Netherlands	1	482	—	—
Pakistan	3	725	1	315
Romania	2	1,300	—	—
Russian Federation	33	23,643	10	8,203
Slovakian Republic	4	1,816	2	782
Slovenia	1	688	—	—
South Africa	2	1,800	—	—
Spain	8	7,567	—	—
Sweden	10	9,313	—	—
Switzerland	5	3,263	—	—
Ukraine	15	13,107	2	1,900
United Kingdom	19	9,920	—	—
United States	104	101,240	1	1,165
Total	**435**	**368,267**	**63**	**61,032**

Source: European Nuclear Society, Nuclear power plants, worldwide, http://www.euronuclear.org/info/encyclopedia/n/nuclear-power-plant-world-wide.htm. With permission.

with the Energy Policy of 2005 [39]. Under those guarantees, the government will provide up to 80% of the cost of construction. If the reactors are operated profitably, the borrower will repay the banks in accordance with the loan and pay a fee to the federal government for having provided the guarantees. However, if the borrowers default, the federal government, that is, the taxpayers, must repay the banks. In the case of VOGTLE 3 and 4, the project may also be eligible for additional loan guarantees from the Japanese government, since the reactors were designed by a Japanese company. The estimated cost for the two units was $14 billion,

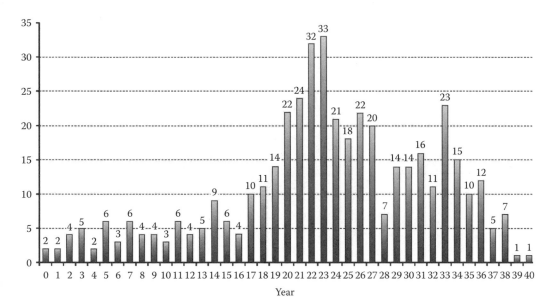

FIGURE 1.21
Number of operating reactors by age (as of June 26, 2007). *Note:* Age of a reactor is determined by its first grid connection. (From IAEA, *Power Reactor Information System (PRIS)*, Vienna, Austria, http://www.iaea.org/programmes/a2/. With permission.)

which does not include the cost of transmission lines that already exist from the previous two units, the future costs of dismantling the plant and safely storing used nuclear fuel, and the cost of security measures in which the U.S. nuclear energy industry has invested more than $2 billion since 2001 [40]. A recent cost estimate for the construction of a new nuclear power plant in the United States was $7 per watt [41], and according to an estimate in [42], dismantling of the plant and safe storage of the waste may substantially add to this cost.

As will be shown later in the book, the large capital investment required for nuclear power and the perceived risk by the public regarding nuclear radiation will make the future of nuclear power in the United States largely dependent on the financial success of nuclear power plants such as AP1000 that are currently being built in many parts of the world. Public acceptance may also depend on how the country's aging nuclear power plants and the large amount of nuclear waste already generated is handled in the next 10 years. It is important to note that most reactors today are more than 25 years old (70%), and about 23% are more than 35 years old. As a result of the age of the reactors, within the next 10 or 20 years, at least 150, or about 30%, will have to be decommissioned. The age distribution of operating reactors is shown in Figure 1.21.

1.7.1 Nuclear Power Conversion Technology

The most common power plant design is the pressurized water reactor (PWR) shown schematically in Figure 1.22. In a PWR, water at a pressure above the critical value (i.e., 255 bar) is heated by nuclear power in the primary loop to about 300°C. The primary loop contains the reactor core with uranium in the form of pellets arranged in rods. The rate of heat generation is carefully controlled by control rods that can absorb neutrons readily (e.g., boron). The secondary loop is a Rankine cycle steam power plant with steam at 70 bar. The PWR is widely considered to be the safest design, and most new reactors are of this type.

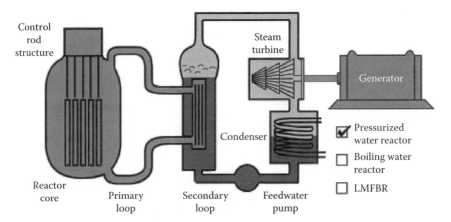

FIGURE 1.22
Schematic diagram of pressurized water reactor (PWR).

The other type of reactor is the boiling water reactor (BWR) shown schematically in Figure 1.23. In a BWR, the same water loop serves as moderator, coolant for the core, and steam source for the turbine. The steam is generated directly in the reactor core where the pressure and heat generation are carefully controlled by a control rod structure. The steam from the reactor core passes directly through a turbine and a condenser, and is returned by a pump to the reactor core as in a conventional Rankine cycle. Up-to-date details of the construction and operation of current as well as proposed next-generation nuclear power plants are presented in [43].

Of all power reactors operating worldwide, about 78% are cooled by light water reactors (LWRs). The choice of water as a coolant dictates that the reactor system environment will be of high pressure (above 255 bar) and medium temperature (roughly 300°C). Light water absorbs neutrons more readily than does heavy water. Consequently, an LWR cannot become critical using natural uranium. Heavy-water cooled/moderated reactors can become critical using natural uranium fuel, as can graphite-moderated, gas-cooled reactors.

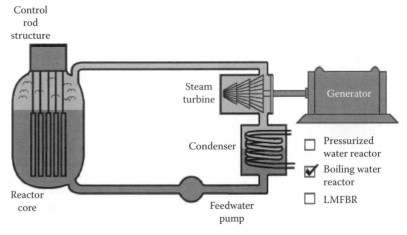

FIGURE 1.23
Schematic diagram of boiling water reactor (BWR).

1.7.2 Nuclear Disasters

The first serious nuclear accident was at Three Mile Island, Pennsylvania, in 1979. The accident was the worst in the history of U.S. commercial power industry. Mechanical problems and human error caused loss of coolant water and led to a partial nuclear meltdown and the release of some radioactive gases. Orders for new nuclear power plants were cancelled in the United States after the accident, and public sentiment shifted to concern about safety and issues of nuclear waste disposal. On April 26, 1986, a catastrophic accident at the Chernobyl Nuclear Power Plant, Ukraine, involved a massive explosion and fire that released large amounts of radioactive contamination across the USSR and Europe. 350,000 people were evacuated and permanently resettled from the most highly contaminated areas of Belarus and Ukraine. The accident killed 31 plant workers and a further 64 deaths from radiation, mostly among emergency and plant workers. The Chernobyl disaster remained the most devastating nuclear accident until 2011.

On March 11, 2011, a massive 9.0 magnitude earthquake struck 109 mi from Fukushima in the Pacific Ocean. The tsunami that followed breached the safety wall that had been built to protect the nuclear reactors at the Fukushima Daiichi nuclear power plant and flooded the pumps necessary for cooling the reactor rods. The 4.7 GWe (gigawatt electric) Fukushima Daiichi Nuclear Plant consists of six reactor units designed by General Electric [44]. All of the reactors came online between 1970 and 1979, which makes them all more than 30 years old. The reactors were of the boiling water type. The number one reactor was slated to be shut down in 2011, but was granted a 10 year extension by Japanese regulators in 2009 although reactor risks had been foretold earlier [45,46].

Three of the units at Fukushima Daiichi Nuclear Power Plant had been shut down prior to the earthquake for maintenance, and the other three reactors shut down automatically after the earthquake, with the remaining decay heat of the fuel rods being cooled with power from emergency generators. The construction of the plant is shown schematically in Figure 1.24. The 5.7 m tsunami wall was breached by waves reaching heights of 14 m. The tsunami not only destroyed the primary cooling system, but also disabled the emergency diesel generators that are required to provide power to the pumps used to cool the reactor core in case of primary system failure [44]. In the following 3 weeks, there was evidence of partial nuclear meltdowns in Units 1, 2, and 3, visible explosions believed to be caused by hydrogen gas in Units 1 and 3, and a suspected explosion in Unit 2 that may have damaged the primary containment vessel. Also due to a loss of water in the pools

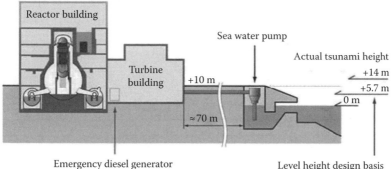

FIGURE 1.24
Schematic diagram of the Fukushima Daiichi power plant layout. (From Kitamura, M. and Shiraki, M., 2011, Japan's reactor risk foretold 20 years ago in U.S. agency report. Bloomberg. http://www.bloomberg.com/news/2011-03-16/japan-s-reactor-risk-fore-told-20-years-ago-in-u.s.-nuclear-agenca-s-report.html)

used to store the spent fuel rods for three of the reactors, some of the rods were uncovered and heated up beyond the safe limit.

Although the events leading up to the nuclear disaster at Fukushima were not the direct result of shortcomings in the nuclear plant operations, they did highlight the vulnerability of nuclear power plants to natural disasters. In response to concerns in the United States about the safety of its own nuclear power plants, the Nuclear Regulatory Commission (NRC) prepared a report outlining the changes in safety precautions recommended for nuclear power plants [47]. The NRC recommended that nuclear facility operating licenses should be denied unless the operators can demonstrate the ability to deal with a complete loss of power for 8 h using backup generators and be able to provide cooling to the radioactive core and spent fuel rods in a water pool for at least 72 h. The pump failures at Fukushima also caused ignition of tightly packed spent fuel rods stored in cooling water ponds and caused the water to evaporate after the pumps failed. Spent fuel rods generate heat for many years' time after they are removed from the reactor. This is important for overall safety because the rods in Japan were less tightly packed than the fuel rods in cooling ponds in the United States. As a precaution to a cooling water accident in the United States, the recommendations from the NRC include requirements for dry storage of the fuel rods after no more than 5 years in a cooling pond. All of these measures will make nuclear reactors safer, but also increase the cost of nuclear power in the United States.

The most significant reaction to the Fukushima disaster occurred in Germany, where the government agreed on a roadmap for phasing out nuclear power entirely. It was decreed that all of the country's 17 nuclear power plants are to go off-line by 2022, with a possible extension for only three reactors, in case of an electric shortfall [41]. As a roadmap for transitioning to a sustainable energy future, it will be important to watch the results of political actions taken in Germany, as well as in other countries such as Switzerland, which will eliminate use of nuclear power in the future energy system of these countries.

1.7.3 Uranium Resources

Nuclear power uses uranium as the energy source. Available resources of uranium are divided into two groups: reasonably assured resources (RAR), and inferred resources (IR). RAR are the uranium in known deposits of sufficient size, grade, and configuration that it could be recovered within reasonable production costs with currently available mining and processing technology. IR are inferred resources whose availability is uncertain. The categories are further divided into various cost classes according to estimated extraction costs. The division of these classes into below $50/kg, below $80/kg, and below $130/kg is widely accepted. According to some experts, RAR for less than $80/kg uranium are called proven resources. It is estimated that of this type, about 2.6 million ton is available globally. When used in LWRs, this corresponds to an energy equivalent of 28 billion tons of hard coal. Figure 1.25 shows the deposits of uranium and the various countries in which they are located. If mining costs up to $130/kg are included, the global uranium sources are increased to 3.3 million tons. The total, irrespective of costs, may be as high as 15 million tons. The amount of energy that can be extracted from these uranium resources depends upon whether or not spent fuel from LWRs can be reprocessed. Reprocessing all of the spent fuels from LWRs and recycling the uranium and plutonium in mixed oxide (MOX) fuel to operate LWRs with 30% MOX and 70% uranium would lead to a cumulative savings of some

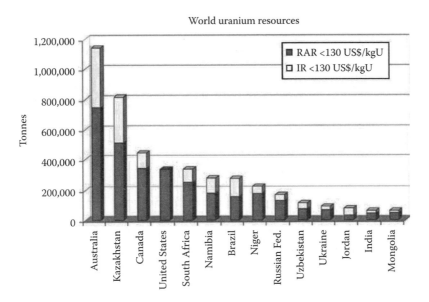

FIGURE 1.25
The deposits of uranium and the various countries in which they are located. (From *Nuclear Power Daily*, Progress energy Florida signs contract for new, advanced-design nuclear plant, January 16, 2009, http://www.nuclearpowerdaily.com/reports/Progress_Energy_Florida_Signs_Contract_For_New_Advanced_Design_Nuclear_Plant_999.html, retrieved on May 18, 2009.)

600,000 ton of natural uranium in the next 40 years. However, current legislation in the United States forbids reprocessing because it produces plutonium that could be used as feedstock for a bomb.

1.8 Renewable Energy

By definition, the term "reserves" does not apply to renewable resources. So we need to look at the annual potential of each resource. Table 1.9 summarizes the resource potential and the present costs and the potential future costs for renewable resources. As in the case of other new technologies, it is expected that cost competitiveness of the renewable energy technologies will be achieved with R&D, scale-up, commercial experience, and mass production. The experience curves in Figure 1.26 show industry-wide cost reductions in the range of 10%–20% for each cumulative doubling of production for wind power, PV technologies, ethanol, and gas turbines [4]. Similar declines can be expected in solar thermal power and other renewable technologies. As seen from Figure 1.26, wind energy technologies have already achieved market maturity, and PV technologies are well on their way. Even though concentrating solar thermal power (CSP) is not shown in Figure 1.26, a Global Environmental Facility report estimates that CSP will achieve the cost target of about $0.05/kWh by the time it has an installed capacity of about 40 GW [48]. As a reference point, wind power achieved that capacity milestone in 2003.

TABLE 1.9

Potential and Status of Renewable Energy Technologies

Technology	Annual Potential	Operating Capacity 2005	Investment Costs U.S.$ per kW	Current Energy Cost	Potential Future Energy Cost
Biomass energy					
Electricity	276–446 EJ total or 8–13 TW	~44 GW	500–6,000$/kWe	3–12 c/kWh	3–10 c/kWh
Heat	MSW ~ 6 EJ	~225 GWth	170–1,000$/kWth	1–6 c/kWh	1–5 c/kWh
Ethanol		~36 bln lit.	170–350$/kWth	25–75 c/L (ge)[a]	6–10 $/GJ
Biodiesel		~3.5 bln lit.	500–1,000$/kWth	25–85 c/L (de)[b]	10–15 $/GJ
Wind power	55 TW Theo. 2 TW practical	59 GW	850–1,700 $/kWe	4–8 c/kWh	3–4 c/kWh
Solar energy					
Photovoltaics	>100 TW	5.6 GW	5,000–10,000$/kWe	25–160 c/kWh	5–25 c/kWh
Thermal power		0.4 GW		12–34 c/kWh	4–20 c/kWh
Heat			2,500–6,000 300–1,700	2–25 c/kWh	2–10 c/kWh
Geothermal electricity	600,000 EJ useful resource base	9 GWe	800–3,000	2–10 c/kWh	1–8 c/kWh
Heat	5000 EJ economical in 40–50 years	11 GWth	200–2,000	0.5–5 c/kWh	0.5–5 c/kWh
Ocean energy					
Tidal	2.5 TW	0.3 GWe	1,700–2,500	8–15 c/kWh	8–15 c/kWh
Wave	2.0 TW		2,000–5,000	10–30 c/kWh	5–10 c/kWh
OTEC	228 TW		8,000–20,000	15–40 c/kWh	7–20 c/kWh
Hydroelectric					
Large	1.63 TW Theo.	690 GW	1,000–3,500	2–10 c/kWh	2–10 c/kWh
Small	0.92 TW Econ.	25 GW	700–8,000	2–12 c/kWh	2–10 c/kWh

Sources: Data from Goldemberg, J. and Johansson, T.B. (eds.), *World Energy Assessment: Energy and the Challenge of Sustainability*, United Nations Development Programme, New York, 2004; UNDP (2004) updated from other sources: Worldwatch Institute, *Bio-Fuels for Transportation—Global Potential and Implications for Sustainable Energy in the 21st Century*, Worldwatch Institute, Washington, DC, 2006; World Wind Energy Association, *Worldwide Wind Energy Bulletin*, World Wind Energy Association, Bonn, Germany, www.wwindea.org (accessed October 12, 2006.); Photovoltaic Barometer, EPIA (www.epia.org); World geothermal power generation 2001–2005; *GRC Bulletin*; International Energy Annual, U.S. DOE-EIA. With permission.

[a] ge, gasoline equivalent liter.
[b] de, diesel equivalent liter.

1.8.1 Geothermal Energy

Geothermal heat originates from two sources: residual heat generated in the formation of the planet by gravitational collapse, and heat generated by decay of radioactive isotopes. The core of the Earth is estimated to have temperatures of 4500°C–6600°C. The solid plates that form the crust of the planet are about 35 km thick and have relatively low thermal conductivity so that the temperature at the surface is only about 10°C on average. The average temperature gradient in the crust is 25°C–30°C/km. Commercially viable geothermal resources occur along spreading or sub-duction zones between plates and at volcanic hot spots where magma from

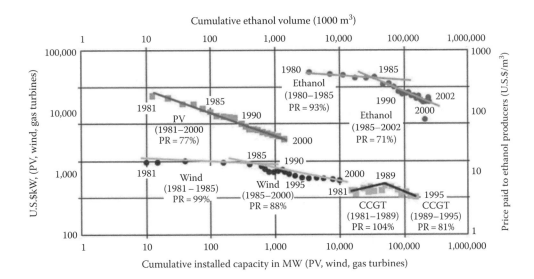

FIGURE 1.26
Experience curves for wind (From Neij, L. et al., The use of experience curves for assessing energy policy programs, Presented at *EU/IEA Workshop*, IEA, Paris, France, January 2003), PV (From Parente, V. et al., *Prog. Photovolt. Res. Appl.*, 10(8), 571, September 2002), ethanol (From Goldemberg, J. et al., *Biomass Bioenergy*, 26(3), 301, March 2004), and gas turbines. (From Claeson Colpier, U. and Cornland, D., *Energy Policy*, 30(4), 309, March 2002.) (Adapted from Goldemberg, J. and Johansson, T.B., eds., *World Energy Assessment: Energy and the Challenge of Sustainability*, United Nations Development Programme, New York, 2004.)

the mantle has pushed up toward the surface. Exploitable geothermal resources depend on the presence of water. These hydrothermal resources are key as the water is heated by convection as it flows through cracks in the hot rock zone. High-quality geothermal resources are tapped by drilling into the high-pressure zone and can produce brine at temperatures around 300°C. The geothermal resource is normally a mixture of liquid and steam with dissolved minerals in the liquid and gases like carbon dioxide and hydrogen sulfide. The geothermal power plant normally is a combination of different types of generators using the different components of the resource. The geothermal fluid is separated and steam is flashed and drives a steam turbine. The majority of the geothermal power extracted is directly from flashed steam. The exhaust from the steam turbine is often condensed in a heat exchanger that boils a binary fluid, usually an organic fluid like pentane or refrigerant, which is in turn used in a Rankine cycle called a bottoming cycle. Most large commercial geothermal power plants utilize resources above 150°C.

Issues with geothermal energy arise from the dissolved minerals and gases in the resource and the nature of the resource. The temperature of the brine cannot be dropped below a point where the dissolved minerals rapidly precipitate out and build up as scale on heat exchangers and pipes. Management of scaling issues is a major part of plant operation and maintenance requirements. Many hydrothermal resources have been used unsustainably and exhausted. The modern practice is to re-inject the brine into the hydrothermal zone at some distance from the production well in order to maintain the water pressure and flow of the resource. Geological investigations and mapping of crack networks are continually developing in sophistication in order to design the best extraction and re-injection locations to improve sustainability of the resource. Nearly all geothermal power plants, even with re-injection, experience substantial changes in the resource enthalpy, usually through reduction of liquid fraction, as the resource is used.

All geothermal power plants have high maintenance rates compared to coal and gas power plants due to corrosion of the steam turbine by aerosols and corrosive gases and scaling of heat exchangers and pipes.

Table 1.10 shows the geothermal energy for direct electric power generation worldwide [50]. Geothermal power development in the United States has experienced slow growth since 1990, but is still the largest geothermal power-producing country with 3187 MW installed. As at 2012, the world geothermal capacity was over 11,000 MW and generated more than twice the amount of energy as solar worldwide. Geothermal power plants have slightly lower utilization factor than coal power plants but are considered base load and are on average available more than 90% of the time. Geothermal development involves drilling and geologic exploration, but in most countries, geothermal development is cheaper than most other renewable resources. In developing countries where diesel generation is the only other viable alternative, geothermal generation, even from low-temperature resources (<150°C), is several times cheaper. It is interesting to note that the largest producer, the United States, generates only about 0.3% of its electricity from geothermal. However, other countries like New Zealand produce over 10% of their power from geothermal.

There are environmental issues with geothermal development. The dissolved minerals and gases in geothermal fluid can include arsenic and other toxic substances. Noncondensable gases (NCG), including CO_2, are vented prior to plant operation. While CO_2 emissions are usually less than 10% of that for coal, the NCG venting may need cleanup to reduce local air pollution. The other major concern is that many geothermal resources, particularly in Japan, are protected in national parks. Geothermal development for a plant in New Zealand caused the dramatic reduction in geyser and other thermal activity of a popular tourist area. The Yellowstone basin in the United States is a massive geothermal resource, but it is difficult to imagine industrial energy development in the nation's first national park.

TABLE 1.10

International Geothermal Power Generation Currently Installed (as of 2012) for Selected Countries and Estimated Development Potential

Country	Installed Capacity (MWe)	Estimated Potential (MW)
African Nations	217	15,000
Indonesia	*5000 Planned by 2025* 1149	27,510
Japan	535	*Includes national parks* 23,000
Philippines	1972	3,447
China	24	887,900,000
Central America	504	13,000
United States	3187	*Currently under development* 5836
Canada	0	13,400
Germany	7.3	*Planned by 2015* 70
Turkey	100	2000
Italy	875	*Planned by 2015* 923
Iceland	660	*Planned by 2015* 890
New Zealand	747	*Planned developments* 1010

Source: GEA, *Geothermal: International Market Overview Report*, May 2012, http://geo-energy.org/reports.aspx

Geothermal fluids are also widely used for direct heating of buildings, water, and swimming pools. The U.S. Geological Survey has prepared an assessment of the geothermal energy potential in the United States [49]. The survey estimated the part of the identified accessible base that could be extracted and utilized in the reasonable future to be 23,000 MWe for 30 years. The undiscovered U.S. resource inferred from knowledge of earth science was estimated to be between 95,000 and 150,000 MWe for 30 years. In 1995, the United States was using over 500 TJ per year of energy from geothermal sources for direct use. Cities such as Boise (Idaho), Elko (Nevada), Clamon Falls (Oregon), and San Bernardino (California) have geothermal district heating systems where a number of commercial and residential buildings are connected to distributed pipelines circulating water at 54°C–90°C. Worldwide, the potential for direct use is estimated to be several times the power generation potential.

Ground source energy is also a source of thermal energy for heat pumps. The heat pump evaporator unit consists of a field of tubes buried in the ground rather than air. Ground-coupled heat pump technology uses the reservoir of constant temperature, shallow ground water, and moist soil as the heat source during the winter and as the heat sink during summer cooling. The energy efficiency of geothermal heat pumps is 30% better than that of air-coupled heat pumps in cold climates and at least 50% better than direct electric heating [51,52]. Application for heat pumps in individual buildings has been widely demonstrated, and their use is entirely dictated by economic considerations [51].

Recently, interest in enhanced geothermal systems (EGS, sometimes referred to as engineered geothermal systems) for primary energy recovery using heat mining technology designed to extract and utilize the earth's stored thermal energy at great depth has received a great deal of attention. A large-scale effort was mounted in 2006, and the report summarizing the results is available on the Internet at http://geothermal. inel.gov and www1.eere.energy.gov/geothermal/egs_technology.html. The EGS technology attempts to drill into the depth of the earth up to 1000 m or more and pump hot water through two interconnected pipes, which would then be used to raise steam in a heat exchanger to drive a conventional turbine. The success of the EGS efforts has so far only been marginal as described in the report referenced earlier, although the study gave projections of potentially very large recoverable energy, with estimated extractable portion of about 200,000 EJ, which is many times the annual consumption of the United States. A demonstration plant, Ogachi project in Japan, has highlighted the main problem as hydrothermal fluid flow [53]. No commercial system has yet been built and put into operation.

1.8.2 Biomass and Biofuel

At least 13.1% of the world's total primary energy supply came from renewable energy in 2009. However, almost 76% of this renewable energy supply was from biomass as illustrated in Figure 1.27. In developing countries, wood and dung are mostly used in traditional open combustion for cooking, which is very inefficient and creates serious health problems. Even a small improvement in conversion efficiency for cooking stoves could reduce wood fuel demand and improve indoor air pollution for millions of people. Such a development was recently made by Professor Bryan Wilson and researchers at Colorado State University resulting in a range of affordable and clean burning wood stoves that have been brought to the world market (http://www.envirofit.org). Biomass provides only 10% of the world's total primary energy, which is much less than its real potential. However, given the immense pressure on forests due to demand for wood, and population pressures for agricultural

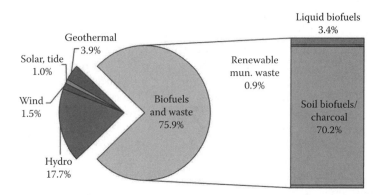

FIGURE 1.27
The year 2009 resource shares in world renewable energy supply. (Data from IEA, *World Energy Outlook*, IEA, Paris, France, 2010.)

land and urban development, extreme caution should be taken before proclaiming biomass as a great source of energy for growth in the future. Although theoretically harvestable biomass energy potential is on the order of 90 TW, the technical potential on a sustainable basis is on the order of 8–13 TW or 270–450 EJ/year [6]. This potential is three to four times the present electrical generation capacity of the world. It is estimated that by 2025, even the municipal solid waste (MSW) could generate up to 6 EJ/year.

Governments of developed countries around the world are viewing development of liquid biofuels from biomass very favorably. Biofuels have the potential to replace as much as 75% of the petroleum fuels in use for transportation in the United States today if new ways to convert nonfood biomass and new programs to increase forest production could be developed [54]. Biofuels for transportation require new supply chain infrastructure for blending and vending, and vehicle fleets can only use low percentage blends for some time in the future. In 2012, the world ethanol production had reached about 7.5 billions of gallons per year while biodiesel production topped 3.5 billion liters during the same year. The United States and Brazil were the largest producers of ethanol, and Germany was by far the largest producer of biodiesel. Producing liquid biofuel from food crops is the most economical route, although production has dropped where government subsidies were removed. Biofuel production depends primarily on agricultural production and therefore on land use as shown in Figure 1.28. So-called second-generation biofuels are the subject of research that seeks to convert wood, grass, algae, or MSW into liquid fuels. The processes being investigated range from chemical dissolution to thermal processing using gasification and steam reforming over catalysts, called the Fischer–Tropsch process that produces methanol. The cost for second-generation biofuels is estimated to be at least five times that of current processes [54].

1.8.3 Hydroelectric Generation

The total share of all renewable sources for electricity production in 2009 was about 13%, a vast majority (72%) of it being from hydroelectric power. Worldwide, the power generation from hydroelectricity is over 3 million GWh. Depending on weather, 7%–10% of U.S. electricity can be generated by hydro. Hydroelectric generation technology is mature as is the civil engineering of earth and concrete dams. The turbine must rotate at a continuous speed so there are several ways to control the flow of water through the

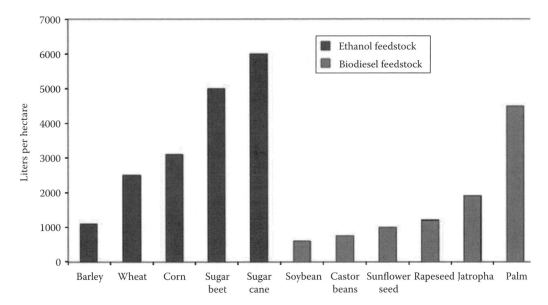

FIGURE 1.28
Biofuel yields of selected ethanol and biodiesel feedstock. (From Worldwatch Institute, *Bio-Fuels for Transportation—Global Potential and Implications for Sustainable Energy in the 21st Century*, Worldwatch Institute, Washington, DC, 2006. With permission.)

turbine to match the power generation to the load on the grid at a constant rotational speed. Sluice gates slide up or down to control the water flow from the reservoir into the penstocks. The water intake to the penstocks can be from the top or the bottom of the reservoir. A scroll wraps around the turbine imparting a circulation flow to the water so that the flow angle into the turbine is optimal. Wicket gates are louvers that can be opened and closed relatively quickly to provide partial flow control. The water flows down through the turbine and into the outlet and surge basin where the flow velocity is dissipated prior to entering the river. Francis turbines are common radial flow turbines with fixed angle turbine blades. The flow enters the outside edge of the turbine from the scroll and flows toward the center of the turbine and downward. Kaplan turbines are also commonly used for lower head reservoirs and for peaking operation because they are axial flow turbines with variable-pitch blades for further speed control. The Kaplan turbine has a higher efficiency at lower flow rates, but hydropower generation typically converts 80%–94% of the kinetic energy in the water flow to shaft work. The turbine is on a vertical shaft with the electric generator mounted above the turbine in the powerhouse. Generators are typically 95%–99% efficient. When the flow rate is lower but the head is very high, a Pelton wheel turbine is used. The water is directed via a nozzle to impact the cup-shaped blades of the turbine at a tangent to the axle.

Hydroelectric power plants require large-scale construction of dams and storage reservoirs. There have been serious environmental consequences of dams, the most costly being the loss of salmon fisheries in the Columbia River basin. The 90% reduction in salmon entering the river system directly attributable to the 130 dams built in the watershed between 1960 and 1980 is estimated to have come at a cost of $6.5 billion by the National Marine Fisheries Service. Hydroelectricity is a renewable resource with long-term sustainability. It is a sizeable, reliable, and low-cost power source. However, whether the resource is developed through damming a river or using a run-of-the-river system, there are long-term and large-scale environmental consequences of all hydro-electric developments. When naturally flowing river water is stilled behind a storage dam, the temperature, oxygen, and other chemical changes can contaminate the reservoir and the river downstream. The release of water from dams coinciding with peak power generation can seriously damage the aquatic ecosystems of the river downstream due to short-cycle scouring. Reservoirs also trap river sediments and nutrients. During warm weather, this can cause algae blooms that can degrade water quality. The high sediment loading can also cause oxygen depletion and acidification of the reservoir that can in turn damage the turbines and penstocks and make the water un-drinkable. Reservoirs produce substantial amounts of methane, a potent greenhouse gas, during the first 10 years after filling or even longer depending on the amount of vegetation submerged.

Floods on rivers can cause property damage and loss of life. Many of the dams built for hydroelectric generation were also built as storage of seasonal precipitation for agriculture and cities, and to control floods. However, the floods also deposited nutrients in the flood-plain soils where people used the natural fertilization for agriculture. Storage of water behind dams in arid locations also increases the salinity of the water due to increased evaporation. For example, it is estimated that about 10% of the water behind the High Aswan Dam in Egypt is lost to evaporation, which is about equal to the withdrawals for agriculture and cities from the Nasser Reservoir.

The worst hydroelectric disaster occurred in August 2009 at the Sayano–Shushenskaya power plant in Russia. The plant was one of the world's largest at 800 ft high with 10 Francis turbines rated at 650 MW each. The plant was built in 1978, and unit 2 was only 2 months short of its maximum service life. Unit 2 had been vibrating outside of allowable limits for some time, but there were no operational requirements to shut down the unit because of excessive vibration. At 8:13 a.m. while more than 300 people were at the plant doing maintenance and taking tours, the bolts on the turbine cover of unit 2 failed. The catastrophic pressure surge lifted the turbine and generator, comprising some 900 ton, from its seat and ejected it into the powerhouse. Water then flooded the powerhouse and the turbine rooms and the transformers exploded. Seventy-five workers were killed, and all but one of the generator units, which was closed for maintenance at the time, were totally destroyed as was the powerhouse. Transformer and other oils spilled into the river causing an environmental disaster many miles downstream.

1.8.4 Wind Energy

Wind energy technology has progressed significantly over the last two decades. The technology has been vastly improved, and capital costs have come down to as low as $1100 per kW. At this level of capital costs, wind power is already economical at locations with fairly good wind resources. The average annual growth in worldwide wind energy capacity from 2000 to 2003 was over 30%, and it continued to grow at that rate in 2004 and 2005. The average growth in the United States over the same period was 37.7%. The total worldwide installed wind power capacity, which was 39 GW in 2003, reached a level of 59 GW in 2005 and 238 GW in 2011 [55]. The total theoretical potential for onshore wind power for the world is around 55 TW with a practical potential of at least 2 TW [6], which is about two-thirds of the entire present worldwide generating capacity. The offshore wind energy potential is even larger.

The biggest issue with wind generation is the intermittency and inability to generate on demand. When the wind slows suddenly, a large drop in wind generation can cause instability on the grid. Thus, wind generation requires the availability of firming capacity. Firming capacity is usually natural gas turbines due to the rapid start-up, but can also be hydroelectricity. A high level of cooperation among power generators and network operators on the grid is another possible way to manage intermittent generation. This is currently a major challenge in the deregulated U.S. power market. In sustainable energy engineering, regulation and management of the energy market are often much bigger challenges and barriers than technology or resources.

1.8.5 Solar Energy

The present state of solar energy technologies is such that solar cell efficiencies have reached over 40% in the laboratory, and solar thermal systems provide efficiencies of 40%–60%. Experience is the most important factor in improving solar technology and reliability of installations. As more experience is gained, the outlook for improving cost and integration for solar power, solar thermal, and passive solar design into existing energy systems is promising.

Solar PV panels have come down in cost from about $30/W to about $1/W in the last three decades.* At $1/W panel cost, the overall system cost is around $3/W, which is still too high to compete with other resources for base-load grid electricity. However, there are many off-grid applications where solar PV is already cost-effective, and peak load shaving for large commercial customers has very short payback times. With net metering and governmental incentives, such as feed-in laws and other policies, even grid-connected applications such as building integrated PV are becoming cost-effective. As a result, the worldwide growth in PV production has averaged over 30% per year from 2000 to 2010, with Germany showing the maximum growth of over 51% (Table 1.11).

Solar thermal power using concentrating solar collectors was the first solar technology that demonstrated its grid power potential. A 354 MW solar thermal power plant has been operating continuously in California since 1988. Progress in solar thermal power stalled after that time because of poor government policy and lack of R&D. However, the last 5 years have seen a resurgence of interest in this technology, and a number of solar thermal power plants around the world are under construction. The cost of power from these plants (which is so far in the range of 12–16 U.S. cents/kWh) has the potential to go down to 5 U.S. cents/kWh. An advantage of solar thermal power is that thermal energy can be stored efficiently and fuels such as natural gas or biogas may be used as backup to ensure continuous operation. If this technology is combined with power plants operating on fossil fuels, it has the potential to extend the time frame of the existing fossil fuels.

Low-temperature solar thermal systems and applications have been well developed for quite some time. They are being actively installed wherever the policies favor their deployment. Just in 2003, over 10 MW_{th} low-temperature solar collectors were deployed around the world, a vast majority of those being in China for domestic hot water.

---

* See e.g., http://thinkprogress.org/climate/2013/01/17/1604661/chinese-companies-projected-to-make-solar-panels-for-42-cents-per-watt-in-2015/

TABLE 1.11

Annual Solar Photovoltaics Production in Megawatts by Country, 1995–2010

Year	China	Taiwan	Japan	Germany	United States	Others	World
				Megawatts			
1995	n.a.	n.a.	16	n.a.	35	n.a.	78
1996	n.a.	n.a.	21	n.a.	39	n.a.	89
1997	n.a.	n.a.	35	n.a.	51	n.a.	126
1998	n.a.	n.a.	49	n.a.	54	n.a.	155
1999	n.a.	n.a.	80	n.a.	61	n.a.	201
2000	3	n.a.	129	23	75	48	277
2001	3	4	171	24	100	70	371
2002	10	8	251	55	121	97	542
2003	13	17	364	122	103	131	749
2004	40	39	602	193	139	186	1,199
2005	128	88	833	339	153	241	1,782
2006	342	170	926	469	178	374	2,459
2007	889	387	938	777	269	542	3,801
2008	2,038	813	1,268	1,399	401	1,207	7,126
2009	4,218	1,411	1,503	1,496	580	2,107	11,315
2010	10,852	3,639	2,169	2,022	1,115	4,248	24,047

Sources: Compiled by Earth Policy Institute (EPI) with 1995–1999 data from Worldwatch Institute, *Signposts 2004*, CD-ROM, Washington, DC, 2005; 2000 data from Prometheus Institute, 23rd annual data collection—Final, *PVNews*, 26(4), pp. 8–9, April 2007; 2001–2006 from Prometheus Institute and Greentech Media, 25th annual data collection results: PV production explodes in 2008, *PVNews*, 28(4), pp. 15–18, April 2009; 2007–2010 from Shyam Mehta, GTM Research, e-mail to J. Matthew Roney, EPI, July 28, 2011.

Note: n.a. = data not available.

1.9 Hydrogen

Although the idea of a hydrogen economy has been promulgated by some energy analysts, as well as by ex-President George W. Bush, this is a somewhat misleading concept because hydrogen is an energy carrier but not an energy source. The concept of a hydrogen economy was proposed back in the 1870s as a fanciful speculation of Jules Verne in his novel *The Mysterious Island* [56]. Hydrogen production was examined extensively in the 1970s by experts for the Institute of Nuclear Energy in Vienna and the Electric Power Research Institute [57,58]. The basic idea was to generate hydrogen by high-temperature nuclear reactions and then use the hydrogen to generate electricity, thereby replacing fossil fuels. The results of this study showed, however, that generating hydrogen with high-temperature thermal–nuclear methods was inferior in cost and efficiency to generating electricity from nuclear reactors and then producing hydrogen by electrolysis [59]. But the study also showed that using the electricity from the nuclear plants directly was preferable in cost and efficiency to the hydrogen path to generate electricity with a fuel cell.

Hydrogen is abundant on Earth, but only in chemically bound form. In order to use hydrogen as a fuel, it is necessary that it be available in unbound form. A substantial energy input is needed to obtain unbound hydrogen. This energy input exceeds the energy released by the same hydrogen when used as a fuel. For example, to split water into hydrogen and oxygen according to the reaction

$$H_2O \rightarrow H_2 + \frac{1}{2}O_2$$

120 MJ/kg-hydrogen is needed (all gases at 25°C), while the reverse reaction of combining hydrogen and oxygen to give water (all gases at 25°C) ideally yields 120 MJ/kg hydrogen. But because no real process can be 100% efficient, more than 120 MJ/kg must be added to the first reaction, while less than 120 MJ/kg of useful energy can be recovered from the recombination. To evaluate the losses, it is, therefore, necessary to examine the "cradle-to-grave" energetics of hydrogen production processes quantitatively.

Figure 1.29 shows all the major pathways to produce hydrogen and to utilize it as an energy carrier. The top row of Figure 1.29 shows the primary energy sources: fossil fuels, nuclear materials, and renewable sources. The next three rows show the major processing steps for conversion of the primary energy into hydrogen. Following the hydrogen row are the two methods of using hydrogen in energy applications: one is to combust hydrogen to produce heat for various applications, and the other is to generate electricity from hydrogen by means of a fuel cell.

Fossil fuels, nuclear energy, solar thermal (including OTEC), biomass, wind, and PVs can all be used to generate electricity. All of these, except PVs and wind, generate electricity by first producing heat, which is then converted to mechanical energy, which, in turn, is finally converted to electricity. PV cells generate electricity directly from solar radiation,

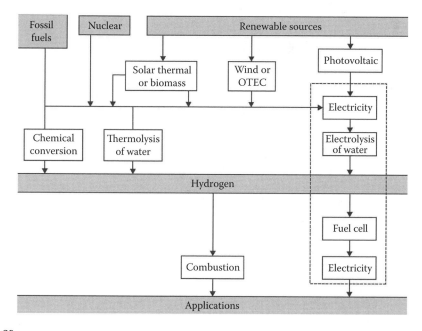

FIGURE 1.29
Pathways for hydrogen production and utilization. (From Kreith, F. and West, R.E., *J. Energy Resources Technol.*, 126, 249, 2004. With permission.)

while wind turbines directly generate mechanical energy and then electricity. In principle, some of the heat-producing technologies can also make hydrogen by thermolysis of water, that is, heating of water to a sufficiently high temperature (greater than 3000 K) to break it into hydrogen and oxygen. However, practical engineering obstacles such as the cost and manufacture of materials to withstand these ultrahigh temperatures make this route impractical.

In Figure 1.29, the dashed box isolates that portion of the pathway in which electricity is used to produce hydrogen by electrolysis of water, and hydrogen subsequently is used to produce electricity via a fuel cell. These steps are common to all energy sources that produce hydrogen by electrolysis. The efficiency can be evaluated by examining the electrolysis and hydrogen utilization steps. These steps are shown in Figure 1.30 with estimated present and highly optimistic future efficiency for the electrolysis and fuel cell steps.

With present technology, it would take 2.9 kWh electricity input to produce 1 kWh of electricity output, and with the most optimistic advanced efficiency, 1.9 kWh would only yield 1 kWh of output. The difference between input and output would be wasted because inefficiencies are involved in each of these sequential steps. Hence, the output of electricity via the hydrogen fuel cell path would cost approximately two to three times as much as the electricity input. Moreover, this cost ratio does not include the capital cost and nonelectrical operating cost of the electrolysis fuel cell and hydrogen storage equipment. Hence, it may be concluded [34] that the conversion of electricity to hydrogen by electrolysis and then the conversion of hydrogen to electricity via a fuel cell are inefficient and not a desirable basis for an economically and environmentally sound, sustainable energy policy.

Similarly, hydrogen produced by electrolysis could be used to produce heat by combustion. However, the efficiency of producing hydrogen by means of electrolysis is only 70%–80%, and burning the hydrogen at an efficiency of 85% yields heat with an overall efficiency of about 60%–70%, while electricity can be converted to heat with essentially no losses. Hence, to use hydrogen made by electrolysis to produce heat is also inefficient and wasteful.

In summary, based upon a cradle-to-grave analysis, any currently available hydrogen production pathway irrespective of whether it uses fossil fuels, nuclear fuels, or renewable technologies as the primary energy source to generate electricity or heat is inefficient

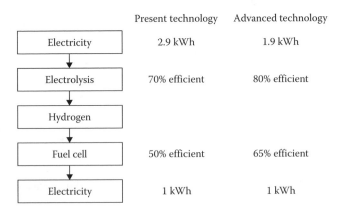

FIGURE 1.30
Generating electricity through hydrolysis to produce hydrogen.

compared to using the electric power or heat from any of these sources directly. Hence, the idea of a hydrogen economy does not appear to be a viable concept with any currently available engineering process and equipment. This conclusion is further elucidated in [59] with appropriate additional references as a basis for the analysis.

There are processes described in the technical literature that can produce hydrogen from renewable or nuclear sources without using either electrolysis or thermolysis of water [60]. For example, biomass may be chemically converted to hydrogen by processes similar to those used with fossil fuels, or it may also be converted to hydrogen by biological conversion processes. Photochemical and photo-electrochemical reactions can produce hydrogen directly with solar radiation input. Thermochemical and hybrid thermochemical/electrochemical cycles use nuclear or solar thermal heat and electricity to drive chemical cycles that produce hydrogen from water. However, none of these ideas is anywhere close to commercialization, and they are topics for future R&D, not viable technologies for a sustainable energy policy [60,61].

1.10 NREL System Advisor Model

Estimating the performance and cost of most renewable energy systems is difficult because the availability of the source, be it wind or solar, varies with time of day, time of year, the local weather, and location. In order to make realistic estimates of performance and cost, it is necessary to understand and be able to simulate the basic theory of the energy generation system, using local environmental data. Fortunately, the NREL has recently developed software to facilitate calculating the performance and economics of virtually any renewable energy system. The System Advisor Model (SAM) makes performance predictions and economic estimates for electric power generation projects in the distributed and central generation markets. The model calculates the performance and cost of generating electricity or heat, based upon data about the project's location, installation, operating costs, financing, and physical system specifications. SAM's performance model makes hour-by-hour calculation for all types of solar power systems' thermal or electric output, generating a set of 8760 hourly values that present the system's heat or energy production during a typical year. SAM then uses the hourly performance data to estimate the system's total annual output and capacity factors for practical performance evaluations. Throughout this text, you will find calculations for each of the major technologies based upon detailed engineering analysis, as well as side-by-side calculations using SAM for the same or a similar system. SAM demonstrates the use of the model based on typical performance calculations at a given location using default parameters based upon experience with and data from previous installations. In order to use the model properly, however, it is necessary to first read and understand the engineering basis for the given technology and then replace the default parameters with the actual parameters for the system being analyzed. The technical details needed are provided in the chapters.

As of 2011, SAM provides performance models for the following technologies:

- Photovoltaic systems (flat and concentrating)
- Parabolic trough concentrating solar power systems
- Power tower concentrating solar power systems

- Dish Stirling concentrating solar power systems
- A simple model for other technologies, including fossil fuel thermal power plants
- Wind power electric turbine
- Flat-plate collectors for domestic or commercial hot water heating

You must have information about the location, the type of equipment used in the system, the various operating parameters related to the resource, the geometry of the energy generation system, the cost of installation, and operating and fuel costs during its life, as well as economic parameters such as interest and inflation rates. Running the model requires a weather file for the project's location, and SAM reads these files in three formats called TMY2, TMY3, and EWP. Data for many locations in the United States and around the world are available on the Internet, and SAM utilizes these data for the performance calculations. SAM provides, for each technology, a set of sample files that contain complete sets of sample cost and performance input data. However, it is your responsibility as an analyst to modify the input data as appropriate for the design of the system you are working on.

Finally, SAM displays modeling results in tables and graphs, displaying first-year annual energy production, LCOE, and other performance data of importance for the analysis. A built-in graphic tool displays a set of default graphs and allows for the creation of custom graphs that fit your system and requirements. Suppose, for example, the default analysis of a flat-plate solar collector in a specific location uses a default tilt angle of 20°, whereas the collector you are planning to use has a tilt value of 40°. In that case, it is necessary for you to replace the default value by the actual tilt angle of the installation. If, in your design, you have an option to select the particular tilt angle, you can run SAM for various angles, compare the results, and then select the value most appropriate for your installation. The availability of SAM makes it possible for the designer or operator to quickly assess the effect of changing parameters and avoid the tedious calculations previously associated with inputting the data and carrying out hour-by-hour calculations in order to obtain the daily, monthly, or yearly performance of the system. It should be emphasized, however, that using SAM without a thorough knowledge of the basic engineering information associated with the system can lead to serious errors. It should also be noted that SAM is an ongoing project, and additional technology performance models are constantly being developed by the NREL in collaboration with Sandia National Laboratories. It is, therefore, important to keep abreast of the website describing the progress of this project (https://sam.nrel.gov/).

New versions of SAM are released every 6 months or so. It is, therefore, likely that a specific solution to a given problem will change a little from version to version. Figure 1.31 compares the SWH system pages in two versions to demonstrate the difference. In version SAM 2011.6.30, the user was required to provide the hourly water draw data for the entire year. In the more recent version, SAM 2011.12.2, there is a new option to scale this hourly hot water draw data to an average daily hot water usage.

Fortunately, all previous versions of SAM are available for download from the NREL website, and it is therefore easy for the student to utilize the appropriate SAM website that corresponds to the particular version used in the book. To access these "legacy versions," it is first necessary to establish an account in their system. After one logs in, these versions can be seen on the right-hand side of the webpage https://sam.nrel.gov/content/downloads.

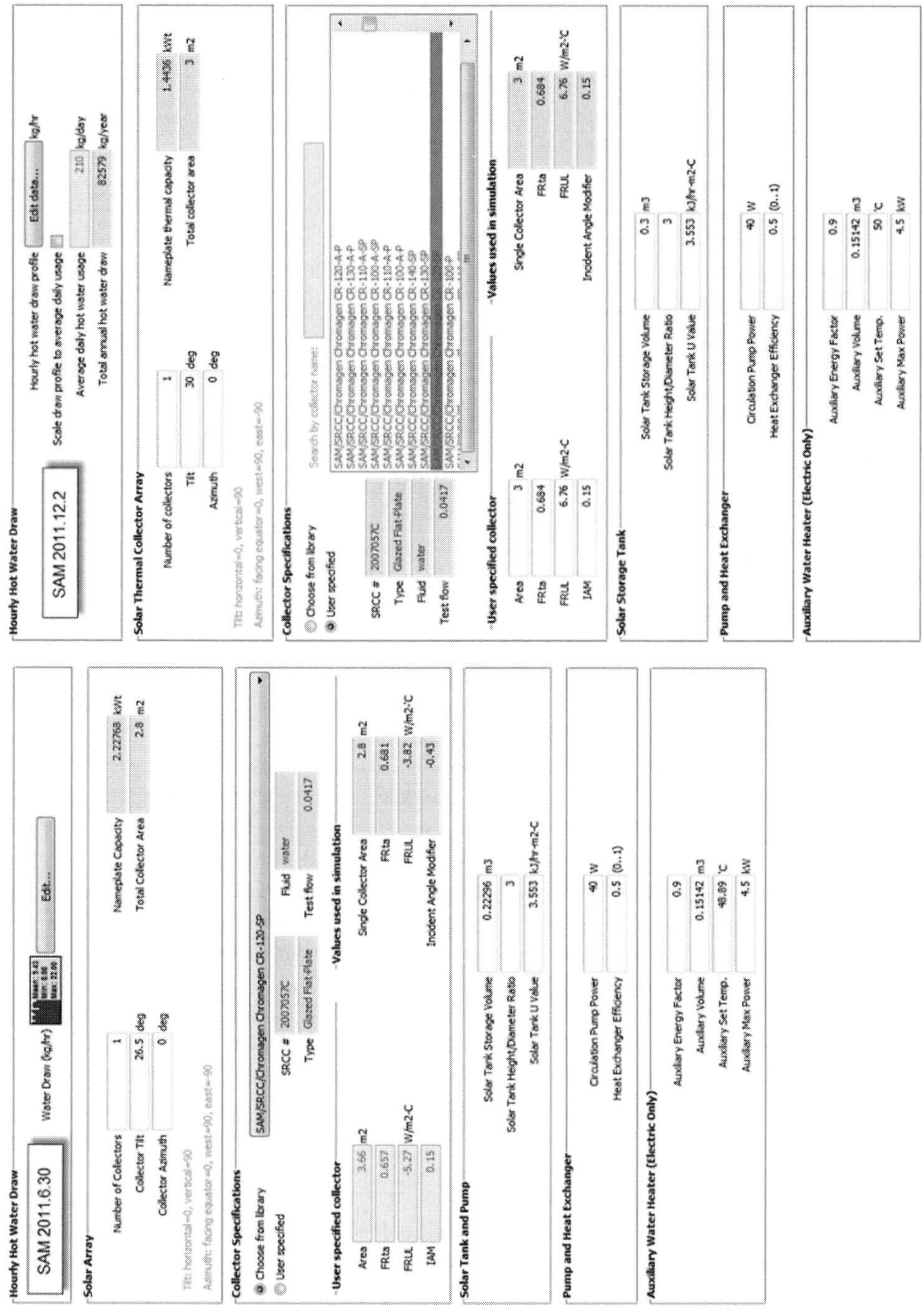

FIGURE 1.31

Comparison of two SAM versions for calculating the performance of a solar hot water heater.

Energy Units and Conversion Factors

In reading the literature pertaining to energy, one encounters an overwhelming number of units that are not usually used in conventional engineering calculations. To assist the reader in handling data from various sources that include unusual units such as barrels of oil equivalent, conversion factors are presented in Table 1.12.

TABLE 1.12

Energy Conversion Factors

	Btus	Quads	Calories	kWh	MWy
Btus	1	10^{-15}	252	2.93×10^{-4}	3.35×10^{-11}
Quads	10^{15}	1	2.52×10^{17}	2.93×10^{11}	3.35×10^{4}
Calories	3.97×10^{-3}	3.97×10^{-18}	1	1.16×10^{-6}	1.33×10^{-13}
kWh	3413	3.41×10^{-12}	8.60×10^{5}	1	1.14×10^{-7}
MWy	2.99×10^{10}	2.99×10^{-5}	7.53×10^{12}	8.76×10^{6}	1
bbls oil	5.50×10^{6}	5.50×10^{-9}	1.38×10^{9}	1612	1.84×10^{-4}
Tonnes oil	4.04×10^{7}	4.04×10^{-8}	1.02×10^{10}	1.18×10^{4}	1.35×10^{-3}
kg coal	2.78×10^{4}	2.78×10^{-11}	7×10^{6}	8.14	9.29×10^{-7}
Tonnes coal	2.78×10^{7}	2.78×10^{-8}	7×10^{9}	8139	9.29×10^{-4}
MCF gas	10^{6}	10^{-9}	2.52×10^{8}	293	3.35×10^{-5}
Joules	9.48×10^{-4}	9.48×10^{-19}	0.239	2.78×10^{-7}	3.17×10^{-14}
EJ	9.48×10^{14}	0.948	2.39×10^{17}	2.78×10^{11}	3.17×10^{4}

	Btus	Quads	Calories	kWh	MWy		
	bbls Oil Equivalent	Tonnes Oil Equivalent	kg Coal Equivalent	Tonnes Coal Equivalent	MCF Gas Equivalent	Joules	EJ
Btus	1.82×10^{-7}	2.48×10^{-8}	3.6×10^{-5}	3.6×10^{-8}	10^{-6}	1055	1.06×10^{-15}
Quads	1.82×10^{8}	2.48×10^{7}	3.6×10^{10}	3.6×10^{10}	10^{9}	1.06×10^{18}	1.06
Calories	7.21×10^{-10}	9.82×10^{-11}	1.43×10^{-7}	1.43×10^{-10}	3.97×10^{-9}	4.19	4.19×10^{-18}
kWh	6.20×10^{-4}	8.45×10^{-5}	0.123	1.23×10^{-4}	3.41×10^{-3}	3.6×10^{6}	3.6×10^{-12}
MWy	5435	740	1.08×10^{6}	1076	2.99×10^{4}	3.15×10^{13}	3.15×10^{-5}
bbls oil	1	0.136	198	0.198	5.50	5.80×10^{9}	5.80×10^{9}
Tonnes oil	7.35	1	1455	1.45	40.4	4.26×10^{10}	4.26×10^{-8}
kg coal	5.05×10^{-3}	6.88×10^{-4}	1	0.001	0.0278	2.93×10^{7}	2.93×10^{-11}
Tonnes coal	5.05	0.688	1000	1	27.8	2.93×10^{10}	2.93×10^{-8}
MCF gas	0.182	0.0248	36	0.036	1	1.06×10^{9}	1.06×10^{-9}
Joules	1.72×10^{-10}	2.35×10^{-11}	3.41×10^{-8}	3.41×10^{-11}	9.48×10^{-10}	1	10^{-18}
EJ	1.72×10^{8}	2.35×10^{7}	3.41×10^{10}	3.41×10^{7}	9.48×10^{8}	10^{18}	1

Notes: To convert from the first column units to other units, multiply by the factors shown in the appropriate row (e.g., 1 Btu = 252 cal). MWy = megawatt-year; bbls = barrels; tonnes = metric tons = 1000 kg = 2204.6 lb; MCF = thousand cubic feet; EJ = exajoule = 10^{18} J. Nominal calorific values assumed for coal, oil, and gas.

Problems

1.1 The hallways of a three-story building in Denver, Colorado, use 600 lights per floor, and each light is an incandescent bulb rated at 60 W.

   a. Estimate the amount of energy that could be saved in August by turning the lights on when the sun sets and turning them off when it becomes light.

   b. Assuming that the power for the lights comes from a 30% efficient coal-fired power plant, estimate the amount of CO_2 that would be saved with this conservation measure. Assume that the chemical composition of the coal is $CH_{0.8}$ and that coal has a heating value of 29 MJ/kg.

   c. Prepare a brief report, making recommendations for reasonable conservation measures.

1.2 On May 27, 2009, U.S. Senator Lamar Alexander proposed to build 100 new nuclear power plants in 20 years "for a rebirth of industrial America while we figure out renewable electricity." In his talk, he stated "This would double America's nuclear plants, which today produce 20% of all of our electricity and 70% of our pollution-free, carbon-free electricity."

   Analyze the proposal of the senator from several perspectives:

   a. Cost of 100 nuclear power plants

   b. Need to replace existing nuclear power plants to generate net power output

   c. Amount of water required for the condensers of the new power plants

1.3 Two different fuels are being considered for a 1 MW (net output) heat engine, which can operate between the highest temperature produced during the burning of the fuels and an atmospheric temperature of 300 K. Fuel A burns at 2500 K, delivering 50 MJ/kg (heating value) and costs $2.00/kg. Fuel B burns at 1500 K, delivering 40 MJ/kg and costs $1.50/kg. Compare the fuel costs per hour of fuel A and fuel B, assuming that the heat engine operates

   a. At Carnot efficiency

   b. At 50% of Carnot efficiency

1.4 A solar energy collector produces a maximum temperature of 100°C. The collected energy is used in a cyclic heat engine that operates in a 5°C environment. Estimate the thermal efficiency if the engine operates at 40% of Carnot efficiency. How would the answer change if the collector were redesigned to focus the incoming solar radiation to enhance the maximum temperatures to 400°C.

1.5 Since meat is higher on the food chain than grains, the more calories you get from meat, the more agricultural land you use. Using the arable land area in your state and data from [30] (which concludes that 5.6 GJ metabolized energy is produced annually per hectare for meat and that 41.4 GJ of metabolized energy is produced annually per hectare for gain), estimate

   a. How many vegetarians could you feed with land available in your state? Determine this as a number (percentage of 2011 population) and percentage of estimated population in 2030.

   b. Repeat this estimate for the average American diet (the typical American gets two-thirds of his or her calories from meat and one-third from grain).

1.6 If 10% of your state would be covered with PV collectors, estimate the power produced in kWh/day per capita on average. If 10% of the people in your state had to drive 100 km/day to commute to work, how many kWh/day would that consume? Discuss the implications of these two estimates. How many collectors of 3 m × 10 m would be required to produce the energy for commuting with an all-electric car with 60% efficient batteries?

1.7 Assuming that a heat pump has a coefficient of performance equal to 4, how much heat could this device deliver to the interior of a building with 1 kWh of electricity?

1.8 If there were a carbon tax of $250/ton of CO_2 on fossil fuels, estimate the increase in the price of a barrel of oil.

1.9 In 2009, the administration of President Obama instituted a so-called Cash for Clunkers program. You are to evaluate the effectiveness of the program, which was intended to stimulate the economy and reduce the import of oil. According to the program announcement, an average of $4000 per car in rebate was offered if a consumer brought in a "clunker" and exchanged it for a new car that would have a fuel efficiency equal to at least 10 mi/gal better than the car they traded. The total amount of money allocated to the program was $3 billion.

    a. What reduction in oil import per year, in terms of both barrels of oil and percentage of total import, was achieved by the program?

    b. Assuming that imported oil cost $55 per barrel, how long did it take for the savings in import to equal the investment in the program?

    c. The energy required to build a new car, called the embedded energy in car construction, has been estimated to be 30,500 kWh/car by Argonne National Laboratory [31]. The embedded energy is mostly coal and nuclear, while the energy saved by the improved efficiency is mostly imported over the life of the car. Estimate how many years it would take to recoup the energy invested in the construction of the car by the savings in energy from the improved mileage of the new vehicle.

    d. What reduction in CO_2 generation did the program achieve?

    Discuss the effectiveness of the program from an energy perspective and what changes in the program you would recommend to make it more effective.

1.10 While completing your homework, imagine the study space you are using employs a space heater with a rating of 30 kW, a refrigerator (for snacks) that is rated for 5 kW, and a light fixture with two 100 W bulbs, all operating at a constant rate for the 1.5 h it takes you to finish.

    a. What is the total power consumed at any given moment?

    b. How much energy do you consume to complete the task?

    c. Assume the utility company charges 6.5 cents per unit of energy in part b, how much did it cost you to complete the assignment?

    d. If you were willing to pay the utility company $5, what would be the power rating of an additional appliance you could add to your study space?

1.11 In 2009, a group of environmentalists approached the Boulder City Council with a request to deny an application to continue operating the local coal-fired 200 MW power plant in order to reduce CO_2 emissions. The group suggested that the coal-fired power plant be replaced by a nonpolluting solar plant.

To evaluate the feasibility of their request, a steel plate is placed in sunlight in Boulder at an angle equal to the latitude. The plate is treated with a coating that has an absorptivity for solar insolation of 0.9 and an emissivity in the long-range spectral range of 0.2.

a. If the average insolation between 9 a.m. and 5 p.m. is equal to 400 W/m² and the convective heat-transfer coefficient from the top surface of the plate is 8 W/m² K, estimate the equilibrium temperature of the plate assuming the bottom is insulated and the effective sky temperature is 50 K.

b. If you could operate a heat engine between the plate temperature calculated in part (a) and the environment at 10°C, estimate the plate surface area required to generate power equal to the coal-fired power plant. Then comment on the result and compare it with a PV system of 12% efficiency.

1.12 A large nuclear power plant has a rated capacity of 1 GW electric. Its actual output is estimated to average about 80% of capacity due to the need for maintenance and imperfect match between supply and demand. Assuming that year-round average residential electricity needs are about 1 kW electric per housing unit and there are three residents per unit, estimate how many power plants would be needed to satisfy the demand of a metropolitan area with a population of 30 million.

1.13 The yearly average solar flux incident on a horizontal surface in the United States is about 180 W/m². The average total energy demand in the United States is approximately 100 quads (100×10^{18} J) per year. Assuming an average efficiency of PV cells of 14% and a biomass conversion efficiency around 1%, estimate the percentage of land required to provide the energy of the country by these two renewable technologies. (The total land area of the United States is about 9.2×10^6 km².)

1.14 Calculate the doubling time for annual percentage increases of 2% and 10%, assuming an exponential growth function.

1.15 In 2011, the United Nations announced that the human population reached 7 billion and was increasing at approximately 1.5%. Estimate the doubling time of the human population and then estimate the available land area on Earth per person at the end of the first and the third doubling times. The Earth is approximately two-thirds water and can be approximated as a sphere of 8000 mi in diameter. Then, discuss the implications for these estimates.

1.16 Experts claim that there is enough coal left for at least another 400 years at current rates of consumption. Estimate how long the coal would last if its annual usage rate increases by 5% and by 10% per year.

1.17 The following equation can be used to solve all problems involving exponential growth and decay:

$$N = N_0 e^{kt}$$

where
N is the number in the population after a time t
N_0 is the initial number
k is the growth constant (if *positive*) or the decay constant (if *negative*)
e is the base of the natural logarithms (approximately 2.72)

1. The population for a town was 100,000 in 2011. To estimate future energy needs, calculate the population in 2030, assuming a growth rate of 3% per year.
2. When would the population reach 1,000,000 at this growth rate?
3. If the city placed a limit of 200,000 people for 2030, what would the allowable growth rate be?
4. Suppose your parents bought a house in 2000 for $200,000 and expected to sell it for their retirement in 2011 for $250,000. What exponential growth rate did they assume?
5. When your parents put the house on the market, the best offer was $170,000. What was the exponential decay rate?
6. Explain the collapse in the U.S. housing market a few years ago in terms of the earlier problems (4 and 5).
7. Obtain data from reliable sources for the global population and oil production between 1850 and 2010. Then plot the data and fit an exponential curve to it. What is the growth constant for the two parameters? Is there a physical explanation for the result?

Discussion Questions

Discussion of sustainability principles

Propose a "Sustainable Development Act" legislation that would embody the Brundtland definition of sustainable development in regulations on land use, energy development, end-user appliances, or the management and use of some resource. Do some research on a resource management regulation in your area. There are fishing regulations in most areas: many national parks have limits on the number of visitors and how they can access sensitive areas; regulations may exist for grazing or logging on public lands. Do all of these existing regulations arise because of previously unsustainable exploitation? Do they all involve limits on what people can take so that the resource can continue into the future?

Discussion on oil supply, depletion effects, and the economy

Consider the short-term impacts on a small company that uses 9 million barrels of oil per year. Companies must make annual budgets that spell out how much they will spend and expected earnings. During most of 2010, the price of oil had remained below $80 per barrel. The company's 2011 budget for oil was $720 million (at a price of $80), with other planned expenditures on new production equipment worth $100 million and $80 million for new workers. Look up the historical price of oil based on U.S. EIA data on a reputable website like www.indexmundi.com, and you will see that the oil price rose to over $100 per barrel for most of 2011. Was the company able to purchase the new equipment, increase production, or hire new workers? Was oil price an issue? What risk did it pose to the company's operation? The 2007–2008 oil price shock had had a negative impact on the company's financial position. What if the company had conducted an energy audit on their oil use and had worked with some energy engineers to devise a plan where they could reduce their oil consumption to 7 million barrels in 2011 (22% reduction) at a cost of $20 million to implement? How would this be a risk management approach? Would the company be able to buy the equipment and hire the new workers if they had implemented this plan? What would happen if the price of oil went back down to $80 per barrel in 2012? Is it fair to say that energy engineering to reduce the amount of energy used in a business is good risk management and good for the economy?

Discussion on the solar energy flow rate and photosynthesis conversion

Put the zip code of your hometown into the solar database PVWatts viewer from NREL (http://gisatnrel.nrel.gov/PVWatts_Viewer/index.html) and find the average daily solar insolation per m^2. The maximum photosynthetic conversion efficiency of plants in direct sunlight is 8%, but the annual average can be assumed around 2%. In Section 1.2.3, a modest estimate of 1 ac per person was used for food production, remembering that we only eat part of the plants. Compare the solar energy flows for food, the ecosystem, and storage as fossil fuels.

Discussion of energy content of corn and transportation

If the American SUV driver rode his bicycle the 225 miles that the tank of E100 would have provided, pedaling at an average power output of 220 W, how many kg of corn would he need to eat to fuel his transportation? Discuss any ideas you have for how market forces would balance the demand for driving low efficiency in the United States with eating tortillas in Costa Rica? Do some research on *green marketing* and *environmental behavior*.

Discussion on CO_2 emissions

Make a spreadsheet of CO_2 emissions to the year 2112. Start with the current figure of 31.6 Gt/year. Think about different scenarios and try plotting them. Make sure to keep track of the sum total of CO_2 emitted. What year would the 565 Gt safety margin be breached if growth in fossil fuel production continues at 3% per year? What if growth were 2% per year? When would the total endowment of hydrocarbons have been exhausted? What if an aggressive demand management program were driven by the engineering profession worldwide to stop growth of fossil fuels and maintain consumption at the current level? The first 10% demand reduction would be the most economical and achievable, but a further 10%–20% could be achieved economically in nearly all end uses. Model several different production rate reduction scenarios. What would it take to keep within the emission limits needed to manage the climate risks? Discuss the benefits to people later in the century if there are still economically recoverable fuel resources available.

Discussion of instability due to growth

Can you think of some examples of civilizations that have faced risks from hitting the limits of growth? What were the implications? Did any of these civilizations know they were on an unsustainable path? What did they do about it? Most engineers are employed in industry or public service. Can you think of any examples of how engineering work has been influenced by these external factors? Can you think of any ideas of how engineering work could help to manage the risks of growth beyond carrying capacity limits?

Online Resources

The field of sustainable energy is changing so rapidly that a single book is not able to capture information in a real-time basis. The following websites will make it possible for the reader to access developments in the sustainable energy field and keep up to date. The list is not exhaustive, but will provide an entry into useful data.

http://www.technologyreview.com—MIT's publications on current technological advances in computing, internet, communications, energy, materials, biomedicine, and business.

http://www.nrel.gov—Publications from the National Renewable Energy Laboratory that focus on innovations for our nation's future energy. The following are some of the numerous tech subjects:

Advanced vehicles and fuels

Basic sciences

Biomass

Buildings

Electric infrastructure systems

Energy analysis

Geothermal

Hydrogen and fuel cells

Solar

Wind

http://www.eere.energy.gov—U.S. Department of Energy efficiency and renewable energy publications focusing on a wide range of programs from biomass, to solar energy, and building technologies.

http://cleanenergysector.com—Profits from the global energy transformation with clean and solar energy technologies.

http://sciencedaily.com—Contains current science advancements and articles of the latest science news from earth to space.

http://www.repp.org—Center for Renewable Energy and Sustainable Technology while building a renewable energy industry by breakdown of each state.

http://www.sciencedirect.com—Current science advancements in physical science and engineering.

http://www.skyfuel.com—Information from a leader in solar thermal technology.

http://www.rangefuels.com—Information from a Colorado company about cellulosic ethanol. It built a cellulosic ethanol plant in South Georgia.

http://www.lignol.ca—Cellulosic ethanol information.

http://www.awea.org—American wind energy association promotes wind energy and has the most up-to-date news on wind power.

http://www.renewableenergyworld.com—Global perspective on renewable energy.

http://www.theoildrum.com—Discussion about renewable energy from reputable sources.

http://www.aceee.org—Information and news about energy efficiency economy on a federal and a state level.

http://www.withouthotair.com—A textbook with careful analysis of the options for a sustainable energy system in Great Britain.

http://www.iea.org—The International Energy Agency provides useful information on global energy topics with important up-to-date data.

References

1. Brundtland, G.H. (1987) *Our Common Future*, World Commission on Environment and Development, Oxford University Press, New York.
2. United Nations (1994) *World Population Prospects: 1994 Revision*, U.S. Census Bureau International Programs Center, International Database, New York.
3. Scanlon, B., Faunt, C., Longuevergne, L., Reedy, R., Alley, W., McGuire, V., and McMahon, P. (June 2012) Groundwater depletion and sustainability of irrigation in the US High Plains and Central Valley, *Proceedings of the National Academy of Sciences*, 109(24), 9320–9325. www.pnas.org/cgi/doi/10.1073/pnas.1200311109
4. World Bank (2003) Development data and statistics, www.worldbank.org/data (accessed on May, 2007).
5. Hoekstra, A.Y. and Chapagain, A.K. (2008) *Globalization of Water: Sharing the Planet's Freshwater Resources*, Blackwell Publishing, Oxford, U.K.
6. Goldemberg, J. and Johansson, T.B. (eds.) (2004) *World Energy Assessment: Energy and the Challenge of Sustainability*, United Nations Development Programme, New York.
7. S. Loupasis (May 2002) Technical analysis of existing RES desalination schemes—RE driven desalination systems REDDES, Report, Contract # 4.1030/Z/01-081/2001.
8. Alliance for Water Efficiency and American Council for an Energy-Efficient Economy (May 2011) *Addressing the Energy-Water Nexus: A Blueprint for Action and Policy Agenda*; ASME, *ETP: Energy–Water Nexus—Cross-Cutting Impacts*.
9. Carney, B., Feeley, T., and McNemar, A. (2012) Power plant–water R&D program, DOE, National Energy Technology Laboratory, Morgantown, WV (accessed on May 25, 2012).
10. Kenny, J., Barber, N., Huston, S., Linsey, K., Lovelace, J., and Maupin, A. (2005) Estimated use of water in the United States in 2005, Circular 1344, U.S. Geological Survey, Washington, DC, pp. 38–41.
11. U.S. DOE (2006) Energy demands on water resources, Report to Congress, U.S. Department of Energy, Washington, DC.
12. U.S. Census Bureau (USCB) (2007). Statistical abstracts of the United States. http://www.census.gov/prod/www/statistical-abstract.html (accessed on March 21, 2008).
13. Wu, M., Mintz, M., Wang, M., and Arora, S. (January 2009) Consumptive water use in the production of ethanol and petroleum gasoline, Argonne National Laboratory, Argonne, IL.
14. Barney, G.O., Blewett, J., and Barney, K.R. (1993) *Global 2000 Revisited: What Shall We Do?* The Millennium Institute, Alexandria, VA, p. 11.
15. Associated Press (2009) U.N.: World's hungry now more than 1 billion, June 19, 2009. http://www.msnbc.msn.com/id/31449307/ (accessed on June 20, 2009).
16. Pimentel, D., Williamson, S., Alexander, C., Gonzalez-Pagan, O., Kontak, C., and Mulkey, S. (2008) Reducing energy inputs in the US food system, *Human Ecology*, 36, 459–471.
17. Schnepf, R. (January 2012) Agriculture-based biofuels: Overview and emerging issues, Congressional Research Service, R41282.
18. Krumdieck, S. and Hamm A. (2009) Strategic analysis methodology for energy systems with remote island case study, *Energy Policy*, 37(9), 3301–3313.
19. U.S. Energy Information Administration, Primary energy production by source, 1949–2010. http://www.eia.gov/totalenergy/data/annual/showtext.cfm?t=ptb0102 (accessed on April 18, 2013).
20. EIA (2011) Annual energy outlook. http://www.eia.gov/forecasts/aeo/ (accessed on April 18, 2013).
21. Metz, B., Davidson, O.R., Bosch, P.R., Dave, R., and Meyer, L.A. (eds.) (2007) *Contribution of Working Group III to the Fourth Assessment Report of the Intergovernmental Panel on Climate Change, 2007—IPCC Fourth Assessment Report (AR4)*, Cambridge University Press, Cambridge, U.K.

22. National Renewable Energy Laboratory Solar Energy Technologies Program (2004) DOE/ GO-102004-1847, January 2004.
23. Murphy, D.J., Hall, C.A.S., and Cleveland, C. (2011) Order from chaos: A preliminary protocol for determining EROI of fuels. *Sustainability*, 3, 1888–1907.
24. Lazard, Ltd. (2009) *Energy Technology Assessment*, Lazard, Ltd., New York. www.lazard.com
25. National Carbon Dioxide Storage Capacity Assessment Act of 2007, and Department of Energy Carbon Capture and Storage Research, Development, and Demonstration Act of 2007. (April 16, 2007). 110th Cong. (testimony of Thomas Shope).
26. Fackler, M. (2012) Japanese nuclear industry near idle, *NYT Asia-Pacific*, March 8, 2012.
27. U.S. DOE (2011) *Transportation Energy Data Book*, 31 edn. http://cta.ornl.gov/data/download31. shtml (accessed on April 18, 2013).
28. Greene, D.L., Roderick, L., Hopson, J.L. (2011) OPEC and the costs to the U.S. economy of oil dependence: 1970–2010, Oak Ridge National Laboratory Memorandum.
29. U.S. DOE (2012) Advanced Manufacturing Office, Energy matters newsletter, http:// www1.eere.energy.gov/manufacturing/tech_deployment/ (accessed on August 2012).
30. Dale, M., Krumdieck, S., and Bodger, P. (2012), Global energy modeling—A biophysical approach (GEMBA) part 2: Methodology and results. *Ecological Economics*, 73, 158–167.
31. Chandler, D. (June 9, 2011) Report: Natural gas can play major role in greenhouse gas reduction, *MIT News Office*. Accessed at: http://web.mit.edu/newsoffice/2011/natural-gas-full-report-0609.html (accessed on April 18, 2013).
32. Hansen, J. (2009). *Storms of My Grandchildren*, Bloomsbury Publishing, London, U.K.
33. Page, S.C., Williamson, A., and Mason, I. (2009) Carbon capture and storage: Fundamental thermodynamics and current technology, *Energy Policy*, 37, 3314–3324.
34. Areva. Finland-Olkiluoto 3. http://www.areva.com/EN/operations-2389/finland—olkiluoto-3. html (accessed on January 18, 2012).
35. European Nuclear Society, Nuclear power plants, worldwide. www.euronuclear.org/info/ encyclopedia/n/nuclear-power-plant-world (accessed on January 18, 2012).
36. Kanter, J. (2009) In Finland, nuclear renaissance runs into trouble, *NYT Energy and Environment*, May 28, 2009.
37. IHS (2011) EDF delays flamanville NPP by another two years, July 21, 2011. http://www.ihs. com/products/global-insight/industry-economic.report.aspx?id=1065930112 (accessed on April 18, 2013).
38. Southern Company, *Plant Vogtle Units 3 and 4 Background*. www.southerncompany.com (accessed on January 25, 2012).
39. Wald, M. (2010) U.S. supports new nuclear reactors in Georgia, *NYT Energy and Environment*, February 16, 2010.
40. Southern Company. Why nuclear (accessed on January 25, 2012). www.southerncompany.com (accessed on April 18, 2013).
41. Siemens. One year after Fukushima—Germany's path to a new energy policy (accessed on May 30, 2012). http://www.siemens.com/press/en/feature/2012/corporate/2012-13-energiewende.php (accessed on April 18, 2013).
42. Storm van Leeuwen, J.W. and Smith, P. (2008 updated), Nuclear power and the energy balance. http://www.stormsmith.nl/ (accessed on August 22, 2009).
43. Kok, K. (2009) *Nuclear Engineering Handbook*, CRC Press, Boca Raton, FL.
44. Dedman, B. (March 13, 2011) General Electric—Designed reactors in Fukushima have 23 sisters in U.S., MSNBC. http://openchannel.msnbc.msn.com/_news/2011/03/13/6256121-general-electric-designed-reactors-in-fukushima-have-23-sisters-in-us (retrieved on March 14, 2011).
45. Yamaguchi, M. and Donn, J. (2011) Japan quake causes emergencies at 5 nuke reactors, Associated Press. http://www.aolnews.com/2011/03/11/japan-quake-causes-emergencies-at-5-nuke-reactors/ (accessed on March 11, 2011).
46. Kitamura, M. and Shiraki, M. (2011) Japan's reactor risk foretold 20 years ago in U.S. agency report, Bloomberg. http://www.bloomberg.com/news/2011-03-16/japan-s-reactor-risk-foretold-20-years-ago-in-u-s-nuclear-agenca-s-report.html (accessed on March 15, 2011).

47. NRC. (n.d.) Implementing lessons learned from Fukushima, http://www.nrc.gov/reactors/operating/ops-experience/japan-info.html (accessed on April 18, 2013).

48. Global Environmental Facility (2005) Assessment of the World Bank/GEF strategy for the market development of concentrating solar thermal power, GEF Report, GEF/c.25/Inf.11, GEF, Washington, DC.

49. Muffler, L.J.P. (ed.) (1979) *Assessment of Geothermal Resources of the United States*, U.S. Geological Survey, Circular 790, Washington, DC.

50. GEA (May 2012) Geothermal: International Market Overview Report, http://geo-energy.org/reports.aspx (accessed on April 18, 2013).

51. Renner, J.L. and Reed, M.J. (2008) Geothermal energy, in: Goswami, D.Y. and Kreith, F. (eds.), *Energy Conversion*, CRC Press, Boca Raton, FL.

52. Kreith, F. and Goswami, Y. (2005) *CRC Handbook of Mechanical Engineering*, CRC Press, Boca Raton, FL.

53. Tester, J.W., Blackwell, D., Petty, S., Richards, M., Moore, M., Anderson, B. et al. (2007) The future of geothermal energy, in: *Proceedings of the 32nd Workshop on Geothermal Reservoir Engineering*, Stanford University, Stanford, CA, January 22–24, 2007, SGP-TR-183.

54. Worldwatch Institute (2006) *Bio-Fuels for Transportation—Global Potential and Implications for Sustainable Energy in the 21st Century*, Worldwatch Institute, Washington, DC.

55. World Wind Energy Association (2006) *Worldwide Wind Energy Bulletin*, World Wind Energy Association, Bonn, Germany. www.wwindea.org (accessed on October 12, 2006).

56. Verne, J. (1888) *The Mysterious Island*, Atheneum, New York.

57. Shinnar, R. (2003) The hydrogen economy, fuel cells, and electric cars, *Technology in Society*, 25(4), 453–476.

58. Shinnar, R., Shapira, D., and Zakai, A. (1981) Thermochemical and hybrid cycles for hydrogen production—A differential comparison with electrolysis, *IEC Process Design Development*, 20, 581.

59. Kreith, F. and West, R.E. (2004) Fallacies of a hydrogen economy: A critical analysis of hydrogen production and utilization, *Journal of Energy Resources Technology*, 126, 249–257.

60. National Research Council (2000) *Renewable Power Pathways—A Review of the U.S. Department of Energy's Renewable Energy Programs*, National Research Council, National Academics Press, Washington, DC.

61. National Research Council, National Academy of Engineering (2000) *The Hydrogen Economy: Opportunities, Costs, Barriers, and R & D Needs Draft*, The National Academies Press, Washington, DC, see www.nap.edu

62. Population Division of the Department of Economic and Social Affairs of the United Nations Secretariat (March 2009) World population prospects: The 2008 revision, http://esa.un.org/unpp (accessed on April 18, 2013).

63. Energy Information Agency (2011) U.S. DOE annual energy review 2010, Washington, DC.

64. IAEA, *Power Reactor Information System (PRIS)*, Vienna, Austria, http://www.iaea.org/programmes/a2/

65. *Nuclear Power Daily* (2009) Progress Energy Florida signs contract for new, advanced-design nuclear plant, January 16, 2009. http://www.nuclearpowerdaily.com/reports/Progress_Energy_Florida_Signs_Contract_For_New_Advanced_Design_Nuclear_Plant_999.html (retrieved on May 18, 2009).

66. Neij, L. et al. (January 2003) The use of experience curves for assessing energy policy programs, Presented at *EU/IEA Workshop*, IEA, Paris, France.

67. Parente, V. et al. (September 2002) Comments on experience curves for PV modules, *Prog. Photovolt. Res. Appl.*, 10(8), 571.

68. Goldemberg, J. et al. (March 2004) Ethanol learning curve—The Brazilian experience, *Biomass Bioenergy*, 26(3), 301.

69. Claeson Colpier, U. and Cornland, D. (March 2002) The economics of the combined cycle gas turbine—an experience curve analysis, *Energy Policy*, 30(4), 309.

70. The World Factbook (May 2, 2012) CIA. https://www.cia.gov/library/publications/the-world-factbook/print/textversion.html

71. BP (June 2011) Statistical review of world energy. www.bp.com/statisticalreview
72. Freeston, D.H. (1995) Worldwide direct uses of geothermal energy, in: *Proceedings of the World Geothermal Congress, 1995*, International Geothermal Association, Florence, Italy, 1995, pp. 15–26.
73. U.S. Energy Information Administration (August 19, 2010) 2010 estimate includes non-conventional (fracking) sources, *Annual Energy Review*, Web: October 15, 2011.
74. Potential Gas Committee (April 27, 2011) *Potential Gas Committee Reports Substantial Increase in Magnitude of U.S. Natural Gas Resource Base*, Web: October 15, 2011.
75. National Petroleum Council (July 18, 2007), 2007 estimate without non-conventional sources, *Facing the Hard Truths about Energy*, Web. October 15, 2011.
76. European Nuclear Society, Nuclear power plants, worldwide, http://www.euronuclear.org/info/encyclopedia/n/nuclear-power-plant-world-wide.htm
77. IEA (2005) *Renewables Information*, IEA, Paris, France.
78. Worldwatch Institute (2005) *Signposts 2004*, CD-ROM, Washington, DC.
79. Prometheus Institute (April 2007) 23rd annual data collection—Final, *PVNews*, 26(4), pp. 8–9.
80. Prometheus Institute and Greentech Media (April 2009) 25th annual data collection results: PV production explodes in 2008, *PVNews*, 28(4) (April 2009), 15–18.
81. Jochem, E. (2000) Energy end-use efficiency, in: *World Energy Assessment, Energy and the Challenge of Sustainability*, UNDP, New York, pp. 173–218, Chapter 6, http://www.undp.org/content/dam/undp/library/Environment%20and%20Energy/Sustainable%20Energy/wea%202000/chapter6.pdf

Suggested Readings

Boyle, G. (2004) *Renewable Energy—Power for a Sustainable Future*, Oxford University Press, Oxford, U.K.
Caputo, R. (2009) *Hitting the Wall—A Vision of a Secure Energy Future*, Morgan & Claypool Publishers, San Rafael, CA.
Goswami, D.Y., Kreith, F., and Kreider, J. (2000) *Principles of Solar Engineering*, 2nd edn., Taylor & Francis, Philadelphia, PA.
Heinberg, R. (2005) *The Party's Over—Oil War and the Fate of Industrial Society*, New Society Publishers, Gabriola Island, British Columbia, Canada.
Kreider, J. and Keith, F. (1981) *Solar Energy Handbook*, McGraw-Hill, New York.
Kreith, F. and West, R.E. (1996) *Handbook of Energy Efficiency*, CRC Press, Boca Raton, FL.
Metz, B., Davidson, O.R., Bosch, P.R., Dave, R., and Meyer, L.A. (eds.) (2007) *Contribution of Working Group III to the Fourth Assessment Report of the Intergovernmental Panel on Climate Change, 2007—IPCC Fourth Assessment Report (AR4)*, Cambridge University Press, Cambridge, U.K.
National Research Council (1982) *Energy in Transition 1985–2010: Final Report of the Committee on Nuclear and Alternative Energy Systems*, National Academies Press, Washington, DC.
Tester, J.W., Drake, E.M., Driscoll, M.J., and Foley, M.W. (2012) *Sustainable Energy: Choosing among Options*, MIT Press, Cambridge, MA.

2

Economics of Energy Generation and Conservation Systems*

> Anyone who believes exponential growth can go on forever in a finite world is either a madman or an economist.
>
> **—Kenneth Boulding (ca. 1980)**

This chapter presents useful analysis tools that engineers should be familiar with. These include a basic process for economic assessment, energy flow analysis, and resource accounting methods for complex energy systems. The objective of economic assessment is to provide useful financial insight and enable a comparison between various energy options. No familiarity with microeconomics is presumed, and given the uncertainty in predicting a future economic environment, one could argue that added complexity may not be warranted. The standard methods presented here are equally applicable to renewable and nonrenewable energy systems, as well as to energy conservation projects. The objectives of energy flow analysis and resource accounting are to conceptualize and compare different energy system investments, to account for resource and external costs, and to identify sustainability issues.

More than 85% of world commercial energy production is currently from fossil resources. Hydropower is the largest renewable electricity resource and has been fully developed in the United States for several decades. Other renewables are experiencing rapid growth, but from a small base and with upper limits on capacity near that for hydro. There are plentiful fossil resources, and fossil fuels will continue to be the most convenient and valuable energy resources for the near future. But fossil resources are exhibiting depletion effects, and the future growth of net energy supply is increasingly unlikely. The approach of sustainable energy engineering is to use quantitative analysis to understand the end use, environmental and social requirements, and the resource availability, and then help to make development decisions based on analysis of the dynamics of energy activity systems into the future. This analysis should lead to adaptive design innovations that meet the environmental, social, and economic requirements, manage risks, and provide benefits. This section explores several different ways to understand the energy/social/economic system and the ways that this understanding can inform investment and development decisions.

2.1 Unit Cost of Energy

The source of energy in renewable energy (RE) systems such as wind turbines and solar collectors is essentially free, but so are oil, coal, and gas in the current economic system that only requires purchase of mineral rights and a small royalty on mining as the price of access to fossil resources. The costs for energy prospecting, extraction, refining, and

---

* Sections in this chapter marked with an asterisk may be omitted in an introductory course.

generation are largely associated with the equipment purchased and the fuels or electricity used [1]. Neglecting interest charges on capital and operating costs, the unit cost of energy produced by a system, C_s dollars, with initial construction and installation cost, C_0 dollars, over the life of the system, t years, and annual energy production, Q units of energy, is given by

$$C_s = \frac{C_0}{Q \cdot t} \tag{2.1}$$

For example, consider the Solartech 40 W solar PV panel that has a list price of $167 for a 0.38 m² panel ($440/m²). The panel has a 10-year warranty for 90% of nominal power output and 25-year warranty for 80% power output. The performance specifications are measured at standard test conditions (25°C, 1000 W/m²) giving conversion efficiency of 10.5%, and the panel is installed in a location where the mean horizontal surface irradiance is 200 W/m². The electricity production can be estimated for a year as $Q = 0.2$ (kW/m²) × 10.5% × 8760 (h/year) = 184 (kWh/year/m²). The cost of solar electricity can be estimated by using Equation 2.1:

$$C_s = \frac{440(\$/m^2)}{184(kWh/year \cdot m^2) \times (0.9 \times 10\, year + 0.8 \times 15\, year)} = \$0.11/kWh$$

Note that we use three significant figures when evaluating energy costs, as the methods do not have accuracy to justify fractions of cents. The sensitivity of energy production cost to the technical factors and resource factors can be explored using Equation 2.1. If the solar PV panel were installed in a location with only 100 W/m² mean solar radiation, although the system cost would be the same, the production cost would increase to $0.22/kWh. If it were possible that research in materials and manufacturing for the solar PV panel could increase efficiency to 20%, reduce cost to $300/m², and improve lifetime and performance degradation to 90% power output over 30 years, then the production cost would be reduced to just $0.03/kWh.

2.2 Payback Period

A common method of evaluating the value of an energy generation or conservation system is to determine the time required for the system to payback the initial costs. The payback period (PP) is determined by comparing the total cost of equipment and installation, C_0, to the annual benefit in either value of net energy produced or net energy saved per year. The value of the energy saved equals the amount of energy saved per year, B_i (kWh/year) times the unit cost of energy, C_i ($/kWh). Once the yearly value of the net saved energy and the cost of installing a conservation measure are known, the simple PP in years is

$$PP = \frac{C_0(\$)}{B_i C_i(\$/year)} \tag{2.2}$$

This approach is acceptable for preliminary estimates if the PP is short, say less than 4 years. For a more precise estimate, or for longer horizons, the time value of money, the inflation rate, and the escalation in fuel costs or maintenance costs must be considered.

2.3 Time Value of Money

If money is deposited in a bank, it earns interest. Similarly, when money (capital) is invested in a project or a business that is financed through a purchase of bonds or stock, a certain rate of return on the investment is expected, to compensate the owners of the invested capital for both the opportunity cost of having their money temporarily unavailable for other uses, and also for the risk of loss, since few investments are "sure things." Conversely, if you borrow money from a bank, you have to pay interest (or dividends) as the cost of the money lent to you. These issues are known as the *time value of money*.

Since it is convenient to conduct economic analyses in a unit of fixed monetary value, a concept known as *present worth* is used. The present worth of a future cash flow or a future payment is the future value of that flow with the time value of the money removed. The opportunity for investing money at a given time with the expectation of a future return is known as the time value of money. For example, the future value F, of a sum of money P, invested today at an annual interest rate i is

$$F = P(1+i)^t \qquad (2.3)$$

where t is the future time in years or other time unit corresponding to the time basis of the interest rate i in percent. This equation essentially states that an initial sum of money P appreciates by a multiplier factor, $(1 + i)$ each year, disregarding monetary inflation.

Stated alternatively, a future sum of money F has a *present worth P* given by

$$P = \frac{F}{(1+i)^t} \qquad (2.4)$$

This indicates that the present worth of a given amount of money in the future is discounted, in constant dollars, by a factor of $(1 + i)^{-t}$ for each year in the future. In this context, i is generally called the discount rate, but is also referred to as the cost of money or rate of return. Note that this approach assumes that i remains constant over the period of concern. In real life, however, it may well not be constant and constitutes a source of uncertainty.

The factor $(1 + i)^{-t}$ is called the present worth factor [PWF(i, t)] and is given by

$$\mathrm{PWF}(i,t) = \frac{1}{(1+i)^t} \qquad (2.5)$$

To obtain the present value of a cash flow in the future at time t, it is multiplied by its PWF(i, t). Table 2.1 is a tabulation of PWFs for various interest rates, i, and number of years, n.

TABLE 2.1

Present Worth Factors (PWF [*i*, *t*])

Year	Interest Rate%													
	0	1	2	3	4	5	6	7	8	9	10	11	12	13
1	1.000	0.990	0.980	0.971	0.962	0.952	0.943	0.935	0.926	0.917	0.909	0.901	0.893	0.885
2	1.000	0.990	0.961	0.943	0.925	0.907	0.890	0.873	0.857	0.842	0.826	0.812	0.797	0.783
3	1.000	0.971	0.942	0.915	0.889	0.864	0.840	0.816	0.794	0.772	0.751	0.731	0.712	0.693
4	1.000	0.961	0.924	0.888	0.855	0.823	0.792	0.763	0.735	0.708	0.683	0.659	0.636	0.613
5	1.000	0.951	0.906	0.863	0.822	0.784	0.747	0.713	0.681	0.650	0.621	0.593	0.567	0.543
6	1.000	0.942	0.888	0.837	0.790	0.746	0.705	0.666	0.630	0.596	0.564	0.535	0.507	0.480
7	1.000	0.933	0.871	0.813	0.760	0.711	0.665	0.623	0.583	0.547	0.513	0.482	0.452	0.425
8	1.000	0.923	0.853	0.789	0.731	0.677	0.627	0.582	0.540	0.502	0.467	0.434	0.404	0.376
9	1.000	0.914	0.837	0.766	0.703	0.645	0.592	0.544	0.500	0.460	0.424	0.391	0.361	0.333
10	1.000	0.905	0.820	0.744	0.676	0.614	0.558	0.508	0.463	0.422	0.386	0.352	0.322	0.295
11	1.000	0.896	0.804	0.722	0.650	0.585	0.527	0.475	0.429	0.388	0.350	0.317	0.287	0.261
12	1.000	0.887	0.788	0.701	0.625	0.557	0.497	0.444	0.397	0.356	0.319	0.286	0.257	0.231
13	1.000	0.879	0.773	0.681	0.601	0.530	0.469	0.415	0.368	0.326	0.290	0.258	0.229	0.204
14	1.000	0.870	0.758	0.661	0.577	0.505	0.442	0.388	0.340	0.299	0.623	0.232	0.205	0.181
15	1.000	0.861	0.743	0.642	0.555	0.481	0.417	0.362	0.315	0.275	0.239	0.209	0.183	0.160
16	1.000	0.853	0.728	0.623	0.534	0.458	0.394	0.339	0.292	0.252	0.218	0.188	0.163	0.141
17	1.000	0.844	0.714	0.605	0.513	0.436	0.371	0.317	0.270	0.231	0.198	0.170	0.146	0.125
18	1.000	0.836	0.700	0.587	0.494	0.416	0.350	0.296	0.250	0.212	0.180	0.153	0.130	0.111
19	1.000	0.828	0.686	0.570	0.475	0.396	0.331	0.277	0.232	0.194	0.164	0.138	0.116	0.098
20	1.000	0.820	0.673	0.554	0.465	0.377	0.312	0.258	0.215	0.178	0.149	0.124	0.104	0.087

As an example, consider a discount rate of 5% and a cash inflow of $100, 1 year from now. Using the earlier formula, the present worth is only $95. Conversely, if $95 was invested at 5% interest today, it would be worth approximately $100 a year from now. The same approach can be used to evaluate cash flows at any future period.

The present worth concept is the key for a discounted cash flow analysis. Its use permits all calculations to be made in terms of *present discounted monies*. Every kind of business or organization has a discount rate that it uses and that reflects the rate of return it expects on investments. Remember, the company could invest capital in making more products and expect a certain rate of return, determined by their experience with past investments. If they instead invest that capital in an energy conservation project, they want to know how that investment compares to other options. The mathematics of the discounted value of money demonstrates why revenues realized in the future are worth less than money earned at the present. For example, $100 of revenue realized 40 years in the future at an annual discount rate i of 10% has a present worth of only $1.83 today. This explains why the far future has little influence on conventional business planning and why many businesses shy away from investments in RE systems or conservation measures that have high up-front costs and long-term benefits.

When future cash flows are fixed in size, F, and regularly occur over a specific number of periods, the situation is known as an annuity. For an annuity, the present worth is given by Equation 2.6.

$$P = F\frac{(1+i)^t - 1}{i(1+i)^t} \tag{2.6}$$

For example, assume a cash flow of $100/year at the end of each year for the next 5 years at a discount rate of 10%. According to Equation 2.6, the present value of the $500 of inflow is only $379.

$$P = \$100 \times \frac{(1+0.10)^5 - 1}{0.10(1+0.10)^5} = \$379$$

The present worth is most commonly used to compare different options for investment in different energy management or energy conservation opportunities as illustrated in the following example.

Example 2.1

A company has a discount rate of 10% for its investments. The company has two different energy conservation opportunities. Assume that opportunity A generates benefits equal to $100, $150, and $200 at the end of years 1, 2, and 3, respectively. Assume that opportunity B yields benefits of $225 in year 2 and $225 in year 3. Over 3 years, both opportunities A and B yield benefits of $450. However, the timing of the benefits received is different in each case. By using the present worth technique, the two benefit flows can be viewed in terms of today's dollar value.

Solution

Step 1
Compute the PWF using Equation 2.5 or Table 2.1 for the discount rate of $i = 10\%$

$$PWF(10,1) = \frac{1}{(1+0.1)^1} = 0.909$$

$$PWF(10,2) = \frac{1}{(1+0.1)^2} = 0.826$$

$$PWF(10,3) = \frac{1}{(1+0.1)^3} = 0.751$$

Step 2
Compute the present worth of each opportunity benefit flow by multiplying the PWF by the annual benefit amount. Note that we use the same significant figures to report results as was given in the cost and benefit estimates.

Year	PWF	Annual Benefit ($)		Present Worth ($)	
		A	B	A	B
1	0.909	100	0	91	0
2	0.826	150	225	124	186
3	0.751	200	225	150	169
Total		450	450	365	355

Note that although the total benefits are the same, the present worth for case opportunity A is higher because the annual benefits occur sooner. Since opportunity A provides cash sooner, that money can be invested for additional financial return, which is more beneficial for the investor.

2.4 Inflation

The cost and revenues of a project can be expressed either in "current dollars" or "constant dollars." Actual cash flows observed in the marketplace at a given time are called current dollar cash flows. They are the actual number of dollars required or acquired in the year the cost or the benefit is incurred. The base year is defined as the year around which an analysis is structured. Most commonly, it is the initial period, which may be denoted by 0. In the base year, constant and current dollar cash flows are the same. The future current dollar equivalent, F_t, of a current amount, F_0 dollars, after t years at inflation rate of j per cent per year is

$$F_t = F_0(1+j)^t \tag{2.7}$$

For example, energy savings due to efficiency improvement of $100 in current 2013 dollars, assuming 4% inflation over the next 10 years, will be worth $66.50 in 2023 dollars. The effect of inflation on the future value is to reduce the value by a factor $(1 + j)$ per year.

Inflation rate in the future will not be constant and will not be known, which constitutes a source of uncertainty in financial analysis (in fact, the major source). The most common way to deal with inflation risk is to add an extra percentage to the future annual inflation rate.

In practice, it is important to distinguish between real and nominal discount rates. A real discount rate excludes inflation, whereas a nominal value includes inflation:

$$i' = i + j \tag{2.8}$$

where

i' is the nominal discount rate
i is the real discount rate in the absence of inflation
j is the inflation rate

For example, given an inflation rate of 6% and a nominal discount rate of 10%, the real discount rate is 4%. For power generation projects the typical real discount rate is usually 3%–4%.

When calculating a present value, it is important to remember "that the real discount rate must be used if cash flows are to be in constant dollars, whereas a nominal discount rate is to be used if the cash flow is in current dollars."

If it is necessary to convert a stream of cash flows from current dollars to constant dollars or vice versa, it is useful to refer to an inflation rate index. There are a number of inflation rate indices, for example, the consumer price index (CPI). The CPI measures the average change in prices of goods and services at the retail end. The GNP is the measure of output supplied by U.S. labor and property. Table 2.2 presents the changes in the GNP, as well as the CPIs from 1970 to 2012. Additional information and up-to-date data can be obtained from the U.S. Department of Commerce, which publishes the *Statistical Abstract of the United States* every year (available from http://www.ntis.gov/products).

The effect of inflation can be integrated into the present value analysis by using the nominal discount rate:

$$PV(F_0) = \frac{F_t}{(1+i')^t} \tag{2.9}$$

It should be noted that if all cash flows are inflated at the same rate and are discounted with the appropriate discount rate, then working with either all constant or all current dollars gives the same result for PV, as demonstrated in Examples 2.2 and 2.3.

2.5 Societal and Environmental Costs

The use of conventional technologies, such as coal-fired power plants, to generate electricity creates environmental degradation and societal costs that are usually considerably higher than those from most renewable technologies. These hidden costs are called *externalities* and are usually not included in conventional energy economics. Estimates of

TABLE 2.2

Selected Inflation Indices

Year	GDP Implicit Price Deflator (2000 = 100)[a]	GNP Implicit Price Deflator (2000 = 100)[a]	Consumer Price Indices (83–84 = 100)[b]	Annual Inflation Rate
1970	27.53	27.5	38.8	5.84
1971	28.91	28.9	40.5	4.3
1972	30.16	30.1	41.8	3.27
1973	31.85	31.8	44.4	6.16
1974	34.73	34.7	49.3	11.03
1975	37.99	38.0	53.8	9.20
1976	40.19	40.2	56.9	5.75
1977	42.74	42.7	60.6	6.50
1978	45.74	45.7	65.2	7.62
1979	49.54	49.5	72.6	11.22
1980	54.04	54.0	82.4	13.85
1981	59.12	59.1	90.9	10.35
1982	62.73	62.7	96.5	6.16
1983	65.19	65.2	99.6	3.22
1984	67.65	67.6	103.9	4.30
1985	69.71	69.7	107.6	3.55
1986	71.25	71.2	109.6	1.91
1987	73.19	73.2	113.6	3.66
1988	75.69	75.7	118.3	4.08
1989	78.55	78.5	124	4.83
1990	81.6	81.6	130.7	5.39
1991	84.4	84.4	136.2	4.25
1992	86.4	86.4	140.3	3.03
1993	88.4	88.4	144.5	2.96
1994	90.3	90.3	148.2	2.61
1995	92.1	92.1	152.4	2.81
1996	93.8	93.9	156.9	2.93
1997	95.4	95.4	160.5	2.34
1998	96.5	96.5	163	1.55
1999	97.9	97.9	166.6	2.19
2000	100.0	100.0	172.2	3.38
2001	102.4	102.4	177.1	2.83
2002	104.2	104.2	179.9	1.59
2003	106.4	106.4	184	2.27
2004	109.5	109.4	188.9	2.68
2005	113.0	113.0	195.3	3.39
2006	116.7	116.7	201.6	3.24
2007	119.8	119.8	207.3	2.85
2008	122.4	122.4	215.3	3.85
2009	123.4	123.4	214.5	−0.34
2010	125.1	125.1	218.1	1.64
2011	127.8	127.8	224.9	3.16

[a] Data from http://www.bea.gov/
[b] Data from http://www.bls.gov/cpi/

societal costs have been made by many different studies. The results vary widely, because they depend on many factors that are difficult to pinpoint. For example, what is the appropriate discount rate for long-term effects? How does one assign a monetary value to items such as human health or life? How does one treat the effects of global warming?

The range of externality estimates, which does not include global warming, is shown in Figure 2.1 [2]. A 1992 study by the American Solar Energy Society Economics of Solar Energy Technologies [3] estimated that the median externality cost of U.S. energy consumption is about $2 per million Btu. For electricity, the estimate of societal cost is about $0.02/kWh, because typically 1 million Btu of heat is used to produce 100 kWh of electricity. Since 8 gal of gasoline contains about 1 million Btus, the societal cost of burning gasoline in automobiles is at least $0.25/gal. While it is difficult to precisely determine these externality costs, they are real, even if they do not appear on a utility bill. A methodology for evaluating the economic cost of the environmental impacts of energy generation is presented in [4].

The major societal cost of renewable technologies comes from building the equipment necessary for the energy conversion. Once built, with the exception of biomass, and to a certain degree, hydro reservoirs and noncondensable gases from geothermal, RE technologies do not emit polluting gases into the atmosphere. Burning biomass, if the combustion engineering and operation are modern and meet clean air standards, produces less pollution than burning coal or natural gas, because biomass has a much smaller sulfur and ash content. Furthermore, burning biomass does not increase the concentration of CO_2 in the atmosphere from fossil sources, because an approximately equivalent amount of CO_2 is absorbed in growing the feedstock. The key goal for biomass utilization is either to grow the feedstock in a sustainable manner that does not impact food production or to utilize waste products such as wood chips or wood industry wastes from sustainably managed forests.

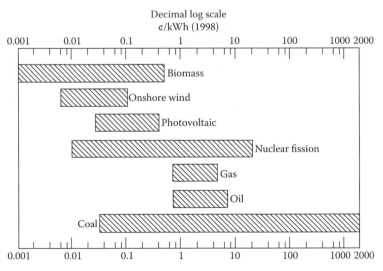

FIGURE 2.1
Externality estimates in ¢/kWh. (Reprinted from *Futures*, 24, Stirling, A. Regulating the electricity supply industry by valuing environmental effects: How much is the emperor wearing?, 1024–1047, Copyright 1992, with permission from Elsevier.)

2.6 Total Life Cycle Costs

A total life cycle cost (TLCC) analysis should consider all significant dollar costs over the life of a project. These costs are then discounted to a base year value using present value analysis. Any revenue generated from the resale of an investment is also discounted to the base year and subtracted from present value costs. TLCC analysis includes many different costs and must use the appropriate form of the cash flows (current or constant dollars) so that the correct discount rate is applied. If the cash flows are not discounted with the appropriate rate, the analysis will be flawed. The TLCC analysis is normally done to evaluate differences in the costs and the timing of these costs between alternative projects. TLCC formula for calculating the present value of TLCC is

$$TLCC = \sum_{t=0}^{N} \frac{C_t}{(1+i')^t} \tag{2.10}$$

where
   C_t is the cost in period t, including finance charges, operating and maintenance (O&M) costs, and energy costs
   N is the analysis period
   i' is the nominal discount rate

Since the major costs associated with most conservation and RE technologies are the initial investment costs and O&M costs (including fuel), the formula for TLCC for residential, government, and nonprofit organizations simplifies to

$$TLCC = C_0 + PVOM$$

where
   C_0 is the initial investment
   PVOM is the present value of all O&M costs

$$PVOM = \sum_{t=0}^{N} \frac{(O\&M)_t}{(1+i)^t} \tag{2.11}$$

If the investment is by a for-profit organization, taxes must be included in the calculations as shown as follows:

$$TLCC = C_0 - (T \times PVDEP) + PVOM(1 - T) \tag{2.12}$$

where
   T is the income tax rate
   PVDEP is the present value of depreciation

Example 2.2

A nonprofit organization invests $10,000 in new double-paned windows and estimates that O&M costs will be $1,300 in the first year for new painting and no salvage value. Assuming an annual inflation rate of 3% and a nominal discount rate of 12%, Table 2.3 demonstrates the TLCC evaluation, both in current dollars and in constant dollars. Note that the TLCC is the same, but the discount rate and cash flows are different.

TABLE 2.3

Current and Constant Dollar TLCC Evaluation

Year	Investment	Discounted Investment	Current Dollar O&M Costs	Discounted O&M Costs (w/12% Nominal)	Constant Dollar O&M Costs	Discounted O&M Costs (w/8.74% Real)
0	$10,000	$10,000	$0	$0	$0	$0
1	0	0	$1,339	$1,196	$1,300	$1,196
2	0	0	$1,379	$1,100	$1,300	$1,100
3	0	0	$1,421	$1,011	$1,300	$1,011
4	0	0	$1,463	$930	$1,300	$930
5	0	0	$1,507	$855	$1,300	$855
NPV		$10,000		$5,091		$5,091

TLCC = $15,091.

TABLE 2.4

After-Tax TLCC Evaluation

Year	Investment	Discounted Investment	Depreciation	Discounted Depreciation	O&M Costs	Discounted O&M Costs
0	$10,000	$10,000			$0	$0
1	0	0	$4,000	$3,571	$1,339	$1,196
2	0	0	$2,400	$1,913	$1,379	$1,100
3	0	0	$1,440	$1,025	$1,421	$1,011
4	0	0	$1,080	$686	$1,463	$930
5	0	0	$1,080	$613	$1,507	$855
NPV		$10,000		$7,809		$5,091

TLCC = $10,000 − (0.34 × $7,809) + ($5,091 × 0.66) = $10,705.

Example 2.3

Suppose that the investor of Example 2.2 is a private company in the 34% federal income bracket but pays no state income tax. The initial investment is $10,000, O&M is $1,300 per year in current dollars in year zero with an inflation rate of 3%, and the nominal discount rate is 12%. The investor uses a "5 year double-declining depreciation schedule" as shown in column 4, and there is no salvage value. The results are shown for each of the steps for evaluating the after-tax TLCC (Table 2.4).

*2.7 Internal Rate of Return

The internal rate of return (IRR) analysis is a convenient procedure to compare a variety of investment activities. It is commonly used for accept/reject decisions by comparing the IRR with a minimal acceptable rate, often referred to as the "hurdle rate." The IRR of an investment that has a series of future cash flows (F_0, F_1, ..., F_t) is that rate of return on investment that causes the NPV of cash flows to equal zero. IRR implicitly assumes reinvestments of any return at the IRR rate and is therefore not recommended for ranking projects. However, it has the advantage of giving an investor a quick comparison of the after-tax return on various financial instruments such as bonds and providing the investor with a quick assessment of energy management project options. The net present value (NPV) can be expressed in equation form as

$$\text{NPV} = \sum_{t=0}^{N} \frac{F_t}{(1+i')^t} \tag{2.13}$$

where
  NPV is the net present value of the capital investment
  F_t is the net cash flow received at time t
  i' is the nominal discount rate

As mentioned before, IRR is the rate of return that will make the NPV equal to zero. In equation form,

$$\text{NPV} = 0 = \sum_{t=0}^{N} \frac{F_t}{(1+D)^t} \tag{2.14}$$

where D is the rate of return that equates the present value of positive and negative cash when used as the discount rate.

Example 2.4

As in Example 2.3, an investment of $10,000 is made initially. The O&M cost in the first year is $1300, and it inflates at a rate of 3% per year thereafter. A 5 year double-declining depreciation schedule is used with a marginal federal income tax rate of 34%. There is no state income tax and no salvage value. The project saves 1000 units of energy in its first year of operation, 950 in the second, 925 in the third, and 900 in the fourth and fifth years. The energy is sold at $10/unit at year zero and inflates at 3% per year thereafter. The results of the IRR analysis calculations are shown in Table 2.5. The last column in

TABLE 2.5

After-Tax IRR

Year	Investment (a)	O&M Costs (b)	Revenue (c)	Depreciation (d)	Not Taxable Income (e) = (c − b − d)	Federal Income Tax (f) = e × tax	After-Tax Cash Flow $F_n(g)$ = (c − a − b − f)	$F_n/(1.61)^n$
0	$10,000	$0	$0	$0	N/A	N/A	−$10,000	
1	0	1,339	10,300	4,000	$4,961	$1,687	7,274	$4,516
2	0	1,379	10,079	2,400	6,300	2,142	6,558	2,530
3	0	1,421	10,108	1,440	7,247	2,464	6,223	1,490
4	0	1,463	10,130	1,080	7,587	2,580	6,087	900
5	0	1,507	10,434	1,080	7,847	2,668	6,259	580
								$10,010

the table demonstrates that if a return of 61% were earned from the after-tax cash flow each year, the total cash flow would equal $10,000, which is the initial investment. A trial-and-error method may be used to obtain the appropriate IRR. For this example, if the hurdle rate were higher than 61%, then an investment in another project would be appropriate, but if the hurdle rate is lower than 61%, the investment in this project represents a good opportunity.

Potential problems with IRR analyses are described in [2]. One of them is that IRR might overstate profitability, because it assumes that all interim proceeds are reinvested at a rate equal to the IRR. Hence, for projects of different scales and lives, it is possible for different ranking criteria such as NPV and IRR to produce conflicting results as a consequence of the differing reinvestment assumptions. The IRR method can, however, be modified, and this modified IRR analysis method can be found in Ref. [2].

Example 2.5

Estimate the payback time and the IRR for a PV system with rated production capacity of 3 kW (@25°C, 1000 W/m²) installed in Boulder, Colorado (40°N latitude, 105°W longitude), in 2008 (see Figure 2.2). The cost of the system was $25,294. The electric utility serving the customer offered a rebate of $4.50 per Watt in order to meet the RE mandate in the Colorado constitution that requires that 20% of all electric power be from renewable sources by 2020. Moreover, the federal government offered a tax credit capped at $2000 for the 2008 tax year.

Use the following assumptions:

Price of electricity: $0.152 per kWh
Average annual rate of power cost increase: 5% per year
Tilt: 20°
Azimuth: 180°
Discount rate: 6%
Inflation rate: 3%

To find the system's annual energy output, input the system parameters at the following website: http://rredc.nrel.gov/solar/calculators/PVWATTS/version1/

FIGURE 2.2
Kreith residence, 3 kW Sanyo high-efficiency photovoltaic system. (See also Figure 6.1 in color insert.)

Solution

To find the value of the annual savings, A, we first multiply the annual energy output by the price of electricity. Since the cost of electricity will increase 5% per year, we must project the cost of energy for the next 20 years. The future cost of energy, F_t, is defined in Equation 2.7, where F_0 is the present cost of energy, r is the rate of increase, and t is the year number.

$$F_t = F_0(1+r)^t$$

So in the second year, $F_2 = 0.152(1 + 0.05)^1 = \$0.16$. Then in the third year, as for each successive year, the value of F_t for the previous year becomes the value of F_{t-1}. So in the 20th year, electricity will cost $\$0.38$ per kWh. Then, the electricity costs for each year are used to calculate annual savings by multiplying by 4207 kWh/year.

The future costs of energy for each of the next 20 years are shown in Table 2.6.

The annual values of electricity produced by the PV system are found by multiplying each year's respective cost of energy by the annual energy output. The results are shown in Table 2.7.

The savings this electricity production represents will be cumulative and must be adjusted for the effective interest rate. To find the cumulative savings, simply add all of the annual savings values. The cumulative savings is $\$21,117$. The effective discount rate, denoted by i', is found in Equation 2.8, giving the effective discount

TABLE 2.6

Future Cost of Energy

Year	1	2	3	4	5	6	7	8	9	10
Price/kWh	$0.15	$0.16	$0.17	$0.18	$0.18	$0.19	$0.20	$0.21	$0.22	$0.24
Year	11	12	13	14	15	16	17	18	19	20
Price/kWh	$0.25	$0.26	$0.27	$0.29	$0.30	$0.32	$0.33	$0.35	$0.37	$0.38

TABLE 2.7

Value of Electricity Produced

Year	1	2	3	4	5	6	7
kWh	4207	4207	4207	4207	4207	4207	4207
Price/kWh	$0.15	$0.16	$0.17	$0.18	$0.18	$0.19	$0.20
Annual savings	$638.62	$670.55	$704.08	$739.29	$776.25	$815.06	$855.82

Year	8	9	10	11	12	13	14
kWh	4207	4207	4207	4207	4207	4207	4207
Price/kWh	$0.21	$0.22	$0.24	$0.25	$0.26	$0.27	$0.29
Annual savings	$898.61	$943.54	$990.71	$1,040.25	$1,092.26	$1,146.87	$1,204.22

Year	15	16	17	18	19	20	
kWh	4207	4207	4207	4207	4207	4207	
Price/kWh	$0.30	$0.32	$0.33	$0.35	$0.37	$0.38	
Annual savings	$1,264.43	$1,327.65	$1,394.03	$1,463.73	$1,536.92	$1,613.77	

TABLE 2.8

Annual Adjusted Net Savings

Year	1	2	3	4	5
Annual savings	$638.62	$670.55	$704.08	$739.25	$776.25
Cumulative savings	$638.62	$1,309.18	$2,013.26	$2,752.54	$3,528.79
Adjusted annual savings	$638.62	$651.02	$663.66	$676.55	$689.69
Adjusted cumulative savings	$638.62	$1,289.65	$1,953.31	$2,629.86	$3,319.55

Year	6	7	8	9	10
Annual savings	$815.06	$855.82	$898.61	$943.54	$990.71
Cumulative savings	$4,343.86	$5,199.67	$6,098.28	$7,041.81	$8,032.53
Adjusted annual savings	$703.08	$716.73	$730.65	$744.84	$759.30
Adjusted cumulative savings	$4,022.63	$4,739.36	$5,470.01	$6,214.85	$6,974.15

Year	11	12	13	14	15
Annual savings	$1,040.25	$1,092.26	$1,146.87	$1,204.22	$1,264.43
Cumulative savings	$9,072.78	$10,165.04	$11,311.91	$12,516.13	$13,780.56
Adjusted annual savings	$774.04	$789.07	$804.39	$820.01	$835.17
Adjusted cumulative savings	$7,748.19	$8,537.26	$9,341.66	$10,161.167	$10,997.61

Year	16	17	18	19	20
Annual savings	$1,327.65	$1,394.03	$1,463.73	$1,536.92	$1,613.77
Cumulative savings	$15,108.21	$16,502.24	$17,965.98	$19,502.90	$21,116.67
Adjusted annual savings	$852.17	$868.72	$885.58	$902.78	$920.31
Adjusted cumulative savings	$11,849.77	$12,718.49	$13,604.07	$14,506.85	$15,427.16

rate of 3%. This value is used in Equation 2.10 to adjust the yearly savings for the NPV of money. After each year's savings are adjusted for the time value of money, they are added together to find the cumulative NPV savings. The cumulative NPV savings of each year is shown in Table 2.8. After 20 years, the total cumulative NPV savings is $15,427.

After the deduction of the utility's rebate and the federal tax credit, the final net capital cost to the customer is $25,294 − ($4.50 × 3000) − $2000 = $9794. After 20 years, the customer has made a profit of $5653. The PP is calculated by comparing the cost of the system to the customer and the value of the energy saved per year. From Table 2.8, we can see that the adjusted cumulative savings (the cost of the energy saved) equals the net cost of the system to the customer between the 13th and 14th years of operation. Therefore the payback time of the system for the customer with the applied incentive and tax rebate is 13.5 years. The cost of the system is not paid back to the owner through energy savings without the incentives.

The IRR, found by using Equation 2.13, represents the summation of the discounted cash flows that will equal the initial investment. It is found by iterating on the discount rate until the summation of the yearly discounted cash flows equals zero. The results are shown in Table 2.9, which shows the calculation for a discount rate of 9.08%, representing the IRR.

TABLE 2.9

Table of Year-by-Year Calculations for IRR

Year	Net Investment $	O&M Cost $/Year	Annual Savings $/Year	Annual Adjusted Cash Flow[a] $/Year	Cash Flow Discounted at 9.08% $/Year
0	9794	0	−9794	−9794	−9794
1	0	0	671	692	634
2	0	0	705	733	616
3	0	0	740	777	599
4	0	0	777	824	582
5	0	0	816	873	565
6	0	0	857	925	549
7	0	0	900	981	534
8	0	0	945	1040	519
9	0	0	992	1102	504
10	0	0	1042	1168	490
11	0	0	1094	1239	476
12	0	0	1148	1313	463
13	0	0	1206	1392	450
14	0	0	1266	1475	437
15	0	0	1329	1564	425
16	0	0	1396	1757	413
17	0	0	1539	1862	401
18	0	0	1539	1862	390
19	0	0	1616	1974	379
20	0	0	1697	2092	368

Note: Summation of discounted cash flows = NPV.
[a] Adjusted for inflation.

2.8 Capital Recovery Factor

In many economic analyses, a *series* of annual or monthly sums is invested or used to pay off a loan. The series of equal periodic payments necessary to pay off such a loan is determined as follows. If a sum P_{sum} is invested each year at interest rate i, the present worth of the *sum* of these payments S is

$$S = \frac{P_{ann}}{1+i} + \frac{P_{ann}}{(1+i)^2} + \frac{P_{ann}}{(1+i)^3} + \cdots, \tag{2.15}$$

or

$$S = P_{ann}\left[(1+i)^{-1} + (1+i)^{-2} + (1+i)^{-3} + \cdots + (1+i)^{-t}\right]$$

The expression in the following brackets is a geometric series with the first term $P_{ann}/(1+i)$ and ratio $(1+i)^{-n}$. It follows from the expression for the sum of such a series that

$$P_{ann} = S\left[\frac{i}{1-(1+i)^{-t}}\right] \tag{2.16}$$

where P_{ann} is the annual payment required to achieve a sum S, after t years. It converts equal payments made at the end of each of t consecutive periods to the equivalent amount in the present using an interest or discount rate i. The term in brackets in Equation 2.16 is called the *capital recovery factor* (CRF(i, t)):

$$CRF(i,t) = \left[\frac{i(1+i)^t}{(1+i)^t - 1} \right]$$
(2.17)

Table 2.10 is a summary tabulation of CRFs. The term P_{ann} is the annual or periodic payment on a self-amortizing loan of value S. Each payment is a mix of interest and principal repayment. Early payments are mostly interest because of the large outstanding balance; later payments are primarily principal repayment. Most solar systems owned by private firms or individuals are purchased with self-amortizing loans. This repayment process is also widely used in paying for the loan of a house.

TABLE 2.10

Capital Recovery Factors (CRF [i, t])

	Interest Rate, i									
t	0%	2%	4%	6%	8%	10%	12%	15%	20%	25%
1	1.00000	1.02000	1.04000	1.06000	1.08000	1.10000	1.12000	1.15000	1.20000	1.25000
2	0.50000	0.51505	0.53020	0.54544	0.56077	0.57619	0.59170	0.16152	0.65455	0.69444
3	0.33333	0.34675	0.36035	0.37411	0.38803	0.40211	0.41635	0.43798	0.47473	0.51230
4	0.25000	0.26262	0.27549	0.28859	0.30192	0.31547	0.32923	0.35027	0.38629	0.42344
5	0.20000	0.21216	0.22463	0.23740	0.25046	0.26380	0.27741	0.29832	0.33438	0.37184
6	0.16667	0.17853	0.19076	0.20336	0.21632	0.22961	0.24323	0.26424	0.30071	0.33882
7	0.14286	0.15451	0.16661	0.17914	0.19207	0.20541	0.21912	0.24036	0.27742	0.31634
8	0.12500	0.13651	0.14853	0.16101	0.17401	0.18744	0.20130	0.22285	0.26061	0.30040
9	0.11111	0.12252	0.13449	0.14702	0.16008	0.17364	0.18768	0.20957	0.24808	0.28876
10	0.10000	0.11133	0.12329	0.13587	0.14903	0.16275	0.17698	0.19925	0.23852	0.28007
11	0.09091	0.10218	0.11415	0.12679	0.14008	0.15396	0.16842	0.19107	0.23110	0.27349
12	0.08333	0.09156	0.10655	0.11928	0.13270	0.14676	0.16144	0.18148	0.22526	0.26845
13	0.07692	0.08812	0.10014	0.11296	0.12652	0.14078	0.15568	0.17911	0.22062	0.26454
14	0.07143	0.08260	0.09467	0.10758	0.12130	0.13575	0.15087	0.17469	0.21689	0.226150
15	0.06667	0.07783	0.08994	0.10296	0.11683	0.13147	0.14682	0.17102	0.21388	0.25912
16	0.06250	0.07365	0.08582	0.09895	0.11298	0.12782	0.14339	0.16795	0.21144	0.25724
17	0.05882	0.06997	0.08220	0.09544	0.10963	0.12466	0.14046	0.16537	0.20944	0.25576
18	0.05556	0.06670	0.07899	0.09236	0.10670	0.12193	0.13794	0.16319	0.20781	0.25459
19	0.05263	0.06378	0.07614	0.08962	0.10413	0.11955	0.13576	0.16134	0.20646	0.25366
20	0.05000	0.06116	0.07358	0.08718	0.10185	0.11746	0.13388	0.15976	0.20536	0.25292
25	0.04000	0.05122	0.06401	0.07823	0.09368	0.11017	0.12750	0.15470	0.20212	0.25095
30	0.03333	0.04465	0.05783	0.07265	0.08883	0.10608	0.12414	0.15230	0.20085	0.25031
40	0.02500	0.03656	0.05052	0.06646	0.08386	0.10226	0.12130	0.15056	0.20014	0.25003
50	0.02000	0.03182	0.04655	0.06344	0.08174	0.10086	0.12042	0.15014	0.20002	0.25000
100	0.01000	0.02320	0.04081	0.06018	0.08004	0.10001	0.12000	0.15000	0.20000	0.25000
∞		0.02000	0.04000	0.06000	0.08000	0.10000	0.12000	0.15000	0.20000	0.25000

Note: For interest rates i from 0% to 25% and for periods of analysis n from 1 to 100 years.

2.9 Levelized Cost of Energy

The levelized cost of energy (LCOE) is widely used for comparing different energy generation technologies. For example, LCOE could be used to compare the cost of electric energy
from a renewable source with that from a nuclear generating unit. The LCOE is defined as
that cost that, if assigned to every unit of energy produced (or saved) by the system over the
analysis period, will equal the TLCC when discounted back to the base year. The LCOE is the
present value of all the investment costs plus operation and maintenance costs plus fuel costs
in each future year per unit power generation, which is also discounted to the present time:

$$LCOE = \frac{\sum_{t=1}^{N} \frac{C_t}{(1+i)^t}}{\sum_{t=1}^{N} \frac{Q_t}{(1+i)^t}} \qquad (2.18)$$

where
  Q_t is the electricity generated by the system (kWh) in year t (or the energy savings from
    conservation measures), calculated by the performance model based on weather data
    and system performance parameters, shown in the Energy row in the project cash
    flow
  N is the analysis period in years as defined on the financing page of SAM
  C_t is the after tax cash flow in the year t, equal to state tax savings + federal tax savings
    + PBI incentives – operating costs – debt total payment + energy value in the project
    cash flow

LCOE can use the real discount rate or the nominal discount rate. Substituting Equation
2.10, we can calculate the LCOE in terms of the total lifecycle cost:

$$LCOE = \frac{TLCC}{\sum_{t=1}^{N} \left(Q_t/(1+i)^t \right)} \qquad (2.19)$$

Note that, if the system energy output, Q_t (or the energy savings), is constant over the
time of analysis, the relation for LCOE simplifies to

$$LCOE = \left(\frac{TLCC}{Q} \right) \times CRF \qquad (2.20)$$

where Q is the annual energy output or energy saving in the case of a conservation system.
Using the data from Example 2.2, and assuming that the system produces 120,000 kWh per
year, the uniform capital recovery factor for $i' = 0.12$ and $t = 5$ is 0.277 and, therefore, the
nominal LCOE is

$$LCOE = (\$15,091/120,000) \times 0.277 = \$0.0348/kWh$$

The next example illustrates the LCOE estimate for a conservation measure.

Example 2.6

Calculate the TLCC and LCOE by replacing a 75 W incandescent light bulb by a 40 W fluorescent bulb. Assume that the incandescent bulb operates 6 h per night and needs to be replaced every year at a cost of $1.00, while the fluorescent bulb lasts for 5 years with an initial cost of $15. The cost of electricity over this period is $0.08/kWh and assume a discount rate of 12%.

Solution

The yearly energy requirement for the incandescent lamp is 164 kWh, while the 5 year discounted O&M cost, that is, the total discounted cost of annual replacement, is

$$\$1\left[1+\left(\frac{1}{1.12}\right)+\left(\frac{1}{1.12}\right)^2+\left(\frac{1}{1.12}\right)^3+\left(\frac{1}{1.12}\right)^4\right]=\$4.03$$

The discounted cost of electricity at the end of the first year is

$$\text{Cost of Electricity} = \frac{164\text{ kWh}}{\text{year}} * \frac{\$0.08}{\text{kWh}} * \left(\frac{1}{1.12}\right)^1 = \$11.71$$

The following table shows the discounted cost of electricity after every year and the total discounted cost of electricity assuming that the electricity is paid at the end of each year.

75 W Incandescent Light Bulb						
Year	1	2	3	4	5	Total
Cost of electricity	$11.71	$10.46	$9.34	$8.34	$7.44	$47.29

The TLCC for the incandescent light bulb can now be calculated to be

$$\text{TLCC} = \$4.03 + \$47.29 = \$51.32$$

The 40 W fluorescent bulb will require 87.6 kWh per year. The discounted electric power cost is $25.26 if the electricity is also paid at the end of each year. The following table shows the same information as the earlier table, but for the fluorescent bulb.

40 W Fluorescent Light Bulb						
Year	1	2	3	4	5	Total
Cost of electricity	$6.26	$5.59	$4.99	$4.45	$3.98	$25.26

Therefore, the TLCC for the fluorescent light bulb is

$$\text{TLCC} = \$15 + \$25.26 = \$40.26$$

Thus, the net 5-year savings is $11.06 and the CRF from Table 2.10 is 0.277. Hence, the levelized cost of the saved energy is

$$\text{LCOE} = \frac{\$11.06}{76.4 \text{ kWh}} \times 0.277 = \$0.04/\text{kWh}$$

Note that LCOE for an energy conservation measure should be based on the incremental costs and savings for the system over the analysis period.

More extensive economic analyses, including taxes, incentives, depreciation, and salvage value, are presented in [2].

Example 2.7

Consider the generation of electricity by a nuclear light water reactor (LWR). The application of the present worth and levelization concept is shown in Table 2.11. The levelized unit cost of electricity produced in cents/kWh at the busbar (the entry to the electric transmission grid) is obtained by equating levelized revenue to levelized expenditures for capital costs, O&M costs, and fuel costs (if applicable). To add realism to this example, as proposed in [5], the cost of money is given by a weighted sum of specified returns on bonds and anticipated return on stocks. The carrying charge rate assumes that bond interest is tax-free, which may not be applicable in all cases. Furthermore, according to *Nuclear Power Daily*, Westinghouse Electric Company and The Shaw Group estimate the total cost for two 1105 MW generating nuclear units to be approximately $14 billion, and therefore plant capital costs at time zero are about $6335/kW for the year 2009 [6]. This figure is computed as an overnight cost, that is, equivalent to hypothetical instantaneous construction, corrected for escalation and interest paid on borrowed funds over a construction period starting n years before operation begins. Additional details about this example can be found in [2].

The capacity factor, L, is the percentage of the hours per year the plant is generating at the nameplate generation capacity. Typically, nuclear units must shut down approximately 1 month for every 2 years of service for refueling. Renewable options are constrained by the diurnal and intermittent nature of sun and wind. Wind turbine capacity factors are typically about 34% for favorable inland locations and up to 50% for offshore locations. Photovoltaic (PV) and solar thermal generation units are mostly on land and may have somewhat lower capacity factors, 18%–26% for PV in favorable locations depending on whether or not tracking is used, and ~30% for solar thermal, most of which must track the sun. The evaluation of capacity factors for RE systems will be treated in later chapters.

The useful life of a plant, T, is a very important factor, but somewhat difficult to specify. The useful life of a nuclear plant was originally specified in the design as 30 years, but has been extended for most operating plants to somewhere between 40 and 55 years, whereas PV solar thermal generation units may last as long as 30–50 years, may reach an even longer life. Once the capital investment for a nuclear plant has been repaid, the cost of electricity produced from nuclear energy drops dramatically because repayment of the capital accounts for more than half of the levelized energy cost. Also note that societal costs have not been included in this example. The cost of externalities is often overlooked in conventional economic analyses. However, given the very long-lived impacts and costs of uranium mining, milling, processing, and safe, secure storage of both low-level and high-level radioactivity materials, a thorough accounting of externalities should be included.

TABLE 2.11

Lifetime-Levelized Busbar Cost of Electrical Energy from a Nuclear LWR[a]

e_b, The cost of electric power in cents per kilowatt hour (0.1 times mills per kilowatt hour) is the sum of

Capital-related costs

$$\frac{100\varphi}{8766 \cdot L}\left(\frac{I}{K}\right)_{-c}\left[1+\frac{x+y}{2}\right]^c$$

Plus O&M costs

$$+\frac{100}{8766 \cdot L}\left(\frac{O}{K}\right)_O\left[1+\frac{yT}{2}\right]$$

Plus fuel costs

$$+\left\{\begin{array}{l}\text{Nuclear}\left[\dfrac{100}{24}\dfrac{F_o}{\eta B}\right]\left[1+\dfrac{yT}{2}\right]\\[1em]\text{or}\\[1em]\text{fossil}\left[\dfrac{0.0034 f_o}{\eta}\right]\left[1+\dfrac{yT}{2}\right]\end{array}\right.$$

where	Typical LWR Value
L is the plant capacity factor: actual energy output + energy if always at 100% rated power	0.90
φ is the annual fixed charge rate (i.e., effective "mortgage" rate) = $x/(1-\tau)$ where x is the discount rate and τ is the tax fraction (0.4)	0.15/year
$X = (1-\tau)b \cdot r_b + (1-b)r_s$ in which b is the fraction of capital raised selling bonds (debt fraction), and r_b is the annualized rate of return on bonds, while r_s is the return on stock (equity)	0.09/year
$\left(\dfrac{I}{K}\right)_{-c}$ = overnight specific capital cost of plant, as of the start of construction, dollars per kilowatt: cost if it could be constructed instantaneously c years before startup in dollars without inflation or escalation	$6,335/kWe
y is the annual rate of monetary inflation (or price escalation, if different)	0.04/year
c is the time required to construct plant, years	5 years
T is the prescribed useful life of plant, years	40 years
$\left(\dfrac{O}{K}\right)_O$ = specific O&M cost as of start of operation, dollars per kilowatt per year	$95/kWe year
η is the plant thermodynamic efficiency, net kilowatt electricity produced per kilowatt of thermal energy consumed	0.33
F_O is the net unit cost of nuclear fuel, first steady-state reload batch, dollars per kilogram of uranium, including financing and waste disposal charges, as of start of plant operation	$2,000/kg
B is the burnup of discharged nuclear fuel, megawatt days per metric ton	45,000
f_o is the fossil fuel costs, at start of operation, cents per million British thermal units = (approximately) dollars per barrel times 16 for residual oil: dollars per ton times 4 for steam coal, cents per thousand standard cubic feet for natural gas	

(continued)

TABLE 2.11 (continued)

Lifetime-Levelized Busbar Cost of Electrical Energy from a Nuclear LWR[a]

Thus, for an LWR nuclear power plant, using the representative values cited earlier

$$Cap + O\&M + Fuel$$

$$e_b = 12.8 + 2.2 + 1.0 = 16 \text{ cents/kWhr}_e$$

Source: Adapted from Tester, J.W., Drake, E.M., Golay, M.W. et al., *Sustainable Energy: Choosing among Options*, MIT Press, Cambridge, MA, 2005, with updated cost figures. With permission from Pluto Press.

[a] Note that these costs represent only the cost of generating the electricity (i.e., excluding transmission and distribution). These costs are lifetime-average (i.e., "levelized") costs for a new plant starting operations today.

The projection of energy cost is always subject to uncertainties resulting from tax incentives, technical innovations, market changes, inflation, and competition. For example, the cost of electricity from nuclear power was the topic of an international symposium of experts at the Keystone Center in June 2007. Their results, published as *Nuclear Power Joint Fact-Finding*, gave a summary of construction cost estimates and levelized energy cost presented in Tables 2.12 and 2.13 [7]. These estimates for low-cost and high-cost scenarios

TABLE 2.12

Summary of Cost Assumptions for Nuclear Power Plant Scenarios

Main Assumptions (2007$)	Low Case	High Case
Overnight cost	$2950/kW	$2,950/kW
Plant life	40 years	30 years
Capital cost, including interest	$3600/kW	$4,000/kW
Capacity factor	90%	75%
Financial	8% debt, 12% equity, 50/50 ratio	8% debt, 15% equity, 50/50 ratio
Depreciation	15 year accelerated	15 year accelerated
Fixed O&M	$100/kW/year	$120/kW/year
Variable O&M	5 mil/kWh	5 mil/kWh
Fuel	1.2 cents/kWh	1.7 cents/kWh
Grid integration	$20/kW/year	$20/kW/year

Sources: Reprinted from *Nuclear Power Joint Fact-Finding*, The Keystone Center, Keystone, CO, June 2007.

TABLE 2.13

Summary of Levelized Cost of Energy for Nuclear Power Plants (Cents/kWh)

Cost Category	Low Case	High Case
Capital costs	4.6	6.2
Fuel	1.3	1.7
Fixed O&M	1.9	2.7
Variable O&M	0.5	0.5
Total (levelized cents/kWh)	8.3	11.1

Sources: Reprinted from *Nuclear Power Joint Fact-Finding*, The Keystone Center, Keystone, CO, June 2007.

are shown in Tables 2.12 and 2.13. Clearly, the levelized cost estimates for nuclear power from [7] or the earlier example, between $0.10 and $0.16 per kWh, are of the same order of magnitude as the projected costs from large solar or wind systems, as shown later in this book. It should be noted that the capital cost estimates between the years 2007 and 2009 increased considerably from $4,000/kWe to $6,333/kWe.

2.10 Input–Output Analysis

Economists use input–output (I/O) analysis to characterize the effect of production in one sector on the production in other sectors. Industries use the products of other industries to produce their own products that are in turn used by other industries. A table is normally constructed to catalog the interdependence of sectors. Energy engineers rarely work on constructing I/O tables, but they do illustrate the "blow-up" result of the causal loops between sectors. For example, let us suppose we want to look at the production of one ton of steel and one ton of coal. Let us say that it takes 3 tons of coal to make a ton of steel. Let us also say that the direct steel requirement for the mining, crushing, and transport of coal is 0.2 tons of steel per ton of coal. Let us say the demand for coal in the economy is 1000 tons and for steel is 10 tons. In order to make the steel for the demand and for the coal mining, we actually need to produce 10 + (1000)(0.2) = 210 tons of steel. But making that extra 200 tons of steel takes 600 tons of coal. Now that extra coal would require another (600) (0.2) = 120 tons more of steel, which would in turn require additional (120)(3) = 360 tons of coal. This causal loop of internal production demand is shown in Figure 2.3, together with 7 rounds of incremental accounting of the internal demand. The I/O matrix shown in Figure 2.3 gives the actual production of coal and steel needed to meet the demand from the economy. These coproduction factors could be found from continuing the looping to a limit or by solving for the causal production requirements for one unit of each good:

$$x_{11} - 3x_{12} = 1 \quad x_{11} - 0.2x_{12} = 0$$

$$x_{21} - 3x_{22} = 0 \quad x_{21} - 0.2x_{22} = 1$$

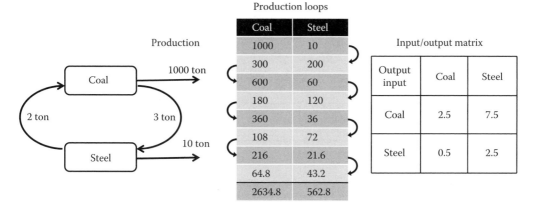

FIGURE 2.3
Causal loop diagram and input–output matrix for coal and steel illustrating the "blow-up" effect for the baseline production rates of different sectors in order to produce a unit of output in another sector.

The implications of this I/O analysis are interesting. If we just looked at the unit production requirement, we might reason that production of 10 tons of steel requires 30 tons of coal. But the I/O causal loop analysis shows that the sector production of 10 tons of steel actually required 75 tons of coal and 25 tons of additional steel! Let us assume that we are charged with figuring out what to do to meet the climate change risk management goal of reducing coal production and consumption by 50%. We can see that this could be accomplished by recession basically reducing steal production by half. There are alternatives to recession such as reducing the amount of steel needed for coal mining by half, or by improving the steel making process so that it only requires 1.3 tons of coal per ton of steel. The first route to reducing emissions is contraction, whereas the second is adaptation.

I/O matrices can be represented in terms of energy or dollars. To convert from dollars to energy units in this approach, one attaches energy flows, which can be obtained from a national energy balance, to the monetary value flows within such an I/O table. Then, an energy matrix based on the physical flows and their respective monetary values can be formulated [9]. This energy matrix finally reveals the amount of primary energy (Btu or kWh) necessary for producing one unit of monetary value in a given economic sector of the I/O matrix, including all causal energy expenditures. The I/O analysis is also used for material relationships.

2.11 Energy System Analysis Methodologies

The objective of energy system analysis is to understand the energy production and supply system and the relationships to consumer energy and to the economy. Energy system analysis is necessary to identify and evaluate sustainability issues and potentially beneficial developments. Figure 2.4 demonstrates the flow of energy and the flow of money

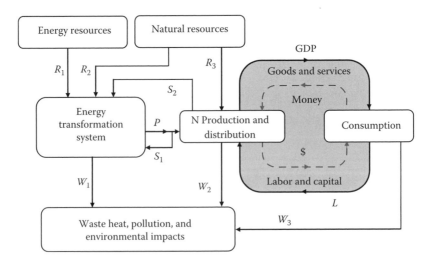

FIGURE 2.4
Schematic diagram of the flows of energy, resources, goods, and money through the economy. R_i are the flows of primary energy and natural resources, P is the consumer energy produced for sale into the economy, S_1 is the energy (electricity or fuel) needed to run the energy transformation system, and S_2 is the embodied energy in equipment and materials needed to extract and process energy. N is the net energy to the economy, GDP refers to dollar or unit values of products and services, L includes the units of labor and investment or taxes, and W_i are the waste flows.

TABLE 2.14

Assessment of Energy and Resource Characteristics of the Whole Economy, a Particular Sector or a Particular Subsystem Using the Structure in Figure 2.1

Indicator	Parameters	Example
Energy conversion efficiency	$\eta_C = P/R_1$	MWh_e/ton coal
Energy resource intensity	P/R_2	gal ethanol/gal water
Energy intensity of the economy	P/GDP	kWh/$
Energy return on energy invested	$EROI = P/(S_1 + S_2)$	MJ/MJ
Embedded energy	R_1/GDP	MJ/kg steel
Net energy to the economy	$N = P - (S_1 + S_2)$	GWh electricity or GJ fuels
Externality loading of conversion	W_1/P	ton-CO_2/MWh
Life cycle assessment	$\Sigma R/GDP$	Resources/unit good or service
External costs	$\Sigma W/GDP$	Impacts/unit good or service
Impact footprint	W/GDP	Carbon footprint
Levelized cost of energy	$LCOE = (\Sigma S + \Sigma R + W_1)/P$	Annualized investment cost
Energy prosperity	$N/P = 1 - 1/EROI$	Rate of return on energy

through the economy. Money, created and circulated in the economy, is not subject to any physical constraints. On the contrary, the energy and resources supplied to the economy by energy transformation systems and the associated environmental impact are subject to both technological and resource constraints. For the development of sustainable energy systems, it is important to evaluate the potential of a proposed energy production system to substitute for fossil fuels. It is also important to evaluate the adaptive potential of the end use and the upstream effects of demand. The energy provided to the end user has to be "economical," meaning that it has to be priced at an acceptable market price, but must also cover the total production cost including any waste disposal or emissions costs.

Figure 2.4 can be used to conceptualize the entire energy system or one particular subsystem. The following definitions listed refer to the figure and can be calculated on an annual, unit, or cumulative basis. Some common characterizations of the energy and resource flows are given in Table 2.14. A flow of energy consumed by an energy transformation plant (e.g., S_1) can be measured in mass, volume, or the heating value. A flow of capital equipment (e.g., S_2) can be measured in dollars or embedded energy. Production (e.g., GDP) can be measured in units of goods or dollars or embedded energy or resources according to the objective of the analysis. Some of the following energy system analysis tools are qualitative or provide frameworks to conceptualize complex energy systems. Others are quantitative, like energy return on energy invested (EROI) and energy intensity.

2.11.1 Process Chain Analysis

Process chain analysis (PCA) is a broad description of the methods that are used to audit any system or subsystem. The purpose of PCA is to take account of the energy and resources needed and the waste produced over different production steps. The PCA approach is used in life cycle cost assessment (LCA), embedded energy calculations, supply chain analysis, EROI calculations, and many other types of analysis. PCA can be used to provide conceptual understanding or to calculate specific energy and material flows and costs. In PCA, the production steps are separated, and for each of them, the relevant inputs and outputs are identified and evaluated according to the specific goals of the analysis.

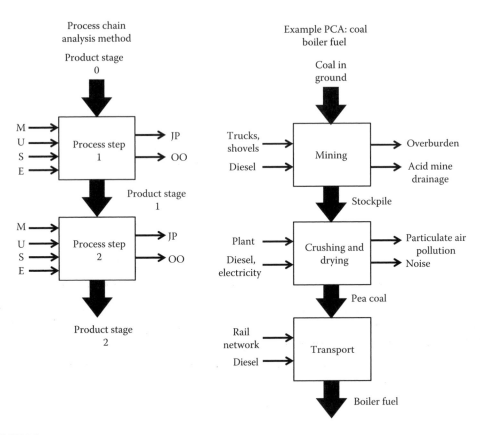

FIGURE 2.5
Illustration of the concept of PCA with an example of qualitative assessment of the transformation of coal deposits to delivered pea coal boiler fuel.

Figure 2.5 shows a basic schematic of the PCA methodology. The individual sequential steps in the process are separated out, and all inputs and outputs are identified for each process in the trajectory. Inputs and outputs include the product as it is transformed, the utilities, U, services, S, materials, M, and equipment, E, as well as possibly joint products, JP, and other outputs, OO. A PCA can be done for any one part of a process chain, for example, from natural resources, R, to consumer products, P, or the PCA can be done for the whole lifespan through the steps of consumer end use and disposal. The example shown for coal fuel production could be compared to the product chain for an alternative fuel like wood. Depending on the regional locations and the grade of coal, the PCA analysis may be helpful in understanding the energy penalty in gathering wood from a plantation, chipping and drying, and transporting a fuel with nominally half the energy density as coal. The technology for wood-fired boilers is available. However, there may be no rail network available to transport wood regionally the way there is for coal. The diesel fuel used to gather and transport wood from the land is much higher per MJ of energy than for coal, and the energy to chip wood may be greater than to crush friable coal. Thus, even a basic conceptual PCA can help to inform energy system investments and policy.

Evaluating the relevant inputs and outputs for each step of the process can be tedious and complicated if exact figures for a sector or specific product are desired. PCA can be

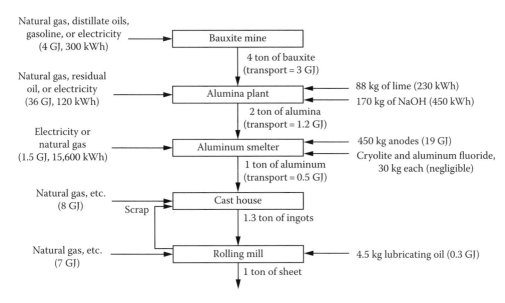

FIGURE 2.6
Quantitative PCA analysis of the energy use in the production of primary aluminum. (From Boercker, S.W., *Energy Use in the Production of Primary Aluminum*, ORAU/IEA-78-14, Oak Ridge Associated Universities, Oak Ridge, TN, 1978. With permission.)

used to evaluate hypothetical systems, like the hydrogen analysis in Chapter 1, but then assumptions must be made about possible technologies where none actually exists for reference. The results can vary widely depending on the assumptions made. PCA can be used to determine embedded energy for materials and products. Figure 2.6 illustrates the PCA for primary aluminum production. For details of the assumptions and possible pitfalls, the reader is referred to [8].

2.11.2 Embedded Energy

Embedded energy is a quantitative measure of the energy cost associated with the processing chain for a material or product. Embedded energy for a product is calculated by using a PCA of all the materials and energy used in the extraction and manufacture. Energy intensity factors (Btu/$ or kWh/$) give the energy required to produce or provide a given product or service and are generally determined through economic analysis by looking at the energy use and activity of whole sectors. By quantifying all the energy and dollars flowing into particular sectors, say the steel industry, it is possible to estimate the amount of energy the steel industry uses in the form of energy embodied in all the inputs it purchases from other sectors. The ratio of all such energies to the dollar value of the steel produced is the energy intensity in Btu/dollar or kWh/dollar of steel production purchases from other sectors. The energy intensity factors in Table 2.15 were calculated for an energy-based I/O model of the U.S. economy in 1977 [9,10] and adapted for 2008 dollars [10]. The ratio of 2008s' CPI to 1977s' CPI is 3.52 [11]. The CPI is a measure of the price of consumer goods and services. The data for 2008 *do not* mean that manufacturing has become more energy/cost productive, rather it shows that the same energy investment produces product value in terms of

TABLE 2.15

Embedded Energy in Various Materials and Commodities

Commodity	1977 Values Energy Intensity (Btu/$)	1977 Values Energy Intensity (kWh/$)	Adjusted 2008 Values Energy Intensity (kWh/$)
Alum castings	70,000	20.5098	6.6821
Air transport	80,846	23.6877	7.7174
Aircraft	21,288	6.2373	2.0321
Aircraft engines	29,238	8.5667	2.7910
Alum rolling	106,723	31.2696	10.1876
Arch metal work	45,552	13.3466	4.3483
Asbestos product	72,140	21.1368	6.8864
Asphalt	184,121	53.9470	17.5759
Auto repair	22,055	6.4621	2.1053
Banking	9,820	2.8772	0.9374
Bearings	42,638	12.4928	4.0702
Blowers	35,077	10.2775	3.3484
Boatbuilding	23,933	7.0123	2.2846
Brass other cast	50,508	14.7987	4.8214
Bricks	127,994	37.5019	12.2181
Canned fruit, vegetable	40,981	12.0073	3.9120
Carbon products	108,949	31.9218	10.4001
Cement	216,631	63.4723	20.6792
Ceramic tile	63,362	18.5649	6.0484
Chem mineral min	128,720	37.7146	12.2874
Clay products	119,323	34.9613	11.3904
Communications	6,887	2.0179	0.6574
Computing mach	21,306	6.2426	2.0338
Concrete blocks	70,374	20.6194	6.7178
Concrete product	50,789	14.8810	4.8482
Const machinery	34,534	10.1184	3.2966
Conveyers	31,042	9.0952	2.9632
Copper mining	86,570	25.3648	8.2638
Copper rolling	70,649	20.7000	6.7440
Doctors, dentists	8,488	2.4870	0.8103
Eat and drink places	23,620	6.9206	2.2547
Elec. h'wares	36,919	10.8172	3.5242
Elec. Ind. Apparat.	31,648	9.2728	3.0211
Elec. meas. instr.	21,561	6.3173	2.0582
Electric lamps	28,390	8.3182	2.7101
Electrical equip.	38,118	11.1685	3.6387
Electronic comp.	31,502	9.2300	3.0071
Elevators	33,497	9.8145	3.1976
Engine elec. eq.	29,349	8.5992	2.8016
Fab. metal prod.	47,163	13.8186	4.5021
Fab. struc. steel	49,469	14.4943	4.7222
Fab. wire product	71,355	20.9068	6.8114
Farm machinery	32,417	9.4981	3.0945

TABLE 2.15 (continued)

Embedded Energy in Various Materials and Commodities

Commodity	1977 Values Energy Intensity (Btu/$)	1977 Values Energy Intensity (kWh/$)	Adjusted 2008 Values Energy Intensity (kWh/$)
Fed govt. enterp.	16,946	4.9651	1.6176
Feed grains	66,423	19.4618	6.3406
Fertilizers	179,710	52.6546	17.1548
Food prod mach.	27,346	8.0123	2.6104
For, grhouse, nurs.	47,697	13.9751	4.5531
Forest fish prod	43,390	12.7132	4.1419
General ind mach	28,793	8.4363	2.7485
Glass containers	80,487	23.5825	7.6832
Glass products	58,274	17.0741	5.5627
Guided missiles	14,272	4.1817	1.3624
Gypsum products	99,292	29.0923	9.4783
Hardware	38,859	11.3856	3.7094
Heating equip.	33,553	9.8309	3.2029
Hoists, cranes	30,233	8.8582	2.8860
Hospitals	23,123	6.7750	2.2073
Hotels	33,939	9.9440	3.2398
Ind. controls	23,412	6.8597	2.2349
Indus. furnaces	27,255	7.9856	2.6017
Industrial truck	34,245	10.0337	3.2690
Inorg–org chem.	171,572	50.2701	16.3780
Int. combust eng	31,345	9.1840	2.9921
Ir. stl. forging	75,110	22.0070	7.1699
Iron ore mining	92,650	27.1462	8.8442
Light fixtures	37,594	11.0149	3.5887
Lime	235,675	69.0522	22.4972
Local transport	32,804	9.6115	3.1314
Mach shop prod.	27,116	7.9449	2.5884
Man-made fibers	123,386	36.1518	11.7782
Manufactured ice	61,718	18.0832	5.8915
Measuring pumps	29,195	8.5541	2.7869
Meat products	42,456	12.4395	4.0528
Meat, animal prod.	46,311	13.5690	4.4208
Medical instr.	29,339	8.5962	2.8007
Met. cutting tool	23,163	6.7867	2.2111
Met. forming tool	28,021	8.2101	2.6748
Met. working mach.	30,373	8.8992	2.8994
Metal stampings	52,502	15.3829	5.0118
Mineral wool	79,474	23.2857	7.5865
Mining machinery	31,541	9.2414	3.0109
Misc. bus service	10,000	2.9300	0.9546
Misc. chem. prod.	95,178	27.8869	9.0855
Misc. leather	32,165	9.4243	3.0704

(continued)

TABLE 2.15 (continued)

Embedded Energy in Various Materials and Commodities

Commodity	1977 Values Energy Intensity (Btu/$)	1977 Values Energy Intensity (kWh/$)	Adjusted 2008 Values Energy Intensity (kWh/$)
Misc. metal work	65,247	19.1172	6.2284
Misc. plastics	63,281	18.5412	6.0407
Misc. rubber prod	53,271	15.6083	5.0852
Motor transport	29,196	8.5544	2.7870
Motor veh. and parts	35,846	10.5028	3.4218
Motor, bicycles	36,373	10.6572	3.4721
Motors, generator	33,556	9.8318	3.2032
NCNST highways	59,333	17.3844	5.6638
NCST dams, resv.	44,509	13.0410	4.2488
NCST elect utility	30,648	8.9798	2.9256
NCST gar., srv., sta.	38,198	11.1919	3.6463
NCST gas utility	28,039	8.2154	2.6766
NCST indust. Bldg	32,103	9.4061	3.0645
NCST railroads	35,264	10.3323	3.3662
Newspapers	30,902	9.0542	2.9499
Nonclay refract.	69,607	20.3947	6.6446
Nonfer. casting	61,395	17.9886	5.8607
Nonfer. forging	64,560	18.9159	6.1628
Nonfer. rolling	81,118	23.7674	7.7434
Nonfer. wire	66,253	19.4120	6.3244
Nonfer. mining	53,225	15.5948	5.0808
Nonmet. min. prod.	56,700	16.6130	5.4125
Oil-bearing crop	23,352	6.8421	2.2291
Oil field mach	30,547	8.9502	2.9160
Optical instr.	28,458	8.3381	2.7166
Organic fibers	99,470	29.1444	9.4952
Paint products	75,217	22.0384	7.1801
Paperboard cont.	64,427	18.8769	6.1501
Paving	181,838	53.2781	17.3580
Personal service	24,405	7.1506	2.3297
Photographic eq.	35,148	10.2983	3.3552
Pipe	37,637	11.0275	3.5928
Pipe line transp.	67,173	19.6815	6.4122
Plastics	126,087	36.9432	12.0361
Plumb fittings	36,093	10.5752	3.4454
Plumbing fixture	60,760	17.8025	5.8001
Power trans eq.	31,596	9.2575	3.0161
Prim nonfer. met.	109,365	32.0437	10.4398
Primary aluminum	158,201	46.3525	15.1016
Primary battery	41,578	12.1822	3.9690
Primary copper	112,850	33.0648	10.7725
Primary lead	130,108	38.1213	12.4199
Primary zinc	137,242	40.2115	13.1009

TABLE 2.15 (continued)

Embedded Energy in Various Materials and Commodities

Commodity	1977 Values Energy Intensity (Btu/$)	1977 Values Energy Intensity (kWh/$)	Adjusted 2008 Values Energy Intensity (kWh/$)
Pumps, compressors	27,731	8.1251	2.6472
Radio, TV sets	24,898	7.2950	2.3767
Railroad	46,041	13.4899	4.3950
Railroad equip.	38,825	11.3756	3.7062
Ready-mix concr.	80,909	23.7061	7.7234
Real estate	6,525	1.9118	0.6229
Refrig. mach.	36,589	10.7205	3.4927
Retail trade	19,433	5.6938	1.8550
R-TV commun. eq.	18,243	5.3452	1.7414
Scien. instr.	24,659	7.2250	2.3539
Screw mach prod.	44,247	12.9643	4.2237
Semiconductors	29,749	8.7164	2.8398
Sheet metal work	56,289	16.4925	5.3733
Shipbuilding	29,048	8.5110	2.7729
St loc govt. entr.	31,395	9.1987	2.9969
Steel prod.	115,724	33.9068	11.0468
Steel springs	59,220	17.3513	5.6530
Stone product	39,667	11.6223	3.7865
Storage battery	69,024	20.2238	6.5889
Switchgear	25,726	7.5377	2.4558
Syn. rubber	143,451	42.0308	13.6936
Tanks	39,964	11.7093	3.8149
Temp. controls	32,331	9.4729	3.0863
Textile goods	65,964	19.3273	6.2968
Textile mach	31,319	9.1764	2.9897
Tires	55,977	16.4011	5.3435
Trans services	17,678	5.1796	1.6875
Transformers	42,591	12.4791	4.0657
Transport equip	44,317	12.9848	4.2304
Truck, bus bodies	35,634	10.4407	3.4016
Vegt, misc. crops	29,798	8.7307	2.8445
Water transport	11,369	3.3311	1.0853
Water, sanit. Ser.	89,974	26.3621	8.5888
Wholesale trade	18,326	5.3695	1.7494
Wiring devices	40,586	11.8916	3.8743
Wood products	52,030	15.2447	4.9667
X-ray equipment	26,598	7.7931	2.5390

money that is worth less than 30 years earlier. This is a good illustration of the destructive effects of inflationary economic system. All producers, whether producing dairy products or ingots of aluminum, must continuously expand their production, and thus their energy and resource consumption, in order to even maintain a particular level of income in real terms.

2.12 Energy Return on Energy Invested

EROI is an important concept for sustainable energy engineers to understand and to communicate to the public and policy makers. In terms of the schematic in Figure 2.4, the EROI is given by the following equation [8,9]:

$$\text{EROI} = \frac{P}{S_1 + S_2} \tag{2.21}$$

where
    P is the rate of energy production (kWh/analysis period) from the transformation system
    S_1 is the conversion energy input (kWh/analysis period)
    S_2 is the embodied energy in the various items used by the production system (kWh/analysis period)

The EROI can be evaluated for the whole energy system in a given year, or it can be assessed for a particular energy production technology over the useful life of the plant. The only reason for building an energy conversion plant (investing embedded energy in capital equipment, S_2) and operating the plant (investing process energy, S_1) is to produce consumer energy for the market (distributing and selling P). The EROI is a measure of the energy profitability of an energy transformation system. Since the energy invested in the transformation system could not be used by the economy, the EROI is also an indication of the availability of energy to meet demand and to provide the surplus needed for maintenance, replacement, and new manufacturing and construction.

The concept of EROI is simple, but exact calculation can be difficult. Companies are not likely to invest in an energy platform with low EROI because it would be bad business. However, misguided government policy and mandates can provide incentives and subsidies for platforms with poor EROI and end up wasting time and money, reducing the overall EROI of the economy, and increasing the risk exposure to supply security issues.

The main point of evaluating the EROI of a given energy supply system is the same as for financial analysis of different investments. EROI lets us compare different energy transformation platforms and make informed investment decisions. The objective of investing is to realize a profit. In the energy system, the "profit" for the society as a whole is the net energy. The net energy yield, N, given by Equation 2.22 is as follows:

$$N = P - (S_1 + S_2) = P\left(1 - \frac{1}{\text{EROI}}\right) \tag{2.22}$$

The net energy yield can be assessed for a particular energy conversion system or for the sector as a whole. The net energy yield can be evaluated on a year-by-year basis or over the lifetime of the energy conversion plant. On a year-by-year basis, the net energy yield may be negative at the early stage of a power plant, and the point at which the cumulative net energy return reaches zero is also known as the playback period (PP).

In order to understand the importance of EROI and net energy yield for a civilization, we will consider a simple model system of the eighteenth-century farming. At this time, more than 85% of all labor would have been engaged in agriculture and primary production. Agriculture was the first energy system developed by people. A preindustrial farm required inputs (S_1) of grains and fodder for animals, plus food for the workers, plus wood for heat and processing. The farm required capital equipment (S_2) like a plough, farm buildings, a thresher, harnesses, tools, and the equipment for the farmhouse that were also part of the farm system. The production from the farm would be the produce taken to the market to fuel the work of other people and animals in the society or to provide the wood for fuel or the fibers for clothes. A prosperous farm (large EROI) would have high-quality, well-maintained equipment, healthy animals, and wealthy farmers with nice houses. A prosperous farm community would have a town with thriving businesses, civic amenities, schools, police, etc. A large net "energy" return in the form of plentiful harvest compared to farm inputs means prosperity. A farming community that is not prosperous would have hunger, unemployment, poor civic services, and if the net return is zero, this implies a total failure of the farm economy system because all of the production from the farm was needed just to run the farm.

EROI is the underlying determining factor for prosperity. The astute reader will notice that we have not yet discussed the price of the energy or resources in this discussion. In the areas of primary production like food, materials, and energy, prosperity is more fundamentally determined by the physical ability to return a positive net product to the economy than by the prices the economy attaches to commodities. In fact, there is an inverse relation between prosperity and price of commodities—scarcity causes price rise for primary goods, but does not result in prosperity as demonstrated in many historical cases, including the energy crisis caused by the OPEC oil embargo.

The implication of low EROI energy investments is demonstrated in Figure 2.7, where the net energy return to the economy as a percentage of the total energy produced, $N/P = 1 - 1/EROI$, is plotted versus EROI. The ratio N/P is essentially a measure of the prosperity for a society. An N/P of 95% implies that all of the energy and resources used by the energy sector were returned to the economy as useful energy production with a 95% rate of return. An N/P of 50% means that the energy transformation sector consumed half of

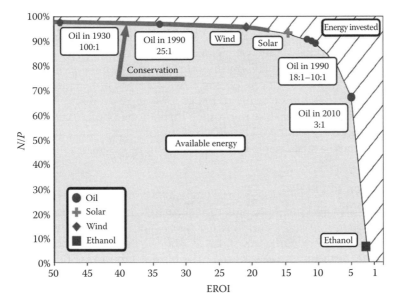

FIGURE 2.7
Ratio of net energy to the rate of energy production vs. EROI.

the resources internally and only returned half to the economy. Inspection of Figure 2.7 shows that once EROI drops below 5, the useful available energy decreases precipitously.

Oil is a good example to consider. We can see that around 1930, EROI was over 80:1, and the surplus energy to the economy was very large. In other words, for 1 Btu invested in finding and producing oil, 80 Btu of oil energy was returned with a rate of return of nearly 99%. By 1956, the EROI had dropped to 25, but the rate of return of 96% was still very favorable. Today the EROI is below 10 for new oil fields giving a rate of return of 90%. This is still pretty good, but if the liquid fuel sector starts developing tar sands that have an EROI of only 2, then there will still be a rate of return, N/P, of 50%, but this is only half the level of prosperity of conventional oil. This would mean that a fuel system based on tar sands would be able to support only half the fuel demand in the economy as conventional oil, even if the tar sand system would have the same production volume as conventional oil. As clearly shown in Figure 2.7, corn ethanol with an EROI less than 2 as clearly shown in Figure 2.7, is not a primary fuel that can provide the same level of prosperity historically provided by oil.

Cross-sector EROI comes into play in electricity generation. Hydroelectricity has the highest EROI of any power generation platform, estimated at over 40 and as high as 100 depending on the site. For example, some hydroelectric facilities do not require construction of a large dam and can be built at the outlet of a natural lake. Most of the hydroelectric dams in the United States were built many decades ago using high EROI diesel fuel (50+) and coal for making the cement and steel (EROI 80+). The water flowing through hydro turbines does not need to be extracted from mines and transported to the power plant; it simply arrives, courtesy of weather, geography, and gravity. Areas of the country where hydro generation was built quickly developed prosperous economies based on plentiful, secure, on-demand electricity. The second highest EROI electricity generation platform is from coal with EROI over 10. Historically, areas of the country with easily extractable coal and navigable waterways also developed prosperous manufacturing and production economies. Today, the energy intensity of primary materials is increasing, and the EROI of fuels and electricity is declining. Wind- and solar-generated electricity have lifetime EROI of 5–15, which is highly dependent on the conversion efficiency, service life, and utilization factor. But manufacture of the wind and solar plant requires investment of fossil fuels and energy-intensive materials, so RE does not offer an "alternative" to fossil fuels in the way often discussed in the popular media. Reduced fossil fuel use means reduction in net energy to the economy. One way to address this is to invest in RE. But another way to address the situation is to invest in conservation—changing existing energy end-use systems to reduce demand. The EROI for conservation systems has the same definition as for energy production. The "low-hanging fruit" conservation investments like improving insulation in buildings have very high lifetime EROI and are also profitable. These factors are of great importance in strategic planning for a transition to a sustainable future, which will be discussed in Chapter 13.

2.12.1 Calculation of EROI

Energy analyses are useful supplements to standard econometric procedures based on monetary assessments of costs and benefits for several reasons. All economic analyses require assumptions about the future, such as interest rates, future price of conventional fuels, and inflation rates. Also assumptions about future government regulation, taxation rates, and political decisions can greatly influence the economic evaluation process. EROI is calculated by putting all inputs and outputs into energetic terms so the evaluation is more straightforward and the result can be interpreted in terms of the general potential for prosperity of the economy.

Researchers and analysts have been increasingly interested in calculating the EROI for different energy platforms. There are different time scales and different assumptions that can be made in the analysis. Determining the embedded energy in the power plant and the exploration and extraction equipment for fossil fuels can be difficult, so back-of-the-envelope calculations using standard embedded energy values for materials and estimates of the weight of the equipment usually provide sufficient accuracy. Another possibility is to use the dollar cost for each item of the energy production system. Then each material input to the entire system is assigned an energy intensity factor from Table 2.15 calculated for the sector that produced the material. The dollar value of the input times its energy intensity is the indirect energy cost of that item. Summing the indirect energy costs for all the items of the system gives the indirect energy cost for the entire energy system. It is also usually necessary to construct a PCA of the energy transformation platform in order to define the energy and materials used for analysis.

In electricity generation from thermal power plants, clarity is needed because the generation is often given in technical data, but the parasitic power used for pumps and fans represent components of S_1. Make sure the parasitic energy requirements are accounted for when using net power plant output. In coal and nuclear power plants, the energy content in the fuel is usually not included in S_1 in the calculation of EROI, but the consumer energy used to extract and process the fuel should be included. For petroleum, the exploration, drilling, and oil well equipment are included in S_2, and the thermal energy, and electricity for pumps and fans used in refining the petroleum into consumer fuel products are counted in S_1.

Figure 2.8 illustrates schematically the EROI calculation method for a hypothetical energy transformation system. During the construction phase, two units of energy are used to build the system; during its operational life, two units of energy are used; finally,

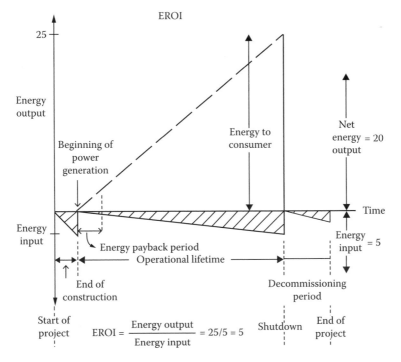

FIGURE 2.8
Cumulative energy input and output for an RE generation system during its lifetime.

one unit of energy is used to decommission the system. The energy produced is 25 units, and the net energy output for this system is 20 units. Thus, the EROI for this example is 5. In Figure 2.8, the shaded area of energy input during the operational lifetime includes the energy of the fuel input for a fossil or nuclear plant. If this were a renewable electricity generation plant that did not require energy fuel or decommissioning, the EROI would be 12.5.

A key difference between the EROI for fossil or nuclear generating systems and renewable technologies, such as solar or wind, is the fact that the renewable systems depend on a primary energy source that not only varies with time of day and time of year, but also with the location at which the system is deployed. Therefore, the evaluation of the energy produced has to be carried out in considerably more detail for a renewable system than for a fossil or nuclear system. In the following chapters, we will present detailed analytical and experimental methods for estimating the energy output of renewable systems. Here, we will simply specify the capacity factor, as well as system lifetime and multiply the plant capacity by that factor to obtain a time-average value of electricity or heat generation over the lifetime of the system.

The EROI for conservation measures and waste heat recovery is calculated using the same approach as for energy transformation systems. The energy being wasted or rejected to the environment is treated as a resource like wind or solar, and the energy recovered or saved is treated as the energy production, P. The following example illustrates the procedure.

Example 2.8: Double-Paned Windows

In 2008, a new building in Boulder, Colorado, is designed to have 696.8 m² of windows. The architect wanted to know if it is worth using double-paned windows or single-paned windows. He asks you to determine the following:

- The annual reduction in heat loss if double-pane (double-glazed) windows are used instead of single-paned windows.
- An estimated payback time for the extra energy used to make double-paned instead of single-paned windows.
- The total reduction in CO_2 emissions if double-paned windows are used (assume electric heating).
- The financial PPs with an estimate that includes the time value of money at an interest rate of 7% and an inflation rate of 3%.

Assume the cost of a single-paned window is $149 and that of a double-paned window is $194, the lifetime of the windows is 30 years, and the average indoor temperature is 20°C. For complete analysis, average outdoor temperatures can be obtained from http://www.wrcc.dri.edu/cgi-bin/cliMONtavt.pl?coboul. For the purposes of this simplified analysis, use 9.13°C for a 9 month average outdoor temperature during the heating season that lasts for 9 months in Boulder. A more sophisticated approach would be to use the heating degree days data, but this example will use a back-of-the-envelope approach.

Price distribution for windows is as follows:

	Single Pane (%)	Double Pane (%)
Glass	16	24
Metal	12	18
Rubber	5	8
Production/installation	67	50

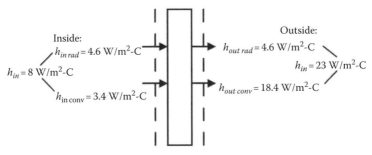

(a)

Pane thickness = 4 mm
k_{glass} = 0.9 W/m-C

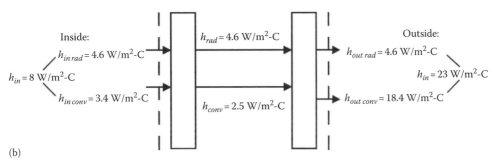

(b)

FIGURE 2.9
Thermal performance of (a) single- and (b) double-paned windows.

Thermal analysis for the windows (Figure 2.9)
First, find the annual reduction in heat loss.
For single-paned windows, assume the heat loss coefficient, U_1 is

$$U_1 = \frac{1}{\dfrac{1}{h_{in}} + \dfrac{d}{k} + \dfrac{1}{h_{out}}} = 5.8\,\text{W/m}^2{}^\circ\text{C}$$

$$Q_1 = \frac{U_1 A_{total} t (T_{in} - T_{out})}{1,000} = 284,700 \text{ kWh}$$

$$A_{total} = 696.8\,\text{m}^2$$

$$(T_{in} - T_{out}) = 20 - 9.13 = 10.87\,^\circ\text{C}$$

$$t = 24\frac{\text{h}}{\text{day}} * 30\frac{\text{days}}{\text{month}} * 9 \text{ months} = 6480\,\text{h/year}$$

Similarly, for double-paned windows, assume U_2 is

$$U_2 = \frac{1}{\dfrac{1}{h_{in}} + \dfrac{1}{h_{mid}} + \dfrac{2d}{k} + \dfrac{1}{h_{out}}} = 3.1\,\text{W/m}^2{}^\circ\text{C}$$

$$Q_2 = \frac{U_2 A_{total} t (T_{in} - T_{out})}{1,000} = 152,150 \text{ kWh}$$

Annual difference in heat loss = 284,700 kWh/year − 152,150 kWh/year = 132,600 kWh/year

Next, find the EROI and PP.

Energy intensity calculations for single-paned windows

Single pane price: $149 per window

Glass

$$\left(58,274\frac{Btu}{1977\$}\right)\left(\frac{60.6}{215.303}\frac{1977\$}{2008\$}\right)\left(149\frac{2008\$}{window}\right)$$

$$\times\left(500\,windows\right)\left(0.16\right)\left(0.00029\frac{kWh}{Btu}\right)=56,700\,kWh$$

Metal

$$\left(115,724\frac{Btu}{1977\$}\right)\left(\frac{60.6}{215.303}\frac{1977\$}{2008\$}\right)\left(149\frac{2008\$}{window}\right)$$

$$\times\left(500\,windows\right)\left(0.12\right)\left(0.00029\frac{kWh}{Btu}\right)=84,500\,kWh$$

Rubber

$$\left(53,271\frac{Btu}{1977\$}\right)\left(\frac{60.6}{215.303}\frac{1977\$}{2008\$}\right)\left(149\frac{2008\$}{window}\right)$$

$$\times\left(500\,windows\right)\left(0.05\right)\left(0.00029\frac{kWh}{Btu}\right)=16,200\,kWh$$

Production and installation labor

$$\left(10,000\frac{Btu}{1977\$}\right)\left(\frac{60.6}{215.303}\frac{1977\$}{2008\$}\right)\left(149\frac{2008\$}{window}\right)$$

$$\times\left(500\,windows\right)\left(0.67\right)\left(0.00029\frac{kWh}{Btu}\right)=40,700\,kWh$$

Total energy intensity for single-paned windows = 198,100 kWh

Energy intensity calculations for double-paned windows

Double pane price: $194 per window

Glass

$$\left(58,274\frac{Btu}{1977\$}\right)\left(\frac{60.6}{215.303}\frac{1977\$}{2008\$}\right)\left(194\frac{2008\$}{window}\right)$$

$$\times\left(500\,windows\right)\left(0.24\right)\left(0.00029\frac{kWh}{Btu}\right)=110,700\,kWh$$

Metal

$$\left(115,724\frac{Btu}{1977\$}\right)\left(\frac{60.6}{215.303}\frac{1977\$}{2008\$}\right)\left(194\frac{2008\$}{window}\right)$$

$$\times\left(500\,windows\right)\left(0.18\right)\left(0.00029\frac{kWh}{Btu}\right)=164,900\,kWh$$

Rubber

$$\left(53,271\frac{\text{Btu}}{1977\$}\right)\left(\frac{60.6}{215.303}\frac{1977\$}{2008\$}\right)\left(194\frac{2008\$}{\text{window}}\right)$$

$$\times(500\,\text{windows})(0.08)\left(0.00029\frac{\text{kWh}}{\text{Btu}}\right)=33,700\,\text{kWh}$$

Production and installation labor

$$\left(10,000\frac{\text{Btu}}{1977\$}\right)\left(\frac{60.6}{215.303}\frac{1977\$}{2008\$}\right)\left(194\frac{2008\$}{\text{window}}\right)$$

$$\times(500\,\text{windows})(0.50)\left(0.00029\frac{\text{kWh}}{\text{Btu}}\right)=39,600\,\text{kWh}$$

Total energy intensity for double-paned windows = 348,900 kWh

Extra energy to produce double-paned windows = 348,900 kWh − 198,100 kWh = 150,800 kWh

The energy saved over a 30-year life is E_g is 132,600 kWh/year × 30 years = 3.98 × 10⁶ kWh.

The ratio of energy saved to energy invested for this saving (E_c) is the EROI, or

$$\text{EROI}=\frac{E_g}{E_c}=\frac{3.98\times10^6\,\text{kWh}}{348,900\,\text{kWh}}=11.4$$

The energy PP is equal to the extra embodied energy in the saving features divided by the yearly energy savings.

$$\text{PP}=\frac{150,800\,\text{kWh}}{132,600\,\text{kWh/year}}=1.14\,\text{years}$$

To calculate CO_2 reduction, make the following assumptions:

Coal plant efficiency = 35%

Thus, the annual extra energy $=\dfrac{132,600}{0.35}=378,857$ kWh

Heating value of coal = 7.75 kWh/kg

%Carbon = 40%

$MW_C = 12$ g/mol; $MW_{CO2} = 44$ g/mol

Extra coal for single pane = 378,857/7.75 = 48,885 kg

$C + O_2 \rightarrow CO_2$

Mass carbon = 48,885 (.4) = 19,554 kg

Moles carbon = 19,554(1,000)/12 = 1,629,493 mol

Moles CO_2 = 1, 629,493 mol

Mass CO_2 = 1, 629,493*44/1,000 = 71,698 kg

Lifetime of windows = 30 years

Total CO_2 reduction = 30(71,698) = 2,150,931 kg

Financial payback of double-paned windows

Extra cost for double pane = 500(194 − 149) = \$22,500

$i = 7\%$ $j = 3\%$

Future value of P = \$22,500 after x years:

$x = 1$, $P((1 + i)/(1 + j))^x = \$23,374$
$x = 2$, $P((1 + i)/(1 + j))^x = \$24,282$
$x = 3$, $P((1 + i)/(1 + j))^x = \$25,224$
Annual electric savings = 132,600 kWh (\$0.08/kWh) = \$10,608

Energy cost savings P = \$11,722 for given year x:
$x = 1$, $P(1 + j)(x - 1) = \$10,608 = S_1$
$x = 2$, $P(1 + j)(x - 1) = \$10,926 = S_2$
$x = 3$, $P(1 + j)(x - 1) = \$11,254 = S_3$

Future value of invested energy cost savings after x years:

$x = 1$, $S1 = \$10,608$
$x = 2$, $S_1((1 + i)/(1 + j)) + S_2 = \$21,946 = F_1$
$x = 3$, $F_1((1 + i)/(1 + j)) + S_3 = \$34,052$
After 2 years, the value of \$22,500 invested is \$24,282
After 2 years, the value of annual energy savings is \$21,946
After 3 years, the value of annual energy savings is \$34,052

PP for double pane = 2 + (24,282 − 21,946)/(34,052 − 21,946) = 2.2 years

2.12.2 EROI and Energy Budgets

Energy budgeting is an important tool for planning a sustainable energy future. In an industrial society, a certain amount of energy is necessary to maintain the so-called energy infrastructure, which includes the pipelines and roads for distributing liquid fuel or natural gas, as well as the hydroelectric dams and power plants that generate electricity. The need to maintain a flow of energy requires drilling for oil or digging for coal and uranium in a fossil-based energy system or building wind turbines and solar PV in a sustainable energy structure. In addition, we must also invest some energy in building and maintaining the capital structure necessary for a modern industrial society. This includes items such as the roads and bridges for transportation, and the electricity and buildings to maintain human comfort. An additional amount must be set aside in order to find more energy to replace the amount of energy being used up. All of these items are mandatory expenses for a steady-state energy budget. They are unavoidable expenses related to the energy necessary for a particular industrial society.

The remaining part of the available energy is divided into two parts: The first part is used for basic needs that support the standard of living such as health, shelter, water, and food. The second part is for discretionary uses associated with lifestyle such as entertainment, vacations, television, and the arts. This latter part can also be used for growth of the economy. Figure 2.10 shows an energy flow diagram that accounts for the energy budget of the system depicted in Figure 2.4.

The energy budget diagram is useful in illustrating the impact on the economy of the trend to low EROI energy resources. Energy budget diagrams for an energy system with EROI = 20 and for EROI = 4 are shown in Figure 2.11a and b. In a simplified fashion, we can divide our energy budget into (a) energy necessary to maintain the energy flows, buildings, roads, and other basic needs, and (b) energy that can be used for discretionary activities. If there is a surplus of energy, we can apply that toward growth and/or prosperity, but the amount is finite and cannot be fully allocated to both simultaneously.

The EROI of oil production has declined over the past 70 years as the most productive resources were initially used, and it was necessary to go further from easily accessible locations, or to drill deeper for increasingly smaller oil fields as time went on. It is estimated

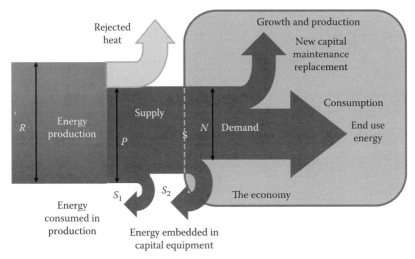

FIGURE 2.10
Energy flow diagram depicting the energy budget from extraction and transformation to end use.

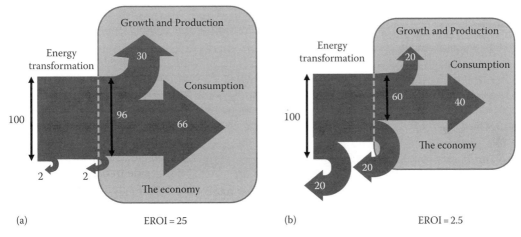

FIGURE 2.11
Energy budget diagrams for two energy transformation systems: (a) has an EROI = 25 and supports a larger economy than (b), which has an EROI = 2.5 even though both have the same energy production.

that by 1970, the EROI of oil production had dropped to 25; and in 1990, it was somewhere between 10 and 18. Today, the EROI of new energy oil production is somewhere between 3 and 4 [12]. This decline in the EROI is the result of having used up the most productive sources such as the Spindle Top in Texas, and about half of the Ghawar Field in Saudi Arabia, which were about 1,000 ft below the surface. In contrast, the fields in the Gulf of Mexico are more than 20,000 ft below the surface and hold only a fraction of the amount of oil that the Saudi Arabian and Texan oil fields once produced [13].

There are some oil economists who claim that as liquid fuels are becoming scarcer, we can turn to the tar sands of Canada and the oil shale in the United States to replace oil as the liquid fuel. It is true that these resources hold an enormous amount of energy, but the energy and financial cost of extracting useful energy from them is much greater than that of the oil fields of the past. It has been estimated that for extracting oil from tar sands, the EROI is less than 5, while for oil shale it may be as low as 2. This means that an enormous amount of energy needs to be invested for a relatively small amount in return. In addition, extraction of oil from both tar sands and oil shale has a deleterious effect on the ecology, as well as requiring large amounts of water for their energy output.

The EROI of RE sources, which are necessary for a sustainable energy system, varies widely. As shown in other parts of the book, the EROI of modern wind systems in a high-quality location is on the order of 20, and solar thermal energy, depending on its location and eventual use, can be on the order of 7–25. On the other hand, the EROI for ethanol from corn is only barely greater than unity, which means that it is not a truly sustainable energy source. It is a somewhat ironic fact that of all the sustainable energy sources, only ethanol from corn has received substantial financial support from the U.S. Government for its production. This is a result of pure politics, since the corn-growing states represent an important political power in the U.S. Congress. The EROI of ethanol from sugar cane in Brazil may be as high as 6–8. Considerable research is going on worldwide to produce ethanol from cellulosic material and algae, but the technology is not sufficiently advanced to make quantitative estimates.

A conclusion to be drawn from these facts is that, as the most productive fossil energy sources are used up, we will have far less energy surplus available for our future basic needs and particularly for discretionary activities that modern man has come to expect to enjoy ever since the fossil fuel era started 150 years ago. The reality of our times is that the energy investment to produce more energy will increase as the energy return on these investments dwindles. This means that the high-energy lifestyles of Western society will have to adapt to lower-energy use. A few years ago, there were some who extolled the potential of a so-called hydrogen economy. However, as shown in Chapter 1, hydrogen is not an energy source. It has to be produced from other forms of energy, such as electricity via hydrolysis or chemical reactions from natural gas. There are no hydrogen energy reservoirs.

The perspective on energy budgeting also has some major implications on the engineering design and optimization of energy production systems. The Electric Power Research Institute (EPRI) released in June of 2011 a 100-page study designed to compare various current and future energy generation systems. The current cost of capital investment and fuel to operate various systems is shown in Table 2.16. As can be seen, the life expectancy in the table for fossil fuels is assumed to be 40 years, whereas for RE systems such as wind or solar, the life expectancy is 20 years. However, since the experimental life for wind and solar systems has been demonstrated to be at least 40 years, the LCOE of the systems has also been calculated for these longer lifetimes to provide a fair comparison. The cost of nuclear power was estimated by EPRI before the Fukushima disaster in Japan and does not include the extra cost for additional safety measures required by updated regulations. The EPRI estimates show that the power from wind and natural gas are the lowest cost, followed closely by nuclear. However, solar thermal and PV are close to competitive in favorable locations.

The levelized costs of the RE systems, as well as the energy return on investment, depend heavily on the assumed lifetime. In fact, since there is no fuel costs associated with any of the renewable systems, the EROI is directly proportional to the lifetime of the system; whereas for fossil fuel generation, the initial capital investment is relatively small, but the operation requires a continuous flow of energy. This flow of energy has its own EROI associated with it, as illustrated in connection with oil. Consequently, the EROI of the energy fuel will decrease with time, and the cost of the fuel is likely to increase along with increases in exploration and delivery costs. Therefore, for future energy systems using RE

TABLE 2.16

Cost and Performance of Power Generation Technologies (2025)

	Capacity Factor (%)	Life (years)	Average Capital Cost ($/kW)	LCOE ($/MWh)
Coal IGCC[c]	80	40	3450	85–101
NG NGCC	80	30	1105	47–74
Nuclear[b]	80	40	5400	68–109
Wind: On-shore	28–40	20	2280	73–134 (37–67)[a]
Solar Thermal	25–49	30	4800	116–173 (87–130)[a]
Solar PV	15–28	20	3750	210–396 (105–198)[a]

Source: EPRI, 2011, http://my.epri.com/portal/server.pt?space=CommunityPage &cached=true&parentname=ObjMgr&parentid=2&control=SetCommuni ty&CommunityID=404&RaiseDocID=000000000001022782 &RaiseDocType=Abstract_id (accessed February 24, 2011).

[a] Recalculated for 40 year life.
[b] Does not include cost of a permanent nuclear waste disposal and storage or cost of dismantling nuclear power plants after decommissioning.
[c] Includes carbon capture.

sources, the key goal of the engineering design should be the highest possible lifetime, which may in many instances be more important for the EROI, as well as probably also for the levelized costs, than the efficiency of the system.

2.13 EROI for a Wind Energy System

A detailed energy analysis for three wind power systems has been made by White and Kulsiwski [14], who calculated the EROI, as well as the air pollution emitted. The project for which the most detailed information was available was built in 1997 and is located near the city of De Pere in Wisconsin and will be referred to as the De Pere Wind Project (DPWP). It consists of two Tacke 600e turbines that have a combined electric power capacity of 1.2 MWe with a capacity factor of 31% (predicted) and a predicted life for each of the nacelles of 20 years. The other two wind projects were designed for a maximum power of 25 MWe (BR-I) with 73 turbines and 107 MWe (BR-II) with 143 turbines. Their respective predicted capacity factors were 33% and 35%, and the predicted nacelle life was 25 years for both.

Table 2.17 shows the energy requirements to construct the two De Pere wind turbines. These requirements are obtained from the cost data shown in column 1, for example, craning cost was $75,000. Then the energy per dollar in 1977 dollars is obtained from the I/O sector data in Table 2.17, for example, 30,233 Btu/1977$ for hoists and cranes. But the time at which the craning cost was $75,000 was 1997, while the I/O tables are for 1977$.

Hence we must adjust the cost figure for inflation, which for these periods can be obtained from Table 2.2. Conversion from 1997 dollars (the year of construction), to 1977 dollars (the year the I/O tables were developed) was done using the CPI ratio 1977:1997 = 60.6:160.5, obtained from the CPI of the Bureau of Labor Statistics [11]. Multiplying the cost times the price index ratio and the Btu/$ gives the embodied energy, that is, 8.56×10^8 Btu. A similar analysis for the

TABLE 2.17

Energy Requirements to Construct Two Turbine Wind Farms at De Pere, Wisconsin

	1997$	I/O Sector[a]	Btu/1977$[b]	Btu	GJth/2 Turbines
Craning	75,000	Hoists, cranes	30,233	8.56E+08	903
Labor	25,000	Misc. business services	10,000	9.44E+07	100
Local equipment rental	3,000	Construction machinery	34,534	3.91E+07	41
Lodging and food for employees	8,000	AVI[c]	28,780	8.69E+07	92
Electrical grounding	12,000	NC, electrical utilities	30,648	1.39E+08	147
				Total: 1283 GJth/2 turbines	

Source: White, S.W. and Kulsiwski, G.L. *Net Energy Payback and CO$_2$ Emissions from Wind-Generated Electricity in the Midwest*, Fusion Technology Institute, University of Wisconsin, Madison, WI. December, 1998.
[a] I/O sector data are from [10].
[b] 1997$ data were calculated from the CPI by the scale 1977/1997 = 60.6/160.5 = 0.3776 [12].
[c] Sector AVI is an average of the hospitality establishments in the area.

TABLE 2.18

Costs and Energy Requirements for On-Site Construction of the De Pere Wind Power Plant

	1997$	I/O Sector	Btu	GJ
Foundation preparation	73,000	Ready mix concrete	2.23E+09	2353
Transformer (2 at $10 K per)	20,000	Transformers	3.22E+08	339
Craning	75,000	Hoists, cranes	8.56E+08	903
Labor	25,000	Misc. business services	9.44E+07	100
Local equipment rental	3,000	Construction machinery	3.91E+07	41
Lodging and food for employees	8,000	AVI	8.69E+07	92
Electrical grounding	12,000	NC, electric utilities	1.39E+08	147
				1282

Source: White, S.W. and Kulsiwski, G.L., *Net Energy Payback and CO$_2$ Emissions from Wind-Generated Electricity in the Midwest*, Fusion Technology Institute, University of Wisconsin, Madison, WI. December, 1998.

other items in column 1 yields the total energy input required to make the turbines. A similar analysis for the energy embodied in on-site construction is shown in Table 2.18.

The lifetime and annual energy investments in the DPWP are shown in Table 2.19. Using these data, the *total energy input* per installed 1.2 MWe is 13,933 GJth or 3.89×10^6 kWhth. The power output is 1.2×10^3 kWe $\times 0.31 = 0.372 \times 10^3$ kWe, and for the assumed lifetime of the system of 20 years, the *total power output* is 0.372×10^3 kWe $\times 20$ year $\times 365$ day/year $\times 24$ h/day $= 65.2$ kWh$_e$. Hence, EROI $= 65.2/3.89 = 16.75$.

TABLE 2.19

Lifetime and Annual Energy Investments for the De Pere Wind
Power Plant

Process	Source	Total Energy per Installed 1.2 MW$_e$ GJth/Power Plant	Annual Energy per GW$_e$y GJth/GW$_e$y
Wind turbine (embodied)			
Blades	PCA	154	
Nacelles	PCA	2,273	
Inverter	I/O	339	
Wiring	PCA	209	
Tower	PCA	4,890	
Foundations	PCA	1,925	
Materials total		9,790	326
Transportation			
Blades		21	1
Nacelles		257	9
Towers		178	6
Concrete		23	
Control cabinets		0.29	0.01
Transportation totals		480	15
Construction	I/O	1,282.28	43
Maintenance	I/O	1,636	55
Decommissioning		641	21
Total required energy		13,933	

If the wind farm were located at a place where new transmission lines would have to be built to deliver the power to a user, additional energy input for grid extension could be included. Currently, energy storage is not used with commercial wind farms. If the wind-generated electricity were to be stored, as is often done for remote power installations, then the energy embedded in batteries would need to be included.

The EROI of wind energy has increased as the technology has improved. Table 2.20 shows the EROI for various wind systems. The average EROI for systems built in the year 1983 was about 2.5, while for systems built in 1998 and 1999, the average was about 23, a ninefold increase. It is therefore not surprising that wind energy has become economically competitive with conventional fossil and nuclear power plants.

*2.14 EROI for Nuclear Power

The EROI for nuclear power today is difficult to evaluate because no nuclear plant has been built in the United States since 1971. An estimate for the average EROI of nuclear power can be obtained by a simplified analysis [7] based on the existing 104 nuclear power plants in the United States, which have a rated capacity of about 100 GW electric [15]. The installed

TABLE 2.20

EROI for Wind Systems

Year of Study	Location	EROI	CO_2 Intensity (g CO_2/kWh)	Power Rating (kW)	Lifetime (Year)	Load Factor (%)	Analysis Type	Turbine Type	Rotor Diameter (m)	Hub Height (m)	Remarks
1999	United States	23	14.4	342.5	30	24	I/O	Kenetech KVS-33	32.9	36.6	25 MW farm
1999	United States	17	20.2	600	20	31	I/O	Tacke 600e	46.0	60.0	1.2 MW farm 6.1 m/s wind spd
1999	United States	39	8.9	750	25	35	I/O	Zond Z-46	46.0	48.5	107 MW farm
1998	Germany	21.7		1500	20	31	PA	Enercon E-66	66	67	3 blades
1998	Germany	23.8		500	20	29.6	PA	Enercon E-40	40.3	44	3 blades
1998	Germany	15.4		500	20	29.6	PA	Enercon E-40	40.3	44	3 blades
1996	Germany	8.3	17	100	20	31.4	PA	3 blades	20	30	3 blades
1998	Germany	14.1		1500	20	31	I/O	Enercon E-66	66	67	3 blades
1994	Germany	14.7	8.1	500	20	36.5	PA	2/3 blades	39	41	3 blades
1991	Japan	4	71.7e	100	20	31.5	I/O				10% aux power
1991	Japan	2.9	95.6e	100	20	31.5	I/O				Downwind
1996	Japan	2.3	123.6e	100	30	20	I/O				
1983	Germany	3.5		6	15	45.7	I/O				Average values
1983	Germany	5		12.5	15	45.7	I/O				Average values
1983	Germany	8.3		32.5	15	45.7	I/O				Average values
1983	Germany	1.3		3000	20	20.5	I/O	2 blades	100	100	GROWNIAN prototype
1997	Denmark	8.3		15	20	19.9	I/O	1980	10	18	Vintage model
1997	Denmark	8.1		22	20	19.9	I/O	1980	10.5	18	Vintage model
1997	Denmark	10		30	20	19	I/O	1980	11	19	Vintage model
1997	Denmark	15.2		55	20	20.69	I/O	1980	16	20	Vintage model
1997	Denmark	27		600	20	26.5	I/O	3 blades	47	50	15 m/s
1991	Germany	11.8		30	20	14.4	PA	Hsw-30	12.5	14.8	2 blades, 13 m/s
1991	Germany	20.4		33	20	29.4	PA	MAN-Aeromann	14.8	22	2 blades 11 m/s

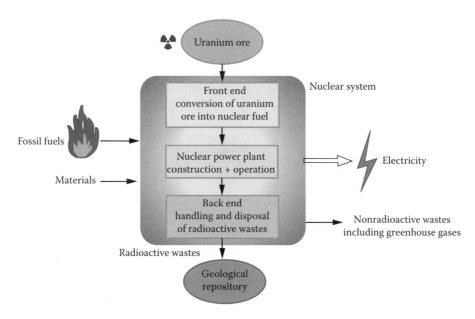

FIGURE 2.12
Basic nuclear fuel cycle. (From World Nuclear Association, www.world-nuclear.org)

new cost of nuclear plants is estimated at $7 billion per 1 GW electric unit [6,7]. The cost of installing 100 GW of nuclear power would therefore be on the order of $700 billion.

A schematic diagram of the basic nuclear fuel cycle is shown in Figure 2.12. Environmental concerns include effects of mining and spent fuel storage. Mining produces tailing piles, which can pose health hazards to people living near them. The storage time for spent fuel rods before they become cooled enough to handle safely has a wide range depending on the remaining U-235 content. The spent fuel rods remain in cooling water ponds at the nuclear reactors in the United States for about 5 years. After that time, a repository for the waste uranium fuel rods must be engineered to provide 10,000 years of safe, secure storage where water and people cannot come in contact with the containers. The only designated U.S. repository at Yucca Mountain, Nevada, has had more than $50 billion invested, but as of 2013, development has been discontinued. Radiation from nuclear power plants is a serious concern to the public, but except for radiation emitted as a result of accident, nuclear power plants do not produce radiation effects more serious than living at very high altitudes.

To estimate the embedded energy in 100 GW of nuclear power, consider that the American industrial sector consumed 33% of the total energy of about 100 quads in 2004, while the GDP in that year equaled $11.7 trillion nominal dollars [15]. Since the American industrial sector constituted 20% of the GDP [16], the average American energy intensity was approximately 8500 Btu/$. However, industrial energy intensity, such as building power plants, is larger than other sectors, estimated by a factor of 33/20, resulting in an energy intensity for nuclear power plants of about 14,000 Btu/$ [7]. Thus, the embedded energy in an expenditure of $700 billion for new nuclear power plants would be 98×10^{14} Btu (9.8 quads).

The energy embedded in the nuclear fuel is a relatively small component of the cost of energy delivered. According to [17], the market price is 0.5 cents per kWh. With a capacity factor of 90%, the electric output of all the nuclear plants working for 8760 h per year would equal about 800 trillion Wh per year. At 0.5 cents per kWh, the cost of the nuclear fuel would be $4 billion/year. At an embedded energy of 14,000 Btu/$, this equates to 56 trillion Btu/year or 0.056 quads per year. Hence, the embedded energy in all the fuel for the nuclear plants during their 40 year life is about 0.056 quads/year × 40 years, for an embedded energy in all nuclear fuels of about 2.25 quads. Therefore, the total embedded energy of the nuclear power production of 100 plants including fuel and construction, which is approximately 10 quads, during 40 years of operation amounts to 12.25 quads.

U.S. primary energy consumption for producing electricity in 2004 was about 41 quads, according to the EIA Annual Energy Review [18]. With an overall efficiency for all power production at about 33% [19], 13.5 quads of electric energy was produced in 2004. Since nuclear plants produced 20% of that, that is, 2.7 quads of electricity per year, the total electricity produced over 40 years was 108 quads, yielding an EROI (electric power out to primary fossil in) of 108/12 or 9. This estimate does not include O&M, storage of spent fuel, the social cost of mining uranium, and dismantling of the plant, which may reduce the EROI substantially. A previous analysis performed by the Oak Ridge National Laboratory in the 1970s [19] gave an EROI for nuclear power, including disposal and dismantling of the plant, of about 6. An extensive analysis of the energy balance of nuclear power was made in 2001 and was updated in February of 2008 by the CEEDATA consultancy in Shaam, the Netherlands [20] and is available at http://www.stormsmith.nl/. This lengthy and detailed report provides a background for the total cost of nuclear power, which includes all the social and environmental costs. It concluded that the EROI on that basis was only about 2.3, but the assumed life of the plant was 30 years, the original design life but which has been extended for all power plants that have applied for permission to continue operation.

A rebuttal to this document was prepared by the nuclear industry, the World Nuclear Association (world-nuclear.org), and a response to the rebuttal was added to the final website. Another important publication regarding the nuclear power prospect was prepared in June 2009 by the Institute for Energy and the Environment of the Vermont Law School [21]. The latter of these two reports approaches the question of nuclear power mostly from an economic perspective. In both cases, the nuclear industry responded with strongly worded rebuttals, and in the absence of hard data, it is difficult to reach a conclusion about the economic and environmental consequences of nuclear power. However, investment in nuclear power requires a large commitment in money and energy, and before such an undertaking on a large scale is initiated, additional careful and unbiased EROI and sustainability analysis for this technology is important. It should also be noted that nuclear power incurs unique financial risks due to the very large capital requirement that are beyond the ability of any single entity to underwrite. Moreover, the very long development time before any revenue is generated incurs specific risk premiums on the cost of capital.

*2.15 Relation between Energy Return on Investment and Monetary Return on Investment

The relation between EROI and monetary return on investment (MROI) is a complex topic of interdisciplinary research by economists and energy analysts. A simplified analysis of this important relationship has been published by King and Hall [22]. The first step in the analysis is to determine all of the different, N types and streams of energy input and M types of energy produced. Then, determine the energy intensity of the direct or indirect energy output, e_i (e.g., MJ/barrel oil), for each of the unit energy production sold in the market, m_i, input or output flows of energy or materials (e.g., short ton coal). Finally, the price per unit of energy or material input or output, p_i, will need to be determined.

$$\text{MROI} = \frac{\sum \$_{out}}{\sum \$_{investment}} \tag{2.23}$$

which can be related to EROI through the energy intensity figures and the price of the fuel or consumer energy flows produced:

$$\text{EROI} = \frac{\sum_{i=1}^{M} m_i e_i * \sum_{i=1}^{N} n_i p_i}{\sum_{i=1}^{M} m_i p_i * \sum_{i=1}^{N} n_i e_i} * \text{MROI} \tag{2.24}$$

If there is only one type of energy production, that is, $M = 1$, one can solve Equation 2.24 for any output variable as a function of the other variables. For example, the required sales price, p_i, of a unit energy production ($/BBL oil) as a function of EROI and MROI is

$$p_i = \frac{\text{MROI}}{\text{EROI}} * \frac{e_i}{e_{investment}} \tag{2.25}$$

and the MROI as a function of EROI is

$$\text{MROI} = \text{EROI} * \frac{(p_i * e_{investment})}{e_i} \tag{2.26}$$

where
  $e_{investment}$ is the energy intensity of the investment (e.g., kWh/$)
  e_i is the energy intensity of direct or indirect energy output (e.g., MJ/$)
  p_i is the unit price of produced energy (e.g., $/gal gasoline)
  m_i is the unit energy production sold in the market (e.g., BBL of oil or ft^3 of natural gas)

Examination of Equations 2.25 and 2.26 shows that as energy gets more expensive, evidenced by a decreasing energy intensity (Energy/$) of investment in energy production ($e_{investment}$), the cost of energy increases even when the EROI is constant. As the energy intensity of investing in energy production increases, the cost of energy to make a constant profit goes down. The reason for this odd result is that higher-intensity purchases represent cheaper energy inputs.

Figure 2.13 shows the oil price as a function of EROI. It is apparent that with the same EROI, the energy production system with a higher intensity of investment can

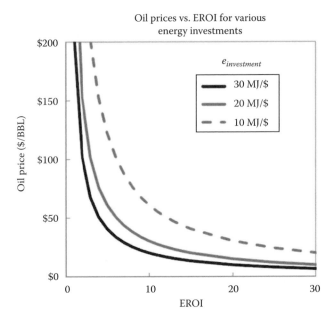

FIGURE 2.13
Oil price as a function of EROI. (From King, C.W. and Hall, C.A.S., *Sustainability*, 3, 1810, 2011).

be sold for less or a greater profit can be obtained at the same price. The graph also shows that when EROI drops below 5, the price begins to increase precipitously. This result, from the monetary view, is in agreement with the result of the energy analysis in Figure 2.7. The drop in net energy when EROI decreases below 5 has been called the energy cliff.

Problems

2.1 The doubling time of any quantity n is related to the continuous growth rate R_{cont} by the relation $2 = \exp n\ R_{cont}$. Derive an expression for n in terms of the annual growth rate and then estimate what the doubling time for a population growth rate is averaging about 2% per year.

2.2 What is the present value of a $10,000 payment in 2015 if the discount rate is 7%?

2.3 Repeat the previous problem in constant dollars for the $10,000 payment if the inflation rate is 4%/year.

2.4 Calculate the annual payments for a solar system whose initial extra cost is $10,000. Assume that the interest rate is 8% and the mortgage term is 15 years.

2.5 Assuming that the price of fuel is $5.00/GJ and is increasing at a rate of 4%/year while the discount rate is 6%, calculate the equivalent levelized price over 20 years.

2.6 A home buyer obtains a mortgage of $100,000 at an interest rate of 8% over 20 years. Estimate the annual payments to repay this loan.

2.7 A rooftop solar hot water heater with collector areas of 8 m² and storage of 640 L is installed in Phoenix, Arizona, on a house that uses 160 L/day of hot water at 60°C. Of the total energy required, 70% is supplied by solar energy (the solar fraction is 70%) and 30% is supplemented with an electric heater. The solar system costs $3000 for installation, whereas an electric water heater and tank for the same supply costs $800 for installation. Assuming that the electricity cost is $0.10/kWh and does not change over time, estimate the simple payback time for the solar hot water heater. The temperature of the water in the main from which it is to be heated is 10°C.

2.8 Repeat the previous problem, but assume that the cost of installation is to be repaid at 7% interest over 15 years and that the electricity costs escalate at 10% per year. Compare the NPV of the solar system in Problem 2.7 to that of an electric heating system over a 15 year period, then discuss whether or not the solar system is a good investment relative to the payback Time. Also prepare a spreadsheet showing the NPV for each year.

2.9 Repeat the previous problem, but assume that the cost of the equipment is reduced from a government grant by 20%.

2.10 Compare the capital costs and the levelized costs of operation for a 1000 MW nuclear power plant and a 1000 MW PV system constructed in the Mojave desert in California. The PV system is to use thin film technology currently available. Neglect the cost of nuclear fuel. Comment on environmental impact and availability of water.

2.11 A solar system is to be installed in Phoenix, Arizona. The estimated amounts of energy from the system as a function of collector area are shown in the following tabulation. If the solar system costs $195/m² and is to be paid off over 25 years at 10% interest, prepare the total cost curve and specify the cost optimal system. Assume that the backup fuel is fuel oil at $7/GJ and that the fuel price is inflating at 8%/year. The total energy demand for which the solar system is constructed is estimated to be 1000 GJ/year.

Collector Area (m²)	Energy Delivered (GJ)
100	336
150	444
200	531
250	612
300	673
400	791
500	856
600	915

2.12 Calculate and tabulate the annual cash flows associated with a solar system for a 10 year period of analysis if the initial solar cost is $6000 and if

Interest rate = 9%

Power cost = $30/year escalating at 10%/year

Property tax = 0

Income tax bracket = 32%

Maintenance = 0.5% of the capital per year (escalating at 10%/year)

Scrap value = 50% of initial cost

If the solar system saves $550/year in conventional fuel (escalating at 10%/year), is it cost-effective? Work the problem in current dollars.

Compute the annual payment of a self-amortizing loan for the solar system with the characteristics tabulated as follows:

Factor	Specification
Expected system lifetime t (year)	20
Effective interest rate (%)	8
Collector area A_c (m²)	20
Collector cost ($/m²)	100
Storage cost ($/m²)	6.25
Cost of control system ($)	100
Miscellaneous costs (e.g., pipes, pumps, motors, heat exchangers) ($)	$200 + (5A_c)$

2.13 If the initial extra cost of a solar system is $5,000, what are the annual payments of a self-amortization loan if the interest rate is 12%, the inflation rate is 4%, and the mortgage term is 15 years.

2.14 A factory uses 700.40 W standard fluorescent lamps for its lighting. The lamps are to be replaced by 34/W high-efficiency lamps that can operate on the existing standard ballasts. The standard and high-efficiency fluorescent lamps cost $1.80 and $2.50 each, respectively. The factory operates 3000 h/year, and all of the lamps operate during that period. Assuming the unit cost of electricity is 6 cents/kWh and the ballasts consume 10% of the rated power of the lamps, calculate how much energy and money will be saved as a result of switching to the high-efficiency fluorescent lamps. Then, estimate the PP assuming an interest rate of 6% and an inflation rate of 3%. Finally, calculate the levelized cost of the saved energy assuming the lamps last for 2 years.

2.15 The hallways of a university are lit by 40 fluorescent lights, each containing two lamps rated at 60 W each. The lights are on all day and night during the year, but the building is only used from 7 a.m. to 7 p.m. 5 days per week during the year. If the price of electricity is 7 cents/kWh, estimate the amount of energy and money that could be saved by installing two motion sensors that turn off the lights when the building is not being used. Then, estimate the simple payback and the payback with a 6% effective interest rate if the price of a sensor is $50 and the cost of installation is $40.

References

1. Rueg, R.T. (1994) Economic methods, in: Kreith, F. and West, R.E. (eds.), *CRC Handbook of Energy Efficiency*, Chap. 3, CRC Press, Boca Raton, FL.
2. Tester, J.W., Drake, E.M., Golay, M.W., Driscoll, M.J., and Peters, W.A. (2005) *Sustainable Energy: Choosing among Options*, MIT Press, Cambridge, MA.
3. Larson, R.W., Vignola, F., and West, R.E. (December 1992) *Economics of Solar Energy Technologies*, ASES, Boulder, CO.

4. Rabl, A. and Spadaro, J.V. (2007) Environmental impacts and cost of energy, in: Kreith, F. and Goswami, D.Y. (eds.), *CRC Handbook of Energy Efficiency and Renewable Energy*, Chap. 4, CRC Press, Boca Raton, FL.

5. Short, W., Packey, D.J., and Holt, T. (1995) *A Manual for the Economic Evaluation of Energy Efficiency and Renewable Energy Technologies*, U.S. Department of Energy, NREL, Golden, CO.

6. Nuclear Power Daily (2009) *Progress Energy Florida Signs Contract for New, Advanced-Design Nuclear Plant*, January 16, 2009. http://www.nuclearpowerdaily.com/reports/ Progress_ Energy_Florida_Signs_Contract_For_New_Advanced_Design_Nuclear_Plant_999.html (retrieved May 18, 2009).

7. *Nuclear Power Joint Fact Finding* (June 2007) The Keystone Center, Keystone, CO.

8. Spreng, D.T. (1988) *Net Energy Analysis and the Energy Requirements of Energy Systems*, Praeger Publishing Company, New York.

9. Ballard, C.W., Penner, P.S., and Pilati, D.A. (November 1978) Net energy analysis—Handbook for combining process and input-output analysis, *Resources and Energy* 1, 267–313.

10. Hannon, B., Casler, D.S., and Blasek, T. (1985) *Energy Intensity for the U.S. Economy-1977*, Energy Research Group, University of Illinois Press, Urbana, IL.

11. Bureau of Labor Statistics (1998) Consumer price index—All urban consumers, March 13, 1998. http://146.142.4.24/cgi-bin/surveymost?cu

12. Cleveland, C.J. (2005) Net energy from oil and gas extraction in the United States, 1954–1997, *Energy*, 30,769–781.

13. Cleveland, C.J. and O'Connor, P. (2010) An assessment of the energy return on investment of oil shale, Western Resource Advocates, www.westernresourceadvocates.org/land/pdf/oseroire-port.pdf.

14. White, S.W. and Kulsiwski, G.L. (December 1998) *Net Energy Payback and CO_2 Emissions from Wind-Generated Electricity in the Midwest*, Fusion Tech Inst., University of Wisconsin, Madison, WI. http://fti.neep.wisc.edu/pdf/fdm1092.pdf

15. Kennedy III, R.G. (2008) Member of ASME Energy Committee, private communication, April 12, 2008.

16. Statistical Abstract of the United States, 2007, accessed at http://www.census.gov/compendia/ statab

17. Economists, *The Economist/Pocket World in Figures*, 2004 edition, Economists Book, London, 2004.

18. *EIA Annual Energy Review*, 2004, accessed at http://www.eiq.doe.gov/aer/

19. Rotty, R.M., Perry, A.M., and Reister, D.B. (May 1976). *Net Energy from Nuclear Power*, Oak Ridge Associated Universities, Inc., Oak Ridge, TN.

20. Storm van Leeuwen, J.W. and Smith, P. (2008, update), Nuclear power and the energy balance. http://www.stormsmith.nl/ (accessed August 22, 2008).

21. Cooper, M., 2009. *The Economics of Nuclear Reactors: Renaissance or Relapse?* Institute for Energy and the Environment, Vermont Law School, South Royalton, VT.

22. King, C.W. and Hall, C.A.S. (2011). Relating financial and energy return on investment. *Sustainability*, 3, 1810–1832.

3

Wind Energy

First, there is the power of the Wind, constantly exerted over the globe.... Here is an almost incalculable power at our disposal, yet how trifling the use we make of it! It only serves to turn a few mills, blow a few vessels across the ocean, and a few trivial ends besides. What a poor compliment do we pay to our indefatigable and energetic servant!

Henry Thoreau, "Paradise, Paradise (To Be) Regained" (1843)

... As yet, the wind is an untamed, and unharnessed force; and quite possibly one of the greatest discoveries hereafter to be made, will be the taming, and harnessing of the wind.

Abraham Lincoln, "Discoveries and Inventions" (1858)

3.1 Wind Power in a Nutshell

Put simply, wind turbines (WTs) convert the kinetic energy of the wind into electrical power. This process begins when incoming wind strikes the rotor. The rotor consists of blades, typically three in a modern large-scale turbine, attached to a hub. The rotor assembly can be seen in Figure 3.1, which shows a large WT manufactured by Clipper Windpower.

The shape of the turbine blades creates a lifting force, which is the same as that of an airplane wing. The lifting force is directed at an angle causing the rotor to rotate. The rotor is typically connected to a low-speed shaft. The spinning low-speed shaft is in turn connected to a gearbox that increases the low-speed shaft rotation speed (50–100 rpm in a "large" WT) to a high-speed rotation (typically 2000–3000 rpm). A high-speed shaft on one end of the gearbox is then connected to an electrical generator that produces the desired electrical energy. The details of the hub and gearbox, as well as the general layout of a wind energy system, can be seen schematically in Figure 3.2.

As one can imagine, it takes quite a bit of engineering work to go from this simple description of a WT to an actual turbine with a hub height of 100 m, producing 5 MW of power. In this chapter, we will explore some of the key elements involved in turning this simple concept into reality by focusing on commercial size (1–5 MW) WTs. These elements include

Fundamentals of power and energy

Frequently asked questions

Turbine types and their difference

Aerodynamics (blade design)

FIGURE 3.1
(See color insert.) 2.5 MW Liberty wind turbine (Clipper Windpower) near Medicine Bow, Wyoming. (Courtesy of DOE/NREL, Golden, CO. With permission.)

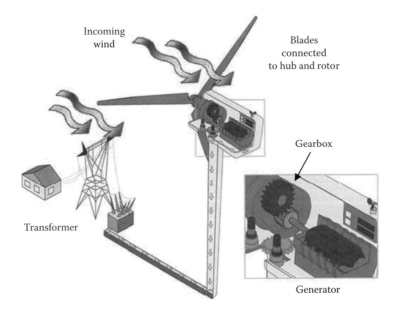

FIGURE 3.2
Schematic of typical wind turbine layout. (From Energy Efficiency & Renewable Energy (EERE) (DOE), How does a wind turbine work? 2010, http://www1.eere.energy.gov/wind/wind_animation.html. With permission.)

Where wind comes from, how it is modeled, and WT performance

How turbine design affects the cost of electricity produced

WT farms and offshore WTs

3.2 Power and Energy

Power and energy are common terms that are often misused. Power is the rate at which energy is used or produced. Light bulbs use power; WTs generate power. Units of power are expressed in work (energy) over an amount of time:

$$\text{Power} = \frac{\text{Work}}{\text{Time}} = \frac{\text{Energy}}{\text{Time}} = \frac{N \cdot m}{s} = \frac{J}{s} = W \tag{3.1}$$

WTs are rated in terms of power. Large commercial WTs have power ratings in the 1–5 MW range while a WT used for a single home is likely rated in the 3–15 kW range. This means commercial-sized WTs produce power at a rate 100 times greater than a residential-sized WT:

$$\frac{\text{Commercial}}{\text{Residential}} = \frac{1 \text{MW}}{10 \text{kW}} = \frac{1 \times 10^6 \text{ W}}{10 \times 10^3 \text{ W}} = 100 \tag{3.2}$$

Energy, on the other hand, is a measure of how much work can be done by a force. Units of energy are the product of force and distance:

$$\text{Energy} = N \cdot m = J \tag{3.3}$$

The link between power and energy comes from the rate of power usage. In other words, energy equals the product of power and time or power is energy divided by time. If a 100 W light bulb is turned on for 24 h, the energy usage is

$$\text{Energy usage} = 100 \text{ W} \cdot 24 \text{ h} = 2400 \text{ Wh} = 2.4 \text{ kWh} = 2.4 \text{ kWh} \tag{3.4}$$

$$\text{In terms of Joules: } 2400 \text{ W h} = \frac{2400 \text{ J} \cdot \text{h}}{s} \times \frac{3600 \text{ s}}{h} = 8.64 \times 10^6 \text{ J} \tag{3.5}$$

Turbines are denoted by their rated power production (e.g., 100 kW, 2.5 MW, etc.), while we pay for electricity based on how much energy we use (e.g., $0.06 per kWh).

In 2006, the per capita U.S. annual energy consumption was approximately 98 MWh [1]. A 10 kW WT running 24 h/day for 365 days (this is not realistic, but rather illustrative) would produce approximately 88 MWh of energy. In comparison, a gallon of gasoline has an energy content of approximately 36 kWh (130 MJ). If our only energy source was gasoline, then the average American would have consumed the equivalent of 2722 gal of gasoline during the year or 7.5 gal of gasoline/day. Creating more sources of renewable energy will expand the options available for the world's energy needs.

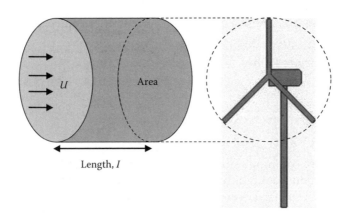

FIGURE 3.3
Incoming air column and wind turbine.

We will now use these fundamental concepts of power and energy to determine the power contained in the wind moving toward a WT. The kinetic energy, KE, of a mass of air, m (in kg), moving at speed, U (in m/s), is

$$KE = \frac{1}{2}mU^2 \tag{3.6}$$

$$\text{Unit check:} \quad kg\left(\frac{m}{s}\right)^2 = \frac{kg \cdot m^2}{s^2} = \left(\frac{kg \cdot m}{s^2}\right) \cdot m = N \cdot m = J \tag{3.7}$$

Now, imagine a horizontal cylinder of air of area A (in m²) and velocity U, moving toward a WT, as seen in Figure 3.3. Note that the area, A, corresponds to the area swept out by the rotating WT blades. The mass m of the air column is volume V times the density, where ρ represents the air density in kg/m³. Mass can then also be expressed as

$$m = \rho V = \rho A l = \rho A U t \tag{3.8}$$

where
  ρ is the air density, kg/m³
  V is the volume, m³
  A is the area, m²
  l is the length, m
  U is the wind speed, m/s
  t is the time, s

where the length of the column of air, l, is equal to the distance the wind travels in a given time interval, t, and is found by multiplying the wind speed, U, by the time. To check the units of the mass expression provided earlier (Equation 3.8),

$$\text{mass} = \frac{kg}{m^3} \cdot m^2 \cdot \frac{m}{s} \cdot s = kg \tag{3.9}$$

Combining Equation 3.6 with Equation 3.8, the kinetic energy becomes

$$KE = \frac{1}{2}mU^2 = \frac{1}{2}(\rho AUt)U^2 = \frac{1}{2}\rho AtU^3 \qquad (3.10)$$

As discussed earlier, power is equal to energy divided by time, so simply by dividing the KE by time yields the power of a moving column of air:

$$\text{Power} = \frac{KE}{t} = \frac{1}{2}\rho AU^3 \qquad (3.11)$$

In other words, the power in a column of wind is linearly proportional to the air density and area of the air column. Of even more significance, the power is also proportional to the cube of the wind speed. We will see later that a key reason WTs have grown in height over time is to place the WT rotor blades higher in the atmosphere where the wind speed is greater and thus harnessing a much higher kinetic energy output.

Since the air density cannot readily be changed to increase power output, this leaves the area of the WT blades as the only variable (besides the WT height) that can be modified on a macroscale to affect the power output. Since the WT power output is linearly proportional to the area, a large-scale commercial WT designer would want to increase the blade length (i.e., radius) to increase the power output. Since the circular area swept out by the turbine blades is proportional to the blade radius squared (R^2), turbine power is also proportional to the blade radius squared (R^2). Turbine blades sweep out a circular area.

$$\text{Power} = \frac{1}{2}\rho AU^3 = \frac{1}{2}\rho \pi R^2 U^3 \qquad (3.12)$$

In summary, longer blade lengths and taller turbines (providing access to higher wind speeds) are options to increasing WT power output. However, this performance increase comes at a price. Longer blades need more sophisticated manufacturing techniques to build blades strong enough to withstand wind gusts and extreme wind speeds (in excess of 100 mph). Taller towers must be strong enough to withstand the forces transmitted from the blades, weight of the nacelle, and other parts while still being cost-effective.

In contrast, why make small and/or shorter WTs? Some applications, for example, remote homes off the electrical grid, require more modest amounts of power. The design challenges here are just as significant: structures (blades, tower, etc.) must be strong enough to withstand wind forces, but also must be light, cost-effective, and reliable.

3.3 Fact or Fiction: Common Questions about Wind Turbines

Other challenges facing WTs beyond common design tasks are strength, reliability, and performance. As in any technology, there may be other unanticipated challenges. Some common questions that can cause controversy about WTs are

How noisy are WTs?

What are the lifetime environmental impacts of WTs?

What about bird and bat kills?

3.3.1 Noise Issues

Noise issues related to wind energy have been reported since the 1980s. Problems were initially found with the mechanical design as well as blade design and operation. For example, the 2 MW MOD-1 turbine developed by NASA in the late 1970s was a two-bladed turbine (61 m rotor diameter) [2]. It operated with the turbine blades downstream of a lattice tower structure. Low-frequency noise was produced as the turbine blades moved in and out of the "tower shadow" caused by wind flowing around and through the lattice tower. Noise of significant magnitude was produced, causing complaints from local residents.

Today, most large commercial WTs have the blades operating upstream of the tower and use a monopole tower. Modern commercial WTs have noise levels of approximately 35–45 dB at 750–1000 ft. In contrast, typical noise or loudness levels in everyday life are given in Table 3.1 [3]. As seen in Table 3.1, the modern turbine (at distances of 750–1000 ft) has a noise level comparable to a library or an average home (40–50 dB). When locating turbines, the distance between residences and WTs needs to be considered. Small WTs can actually have higher noise levels due to higher rates of rotation.

The sound typically heard from a WT comes from the blades. Careful aerodynamic design of the blades, as well as understanding the maximum allowable rotational speed of the blades, has reduced the aerodynamic noise level from previous levels. Similarly, mechanical components have been redesigned to virtually eliminate mechanical noise, leaving blade noise as the predominant noise source, albeit at a low noise level [4].

3.3.2 Lifetime Environmental Impact

An important consideration for any technology is the greenhouse gas emissions for the total fuel cycle. Each phase of the fuel cycle, including resource extraction, facility construction, and facility operation, must be evaluated in order to quantify and compare emissions from different technologies [5]. When the operations needed to manufacture WTs and build wind farms are included in a CO_2 emission analysis, the use of wind power compared to coal power reduces CO_2 emissions by 99%. Similarly, using wind power instead of natural gas reduces CO_2 emissions by 98% [6].

Other environmental impacts from WTs do exist. For example, construction of a wind farm could cause erosion due to the construction and use of roads needed to build the

TABLE 3.1

Loudness Levels of Common Noises

Noise Level (dB)	Equivalent Activity
120	Jet plane takeoff, amplified rock music at 4–6 ft, car stereo, band practice
110	Rock music, model airplane
100	Snowmobile, chain saw, pneumatic drill
90	Lawnmower, shop tools, truck traffic, subway
80	Alarm clock, busy street
70	Vacuum cleaner
60	Conversation, dishwasher
50	Average home
40	Quiet library
30	Quiet bedroom

farm (truck and crane traffic, electrical line/equipment installation, etc.), but this impact can be mitigated using standard environmental reclamation techniques.

In Europe, "shadow flicker" created by the shadow of rotating WT blades on residences has occurred due to placing WTs too close to the residences. The resulting flicker can be very annoying. Proper planning and accounting for latitude allow the necessary distance between turbine and residence to be calculated so no turbine shadow falls on the residence [7]. This distance should also take into account the noise level of the operating turbine (see the earlier text). The turbine should be located a sufficient distance away (based on sun angle) to eliminate any shadow flicker or noise concerns.

3.3.3 Bird and Bat Kills

As can be seen in Figure 3.4, the largest single cause of bird deaths are buildings and/or windows. A recent study [8] indicated that about 7000 bird fatalities occurred in 2006 due to wind farms, while nuclear plants and fossil-fueled power stations are estimated to kill 46 and 2070 times more birds, respectively. A study by Erickson [9] provides an annual estimate of the sources of bird fatalities (see Figure 3.4) including deaths from WTs. Bird deaths from WTs are estimated to be much less than 0.0001% of the total number of projected deaths, on the order of 30,000/year. In contrast, bird deaths from cats or buildings are projected to be 3.5 and 19 million times, respectively, more likely than that from WTs. If WTs are sited properly (e.g., away from migratory paths) or even shut down briefly with the help of radar to determine when migratory birds are approaching, the fatalities can be

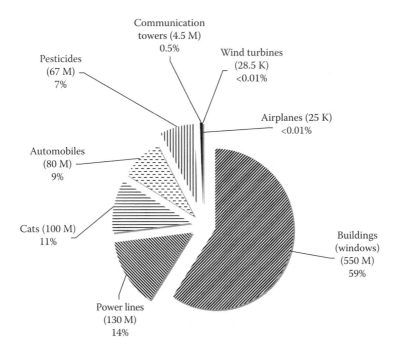

FIGURE 3.4
Most common causes of bird fatalities. Bird deaths are shown in millions (M) or thousands (K) with an annual estimated total near 1 billion birds. (From Erickson, W. et al., A summary and comparison of bird mortality from anthropogenic causes with an emphasis on collisions, USDA Forest Service General Technical Report PSW-GTR-191, 2002, http://www.fs.fed.us.psw/publications/documents/psw_gtr191/Asilomar/pdfs/1029-1042. With permission.)

minimized [55]. Increasing the number of wind farms would likely increase the number of deaths, and this topic remains an area of research in order to minimize the effect on avian mortality.

While bird fatalities occur due to birds being struck by blades or running into towers, bat fatalities have been found to be due to a different cause. Researchers proposed that decompression killed bats by barotraumas, which is a rapid reduction in air pressure near rotating turbine blades. Study results showed that, of the bats affected by WTs, "90% of bat fatalities involved internal hemorrhaging consistent with barotrauma, and that direct contact with turbine blades only accounted for about half of the fatalities" [54]. Studies have shown a higher rate of fatalities of bats with WTs than birds.

Bird fatalities per turbine ranged from 0.6 to 7.7 while bat fatalities ranged from 3.4 to 47.5 per turbine [10]. Research is underway to understand the higher fatality rate as well as ways to stop/minimize bat fatalities as well as predict and avoid high-risk sites [11].

One wind power project conducted a study to determine the effect of turning off the WTs at night during a 11 week bird migratory period corresponding to low winds. From this project, fatalities declined by approximately 50%. However, if used, this method would result in an annual loss in generating capacity of 1% if the turbines were turned off when they could have been generating electricity [12]. Further study and work is needed to ensure that WT technology is developed to coexist with environmental issues.

3.4 History of Wind Turbine Development: HAWTs and VAWTs

Much work has occurred over the last three decades leading to optimization of performance and cost in WT design. Currently, the exclusive commercial WT design (e.g., 1 MW and larger) is a horizontal axis wind turbine (HAWT) in which the axis of the rotor is horizontal to the ground, as shown in Figure 3.5, with the turbine blades spinning in the same orientation as if they are on a plane. Conversely, vertical axis wind turbines (VAWTs) have blades that spin about a vertical axis, as shown in Figure 3.6. The biggest advantage of this design is that it does not have to be pointed in the direction of the wind.

Earliest indication of wind devices (anything that converts wind energy to other forms of energy, e.g., rotational and mechanical) was in Persia around 900–1000 AD or earlier. The first practical vertical axis windmills arose in the seventh century in Sistan, Afghanistan, to process grain and draw up water. Horizontal axis wind machines like that shown in Figure 3.5 appeared in the eleventh and twelfth centuries in Northwest Europe to grind grain. The Dutch built thousands of windmills over the twelfth and nineteenth centuries to move water. Clearly, early development of wind machines occurred all over the world, and this trend continued as WT technology continued to mature.

In Scotland, Professor James Blyth built the first wind turbine for electricity production and successfully generated electricity in July 1887 [13,14]. Shortly afterward, the first automatically operating WT in the world was made by Charles F. Brush in Cleveland, Ohio, in the winter of 1887–1888. The HAWT turbine produced 12 kW and had a 17 m (50 ft) rotor diameter [15].

The first MW-sized turbine to supply electricity to the local electrical grid occurred in 1941 on a hill Grandpa's Knob in Vermont. The 1.25 MW turbine was designed by Palmer Cosslett Putnam and was 53 m in diameter and had two blades. A failure of a metal blade in 1945 after only 1100 h of operation brought the project to an end [16,17].

FIGURE 3.5
An example of a horizontal axis wind turbine.

FIGURE 3.6
An example of a Darrieus VAWT, Quebec Canada.

The modern era in WTs began in the late 1970s as a result of the oil crises earlier in the decade. An increase in turbine development occurred in multiple countries around the world. For example, Danish manufacturers Kuriant, Vestas, Nordtank, and Bonus began producing 20–30 kW machines, while NASA, at the Lewis (now Glenn) Research Center in Cleveland, Ohio, developed a government research and development program producing the first of the "Mod" series of WTs ranging from 0.2 to 2 MW [18].

As we have seen, VAWTs have a long history, while HAWTs have a shorter history. This raises the question of why today's commercial WTs are almost exclusively horizontal axis, two- or three-bladed, upwind turbines. To help answer this question, we first look at a summary of the qualitative differences between the two technologies and then have a discussion of quantitative differences in power-generating ability between the two designs.

3.4.1 VAWTs

Strengths

Turbine always aligned with the wind direction

May start producing electricity at lower speeds

Simpler design without large structures in the air (house mechanical and electrical components at ground level)

May be lower noise than HAWT

Disadvantages

Typically produce 50% less energy than HAWT (due to drag of blades through full rotation)

Wind speeds are lower at lower heights meaning the energy potential of a VAWT at a given location is lower than a HAWT having a taller tower height at the same location

Typically not self-starting

Difficult to maintain parts that are under weight of tower

Cannot control power output or rotational speed through blade settings

Can require high tip-speed ratios (TSRs) causing increased noise (Darrieus)

3.4.2 HAWTs

Advantages

Highest operating efficiency (see Figure 3.7)

Turbine blades can be placed in a higher wind speed regime (more power) on top of a tower

Power output and blade speed can be controlled and maximized by varying the angle of the blades

Disadvantages

Transportation costs of long (50–90 m) blades and tall towers can be a significant portion (20%) of the installation cost

Installing tall HAWTs (tower, nacelle, blades, etc.) require skilled operators, expensive cranes

Tall height means proximity to residences must be considered for visual impact

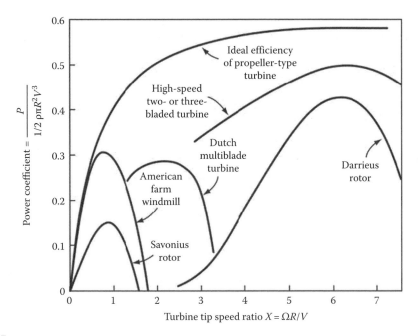

FIGURE 3.7
Various turbine power coefficients versus TSR. (From Wilson, R.E. and Lissaman, P.B.S., *Applied Aerodynamics of Wind Power Machines*, Oregon State University, Corvallis, OR, 1974, http://ir.library,oregonstate.edu/jspui/handle/1957/8140. With permission.)

Sophisticated control systems required to point turbine blades into wind, prevent overspeed conditions, etc.

Readers interested in more information on a variety of WT concepts are directed to [19,20].

3.5 Introduction to Wind Turbine Performance

A key performance measure in WT design is the coefficient of performance, C_p, which is a ratio of the aerodynamic power extracted by the WT, P_{aero}, divided by the total power in the wind, P_{wind}, and is calculated as

$$C_p = \frac{P_{aero}}{P_{wind}} = \frac{P_{aero}}{1/2\rho A U^3} = \frac{P_{aero}}{1/2\rho\pi R^2 U^3} \tag{3.13}$$

The theoretical maximum turbine efficiency was derived by Albert Betz in 1920 [21], as well as others, see [22], and is

$$C_p = \frac{16}{27} \approx 0.59 = \text{Betz limit} \tag{3.14}$$

In addition, C_p is a function of how fast the tips of the WT blades are rotating compared to the incoming wind. The ratio of the blade's rotational speed divided by the incoming wind speed is known as the TSR:

$$\text{Turbine tip speed ratio} = \frac{\Omega R}{U} \qquad (3.15)$$

where
   Ω is the turbine blade rotational speed, rad/min
   R is the rotor radius, m
   U is the wind speed, m/s

In Figure 3.7 [23], the power coefficient is plotted as a function of TSR for different WT configurations. Several key points can be seen in Figure 3.7, starting from the top. The highest line shows how the theoretical maximum power coefficient increases with increasing TSR. The Betz limit value of 0.59 is approached asymptotically as the TSR increases toward a value of seven (TSR = 7). The next highest curve denotes the performance of today's modern high-speed horizontal axis two- or three-bladed turbines. The maximum theoretical performance coefficient for these turbines is approximately $C_p = 0.5$, while the Darrieus vertical axis rotor has a maximum C_p of approximately 0.40–0.45. Thus the modern two- and three-bladed HAWTs are approximately 10% more efficient than a Darrieus (VAWT) machine and thus have the potential to make 10% more energy and money.

Not only are the large commercial WTs configured horizontally, they are also predominantly three-bladed upwind (blades rotate in front of the tower). Why are HAWT mainly upwind? The blades on a downwind turbine experience cyclic loading (alternating higher forces and lower forces) due to the blades passing behind the tower which may cause fatigue failure of the blades as well as noise. Three-bladed turbines have also been found to be more aesthetically pleasing than two-bladed turbines due to a blade "disappearing" when one of the two blades is aligned with the tower.

3.6 Aerodynamics

Examining the aerodynamics of WT blades helps explain why different WT configurations have such different performance coefficients, C_p [24,25]. Figure 3.8 shows a WT blade rotating (in the plane of the paper) in a clockwise direction. The atmospheric wind, U_w, is blowing toward the blade (i.e., into the plane of the paper). A section view of the blade is then taken near the tip of the blade looking in toward the root of the blade (Figure 3.8). The cross-sectional view of the blade airfoil is shown in the right-hand side of the figure. The two airflows acting on the blade are the incoming wind speed, U_w, which is perpendicular to the plane of the rotating blades, and the rotational wind speed, $U_{Rot} = \Omega R$, created by the turbine blade rotating at an angular speed, Ω, at the radius of the cross section, R. Note that the side of the airfoil facing the wind (designated the front side) is relatively flat compared to the curved rear side of the airfoil. The leading edge of the airfoil hits the airflow arising from the blade rotation first causing the airflow to separate into two streams over the sides of the airfoil. Finally, the trailing edge is where the airflow over the two sides of the airfoil is rejoined.

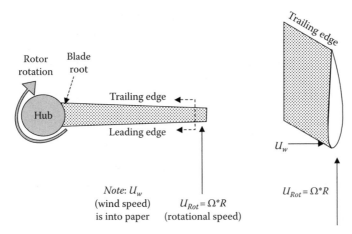

FIGURE 3.8
Rotor blade diagram.

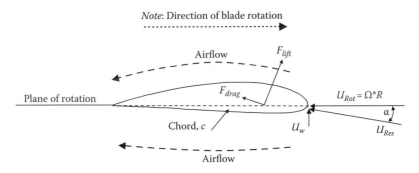

FIGURE 3.9
Lift and drag forces on a blade cross section near tip.

In Figure 3.9, a blade cross section near the tip of the blade is shown with the resultant forces created by the passing wind, the lift force, F_{Lift}, and the drag force, F_{Drag}. The straight line connecting the front (leading) and rear (trailing) edges of the airfoil is known as the chord line, and the distance between the two points is designated as the chord, c. Also shown are the resultant velocity component, U_{Res}, between the wind speed, U_w, and the wind arising from the blade's rotational speed, U_{Rot}. Finally, the angle between the resultant wind speed, U_{Res}, and the chord line, c, is the angle of attack α.

Via Bernoulli's law, Equation 3.16, the velocity is highest across the curved (upper) surface of any airfoil (e.g., airplane wing). In addition, Bernoulli's law tells us that the pressure is the lowest where the velocity is the highest. In the case of a WT blade, the pressure is lower over the rear (i.e., upper) surface of the airfoil compared to the front (i.e., lower) airfoil surface.

$$p + \frac{1}{2}\rho U^2 = \text{const.} \tag{3.16}$$

In summary, the lowest pressure is near the leading edge ("nose") of the airfoil, where the wind velocity is the highest, and over the rear (or upper) curved surface. This results in a net force, F_{Lift}, on the airfoil, causing it to be drawn forward, in the direction of the "nose" of the airfoil and hence rotating. By convention, the resultant forces on the airfoil are drawn perpendicular, F_{Lift}, and parallel, F_{Drag}, to the resultant velocity, U_{Res}. The higher the lift force (and conversely the lower the drag force) along a WT blade, the greater the rotational force, which means the WT is capturing and converting more of the wind's energy and converting it to electricity. In other words, the higher the lift-to-drag ratio (L/D) is, the more efficient the blade will be at capturing the wind's energy.

NOTE: It is now possible to better understand the performance of the different WT configurations shown in Figure 3.7. It turns out that a Savonius VAWT has a poor L/D ratio and thus a very low maximum C_p, as shown in Figure 3.7. The American farm windmill and Dutch multiblade turbines have higher L/D ratios and hence higher maximum C_p values. Lastly, machines with high L/D ratios are the Darrieus VAWT rotor and the modern high-speed two- or three-bladed turbines, with the modern two- and three-bladed turbines having the highest C_p, as noted earlier.

WT blades have very complicated shapes that vary along the blade. The shape variation is needed not only to help start the WT rotating but also to operate very efficiently as the wind speed increases. The airfoil is twisted near the blade root to provide the required torque to start the turbine rotating at low wind speeds. Once rotating, the optimized blade profiles in the outer portions of the blade provide the necessary aerodynamic forces to keep the blade rotating at design speeds.

Recall that Figure 3.9 illustrates a cross section of a blade in the outer portion of the blade. Figure 3.10 illustrates the lift and drag forces on a blade cross section near the root of the blade. Note that the blade is twisted, and the chord line now forms an angle θ with the plane of rotation of the blade. This twist angle, θ, is a function of turbine blade radius and has a larger value near the blade root and a decreasing value with increasing blade radius. Since the resultant velocity, U_{Res}, has decreased, this increases the angle between the chord line and U_{Res}, and results in an increased angle of attack approaching the blade root.

While the root portion of the blade encounters the same incoming wind speed, U_w, the rotational speed, $U_{Rot} = \Omega R_{Root}$, is lower. This occurs because, while the angular rate of rotation, Ω, is the same along the blade, the radius at the blade root, R_{Root}, is of course smaller. The resulting lift force, F_{Lift}, per unit area is lower at the blade root compared to the blade tip. However, the direction of the force is better aligned to apply a torque to begin rotating the blades. In addition, the surface area of the blade is considerably larger at the root. This provides sufficient area for the lower magnitude aerodynamic forces to act over, resulting in sufficient torque to start rotation at low wind speeds. Finally, the larger surface area is needed to accommodate the necessary structural strength to handle the cantilevered loads arising from wind forces on the blade, weight of the blade, etc.

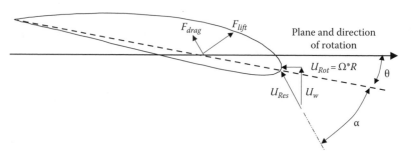

FIGURE 3.10
Lift and drag forces on a blade cross section near the blade root.

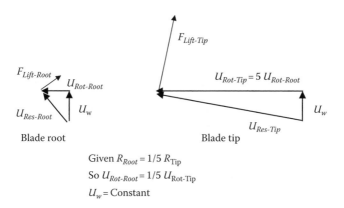

FIGURE 3.11
Comparison of velocity vector diagrams for blade root and tip cross sections.

Figure 3.11 shows the vector diagrams from Figures 3.9 and 3.10 side by side. Assuming a constant wind speed at both the blade root and tip cross sections, we will consider the case where the root cross section is located at a radial location one-fifth of the blade tip cross section. This last assumption means that the rotation speed at the blade root, $U_{Rot\text{-}Root}$, is one-fifth the speed at the blade tip, $U_{Rot\text{-}Tip}$. Note that $U_{Rot\text{-}Root}$ and $U_{Rot\text{-}Tip}$ are drawn to scale. The resultant velocity vector at the tip is much larger due to the contribution of the rotational wind speed. This results in a larger lift force per unit area at the tip. At the blade root, the lift force is better aligned to provide a start-up torque. While the blade root lift force is lower per unit area, it is applied over a large surface area. Interestingly, at the tip, the contribution of the wind speed to the resultant velocity is small compared to the contribution from the rotational wind speed when the turbine is at operational conditions.

3.7 Wind Characteristics

3.7.1 Wind Generation

Now we will turn our attention to the atmospheric wind: generation, distribution, characteristics, etc. The motion of air, otherwise known as wind, is caused by uneven heating of the Earth by solar radiation and the rotation of the Earth. The predominant wind direction varies with latitude and altitude above the Earth's surface, as shown in Figure 3.12 [25–27]. Energy is transferred from the equator to the poles via wind. As the wind moves, the Earth's rotation influences the wind direction in two ways. First, Coriolis forces accelerate a particle of air to the right in the Northern Hemisphere (and to the left in the Southern Hemisphere). Secondly, each air particle has an angular momentum from west to the east due to the Earth's rotational direction. As the air particle moves toward the pole, conservation of angular momentum requires its west-to-east velocity component to increase. This effect is pronounced in the midlatitudes, in the region called the westerlies, near the ground [25]. In general, the westerlies are the best-suited region for wind energy applications. In addition, the polar jet stream (7–12 km or 20,000–40,000 ft above sea level) can have a profound impact on the westerlies, moving them north or south, resulting in lower or higher energy production from year to year.

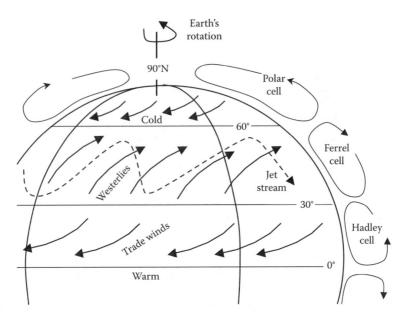

FIGURE 3.12
Wind flow in the Northern Hemisphere.

Of course, the most critical part of the atmosphere for a WT is relatively close to the ground. The atmospheric boundary layer (ABL) is the lowest part of the atmosphere and varies from the ground up to 500–2000 m in height depending on the time of day, temperature, and pressure. In the ABL, the mean wind speed varies from zero at or near the ground up to the free stream velocity of the atmosphere at the top of the ABL. In addition, the velocity field exhibits turbulent fluctuations. In other words, each of the three velocity components for an air particle has a mean as well as fluctuating component. The turbulent velocity component can be quite large and by nature is time varying. This results in unanticipated fluctuating forces (high velocity causes high blade loading), which can lead to fatigue failures of WT components such as blades, gearboxes, and tower. The turbulent portion of the flowfield is typically described statistically. Methods of accurately simulating turbulent flowfields for a WT design are an active field of research [28].

3.7.2 Distribution of Wind

Now we transition from a global view of wind to a local area view of the wind resource. The wind resource is the mean wind speed as well as how many hours per year the wind blows at a given speed. A wind farm developer needs high confidence that the wind speeds at a given location are "high" and consistent from year to year. How much of a concern is this? If the average wind speed is only 5% lower than predicted, the maximum possible power will be 14% lower than anticipated! High confidence in the wind resource allows the developer to estimate the probable energy output and compare that to the investment needed to lay out a wind site and purchase or finance, install, and operate the wind farm (consisting of tens of turbines).

Wind resource maps have been developed for the entire world. For example, the U.S. wind resource map is shown in Figure 3.13 [29]. Smaller-scale maps are also available, including each state in the United States [29]. Similar maps are available for all parts of the world.

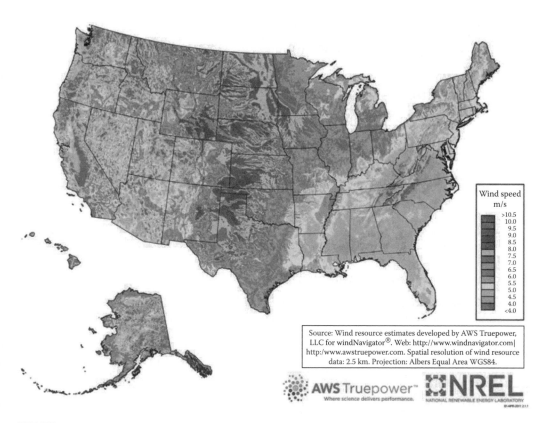

Wind speed m/s

>10.5
10.0
9.5
9.0
8.5
8.0
7.5
7.0
6.5
6.0
5.5
5.0
4.5
4.0
<4.0

Source: Wind resource estimates developed by AWS Truepower, LLC for windNavigator®. Web: http://www.windnavigator.com| http://www.awstruepower.com. Spatial resolution of wind resource data: 2.5 km. Projection: Albers Equal Area WGS84.

FIGURE 3.13
(See color insert.) Wind resource map of the United States. This map shows the annual average wind power estimates at a height of 50 m. It is a combination of high-resolution and low-resolution datasets produced by NREL and other organization. The data were screened to eliminate areas unlikely to be developed onshore due to land-use or environmental issues. In many states, the wind resource on this map is visually enhanced to better show the distribution on ridge crests and other features. (Courtesy of U.S. Department of Energy, National Renewable Energy Laboratory, Washington, DC.)

Figure 3.13 color codes the estimated annual average wind power 50 m above the ground, as shown in Table 3.2. For an enlarged view of the relationship between wind power class (denoted by color code in Figure 3.13) and mean wind speed (in m/s), consult Table 3.2. Several items stand out in the U.S. map. First, the highest-class wind power (or highest mean wind speeds) regions are offshore along the East and West coasts, and are Classes 6 and 7 (which represents Outstanding and Superb wind resources, respectively). Second, the middle of the country has a wind power classification of Fair to Good running from the Canadian border into Northern Texas.

A small but significant region of Outstanding/Superb wind power can be seen in southeastern Wyoming. Another valuable "offshore" resource exists in the Great Lakes, which have a power classification of Good to Outstanding. Similar regions of high annual wind power can be found in many areas of the country but are not visible at this resolution. This highlights a key point, not surprisingly, that the wind resource is highly dependent on local conditions such as vegetation (forests), terrain (e.g., hills, mountains), and water (ponds, rivers, lakes, oceans). Given sufficient resolution, one can search for and decide on a site to locate a WT. In order to predict the annual power output of a turbine, we must understand the variation of wind speed (and thus power) with height.

TABLE 3.2

Rating Wind Energy Resource Potential

Wind Power Class	Resource Potential	Wind Power Density at 50 m (W/m²)	Mean Wind Speed[a] at 50 m (m/s)	Mean Wind Speed[a] at 50 m (mph)
1	Poor	0–200	0.0–5.9	0.0–13.2
2	Marginal	200–300	5.9–6.7	13.2–15.0
3	Fair	300–400	6.7–7.4	15.0–16.6
4	Good	400–500	7.4–7.9	16.6–17.7
5	Excellent	500–600	7.9–8.4	17.7–18.8
6	Outstanding	600–800	8.4–9.3	18.8–20.8
7	Superb	>800	>9.3	>20.8

Note: Enlarged from Figure 3.13.

[a] Wind speeds are based on a Weibull *k* value of 2.0 (Rayleigh).

3.7.3 Wind Speed Increasing with Height

The wind resource maps in Figure 3.13 show, on a large scale, where winds suitable for commercial WTs exist. The next step is to develop a wind speed profile as a function of height. As mentioned earlier, wind speed increases with height and thus power increases with height.

Using local data from airports and weather stations, developers gather the data needed to get site-specific wind speed data. The data are then processed to arrive at an equation expressing the wind speed as a function of height. Two common approaches are the log law and the power law [30].

3.7.4 Log Law Wind Speed Profile

The log law wind speed profile has a physical basis. It is assumed that the shear stress in the wind is constant with height, the pressure gradient near the Earth's surface is small, and the Earth's rotation is ignored. The height of this constant shear stress layer is 100–200 m and is known as the "surface layer" of the ABL (see, for example, [31]). For a smooth surface, the mixing length, *l*, of the turbulent wind is given as

$$l = \kappa z \tag{3.17}$$

where

κ is von Karman's constant ($\kappa = 0.4$)

z is the height above the ground

The velocity profile as a function of height can be found to be

$$U(z) = \frac{U^*}{\kappa} \ln\left(\frac{z}{z_0}\right) \tag{3.18}$$

where

$U(z)$ is the wind speed at a height, z

U^* is the friction velocity arising from shear stress

z_0 is the surface roughness length

Values for z_0 are shown in Table 3.3.

TABLE 3.3

Surface Roughness for Various Outdoor Surfaces

Terrain Type	Roughness Length, z_0 (m)
Cities, forests	0.7
Suburbs, wooded countryside	0.3
Countryside with trees and hedges	0.1
Open farmland	0.03
Flat grassy plains	0.01
Flat desert, rough sea	0.001

Sources: Burton, T. et al., *Wind Energy Handbook*, Wiley, Chichester, U.K., 2001; Manwell, J.F. et al., *Wind Energy Explained, Theory, Design and Application*, Wiley, Chichester, U.K., 2002.

There are two unknowns in this equation: U^* and z_0. Substituting typical values for U^* and z_0, the shape of the resulting atmospheric velocity profile is illustrated in Figure 3.14 for a neutrally stable atmosphere. One sees a predicted gradual change in wind speed through the first 20 m or so of the ABL and then an exponential growth approaching a steady-state value above 200 m.

There are three stability states of the atmosphere: neutral, stable, and unstable [32]. Neutral stability occurs when there is no change in temperature with height through the ABL due to adiabatic cooling. This state is typically accompanied by high winds. A stable atmosphere occurs when the air temperature on the ground is cold and increases with increasing altitude. This results in a steep increase in the wind speed with height (i.e., large wind shear). An unstable atmosphere occurs when the air temperature near the ground is warmer than the temperature at higher altitudes. The air on the ground rises while the colder air above sinks creating turbulent wind conditions.

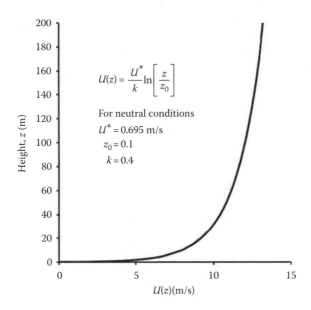

$$U(z) = \frac{U^*}{k} \ln\left[\frac{z}{z_0}\right]$$

For neutral conditions
$U^* = 0.695$ m/s
$z_0 = 0.1$
$k = 0.4$

FIGURE 3.14
Log law velocity distribution.

If one wants to determine an unknown wind speed at a given height, it can be estimated from a known wind speed at another height using this equation. This approach is especially valuable because wind speed data are typically known at heights of 50 m or less (reference height); however, turbines operate at much higher heights where wind speed data are unknown. Using this method, wind speeds can be estimated at hub height from data available at lower heights.

So, taking Equation 3.18 at a height z, for the wind speed $U(z)$ and then dividing it by the same equation for a reference height z_r at the reference wind speed $U(z_r)$, eliminates the unknown friction velocity U^* and yields

$$\frac{U(z)}{U(z_r)} = \frac{\ln(z/z_0)}{\ln(z_r/z_0)} \tag{3.19}$$

This equation now has only one unknown, z_0, the roughness length, which is estimated from Table 3.3. Consider a meteorological tower taking measurements at 40 m ($z_r = 40$ m), measuring a wind speed of $U(z_r) = 7.2$ m/s at z_r, located in a wooded countryside, $z_0 = 0.3$, then the wind speed, $U(z)$, at a hub height of $z = 95$ m is

$$U(z) = 7.2 \cdot \frac{\ln(95/0.3)}{\ln(40/0.3)} = 8.5\,\text{m/s} \tag{3.20}$$

In other words, the wind speed is predicted to increase by approximately 18% from 72 to 95 m if the atmospheric assumptions in the model hold. In this case, the wind power would increase by $(9.7/7.2)^3 = 164\%$. Height does indeed matter when it comes to wind turbines.

3.7.5 Power Law Wind Speed Profile

Another model commonly used in engineering wind studies is the power law profile. It is used to estimate wind speeds given a reference height, z_r, to some desired height, z. The power law profile is used when one does not know the surface roughness, z_0, and/or stability information (i.e., earlier defined friction velocity, U^*) is unknown. The power law profile is given in Equation 3.21 as

$$U(z) = U(z_r) \cdot \left[\frac{z}{z_r}\right]^{\alpha} \tag{3.21}$$

where α is the power law or shear exponent.

Typical values for α are

$$\text{For neutral flows:} \quad \alpha \cong \frac{1}{7} = 0.143$$

$$\text{For highly sheared flows:} \quad \alpha > 1$$

A comparison between the log law and power law wind speed profiles is shown in Figure 3.15. This figure shows that the wind speed profile of the power law profile for a value of $\alpha = 0.17$ is similar in shape to the log wind profile plotted earlier for neutral atmospheric conditions. In any case, it should be noted that the value of α is highly dependent

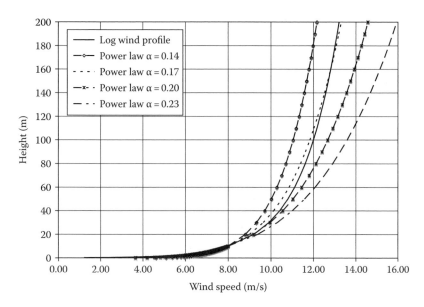

FIGURE 3.15
Log law and power law velocity profile comparison.

TABLE 3.4

Variation in Wind Speed with α
at a Height of 100 m

α	Wind Speed $U(z)$ (m/s)
0.14	11.0
0.23	13.6

on temperature, weather, geographic location (elevation and terrain), and time (of day and season). The power law profile is typically used in engineering studies.

Consider Table 3.4 that shows the calculated wind speeds for two different values of α at a height of 100 m, which is the typical hub height for modern commercial WTs. The change in predicted power based on these wind speeds is the ratio $(13.6/11)^3 = 1.89$. This indicates that the predicted power can nearly double at 100 m if the value of α increases from 0.14 to 0.23. This has large economic implications. In other words, determining the correct value of α for the power law profile (or using an entirely different method to determine wind speed as a function of height) for a given location and time period is critical when making choices about where to build wind farms. The best option is to gather long-term meteorological data at hub height in a sufficient number of locations at a potential WT site (if the cost can be afforded).

Consequently, methods have been developed, and continue to be developed, to choose the best value of α given a set of wind data and environmental wind factors [30].

3.7.6 Probability of Observing a Given Wind Speed

As has been discussed, the wind speed (and hence power available) varies. The next step is to determine the probability of a given wind speed occurring over a desired time period using a probability distribution function (PDF). A common PDF is the normal or

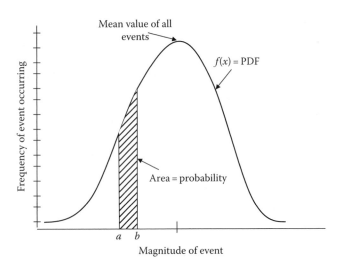

FIGURE 3.16
Illustration of a Gaussian or normal probability density function.

Gaussian PDF, which can be seen in Figure 3.16. Many natural events can be described by a normal PDF, for example, the height distribution of 20 year old students in the United States. When graphed, this Gaussian PDF yields the commonly known "bell-shaped" curve. For all PDFs, the vertical axis is the observed frequency of a given event within a certain range or magnitude, for example, number of students between the heights of 5'0" and 5'1". The horizontal axis contains the range of all events, for example, 4'0"–6'10". The peak of any PDF curve (where the curve is denoted by $f(x)$) is the mean magnitude of all occurring events.

The area under any section of a PDF curve corresponds to the probability, P, of a range of events occurring. As shown in Figure 3.16, the area represents the probability of having an event occur with a magnitude between a and b. The probability, P, of a range of events occurring is obtained by integrating the PDF, $f(x)$, between the range of the events, a and b.

$$P = \int_a^b f(x)\,dx \qquad\qquad (3.22)$$

It should also be noted that integrating the PDF between the limits of 0 and ∞ yields a value of 1. In other words, the probability of having an event occur of any magnitude between the limits of all events is 100%.

$$P = \int_0^{\infty} f(x)\,dx = 1 \qquad\qquad (3.23a)$$

Two commonly used PDFs, Weibull and Rayleigh, are used to describe the probability of observing a given wind speed. These distributions are used because they more closely represent the distribution of wind speeds at given location. These PDFs are NOT symmetric as is the Gaussian distribution (see Figure 3.16) but are skewed to the "left"

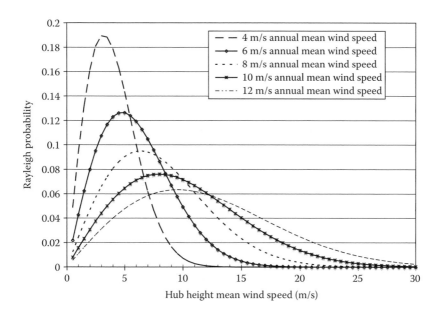

FIGURE 3.17
Rayleigh wind distributions.

(see Figure 3.17). A wind speed distribution skewed to the "left" means that the probability of seeing a lower wind speed is higher than the probability of seeing a higher wind speed (i.e., there is a higher probability of a 3 m/s wind speed than a 25 m/s wind speed at a typical WT site). These PDFs are typically determined at a desired height, for example, hub height. These probability distributions are, of course, only defined for values greater than zero. Integrating the PDF will provide the probability, $P(U)$ of wind speed, U, occurring between two wind speeds U_a and U_b.

$$P = \int_{U_a}^{U_b} f(U)dU \tag{3.23b}$$

Several other important statistical relationships can be found given a wind speed PDF, $f(U)$, including

$$\text{Mean wind speed:} \quad \bar{U} = \int_0^\infty Uf(U)dU \tag{3.24}$$

$$\text{Standard deviation:} \quad \sigma = \sqrt{\int_0^\infty (U - \bar{U})^2 f(U)dU} \tag{3.25}$$

Lastly, one is likely to want to know the probability (or the percent of time) that the wind will be equal to or less than a given wind speed. For example, what percent of a year will the wind speed be too low for the turbine to operate? This is known as the cumulative distribution function, $C(U)$, and is found by integrating from 0 m/s to a desired wind speed, U m/s:

$$C(U) = \int_0^U f(U)dU \tag{3.26}$$

The Weibull PDF and cumulative distribution function are given by

$$f(U) = \left(\frac{k}{c}\right)\left(\frac{U}{c}\right)^{k-1} \exp\left[-\left(\frac{U}{c}\right)^k\right] \tag{3.27}$$

$$C(U) = 1 - \exp\left[-\left(\frac{U}{c}\right)^k\right] \tag{3.28}$$

where
  k is the shape parameter
  c is the scale parameter

Both k and c are functions of the mean wind speed, $\overline{U}$, and the standard deviation of the wind speed, σ. The Rayleigh PDF and cumulative distribution function are given by

$$f(U) = \frac{\pi}{2}\left(\frac{U}{\overline{U}^2}\right)\exp\left[-\frac{\pi}{4}\left(\frac{U}{\overline{U}}\right)^2\right] \tag{3.29}$$

$$C(U) = 1 - \exp\left[-\frac{\pi}{4}\left(\frac{U}{\overline{U}}\right)^2\right] \tag{3.30}$$

The Rayleigh PDF only depends on one parameter, the mean wind speed, $\overline{U}$, making it much easier to use (but not necessarily more accurate!). Note that the Rayleigh distribution is a special case of the Weibull distribution when $k = 2$ and $c = \overline{U}$. An example of the Rayleigh distribution for different annual mean wind speeds is shown in Figure 3.17. Figure 3.17 shows that as the mean speed increases, there is a larger probability of having higher wind speeds at a given site, and hence the annual power production has the potential to be higher.

Example 3.1: Rayleigh Wind Speed Distribution Calculations

An analysis of wind speed data (10 min interval average, taken over a 1 year period) has yielded an average speed of $\overline{U} = 6$ m/s for a potential WT site. It has been determined that a Rayleigh wind speed distribution gives a good fit to the wind data. Recall that the Rayleigh PDF is given by Equation 3.29.

A. Estimate the number of hours per year that the wind speed will be between $U_B = 10.5$ and $U_A = 9.5$ m/s during the year.

To do this, we will find the probability the wind speed will be in a given range during the year, $P(U_A \leq U \leq U_B)$, and multiply the probability by the number of hours in a year.

$$P(U_A \leq U \leq U_B) = \int_{U_A}^{U_B} f(U)dU = \int_0^{U_B} f(U)dU - \int_0^{U_A} f(U)dU = C(U_B) - C(U_A)$$

For a Rayleigh distribution $C(U) = 1 - \exp\left[-\frac{\pi}{4}\left(\frac{U}{\overline{U}}\right)^2\right]$
So

$$C(U_B) - C(U_A) = \exp\left[-\frac{\pi}{4}\left(\frac{U_B}{\overline{U}}\right)^2\right] - \exp\left[-\frac{\pi}{4}\left(\frac{U_A}{\overline{U}}\right)^2\right] = P(U_A \leq U \leq U_B)$$

Substituting $\overline{U} = 6$ m/s, $U_B = 10.5$ m/s, $U_A = 9.5$ m/s yields

$$P(U_A \leq U \leq U_B) = 0.049$$

In other words, the wind speed will be between U_B and U_A m/s 4.9% of the year. Thus the total number of hours the wind speed is between u_B and u_A in a year is

$$\frac{\#h}{\text{year}} = 0.0494 \cdot \frac{24\,h}{\text{day}} \cdot \frac{365\,\text{day}}{\text{year}} = 432\,h/\text{year}$$

B. Estimate the number of hours per year that the wind speed is above 16 m/s. This problem is the same as part A with the wind speed ranges being

$$U_B = \infty \quad \text{and} \quad U_A = 16\,\text{m/s}.$$

$$P(\infty \leq U \leq 16) = \int_{16}^{\infty} P(U)dU = C(\infty) - C(16)$$

$$C(\infty) - C(16) = 1 - \exp[-\infty] - \left[1 - \exp\left[-\frac{\pi}{4}\left(\frac{U_A}{\overline{U}}\right)^2\right]\right] = \exp\left[-\frac{\pi}{4}\left(\frac{U_A}{\overline{U}}\right)^2\right] = 0.0038$$

Thus, the total number of hours the wind speed is greater than 16 m/s (i.e., between $U_B = \infty$ and $U_A = 16$ m/s) in a year is

$$\frac{\#h}{\text{year}} = 0.0038 \cdot \frac{24\,h}{\text{day}} \cdot \frac{365\,\text{day}}{\text{year}} = 33\,h/\text{year}$$

3.8 Turbine Performance

Given a WT site, selected by examining long-term wind speed data, a turbine can be picked and/or designed for the site. In order to do this, detailed engineering work has to be conducted to design the blades, tower, electrical systems, etc. In this section, we will

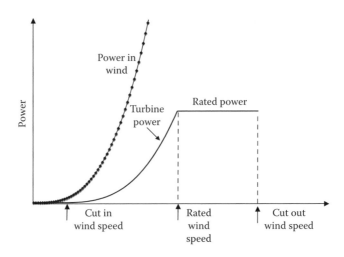

FIGURE 3.18
Wind power and turbine power curves versus wind speed.

focus on understanding the basics of the blade aerodynamics to design an efficient WT. Aerodynamics is the study of bodies moving through gases. Its use allows the blade and tower shape to be optimized as well as optimizing blade rotational speed. The goal is to maximize the extraction of power from the incoming wind stream and minimize the effect of wind loads (both steady and unsteady).

The efficiency of a WT is the power extracted by a WT from the power contained in the wind flowing through the area of the blades. Figure 3.18 contains a schematic comparing this relationship. Note that the power contained in the wind is always greater than the power extracted by the turbine.

The turbine power curve contains several interesting points. The turbine does not start producing power until the wind speed reaches the cut-in wind speed. This is the lowest wind speed where the rotor's aerodynamic power is greater than the sum of equipment loads and mechanical and electrical inefficiencies. The cut-out wind speed is the speed at which additional energy production is less than the cost of structural strength to withstand incremental fatigue damage. Lastly, the rated wind speed is the speed at which the turbine is operating near its peak efficiency. The goal is to operate the WT near or at its maximum efficiency.

Earlier, we saw that the standard measure of a WT's efficiency is known as the power coefficient, C_p, and is defined as

$$C_p = \frac{P_{aero}}{P_{wind}} = \frac{Q\Omega}{1/2\rho A U^3} \tag{3.31}$$

where
Q is the aerodynamic torque
Ω is the rotor rotational speed

A homework problem will show that the maximum theoretical limit of C_p is given by the Betz limit, $C_{p,max} = C_{p,Betz} = 16/27 \approx 0.5926$. In other words, the maximum power a WT should be able to extract from the wind is approximately 59% of the power crossing the area swept by the WT blades.

Another heavily used design metric is the TSR, which is defined as

$$\text{TSR} = \lambda = \chi = \frac{\text{Tip speed}}{U_{wind}} = \frac{\Omega R}{U_{wind}} \tag{3.32}$$

where
  R is the rotor radius
  Ω is the rotor rotational speed
  U_{wind} is the instantaneous incoming wind speed

Design guidelines currently suggest limiting the tip speed to 90 m/s to minimize noise issues.

Example 3.2: Tip-Speed Ratio Calculation

What is the TSR for a 90 m diameter turbine rotating at 15 rpm at a wind speed of 10 m/s?

$$R = \frac{D}{2} = 45\,\text{m} \quad \text{and} \quad U_{wind} = 10\,\text{m/s}$$

$$\Omega = 15\,\text{rpm} = 15\frac{\text{rev}}{\text{min}} \cdot 2\pi \frac{\text{rad}}{\text{rev}} \cdot \frac{\text{min}}{60\,\text{s}} = 1.57\,\text{rad/s}$$

$$\text{TSR} = \frac{\Omega R}{U_{wind}} = \frac{1.57\,\text{rad/s} \cdot 45\,\text{m}}{10\,\text{m/s}} = 7.07$$

Actuator disk theory allows one to find simple relationships for power and thrust. Another more comprehensive theory beyond the scope of this chapter allows one to design the blade shape to optimize power and thrust. This theory, called blade element momentum theory (BEM), is covered in detail in many of the comprehensive wind energy textbooks in Refs. [25,32].

  The first assumption in actuator disk theory is that the area swept by the rotor blades is a permeable disk, shown in Figure 3.19, with an area swept out by the turbine blades. As the wind flows through the disk, momentum (i.e., power and thrust) is extracted.

Figure 3.20 illustrates a top view of an actuator disk (the outlines of a tower and nacelle are omitted). The wind flows from the left to right. As the wind passes through the actuator disk (area swept by the turbine blades), momentum is extracted and the wind speed is reduced, hence

$$U_{wind} > U_{disk} > U_{wake}$$

From the conservation of mass, one can see that as the velocity decreases, the affected area of the column of wind increases (i.e., the mass flow rate remains constant).

  Note that the area of the wind column increases after passing through the actuator disk.

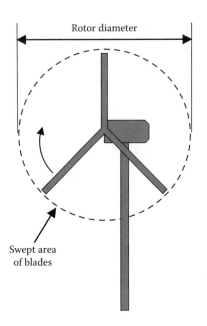

FIGURE 3.19
Actuator disk.

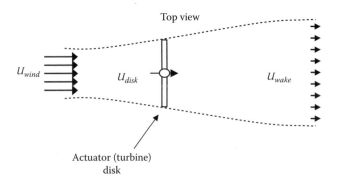

FIGURE 3.20
Actuator disk (top view), showing incoming wind passing through actuator disk. (*Note:* Wind speed is denoted by arrow length.)

Additional assumptions are used to develop this theory. Key ones include the following:

Homogeneous, incompressible, inviscid, and steady-state flow.
Static pressure far upstream and downstream are equivalent.
Velocity on either side of the actuator disk is the same.
Infinite number of blades that are infinitesimally thin.
Uniform thrust over actuator disk.
Nonrotating wake.
Velocity across the disk is constant.

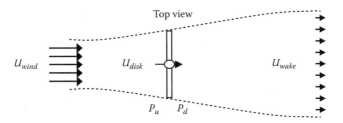

FIGURE 3.21
Actuator disk (top view) with quantities needed for momentum balance. (Note: ρ, air density; A, area of actuator (turbine) disk; P_u, static pressure immediately upstream of disk; P_d, static pressure immediately downstream of disk.)

Bernoulli's theory is valid:

$$P + \frac{1}{2}\rho U^2 = \text{constant} \tag{3.33}$$

More complicated theories such as BEM ease some of these restrictions.

Next, a momentum balance will be made on the actuator disk (see Figure 3.21) to find the power coefficient, C_p.

Defining x as the wind flow direction and using momentum theory,

$$\sum F_x = (\rho A U_{disk})(U_{wind} - U_{wake}) \tag{3.34}$$

$$A(P_{up} - P_{down}) = (\rho A U_{disk})(U_{wind} - U_{wake}) \tag{3.35}$$

Defining the axial induction factor, a, to be

$$a = \frac{U_{wind} - U_{disk}}{U_{wind}} \tag{3.36}$$

or

$$U_{disk} = (1-a)U_{wind} \tag{3.37}$$

In other words, the actuator disk induces a reduction in the wind speed at the disk, U_{disk}, compared to the free stream wind speed, U_{wind}.

From the assumptions, one can find

$$U_{wake} = (1-2a)U_{wind} \tag{3.38}$$

To find the power coefficient, C_p, we first find the aerodynamic power P_{aero}:

$$P_{aero} = \sum F_{disk} U_{disk} = \left(2\rho A U_{wind}^2\, a(1-a)\right)U_{disk}$$

$$= \left(2\rho A U_{wind}^2\, a(1-a)\right)U_{wind}(1-a)$$

$$P_{aero} = 2\rho A U_{wind}^3\, a(1-a)^2 \tag{3.39}$$

And substituting P_{aero} into the relationship for C_p (see Equation 3.13),

$$C_p = \frac{P_{aero}}{P_{wind}} = \frac{2\rho A V_{wind}^3 \, a(1-a)^2}{1/2\rho A U_{wind}^3} \tag{3.40}$$

Simplifying yields

$$C_p = \frac{P_{aero}}{P_{wind}} = 4a(1-a)^2 \tag{3.41}$$

It is left as problem to find the Betz limit of $C_{pmax} = 16/27 = 0.59$ from this relationship.

In a similar fashion, a thrust coefficient, C_T, can be found that characterizes the axial thrust on the actuator disk. Summing forces on the disk yields the thrust, T:

$$T = \sum F_{disk} \tag{3.42}$$

$$T = 2\rho A U_{wind}^2 \, a(1-a) \tag{3.43}$$

Nondimensionalizing the thrust by P_{wind} yields the thrust coefficient

$$C_T = \frac{2\rho A U_{wind}^2 \, a(1-a)}{1/2\rho A U_{wind}^2} \tag{3.44}$$

and simplifying yields

$$C_T = 4a(1-a) \tag{3.45}$$

The power and thrust coefficients are plotted in Figure 3.22 as functions of the axial induction factor, a. Note that the maximum power and thrust occur at different axial induction factors.

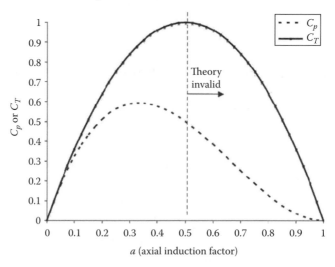

FIGURE 3.22
Power and thrust coefficients versus induction factor showing maximum power COEFFICIENT, $C_{p,max}$.

FIGURE 3.23
Flow visualization tests conducted in the NASA Ames wind tunnel using smoke emitted from the tips of the turbine helped researchers determine the extent of the wake under a controlled set of conditions. (Courtesy of DOE/NREL, Golden, CO.)

The maximum power output of $C_{pmax} = C_{pBetz} = 0.59$ occurs at $a = 1/3$, while the maximum nondimensional thrust of 1.0 occurs at a value of $a = 0.5$. When power is maximum at $a = 1/3$, $C_T = 8/9$.

Also note that the axial induction factor theory is not valid for $a > 0.5$. Equation 3.38 shows that the speed of the wake, U_{wake}, behind the turbine decreases as the axial induction factor increases from 0.0 to 0.5. At $a = 0.5$, the wake speed equals zero. For values of $a > 0.5$, the wake speed would be negative (Figure 3.23), violating the assumptions of this simple theory. Figure 3.23 illustrates the wake tip vortex produced by a rotating WT located in a wind tunnel.

Actuator disk theory provides a theoretical limit for the maximum power coefficient of a WT, the Betz limit. In practice, C_p is actually a function of many variables including wind speed, pitch angle, and TSR. Values of C_p are obtained through experiment or numerical modeling.

The aerodynamic power of the wind is determined by rearranging

$$C_p = \frac{P_{aero}}{P_{wind}} \tag{3.46}$$

to yield

$$P_{aero} = C_p P_{wind} = \frac{1}{2} C_p \rho A U_{wind}^3 \tag{3.47}$$

Given C_p and the expected probability of wind distribution, the annual energy production can be estimated. This in turn will allow the cost of energy (COE) to be calculated, as shown in Section 3.9.

Example 3.3: Determine Annual Energy Production

Given a WT with the following parameters, calculate the annual energy production: $C_p = C_{pmax} = 0.48$ (C_p value assumed maximum and constant for all wind speeds)

Rated power = 5000 kW = 5.0 MW
Annual average wind speed = 8.0 m/s
Rated wind speed = 13.0 m/s
Cut-out wind speed = 25 m/s
Rotor radius = 48.0 m
Area = 7238.2 m²
Air density = 1.20 kg/m³

The annual energy production for all wind speeds will be illustrated by first calculating the annual energy production at a single wind speed. Then, the methodology will be applied to all wind speeds to arrive at the annual energy production for all wind speeds.

First, determine the power production at a single wind speed, say 10 m/s. Then a Rayleigh distribution will be calculated to determine the number of hours the wind speed is at 10 m/s. The power and number of hours are then multiplied to determine the yearly energy production for a 10 m/s wind speed.

Power production: For a wind speed of 10 m/s, the power produced for the values shown earlier is

$$P_{aero} = \frac{1}{2}C_P \rho A U_{wind}^3 = \frac{1}{2}(0.48 \cdot 1.2 \cdot 7238.2 \cdot 10^3) = 2084.6\,kW$$

A plot of the power produced for all wind speeds is shown in Figure 3.24.

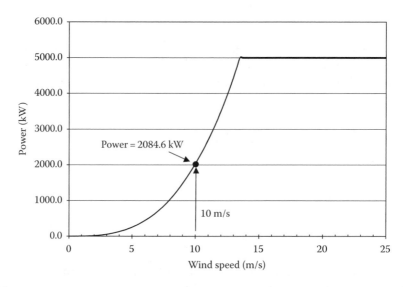

FIGURE 3.24
Power versus wind speed for a 5000 kW turbine with $C_p = C_{pmax} = 0.48$.

Number of hours at 10 m/s: Assuming a Rayleigh PDF with an average wind speed of 8 m/s, the annual probability of a 10 m/s wind is

$$f(10) = \frac{\pi}{2}\left(\frac{U}{\bar{U}^2}\right)\exp\left[-\frac{\pi}{4}\left(\frac{U}{\bar{U}}\right)^2\right] = \frac{\pi}{2}\left(\frac{10}{8^2}\right)\exp\left[-\frac{\pi}{4}\left(\frac{10}{8}\right)^2\right] = 0.0719 = 7.19\%$$

Thus, in a given year, the number of hours the wind will blow at 10 m/s at this site is the probability times the number of hours in a year:

$$0.0719 \cdot \frac{24\,\mathrm{h}}{\mathrm{day}} \cdot \frac{365\,\mathrm{day}}{\mathrm{year}} = 629.8\,\mathrm{h}$$

A plot of the number of hours the wind blows at a given wind speed for all wind speeds is shown in Figure 3.25.

Finally, the annual energy production for this WT, at this site, at a wind speed of 10 m/s is found by multiplying the power production times the number of hours the power is produced, in this case:

Annual energy production at 10 m/s = 2084.6 kW · 629.8 h = 1312.9 MWh

In a similar fashion, this is done for all wind speeds up to cut out. The energy production from all wind speeds is then summed to determine the annual energy production estimate. A summary table of the calculations is shown in Table 3.5.

Examination of Table 3.5 shows that the annual energy produced at this site is calculated to be 13,873 MWh. If electricity costs 10 cents/kWh ($0.10/kWh), the annual value of the energy produced in this example would be

$$\text{Annual value} = 13{,}873{,}070\,\mathrm{kWh} \cdot \frac{\$0.10}{\mathrm{kWh}} = \$1.4\,\mathrm{million}$$

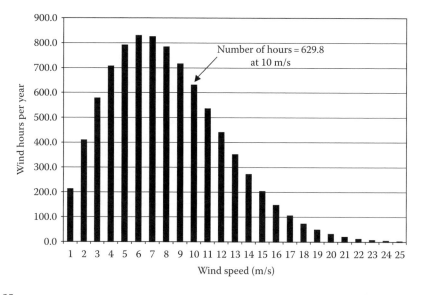

FIGURE 3.25
Number of hours at a given wind speed for a Rayleigh distribution with an annual average wind speed of 8.0 m/s.

TABLE 3.5

Power and Annual Energy Calculation

Wind Speed (m/s)	Power (kW)	Probability (%)	# of Hours/Year	Energy (kWh)
1	2.1	2.42	212.4	442.7
2	16.7	4.67	409.4	6,827.6
3	56.3	6.59	577.6	32,507.8
4	133.4	8.07	706.7	94,283.4
5	260.6	9.03	791.0	206,114.7
6	450.3	9.47	829.3	373,429.3
7	715.0	9.42	824.9	589,808.3
8	1067.3	8.95	784.2	837,017.8
9	1519.7	8.17	716.1	1,088,282.6
10	2084.6	7.19	630.2	1,313,739.7
11	2774.6	6.12	535.7	1,486,478.3
12	3602.2	5.03	440.7	1,587,567.2
13	4579.9	4.01	351.3	1,608,943.1
14	5000.0	3.10	271.6	1,358,130.2
15	5000.0	2.33	203.9	1,019,404.1
16	5000.0	1.70	148.7	743,288.9
17	5000.0	1.20	105.4	526,756.5
18	5000.0	0.83	72.6	362,992.5
19	5000.0	0.56	48.7	243,323.3
20	5000.0	0.36	31.7	158,710.6
21	5000.0	0.23	20.2	100,758.5
22	5000.0	0.14	12.5	62,274.8
23	5000.0	0.09	7.5	37,478.8
24	5000.0	0.05	4.4	21,967.4
25	5000.0	0.03	2.5	12,541.7
		Total: 99.76		Total: 13,873,070.1 kWh

Now the rationale to build wind farms comprised of dozens of WTs with a similar energy (cash) output can be seen. The wind is literally producing money!

Note that in Table 3.5 the power produced above 13 m/s (rated wind speed) is a constant value of 5000 kW. Control schemes are typically implemented to limit the maximum amount of power produced by a turbine to the rated turbine power (5 MW in this case) at the rated wind speed (13 m/s). Limiting the power production is important from both a structural standpoint and an electrical generation standpoint. In both cases, WT components must be sized to operate under maximum defined loads and thus cannot handle larger loads.

Another measure of a WT's performance is the capacity factor. Since the wind never blows continuously, a turbine will never be able to operate continuously at full power. The capacity factor is the ratio of the energy produced to the theoretical maximum the turbine could produce (i.e., rated power production, 24 h/day, 7 day/week for a year):

$$\text{Capacity factor} = \frac{\text{Energy generated}}{\text{Rated power} \times (\# \text{h/year})} = \frac{\text{Energy generated}}{\text{Rated power} \times 8760\,\text{h}} \qquad (3.48)$$

For this example, the capacity factor is found to be

$$\text{Capacity factor} = \frac{13,880\,\text{MWh}}{5\,\text{MW} \times 8760\,\text{h}} = 31.7\%$$

For comparison, typical WT capacity factors are 20%–40% [33].

Finally, the calculations shown in Table 3.5 would typically be done at wind speed increments of 0.10 or 0.25 m/s. Note that slight differences in values between the spreadsheet table and the example calculations are due to round-off errors in the example calculations.

3.8.1 Control Schemes

As a comparison, we will briefly examine the major control schemes currently used in large commercial WTs. In the first case, the rotor speed is held fixed or constant while the blade pitch is variable. In the second case, both the rotor speed and the blade pitch are variable. The fixed rotor speed turbine has a simpler mechanical design, the control scheme is easier to implement, and it also offers advantages by simplifying the electrical power generation. The variable-speed and variable-pitch (VSVP) control scheme is more complicated, both mechanically and electrically. Is the more complicated VSVP turbine worth it from an energy production standpoint?

Figure 3.26 illustrates the power output versus wind speed for both WT control methods. The operating parameters cited in Example 3.3 were also used for Figure 3.26. In addition, the results in Example 3.3 are the same shown here for the VSVP turbine.

The key point is that the VSVP turbine has a higher power output. Two regions can be seen where the VSVP turbine power output is higher than the constant-speed/variable-pitch (CSVP) turbine. This increased power output will lead to a higher annual energy production and capacity factor. This is due to being able to operate the VSVP turbine at

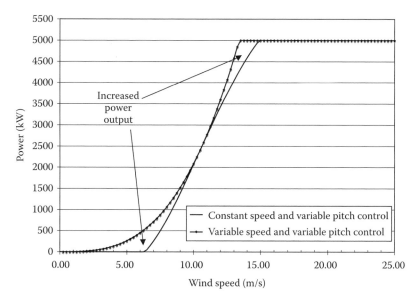

FIGURE 3.26
Power output for constant-speed/variable-pitch and variable-speed and pitch-controlled schemes.

TABLE 3.6

Energy Production and Capacity Factor for VSVT and CSVP
Wind Turbines

	CS/VP	VS/VP	Percent Increase from CS/VP to VS/VP (%)
Energy (MWh)	12,000	13,880	15.3
Capacity factor	27.5%	31.7%	15.3

$C_p = C_{pmax}$ at all wind conditions. The CSVP runs at a variable C_p. Since the turbine cannot vary rotor speed, it cannot run at $C_p = C_{pmax}$ at all conditions.

Note that both turbines eventually reach a rated power of 5 MW. However, the CSVP reaches the 5 MW turbine rating at a higher wind speed, 15 m/s, while the VSVP turbine reaches 5 MW rated power at 13.5 m/s. This is because the CSVP turbine is operating at a lower C_p and thus needs a higher wind speed to reach the higher 5 MW power rating. For the purposes of this illustration, C_p for the CSVP was assumed to be a function of TSR ($\Omega R/$ (wind speed)) and a fixed pitch angle of 0°:

$$C_p(\lambda, \beta) = 0.5176\left(\frac{116}{\lambda_i} - 0.4\beta - 5\right)e^{-21/\lambda_i} + 0.0068\lambda \tag{3.49}$$

where

$$\lambda = \text{Tip speed ratio} = \frac{\Omega R}{\text{Wind speed}}$$

$$\frac{1}{\lambda_i} = \frac{1}{\lambda + 0.08\beta} - \frac{0.035}{\beta^3 + 1}$$

β is the blade pitch angle.

Examining Table 3.6, we can see that the VSVP-controlled turbine produces over 15% more energy (i.e., 15% more yearly revenue), and thus also has a capacity factor 15% higher, than the CSVP-controlled turbine. The goal of WT manufacturers is to design and build turbines, such as the VSVP turbine, that produce more power while minimizing the cost to design and build these turbines, to achieve the highest level of performance possible.

3.9 Cost of Energy

A successful WT design must be strong, reliable, and produce energy at a cost competitive with other technologies. Among others, the U.S. National Renewable Energy Laboratory's (NREL) National Wind Technology Center has developed a wind energy electricity cost–estimating tool. The tool focuses on WT configurations that are currently the most common: three-bladed, upwind, pitch-controlled, variable-speed, and for both land-based and offshore WTs [34]. Results are in 2002 dollars, and cost data are based on a mature technical design and mature component production for a 50 MW wind farm installation.

The Cost of Energy (COE) of a single WT is found using

$$COE = \frac{(FCR \cdot ICC)}{AEP_{net}} + AOE$$

where
  COE is the levelized cost of energy (i.e. electricity) (LCOE) (in constant $) ($/kWh)
  FCR is the fixed charge rate (in constant $. Interest rate in%/year expressed as decimal value/year)

$$ICC = initial\, capital\, cost\, (\$) = Turbine\, cost + balance\, of\, station$$

$$AEP_{net} = Net\, annual\, energy\, production\, (kWh/year)$$

$$= Capacity\, factor \cdot rated\, turbine\, power\, (kW) \cdot 8766\, (h/year)$$

where

$$Capacity\, factor = \frac{Energy\, capture\, (kWh)}{Rated\, turbine\, power\, (kW) \times 8766\, h/year}$$

NOTE: The number of hours in a year is typically defined in one of two ways:

  1. 24 h/day × 365 day/year = 8760 h/year
or
  2. To take into account a leap year every fourth year, 6 h/year is added to account for one-fourth of the leap day: 24 h/day × 365 day/year + 6 h/year = 8766 h/year

The NREL COE tool uses the second definition, 8766 h/year, for number of hours in a year.

AOE = annual operating expenses ($/kWh)

$$AOE = LLC + \frac{(O\&M + LRC)}{AEP_{net}}$$

where
  LLC is the land lease cost ($/kWh)
  O&M is the levelized operation and maintenance cost ($/year)
  LRC is the levelized replacement/overhaul cost ($/year)

3.9.1 Definitions

Fixed charge rate (FCR) is the interest rate needed to cover the capital cost, a return on debt and equity, and various other fixed charges.
  The initial capital cost (ICC) is the sum of the turbine system cost and the balance of station (BOS) cost: ICC = turbine system cost + BOS

The components in these systems are

Turbine system (rotor, drive train, nacelle) cost

Rotor: blades, hub, pitch mechanisms and bearings, spinner, nose cone

Drive train and nacelle: Low-speed shaft, bearings, gearbox, mechanical brake, high-speed coupling, and associated components, generator, variable-speed electronics, yaw drive and bearing, main frame, electrical connections, hydraulic and cooling systems, nacelle cover

Control, safety system, and condition monitoring

Tower

NOTE: Total WT cost = turbine system + control + tower

The total turbine cost is often expressed as a turbine cost per kW ($/kW) multiplied by the turbine rating (kW).

For example: Given a typical $1100/kW turbine cost per kW for a 5 MW turbine

$$\text{Total turbine cost} = \$1100 \times 5000 = \$5.5 \text{ million}$$

BOS

Foundation/support structure

Transportation

Roads, civil work

Assembly and installation

Electrical interface/connections, cables, transformers

Engineering permits

Offshore turbines have additional costs

Marinization (added cost to handle marine environments)

Port and staging equipment

Personal access equipment

Scour protection

Surety bond (to cover decommissioning)

Offshore warranty premium

NOTE: Currently, a typical BOS cost for a land-based turbine is approximately $700,000 per turbine.

LLCs, one component of the annual operating expense, are the rental or lease fees charged (e.g., land or ocean bottom lease cost) for the turbine installation.

Operation and maintenance (O&M) costs are a component of the annual operating expense that is larger than the LLC. The O&M cost normally includes

Labor, parts, and supplies for scheduled and unscheduled turbine maintenance

Parts and supplies for equipment and facilities maintenance

Labor for administration and support

Levelized replacement/overhaul cost (LRC) distributes the cost of major replacements and overhauls over the life of the WT.

Finally, the AOE includes land or ocean bottom lease cost (LLC), levelized O&M cost, and LRC.

COE can then be expanded to show that the annual cost is due to three major components:

$$COE = \frac{(FCR \ ICC)}{AEP_{net}} + AOE = LLC + \frac{(FCR \times ICC)}{AEP_{net}} + \frac{(O\&M + LRC)}{AEP_{net}}$$

In other words, the COE is due to

1. Costs of leasing land (if land is purchased, this cost is $0)
2. ICCs to buy/construct all WT parts
3. Ongoing costs due to yearly operations, maintenance, overhauls, and replacements

Typical values used to calculate COE are

Fixed charge rate (FCR) = 10%–15%

Turbine cost/kW = $800–1100/kW (for MW size turbines)

Capacity factor = 25%–40%

O&M costs (scheduled maintenance) = $0.005–$0.010/year

Finally, typical component costs for a MW-scale WT are shown in Figure 3.27. Figure 3.27 shows BOS (21%), O&M (19%), and LRC (11%) sum to 51% of the total COE cost, while the drive train (27%) has the largest single cost of the turbine components. The rotor (12%), tower (9%), and control system (1%) sum to only 22% of the total COE. When the drive train is added to these components, the sum of all WT components is 49% of the total

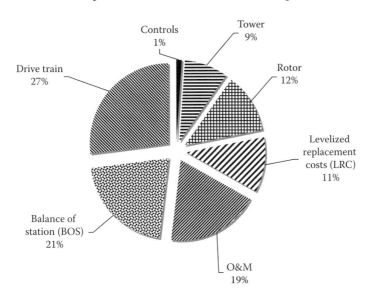

FIGURE 3.27
Individual component costs as % of COE for a land-based, MW-scale wind turbine.

COE cost. In summary, installing, maintaining, and overhauling/replacing components is approximately half (51%) of the cost of electricity for a WT, which illustrates the need for an efficient installation method and a well-designed, reliable WT.

Example 3.4: COE Calculation

What is the COE for a WT with the following characteristics?

> Turbine rating = 5 MW = 5,000 kW
> CF = capacity factor = 40.00%
> TC = turbine cost = $1100/kW
> O&M cost/year = $87,600/year
> FCR = fixed charge rate = 10.00%
> BOS = balance of station = $2,360,000
> LRC = levelized replacement costs = $123,000/year
> LLC = land lease costs = $0/year

Answer:

AEP net = Annual energy production, Net (kWh/year)
$$= CF \cdot \text{turbine rating} \cdot 8760 = 17{,}520{,}000$$

$$ICC = \text{Initial capital cost} = \text{Turbine cost} + \text{Balance of station}$$

$$= \text{Rating} \cdot TC + BOS = \$7{,}860{,}000$$

$$AOE = LLC + (O\&M + LRC)/AEP \text{ net} = 0 + (87{,}600 + 123{,}000)/17{,}520{,}000$$

$$= \$0.012/kWh$$

$$COE = \frac{(FCR \cdot ICC)}{AEP_{net}} + AOE = \frac{(0.1 \cdot 7{,}860{,}000)}{17{,}532{,}000} + 0.012 = 0.045 + 0.012 = \$0.057/kWh$$

% of total COE due to ICC = 79%
% of total COE due to AOE = 21%

In summary, for this example, 79% of the yearly cost is due to the actual WT and installation of the turbine. The remaining 21% of the yearly cost is due to expenses needed to keep the turbine running.

This analysis assumes that costs are known for the turbine of interest. If one is planning to design a turbine, the NREL report contains cost relationships for turbine components, for example, estimated cost of a WT blade as a function of length. This allows one to estimate the turbine, control system, tower, BOS, and offshore turbine costs. Knowledge of maintenance costs, replacement costs, and actual WT site information (for capacity factor) allows the COE for a WT at a site to be determined and evaluated for possible commercial development.

3.10 Wind Farms

As noted earlier, a wind farm is a group of WTs producing electricity in a single location (see Figure 3.28). In order to have a successful wind farm, selecting a proper location is key [35,36]. The most important item is selecting a site with the desired wind resource. The winds must

FIGURE 3.28
Wind farm located in Tehachapi, California, operational since early 1980s. (Courtesy Ian Kluft.)

be consistent and have a sufficient speed to generate the desired energy. Wind resource maps (see Section 3.7.2) are a typical starting point. Meteorological data from "met" towers erected by companies interested in a site give a detailed view of the wind resource at a specific site.

Potential impacts of WT farms also need to be considered. As noted earlier, bird/bat activity and avian migration need to be considered to prevent negative effects on the environment. Noise effects can also have an impact in the area local to a proposed wind farm site. In addition, larger questions are also studied, for example: Will the turbines create interference for local air traffic? These additional factors must be weighed in addition to the wind resource when selecting a WT site.

Besides wind resource and environmental effects, other factors that must be considered include land access (lease or purchase, zoning, and permits must be obtained) and sufficient capital to fund the project (approximately $1 million per MW of installed generating capacity) (AWEA [31]). The wind farm would ideally be located near existing transmission lines to limit the cost (up to $1 million/mile) of installing large lengths of new transmission lines that drive the BOS costs up. In addition, there has to be a market available and an agreement in place to buy the electric power produced. Assuming that other factors are favorable, based on wind resource and turbine costs, the wind farm developer selects a WT model. Additional steps include checking with the manufacturer for availability and considering a service provider for annual maintenance.

Another consideration is the issue of intermittency. The power output from a single WT can be intermittent since the wind speed can fluctuate below and above rated wind speed. On the other hand, a wind farm, with WTs distributed over a large, geographically diverse area, will have less chance of the wind speed being below rated wind speed for all WTs in the wind farm. A study by Archer and Jacobson [37] indicates that when multiple wind farms are interconnected together in an array through the electric transmission grid, the correlation between the wind speeds at the sites decreases. Consequently, the array of wind farms behaves more like a single wind farm experiencing a steady wind speed and

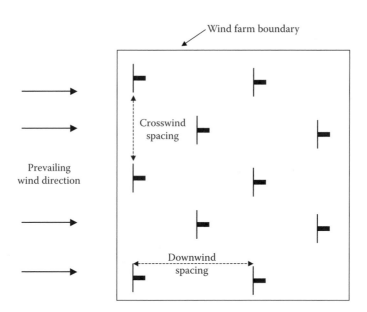

FIGURE 3.29
Schematic (top view) of typical wind farm layout.

thus producing a constant power output. In addition, the benefits of interconnecting wind farms continued to increase as the number of connected wind farms increased.

On the other hand, distributing WTs in a wind farm over a large geographic area drives up BOS costs due to the increased cost of electrical interconnects and the potential need for new transmission lines. Wind farms are typically arranged with a rectangular grid pattern of turbines with each successive row staggered, as shown in Figure 3.29. Turbines are staggered by a specified number of rotor diameters apart, both side to side (crosswind spacing) and between rows (downwind spacing) of turbines. This distance varies with the wind farm owner, site conditions, and wind resource. If only BOS costs are considered, the turbines should be placed closer together. From a power output standpoint, if 100 turbines rated at 2 MW each are to be placed at this site, the total power output could be anticipated to be 200 MW. Unfortunately, this is not the case.

Consider a typical wind farm development scenario. Once transmission, land-use issues, and a power purchase agreement are in place, the wind farm development can proceed. The wind resource estimates from the met tower for a potential site are anticipated to be sufficient for power production; however, wind farms do not always produce the expected power. Several factors could be responsible for this unexpected reduction in power output.

For example, the met tower used to collect data to make the wind resource predictions can only measure wind at one location. The wind resource could change throughout the site depending on terrain such as ridges and hills or on surface features like a forest. As turbines are built farther away from the met tower, the resource prediction becomes less accurate.

Another suspect for producing less than expected power is wake effects or array effects. Depending on the locations of the turbines positioned on the site, and the direction of the wind, some turbines can be located downwind of other turbines. As a turbine extracts energy from the wind, the flow downstream becomes slower and more turbulent. This slower, more turbulent flow persisting downwind of a turbine is known as a wake. Turbines built in the wakes of other turbines have a diminished wind resource and do not produce the power levels of the upwind turbines. This will result in the production of less energy, which means a reduced monetary return for investors. Downstream of a WT (perhaps 15 diameters or much more depending on conditions), the wind speed in the wake will begin to approach the free stream velocity allowing higher energy production by downstream turbines. Unfortunately, increasing the spacing between turbines increases BOS costs (e.g., more construction and materials) resulting in a lowered return on investment. Several projects into studying wake effects strive to quantify wake losses to better design wind farms.

Extensive turbine wake modeling has been conducted by efforts such as the Efficient Development of Offshore Wind farms (ENDOW) project and UpWind projects. As part of the ENDOW project, Rados [38] reviews six wake models and compares the results to experimental data for the decease in wind speed behind the turbine (wake velocity deficit) and turbulence intensity (a measure of the variability of the wind speed). The results show that the models consistently overpredict the wake velocities and turbulence intensity. In a review of current wind farm models, Barthelmie [39] recognizes that "wind farm models are lacking one or more components which account for the modification of the overlying boundary layer by the reduced wind speed, high turbulence atmosphere generated by large wind farms."

A study by Stovall [40] using computational fluid dynamics (CFD) and large eddy simulations looked at the overlying boundary layer effects on wake propagation. Two turbines, with 100 m rotor diameters, are modeled as being inline. The incoming wind speed to the upstream turbine is approximately 8–10 m/s. The objective is to determine the effect the wake has on the power output of the downstream turbine compared to power output of the upstream turbine. In this study, the turbines have a 9 diameter downwind spacing. The power coefficient for the turbines in this simulation is approximately $C_p = 0.50$.

Prior to modeling the power output of the WTs, a simulation of the ABL is created. Once the ABL simulation is validated, turbines are imposed into the domain with the turbine rotors modeled via actuator disks. The actuator disks impose a pressure drop on the flow-field proportional to the power extracted by a turbine.

Results from the numerical simulation show the diminished wind resource caused by the wake-forming downstream of a WT. Figure 3.30 illustrates the power in the wind

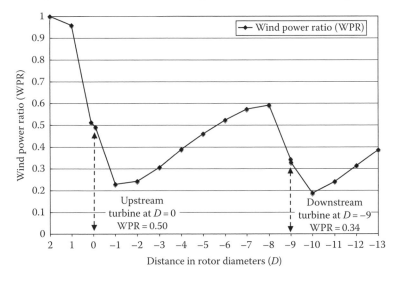

FIGURE 3.30
Wind power wake reduction: Local wind power normalized by wind power level at inlet. (From Stovall, T.D., Simulations of wind turbine wake interactions in OpenFOAM, MS thesis, University of Colorado, Boulder, CO, 2009. With permission.)

calculated as the area integral of velocity cubed (Equation 3.11). Distances between the turbines are shown in rotor diameters (D). The inlet of the computational domain used in this numerical simulation is located at $D = +2$. The first turbine is located at $D = 0$ while the second turbine is located 9 diameters downstream of the first turbine at $D = -9$. All wind powers are expressed as a wind power ratio (WPR) of the wind power at a given location divided by the inlet wind power (located at $D = +2$). This results in a WPR = 1.0 at the inlet ($D = +2$).

Notice that the nondimensional wind power extracted by the first turbine is half of the wind power at the inlet, yielding a WPR = 0.50, as would be expected since the power coefficient is $C_p = 0.50$.

Downstream of the first turbine, the wake is represented by a further reduction in wind power due to turbulent losses. As distance increases downstream of the first turbine, the wake begins to recover (D increases from $-1 \rightarrow -8$) as surrounding higher speed (i.e., higher energy) wind mixes with the wake. The power in the wake, however, does not achieve free stream power levels prior to reaching the second turbine.

Since the incoming power of the wind is lower immediately upstream of the second turbine, the power output of the second turbine should be lower than the power output of the first turbine. Indeed, the WPR of the second turbine in this simulation is WPR = 0.34, indicating that the second turbine was only able to extract 34% of the energy available in the incoming stream at $D = +2$. The wake from the first turbine adversely impacts the power output of the second turbine. Another way to view this is that the ratio of the power extracted by a second turbine (located $9D$ downstream of the first turbine in this study) to the first turbine is approximately 67% (i.e., $0.34/0.50 = 0.67\%$). In economic terms, the second turbine will only produce 67% of the income of the first turbine in this scenario.

Increasing the spacing between the turbines in the downwind direction reduces the wake effect but increases the BOS costs. A trade-off between these two competing economic factors remains an active area of research. Although the results from the Stovall study match up with experimental data from wind farms, the simulation is not yet commercially viable for developing wind farms. The computational resources required for modeling an entire wind farm (dozens of turbines) are currently too high for industry to optimize wind farm layouts. Existing models such as Riso DTU [41] and Garrad Hassan [42] are designed for wind farm optimization, but lack the accuracy of full CFD simulations. Significant work still remains to develop a robust, accurate, and fast optimization tool for wind farm wake effects. With such a tool, BOS costs can be optimized simultaneously with turbine location resulting in maximum energy (income) production.

3.11 Offshore Wind Energy

Offshore WT farms (see Figure 3.31) address limits of land-based wind farms such as transmission line access and capacity [43], space for WTs (more of an issue in higher-density population areas such as Europe), while providing access to a vast wind resource. Significant development still remains for wind farms in water deeper than 30 m while BOS costs can double in an offshore installation compared to land-based installation. Still, the complexity of offshore technology development is more than offset by the need and desire to harness the higher wind resource available offshore (ocean as well as large inland lakes).

FIGURE 3.31
Offshore wind farm. (Courtesy of Hans Hillewaert.)

Offshore wind energy has a number of benefits including

Better wind resources

Reduced turbulence resulting in steadier winds

Higher wind speed resulting in higher energy production

Higher-capacity factors resulting in better electric load matching

Closer proximity to electric load centers resulting in

Lower transmission costs and constraints

Ability to serve high-cost regions

Avoids land-based size limits

Shipping—no land-based roadway limits

Assembly—sea-based cranes are taller

Larger machines are more economical

These benefits are significant throughout the world. For example, the United States has no offshore wind farms as of 2009 while nine farms are proposed for the future [36]. Of the 48 contiguous United States, 28 have a coastal boundary, and these 28 states use 78% of the electricity in the United States [44]. Since these coastal load centers are typically located hundreds if not thousands of miles from land-based wind energy sources, offshore wind power provides a nearby source of energy. Of these 28 states, only 6 have a sufficient land-based wind energy resource to meet more than 20% of their electric requirements through wind power. If shallow water offshore potential (less than 30 m in depth) is included in the wind resource mix, 26 of the 28 states would have the wind resources to meet at least 20% of their electric needs. Several of these states would have sufficient offshore wind resources to meet 100% of their electric needs [45]. The goal of the United States achieving 20% of its energy from wind energy by 2030 [36] assumes significant contributions from offshore turbines to meet this goal.

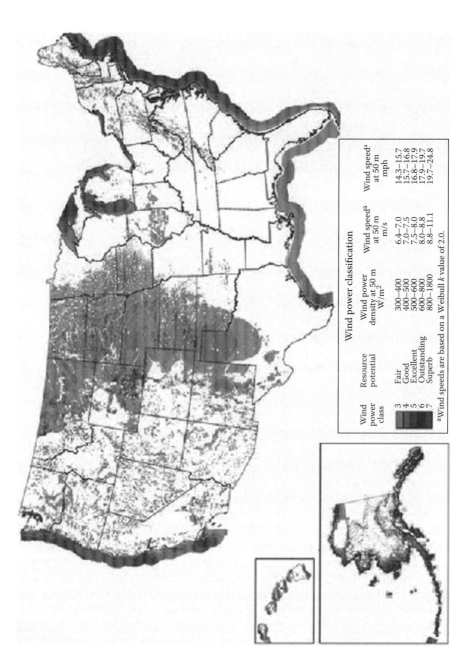

FIGURE 3.32
Wind resource map of the United States. This map shows the annual average wind power estimates at a height of 50 m. It is a combination of high-resolution and low-resolution datasets produced by NREL and other organizations. The data were screened to eliminate areas unlikely to be developed onshore due to land-use or environmental issues. In many states, the wind resource on this map is visually enhanced to better show the distribution on ridge crests and other features. (Courtesy of U.S. Department of Energy, National Renewable Energy Laboratory, Washington, DC. With permission.)

TABLE 3.7

Offshore U.S. Wind Power (GW) for Wind Class 5 or Greater by Water Depth

Coastline	0–30 m	30–60 m	60–900 m	>900 m
New England	59.2	127.7	273.4	0
Mid-Atlantic	165.6	181.6	59.7	56.6
South Atlantic Bight	28.4	58.2	13.7	0
Gulf of Mexico	0	12.3	54.7	0
California	2.3	4.8	130.5	277.9
Pacific Northwest	7.5	19.2	188.1	121
Great Lakes	166.6	137	813.2	0
Total	429.6	540.8	1533.30	455.5
Total power all depths = 2,959.2 GW (= 2,959,200 MW = 2.96 TW)				
% Power of total power	14.50	18.30	51.80	15.40
Hawaii (not included in totals)	0.8	1.4	24.9	123.6

Source: Musial, W., *Mar. Technol. Soc. J.*, 41(3), 24, 2007.
Note: Power is in GW, not MW!

Studies have been conducted [46], which illustrate the potential available offshore wind power as a function of depth and coastal location for the United States. Figure 3.32 shows the location of the offshore U.S. wind power (Atlantic and Pacific Coasts, Gulf of Mexico, Great Lakes, and Hawaii). Table 3.7 quantifies the amount of power available by water depth. For example, approximately 430 GW of power is available in shallow waters (depth of 30 m or less). Similarly, the mid-Atlantic region has a potential of 347.2 GW of power available in waters 60 m or less in depth. Considering all seven regions (except Hawaii), 970 GW or nearly 1 TW (where 1 TW = 1,000 GW = 1,000,000 MW) of power is available in water depths of 60 m or less.

In contrast, countries in the European Union, as of September 2009, lead the world in developing offshore wind farms with over 25 operational offshore wind farms (most installed in water less than 22 m deep) and more than 15 additional offshore wind farms under construction. Dozens of offshore wind farms are proposed in countries throughout the world including Canada, China, Germany, Netherlands, United Kingdom, and the United States, among others.

The European Wind Energy Association's (EWEA) statistics [47] for January 2009 cite an operational wind farm capacity of 1.4 GW for Europe, with the United Kingdom and Denmark combining for over 67% of the total capacity, as shown in Figure 3.33.

Proposed European offshore wind farms for 2015, cited in the same EWEA report, show an enormous growth in proposed capacity (see Figure 3.34) from 1471 MW in 2009 to a proposed capacity of 37,444 MW, a 25 times increase! If plans come to fruition, Germany would jump to number one in offshore capacity with a 900 times increase from 24 to 10,928 MW, resulting in a 30% share of the total proposed offshore capacity. The United Kingdom would have the next largest share with 23%, resulting from a 14.8 times increase in offshore capacity (590.8–8755.8 MW). The number of European countries with offshore capacity would likewise increase from 8 countries in 2009 to 13 countries in 2015.

As noted earlier, benefits of offshore turbines include using larger blades since the wind speeds are higher yet steadier offshore, allowing for better economics. Significant technical challenges remain, especially in water deeper than 30 m. From Table 3.7, we can see that in waters 0–30 m and 30–60 m deep, the wind power potential off the coast of California is 2.3 and 4.8 GW, respectively, while in water 60–900 m deep, the potential wind power is 130.5 GW. Significant technology developments remain to make offshore wind power viable to the state of California, which contained 12% of the U.S. population in 2008 [48]

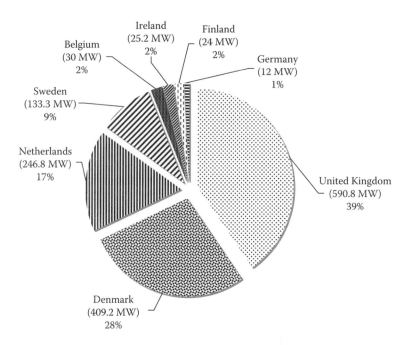

FIGURE 3.33
Operational offshore wind power in Europe as of January 2009. Total capacity of 1471 MW. (From European Wind Energy Association (EWEA), Offshore statistics for January, 2009, http://www.ewea.org/fileadmin/ewea_documents/documents/statistics/Offshore_Wind_Farms_2008.pdf. With permission.)

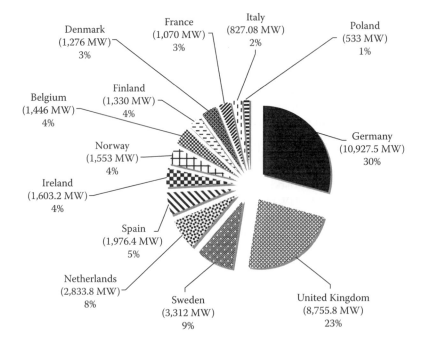

FIGURE 3.34
Proposed offshore wind power in Europe in 2015. Total capacity of 37,444 MW. (From European Wind Energy Association (EWEA), Offshore statistics for January, 2009, http://www.ewea.org/fileadmin/ewea_documents/documents/statistics/Offshore_Wind_Farms_2008.pdf. With permission.)

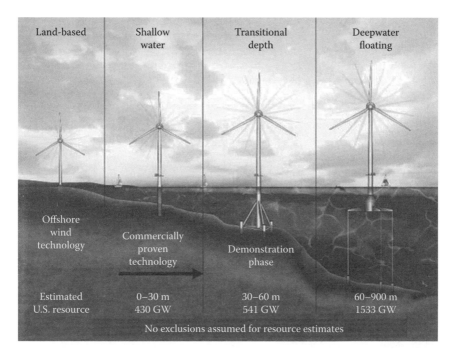

Land-based	Shallow water	Transitional depth	Deepwater floating

Offshore wind technology | Commercially proven technology → | Demonstration phase |

| Estimated U.S. resource | 0–30 m 430 GW | 30–60 m 541 GW | 60–900 m 1533 GW |

No exclusions assumed for resource estimates

FIGURE 3.35

The U.S. offshore wind power resource estimate as a function of depth and status of the needed wind turbine technology. (From Robinson, M. and Musial, W., Offshore wind technology overview, NREL/PR-500-40462, 2006, http://www.nrel.gov/docs/gen/fy07/40462.pdf. With permission.)

and used 7% of the electricity in the United States in 2007 [49], and other parts of the world with deep water close to shore.

Figure 3.35 illustrates the depths for which WT technology is needed. In waters of 30 m or less, proven technology is available. In waters between 30 and 60 m, investigations are currently beginning to develop the generation of technology. In waters deeper than 60 m, significant technology challenges exist to develop floating WTs.

In Figure 3.36, three concepts of different technology approaches for developing floating WTs are shown. Modeling a floating WT is a challenging task because it must include effects from wind and wave interaction with the turbine and floating structure, a nonstationary and rotating turbine, and a control system, which monitors and controls the turbine for these effects. State-of-the-art research is underway using codes such as FAST [50].

The winning technology design will likely include elements from these three concepts as well as other designs.

3.12 System Advisory Model

The System Advisory Model, generally referred to as SAM, is a numerical performance and economic model designed by NREL to make performance predictions and economic estimates for stand-alone or grid-connected electric power projects in the distributed and central generation markets. The model calculates the cost of generating electricity based

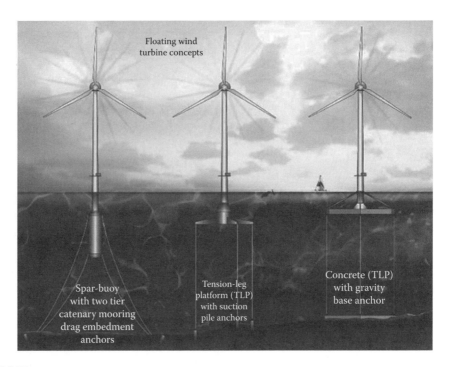

FIGURE 3.36
Floating wind turbine concepts for deepwater (>60 m) wind resource locations. (Butterfield, S. et al., Engineering challenges for floating offshore wind turbines, NREL/CP-500-38776, 2007, http://www.nrel.gov/wind/pdfs/38776.pdf. With permission.)

upon information about a project's desired power output location, WT type, installation, operating costs, and various financial data such as interest rates, incentives, and system specifications. The performance model makes hour-by-hour calculations for the power output of wind systems generating a set of 8760 hourly values that represent the system's electricity production during a single typical year.

For wind energy systems, SAM reads weather files provided by Energy Plus. Running the model requires a file of the wind at the project location. SAM comes with a set of sample files that contain complete sets of sample costs and performance input data called default values, but it is the analyst's responsibility to modify the input data as appropriate for the project being evaluated.

To utilize SAM, begin by downloading and installing NREL's free system from https://www.nrel.gov/analysis/SAM/. From the SAM home page, you can create a new utility wind energy project as follows. The following instructions are for the "Version 2012.5.11":

1. Click on the blue "Create a new file…" button.
2. Choose technology: Wind Power.
3. Choose financing option: Utility Independent Power Producer (IPP).
4. Click on OK at the bottom of the box.

On a typical SAM page, you will find values in black and blue. The blue numbers cannot be changed. They are either given or calculated from other input values. To assist you, the specified input values that have been changed to fit the problem specification are encircled in the solutions. Unless you have specific data for your project, you can use the so-called default options. Note that all of the default inputs can be modified and would need to be changed for a real project. The following example demonstrates the application of SAM for a hypothetical wind farm project using some of the default options.

Example 3.5: SAM Method to Determine Cost of Energy (COE)

Using the NREL's SAM software, model a 50 MW utility wind farm using WTs of 2000 kW nameplate capacity. Assume the cost of each turbine is 1500 $/kWh, a 20 year lifetime, an inflation rate of 2%, an interest rate of 6%, a 25% federal tax credit, and a 1¢/kWh production incentive credit. Run this model for 2005 in three different locations: Cheyenne, WY; Warren, CA; and Long Island, NY. How many turbines are needed to build this wind farm? For each location, find the wind farm's capacity factor and real LCOE with and without incentives.

Solution

After starting up the SAM program (which you can install using the steps detailed earlier), from the SAM home page, select "Create a new file." Under "Select a technology," choose "Wind Power." Under "Select a financing option," choose "Utility Independent Power Producer" and click Next. You are now given a number of system parameters to modify including financing, incentives, wind resource, etc. SAM allows users to create a model as simple or complex as needed. For this example, we will only be modifying the model to fit the assumptions in the problem statement while leaving all other options as the default.

Under the "Financing" tab on the left hand sidebar in the box labeled "General," change the analysis period to 20 years, the inflation rate to 2%, and the real discount rate to 6%. The circled fields in the following solution (see Figure 3.37) are the parameters that have been changed from their default values. In "Taxes and Insurance," change the federal tax rate to 28% and insurance to 1.0%. Leave all other values at the defaults shown in Figure 3.37.

Now, select the "Tax Credit Incentives" tab on the left-hand side bar. Under "Investment Tax Credit" (see Figure 3.38), change the federal percentage to 25%.

Then under the "Payment Incentives" tab on the left-hand side of the screen, inside the "Production Based Incentive (PBI)" box (see Figure 3.39), change the state incentive to 0.01 $/kWh for 20 years. These fields are circled in the following solution.

Now select the "Annual Performance" tab and change the availability to 92% as shown in Figure 3.40.

Now, select the "Wind Resource" tab. The wind farm location can be selected through the "Wind Data File" drop-down menu, which has been circled in the solution in Figure 3.41. SAM has pre-loaded wind data files for Cheyenne, Wyoming, and Warren, California, for various years. For this example, select data from 2005 to match the problem statement.

For utility-scale wind farm simulations, SAM only has data for the western part of the United States as shown in Figure 3.42. Data for any location in this region can be found through the "Location Lookup" tool, which allows the user to input the longitudinal and latitudinal coordinates of the site and obtain accurate wind data for that site. The current version of SAM does not contain wind data for locations outside of the region shown in Figure 3.42.

Until a future version of SAM is created to provide wind resource data for the entire United States, several options exist to create custom wind resource files to use in SAM.

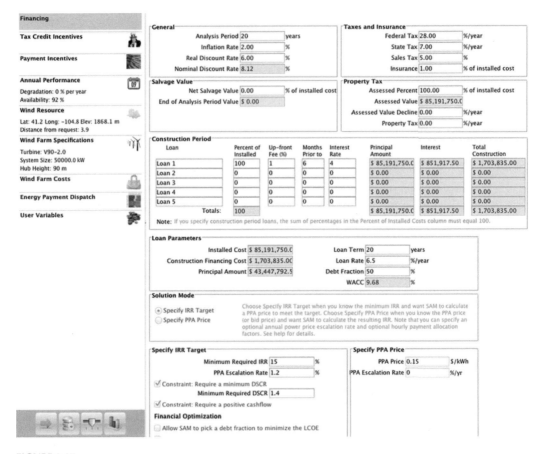

FIGURE 3.37
NREL System Advisory Model (SAM), FINANCING tab settings for utility-scale wind turbine farm example.

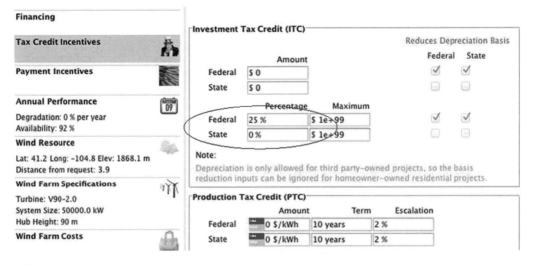

FIGURE 3.38
NREL SAM program, TAX CREDIT INCENTIVES tab settings for utility-scale wind turbine farm example.

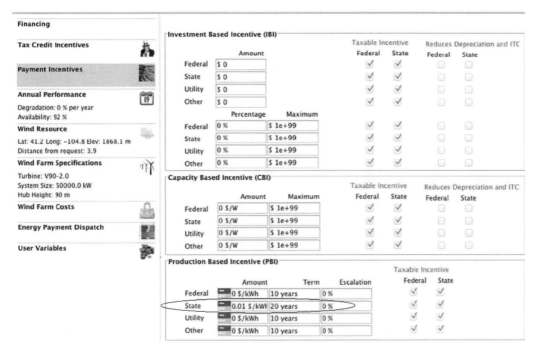

FIGURE 3.39
NREL SAM program, PAYMENT INCENTIVES tab settings for utility-scale wind turbine farm example.

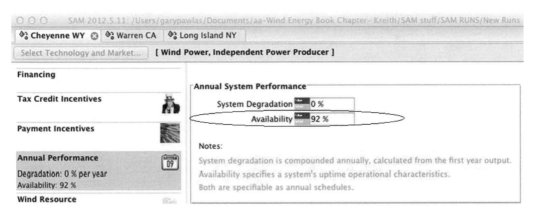

FIGURE 3.40
NREL SAM program, ANNUAL PERFORMANCE tab settings for utility-scale wind turbine farm example.

One method is to modify an existing wind resource file (file extension.swrf) by creating a power law profile for a given location and placing that data into the file (see SAM support forums and search for "SWRF wind resource file creation" [56]). Another method is to download data from NREL's Wind Integration Datasets (WIDs) for the Eastern United States [57]. These data are modeled and are not actually measured. Full details on the methodology used to create the data are contained at this website [57]. Once a desired site is located, for example, Long Island, NY, a data file can be downloaded. SAM's data files

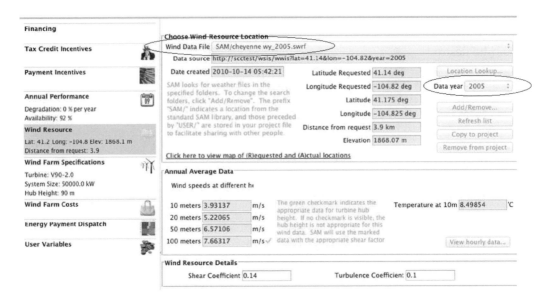

FIGURE 3.41
NREL SAM program, WIND RESOURCE tab settings for utility-scale wind turbine farm example.

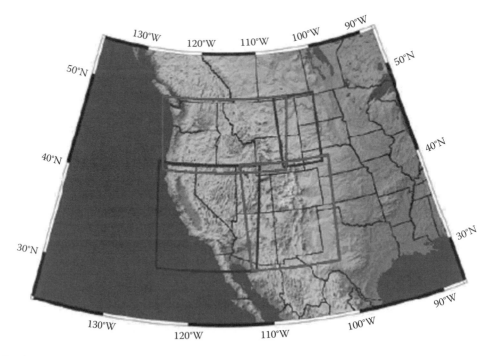

FIGURE 3.42
The region of the western United States used in the SAM program for utility wind models.

contain hourly data for the year (24 h/day*365 days = 8760 wind speeds). The WID files contain 10 min wind speed data at a variety of heights (200, 100, 50 m, etc). Using a text editor, the desired wind resource file must be edited to provide only hourly wind speed data for the desired year, in this case 2005. Since this is a utility-scale WT simulation, only the wind speed data at 100 m will be needed. Then using an existing SAM data file (extension.swrf), the desired wind speed data can be pasted into the file in the 100 m height data column. All other columns can be set to zero except for pressure that is used in conjunction with the desired wind speed data to simulate the performance at the new location. This second method of importing modeled data from the NREL WID was used in this solution to determine LCOE for Long Island, NY. The Cheyenne, WY, and Warren, CA, sites were modeled using the available data in SAM.

Finally, if a project is located in the parts of the United States for which SAM or NREL does not provide a solution, it would be necessary to install a data acquisition system and then use the procedure outlined in the chapter to estimate the cost and performance as a first rough approximation. One could use the estimates for the small-scale version of SAM, which cover the entire country. However, the accuracy of that approach is limited.

Once the desired wind resource file is loaded in the "Wind Resource" tab, move to the next tab, "Wind Farm Specifications." Select the V90-2.0 WT via the "Model Name" drop-down menu. This model has a 2000 kW nameplate capacity. Under "Turbine Farm Layout," change the number of rows to 5 and turbines per row to 5. This gives a total wind farm nameplate capacity of 50 MW. These fields are circled in Figure 3.43.

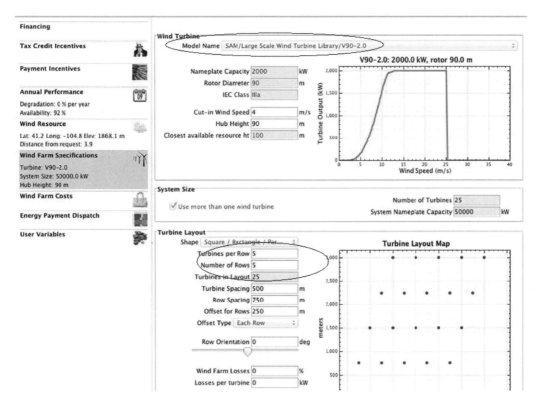

FIGURE 3.43
NREL SAM program, WIND FARM SPECIFICATIONS tab settings for utility-scale wind turbine farm example.

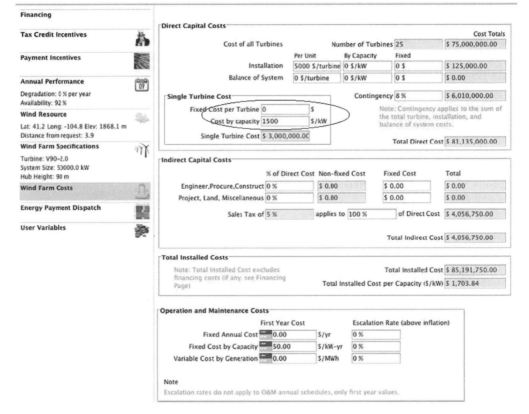

FIGURE 3.44
NREL SAM program, WIND FARM COSTS tab settings for utility-scale wind turbine farm example.

Under the "Wind Farm Costs" tab in the Single Turbine Cost box, change the cost by capacity to 1500 $/kW. This field is circled in the Figure 3.44. Leave the other fields at their default values as shown.

The settings in the "Energy Payment Dispatch" and "User Variables" tabs were not used in this solution and are left unchanged. The wind farm model has now been modified to fit the assumptions in the problem statement. To run the simulation, select the icon with the green arrow at the bottom left corner of the screen. SAM will output a graph displaying different measures of LCOE as seen in Figure 3.45 for Cheyenne, WI, output. This output screen also contains a table of other useful metrics on the bottom left corner. These metrics include the capacity factor, annual energy production, the real LCOE without incentives, etc. Repeat this process for the sites in Warren, California, and Long Island, NY.

Table 3.8 summarizes the LCOE results for all three locations listed in the problem statement. In 2005, the state average residential electricity price [58] in California was about 13 cents/kWh, and the installation of a WT farm in Warren, CA, would be economically advantageous given the Real LCOE w/o incentives of 6.8 cents/kwh. However, given the average residential electricity price in Wyoming of 7.5 cents/kWh in 2005, the

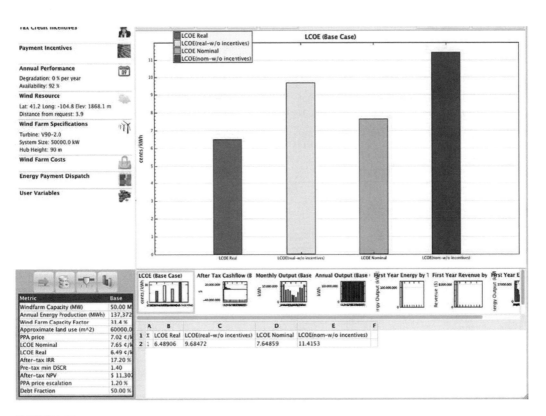

FIGURE 3.45
SAM output for LCOE for Cheyenne, WY.

TABLE 3.8

Wind Farm Metrics in Three Geographic Regions

Location	Real LCOE w/o Incentives (cents/kWh)	Real LCOE w/o Incentives (cents/kWh)	Capacity Factor (%)
Cheyenne, WY	6.49	9.68	31.4
Warren, CA	4.57	6.82	44.5
Long Island, NY	7.34	10.96	27.7

installation of a WT would be only marginally advantageous since the Real LCOE w/o incentives is 9.7 cents/kwh. Finally in Long Island, NY, the installation of a WT farm would be advantageous since the residential electricity price was about 16 cents/kwh in 2005 while the Real LCOE w/o incentives is approximately 11 cents/kwh. Finally, incentives obviously decrease the LCOE in all cases, improving the economic advantage. This example demonstrates the usefulness of determining the economic viability of WT farms in a given geographic location. It also demonstrates the need for accurate wind resource data for use in determining the yearly power output.

3.13 Additional Topics for Study

A single chapter cannot, of course, completely cover a topic as rich, diverse, and rapidly changing as WTs. Topics omitted for brevity include (but are not limited to!) statistical representation of wind, vibration and fatigue, strength of materials, manufacturing (blades, towers, and nacelles), detailed aerodynamic design, gearbox, generators, power converters, control systems, connection to electrical grid, and transmission. Discussion of all these topics and more can be found in wind energy texts such as Manwell et al. [30], Burton et al. [32], or Hau [25].

Acknowledgment

My thanks to Tim Stovall for his contributions to Section 3.10.

Problems

3.1 While completing this homework, imagine the study space you are using employs a space heater with a rating of 30 kW, a refrigerator (for snacks) that is rated for 5 kW, and a light fixture with 2100 W bulbs, all operating at a constant rate for the 1.5 h it takes you to finish.

    a. What is the total power consumed at any given moment?

    b. How much energy do you consume to complete the task?

    c. Assume your local power company, XEnergy, charges 6.5 cents per unit of energy in part b, how much did it cost you to complete the assignment?

    d. If you wanted to pay XEnergy $5, what would be the power rating of an additional appliance you would need to turn on?

3.2 A residential size WT has a diameter of 1 meter. Due to the size, its efficiency is only 25% of the Betz limit. What is the expected power output at a wind speed of 20 miles per hour? Assume the air density is 1.225 kg/m³.

3.3 You have assembled a modern three-bladed upwind WT with a rotor radius of 5 m at sea level where the density is 1.225 kg/m³.

    a. If the wind is blowing at an average speed of 10 m/s, what is the power of the wind traveling through the swept area of the rotor?

    b. What is the average actual power produced by a turbine of this type at this wind speed? (*Hint:* Find C_p)

Hint: Find the maximum power coefficient, C_p, based on the following equation for axial induction.

$$C_p = 4a(1-a)^2$$

3.4 An investor comes to you and tells you that a company wants to invest in a new turbine design with the following performance characteristics:

112 m rotor diameter

13 m/s wind speed

2.1 MW

1.223 kg/m³ air density

a. What is the claimed rotor power coefficient, C_p, for this turbine design?

b1. Do you encourage the investor to invest the company's $1 million? Answer in complete sentence(s).

b2. Why? Answer in complete sentence(s).

3.5 Weibull distribution: From an analysis of wind speed data (hourly interval average, taken over a 1 year period), the Weibull parameters are determined to be $c = 6$ m/s and $k = 1.8$ at a potential WT site.

The Weibull PDF is given by

$$p(U) = \left(\frac{k}{c}\right)\left(\frac{U}{c}\right)^{k-1} \exp\left[-\left(\frac{U}{c}\right)^k\right]$$

a. Estimate the number of hours per year that the wind speed will be between $u_B = 6.5$ and $u_A = 7.5$ m/s during the year.

b. Estimate the number of hours per year that the wind speed is above 16 m/s.

3.6 Given a WT with the following parameters, calculate the annual energy production assuming a Rayleigh PDF:

$C_p = C_{pmax} = 0.40$ (C_p value assumed maximum and constant for all wind speeds)

Rated power = 3000 kW = 3.0 MW

Annual average wind speed = 7.0 m/s

Rated wind speed = 12.5 m/s

Cut-out wind speed = 25 m/s

Rotor radius = 45.0 m

Area = 6361.73 m²

Air density = 1.20 kg/m³

Wind Speed (m/s)	Power (kW)	Probability (%)	# of Hours/Year	Energy (kWh)
1	1.5			
2	12.2			
3	41.2			
4	97.7			
5	190.9			
6	329.8			
7	523.7			
8	781.7			
9	1113.0			
10	1526.8			
11	2032.2			
12	2638.3			
13	3000.0			
14	3000.0			
15	3000.0			
16	3000.0			
17	3000.0			
18	3000.0			
19	3000.0			
20	3000.0			
21	3000.0			
22	3000.0			
23	3000.0			
24	3000.0			
25	3000.0			
	Total:		Total:	kWh
				MWh

3.7 (a) What is the COE for a turbine with the following assumptions:

5 MW rating

31% capacity factor

$1250/kW turbine cost

$91,000 operations and maintenance cost/year

Fixed charge rate of 9%

BOS costs of $1,999,999

Levelized replacement costs of $135,000/year

(b) If a gearbox replacement was required every 5 years at a cost of $600,000, what would the COE be?

How much (in percent) does the COE increase due to the gearbox replacement?

3.8 Power deficit inside wind farms—Barthelmie et al. [59], present results of modeling the behavior of WT wakes in order to improve power output predictions. The following figure shows the predicted power deficit in the Danish offshore wind farm Horns Rev.

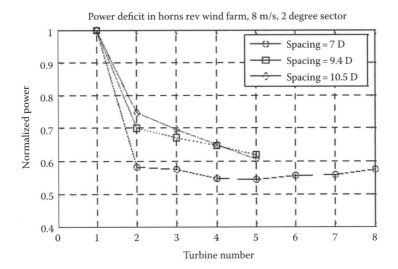

Power deficit in horns rev wind farm, 8 m/s, 2 degree sector

Example: Two 2.3 MW turbines are spaced 9.4D apart directly in line with the wind. What is the total power produced?

Front turbine, normalized power is 1.0, power = >1.0*2.3 MW = 2.3 MW

Second turbine is "waked," normalized power is 0.7, waked power = >0.7*2.3 MW = 1.61 MW

$$\text{Total power} = 2.3 + 1.61 \text{ MW} = 3.91 \text{ MW}$$

(a)   What is the power of three turbines, 9.4D apart, directly in line with the wind?

(b)   What is the power of three turbines, 9.4D apart, perpendicular with the wind?

References

1. Energy Information Administration (EIA). (2008) Per capita (per person) total primary energy consumption (million Btu per person), United States 1949–2007, http://www.eia.doe.gov/emeu/international/energyconsumption.html

2. Winds of Change. (2006) American Federal Projects 1975–1985. http://www.windsofchange.dk/WOC-usastat.php

3. American Speech-Language-Hearing Association (ASHA). (1997) Noise and hearing loss, http://www.asha.org/public/hearing/disorders/noise.htm http://www.sengpielaudio.com/TableOfSoundPressureLevels.htm

4. American Wind Energy Association (AWEA). (2010) Wind power myths vs. facts, http://www.awea.org/pubs/factsheets/MythsvsFacts-FactSheet.pdf

5. San Martin, R.L. (1989) Environmental emissions from energy technology systems: The total fuel cycle. Energy Citations Database: http://www.osti.gov/energycitations/product.biblio.jsp?osti_id = 860715

6. American Wind Energy Association (AWEA). (1996) Wind web tutorial. http://www.awea.org/faq/wwt_environment.html#How%20does%20wind%20stack%20up%20on%20greenhouse%20gas%20emissions

7. Danish Wind Industry Association (DWIA). (1997) Shadow casting from wind turbines, http://www.talentfactory.dk/en/tour/env/shadow/index.htm

8. Sovacool, B.K. (2009) Contextualizing avian mortality: A preliminary appraisal of bird and bat fatalities from wind, fossil-fuel, and nuclear electricity, *Energy Policy* 37(6), 2241–2248.

9. Erickson, W., Johnson, G., and Young, D. Jr. (2002) A summary and comparison of bird mortality from anthropogenic causes with an emphasis on collisions. USDA Forest Service General Technical Report PSW-GTR-191, http://www.fs.fed.us/psw/publications/documents/psw_gtr191/Asilomar/pdfs/1029–1042

10. National Wind Coordinating Committee (NWCC). (2004) Wind turbine interactions with birds and bats: A summary of research results and remaining questions, http://www.nationalwind.org/publications/wildlife/wildlife_ factsheet.pdf

11. Cryan, P.M. (2008) Overview of issues related to bats and wind energy: Web version of presentation to the *Wind Turbine Guidelines Advisory Committee Technical Workshop & Federal Advisory Committee Meeting*, Washington, DC, February 26, 2008.

12. Vestel, L.B. (2009) Study finds reduction in turbine bat kills, http://greeninc.blogs.nytimes.com/2009/05/18/study-finds-reduction-in-turbine-bat-kills/

13. Nebraska Wind and Solar. (2008) History of wind power, http://nebraskawindandsolar.com/history.aspx

14. Price, T.J. (2005) James Blyth—Britain's first modern wind power pioneer, *Wind Engineering* 29(3), 191–200.

15. Danish Wind Industry Association (DWIA). (1997) A wind energy pioneer: Charles F. Brush, http://www.talentfactory.dk/en/pictures/brush.htm

16. Gipe, P. (2003) Smith-Putnam industrial photos, http://www.wind-works.org/photos/Smith-PutnamPhotos.html

17. Renewable Energy Vermont, The story of Grandpa's Knob: How Vermont made wind energy history, http://www.revermont.org/grandpa_knob.pdf

18. Gipe, P. (1995) *Wind Energy Comes of Age*, John Wiley & Sons, Inc., New York.

19. Eldridge, F.R. (1980) *Wind Machines*, 2nd edn., Van Nostrand Reinhold, New York.

20. Nelson, V. (1996) *Wind Energy and Wind Turbines*, Alternative Energy Institute, Canyon, TX.

21. Betz, A. (1920) Das maximum der theoretisch möglichen ausnützung des windes durch windmotoren, *Zeitschrift für das gesamte Turbinenwesen* 26, 307–309.

22. Gijs, A.M. van Kuik. (2007) The Lanchester–Betz–Joukowsky limit, *Wind Energy* 10, 289–291.

23. Wilson, R.E. and Lissaman, P.B.S. (1974) *Applied Aerodynamics of Wind Power Machines*, Oregon State University, Corvallis, OR, http: ir.library.oregonstate.edu/jspui/handle/1957/8140.

24. Stiesdal, H. (1999) The wind turbine: Components and operation, Bonus Energy A/S, *Bonus Info Newsletter* (autumn).

25. Hau, E. (2006) *Wind Turbines—Fundamentals, Technologies, Application, Economics*, Springer, Berlin, Germany.

26. Frost, W.D. and Asphiden, D. (1994) Characteristics of the wind, in: Spera, D.A. (ed.), *Wind Turbine Technology*, ASME Press, New York.

27. Pidwirny, M. (2006) *Fundamentals of Physical Geography*, 2nd edn., Chap. 7(p), http://www.physicalgeography.net/fundamentals/7p.html

28. Kelley, N.D. and Jonkman, B.J. (2007) Overview of the TurbSim stochastic inflow turbulence simulator version 1.21 (revised February 1, 2007), National Renewable Energy Laboratory Technical Report, NREL/TP-500-41137, http://wind.nrel.gov/designcodes/preprocessors/turbsim/TurbSimOverview.pdf

29. Wind Powering America. (2009) Wind resource maps, http://www.windpoweringamerica.gov/wind_maps.asp

30. Manwell, J.F., McGowan, J.G., and Rogers, A.L. (2002) *Wind Energy Explained, Theory, Design and Application*, Wiley, Chichester, U.K.

31. Holton, J. (2004) The planetary boundary layer, in: *Dynamic Meteorology*, International Geophysics Series, vol. 88, Chap. 5, Elsevier Academic Press, Burlington, MA.

32. Burton, T., Sharpe, D., Jenkins, N., and Bossanyi, E. (2001) *Wind Energy Handbook*, Wiley, Chichester, U.K.

33. Renewable Energy Research Laboratory: U. Massachusetts at Amherst. (2006) Capacity factor, intermittency, and what happens when the wind doesn't blow, http://www.ceere.org/rerl/about_wind/RERL_Fact_Sheet_2a_Capacity_Factor.pdf

34. Fingersh, L., Hand, M., and Laxson, A. (2006) Wind turbine design cost and scaling model, Technical Report NREL/TP-500-40566, www.nrel.gov/wind/pdfs/40566.pdf

35. American Wind Energy Association (AWEA). (2010) 10 steps in building a wind farm, http://www.awea.org/pubs/factsheets/10stwf_fs.pdf

36. Department of Energy (DOE). (2008) 20% Wind energy by 2030 increasing wind energy's contribution to U.S. electricity supply, http://www.nrel.gov/docs/fy08osti/41869.pdf

37. Archer, C.L. and Jacobson, M.Z. (2007) Supplying baseload power and reducing transmission requirements by interconnecting wind farms. *Journal of Applied Meteorology and Climatology* 46(11), 1701–1717.

38. Rados, K., Larsen, G., Barthelmie, R., Schlez, W., Lange, B., Schepers, G., Hegberg, T., and Magnisson, M. (2001) Comparison of wake models with data for offshore windfarms, *Wind Engineering* 25, 271–280.

39. Barthelmie, R. et al. (2007) Modeling and measurements of wakes in large wind farms, *Journal of Physics: Conference Series* 75, 012049.

40. Stovall, T.D. (2009) Simulations of wind turbine wake interactions in OpenFOAM, MS thesis, University of Colorado, Boulder, CO.

41. Riso DTU. (2003) Wind Atlas Analysis and Application Program (WAsP), www.wasp.dk.

42. Garrad Hassan & Partners Ltd. (2009) GH WindFarmer 4.0 www.garradhassan.com/products/ghwindfarmer/

43. Piwko, R., Osborn, D., Gramlich, R., Jordan, G., Hawkins, D., and Porter, K. (2005) Wind energy delivery issues: Transmission planning and competitive electricity market operation, *IEEE Power and Energy Magazine* 3(6), 47–56.

44. Energy Information Administration (EIA). (2006) Spread sheet state sales. www.eia.doe.gov/cneaf/electricity/epa/sales_state.xls

45. Musial, W. and Ram, B. (2007) Large scale offshore wind deployments: Barriers and opportunities. NREL, Golden, CO.

46. Musial, W. (2007) Offshore wind electricity: A viable energy option for the coastal United States, *Marine Technology Society Journal* 41(3), 24–35.

47. European Wind Energy Association (EWEA). (2009) Offshore statistics for January 2009, www.ewea.org/fileadmin/ewea_documents/documents/statistics/Offshore_Wind_Farms _2008.pdf

48. U.S. Census Bureau. (2009) National and state population estimates: Annual population estimates 2000 to 2008, http://www.census.gov/popest/states/NST-ann-est.html

49. Energy Information Administration (EIA). (2008) Electricity consumption estimates by sector, 2007, http://www.eia.doe.gov/states/sep_fuel/html/pdf/fuel_use_es.pdf

50. Jonkman, J.M. and Buhl, M.L. Jr. (2007) Development and verification of a fully coupled simulator for offshore wind turbines, NREL/CP-500-40979, www.nrel.gov/wind/pdfs/40979.pdf

51. Energy Efficiency & Renewable Energy (EERE) (DOE). (2010), How does a wind turbine work? http://www1.eere.energy.gov/windandhydro/wind_animation.html

52. Robinson, M. and Musial, W. (2006) Offshore wind technology overview, NREL/PR-500-40462, http://www.nrel.gov/docs/gen/fy07/40462.pdf

53. Butterfield, S., Musial, W., Jonkman, J., and Sclavounos, P. (2007) Engineering challenges for floating offshore wind turbines, NREL/CP-500-38776, www.nrel.gov/wind/pdfs/38776.pdf

54. Baerwald, E., D'Amours, G., Klug, B., and Barclay, R. (2008) Barotrauma is a significant cause of bat fatalities at wind turbines, *Current Biology* 18(16), R695–R696 (August).

55. Goldenberg, S. (2009) Texas wind farm pioneers radar technology to protect migrating birds. guardian.co.uk, http://www.guardian.co.uk/environment/2009/may/01/wind-farm-bird-radar
56. NREL System Advisor Model (SAM) (2012) Support Forum link covering creation of swrf files (accessed August 1, 2012). https://sam.nrel.gov/content/swrf-wind-resource-file-creation.
57. NREL wind integration datasets, http://www.nrel.gov/wind/integrationdatasets/eastern/methodology.html
58. Data.gov-Average price of retail electricity customers by state by provider annually 1990–2009, https://explore.data.gov/Energy-and-Utilities/Annual-1990-2009-Average-Electricity-Price-by-Stat/pt8p-pi7b (accessed July 23, 2012)
59. Barthelmie, R.J., Rathmann, O., Frandsen, S.T., et al. (2007) Modelling and measurements of wakes in large wind farms, *Journal of Physics: Conference Series* 75, 012049, The Science of Making Torque from Wind, IOP Publishing.

4

Capturing Solar Energy through Biomass

Biorenewable resources, sometimes referred to as biomass, are organic materials of recent biological origin [8]. This definition is deliberately broad with the intent of only excluding fossil fuel resources from the wide variety of organic materials that arise from the biotic environment.

We have the technology to convert biorenewable resources into a variety of gaseous and liquid fuels using both thermal and biological processes. Which process and product we choose depends upon the nature of available biomass feedstock and the target market. Gaseous bioenergy products include biogas from anaerobic digestion and producer gas or synthetic gas (syngas) from thermal gasification; and hydrogen, methane, ammonia, and dimethyl ether (DME) from the upgrading of the primary products of anaerobic digestion or gasification. Liquid bioenergy products are conveniently classified according to the intermediate substrate derived from physically, chemically, or thermally processing biomass: carbohydrates, triglycerides, syngas, and bio-oil/bio-crude.

The majority of carbohydrate conversion processes focus on ethanol production although we can also produce butanol, isoprenes, furans, and even alkanes. Similarly, refiners convert triglycerides primarily into methyl esters (biodiesel) even though conversion to alkanes via hydrotreating/hydrocracking is receiving increasing industrial attention. Researchers are investigating promising processes to convert syngas into a variety of liquid fuels: Fischer–Tropsch liquids, methanol, ethanol, and methanol-to-gasoline; and to upgrade bio-oil or bio-crude into a range of hydrocarbons such as gasoline, diesel fuel, or aviation (jet) fuel. Finally, we can convert biomass into electricity, which offers an alternative approach to providing energy for transportation.

This chapter provides an introduction to biomass feedstocks, availability, and conversion to valuable energy products. It also includes sections on pressing topics such as land use, environmental impact, and energy efficiency.

4.1 Biomass Production and Land Use

Biomass is a term that encompasses a wide range of materials. Scientists generally classify biorenewable resources as either wastes or dedicated energy crops. A waste is a material that has been traditionally discarded because it has no apparent value or represents a nuisance or even a pollutant to the local environment. Dedicated energy crops are plants grown specifically for the production of biobased products, that is, for purposes other than food or feed. These resources are typically the product of converting solar energy into organic compounds. The properties of these organic compounds, and the means by which they are formed, are subject to numerous factors.

This section describes the different types of biomass and their properties, the use of land for crop production, and the principles of solar energy conversion. These are important considerations in order to understand the potential of biorenewable resources for capturing solar energy.

4.1.1 Waste Materials

Categories of waste materials that qualify as biorenewable resources include municipal solid wastes (MSWs), agricultural and forest residues and their by-products, and manure. MSWs are whatever is thrown out in the garbage and clearly include materials that do not qualify as biorenewable resources, such as glass, metal, and plastics. MSW includes food processing waste that is the effluent from a wide variety of industries ranging from breakfast cereal manufacturers to alcohol breweries. Another category of waste product is agricultural residues. Agricultural residues (see Figure 4.1) are simply that part of a crop discarded by farmers after harvest such as corn stover (husks and stalks), rice hulls, wheat straw, and bagasse (fibrous material remaining after the milling of sugar cane). Modern agriculture continues to heavily employ animals. The recent concentration of animals into giant livestock facilities has led to calls to treat animal wastes in a manner similar to that for human wastes. Table 4.1 shows the potential quantities of agricultural and forest residue available in the United States.

Waste materials share few common traits other than the difficulty of characterizing them because of their variable and complex composition. Thus, waste biomass presents special problems to engineers who are tasked with converting this sometimes unpredictable feedstock into reliable power or high-quality fuels and chemicals. The major virtue of waste materials is their low cost. By definition, waste materials have little apparent economic value and can often be acquired for little more than the cost of transporting

(a)   (b)

(c)

FIGURE 4.1
Corn fields during the early growing season (a), middle growing season (b), and baled corn stover after harvest (c).

TABLE 4.1

Potential Agricultural, Forest, and Process Waste Supply in the United States

	Annual Biomass Supply (Million Dry Mg/yr)
Logging and other residue	58
Fuel treatments	54
Urban wood residues	43
Wood processing residues	64
Pulping liquor	67
Fuelwood	47
Crop residues	405
Process residues	79

Source: Perlack, R. et al., Biomass as feedstock for a bioenergy and bioproducts industry: The technical feasibility of a billion-ton annual supply, Technical Report A357634, Oak Ridge National Laboratory, 2005. With permission.

the material from its point of origin to a processing plant. In fact, it is possible to acquire wastes at a negative cost because of the rising costs for solid waste disposal and sewer discharges and restrictions on landfilling certain kinds of wastes; that is, a biorenewable resource processing plant is paid by a company seeking to dispose of a waste stream. For this reason, many of the most economically attractive opportunities in biorenewable resources involve waste feedstocks.

Clearly, a waste material that can be used as feedstock for an energy conversion process is no longer a waste material. As demand for these new-found feedstocks increases, those that generate it come to view themselves as suppliers and may demand payment for the one-time waste: a negative feedstock cost becomes a positive cost. Such a situation developed in the California biomass power industry during the 1980s [35]. Concerns about air pollution in California led to restrictions on open-field burning of agricultural residues, a practice designed to control infestations of pests. With no means for getting rid of these residues, an enormous reserve of biomass feedstocks materialized. These feedstocks were so inexpensive that independent power producers recognized that even small, inefficient power plants using these materials as fuel would be profitable. A number of plants were constructed and operated on agricultural residues. Eventually, the feedstock producers had plant operators bidding up the cost of their once nuisance waste material. In the end, many of these plants were closed because of the escalating cost of fuel.

4.1.2 Energy Crops

Energy crops are defined as plants grown specifically as an energy resource. We should note that firewood obtained from cutting down an old-growth forest does not constitute an energy crop. An energy crop is planted and harvested periodically. Harvesting may occur on an annual basis, as with sugar beets or switchgrass, or on a 5–7 year cycle, as with certain strains of fast-growing trees such as hybrid poplar or willow. The cycle of planting and harvesting over a relatively short time period assures that the resource is used in a sustainable fashion; that is, the resource will be available for future generations.

Energy crops contain significant quantities of one or more of four important energy-rich components: oils, sugars, starches, and lignocellulose (fiber). Farmers historically cultivated crops rich in the first three components for food and feed: oils from soybeans and nuts; sugars from sugar beets, sorghum, and sugar cane; and starches from corn and cereal crops. Oil, sugars, and starches are easily metabolized. On the other hand, humans find it hard to digest lignocellulose. Certain domesticated animals with specialized digestive tracts are able to break down the polymeric structure of lignocellulose, and use it as an energy source. From this discussion, it might appear that the best strategy for developing biomass resources is to grow crops rich in oils, sugars, and starches. However, even for "oil crops" or "starch crops," the largest single constituent is invariably lignocellulose (Table 4.2), which is the structural (fibrous) material of the plant: stems, leaves, and roots. If we harvest oils, sugars, and starches and leave the lignocellulose behind as an agricultural residue rather than used as fuel, we will waste the greatest portion of the biomass crop.

Research has shown that energy yields (joules per km² per year) are usually greatest for plants that are mostly "roots and stems"; in other words, plant resources are directed toward the manufacture of lignocellulose rather than oils, sugars, and starches. As a result, there has been a bias toward the development of energy crops that focus on lignocellulosic biomass, which is reflected in the discussion that follows.

Dedicated energy crops are typically high-fiber crops grown specifically for their high productivity of holocellulose (cellulose and hemicellulose). Harvesting may occur on an annual basis, as with switchgrass, or on a 5–7 year cycle, as with certain strains of fast-growing trees such as hybrid poplar. Lignocellulosic crops are conveniently divided into herbaceous energy crops (HECs) and short rotation woody crops (SRWCs) [55].

Herbaceous crops are plants that have little or no woody tissue. The above-ground growth of these plants usually lives for only a single growing season. However, herbaceous crops include both annuals and perennials. Annuals die at the end of a growing season and must be replanted in the spring. Perennials die back each year in temperate climates but reestablish themselves each spring from rootstock. Both annual and perennial HECs are harvested on at least an annual basis, if not more frequently, with yields averaging 550–1100 Mg/km²/year, with maximum yields between 2000 and 2500 Mg/km²/year in temperate regions [55]. As with trees, yields can be much higher in tropical and subtropical regions.

Herbaceous crops more closely resemble hardwoods in their chemical properties than they do softwoods. Their low lignin content makes them relatively easy to delignify, which improves accessibility of the carbohydrate in the lignocellulose. The hemicellulose contains mostly xylan, which is highly susceptible to acid hydrolysis, compared to the cellulose. As a result, microbes can easily degrade agricultural residues, destroying their processing potential in a matter of days if exposed to the elements. Herbaceous crops have relatively high silica content compared to woody crops, which can present problems during processing.

TABLE 4.2

Typical Woody Biomass Compositions

Component	Weight (%)
Cellulose	44 ± 6
Hemicellulose	28 ± 4
Lignin	20 ± 5

SRWC is used to describe woody biomass that is fast growing and suitable for use in dedicated feedstock supply systems. Desirable SRWC candidates display rapid juvenile growth, wide site adaptability, and pest and disease resistance. Woody crops grown on a sustainable basis are harvested on a rotation of 3–10 years. Annual SRWC yields range between 500 and 2400 $Mg/km^2/year$.

Woody crops include hardwoods and softwoods. Hardwoods are trees classified as angiosperms, which are also known as flowering plants. Examples include willow, oak, and poplar. Hardwoods can be regrown from stumps, a process known as coppicing, which reduces their production costs compared to softwoods. Advantages of hardwoods in processing include high density for many species; relative ease of delignification and accessibility of wood carbohydrates; the presence of hemicellulose high in xylan, which can be removed relatively easily; low content of ash, particularly silica, compared to softwoods and herbaceous crops; and high acetyl content compared to most softwoods and herbaceous crops, which is an advantage in the recovery of acetic acid.

Softwoods are trees classified as gymnosperms, which encompass most trees known as evergreens. Examples include pine, spruce, and cedar. Softwoods are generally fast growing, but their carbohydrate is not as accessible for chemical processing as the carbohydrates in hardwood. Since softwoods have considerable value as construction lumber and pulpwood, they are more readily available as waste material in the form of logging and manufacturing residues compared to hardwoods. Logging residues, consisting of a high proportion of branches and tops, contain considerable high-density compression wood, which is not easily delignified. Therefore, logging residues are more suitable as boiler fuel or other thermochemical treatments than as feedstock for chemical or enzymatic processing.

4.1.3 Algae

Algae is a broad term that encompasses several eukaryotic organisms. Eukaryotic organisms are characterized by complex structures enclosed within their cell membranes. Although algae do not share many of the structures that define terrestrial biomass, they are capable of photosynthesis and capturing carbon. Algae's affinity to convert CO_2 into lipids has drawn academic and industrial attention as a means to simultaneously lower carbon emissions and produce biofuels.

Algal biomass uses CO_2 as its carbon source and sunlight as its energy source. About 1.8 kg of CO_2 is fixed for every kg of algal biomass, which contains up to 50% carbon by dry weight. Controlled production of renewable fuels from algae has been proposed in either raceway ponds or photobioreactors. Raceway ponds consist of open, shallow recirculation channels with mechanical flow control and surfaces that enhance light retention. Raceway ponds are inexpensive, but relatively inefficient when compared to photobioreactors. There are various photobioreactor designs with the common goal of maintaining a monoculture of algae that is efficiently exposed to sunlight and carbon dioxide. A common design employs arrays of tubes arranged vertically to minimize land use and oriented north–south to maximize light exposure.

Given that algae do not require fresh water or fertile soils, waste lands have been suggested as potential locations to grow algae. One suggestion is to build algae ponds in the desert Southwest United States, where inexpensive flat land, abundant sunlight, water from alkaline aquifiers, and CO_2 from power plants could be combined to generate renewable fuels. Algae's potential for yields of 1.12–9.40 million liters of oil/km^2/year promises significant reductions in the land footprint required to produce biofuels.

4.1.4 Land Use for Biomass Production

Global land use is broadly defined by five categories: pasture, crop, forest, urban, and abandoned. Pasture is land devoted primarily to animal grazing; crop lands are areas actively cultivated for food production; forest land contains primarily large trees; urban areas are heavily populated regions; and abandoned lands are territories that formerly fit one of the previous categories but are no longer employed for human activities. Humans, because of population migrations or land-use change, alter the portions of land devoted to each of these categories over time.

Researchers estimate that 14.5 and 33.2 million km² of global land area were devoted to crops and pasture respectively in 2000 [15]. These land-use groups can coexist within the same region. For example, the U.S. Midwest and parts of the Southeast include regions with more than 70% of the land devoted to crops, and the western sides of the Midwest and Southern U.S. states have a high concentration of land for pasture.

Modern-day farmers devote their production to a small selection of crops depending on socioeconomic factors. Table 4.3 shows a sample of biomass crops grown in various geographical regions and their annual yields. Crops such as corn and sugar cane can serve both food and energy needs due to their high yields of sugar-rich biomass and biomass residue (stover and bagasse respectively).

We can estimate the amount of biomass available in a given region by assuming nominal values for crop productivity and available land-use data using the following equation:

$$\text{Total Biomass}\left[\frac{\text{kg}}{\text{year}}\right] = f \times \text{Crop}_{yield}\left[\frac{\text{kg}}{\text{km}^2 \times \text{year}}\right] \times \text{Land}_{area}[\text{km}^2] \qquad (4.1)$$

TABLE 4.3

Nominal Annual Yields of Biomass Crops

Biomass Crop	Geographical Location	Annual Yield (Mg/km²)
Corn: grain	North America	700
Corn: cobs	North America	130
Corn: stover	North America	840
Jerusalem artichoke: tuber	North America	4500
Jerusalem artichoke: sugar	North America	640
Sugar cane: crop	Hawaii	5500
Sugar cane: sugar	Hawaii	720
Sugar cane: bagasse (dry)	Hawaii	720
Sweet sorghum: crop	Midwest U.S.	3800
Sweet sorghum: sugar	Midwest U.S.	530
Sweet sorghum: fiber (dry)	Midwest U.S.	490
Switchgrass	North America	1400
Hybrid poplar	North America	1400
Wheat: grain	Canada	220
Wheat: straw	Canada	600

Source: Wayman, M. and Parekh, S., *Biotechnology of Biomass Conversion: Fuels and Chemicals from Renewable Resources*, Open University Press, Philadelphia, PA, 1990. With permission.

In Equation 4.1, f is a factor that accounts for crop rotations, farmer participation, and land conservation among other considerations that restrict the land use. As an example, Iowa has a total land area of 144,700 km² that is predominantly covered by corn and soybeans. In 2010, farmers planted 37.5% of Iowa land with corn netting an average yield of 165 bushels per acre (1035 Mg/km²). Thus, the total amount of corn grown in Iowa that year was 56.2 million Mg.

Farmers and seed companies have managed to increase crop yields every year for the past couple of decades. Crop yield increases follow the exponential growth formula:

$$\text{Crop}_{yield}(t)\left[\frac{\text{kg}}{\text{km}^2}\right] = \text{Crop}_{yield,0}e^{kt} \qquad (4.2)$$

where k is the growth rate and t is the period of time since the initial value $\text{Crop}_{yield,0}$. Iowa corn yields have increased from an average of 1162 Mg/km² in 1990 to 1533 Mg/km² in 2010. This improvement represents a 1.32% annual growth rate as shown in Equation 4.3:

$$k = \frac{\ln(1533/1162)}{21} = 0.0132 \qquad (4.3)$$

The U.S. Department of Agriculture (USDA) maintains a comprehensive database of agricultural statistics (available online for free at http://quickstats.nass.usda.gov/). The data span several years and include county-level data for crops, demographics, economics, animals and products, and environmental impacts.

Example 4.1: Biomass Collection and Transportation Distance Calculations

A facility would like to collect 10,000 metric tons of corn stover per day for 330 days per year into the outskirts of Springfield, Illinois. We can estimate how far, on average, corn stover would need to travel if it is all delivered by truck.

Springfield is centrally located in Sangamon county and surrounded by the counties listed in Table 4.4. The total amount of stover produced in the area is 6,306,000 tonnes per year or enough to supply 19,110 tonnes per day to the facility. However, let us assume that only 70% of the stover is sustainably recoverable ($f = 0.7$). The distance between county seats and Springfield is the travel distance estimated from online mapping tools.

1. The facility has decided to collect the stover from farm gates. They approximated the transport distance by assuming that the facility would need to collect stover from a circular area surrounding the facility. The total area for the seven counties is approximately 10,860 km. The average distance for transport within a circle with this total area is

$$\tau \times \frac{2}{3} \times \sqrt{\frac{\text{Area}}{\pi}} = 50.95 \text{ km} \qquad (4.4)$$

where τ is known as the tortuosity factor—a way to account for windings and road degradation in transport networks. Here we assumed that τ is 1.3, and it varies between 1 (straight line) and 1.5 (square grid network).

TABLE 4.4

Counties Surrounding Springfield, IL, and Their County Seats, Distance to Springfield, and 2010 Corn Stover Yields

County	County Seat	Distance to Springfield, IL (km)	Stover Production (Mg/year)	Planted Area (km²)	County Area (km²)
Sangamon	Springfield	0	969,347	1000	2271
Menard	Petersburg	38.3	310,667	340	816
Christian	Taylorville	43.1	913,248	927	1854
Cass	Virginia	54.4	351,039	376	995
Morgan	Jacksonville	56.8	580,518	631	1481
Logan	Lincoln	61	838,452	858	1603
Montgomery	Hillsboro	78.2	746,398	761	1839
Macon	Decatur	80.5	749,017	708	1515
Macoupin	Carlinville	83.5	847,035	919	2248

Source: United States Department of Agriculture, Agricultural statistics database, July 2012, 2012, http://quickstats.nass.usda.gov

2. In order to save costs, the facility has convinced farmers to deliver stover to satellite depots in their respective county seats. For this case, the average stover transport distance for the facility is the weighted average of stover transport distance to Springfield (Equation 4.5) for the amount required (Equation 4.6).

$$\text{Distance (ave. km)} = \frac{\sum_{i=0}^{n} D_i \times S_i}{\sum_{i=0}^{n} S_i} \quad \forall i \in \text{County Seats} \quad (4.5)$$

$$\sum_{i=0}^{n} S_i \leq 10{,}000 \text{ tonnes/day} * 330 \text{ days} \quad (4.6)$$

For this case, the seven nearest counties (Sangamon, Menard, Christian, Cass, Morgan, Logan, and Montgomery) can sustainably provide 3,300,000 tonnes per year to the facility. The average distance that stover would ship from the county seat depots to Springfield is about 45.2 km.

Every year, 5.6 million exajoules (EJ) of solar radiation impacts the top of our planet's atmosphere. This is enough energy to supply the world's energy consumption for several thousand years as the world consumes about 570 EJ per year. Unfortunately, this energy is very diffuse and difficult to convert efficiently. A majority of the energy is unreachable on land surfaces, and only a minuscule amount is converted into biomass.

Only 70% of solar radiation makes it through the planet's atmosphere, and the remaining 30% is absorbed, reradiated, or reflected in the atmosphere—the earth's surface area is approximately 29.2% land. Furthermore, biomass covers approximately 21% of the earth's surface. This leaves 6.1% of the solar energy incident on the earth's surface available for crops.

Biomass captures a small amount of the incident solar irradiation energy via photosynthesis. The chemical formula 4.7 can summarize the overall reaction:

$$6CO_2 + 6H_2O + (\text{sunlight}) \rightarrow C_6H_{12}O_6 + 6O_2 \quad (4.7)$$

The efficiency depends on the plant's carbon fixation pathway. Plants employ carbon fixation as a biochemical mechanism to capture carbon from airborne CO_2 and convert it to organic forms. Scientists categorize carbon fixation pathways after the carbon chain length of the first carbohydrate formed. C3 and C4 carbon fixation pathways proceed from initial molecules with three and four carbon atoms respectively. The vast majority of plants employ C3 carbon fixation whereas about 3% of known species employ the C4 pathway. Corn, sugar cane, and sorghum are common C4 plants.

Theoretical estimates for solar energy to biomass conversion are 4.6% and 6% for C3 and C4 plants respectively. C3 plants consume energy during carbohydrate synthesis, photorespiration, and respiration. C4 plants employ more energy for carbohydrate synthesis and respiration, but do not suffer from photorespiration energy penalties resulting in higher overall efficiencies. Table 4.5 summarizes each step involved in converting solar energy at the plant's surface into chemical energy (biomass).

In practice, the most efficient conversion measured in C4 plants is 3.7% of the incident solar energy, and C3 plants managed 2.4% conversion efficiency.

Corn grain and soybeans are C4 and C3 plants grown throughout the U.S. Midwest, where global irradiation on the horizontal surface averages around 4 kWh/m²/day. We can estimate the efficiency of solar energy collection in this region. First, we estimate the amount of solar irradiation in the area:

$$\text{Iowa irradiation energy} = 4\frac{\frac{kWh}{m^2}}{day} \times 365\frac{days}{year} \times 3600\frac{s}{h} \times 100\frac{J}{kW}$$

$$= 5.25 \times 10^9 \frac{MJ}{km^*year} \tag{4.8}$$

Second, we estimate how much biomass grows on a plot of land. Total biomass includes above- and below-ground material. The ratio of below-ground biomass to above ground is known as the root-to-shoot ratio.

$$\frac{Root}{Shoot} = \frac{root + exudates}{crowne \text{ or leaves or branches} + trunk \text{ or stalk}} \tag{4.9}$$

TABLE 4.5

Photosynthesis Steps and Efficiencies

Photosynthesis Step	% Total Energy	
Incident solar energy (on leaf surface)	100%	
Energy in photosynthetically active spectrum	48.7%	
Absorbed energy	43.8%	
Photochemically converted energy	37.2%	
Carbon fixation pathway	C3	C4
Energy in synthesized carbohydrates	12.6%	8.5%
Energy available after photorespiration	6.5%	8.5%
Energy available after respiration	4.6%	6.0%

Source: Zhu, X.G. et al., *Curr. Opin. Biotechnol.*, 19(2), 153, 2008. With permission.

Corn has a root-to-shoot ratio of about 0.55. Therefore, the total above- and below-ground biomass in land growing corn includes corn grain (700 Mg/km²), corn cobs (130 Mg/km²), corn stover (840 Mg/km²), and root and exudates (918 Mg/km²). Based on an assumed energy value of 17.58 MJ/kg of corn [20], the total corn energy is

$$
\text{Corn Energy}\left[\frac{\text{MJ}}{\text{km}^2}\right] = 17.58\left[\frac{\text{MJ}}{\text{kg}}\right] \times 2,588,000\left[\frac{\text{kg}}{\text{km}^2}\right] = 45.5e6\left[\frac{\text{MJ}}{\text{km}^2}\right] \tag{4.10}
$$

Based on these estimates, corn has an annual solar energy conversion efficiency of 0.62%. Farmers plant corn seeds between April and June, and they harvest in October. Thus, corn has a growth period of about 180 days yielding a 1.25% growing season efficiency.

   Although biomass has low solar energy conversion efficiency, they require minimal inputs or maintenance. This low efficiency is sufficient for nature's purposes, but scientists are investigating how to increase it. We could dramatically increase the size of our biomass resource base with incremental improvements in biomass efficiency yielding more food and energy for human use.

4.1.5 Important Properties of Biomass

Scientists need information about plant composition, heating value, bulk density, and production yields to evaluate the potential resource of biomass feedstock. Studies report composition in terms of organic components, proximate analysis, or ultimate analysis. Table 4.6 shows properties of representative fibrous biomass feedstock. Compilations of relevant biomass properties are found in Brown [8]. Excellent online databases of biomass properties include the U.S. Department of Energy "Biomass Feedstock Composition and Property Database," and the ECN Phyllis database.

   Organic component reports describe types and quantities of biomass compounds including proteins, oils, sugars, starches, and lignocellulose (fiber). These components

TABLE 4.6

Physical and Thermochemical Properties of Selected Biomass

	Feedstock	Corn Stover	Herbaceous Crop	Woody Crop
Organic composition	Cellulose	53	45	50
(wt-%)	Hemicellulose	15	30	23
	Lignin	16	15	22
	Other	16	10	5
Elemental Analysis	C	44	47	48
(dry wt-%)	H	5.6	5.8	5.9
	O	43	42	44
	N	0.6	0.7	0.5
	Ash	6.8	4.5	1.6
Proximate Analysis	Volatile matter	75	81	82
(dry wt-%)	Fixed C	19	15	16
	Ash	6	4	1.3
HHV (MJ/kg)		17.7	18.7	19.4
Bulk density (kg/m³)		160–300	160–300	280–480
Yield (Mg/ha)		8,400	14,000	14,000

vary widely among plant parts. For example, corn grain is mostly starch (72 wt.%) with a relatively small amount of fiber (13 wt.%) while corn stover, that part of the crop left on the field, is mostly fiber (84 wt.%) with very little starch content. Often the fiber, which is a polymeric composite of cellulose, hemicellulose, and lignin, is reported in terms of these three constituents. Biochemical and thermochemical process design requires knowledge of the organic components because they influence the final product composition.

Proximate analysis is important in developing thermochemical conversion processes for biomass. Proximate analysis reports the yields (% mass basis) of various products obtained upon heating the material under controlled conditions; these products include moisture, volatile matter, fixed carbon, and ash. Since moisture content of biomass is so variable and can be easily determined by gravimetric methods (weighing, heating at 100°C, and reweighing), the proximate analysis of biomass is commonly reported on a dry basis. Volatile matter is that fraction of biomass that decomposes and escapes as gases upon heating a sample at moderate temperatures (about 400°C) in an inert (nonoxidizing) environment. Knowledge of volatile matter is important in designing burners and gasifiers for biomass. The remaining fraction is a mixture of solid carbon (fixed carbon) and mineral matter (ash), which can be distinguished by further heating the sample in the presence of oxygen: the carbon is converted to carbon dioxide leaving only the ash.

Ultimate analysis is simply the (major) elemental composition of the biomass on a gravimetric basis: carbon, hydrogen, oxygen, nitrogen, sulfur, and chlorine along with moisture and ash. Sometimes this information is presented on a dry, ash-free (daf) basis. Compared to fossil fuels, biomass has relatively high oxygen content (typically 40–45 wt.%), which detracts from its heating value and represents new challenges in converting these compounds into substitutes for the hydrocarbons that currently dominate our economy. In many instances, generic molecular formula based on one mole of carbon is convenient for performing mass balances on a process. For example, cellulose and starch have the generic molecular formula $CH_{1.7}O_{0.83}$, hemicellulose can be represented by $CH_{1.6}O_{0.8}$, and wood is $CH_{1.4}O_{0.66}$ [22].

Efficient processing facilities keep close track of energy use. They usually employ heating value as a measure of energy content. Heating value is the net enthalpy released upon the reaction of a particular fuel with oxygen under isothermal conditions (the starting and ending temperatures are the same). If water vapor formed during reaction condenses at the end of the process, the latent enthalpy of condensation contributes to what is known as the higher heating value (HHV). Otherwise, the latent enthalpy does not contribute and the lower heating value (LHV) prevails. These measurements are typically performed in a bomb calorimeter and yield the HHV for the fuel. Heating values of biomass are conveniently estimated from the percent of carbon in the biomass on a dry basis using the empirical relationship [22]:

$$HHV \text{ (dry)} \left[\frac{MJ}{kg} \right] = 0.4571(\% \text{ C on dry basis}) - 2.70 \qquad (4.11)$$

For estimating purposes, assuming an HHV of 18 MJ/kg is a good approximation for many kinds of (dry) biomass. Enthalpies of formation are very useful in thermodynamic calculations, but these data are rarely tabulated for biomass because of the wide variability in its composition. However, if HHV for a biomass fuel has been determined in a bomb calorimeter, its enthalpy of formation can be determined by summing the enthalpies of formation of the products of combustion and subtracting the HHV from this sum.

Biomass transport properties such as bulk density and volumetric energy content are important for transport and storage applications. Bulk density is determined by weighing a known volume of biomass that is packed or baled in the form anticipated for its transportation or use. Clearly, solid logs will have higher bulk density than the same wood chipped. Bulk density will be an important determinant of transportation costs and the size of fuel storage and handling equipment. Volumetric energy content is also important. Volumetric energy content, which is simply the enthalpy content of fuel per unit volume, is calculated by multiplying the HHV of a fuel by its bulk density.

Biomass energy projects need knowledge of appropriate biomass properties to be successful. These include knowledge of the types and quantities of biomass grown in a given area, which depend on many factors including plant variety, crop management (fertilization and pest control), soil type, landscape, climate, weather, and water drainage. Thus, each project will require site-specific information obtained through discussions with state extension agents and local agronomists in combination with field trials in advance of detailed manufacturing plant design.

Example 4.2: Enthalpy of Formation Calculations for Switchgrass

Assume that a sample of switchgrass has an elemental analysis that gives a generic molecular formula of $CH_{1.4}O_{0.8}$ (molecular weight of 26.2 kg/kmol) and its HHV is measured to be 18.1 MJ/kg. The complete combustion of one kilomole of switchgrass can be represented by

$$CH_{1.4}O_{0.8} + 0.95O_2 \rightarrow CO_2 + 0.7H_2O(\text{liquid}) \tag{4.12}$$

The enthalpy of reaction, ΔH_R, is calculated from the various enthalpies of formation, h_f° (note: molecular elements like O_2 have an h_f° equal to zero):

$$\Delta H_R = h_f^{\circ}{}_{CO_2} + 0.7 h_f^{\circ}{}_{H_2O(\text{liquid})} - \left(h_f^{\circ}{}_{CH_{1.4}O_{0.8}} + 0.95 h_f^{\circ}{}_{O_2} \right)$$

$$= -18.1 \left[\frac{MJ}{kg} \right] \times 26.2 \left[\frac{kg}{kmol} \right]$$

$$= -393.5 \left[\frac{MJ}{kmol} \right] + 0.7 \times \left(-285.8 \left[\frac{MJ}{kmol} \right] \right) - \left(h_f^{\circ}{}_{CH_{1.4}O_{0.8}} + 0 \right) \tag{4.13}$$

Solving for the enthalpy of formation for $CH_{1.4}O_{0.8}$ yields

$$h_f^{\circ}{}_{CH_{1.4}O_{0.8}} = -119.3 \left[\frac{MJ}{kmol} \right] \tag{4.14}$$

This value can be used in various thermodynamic calculations for this switchgrass sample, whether gasification to hydrogen or hydrolysis to fermentable sugars. Heat of formation data is available from most engineering thermodynamic textbooks such as Moran and Shapiro's [33].

4.2 Biomass Process Economics and Technology

There are many ways to convert biomass into valuable chemicals, fuels, and electricity. This section will discuss biomass conversion technologies for gaseous and liquid fuel production as well as power generation. Gaseous fuels include biogas from anaerobic digestion and synthetic gas from gasification, and liquid fuels encompass a wide range of products such as ethanol, gasoline, diesel, and Fischer–Tropsch liquids. Combinations of these technologies can improve the overall conversion efficiency, and we will show how to compare these technologies from an energy efficiency perspective.

4.2.1 Biomass Process Economics

Conversion costs are a major factor that determines whether a technology becomes a commercial success. Conversion costs can be grouped into three major categories: capital, operating, and feedstock costs.

Capital costs include equipment purchasing and installation costs as well as interests from financial loans. They are difficult to estimate. The accuracy for most analyses ranges between +100% and −30% (meaning that the project could be twice as expensive or 30% cheaper). The accuracy can be improved with detailed engineering analysis and/or cost databases.

Operating costs consist of all expenses required to maintain the operation of a given facility such as labor, maintenance, and utilities. Operating costs include both fixed and variable cost components. Fixed costs, like labor and maintenance, are independent of the plant throughput capacity. On the other hand, variable costs are typically related to the amount of feedstock input. For example, a facility would have higher heating costs from drying increasing quantities of biomass.

Finally, feedstock costs are the costs of purchasing and delivering biomass to the facility. Facilities may choose to pay a fixed, plant-gate cost in which case the supplier carries the burden of delivering the feedstock. However, delivery costs are an important factor in how much a biorefinery pays for biomass.

The summation of these cost components allows us to estimate the cost of a biomass product. These are typically reported on a normalized basis such as Equation 4.15:

$$\text{Fuel Cost}\left[\frac{\$}{\text{unit output}}\right] = \frac{\text{Capital} + \text{Operating} + \text{Feedstock}}{\text{Product Output}} \qquad (4.15)$$

Fuel costs typically depend on the scale of operation. Large-scale facilities can optimize their operations in order to take advantage of economies-of-scale. The impact of economies-of-scale is greatest for capital costs, and research has shown a strong relationship between the plant capacity and unit capital costs. This relationship is captured by the power law:

$$\text{Capital Cost} = \text{Capital Cost}_0 \times \left(\frac{\text{Capacity}}{\text{Capacity}_0}\right)^n \qquad (4.16)$$

where
n is the scale factor
the index 0 indicates the base or known value (cost or capacity)

Typical values for the scale factor are 0.63 for biochemical facilities and 0.7 for thermo-chemical plants. The higher value is due to the larger capital investment required for most thermochemical projects. Economies-of-scale allow us to estimate the costs of scaling up a facility. For example, if a 40 million gallon per year (MMGPY) ethanol plant costs $46.7 million, the cost for an equivalent 200 MMGPY facility is

$$\text{Capital Cost}_{200\text{MMGPY}} = \$46.7e6 \times \left(\frac{200e6}{40e6}\right)^{0.63} = \$128.7 \text{ million} \tag{4.17}$$

In this case, per unit capital costs decrease from $1.17 per gallon of annual capacity to $0.64. This shows the potential cost reductions from scaling-up operations. In reality, these benefits are not fully realized because of diseconomies of scale. In other words, additional considerations become important at different plant capacities. However, several industries have successfully scaled refineries from a few thousand barrels per day of capacity to over 400,000 barrels per day and benefit from dramatic reductions in per unit costs.

Most operating costs scale linearly with capacity ($m = 1$). However, large facilities can also reduce their per unit operating costs with labor management and equipment mainte-nance strategies. Thus, operating costs are sometimes scaled with a 0.9 scale factor.

Feedstock costs for biorefinery are notable for their diseconomies-of-scale—their per unit costs increase with capacity. Unlike fossil fuels that travel from a single source (coal mine or oil well) to a facility, biomass must be collected from a wide area. The larger the facility, the further away it must seek feedstock. This relationship can be captured simply using a scale factor greater than 1 ($p \approx 1.5$).

From these relationships, we can roughly estimate the costs of scaling-up a known tech-nology using Equation 4.18:

$$\text{Fuel Cost}_M = C_0 \times \left(\frac{\text{Capacity}}{\text{Capacity}_0}\right)^n + O_0 \times \left(\frac{\text{Capacity}}{\text{Capacity}_0}\right)^m + F_0 \times \left(\frac{\text{Capacity}}{\text{Capacity}_0}\right)^p \tag{4.18}$$

where C, O, and F stand for capital, operating, and feedstock costs. Per unit costs in units of $/gal can be calculated by dividing total costs by the biorefinery output capacity.

Inflation should be taken into account when comparing estimates published in different years. There are several indices available to account for inflation: consumer price index (CPI), chemical engineering plant cost index, and the Marshall and Swift index. The CPI is maintained by the U.S. government Bureau of Labor Statistics and is a general measure of the time value of money. The other indices are more relevant for industrial estimates, but are not publicly available. The CPI is available at http://www.bls.gov/cpi.* Table 4.7 shows the CPI values for 2006–2010.

Adjusting for inflation, the $46.7 million corn ethanol facility estimate, developed in 2006, would be more expensive. Based on the annual average CPI value for the years 2006 and 2010, the cost would be

$$\$46.7e6 * \frac{218.06}{201.6} = \$50.5e6 \tag{4.19}$$

---

* Direct link: ftp://ftp.bls.gov/pub/special.requests/cpi/cpiai.txt

TABLE 4.7

Consumer Price Index Factors
for 2006–2010

Year	CPI Factor
2006	201.60
2007	207.34
2008	215.30
2009	214.54
2010	218.06

The previous approach is suitable for quick comparisons of biorefineries operating at different scales. However, it would not help us determine the costs for a new facility. To determine the costs of a new biorefinery, we would need detailed process and cost data of the unit operations. These data can be acquired from process modeling tools and cost databases such as Aspen Plus™ and Aspen Process Economic Analyzer™.

The following sections include brief descriptions of mature biorefinery technologies and their costs. This information can help us conduct comparisons such as the one described here.

4.2.2 Conversion of Biomass to Gaseous Fuels

There are two main approaches to converting biomass into gaseous fuels: anaerobic digestion and gasification. Both of these processes generate a combustible gas suitable for power generation applications. Anaerobic digestion gas is commonly known as biogas, and biomass gasification products are called synthetic gas (syngas) or producer gas depending on the gas composition. The key difference between biogas and syngas is that syngas is better suited for liquid fuel production. Therefore, biomass gasification is a pathway to both power generation and liquid fuel synthesis.

4.2.2.1 Biomass to Biogas

Anaerobic digestion is the decomposition of waste to gaseous products by bacteria in an oxygen-free environment. Bacteria decompose organic wastes into a mixture of methane, carbon dioxide, and trace gases. Although bacteria can convert a wide range of waste materials, their productivity is sensitive to feedstock properties such as moisture and ash content. Anaerobic digestion yields vary from 23 to over 250 m^3/Mg (wet basis) with a methane content of 55%–75% by volume. The thermodynamic efficiency of this process is about 60%.

Biogas with a high methane content can substitute for natural gas in many applications. Clean biogas is suitable for engine generator sets, small gas turbines, and some kinds of fuel cells. Biogas contaminants, sulfur in particular, must be removed before use in some equipment.

The biological processes within an anaerobic digester that lead to biogas, as summarized in Figure 4.2, are relatively complicated [23]. They involve a series of steps through which several bacteria species break down proteins, carbohydrates, and fats into simple acids, alcohols, and gaseous compounds. Anaerobic digester systems, on the other hand, are simple as shown in Figure 4.3. These systems are typically operated as either batch or steady flow units although semi-batch operation is also an option. The effluents from these systems are biogas and sludge. The biogas may undergo scrubbing to remove hydrogen sulfide that would otherwise yield the pollutant sulfur dioxide.

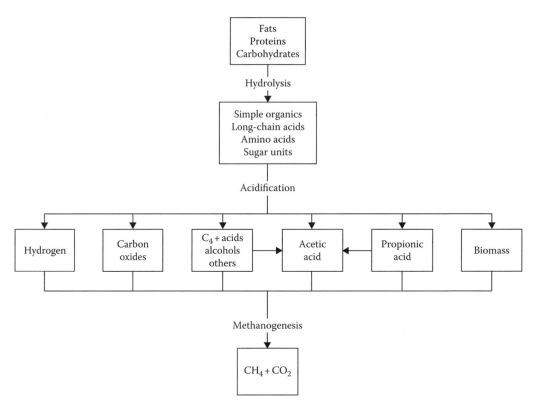

FIGURE 4.2
Microbial phases in anaerobic digestion. (Adapted From Klass, D.L., *Biomass for Renewable Energy, Fuels, and Chemicals*, Academic Press, San Diego, CA, 1998c, p. 356.)

Anaerobic digester designs include batch, plug-flow, and continuously stirred tank, upflow, and two-tank reactors. The batch reactor is a single vessel design in which all steps of the digestion process take place. Advanced reactors aim to improve waste contact with active bacteria and/or to separate and control the environments for acid-forming and methane-forming bacteria. Modifying multiple variables can optimize anaerobic digestion: waste pretreatment, heating, mixing, nutrient addition, specialized bacteria addition, and pH among others.

4.2.2.2 Biomass to Synthetic Gas

Gasification is the high-temperature (750°C–850°C) conversion of solid, carbonaceous fuels into flammable gas mixtures consisting of carbon monoxide (CO), hydrogen (H_2), methane (CH_4), nitrogen (N_2), carbon dioxide (CO_2), and smaller quantities of higher hydrocarbons. When the gas mixture has a high nitrogen content, it is known as producer gas; otherwise, it is known as synthetic gas. The overall process is endothermic and requires either the simultaneous burning of part of the fuel or the delivery of an external heat source to drive the process [40]. Gasification converts over 90% of biomass feed into gas with the remaining material ending as solid char and ash, or viscous tar in some applications. Typical biomass-to-syngas efficiencies range between 70% and 90%.

Figure 4.4 illustrates the steps involved in gasification: heating and drying, pyrolysis, gas–solid reactions that consume char, and gas-phase reactions that adjust the final composition. Solid particles are initially heated and dried at temperatures below 100°C. Pyrolysis takes

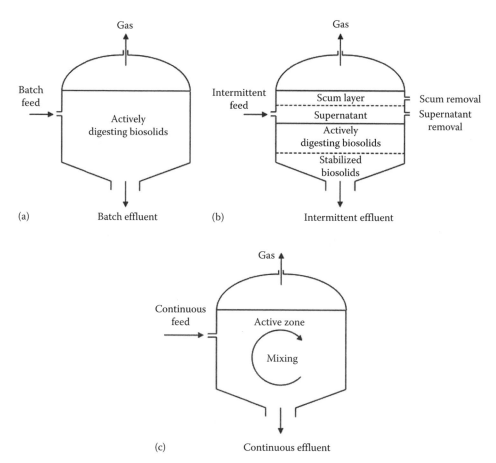

FIGURE 4.3
Types of anaerobic digesters. (a) Batch fed, (b) intermittently fed, and (c) continuously stirred and fed.

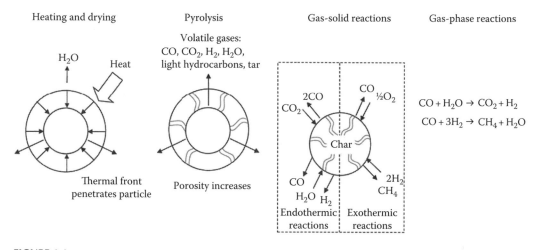

FIGURE 4.4
Processes of thermal gasification.

place as the particle temperature reaches 600°C; pyrolysis produces the intermediate products of char, gases, and condensable vapors (including water, methanol, acids, and heavy hydrocarbons). At high enough temperatures (above 1000°C), it is possible to convert over 95% of the feed carbon with char-consuming reactions. Gas-phase reactions achieve equilibrium in high-temperature gasification; however, gasification is sometimes conducted at low temperatures (approximately 850°C) yielding nonequilibrium products (mostly tars and light hydrocarbons).

Example 4.3: The Role of Carbon Conversion in Gasification

Carbon conversion is a key measure of gasification efficiency. Commercial high-temperature gasifiers can achieve carbon conversions of over 97%. Low-temperature gasifiers commonly employed with biomass operate at lower carbon conversion rates of 90%–95%. Solid residue from the gasifier contains most of the unconverted carbon otherwise known as char. Even though char material yields are low (10 wt.%), char energy content can be high. This example illustrates how much energy can remain unconverted in the relatively small fractions of char and tar.

Based on Table 4.8, we can calculate the percentage of biomass energy remaining as char and tar:

$$\text{Char}_{Energy}\left[\frac{MJ}{MJ_{Biomass}}\right] = \frac{0.12[kg/kg] * 22.7[MJ/kg]}{17.4[MJ/kg]} = 15.6\% \qquad (4.20)$$

$$\text{Tar}_{Energy}\left[\frac{MJ}{MJ_{Biomass}}\right] = \frac{0.07[kg/kg] * 12.1[MJ/kg]}{17.4[MJ/kg]} = 4.87\% \qquad (4.21)$$

These results show that 20.5% of the biomass energy remains unconverted as char and tar after low-temperature gasification. Higher conversion rates are possible by recirculating the char into the gasifier, and some of the tar energy can be recovered by combustion downstream from the gasifier.

Gasification requires an oxygen source for exothermic processes that help with drying, pyrolysis, and endothermic gas-phase reactions. Air, pure oxygen, and steam are the main oxygen sources employed in gasification. Air gasification dilutes the gasification gas (producer gas) with nitrogen lowering the product's energy density. Thus, there are advantages in using oxygen or steam despite the additional costs. Oxygen requirement is an important process parameter that can be estimated using combustion equations. The stoichiometric requirement for complete combustion can be calculated using the combustion formula:

$$C_a H_b O_c N_d S_e + f O_2 + 3.76 f N_2 \rightarrow a CO_2 + \frac{b}{2} H_2 O + e SO_2 + \left(3.76f + \frac{d}{2}\right) N_2 \qquad (4.22)$$

TABLE 4.8

Biomass Gasification Product Distribution and Energy Content

Compound	Yield (wt.%)	Energy Content (MJ/kg)
Biomass	—	17.4
Syngas	81	17.1
Char	12	22.7
Tar	7	12.1

For full combustion, the moles of oxygen required are f:

$$f = a + \frac{b}{4} + e - \frac{c}{2}$$ (4.23)

For example, the moles of oxygen required to fully combust coal are

$$\text{Coal composition} = CH_{0.739}O_{0.091}N_{0.014}S_{0.003}$$ (4.24)

$$f = 1 + \frac{0.739}{4} + 0.003 - \frac{0.091}{2} = 1.14 \text{ moles of } O_2$$ (4.25)

We can calculate the stoichiometric oxygen requirement on a mass basis using

$$\frac{1.14 \text{ moles of } O_2}{\text{mole of Coal}} \frac{32.0 \text{g of } O_2}{\text{mole of } O_2} \frac{\text{mole of Coal}}{14.49 \text{g of Coal}} = 2.52 \frac{\text{g of } O_2}{\text{g of Coal}}$$ (4.26)

Therefore, a kilogram of coal with this composition consumes 2.52 kg of oxygen for full combustion. Note that it is common practice to burn fuel with about 25% excess oxygen than the stoichiometric amount. We can convert the oxygen requirement into an equivalent amount of air. Since air is 23.2 wt.% oxygen, the total amount of air needed for complete combustion of this fuel is 10.87 kg (oxygen required/0.232). However, gasification requires that oxygen levels remain below 25% of the stoichiometric amount for full combustion. Therefore, gasification of this coal feedstock would require less than 0.63 kg of oxygen (2.72 kg of air) per kilogram of coal.

There are four main types of gasifiers: updraft (countercurrent), downdraft (cocurrent), fluidized bed, and entrained flow [6]. These are illustrated in Figure 4.5, and their performance characteristics are summarized in Table 4.9. The most common types of gasifiers are variations of the entrained flow and fluidized bed designs.

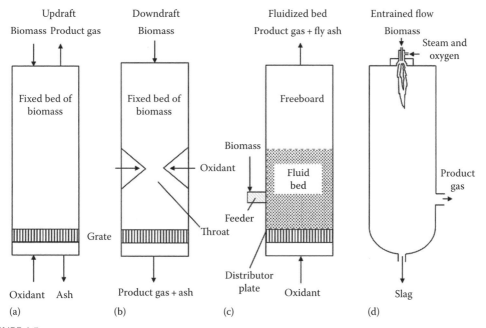

FIGURE 4.5
Common types of biomass gasifiers: (a) updraft, (b) downdraft, (c) fluidized bed, and (d) entrained flow.

TABLE 4.9

Producer Gas Composition from Various Kinds of Gasifiers

Gasifier Type	Gaseous Constituents (vol.% dry)					HHV (MJ/m³)	Gas Quality	
	H₂	CO	CO₂	CH₄	N₂		Tars	Dust

Gasifier Type	H_2	CO	CO_2	CH_4	N_2	HHV (MJ/m³)	Tars	Dust
Air-blown updraft	11	24	9	3	53	5.5	High ($\approx$10 g/m³)	Low
Air-blown downdraft	17	21	13	1	48	5.7	Low ($\approx$1 g/m³)	Medium
Air-blown fluidized bed	9	14	20	7	50	5.4	Medium ($\approx$10 g/m³)	High
Oxygen-blown downdraft	32	48	15	2	3	10.4	Low ($\approx$1 g/m³)	Low
Indirectly heated fluidized bed	31	48	0	21	0	17.4	Medium ($\approx$10 g/m³)	High

The entrained flow gasifier is a highly efficient reactor. This reactor was developed for steam-oxygen gasification of coal at temperatures of 1200°C–1500°C. These high temperatures assure excellent char conversion (approaching 99%) and low tar production and convert the ash to molten slag, which drains from the bottom of the reactor. The technology is attractive for advanced coal power plants but has not been widely explored for biomass because of the expense of finely dividing biomass, the difficulty of reaching high temperatures with biomass, and the presence of alkali in biomass, which leads to severe ash sintering. Despite these difficulties, entrained flow reactors are being considered for gasification of pretreated biomass (such as torrefied biomass and pyrolysis liquids) [19].

Torrefied biomass is the product of heating biomass to between 200°C and 350°C—a process known as torrefaction. At these temperatures, biomass undergoes drying and minor loss of volatile content. The remaining product has a closer resemblance to coal: high energy density and improved grindability. It is also hydrophobic, which improves its storage properties by reducing water-related degradation. Torrefaction has attracted recent interest as an approach to introduce biomass into coal-based facilities.

Fluid bed gasifiers are markedly different from entrained flow designs. In a fluidized bed gasifier, gas streams from the bottom of the reactor through a bed of particulate material (typically sand) to form a turbulent mixture of gas and solids. Feed is added at a small enough rate to maintain a low concentration in the bed. Typically a fluidized gasifier operates in the range of 700°C–850°C. By injecting fuel in the base of the bed, much of the tar can be cracked within the fluidized bed. However, a large insulated space above the bed, known as the freeboard, is usually included to promote additional tar cracking as well as more complete conversion of char. However, further conditioning is required to protect downstream equipment. Fluidized beds are attractive for biomass gasification because of their feedstock flexibility and ability to scale.

4.2.3 Conversion of Biomass to Liquid Fuels

Almost 25% of energy consumption in the United States goes toward transportation. Approximately half of this amount comes from imported petroleum. Thus, development of transportation fuels from biorenewable resources (biofuels) is a priority if decreased dependence on foreign sources of energy is to be achieved.

There are two major platforms for biofuel production: biochemical and thermochemical. The biochemical platform employs microorganisms to convert primarily sugars into mostly alcohols, and it is best known for the corn- or sugarcane-to-ethanol process. The thermochemical platform relies on heat and catalysts (precious metals) to synthesize a wide range of compounds that include alcohols and hydrocarbons; this technology builds upon research from the coal, gas, and petroleum industries.

This section will discuss how to convert biomass into transportation fuels (biofuels) via the biochemical and thermochemical platforms. We will discuss the overall efficiencies, the various products generated, and overall process costs.

4.2.3.1 Corn Ethanol

Ethanol is the dominant biofuel with global production exceeding 50 billion liters. The majority of this ethanol comes from Brazilian sugarcane and U.S. corn grain. Microbes convert these sugar-rich crops into alcohols with relative ease and at low cost. Two major corn-to-ethanol facilities have been operated in the United States: dry-grind and wet-milling.

Dry-grind plants, shown in Figure 4.6, grind the whole kernel while wet-milling plants soak the grain with water and acid to separate the corn germ, fiber, gluten, and starch components before mechanical grinding [3]. The capital investment for dry-grind is less

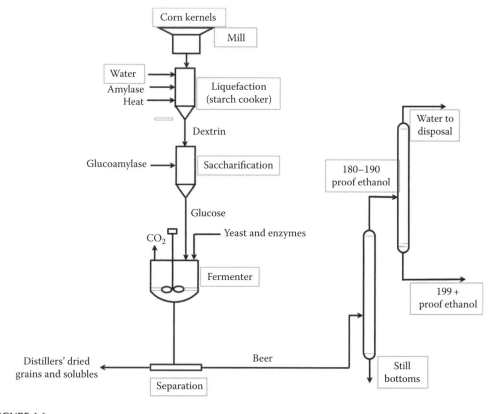

FIGURE 4.6
Dry-grind corn to ethanol.

TABLE 4.10

Ethanol Yields from Various Biorenewable Resources

Feedstock	Yield (L/Mg)
Apples	64
Barley	330
Cellulose	259
Corn	355–370
Grapes	63
Jerusalem artichoke	83
Molasses	280–288
Oats	265
Potatoes	96
Rice (rough)	332
Rye	329
Sorghum (sweet)	44–86
Sugar beets	88
Sugarcane	160–187
Sweet potatoes	125–143
Wheat	355

Source: Klass, D.L., *Biomass for Renewable Energy, Fuels, and Chemicals*, Academic Press, San Diego, CA, 1998e, p. 416. With permission.

than that for a comparably sized wet-milling plant. However, the higher value of its by-products, greater product flexibility, and simpler ethanol production can make a wet-milling plant a more profitable investment.

Dry-grind ethanol takes place in four major steps: pretreatment, cooking, fermentation, and distillation. Pretreatment consists of grinding the corn kernel into flour "meal," which is mixed with water, enzymes, and ammonia. This mixture ("mash") is then "cooked" to reduce bacteria levels. After cooling, the mash is sent to the fermenter where it remains for 40 h or more. The beer resulting from fermentation consists of a mixture of ethanol and stillage. Energy-intensive distillation of the beer is necessary to separate the stillage and water content from the ethanol and achieve maximum concentrations of 95%. This is followed by further purification to 100% ethanol using molecular sieves.

A modern dry-grind plant will produce over 2.7 gal of ethanol per bushel of corn processed. Yields of coproducts per bushel of corn are 7.7–8.2 kg (17–18 lb) of dry distillers grain with solubles and 7.3–7.7 kg (16–17 lb) of carbon dioxide evolved from fermentation, the latter of which can be sold to the carbonated beverage industry. As a rule of thumb, the three products are produced in approximately equal weight per bushel. Table 4.10 shows ethanol yields for various feedstock [63].

Costs for ethanol from corn grain have been developed by the USDA [28] for the dry mill process. Capital costs for a 40 MMGPY ethanol plant were estimated at $46.7 million with fuel production costs of $1.03 per gallon (2006).

4.2.3.2 Cellulosic Ethanol

Much of the carbohydrate in plant materials is structural polysaccharides, providing shape and strength to the plant. This structural material, known as lignocellulose, is a composite of cellulose fibers embedded in a cross-linked lignin–hemicellulose matrix [42].

Depolymerization to basic plant components is difficult because lignocellulose is resistant to both chemical and biological attacks [45]. However, depolymerization is necessary for microbes to efficiently convert cellulosic biomass into alcohols.

Cellulose to ethanol consists of four steps: pretreatment, enzymatic hydrolysis, fermentation, and distillation [3]. Of these, pretreatment is the most costly step, accounting for about 33% of the total processing costs [31]. An important goal of all pretreatments is to increase the surface area of lignocellulosic material, making the polysaccharides more susceptible to hydrolysis. Thus, comminution, or size reduction, is an integral part of all pretreatments.

Enzymatic hydrolysis was developed to better utilize both cellulose and hemicellulose from lignocellulosic materials. Three basic methods for hydrolyzing structural polysaccharides in plant cell walls to fermentable sugars are available: concentrated acid hydrolysis, dilute acid hydrolysis, and enzymatic hydrolysis [45,54]. The two acid processes hydrolyze both hemicellulose and cellulose with very little pretreatment beyond comminution of the lignocellulosic material to particles of about 1 mm size. The enzymatic process must be preceded by extensive pretreatment to separate the cellulose, hemicellulose, and lignin fractions.

Although thermodynamic efficiencies for conversion of carbohydrates to ethanol can be calculated, it is more typical to report the volumetric yield of ethanol per unit mass of feedstock. The yield of ethanol from energy crops varies considerably. Among sugar crops, sweet sorghum yields 80 L/ton, sugar beets yield 90–100 L/ton, and sugar cane yields 75 L/ton. Among starch and inulin crops, the ethanol yield is 350–400 L/ton of corn, 400 L/ton of wheat, and 90 L/ton of Jerusalem artichoke. Among lignocellulosic crops, the potential ethanol yield is 400 L/ton of hybrid poplar, 450 L/ton for corn stover, 510 L/ton for corn cobs, and 490 L/ton for wheat straw.

Researchers at the National Renewable Energy Laboratory (NREL) [1] developed a design report showing capital costs of $114 million and operating costs of $1.07 per gallon (2000) of ethanol for a 69 MMGPY ethanol plant.

Example 4.4: Cellulosic Ethanol Energy Return on Investment Calculations

There is growing interest in calculating the Energy Return on Investment (EROI) of biofuel technologies and ethanol in particular [18]. The process of estimating EROI can be summarized into three steps: identifying system inputs and outputs, calculating energy transfer quantities, and determining the EROI following a systematic approach.

1. System Identification: The cellulosic ethanol system involves inputs of fuel, electricity, fertilizer, and herbicides. The primary output is ethanol, but excess heat can be considered a coproduct. Energy inputs/costs can be distributed among coproducts using different methods.
2. Energy Transfer: The life cycle analysis literature includes reports of energy input/output in biofuel systems. Hall et al. reported the values shown in Table 4.11 for cellulosic ethanol production. These values are based on converting switchgrass to ethanol. This process employs a portion of the switchgrass feed to provide process heat and electricity.
3. EROI Calculations: In addition to ethanol, the biorefinery supporting Table 4.11 generates 4.79 MJ/L of excess electricity, which displaces three times the amount of energy in the form of fuel input (based on fuel-to-power efficiency of ≈33%). Thus, the authors concluded that cellulosic ethanol has an energy output of 21.2 MJ/L of ethanol plus 3*4.79 or 14.4 MJ/L for a total of 35.6 MJ/L. The net EROI is therefore 35.7 MJ/L of ethanol.

TABLE 4.11

Cellulosic Ethanol Energy Input/Output

Input (MJ/L Ethanol)	Value
Agriculture: fuel	0.19
Agriculture: electricity	0.00
Fertilizer	0.33
Pesticides/herbicides	0.10
Feedstock transport	0.29
Biorefinery: fuel	0
Biorefinery: electricity	0
Ethanol distribution	≈0.00
Biorefinery: coproducts	0
Total direct	0.91
Indirect	0.13
Total input	1.04

Source: Hall, C.A.S. et al., *Sustainability*, 3(12), 2413, 2011. With permission.

EROI calculations are subject to numerous assumptions and methodologies. Differences in energy input/output values and calculation methods explain some of the large differences in reported EROI estimates.

4.2.3.3 Biomass Fermentation to Alternative Fuels

Ethanol has several limitations as a transportation fuel, including its affinity for water, which prevents it from being fully compatible with the existing fuel infrastructure, and its low volumetric heating value, which is only two-thirds that of gasoline. For this reason, fermentations that produce metabolites other than ethanol have been proposed. Alternative fermentations could produce hydrophobic molecules that are less oxidized than ethanol including higher alcohols (most prominently butanol), fatty acids, fatty alcohols, esters, alkanes, alkenes, and isoprenes.

Alternative biochemical biofuels could address the energetic and fuel compatibility challenges faced by ethanol. Metabolic engineering, reactor design, and hybrid approaches are some of the approaches employed by researchers to identify and improve microbes' selectivity and performance for biofuel production.

4.2.3.4 Biomass to Fischer–Tropsch Liquids

Fischer–Tropsch liquids from biomass have antecedents in the coal-to-liquids industry. Germany extensively developed the Fischer–Tropsch process during World War II when it was denied access to petroleum-rich regions of the world. Likewise, when South Africa faced a world oil embargo during their era of apartheid, they employed Fischer–Tropsch technology to sustain its national economy. A comprehensive bibliography of Fischer–Tropsch literature can be found on the web [2].

Fischer–Tropsch catalysis produces a large variety of hydrocarbons including light hydrocarbon gases, paraffinic waxes, and alcohols according to the generalized reaction 4.27:

$$CO + 2H_2 \rightarrow -CH_2 - + H_2O \tag{4.27}$$

Fischer–Tropsch liquids composition depends on the process selectivity. Process selectivity is affected by various factors including catalyst and feed gas properties. The Anderson–Schulz–Flory (ASF) distribution (Equation 4.28) describes the probability of hydrocarbon chain growth where the molar yield for a carbon chain can be calculated using the following equation [46]:

$$C_n = \alpha^{n-1}(1-\alpha)$$ (4.28)

where α is the chain growth probability of a hydrocarbon of length n. Light hydrocarbons (mostly methane) can be fed into a gas turbine to provide power. Fischer–Tropsch liquids can be separated into various products in a process similar to petroleum distillation. Product distribution is a function of temperature, pressure, feed gas composition (H_2/CO), catalyst type, and composition [50]. Depending on the types and quantities of Fischer–Tropsch products desired, either low-(200°C–240°C) or high-temperature (300°C–350°C) synthesis at pressures ranging between 10 and 40 bar is used. For example, high gasoline yield can be achieved using high process temperatures and an iron catalyst. Fischer–Tropsch synthesis requires careful control of the H_2/CO ratio to satisfy the stoichiometry of the synthesis reactions as well as avoid deposition of carbon on the catalysts (coking). The optimal H_2/CO ratio for the production of naphtha and diesel range fuels sold in Western markets is 2:1.

Swanson et al. developed an analysis of Fischer–Tropsch liquid fuels from biomass [48]. Their estimates of a 2000 Mg per day corn stover facility found capital costs of $498–$606 million with minimum fuel selling prices of $4.27 and $4.83 per gallon of gasoline equivalent depending on whether the process was based on a fluid bed or entrained flow gasifier.

4.2.3.5 Biomass Pyrolysis Oil to Gasoline and Diesel

Pyrolysis is the thermal decomposition of organic compounds in the absence of oxygen [7]. The resulting product streams depend on the rate and duration of heating. Liquid yields exceeding 70% are possible under conditions of fast pyrolysis, which is characterized by rapid heating rates (up to 1000°C/s), moderate reactor temperatures (450°C–600°C), short vapor residence times (<0.5 s), and rapid cooling at the end of the process. Rapid cooling is essential if high-molecular-weight liquids are to be condensed rather than further decomposed to low-molecular-weight gases.

Pyrolysis liquid, also known as bio-oil, is a low-viscosity, dark-brown fluid with up to 15%–20% water, which contrasts with the black, tarry liquid resulting from slow pyrolysis [11]. Fast pyrolysis liquid is a complicated mixture of organic compounds arising from thermal degradation of carbohydrate and lignin polymers in biomass [39]. The liquid is highly oxygenated, approximating the elemental composition of the feedstock, which makes it highly unstable. The HHV of pyrolysis liquids ranges between 17 and 20 MJ/kg with liquid densities of about 1280 kg/m. Assuming conversion of 72% of the biomass feedstock to liquid on a weight basis, yields of pyrolysis oil are about 135 gal/ton.

Example 4.5: Energy Content in Pyrolysis Oils

Producing bio-oils with a high energy content are one of the main challenges for biomass pyrolysis. Bio-oil material yields can exceed 70 wt.% but suffer from high moisture content. In this example, we calculate the energy conversion efficiency for a typical pyrolysis experiment.

TABLE 4.12

Pyrolysis Mass and Energy Yield Values
for Problem Set

Component	Yield (wt.%)	Energy Density (MJ/kg)
Biomass	—	15.8
Pyrolysis oil	65	16.5
Pyrolysis gas	21	5.7
Pyrolysis char	14	27.5

Based on Table 4.12, we can estimate the biomass to bio-oil energy efficiency using Equation 4.29:

$$\text{Efficiency}_{Bio\text{-}oil} = \frac{0.65[kg/kg] * 16.5[MJ/kg]}{15.8[MJ/kg]} = 68.0\% \tag{4.29}$$

Production of pyrolysis oils and its coproducts involves several steps [56] that are illustrated in Figure 4.7. Lignocellulosic feedstock, such as wood or agricultural residues, is milled to a fine powder to promote rapid reaction. The particles are augured into the pyrolysis reactor where they are rapidly heated and converted into condensable vapors, liquid aerosols, noncondensable gases, and charcoal. These products are transported out of the reactor into a cyclone operating above the condensation point of pyrolysis vapors where the charcoal is removed. Vapors and gases are transported to a quench vessel where a spray of pyrolysis liquid cools vapors sufficiently for them to condense. The noncondensable gases, which include flammable carbon monoxide, hydrogen, and methane, are burned in air to provide heat for the pyrolysis reactor. The condensable liquids consist of a mixture of hundreds of organic compounds commonly known as bio-oil.

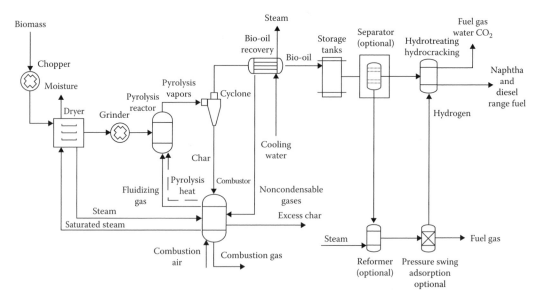

FIGURE 4.7

Biomass fast pyrolysis and bio-oil hydrotreating/hydrocracking to gasoline and diesel.

Bio-oil can be upgraded using conventional oil refinery processes. The most common of these processes are hydrotreatment and hydrocracking. Hydrotreatment can remove most of the bio-oil impurities such as nitrogen, alkali metals, and oxygen carried over from the original biomass. The purpose of hydrocracking is to break down heavy hydrocarbons with long carbon chain lengths into compounds within the naphtha and diesel range (7–20 carbon atoms). Bio-oil contains organic compounds with molecular weights in the hundreds. These compounds can be cracked into lighter hydrocarbons increasing the yield of naphtha and diesel range fuels.

Gasoline and diesel from corn stover fast pyrolysis followed by bio-oil upgrading could cost between $2.00 and $3.10/gal for a 2000 Mg/day biorefinery [56]. This facility would generate 35.4 million gallons per year.

4.2.3.6 Compressed Gases as Transportation Fuel

The ideal transportation fuel is a stable liquid at ambient temperature and pressure that can be readily vaporized and burned within an engine. However, some gaseous compounds are potential transportation fuels by increasing their density through compression. Among these gaseous transportation fuels are hydrogen, methane, ammonia, and DME.

Hydrogen can be manufactured from syngas via the water–gas shift reaction. This moderately exothermic reaction is best performed at relatively low temperatures in one or more stages with the aid of catalysts. Biomass to hydrogen processes face the same fuel delivery challenges as hydrogen from fossil sources in addition to the increased costs associated with using biomass.

Methane can be the main product of gasification under conditions known as hydrogasification [40]:

$$C + 2H_2 \rightarrow CH_4$$

$$CO + 3H_2 \rightarrow CH_4 + H_2O \tag{4.30}$$

Although methane is more easily pressurized or liquefied than hydrogen, its density is still too low to be an attractive transportation fuel except in some urban mass transit applications [53].

Ammonia is produced by the Haber process at 200 bar and 500°C [44]:

$$N_2 + 3H_2 \rightarrow 2NH_3 \tag{4.31}$$

As a widely employed agricultural fertilizer, the United States already has in place production, storage, and distribution infrastructure for its use.

DME, like liquefied petroleum gas, is a nontoxic, flammable gas at ambient conditions that is easily stored as a liquid under modest pressures [47]. It can be produced from syngas and can substitute for diesel after minor engine modifications.

4.2.3.7 Modern Concepts in Biofuel Conversion

There are a wide range of technologies under development that could have a dramatic impact on how we convert biomass to transportation fuels. These technologies include syngas to alcohols, lipid to fuels, hydrothermal processing (HTP), catalytic methylated furan synthesis, and compressed gases.

Syngas affords us with the possibility of generating many different types of alcohols. Efforts in Germany during World War II to develop alternative motor fuels discovered that iron-based catalysts could yield appreciable quantities of water-soluble alcohols from syngas, especially ethanol [24]. The overall conversion can be described by Equation 4.32:

$$CO + 3H_2 \rightarrow CH_3CH_2OH \tag{4.32}$$

These early efforts yielded liquids containing as much as 45%–60% alcohols of which 60%–70% was ethanol. Working at pressures of around 50 bar and temperatures in the range of 220°C–370°C, researchers have developed catalysts with selectivity to alcohols of over 95%, but production of pure ethanol has been elusive. Researchers have proposed several alternative approaches: direct carbonylation of methanol and syngas fermentation.

Methanol, which can be synthesized from syngas, is readily converted into ethanol with direct carbonylation (Equation 4.33). Direct carbonylation of methanol has the advantage of yielding ethanol without coproduct water, which would eliminate energy-intensive distillations. However, the cost-effectiveness of this approach to ethanol synthesis has not been proven.

$$CH_3OH + 2CO + H_2 \rightarrow CH_3CH_2OH + CO_2 \tag{4.33}$$

Methanol can not only be converted into ethanol, but it can also be converted to alcohols with a wide range of carbon chain lengths. The overall reaction (Equation 4.34) can be described by the conversion of CO and H into a mixture of alcohols and water as the main by-product.

$$nCO + 2nH_2 \rightarrow C_nH_{2n} + {}_1OH + (n-1)H_2O \tag{4.34}$$

where n ranges between 1 and 8 [16]. Methanol synthesis is favored at low temperatures and high pressures, and higher alcohols are produced as the temperature is increased. Process conditions for high-temperature, high-pressure synthesis catalysts range between 300 and 425 C, and 12.5 and 30 MPa using modified methanol catalysts.

NREL researchers suggest that ethanol and mixed-alcohol synthesis could be cost competitive with corn ethanol by 2012 [38]. Their estimates show capital costs of $137 million and operating costs of $1.01 per gallon for a 72.6 MMGPY mixed-alcohol biomass plant.

Syngas fermentation is a hybrid thermochemical and biochemical biofuel synthesis approach. In this process, microorganisms consume the CO, CO_2, and H_2 in syngas to form a variety of products including carboxylic acids, alcohols, and esters [9]. The advantages of this approach include ambient temperature and pressure operating conditions, and improved resistance to contaminants such as sulfur. The disadvantages include product inhibition and mass transfer limitations.

Lipids are a large group of hydrophobic, fat-soluble compounds produced by plants and animals for high-density energy storage. Triglycerides of fatty acids, commonly known as fats and oils depending upon their melting points, are among the most familiar form of lipids and have been widely used in recent years for the production of diesel fuel substitutes. The solution to this problem is to convert the triglycerides into methyl esters or ethyl esters of the fatty acids, known as biodiesel, and the by-product 1,2,3-propanetriol (glycerol).

A wide variety of plant species produce triglycerides in commercially significant quantities, most of it occurring in seeds [28]. Average oil yields range from 15,000 L/km² for cottonseed to 81,400 L/km² for peanut oil although intensive cultivation might double these numbers. Soybeans are responsible for more than 50% of world production of oilseed, representing 48–82 million bbl/year. The average oil yield for soybeans is 38,300 L/km² [28]. A higher yielding crop is oil palm, already grown in plantations for vegetable oil production [36].

Although oil palm yields are 10 times higher than soybeans, some environmentalists are concerned that its cultivation for fuel production will encourage rainforest destruction. However, several oil seed crops have been identified that could be grown on waste land or even saline soils, which reduces concerns about competition for food crops and rainforest destruction. These alternative oil seed crops include jatropha, Chinese tallow tree, and salicornia [17,26].

Triglycerides can be converted into transportation fuels via transesterification [25]. The process, described in Figure 4.8, has been commercialized for the production of biodiesel. Although biodiesel can substitute for diesel fuel, it has some shortcomings. Fatty acid methyl esters found in biodiesel are subject to microbial or oxidative attack, making them unsuitable in applications requiring long-term fuel storage. Low-temperature performance of biodiesel is sometimes problematic.

For these reasons, hydrogenation is being evaluated as a replacement for transesterification in the production of lipid-based biofuels [25]. Hydrogenation includes a number of reactions: large molecules are broken into smaller molecules; carbon–carbon double bonds are converted into more stable single bonds; molecular structures are rearranged; and undesirable atoms such as sulfur, nitrogen, and oxygen are removed from the hydrogenated compounds. In the case of lipids, hydrogenation yields alkanes, which are highly desirable fuel molecules.

Finally, lipid-rich microalgae might be grown in brackish water or even in seawater [21]. Algae can produce as much as 60% of their body weight as lipids when deprived of key

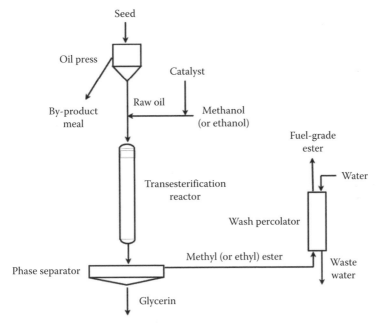

FIGURE 4.8
Conversion of triglycerides to methyl (or ethyl) esters and glycerol (biodiesel synthesis).

TABLE 4.13

Capital and Operating Costs of Biomass to Transportation Fuel Pathways

	Grain Ethanol	Cellulosic Ethanol	Butanol	Methanol	Fischer–Tropsch Liquids	Gasoline from Bio-oil
Publication Date	2006	2005	2000	2002	2010	2010
Plant size (MM L/year)	151	190	136	330	158	134
Capital cost (MM $/year)	46.7	294	110	224	606	$287
Operating and feedstock cost (MM $/year)	41.3	76.0	68.4	60.6	145	$123
Fuel cost ($/L)	$0.27	$0.40	$0.50	$0.18	$1.13	$0.82
Fuel cost ($/lge[a])	$0.44	$0.65	$0.55	$1.13	$0.34	$0.82

[a] Liters of gasoline equivalent.

nutrients such as silicon for diatoms or nitrogen for green algae. They employ relatively low substrate concentrations, on the order of 10–40 g/L. However, algae require the proper combination of brackish water, CO_2, and sunlight, which could limit the number of sites where this process would be profitable. Berkeley researchers have estimated that biofuels from algae would require break-even oil prices of $332/barrel [30].

Researchers have pursued hydrothermal biomass processing as an alternative to producing a variety of biofuels [37]. HTP describes the thermal treatment of wet biomass to produce carbohydrate, liquid hydrocarbons, or gaseous products depending upon the reaction conditions [13,14]. HTP liquids are commonly referred as bio-crude. Bio-crude contains a wide range of organic compounds including hydrocarbons. Unlike pyrolysis oil, bio-crude contains a much smaller amount of oxygenated organic compounds. Unfortunately, HTP requires severe operating conditions: temperatures of 200°C–600°C and pressures of 5–40 MPa to prevent water from boiling. The metallurgy required for this process has so far limited adoption of this technology.

Destructive distillation of wood is one of several routes to produce methylated furans [34]. Methylated furans have heating values and octane numbers comparable to gasoline making them potential transportation fuel [10]. 2,5-Dimethyl furan in particular has received recent interest because new catalytic synthesis routes from sugars have been developed [41,57]. Furfural can be methylated by reaction with methanol over zeolite catalyst to yield methyl furan, dimethyl furan, trimethyl furan, and tetramethyl furan [10,12].

Table 4.13 summarizes the costs for common biofuel conversion processes. These costs are on different basis, but they can be easily compared by scaling the plant capacity and adjusting for inflation.

4.2.4 Conversion of Biomass to Electricity

We can convert the chemical energy in biomass into electric power by a number of different routes. Direct combustion of biomass releases heat that can be used in Stirling engines or Rankine steam power cycles. Fast pyrolysis and thermal gasification, described in earlier sections, yield bio-oil and syngas, respectively, which are suitable for firing in gas turbines or even fuel cells.

4.2.4.1 Direct Combustion

Combustion is the rapid oxidation of fuel to obtain energy in the form of heat. Since biomass fuels are primarily composed of carbon, hydrogen, and oxygen, the main oxidation products are carbon dioxide and water although fuel-bound nitrogen and sulfur can be significant sources of sulfur oxide and nitrogen oxide emissions. Depending on the heating value and moisture content of the fuel, the amount of air used to burn the fuel, and the construction of the furnace, flame temperatures can exceed 1650°C.

Solid-fuel combustion consists of four steps, illustrated in Figure 4.9: heating and drying, pyrolysis, flaming combustion, and char combustion [51]. Heating and drying of the fuel particle is normally not accompanied by chemical reaction. Water is driven from the fuel particle as the thermal front advances into the interior of the particle. As long as water remains, the temperature of the particle does not raise high enough to initiate pyrolysis, the second step in solid-fuel combustion.

Pyrolysis is a complex series of thermally driven chemical reactions that decompose organic compounds in the fuel [27]. Pyrolysis proceeds at relatively low temperatures, which depend on the type of plant material. Hemicellulose begins to pyrolyze at temperatures between 150°C and 300°C, cellulose pyrolyzes at 275°C–350°C, and lignin pyrolysis is initiated between 250°C and 500°C.

The resulting decomposition yields a large variety of volatile organic and inorganic compounds, the types and amounts dependent on the fuel and the heating rate of the fuel. Pyrolysis products include carbon monoxide (CO), carbon dioxide (CO_2), methane (CH_4), and high-molecular-weight compounds that condense to a tarry liquid if cooled before they are able to burn. Fine droplets of these condensable compounds represent much of the smoke associated with smoldering fires. Pyrolysis follows the thermal front through the particle, releasing volatile compounds and leaving behind pores that penetrate to the surface of the particle.

Both the volatile gases and the char resulting from pyrolysis can be oxidized if sufficient oxygen is available to them. Oxidation of the volatile gases above the solid fuel results in flaming combustion. The ultimate products of volatile combustion are CO_2 and H_2O although a variety of intermediate chemical compounds can exist in the flame, including CO, condensable organic compounds, and long chains of carbon known as soot. Indeed, hot glowing soot is responsible for the familiar orange color of wood fires.

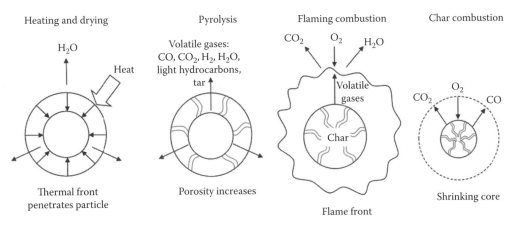

FIGURE 4.9
Processes of solid-fuel combustion.

Combustion intermediates will be consumed in the flame if sufficient temperature, turbulence, and time are allowed. High combustion temperature assures that chemical reactions will proceed at high rates. Turbulent or vigorous mixing of air with the fuel makes certain that every fuel molecule comes into contact with oxygen molecules. Long residence times for fuel in a combustor allow the fuel to be completely consumed. In the absence of good combustion conditions, a variety of noxious organic compounds can survive the combustion process including CO, soot, polycyclic aromatic hydrocarbons, and the particularly toxic families of chlorinated hydrocarbons known as furans and dioxins. In some cases, a poorly operated combustor can produce pollutants from relatively benign fuel molecules.

Both CO and CO_2 can form at or near the surface of burning char [51]:

$$C + \frac{1}{2}O_2 \rightarrow CO$$

$$CO + \frac{1}{2}O_2 \rightarrow CO_2 \tag{4.35}$$

These gases escape the immediate vicinity of the char particle where CO is oxidized to CO_2 if sufficient oxygen and temperature are available; otherwise, it appears in the flue gas as a pollutant.

The next step in the combustion of solid fuels is solid–gas reactions of char, also known as glowing combustion, familiar as red-hot embers in a fire. Char is primarily carbon with a small amount of mineral matter interspersed. Char oxidation is controlled by mass transfer of oxygen to the char surface rather than by chemical kinetics, which is very fast at the elevated temperatures of combustion. Depending on the porosity and reactivity of the char and the combustion temperature, oxygen may react with char at the surface of the particle or it may penetrate into the pores before oxidizing char inside the particle. The former situation results in a steadily shrinking core of char whereas the latter situation produces a constant-diameter particle of increasing porosity.

4.2.4.2 Combustion Equipment

A combustor is a device that converts the chemical energy of fuels into high-temperature exhaust gases. Heat from the high-temperature gases can be employed in a variety of applications, including space heating, drying, and power generation. However, with the exception of kilns used by the cement industry, most solid-fuel combustors today are designed to produce either low-pressure steam for process heat or high-pressure steam for power generation. Combustors integrated with steam-raising equipment are called boilers. In some boiler designs, distinct sections exist for combustion, high-temperature heat transfer, and moderate-temperature heat transfer: these are called the furnace, radiative, and convective sections of the boiler, respectively. In other designs, no clear separation between the processes of combustion and heat transfer exists.

Solid-fuel combustors, illustrated in Figure 4.10, can generally be categorized as grate-fired systems, suspension burners, or fluidized beds [4]. Grate-fired systems were the first burner systems to be developed, evolving during the late nineteenth century and early twentieth century into a variety of automated systems. The most common system is the

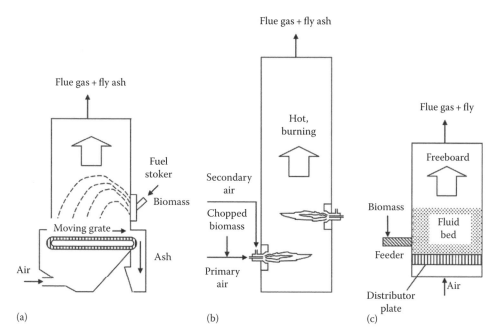

FIGURE 4.10
Common types of combustors: (a) grate-fired, (b) suspension, and (c) fluidized.

spreader-stoker, consisting of a fuel feeder that mechanically or pneumatically flings fuel onto a moving grate where the fuel burns. Much of the ash falls off the end of the moving grate although some fly ash appears in the flue gas. Grate systems rarely achieve combustion efficiencies exceeding 90%.

Suspension burners suspend the fuel as fine powder in a stream of vertically rising air. The fuel burns in a fireball and radiates heat to tubes that contain water to be converted into steam. Suspension burners, also known as pulverized coal boilers, have dominated the U.S. power industry since World War II because of their high volumetric heat release rates and their ability to achieve combustion efficiencies, often exceeding 99%. However, they are not well suited to burning coarse particles of biomass fuel, and they are notorious generators of nitrogen oxides. Biomass is fed from a bunker through pulverizers designed to reduce fuel particle size enough to burn in suspension. The fuel particles are suspended in the primary airflow and fed to the furnace section of the boiler through burner ports where it burns as a rising fireball. Secondary air injected into the boiler helps complete the combustion process. Heat is absorbed by steam tubes arrayed in banks of heat exchangers (waterwall, superheaters, and economizer) before exiting through a bag house designed to capture ash released from the fuel. Steam produced in the boiler is part of a Rankine power cycle.

Fluidized bed combustors are a recent innovation in boiler design. Air injected into the bottom of the boiler suspends a bed of sand or other granular refractory material producing a turbulent mixture of air and sand. The high rates of heat and mass transfer in this environment are ideal for efficiently burning a variety of fuels. Furthermore, the large thermal mass of the sand bed allows the unit to be operated as low as 850°C, which lowers the emission of nitrogen compounds. A commercial market for fluidized bed boilers developed during the 1980s, especially for industrial applications.

4.2.4.3 Biomass Cofiring

As an alternative to completely replacing coal with biomass fuel in a boiler, mixtures of biomass and coal can be burned together in a process known as cofiring [43]. Cofiring offers several advantages for industrial boilers. Industries that generate large quantities of biomass wastes, such as lumber mills or pulp and paper companies, can use cofiring as an alternative to costly landfilling of wastes.

The best wood-fired power plants, which are typically 20–100 MW$_e$ in capacity, have heat rates exceeding 12,500 Btu/kWh. In contrast, large, coal-fired power plants have heat rates of only 10,250 Btu/kWh. The relatively low thermodynamic efficiency of steam power plants at the sizes of relevance to biomass power systems may ultimately limit the use of direct combustion to convert biomass fuels to useful energy.

4.2.5 Fossil and Biomass Fuel Properties

Fossil-based transportation fuels include gasoline, diesel, and jet fuel. Gasoline is intended for spark-ignition (Otto cycle) engines; thus, it is relatively volatile but resistant to autoignition during compression. Diesel fuel is intended for use in compression-ignition (diesel cycle) engines; thus, it is less volatile compared to gasoline and more susceptible to autoignition during compression. Jet fuel is designed for use in gas turbine (Brayton cycle) engines, which are not limited by autoignition characteristics but otherwise have very strict fuel specifications for reasons of safety and engine durability (aviation for example). Biomass can be converted into gasoline, diesel, and jet fuel as well as alternative fuels such as alcohols, DME, and Fischer–Tropsch liquids.

Combinations of biochemical and thermochemical technologies could generate biofuels that meet or even exceed the combustion properties of conventional fuels. Table 4.14 [5,29,32,49] shows key properties of various fossil- and biomass-derived transportation fuels. These properties help identify biofuel substitutes for conventional fuels.

Transportation fuels are characterized by several properties based on broadly recognized standards and their combustion behavior. Specific gravity is a measure of the fuel's density. Kinematic viscosity describes the fuel's ability to flow at a given temperature—a high fuel viscosity could be detrimental to fuel delivery to an engine. Boiling point range and flash point temperature are important because they are key factors during engine start-up. Flash point is the lowest temperature at which enough fuel vaporizes and mixes with air to form a combustible gas. Below this temperature, the fuel would fail to ignite even when exposed to a spark. Flash point differs slightly from the autoignition temperature: autoignition occurs when the fuel combusts without a spark or ignition source.

The octane number is an important transportation fuel figure of merit. The octane number indicates the tendency of a fuel to undergo premature detonation within the combustion cylinder of an internal combustion engine. The higher the octane number, the less likely a fuel will detonate until exposed to an ignition source (electrical spark). Premature denotation is responsible for the phenomenon known as engine knock, which reduces fuel economy and can damage an engine. Various systems of octane rating have been developed, including research octane and motor octane numbers. Federal regulation in the United States requires gasoline sold commercially to be rated using an average of the research and motor octane numbers. Gasoline rated as "regular" has a commercial octane number of about 87 while premium grade is 93.

TABLE 4.14

Comparison of Ignition and Combustion Properties of Transportation Fuels

	Fossil Fuel-Derived				Biomass-Derived					
Fuel Type	Gasoline	No. 2 Diesel Fuel	Methanol	Ethanol	Methyl Ester (from Soybean Oil)	Fischer–Tropsch A	Hydrogen	Methane	Dimethyl Ether
Specific gravity[a]	0.72–0.78	0.85	0.796	0.794	0.886	0.770	0.071 (liq)	0.422 (liq)	0.660
Kinematic viscosity at 20°C–25°C (mm²/s)	0.8	2.5	0.75	1.51	3.9	2.08	105 [61]	16.5 [61]	0.227
Boiling point range (°C)	30–225	210–235	65	78	339	164–352	−253	−162	−24.9
Flash point (°C)	−43	52	11	13	188	58.5	−184	—	—
Autoignition temperature (°C)	370	254	464	423	—	—	566–582	540	235
Octane no. (research)	91–100	—	109	109	—	—	>130	>120	—
Octane no. (motor)	82–92	—	89	90	—	—	—	—	—
Cetane no.	<15	37–56	<15	<15	55	74.6	—	—	>55
Heat of vaporization (kJ/kg)	380	375	1185	920	—	—	447	509	402 [62]
Lower heating value (MJ/kg)	43.5	45	20.1	27	37	43.9	120	49.5	28.88

[a] Measured at 16°C except for liquefied gases, which are saturated liquids at their respective boiling points.

These properties specify the potential uses for any type of transportation fuels. Therefore, biofuel development requires careful consideration of these properties. An extended discussion of this topic is beyond the scope of this chapter, but readers are encouraged to consult engine and combustion texts on this matter.

4.3 Conclusions

Economic and political concerns have renewed interest in biomass development as a domestic source of clean and renewable fuels and chemicals. Initial efforts have unfortunately competed with food crops for low-cost sugars. A sustainable biomass industry will have to provide food to the world's growing population before supporting future energy needs. Thus, researchers are investigating how to convert crop residues and dedicated energy crops into valuable fuels and chemicals. These feedstocks could provide a bridge from first- to second-generation biofuels and beyond.

First-generation biofuels based on grain ethanol and soy diesel will continue to be important sources of alternative transportation fuels for several years because significant investment has been made in their infrastructure and continuing political support of government incentives. However, significant expansion of the biofuel industry will require alternatives to corn and soybean crops as feedstocks. These possibilities include both lignocellulose and lipids from alternative crops.

Each of the biofuel and bioenergy options considered in this chapter has its advantages and disadvantages in terms of market, opportunities, environmental benefits, and process economics. Accordingly, no single biofuel is likely to dominate transportation in the same manner as currently exists for petroleum-based gasoline and diesel. Niche markets and regional considerations will heavily influence the best choice of fuel for specific applications. Finally, increased research efforts could raise new possibilities that change the outlook for biofuels.

Although biochemical processing dominates current-generation biofuels, thermochemical processing is likely to play an important role in advanced biofuels because of its technical maturity and potential to efficiently produce hydrocarbon fuels. However, it is too early to designate winners and losers among the many biofuel options. What is certain is that biofuels will evolve as a result of several forces including new technologies, market demand, environmental concerns, and government policies.

Problems

4.1 Wisconsin produced 13,260,000 bushels (25.4 kg per bushel) of oats on 195,000 ac (4046.85 m^2 per acre) of land. What is the corresponding yield of oat residue in tonnes per km^2 assuming a residue to grain ratio of 2:1?

4.2 Use the USDA Quick Stats service (http://quickstats.nass.usda.gov//) to compare the 2009 oat yields between New York and California. What additional land would California require to meet the same amount of oat production as New York?

4.3 We can estimate future crop yields using the exponential growth formula

$$\text{Yield }(t) = \text{Yield}_0 e^{kt} \tag{4.36}$$

where k is the growth rate and t is the period of time since the initial value Yield_0. Collect the required data from Quick Stats to calculate the rate at which corn yields have increased in Minnesota between 1990 and 2009. Given this growth rate, what is the predicted 2020 corn yield for Minnesota?

4.4 The U.S. total land area is almost 9,307,800 km². Thirty percent of the land is forestland. How much above-ground biomass is available in the U.S. forestland assuming a 15,240 Mg/km² biomass density? How much land would we need to grow an equivalent amount of biomass with fast-growing trees (hybrid poplar), or energy crops (switchgrass)?

4.5 A common assumption for biorefinery capacities is 2000 Mg/day of feedstock input capacity. What is the land area required to grow this quantity of biomass from corn stover with land availability (f) of 40%? How does this compare to switchgrass?

4.6 Given the enthalpy of formation for food waste to be −394.7 MJ/kmol, use the combustion reaction from Equation 4.11 to calculate its HHV (assume that the molecular formula is the same as switchgrass). Repeat this calculation for redwood (assume an enthalpy of formation of −50.7 MJ/kmol).

4.7 Solar irradiation in parts of California exceeds 9 kWh/m²/day (32 million MJ/km²/day). What are theoretical maximum C3 and C4 biomass energy yields assuming a 150 day crop growth period? In Florida, solar irradiation is about 3.5 kWh/m²/day; what would the theoretical maximum yield be if the crop had a maturity period of 180 days?

4.8 How many people can Hawaii feed? Different diets require different energy inputs. Since meat is higher in the food chain than grains, the more calories you get from meat, the more agricultural land you use. About 1 GJ metabolizable energy is produced annually/hectare for meat and 30 GJ metabolizable energy produced annually/hectare for grain.

Hawaii has about 1.3 million acres of zoned agricultural lands and forests. In that, 675,000 ac are designated as prime agricultural lands of importance to the State of Hawaii of which less than 200,000 ac are under cultivation [58].

Assuming all 675,000 ac of prime agricultural land is used for food production, and the average daily energy requirement for a human is about 2000 kilocalories/day (kcal/day), answer the following questions:

- How many vegetarians could Hawaii feed? Determine this as an absolute number, percent of 2010 population, and percent of estimated 2030 population.
- How many typical Americans could Hawaii feed? (The typical American gets two-third of his/her calories from meat and one-third from grain)? Again, determine this as an absolute number, percent of 2010 population, and percent of estimated 2030 population.

Clearly state all assumptions made and sources of information used. Show your calculations. Take care to convert units as needed (acre-to-hectare and kcal/day-to-GJ for example). Note that 1 food Calorie (with a capital "C") = 1 kcal. Here is an excellent unit conversion website: http://www.onlineconversion.com/

4.9 A gasifier can convert 1 kg of wood (19.5 MJ/kg) into 1.5 m³ of syngas (5.2 MJ/m³). If we define energy conversion efficiency as the ratio of syngas energy to biomass energy, what is the energy conversion efficiency of wood to syngas?

4.10 A lactating cow can produce more than 60 kg of manure (about 7 MJ/kg on a dry basis) per day. What is the energy efficiency of converting cow manure to biogas if it yields 25 m³ per Mg with the methane content of 55% (35 MJ/m³)?

4.11 Gasification uses an oxygen-to-fuel ratio that is 25% of the required oxygen for complete combustion. What is the amount of oxygen required to gasify 1 kg of wood ($CH_{1.4}O_{0.66}$)?

4.12 Steam gasification employs steam as the oxygen carrier. How much steam would be needed to gasify 1 kg of wood ($CH_{1.4}O_{0.66}$)?

4.13 How much air is needed to fully combust fuel oil ($CH_{1.67}O_{0.006}N_{0.005}S_{0.002}$) if 25% excess air is required?

4.14 Calculate the energy conversion efficiency of biomass to gasification products assuming the yields and energy contents from Table 4.8.

4.15 In 2005, California had almost 440 MW of coal power-generating capacity. What amount of biomass would be required to produce the same amount of electricity? Assume a biomass heating value of 17 MJ/kg, a biomass-to-syngas thermal efficiency of 85%, and a syngas-to-electricity efficiency of 38%.

4.16 Assuming a chain growth probability of 0.8, calculate the mole yields for hydrocarbons with 1–24 carbon atoms using the ASF distribution. How does this distribution compare to that of a 0.9 chain growth probability.

4.17 Calculate the heat of reaction for ethanol (heat of formation −277.7 MJ/kmol) synthesis with the mixed-alcohol process and with the methanol (heat of formation −238.4 MJ/kmol) conversion process.

4.18 Assuming nominal annual yields of 700,000, 5,500,000, and 3,800,000 kg/km² for corn grain, sugar cane, and sweet sorghum respectively, which conversion pathway is the most productive use of land? Use conversion yield values given in ethanol section and units of L/km².

4.19 A group proposes a 2000 barrel-per-day ethanol output biorefinery using Jerusalem artichokes. What is the daily feedstock requirement for this biorefinery? What is the minimum amount of land required to grow the feedstock ($f = 100\%$)?

4.20 If the yield of bio-oil to liquid fuels (gasoline- and diesel-like fuels) is 42 wt.%, what is the overall energy efficiency? Use the values in Table 4.12 for energy content and product yields. First, calculate the biomass to liquid fuel yield based on the biomass to bio-oil and bio-oil to liquid fuel yields. Finally, assume a liquid fuel energy density of 41.2 MJ/kg and calculate the energy efficiency.

4.21 How many liters of bio-oil can be generated with Iowa's 2009 corn stover yield? Use appropriate data from Quick Stats, a 65 wt.% yield of bio-oil from corn stover, and a bio-oil density of 1201.9 kg/m³.

4.22 Using an ethanol fuel content of 21.2 MJ/L and TTW efficiency of 3.08 MJ per km, calculate the theoretical travel distance (units of km per km² of land) for an E-85 vehicle powered by corn grain ethanol. How does this number compare to sugar cane ethanol and cellulosic ethanol? Employ appropriate assumptions for crop yields and process conversion efficiency.

4.23 Assuming a 10% improvement in the production of H_2 from coal, what is the new WTW efficiency for a H_2 ICE and what is the reduction in CO_2 emissions per kilometer?

References

1. Aden, A., Ruth, M., Ibsen, K., Jechura, J., Neeves, K., Sheehan, J., Wallace, B., Montague, L., and Slayton, A. (2002) Lignocellulosic biomass to ethanol process design and economics utilizing co-current dilute acid prehydrolysis and enzymatic hydrolysis for corn stover, Technical Report NREL//TP-510-32438, National Renewable Energy Laboratory.
2. Anon. (2005) Fischer–Tropsch archive, URL http://www.fischer-tropsch.org/
3. Bailey, B.K. (1996) Performance of ethanol as a transportation fuel, in: Charles Wyman, ed., *Handbook on Bioethanol*, Vol. 1, pp. 37–58.
4. Bain, R.L., Overend, R.P., and Craig, K.R. (1998) Biomass-fired power generation, *Fuel Processing Technology* 54(1–3):1–16.
5. Borman, G.L. and Ragland, K.W. (1998) *Combustion Engineering*, McGraw-Hill Company, New York.
6. Bridgwater, A.V. (1995) The technical and economic feasibility of biomass gasification for power generation, *Fuel* 74(5):631–653.
7. Bridgwater, A.V. and Peacocke, G.V.C. (2000) Fast pyrolysis processes for biomass, *Renewable and Sustainable Energy Reviews* 4(1):1–73.
8. Brown, R.C. (2003) *Biorenewable Resources: Engineering New Products from Agriculture*, Iowa State Press, A Blackwell Publishing Company, Ames, IA, pp. 59–75.
9. Brown, R.C. (2007) Hybrid thermochemical/biological processing, *Applied Biochemistry and Biotechnology* 137(1):947–956.
10. Carlson, T.R., Vispute, T.P., and Huber, G.W. (2008) Green gasoline by catalytic fast pyrolysis of solid biomass derived compounds, *ChemSusChem* 1(5):397.
11. Czernik, S. and Bridgwater, A.V. (2004) Overview of applications of biomass fast pyrolysis oil, *Energy and Fuels* 18(2):590–598.
12. Diebold, J.P. and Evans, R.J. (1988). U.S. Patent No. 4,764,627. Washington, DC: U.S. Patent and Trademark Office.
13. Elliott, D.C., Beckman, D., Bridgwater, A.V., Diebold, J.P., Gevert, S.B., and Solantausta, Y. (1991) Developments in direct thermochemical liquefaction of biomass: 1983–1990, *Energy & Fuels* 5(3):399–410.
14. Elliott, D.C., Neuenschwander, G.G., Hart, T.R., Butner, R.S., Zacher, A.H., Engelhard, M.H., Young, J.S., and McCready, D.E. (2004) Chemical processing in high-pressure aqueous environments. 7. Process development for catalytic gasification of wet biomass feedstocks, *Industrial and Engineering Chemistry Research* 43(9):1999–2004.
15. Field, C.B., Campbell, J.E., and Lobell, D.B. (2008) Biomass energy: The scale of the potential resource, *Trends in Ecology & Evolution* 23(2):65–72.
16. Forzatti, P., Tronconi, E., and Pasquon, I. (1991) Higher alcohol synthesis, *Catalysis Reviews*, 33(1):109–168.
17. Glenn, E.P., O'Leary, J.W., Watson, M.C., Thompson, T.L., and Kuehl, R.O. (1991) *Salicornia bigelovii* torr.: An oilseed halophyte for seawater irrigation, *Science* 251(4997):1065–1067.
18. Hall, C.A.S., Dale, B.E., and Pimentel, D. (2011) Seeking to understand the reasons for different energy return on investment (eroi) estimates for biofuels, *Sustainability* 3(12):2413–2432.
19. Henrich, E. and Weirich, F. (2004) Pressurized entrained flow gasifiers for biomass, *Environmental Engineering Science* 21(1):53–64.
20. Jenkins, B.M., Baxter, L.L., and Miles, T.R. (1998) Combustion properties of biomass, *Fuel Processing Technology* 54(1–3):17–46.
21. Klass, D.L. (1998a) *Biomass for Renewable Energy, Fuels, and Chemicals*, Academic Press, San Diego, CA, pp. 341–344.
22. Klass, D.L. (1998b) *Biomass for Renewable Energy, Fuels, and Chemicals*, Academic Press, New York, pp. 72–90.
23. Klass, D.L. (1998c) *Biomass for Renewable Energy, Fuels, and Chemicals*, Academic Press, San Diego, CA, p. 356.

24. Klass, D.L. (1998d) *Biomass for Renewable Energy, Fuels, and Chemicals*, Academic Press, San Diego, CA, pp. 274–276, (Online service).
25. Kram, J.W. (2013) Aviation Alternatives, *Biodiesel Magazine*, Grand Forks, ND58203, viewed 7 April, 2013 https://www.biodieselmagazine.com/articles/3071/aviation-alternatives
26. Kumar, A. and Sharma, S. (2008) An evaluation of multipurpose oil seed crop for industrial uses (*Jatropha curcas* l.): A review, *Industrial Crops & Products* 28(1):1–10.
27. Kumar, J.V. and Pratt, B.C. (1996) Compositional analysis of some renewable biofuels, *American Laboratory* 28(8):15–20.
28. Kwiatkowski, J.R., McAloon, A.J., Taylor, F., and Johnston, D.B. (2006) Modeling the process and costs of fuel ethanol production by the corn dry-grind process, *Industrial Crops & Products* 23(3):288–296.
29. Laboratory National Renewable Energy Laboratory. Alternative fuels, general table of fuel properties. URL http://www.eere.energy.gov/afdc/altfuel/fuelsdo5(p)roperties.html.
30. Lundquist, T.J., Woertz, I.C., Quinn, N.W.T., and Benemann, J.R. (2010) A realistic technology and engineering assessment of algae biofuel production, in: *Energy Biosciences Institute*, Berkeley, CA, p. 1.
31. Lynd, L.R. (1996) Overview and evaluation of fuel ethanol from cellulosic biomass: Technology, economics, the environment and policy, *Annual Reviews in Energy and the Environment* 21(1):403–465.
32. MacKenzie, J.J. and Avery, W.H. (1996) Ammonia fuel: The key to hydrogen-based transportation, in: *Energy Conversion Engineering Conference, IECEC 96*, Vol. 3.
33. Moran, M.J., Shapiro, H.N., Boettner, D.D., and Bailey, M. (2010) *Fundamentals of Engineering Thermodynamics*, Wiley, New York.
34. Moreau, C., Naceur Belgacem, M., and Gandini, A. (2004) Recent catalytic advances in the chemistry of substituted furans from carbohydrates and in the ensuing polymers, *Topics in Catalysis* 27(1–4):11–30.
35. Morris, G. (2003) The status of biomass power generation in California, July 31, 2003, Technical Report NREL/SR-510-35114, National Renewable Energy Laboratory.
36. Openshaw, K. (2000) A review of *Jatropha curcas*: An oil plant of unfulfilled promise, *Biomass and Bioenergy* 19(1):1–15.
37. Peterson, A.A., Vogel, F., Lachance, R.P., Fröling, M., Antal Jr., M.J., and Tester, J.W. (2008) Thermochemical biofuel production in hydrothermal media: A review of sub-and supercritical water technologies, *Energy & Environmental Science* 1(1):32–65.
38. Phillips, S., Aden, A., Jechura, J., Dayton, D., and Eggeman, T. (2007) Thermochemical ethanol via indirect gasification and mixed alcohol synthesis of lignocellulosic biomass, Technical Report NREL//TP-510-41168, National Renewable Energy Laboratory.
39. Piskorz, J. and Scott, D.S. (1987) The composition of oils obtained by the fast pyrolysis of different woods, *Preprint Paper, American Chemical Society, Division of Fuel Chemistry*, 32(2).
40. Reed, T.B. (1981) *Biomass Gasification: Principles and Technology*, Noyes Data Corp., Park Ridge, NJ.
41. Roman-Leshkov, Y., Barrett, C.J., Liu, Z.Y., and Dumesic, J.A. (2007) Production of dimethylfuran for liquid fuels from biomass-derived carbohydrates, *Nature* 447:982–985.
42. Rowell, R.M., Young, R.A., and Rowell, J.K. (1997) *Paper and Composites from Agro-Based Resources*, CRC Press, Boca Raton, FL.
43. Sami, M., Annamalai, K., and Wooldridge, M. (2001) Cofiring of coal and biomass fuel blends, *Progress in Energy and Combustion Science* 27(2):171–214.
44. Satterfield, C.N. (1991) *Heterogeneous Catalysis in Industrial Practice*, McGraw-Hill Book Co., New York.
45. Schell, D.J., McMillian, J.D., and Philippidis, G.P. (1992) Ethanol from lignocellulosic biomass, *Advances in Solar Energy* 7:373–448.
46. Schulz, H. (1999) Short history and present trends of fischer–tropsch synthesis, *Applied Catalysis A, General* 186(1–2):3–12.
47. Sorenson, S.C. (2001) Dimethyl ether in diesel engines: Progress and perspectives, *Journal of Engineering for Gas Turbines and Power* 123:652.

48. Swanson, R.M., Platon, A., Satrio, J.A., and Brown, R.C. (2010) Techno-economic analysis of biomass-to-liquids production based on gasification, *Fuel* 89:S11–S19.

49. Teng, H., McCandless, J.C., and Schneyer, J.B. (2001) Thermochemical characteristics of dimethyl ether-an alternative fuel for compression-ignition engines [r], SAE technical paper, 2001-01-0154.

50. Tijmensen, M.J.A., Faaij, A.P.C., Hamelinck, C.N., and van Hardeveld, M.R.M. (2002) Exploration of the possibilities for production of fischer tropsch liquids and power via biomass gasification, *Biomass and Bioenergy* 23(2):129–152.

51. Tillman, D.A. (1991) *The Combustion of Solid Fuels and Wastes*, Academic Press, San Diego, CA.

52. United States Department of Agriculture. (2012) Agricultural statistics database, July 2012, URL http://quickstats.nass.usda.gov.

53. Vieira de Carvalho, A.J. (1982) Natural gas and other alternative fuels for transportation purposes, *Energy* 10(2):187–215.

54. Wayman, M. and Parekh, S.R. (1990) *Biotechnology of Biomass Conversion: Fuels and Chemicals from Renewable Resources*, Open University Press, Philadelphia, PA.

55. Wright, L.L. and Hohenstein, W.G. (1994) Dedicated feedstock supply systems: Their current status in the USA, *Biomass and Bioenergy* 6(3):159–241.

56. Wright, M.M., Daugaard, D.E., Satrio, J.A., and Brown, R.C. (2010) Techno-economic analysis of biomass fast pyrolysis to transportation fuels, *Fuel* 89(Supplement 1(0)):S2–S10, ISSN 0016-2361, URL http://www.sciencedirect.com/science/article/pii/S0016236110003765, Techno-economic Comparison of Biomass-to-Biofuels Pathways.

57. Zhao, H., Holladay, J.E., Brown, H., and Zhang, Z.C. (2007) Metal chlorides in ionic liquid solvents convert sugars to 5-hydroxymethylfurfural, *Science* 316(5831):1597.

58. Turano, B., Ogoshi, R., and Uehara, G. (2009) Can biofuels from non-food sources end the food versus fuel debate? Hawaii Energy Policy Forum, http://www.hawaiienergypolicyforum.blogspot.com/

59. Perlack, R., Wright, L., Turhollow, A., Graham, R., Stokes, B., and Erbach, D. (2005) Biomass as feedstock for a bioenergy and bioproducts industry: The technical feasibility of a billion-ton annual supply, Technical Report A357634, Oak Ridge National Laboratory.

60. Zhu, X.G., Long, S.P. et al. (2008) What is the maximum efficiency with which photosynthesis can convert solar energy into biomass? *Current Opinion in Biotechnology* 19(2):153–159.

61. Munson, B.R., Young, D.F., and Okiishi, T.H. (1994) *Fundamentals of Fluid Mechanics*, 2nd edn., John Wiley & Sons, New York, Table 1.6.

62. Kajitani, S., Chen, Z.L., Konno, M., and Rhee, K.T. (1997) *Engine Performance and Exhaust Characteristics of Direct-Injection Diesel Engine Operated with DME*, Society of Automotive Engineers Inc., Warrendale, PA, p. 35.

63. Klass, D.L. (1998e) *Biomass for Renewable Energy, Fuels, and Chemicals*, Academic Press, San Diego, CA, p. 416.

5

Fundamentals of Solar Radiation*

> Having been admonished by the Holy Office [the Inquisition] entirely to abandon the false option that the Sun was the center of the universe and immovable, and that the Earth was not the center of the same and that it moved ... I abjure ...
>
> —Galileo

5.1 Physics of the Sun and Its Energy Transport

The nature of energy generation in the sun is still an unanswered question. Spectral measurements have confirmed the presence of nearly all the known elements in the sun. However, 80% of the sun is hydrogen and 19% helium. Therefore, the remaining 100-plus observed elements make up only a tiny fraction of the composition of the sun. It is generally accepted that a hydrogen-to-helium thermonuclear reaction is the source of the sun's energy. Yet because such a reaction has not been duplicated in the laboratory, it is unclear precisely what the reaction mechanism is, what role the turbulent flows in the sun play, and how solar prominences and sunspots are created.

The nature of the energy-creation process is of no importance to terrestrial users of the sun's radiation. Of interest is the amount of energy, its spectral and temporal distribution, and its variation with time of day and year. These matters are the main subject of this chapter.

The sun is a 13.9×10^5 km diameter sphere comprised of many layers of gases, which are progressively hotter toward its center. The outermost layer, from which energy is radiated into the solar system, is approximately at an equivalent black-body temperature of 5760 K ($10,400°R$). The center of the sun, however, may be at 20×10^6 K. The rate of energy emission from the sun is 3.8×10^{23} kW, which results from the conversion of 4.3×10^9 g/s (4.7×10^6 ton/s) of mass to energy. Of this total, only a tiny fraction, approximately 1.7×10^{14} kW, is intercepted by the earth, which is located about 150 million km from the sun (Figure 5.1).

Solar energy is the world's most abundant permanent source of energy. The amount of solar energy intercepted by the planet earth is 5000 times greater than the sum of all other inputs (terrestrial, nuclear, geothermal, and gravitational energies, and lunar gravitational energy). Of this amount, 30% is reflected to space, 47% is converted to low-temperature heat and reradiated to space, and 23% powers the evaporation/precipitation cycle of the bio-sphere. Less than 0.5% is represented in the kinetic energy of the wind and waves and in photosynthetic storage in plants.

Total terrestrial radiation is only about one-third of the extraterrestrial total during a year, and 70% of that falls on the oceans. However, the remaining 1.5×10^{17} kWh that falls on land is a prodigious amount of energy—about 6000 times the total energy usage of the United States in 2000. However, only a small fraction of this total can be used because of physical and socioeconomic constraints, as described in Chapter 1.

---

* Sections in this chapter marked with an asterisk may be omitted in an introductory course.

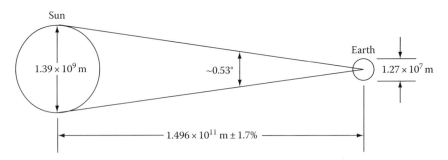

FIGURE 5.1
Relationship between the sun and the earth.

5.2 Thermal Radiation Fundamentals

The material presented in this section has been selected from textbooks on heat transfer and radiation (e.g., Refs. [1–4]). It provides the background needed to understand the nature of solar radiation for the engineering analysis of solar energy systems.

To begin with, all radiation travels at the speed of light, which is equal to the product of the wavelength and the frequency of radiation. The speed of light in a medium equals the speed of light in a vacuum divided by the refractive index of the medium through which it travels:

$$c = \lambda v = \frac{c_0}{n} \tag{5.1}$$

where
λ is the wavelength (m) (or μm, 1 μm = 10^{-6} m)
v is the frequency (s^{-1})
c is the speed of light in a medium (m/s)
c_0 is the speed of light in a vacuum (m/s)
n is the index of refraction of the medium

Thermal radiation is one kind of electromagnetic energy, and all bodies emit thermal radiation by virtue of their temperature. When a body is heated, its atoms, molecules, or electrons are raised to higher levels of activity called excited states. However, they tend to return to lower energy states, and in this process, energy is emitted in the form of electromagnetic waves. Changes in energy states result from rearrangements in the electronic, rotational, and vibrational states of atoms and molecules. Since these rearrangements involve different amounts of energy changes and these energy changes are related to the frequency, the radiation emitted by a body is distributed over a range of wavelengths. A portion of the electromagnetic spectrum is shown in Figure 5.2. The wavelengths associated with the various mechanisms are not sharply defined; thermal radiation is usually considered to fall within the band from about 0.1 to 100 μm, whereas solar radiation has most of its energy between 0.1 and 3 μm.

For some problems in solar energy engineering, the classical electromagnetic wave theory is not suitable. In such cases, for example, in photovoltaic or photochemical processes,

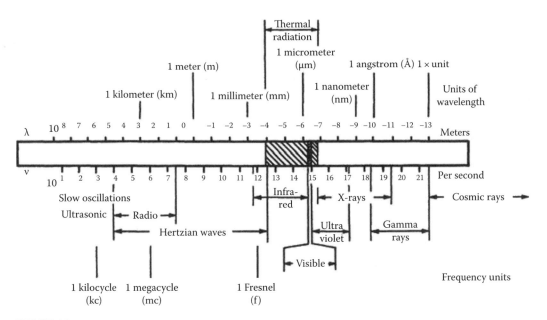

FIGURE 5.2
Electromagnetic radiation spectrum.

it is necessary to treat the energy transport from the point of view of quantum mechanics. In this view, energy is transported by particles or *photons*, which are treated as energy units or quanta rather than waves. The energy of a photon, E_p, of frequency v_p is

$$E_p = h v_p \tag{5.2}$$

where h is the Planck's constant (6.625×10^{-34} J · s).

5.2.1 Black-Body Radiation

The energy density of the radiation emitted at a given wavelength (monochromatic) by a perfect radiator, usually called a black body, is given according to the following relation:

$$E_{b\lambda} = \frac{C_1}{(e^{C_2/\lambda T} - 1)\lambda^5 n^2} \tag{5.3}$$

where
$C_1 = 3.74 \times 10^8$ W $\mu m^4/m^2$ (1.19×10^8 Btu $\mu m^4/h$ ft^2)
$C_2 = 1.44 \times 10^4$ μm K (2.59×10^4 μm °R)
n = refractive index of the medium = 1.0 for vacuum; n is taken to be approximately equal to 1 for air

The quantity $E_{b\lambda}$ has the units of W/m^2 μm (Btu/h ft^2 μm) and is called the monochromatic emissive power of a black body, defined as the energy emitted by a perfect radiator per unit wavelength at the specified wavelength per unit area and per unit time at the temperature *T*.

The total energy emitted by a black body, E_b, can be obtained by integration over all wavelengths. This yields the Stefan–Boltzmann law:

$$E_b = \int_0^\infty E_{b\lambda}d\lambda = \sigma T^4 \tag{5.4}$$

where
   σ is the Stefan–Boltzmann constant = 5.67×10^{-8} W/m² K⁴ (0.1714×10^{-8} Btu/h ft² R⁴)
   T is the absolute temperature (K) (or R = 460 +°F)

The concept of a black body, although no such body actually exists in nature, is very convenient in engineering because its radiation properties can readily be related to those of real bodies.

5.2.2 Radiation Function Tables

Engineering calculations of radiative transfer are facilitated by the use of radiation function tables, which present the results of Planck's law in a more convenient form than Equation 5.3. A plot of the monochromatic emissive power of a black body as a function of wavelength as the temperature is increased is given in Figure 5.3. The emissive power shows a maximum at a particular wavelength.

These peaks, or infection points, are uniquely related to the body temperature. By differentiating Planck's distribution law (Equation 5.3) and equating to zero, the wavelength corresponding to the maximum value of $E_{b\lambda}$ can be shown to occur when

$$\lambda_{max}T = 2897.8 \text{ μm K } (5215.6 \text{ μm R}) \tag{5.5}$$

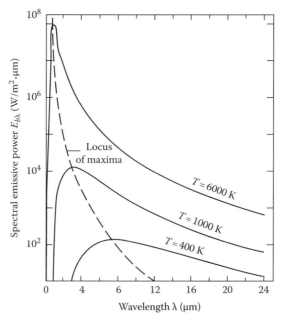

FIGURE 5.3
Spectral distribution of black-body radiation.

Frequently, one needs to know the amount of energy emitted by a black body within a specified range of wavelengths. This type of calculation can be performed easily with the aid of the radiation functions mentioned previously. To construct the appropriate radiation functions in dimensionless form, note that the ratio of the black-body radiation emitted between 0 and λ and between 0 and ∞ can be made a function of the single variable (λT) by using Equation 5.3 as shown (for $n = 1$) in the following equation:

$$\frac{E_{b,0-\lambda}}{E_{b,0-\infty}} = \frac{\int_0^\lambda E_{b\lambda} d\lambda}{\sigma T^4} = \int_0^{\lambda T} \frac{C_1 d(\lambda T)}{\sigma (\lambda T)^5 (e^{C_2/\lambda T} - 1)} \tag{5.6}$$

The earlier relation is plotted in Figure 5.4, and the results are also shown in tabular form in Table 5.1.

In Table 5.1, the first column is the ratio of λ to λ_{max} from Equation 5.5, and the third column the ratio of $E_{b,0-\lambda}$ to σT^4 from Equation 5.6. For use on a computer, Equation 5.6 can be approximated by the following polynomials:

$$v \geq 2 \frac{E_{b,0-\lambda}}{\sigma T^4} = \frac{15}{\pi^4} \sum_{m=1,2,\ldots} \frac{E^{-mv}}{m^4} \left\{ [(mv+3)mv+6]mv+6 \right\} \tag{5.7a}$$

and

$$v \geq 2 \frac{E_{b,0-\lambda}}{\sigma T^4} = \frac{15}{1-\pi^4} v^3 \left(\frac{1}{3} - \frac{v}{8} - \frac{v^2}{60} - \frac{v^4}{5,040} + \frac{v^6}{272,160} - \frac{v^8}{13,305,600} \right) \tag{5.7b}$$

where $v = C_2/\lambda T$.

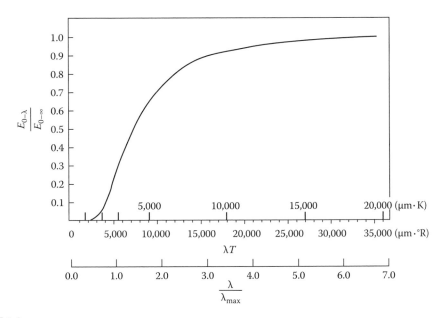

FIGURE 5.4
Fraction of total emissive power in spectral region between $\lambda = 0$ and λ as a function of λT and λ/λ_{max}.

TABLE 5.1

Thermal Radiation Functions[a]

λ/λ_{max}	$E_{b\lambda}/E_{b\lambda,max}$	$E_{b\lambda,0-\lambda}/\sigma T^4$
0.00	0.0000	0.0000
0.20	0.0000	0.0000
0.30	0.0038	0.0001
0.40	0.0565	0.0015
0.50	0.2217	0.0101
0.60	0.4664	0.0325
0.70	0.7042	0.0712
0.80	0.8776	0.1236
0.90	0.9725	0.1849
1.00	1.0000	0.2501
1.10	0.9791	0.3153
1.20	0.9277	0.3782
1.30	0.8600	0.4370
1.40	0.7854	0.4911
1.50	0.7103	0.5403
1.60	0.6382	0.5846
1.70	0.5710	0.6243
1.80	0.5098	0.6598
1.90	0.4546	0.6915
2.00	0.4054	0.7197
2.10	0.3616	0.7449
2.20	0.3229	0.7674
2.30	0.2887	0.7875
2.40	0.2585	0.8054
2.50	0.2318	0.8215
2.60	0.2083	0.8360
2.70	0.1875	0.8490
2.80	0.1691	0.8607
2.90	0.1528	0.8713
3.00	0.1384	0.8809
3.10	0.1255	0.8895
3.20	0.1141	0.8974
3.30	0.1038	0.9045
3.40	0.0947	0.9111
3.50	0.0865	0.9170
3.60	0.0792	0.9225
3.70	0.0726	0.9275
3.80	0.0667	0.9320
3.90	0.0613	0.9362
4.00	0.0565	0.9401
4.20	0.0482	0.9470
4.40	0.0413	0.9528
4.60	0.0356	0.9579
4.80	0.0308	0.9622
5.00	0.0268	0.9660
6.00	0.0142	0.9790
7.00	0.0082	0.9861

TABLE 5.1 (continued)

Thermal Radiation Functions[a]

λ/λ_{max}	$E_{b\lambda}/E_{b\lambda,max}$	$E_{b\lambda,0-\lambda}/\sigma T^4$
8.00	0.0050	0.9904
9.00	0.0033	0.9930
10.00	0.0022	0.9948
20.00	0.0002	0.9993
40.00	0.0000	0.9999
50.00	0.0000	1.0000

[a] λ = wavelength in μm.

λ_{max} = wavelength at $E_{b\lambda,max}$ in μm = $2898/T$.

$E_{b\lambda}$ = monochromatic emissive power in W/m² · μm
$\quad = 374.15 \times 10^6/\lambda^5 \, [\exp(14{,}387.9/\lambda T) - 1]$,

$E_{b\lambda,max}$ = maximum monochromatic emissive power in W/m² · μm
$\quad = 12.865 \times 10^{-12} T^5$,

$E_{b\lambda,0-\lambda} = \displaystyle\int_0^\lambda E_{b\lambda} d\lambda$,

$\sigma T^4 = E_{b\lambda,0-\infty} = 5.670 \times 10^{-8} T^4$ W/m²

T = absolute temperature in K.

5.2.3 Intensity of Radiation and Shape Factor

The emissive power of a surface gives the total radiation emitted in all directions. To determine the radiation emitted in a given direction, we must define another quantity, the radiation intensity I. This quantity is defined as the radiant energy passing through an imaginary plane in space per unit area per unit time and per unit solid angle perpendicular to the plane, as shown in Figure 5.5. I is defined by the relation

$$I = \lim_{\substack{dA' \to 0 \\ d\omega \to \infty}} = \frac{dE}{dA' d\omega} \tag{5.8}$$

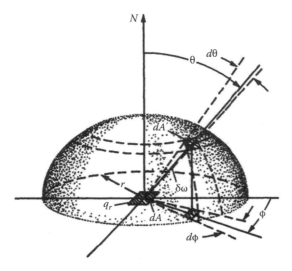

FIGURE 5.5
Schematic diagram illustrating radiation intensity and flux.

Radiation intensity has both magnitude and direction. It can be related to the radiation flux, defined as the radiant energy passing through an imaginary plane per unit area per unit time in all directions.

Note that, whereas for the intensity, the area dA' is perpendicular to the direction of the radiation, for the flux, the area dA is at the base in the center of a hemisphere through which all of the radiation passes. Recalling that the definition for the solid angle between dA' and dA is $d\omega = dA'/r^2$, the radiation flux q_r emanating from dA can be obtained by integrating the intensity over the hemisphere. As shown in Figure 5.5, the unit projected area for I is $dA \cos \theta$ and the differential area dA' on the hemisphere is $r^2 \sin \theta\, d\theta\, d\phi$; thus

$$q_r = \int_0^{2\pi} \int_0^{\pi/2} I \cos \theta \sin \theta\, d\theta\, d\phi \tag{5.9}$$

If the area dA is located on a surface, the emissive power E can also be obtained from Equation 5.9. For the special case of a diffuse surface, for which I is the same in all directions, Equation 5.9 gives

$$q_r = \pi I \tag{5.10}$$

Since all black surfaces are diffuse,

$$E_b = \pi I_b \tag{5.11}$$

Equation 5.11 can, of course, also be written for monochromatic radiation as

$$E_{b\lambda} = \pi I_{b\lambda} \tag{5.12}$$

In the evaluation of the rate of radiation heat transfer between two surfaces, not only their temperatures and their radiation properties but also their geometric configurations and relationships play a part. The influence of geometry in radiation heat transfer can be expressed in terms of the *radiation shape factor* between any two surfaces 1 and 2 defined as follows:

$F_{1\text{-}2}$ = fraction of radiation leaving surface 1 that reaches surface 2

$F_{2\text{-}1}$ = fraction of radiation leaving surface 2 that reaches surface 1

In general, F_{m-n} = fraction of radiation leaving surface "m" that reaches surface "n." If both surfaces are black, the energy leaving surface "m" and arriving at surface "n" is $E_{bm}A_mF_{m-n}$ and the energy leaving surface "n" and arriving at "m" is $E_{bn}A_nF_{n-m}$. If both surfaces absorb all the incident energy, the net rate of exchange $q_{m\Leftrightarrow n}$ will be

$$q_{m\Leftrightarrow n} = E_{bm}A_mF_{m-n} - E_{bn}A_nF_{n-m} \tag{5.13}$$

If both surfaces are at the same temperature, $E_{bm} = E_{bn}$ and the net exchange is zero $q_{m\Leftrightarrow n} = 0$. This shows that the geometric radiation shape factor must obey the reciprocity relation

$$A_mF_{m-n} = A_mF_{n-m} \tag{5.14}$$

The net rate of heat transfer can therefore be written in two equivalent forms:

$$q_{m \Leftrightarrow n} = A_m F_{m-n}(E_{bm} - E_{bn}) = A_n F_{n-m}(E_{bm} - E_{bn}) \tag{5.15}$$

The evaluation of geometric shape factors is in general quite involved. For a majority of solar energy applications, however, only a few special cases are of interest. One of these is a small convex object of area A_1 surrounded by a large enclosure A_2. Since all radiation leaving A_1 is intercepted by A_2, $F_{1-2} = 1$ and $F_{2-1} = A_1/A_2$.

Another case is the exchange of radiation between two large parallel surfaces. If the two surfaces are near each other, almost all of the radiation leaving A_1 reaches A_2 and vice versa. Thus, $F_{1-2} = F_{2-1} = 1.0$, according to the definition of the shape factor. A third case of importance is the exchange between a small surface ΔA_1 and a portion of space A_2, for example, the exchange between a flat-plate solar collector tilted at an angle β from the horizontal and the sky it can see. For this situation, we refer to the definition of radiation flux (see Figure 5.5). The portion of the radiation emitted by ΔA_1 that is intercepted by the surrounding hemisphere depends on the angle of tilt. When the surface is horizontal, $F_{1-2} = 1$; when it is vertical, $F_{1-2} = 1/2(\beta = 90°)$. For intermediate values, it can be shown that [4]

$$F_{1-2} = \frac{1}{2}(1 + \cos \beta) = \cos^2\left(\frac{\beta}{2}\right) \tag{5.16}$$

If the diffuse sky radiation is uniformly distributed and assumed to be black, then a small black area A_1 receives radiation at the rate

$$A_1 F_{1-sky} E_{sky} = \frac{A_1}{2}(1 + \cos \beta)\sigma T_{sky}^4 \tag{5.17}$$

whereas the net radiation heat transfer is given by

$$q_{sky \Leftrightarrow 1} = A_1 F_{1-sky} \sigma \left(T_{sky}^4 - T_1^4\right) \tag{5.18}$$

If the receiving area is gray with an absorptance $\bar{\alpha}$ equal to the emittance $\bar{\varepsilon}$, the net exchange is given by

$$q_{sky \Leftrightarrow 1} = A_1 F_{1-sky} \bar{\alpha} \sigma \left(T_{sky}^4 - T_1^4\right) \tag{5.19}$$

5.2.4 Transmission of Radiation through a Medium

When radiation passes through a transparent medium such as glass or the atmosphere, the decrease in intensity can be described by Bouger's law, which assumes that the attenuation is proportional to the local intensity in the medium. If $I_\lambda(x)$ is monochromatic intensity after the radiation has traveled a distance x, the law is expressed by the equation

$$-dI_\lambda(x) = I_\lambda(x) K_\lambda dx \tag{5.20}$$

where K_λ is the monochromatic extinction coefficient assumed to be a constant of the medium. If the transparent medium is a slab of thickness L and the intensity at $x = 0$ is designated by the symbol $I_{\lambda,0}$, the monochromatic transmittance τ_λ is equal to the ratio of the intensity at $x = L$ to $I_{\lambda,0}$. An expression for $I_\lambda(L)$ can be obtained by integrating Equation 5.20 between 0 and L, which gives

$$\ln\frac{I_\lambda(L)}{I_{\lambda,0}} = -K_\lambda L \quad \text{or} \quad I_\lambda(L) = I_{\lambda,0}e^{-K_\lambda L} \tag{5.21}$$

Then

$$\tau_\lambda = \frac{I_\lambda(L)}{I_{\lambda,0}} = e^{-K_\lambda L} \tag{5.22}$$

The extinction coefficient K_λ is a complex property of the medium since it combines the effects of absorption, emission, and scattering by the molecules and particles that make up the medium. Fortunately, for materials such as glass and plastics with known compositions, this coefficient can be determined accurately. Transmission of radiation through such materials is discussed further in Chapter 7. In this chapter, we are concerned about the transmission of solar radiation through the atmosphere. The atmosphere consists of the molecules of gases in it, such as N_2, O_2, CO_2, H_2O, etc., and aerosols such as dust particles, water droplets, and ice crystals. The extinction processes of the atmosphere consist of (a) absorption and emission by the molecules and aerosols, (b) scattering by the molecules, and (c) scattering by aerosols.

Since the atmosphere consists of a large number of components whose concentration changes as a function of time and location, determining the extinction coefficient of the atmosphere presents a formidable challenge. A major research effort is underway by scientists trying to predict global climate change. Some early attempts for the estimation of extinction coefficient for "average atmospheric conditions" were combined with an empirical approach [5] to use the earlier equation for the estimation of terrestrial solar radiation resource. This approach is described later in this chapter.

5.3 Sun–Earth Geometric Relationship

Figure 5.6 shows the annual orbit of the earth around the sun. The distance between the earth and the sun changes throughout the year, the minimum being 1.471×10^{11} m at winter solstice (December 21) and the maximum being 1.521×10^{11} m at summer solstice (June 21). The year-round average earth–sun distance is 1.496×10^{11} m. The amount of solar radiation intercepted by the earth, therefore, varies throughout the year, the maximum being on December 21 and the minimum on June 21.

The axis of the earth's daily rotation around itself is at an angle of 23.45° to the axis of its ecliptic orbital plane around the sun. This tilt is the major cause of the seasonal variation of the solar radiation available at any location on the earth. The angle between the earth–sun line (through their centers) and the plane through the equator is called the *solar declination*, δ_s. The declination varies between −23.45° on December 21 to +23.45° on June 21. Stated another way, the declination has the same numerical value as the latitude at which the sun is directly overhead at solar noon on a given day. The tropics of Cancer (23.45°N) and

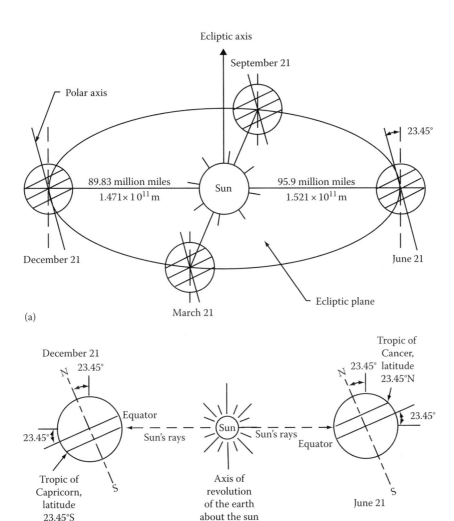

FIGURE 5.6
(a) Motion of the earth about the sun. (b) Location of tropics. Note that the sun is so far from the earth that all the rays of the sun may be considered as parallel to one another when they reach the earth.

Capricorn (23.45°S) are at the extreme latitudes where the sun is overhead at least once a year, as shown in Figure 5.6. The Arctic and Antarctic circles are defined as those latitudes above which the sun does not rise above the horizon plane at least once per year. They are located, respectively, at 66½°N and 66½°S. Declinations north of the equator (summer in the northern hemisphere) are positive; those south, negative. The solar declination may be estimated by the following relation*:

$$\delta_s = 23.45° \sin\left[\frac{360(284+n)}{365}\right]° \tag{5.23}$$

---

* A more accurate relation is sin δ_s = sin(23.45°) sin[360(284 + n)/365]°. Because the error is small, Equation 5.23 is generally used.

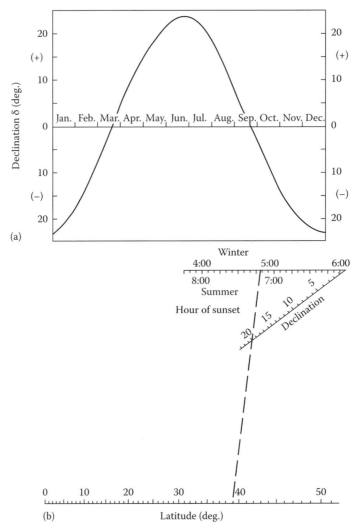

FIGURE 5.7
(a) Graph to determine the solar declination. (b) Sunset nomograph example. Example (b) shows determination of sunset time for summer (7:08 p.m.) and winter (4:52 p.m.) when the latitude is 39°N and the solar declination angle is 20°.

where *n* is the day number during a year with January 1 being *n* = 1. Approximate values of declination may also be obtained from Figure 5.7. For most calculations, the declination may be considered constant during any given day. A summary table of solar ephemeris can be found in Table W.5.1 on the website for the book, http://www.crcpress.com/product/isbn/9781466556966.

For the purposes of this book, the Ptolemaic view of the sun's motion provides a simplification to the analysis that follows. It is convenient to assume the earth to be fixed and to describe the sun's apparent motion in a coordinate system fixed to the earth with its origin at the site of interest. Figure 5.8 shows an apparent path of the sun to an

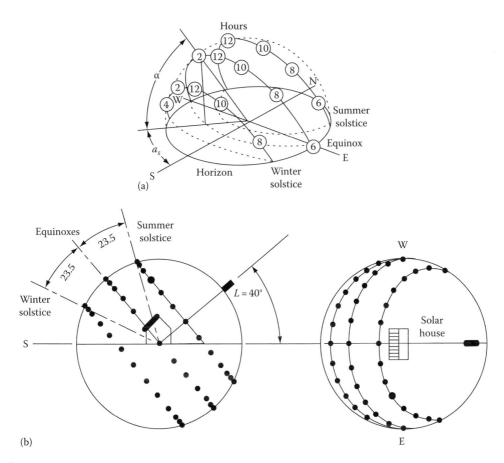

FIGURE 5.8
Sun paths for the summer solstice (6/21), the equinoxes (3/21 and 9/21), and the winter solstice (12/21) for a site at 40°N: (a) isometric view; (b) elevation and plan views.

observer. The position of the sun can be described at any time by two angles, the altitude and azimuth angles, as shown in Figure 5.8. The *solar altitude angle*, α, is the angle between a line collinear with the sun's rays and the horizontal plane. The *solar azimuth angle*, α_s, is the angle between a due south line and the projection of the site to sun line on the horizontal plane. The sign convention used for azimuth angle is positive west of south and negative east of south. The *solar zenith angle*, z, is the angle between the site to sun line and the vertical at the site:

$$z = 90° - \alpha \qquad (5.24)$$

The solar altitude and azimuth angles are not fundamental angles. Hence, they must be related to the fundamental angular quantities *hour angle*, *latitude*, and *declination*. The three angles are shown in Figure 5.9. The solar hour angle h_s is based on the

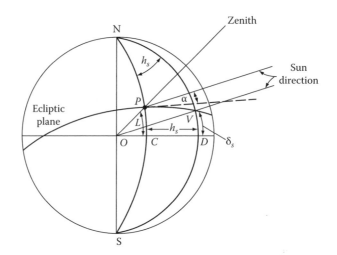

FIGURE 5.9
Definition of solar hour angle h_s (CND), solar declination δ_s (VOD), and latitude L (POC); p, site of interest. (Modified from Kreider, J.F. and Kreith, F., *Solar Heating and Cooling*, Hemisphere Publishing Corp., Washington, DC, 1982.)

nominal time of 24 h required for the sun to move 360° around the earth or 15° per hour. Therefore, h_s is defined as

$$h_s = 15°/\text{h} \cdot (\text{hours from local solar noon})$$

$$= \frac{\text{Minutes from local solar noon}}{4\,\text{min}/\text{degree}} \tag{5.25}$$

Again, values east of due south, that is, morning values, are negative, and values west of due south are positive.

The latitude angle L is the angle between the line from the center of the earth to the site and the equatorial plane. The latitude may be read from an atlas and is considered positive north of the equator and negative south of the equator.

5.3.1 Solar Time and Angles

The sun angles are obtained from the local solar time, which differs from the local standard time (LST). The relationship between the local solar time and the LST is

$$\text{Solar time} = \text{LST} + \text{ET} + (l_{st} - l_{local}) \cdot 4\,\text{min}/\text{degree} \tag{5.26}$$

where
  ET is the equation of time, which is a correction factor that accounts for the irregularity
    of the speed of earth's motion around the sun
  l_{st} is the standard time meridian
  l_{local} is the local longitude

ET may be estimated from Table W.5.1 or calculated from the following empirical equation:

$$ET(min) = 9.87 \sin 2B - 7.53 \cos B - 1.5 \sin B \tag{5.27}$$

where $B = 360(n - 81)/364°$.

The solar altitude angle, α, can be found from the application of the law of cosines to the geometry of Figure 5.9 and can be simplified as

$$\sin \alpha = \sin L \sin \delta_s + \cos L \cos \delta_s \cos h_s \tag{5.28}$$

Using a similar technique, the solar azimuth angle, a_s, can be found as

$$\sin a_s = \frac{\cos \delta_s \sin h_s}{\cos \alpha} \tag{5.29}$$

At local solar noon, $h_s = 0$; therefore, $\alpha = 90 - |L - \delta_s|$, and $a_s = 0$.

In calculating the solar azimuth angle from Equation 5.29, a problem occurs whenever the absolute value of a_s is greater than 90°. A computational device usually calculates the angle as less than 90° since $\sin a_s = \sin(180 - a_s)$. The problem can be solved in the following way:

For $L > \delta_s$, the solar times when the sun is due east (t_E) or due west (t_W) can be calculated by t_E or t_W = 12:00 Noon ± cos^{-1}[tan δ_s/tan$_L$]°(15°/h) (−for t_E, +for t_W).

For solar times earlier than t_E or later than t_W, the sun would be north (south in the southern hemisphere) of the east–west line and the absolute value of a_s would be greater than 90°. Then, a_s may be calculated as $a_s = ±(180° - |a_s|)$.

For $L \leq \delta_s$, the sun remains north (south in the southern hemisphere) of the east–west line and the true value of a_s is greater than 90°.

Sunrise and *sunset* times can be estimated by finding the hour angle for $\alpha = 0$. Substituting $\alpha = 0$ in Equation 5.28 gives the hour angles for sunrise (h_{sr}) and sunset (h_{ss}) as

$$h_{ss} \text{ or } h_{sr} = ±\cos^{-1}[-\tan L \cdot \tan \delta_s] \tag{5.30}$$

It should be emphasized that Equation 5.30 is based on the center of the sun at the horizon. In practice, sunrise and sunset are defined as the times when the upper limb of the sun is on the horizon. Because the radius of the sun is 16′, the sunrise would occur when $\alpha = -16′$. Also, at lower solar elevations, the sun will appear on the horizon when it is actually 34′ below the horizon. Therefore, for apparent sunrise or sunset, $\alpha = -50′$.

Example 5.1

Find the solar altitude and azimuth angles at solar noon in Gainesville, Florida, on February 1. Also find the sunrise and sunset times in Gainesville on that day.

Solution
For Gainesville,

$$\text{Latitude } L = 29° + 41'\text{N or } 29.7°\,\text{N}$$

$$\text{Longitude } l_{local} = 82° + 16'\text{W or } 82.3°\,\text{W}$$

On February 1, day number, $n = 32$. Therefore, from Equation 5.23, the declination is

$$\delta_s = 23.45 \sin\left[\frac{360(284+32)}{365}\right]^{\circ} = -17.5^{\circ}$$

At solar noon, $h_s = 0$. Therefore,

$$\sin\alpha = \cos L \cos\delta_s \cosh_s \sin\alpha = \cos L \cos\delta_s \cosh_s + \sin L \sin\delta_s$$

$$= \cos(29.7^{\circ})\cos(-17.5^{\circ})\cos(0) + \sin(29.7^{\circ})\sin(-17.5^{\circ})$$

or $\alpha = 42.8^{\circ}$

$$\sin a_s = \frac{\cos(-17.5^{\circ})\sin(0)}{\cos(42.8^{\circ})} = 0$$

or $a_s = 0$

At solar noon, α can also be found as

$$\alpha = 90 - |L - \delta_s|^{\circ}$$

$$= 90 - |29.7 + 17.5| = 42.8^{\circ}$$

$$h_{ss} \text{ or } h_{sr} = \pm\cos^{-1}[-\tan L \cdot \tan\delta_s]$$

$$= \pm\cos^{-1}\left[-\tan(29.7^{\circ}) \cdot \tan(17.5^{\circ})\right] = \pm79.65$$

Time from solar noon $= \pm(79.65^{\circ})(4 \text{ min/degree})$

$$= \pm 319 \text{ min or } \pm (5 \text{ h } 19 \text{ min})$$

Sunrise time $= 12:00 \text{ noon} - (5 \text{ h } 19 \text{ min})$

$$= 6 \text{ h } 41 \text{ min a.m. (solar time)}$$

Sunset time $= 12:00 \text{ noon} + (5 \text{ h } 19 \text{ min})$

$$= 5 \text{ h } 19 \text{ min p.m. (solar time)}$$

To convert these times to local times, we need to find ET:

$$ET = (9.87 \sin 2B - 7.53 \cos B - 1.5 \sin B) \text{ min}$$

$$B = \frac{360}{364}(n-81) = \frac{360}{364}(32-81) = -48.46^{\circ}$$

Therefore,

$$ET = -13.67 \text{ min}$$

$$LST = \text{solar time} - ET - 4(l_{st} - l_{local})$$

Gainesville, Florida, is in the eastern standard time (EST) zone, where $l_{st} = 75°\text{W}$. Therefore,

$$LST = \text{Solar time} - (-13.67 \text{ min}) - 4(75 - 82.27) \text{ min}$$

$$= \text{Solar time} + 42.75 \text{ min}$$

Therefore,

$$\text{Sunrise time} = 6:41 \text{ a.m.} + 43 \text{ min} = 7:24 \text{ a.m. EST,}$$

and

$$\text{Sunrise time} = 5:19 \text{ p.m.} + 43 \text{ min} = 6:02 \text{ p.m. EST.}$$

NOTE: Since the sunrise and sunset times are calculated when the center of the sun is at the horizon, they differ from the apparent times. If we use $\alpha = -50'$, the apparent sunrise and sunset times would be 7:20 a.m. EST and 6:06 p.m. EST, respectively.

Knowledge of the solar angles is helpful in the design of passive solar buildings, especially the placement of windows for solar access and the roof overhang for shading the walls and windows at certain times of the year. The following example illustrates this point.

Example 5.2

Find the roof overhang L of a south-facing window of height $H = 1$ m, such that the window is completely shaded at solar noon on April 1 and not shaded at all at noon on November 1. Assume that the roof extends far beyond the window on either side. Location: Gainesville, Florida. Also find the overhang if $S = 1.3$ m.

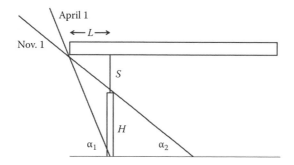

Solution
From the geometry of the figure, we can derive the following two equations with four unknowns, L, S, α_1, and α_2:

$$L = \frac{S}{\tan \alpha_2}$$

$$L = \frac{(S+H)}{\tan \alpha_1}$$

From the following figure, it can be determined that $\alpha + |\text{Lat.} - \delta_s| = 90°$ from simple geometry. Therefore, α_1 and α_2 are dependent on the seasonal declination angle δ_s and the latitude of Gainesville, FL, which is $29.68°$ from Example 5.1.

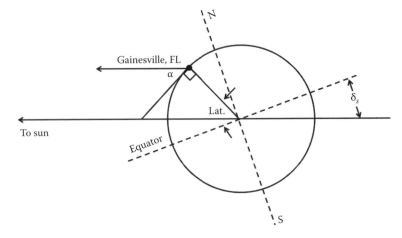

To find the declination angle δ_s on April 1, $n = 91$, we use Equation 5.23:

$$n = 91.$$

$$\delta_s = 23.45° \sin\left[\frac{360(284 + 91)}{365}\right] = 4.02°$$

Therefore, at solar noon,

$$\alpha_1 = 90 - |29.68 - 4.02| = 64.34°$$

On November 1, $n = 305$ and

$$n = 305$$

$$\delta_s = 23.45° \sin\left[\frac{360(284 + 305)}{365}\right]^\circ = -15.4°$$

Therefore,

$$\alpha_2 = 90 - |29.68 + 15.4| = 44.9°$$

$$L = \frac{H}{\tan \alpha_1 - \tan \alpha_2} = \frac{1}{\tan(64.34°) - \tan(44.9°)} = 0.92 \text{ m}$$

$$S = L \tan \alpha_2 = 0.92 \times \tan(44.9°) = 0.92 \text{ m}$$

If $S = 1.3$ m, then

$$L = \frac{1.3}{\tan 44.9°} = 1.3 \text{ m}$$

Also

$$L = \frac{2.3}{\tan 64.34°} = 1.1 \text{ m}$$

Therefore, 1.1 m $\leq L \leq$ 1.3 m.

5.3.2 Sun-Path Diagram

The projection of the sun's path on the horizontal plane is called a *sun-path diagram*. Such diagrams are very useful in determining shading phenomena associated with solar collectors, windows, and shading devices. As shown earlier, the solar angles (α, a_s) depend upon the hour angle, declination, and latitude. Since only two of these variables can be plotted on a two-dimensional graph, the usual method is to prepare a different sun-path diagram for each latitude with variations of hour angle and declination shown for a full year. Sun-path diagrams for different latitudes are shown in Figure 5.10.

The altitude and azimuth of the sun are given by

$$\sin \alpha = \sin L \sin \delta_s + \cos \phi \cos \delta_s \cosh_s \tag{5.31}$$

and

$$\sin a_s = \frac{-\cos \delta_s \sinh_s}{\cos \alpha} \tag{5.32}$$

where
α is the altitude of the sun (angular elevation above the horizon)
L is the latitude of the observer
δ_s is the declination of the sun
h_s is the hour angle of sun (angular distance from the meridian of the observer)
a_s is the azimuth of the sun (measured eastward from north)

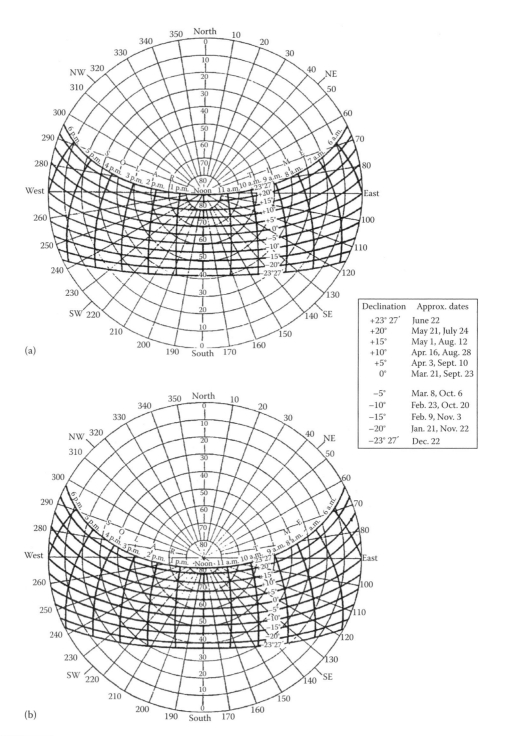

Declination	Approx. dates
+23° 27′	June 22
+20°	May 21, July 24
+15°	May 1, Aug. 12
+10°	Apr. 16, Aug. 28
+5°	Apr. 3, Sept. 10
0°	Mar. 21, Sept. 23
−5°	Mar. 8, Oct. 6
−10°	Feb. 23, Oct. 20
−15°	Feb. 9, Nov. 3
−20°	Jan. 21, Nov. 22
−23° 27′	Dec. 22

FIGURE 5.10

Description of method for calculating true solar time, together with accompanying meteorological charts, for computing solar altitude and azimuth angles. (a) chart, 25°N latitude; (b) chart, 30°N latitude. (Reproduced from List, R.J., *Smithsonian Meteorological Tables*, 6th edn., Smithsonian Institution Press, Washington, DC, pp. 442–443, 1949.)

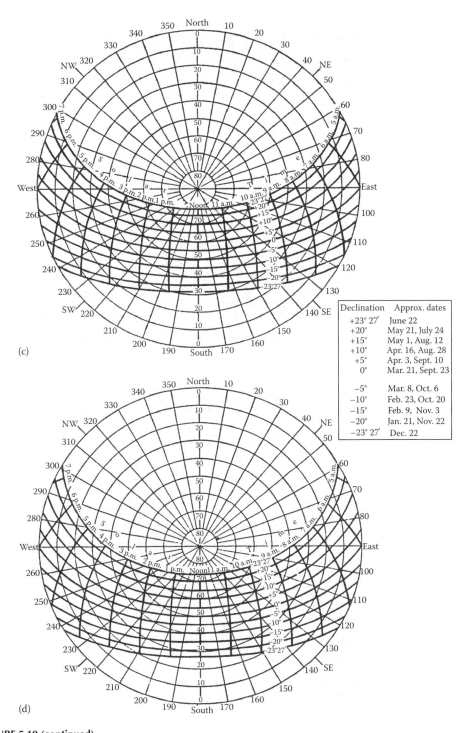

FIGURE 5.10 (continued)
Description of method for calculating true solar time, together with accompanying meteorological charts, for computing solar altitude and azimuth angles. (c) chart, 35°N latitude; (d) chart, 40°N latitude. (Reproduced from List, R.J., *Smithsonian Meteorological Tables*, 6th edn., Smithsonian Institution Press, Washington, DC, pp. 442–443, 1949.)

(*continued*)

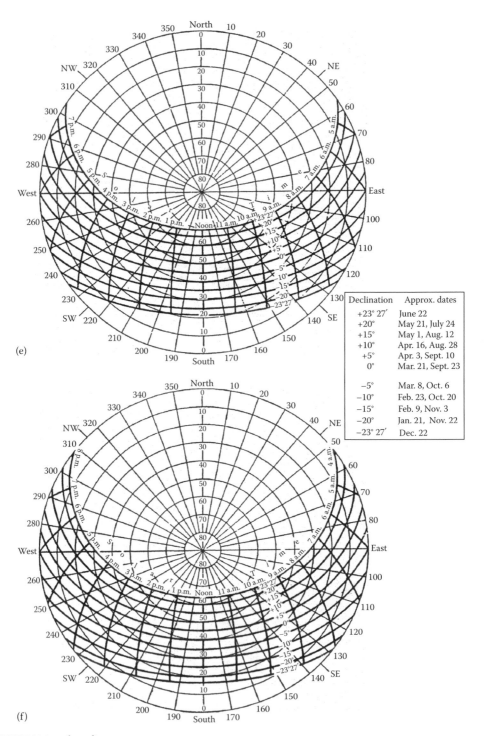

Declination	Approx. dates
+23° 27′	June 22
+20°	May 21, July 24
+15°	May 1, Aug. 12
+10°	Apr. 16, Aug. 28
+5°	Apr. 3, Sept. 10
0°	Mar. 21, Sept. 23
−5°	Mar. 8, Oct. 6
−10°	Feb. 23, Oct. 20
−15°	Feb. 9, Nov. 3
−20°	Jan. 21, Nov. 22
−23° 27′	Dec. 22

(e)

(f)

FIGURE 5.10 (continued)
Description of method for calculating true solar time, together with accompanying meteorological charts, for computing solar altitude and azimuth angles. (e) chart, 45°N latitude; (f) chart, 50°N latitude. (Reproduced from List, R.J., *Smithsonian Meteorological Tables*, 6th edn., Smithsonian Institution Press, Washington, DC, pp. 442–443, 1949.)

From Equations 5.31 and 5.32, it can be seen that the altitude and azimuth of the sun are functions of the latitude of the observer, the time of day (hour angle), and the date (declination).

Figure 5.10a through f provides a series of charts, one for each 5° of latitude (except 5°, 15°, 75°, and 85°) giving the altitude and azimuth of the sun as a function of the true solar time and the declination of the sun in a form originally suggested by Hand. Linear interpolation for intermediate latitudes will give results within the accuracy to which the charts can be read.

On these charts, a point corresponding to the projected position of the sun is determined from the heavy lines corresponding to declination and solar time.

To find the solar altitude and azimuth

1. Select the chart or charts appropriate to the latitude.

2. Find the solar declination δ corresponding to the date.

3. Determine the *true solar time* as follows:

   a. To the *LST* (zone time), add 4° for each degree of longitude the station is east of the standard meridian or subtract 4° for each degree west of the standard meridian to get the *local mean solar time*.

   b. To the *local mean solar time*, add algebraically the equation of time; the sum is the required *true solar time*.

4. Read the required altitude and azimuth at the point determined by the declination and the true solar time. Interpolate linearly between two charts for intermediate latitudes.

It should be emphasized that the solar altitude determined from these charts is the true geometric position of the center of the sun. At low solar elevations, terrestrial refraction may considerably alter the apparent position of the sun. Under average atmospheric refraction, the sun will appear on the horizon when it actually is about 34° below the horizon; the effect of refraction decreases rapidly with increasing solar elevation. Since sunset or sunrise is defined as the time when the upper limb of the sun appears on the horizon, and the semidiameter of the sun is 16°, sunset or sunrise occurs under average atmospheric refraction when the sun is 50° below the horizon. In polar regions especially, unusual atmospheric refraction can make considerable variation in the time of sunset or sunrise.

The 90°N chart is included for interpolation purposes; the azimuths lose their directional significance at the pole.

Altitude and azimuth in southern latitudes: To compute solar altitude and azimuth for southern latitudes, change the sign of the solar declination and proceed as earlier. The resulting azimuths will indicate angular distance from *south* (measured eastward) rather than from north.

Sun-path diagrams for a given latitude are used by entering them with appropriate values of declination δ_s and hour angle h_s. The point at the intersection of the corresponding δ_s and h_s lines represents the instantaneous location of the sun. The solar altitude can then be read from the concentric circles in the diagram, and the azimuth, from the scale around the circumference of the diagram.

Example 5.3

Using Figure 5.10d, determine the solar altitude and azimuth for March 8 at 10 a.m. Compare the results to those calculated from the basic equations (Equations 5.28 and 5.29).

Solution
On March 8, the solar declination is −5°; therefore the −5° sun path is used. The intersection of the 10 a.m. line and the −5° declination line in the diagram represents the sun's location; it is marked with a heavy dot in Figure 5.10b.
The sun's position lies midway between the 40° and 50° altitude circles, say at 45°, and midway between the −40° and −50° azimuth radial lines, say at −45°. So $\alpha \cong 45°$ and $a_s \cong -45°$. Equations 5.28 and 5.29 give precise values for a and a_s:

$$\sin\alpha = \sin(30°)\sin(-5°) + \cos(30°)\cos(-5°)\cos(-30°)$$

$$\alpha = 44.7°$$

$$\sin a_s = \frac{\cos(-5°)\sin(-30°)}{\cos(44.7°)}$$

$$a_s = -44.5°$$

Therefore, the calculated values are within ±0.5° (1%) of those read from the sun-path diagram.

5.3.3 Shadow-Angle Protractor

The shadow-angle protractor used in shading calculations is a plot of solar altitude angles, projected onto a given plane, versus solar azimuth angle. The projected altitude angle is usually called the *profile angle* γ. It is termed as the angle between the normal to a surface and the projection of the sun's rays on a vertical plane normal to the same surface. The profile angle is shown in Figure 5.11a with the corresponding solar altitude angle. The profile angle, which is always used in sizing shading devices, is given by

$$\tan\gamma = \sec a \tan\alpha \tag{5.33}$$

where a is the solar azimuth angle with respect to the wall normal.
Figure 5.11b shows the shadow-angle protractor to the same scale as the sun-path diagrams in Figure 5.10d. It is used by plotting the limiting values of profile angle γ and azimuth angle a, which will start to cause shading of a particular point. The shadow-angle protractor is usually traced onto a transparent sheet so that the shadow map constructed on it can be placed over the pertinent sun-path diagram to indicate the times of day and months of the year during which shading will take place. The use of the shadow-angle protractor is best illustrated by an example.

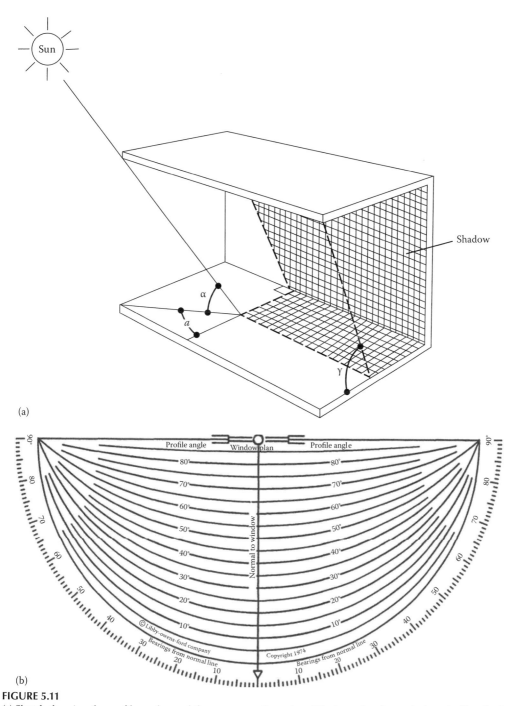

(a)

(b)

FIGURE 5.11
(a) Sketch showing the profile angle γ and the corresponding solar altitude angle α for a window shading device;
(b) the shadow-angle protractor. (Courtesy of Libby-Owens-Ford Glass Co., Toledo, OH.)

Example 5.4

A solar building with a south-facing collector is sited to the north–northwest of an existing building. Prepare a shadow map showing what months of the year and what part of the day point C at the base of the solar collector will be shaded. Plan and elevation views are shown in Figure 5.12. Latitude = 40°N.

Solution

The limiting profile angle for shading is 40° and the limiting azimuth angles are −45° and +10°, as shown in Figure 5.12. These values are plotted on the shadow-angle protractor (Figure 5.13a). The shadow map, when superimposed on the sun-path diagram (Figure 5.13b), shows that point C will be shaded during the following times of day for the periods shown:

Declination	Date	Time of Day
−23°27′	December 22	8:45 a.m.–12:40 p.m.
−20°	January 21, November 22	8:55 a.m.–12:35 p.m.
−15°	February 9, November 3	9:10 a.m.–12:30 p.m.

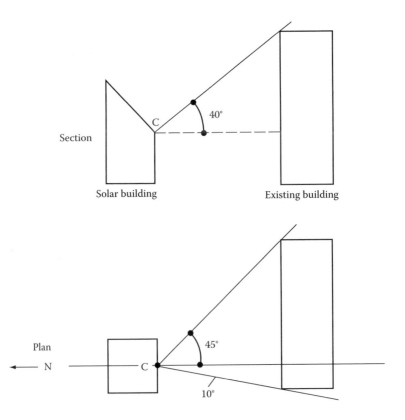

FIGURE 5.12
Plan and elevation views of proposed solar building and existing building, which may shade solar collector at point C.

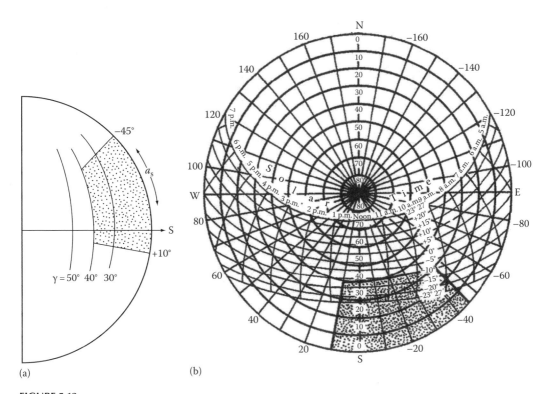

FIGURE 5.13
(a) Shadow map constructed for the example shown in Figure 2.13; (b) shadow map superimposed on sun-path diagram. The degrees on the circumference in this version are shown as increasing or decreasing from south.

In summary, during the period from November 3 to February 9, point C will be shaded between 3 and 4 h. It will be shown later that this represents about a 50% loss in collector performance for point C, which would be unacceptable for a collector to be used for heating a building in winter.

5.4 Solar Radiation

Detailed information about solar radiation availability at any location is essential for the design and economic evaluation of a solar energy system. Long-term measured data of solar radiation are available for a large number of locations in the United States and other parts of the world. Where long-term measured data are not available, various models based on available climatic data can be used to estimate the solar energy availability. Solar energy is in the form of electromagnetic radiation with the wavelengths ranging from about 0.3 μm (10^{-6} m) to over 3 μm, which correspond to ultraviolet (less than 0.4 μm), visible (0.4 and 0.7 μm), and infrared (over 0.7 μm). Most of this energy

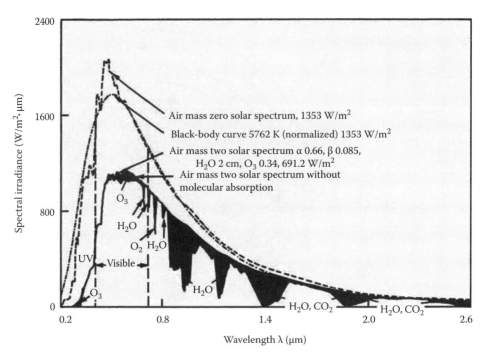

FIGURE 5.14
Extraterrestrial solar radiation spectral distribution. Also shown are equivalent black-body and atmosphere-attenuated spectra.

is concentrated in the visible and the near-infrared wavelength range (see Figure 5.14). The incident solar radiation, sometimes called *insolation*, is measured as irradiance, or the energy per unit time per unit area (or power per unit area). The units most often used are Watts per square meter (W/m^2), British thermal units per hour per square foot ($Btu/h\ ft^2$), and Langleys per minute (calories per square centimeter per minute, cal/cm^2 min).

5.4.1 Extraterrestrial Solar Radiation

The average amount of solar radiation falling on a surface normal to the rays of the sun outside the atmosphere of the earth (extraterrestrial) at mean earth–sun distance (D_0) is called the *solar constant*, I_0. Measurements by NASA indicated the value of the solar constant to be $1353\ W/m^2$ ($\pm 1.6\%$), $429\ Btu/h\ ft^2$ or $1.94\ cal/cm^2$ min (Langleys/min). This value was revised upward by Frohlich et al. [6] to $1377\ W/m^2$ or $437.1\ Btu/h\ ft^2$ or 1.974 Langleys/min, which was the value used in compiling SOLMET data in the United States [7,8]. At present, there is no consensus on the value of the solar constant. However, considering that the difference between the two values is about 1.7% and the uncertainties in estimation of terrestrial solar radiation are 10% or higher, either value may be used. A value of $1367\ W/m^2$ is also used by many references.

   The variation in seasonal solar radiation availability at the surface of the earth can be understood from the geometry of the relative movement of the earth around the sun. Since the earth's orbit is elliptical, the earth–sun distance varies during a year, the variation being ±1.7% from the average. Therefore, the extraterrestrial radiation, I, also varies by the inverse square law as follows:

$$I = I_0 \left(\frac{D_0}{D} \right)^2 \tag{5.34}$$

where
   D is the distance between the sun and the earth
   D_0 is the yearly mean earth–sun distance (1.496×10^{11} m)

The $(D_0/D)^2$ factor may be approximated as [9]

$$\left(\frac{D_0}{D} \right)^2 = 1.00011 + 0.034221\cos(x) + 0.00128\sin(x)$$

$$+0.000719\cos(2x) + 0.000077\sin(2x) \tag{5.35}$$

where

$$x = \frac{360(n-1)}{365°} \tag{5.36}$$

and n = day number (starting from January 1 as 1). The following approximate relationship may also be used without much loss of accuracy:

$$I = I_0 \left[1 + 0.034\cos\left(\frac{360n}{365.25} \right)° \right] \tag{5.37}$$

Figure 5.15 also shows the relationship of the extraterrestrial solar radiation to the solar constant. For many solar energy applications, such as photovoltaics and photocatalysis, it is necessary to examine the distribution of energy within the solar spectrum. Figure 5.15 shows the spectral irradiance at the mean earth–sun distance for a solar constant of 1353 W/m² as a function of wavelength according to the standard spectrum data published by NASA in 1971. The data are also presented in Table 5.2, and their use is illustrated in the following example.

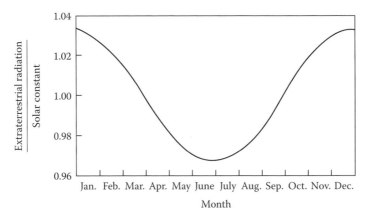

FIGURE 5.15
Effect of the time of year on the ratio of extraterrestrial radiation to the nominal solar constant.

TABLE 5.2

Extraterrestrial Solar Irradiance[a]

λ (μm)	$E\lambda^b$ (W/ m²·μm)	$E\lambda^b$ (Btu/ h·ft²·μm)	$D\lambda^c$ (%)	λ (μm)	$E\lambda$ (W/ m²·μm)	$E\lambda$ (Btu/ h·ft²·μm)	$D\lambda$ (%)	λ (μm)	$E\lambda$ (W/ m²·μm)	$E\lambda$ (Btu/ h·ft²·μm)	$D\lambda$ (%)
0.12	0.007	0.002	1×10^{-4}	0.43	1639	520	12.47	0.90	891	283	63.37
0.14	0.03	0.010	5×10^{-4}	0.44	1810	574	13.73	1.00	748	237	69.49
0.16	0.23	0.073	6×10^{-4}	0.45	2006	636	15.14	1.2	485	154	78.40
0.18	1.25	0.397	1.6×10^{-3}	0.46	2066	655	16.65	1.4	337	107	84.33
0.20	10.7	3.39	8.1×10^{-3}	0.47	2033	645	18.17	1.6	245	77.7	88.61
0.22	57.5	18.2	0.05	0.48	2074	658	19.68	1.8	159	50.4	91.59
0.23	66.7	21.2	0.10	0.49	1950	619	21.15	2.0	103	32.7	93.49
0.24	63.0	20.0	0.14	0.50	1942	616	22.60	2.2	79	25.1	94.83
0.25	70.9	22.5	0.19	0.51	1882	597	24.01	2.4	62	19.7	95.86
0.26	130	41.2	0.27	0.52	1833	581	25.38	2.6	48	15.2	96.67
0.27	232	73.6	0.41	0.53	1842	584	26.74	2.8	39	12.4	97.31
0.28	222	70.4	0.56	0.54	1783	566	28.08	3.0	31	9.83	97.83
0.29	482	153	0.81	0.55	1725	547	29.38	3.2	22.6	7.17	98.22
0.30	514	163	1.21	0.56	1695	538	30.65	3.4	16.6	5.27	98.50
0.31	689	219	1.66	0.57	1712	543	31.91	3.6	13.5	4.28	98.72
0.32	830	263	2.22	0.58	1715	544	33.18	3.8	11.1	3.52	98.91
0.33	1059	336	2.93	0.59	1700	539	34.44	4.0	9.5	3.01	99.06
0.34	1074	341	3.72	0.60	1666	528	35.68	4.5	5.9	1.87	99.34
0.35	1093	347	4.52	0.62	1602	508	38.10	5.0	3.8	1.21	99.51
0.36	1068	339	5.32	0.64	1544	490	40.42	6.0	1.8	0.57	99.72
0.37	1181	375	6.15	0.66	1486	471	42.66	7.0	1.0	0.32	99.82
0.38	1120	355	7.00	0.68	1427	453	44.81	8.0	0.59	0.19	99.88
0.39	1098	348	7.82	0.70	1369	434	46.88	10.0	0.24	0.076	99.94
0.41	1751	555	9.92	0.75	1235	392	51.69	20.0	0.0015	0.005	99.99
0.42	1747	554	11.22	0.80	1109	352	56.02	50.0	0.0004	0.0001	100.00

Sources: Adapted from Iqbal, M., *Sol. Energy*, 23, 169, 1979; Thekaekara, M.P., *Solar Energy*, 14, 109, 1973.

[a] Solar constant = 429 Btu/h ft² = 1353 W/m².

[b] $E\lambda$ is the solar spectral irradiance averaged over a small bandwidth centered at λ.

[c] $D\lambda$ is the percentage of the solar radiation associated with wavelengths shorter than λ.

Example 5.5

Calculate the fraction of solar radiation within the visible part of the spectrum, that is, between 0.40 and 0.70 μm.

Solution

The first column in Table 5.2 gives the wavelength. The second column gives the averaged solar spectral irradiance in a band centered at the wavelength in the first column. The fourth column, $D\lambda$, gives the percentage of solar total radiation at wavelengths shorter than the value of λ in the first column. At a value of 0.40 μm, 8.7% of the total radiation occurs at shorter wavelengths. At a wavelength of 0.70, 46.88% of the radiation occurs at shorter wavelength. Consequently, 38% of the total radiation lies within the band between 0.40 and 0.70 μm, and the total energy received outside the earth's atmosphere within that spectral range is 517 W/m² (163 Btu/h ft²).

5.5 Estimation of Terrestrial Solar Radiation

As extraterrestrial solar radiation, I, passes through the atmosphere, a part of it is reflected back into space, a part is absorbed by air and water vapor, and some gets scattered by molecules of air, water vapor, aerosols, and dust particles (Figure 5.16). The part of solar radiation that reaches the surface of the earth with essentially no change in direction is called *direct* or *beam radiation*. The scattered diffuse radiation reaching the surface from the sky is called the *sky diffuse radiation*.

Although extraterrestrial radiation can be predicted with certainty,* radiation levels on the earth are subject to considerable uncertainty resulting from local climatic inter-actions. The most useful solar radiation data are based on long-term (30 years or more) measured average values at a location, which unfortunately are not available for most locations in the world. For such locations, an estimation method (theoretical model) based on some measured climatic parameters may be used. This chapter describes several ways of estimating terrestrial solar radiation; all have large uncertainties (as much as ±30%) associated with them.

5.5.1 Atmospheric Extinction of Solar Radiation

As solar radiation I travels through the atmosphere, it is attenuated due to absorption and scattering. If K is the local extinction coefficient of the atmosphere, the beam

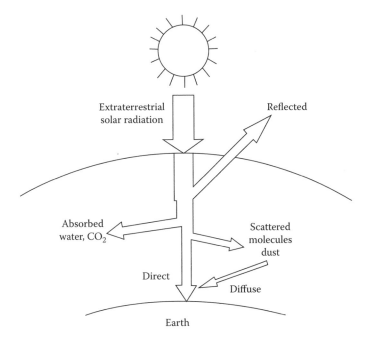

FIGURE 5.16
Attenuation of solar radiation as it passes through the atmosphere.

---

* The effect of sunspots, which may cause up to 0.5% variation, is neglected.

solar radiation at the surface of the earth can be written according to Bouger's law Equation 5.21 as

$$I_{b,N} = Ie^{-\int Kdx} \tag{5.38}$$

where
$I_{b,N}$ is the instantaneous beam solar radiation per unit area normal to the sun's rays
x is the length of travel through the atmosphere

If L_o is the vertical thickness of the atmosphere and

$$\int_0^{L_o} Kdx = k \tag{5.39}$$

the beam normal solar radiation for a solar zenith angle of z will be

$$I_{b,N} = Ie^{-k \, \sec z} = Ie^{-k \, \sin \alpha} = Ie^{-km} \tag{5.40}$$

where m is a dimensionless path length of sunlight through the atmosphere, sometimes called the *air mass ratio* (Figure 5.17). When solar altitude angle is 90° (sun is overhead), $m = 1$.

Threlkeld and Jordan [5] estimated values of k (also known as optical depth) for average atmospheric conditions at sea level with a moderately dusty atmosphere and the amount of precipitable water vapor equal to the average value for the United States for each month. These values are given in Table 5.3.

To account for the differences in local conditions from the average sea-level conditions, Equation 5.41 is modified by a parameter called clearness number C_n, introduced by Threlkeld and Jordan [5]:

$$I_{b,N} = C_n Ie^{-k/\sin \alpha} \tag{5.41}$$

Note that $C_n = 1$ in clear-sky conditions.

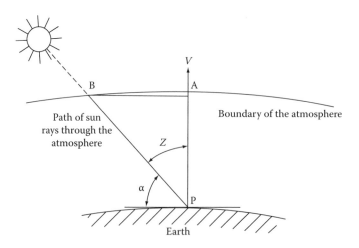

FIGURE 5.17
Air mass definition: air mass $m = BP/AP = \csc \alpha$, where α is the altitude angle. The atmosphere is idealized as a constant thickness layer.

TABLE 5.3

Average Values of Atmospheric Optical Depth (k) and Sky Diffuse Factor (C) for 21st Day of Each Month, for Average Atmospheric Conditions at Sea Level for the United States

Month	1	2	3	4	5	6	7	8	9	10	11	12
k	0.142	0.144	0.156	0.180	0.196	0.205	0.207	0.201	0.177	0.160	0.149	0.142
C	0.058	0.060	0.071	0.097	0.121	0.134	0.136	0.122	0.092	0.073	0.063	0.057

Sources: Threlkeld, J.L. and Jordan, R.C., *ASRAE Trans.*, 64, 45, 1958; Islam, M.R., *RERIC Int. Energy J.*, 16(2), 103113, 1994.

5.5.2 Solar Radiation on Clear Days

Total instantaneous solar radiation on a horizontal surface (see Figure 5.18), I_h, is the sum of the beam or direct radiation, $I_{b,h}$, and the sky diffuse radiation $I_{d,h}$:

$$I_h = I_{b,h} + I_{d,h} \tag{5.42}$$

According to the Threlkeld and Jordan model [5], the sky diffuse radiation on a clear day is proportional to the beam normal solar radiation and can be estimated by using an empirical sky diffuse factor C. Therefore, I_h can be estimated as

$$I_h = I_{b,N} \cos z + C I_{b,N}$$

$$= I_{b,N} \sin \alpha + C I_{b,N}$$

$$= C_n I e^{-k/\sin \alpha}(C + \sin \alpha) \tag{5.43}$$

Values of C are given in Table 5.3.

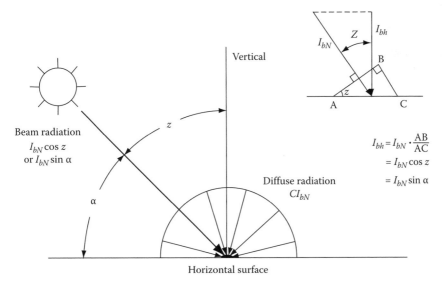

FIGURE 5.18
Solar radiation on a horizontal surface.

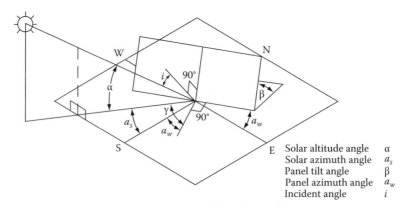

FIGURE 5.19
Definitions of solar angles for a tilted surface. *Note:* α, solar altitude angle; a_s, solar azimuth angle; β, panel tilt angle; a_w, panel azimuth angle; i, incident angel.

5.5.3 Solar Radiation on a Tilted Surface

Solar radiation on an arbitrary tilted surface having a tilt angle of β from the horizontal and an azimuth angle of a_w (assumed + west of south), as shown in Figure 5.19, is the sum of components consisting of beam ($I_{b,c}$), sky diffuse ($I_{d,c}$), and ground-reflected solar radiation ($I_{r,c}$):

$$I_c = I_{b,c} + I_{d,c} + I_{r,c} \tag{5.44}$$

If i is the *angle of incidence* of the beam radiation on the tilted surface, it is simple to show that the instantaneous beam radiation on the surface per unit area is

$$I_{b,c} = I_{b,N} \cos i \tag{5.45}$$

From the geometry in Figure 5.19, it can be shown that the angle of incidence i for the surface (angle between the normal to the surface and a line collinear with the sun's rays) is related to the solar angles as

$$\cos i = \cos \alpha \cos(a_s - a_w) \sin \beta + \sin \alpha \cos \beta \tag{5.46}$$

The diffuse radiation on the surface ($I_{d,c}$) can be obtained by multiplying the sky diffuse radiation on a horizontal surface by the view factor between the sky and the surface[*]:

$$I_{d,c} = I_{d,h} \frac{(1 + \cos \beta)}{2}$$

$$= CI_{b,N} \frac{(1 + \cos \beta)}{2}$$

$$= CI_{b,N} \cos^2\left(\frac{\beta}{2}\right) \tag{5.47}$$

The ground-reflected solar radiation can be found from the total solar radiation incident on a horizontal surface and the ground reflectance ρ as

$$I_{r,c} = I_h \rho \tag{5.48}$$

---

[*] The surface has been assumed infinitely large for this view factor. See Section 5.2.3.

The part of $I_{r,c}$ intercepted by the tilted surface can be found by multiplying the ground-reflected radiation by the view factor between the surface* and the ground:

$$I_{r,c} = \rho I_h \frac{(1-\cos\beta)}{2} = \rho I_h \sin^2\left(\frac{\beta}{2}\right)$$

$$= \rho I_{b,N}(\sin\alpha + C)\sin^2\left(\frac{\beta}{2}\right) \qquad (5.49)$$

For ordinary ground or grass, ρ is approximately 0.2, and for snow-covered ground, it can be taken as approximately 0.8.

Example 5.6

Find the instantaneous solar radiation at 12:00 noon EST on a solar collector surface ($\beta = 30°$, $a_w = +10°$) on February 1 in Gainesville, Florida.

Solution
From Example 5.1, for February 1

$$n = 32, \quad \delta_s = -17.5°, \quad \text{and} \quad ET = -13.7\,\text{min}$$

For finding the values of solar radiation on the collector, we will need to calculate angles α, a_s, h_s, and i.

$$\text{Solar time} = LST + ET + 4(I_{st} - I_{local})$$

$$= 12:00 - 13.7\,\text{min} + 4(75° - 82.27°)$$

$$= 11:17.2\,\text{a.m.,}$$

$$h_s = \frac{\text{Minutes from solar noon}}{4\,\text{min/degree}}$$

$$= \frac{-42.8}{4} = -10.7°(-\text{before noon})$$

From Equation 5.28,

$$\alpha = \sin^{-1}(\sin(29.68°)\sin(-17.5°) + \cos(29.68°)\cos(-17.5°)\cos(-10.7°)) = 41.7°$$

From Equation 5.29,

$$a_s = \frac{\sin^{-1}(\cos(-17.5°)\sin(-10.7°))}{\cos(41.7°)}$$

$$= -13.7°$$

---

* The tilted surface and the ground in front of it have been assumed to be infinitely large for this view factor.

Angle of incidence i for the solar collector is given by Equation 5.46:

$$\cos i = \cos(41.7°)\cos(-13.7° - 10°)\sin(30°) + \sin(41.7°)\cos(30°) \quad i = 23.4°$$

To calculate the solar radiation using Equations 5.44, 5.45, 5.47, and 5.49, we need to find $I_{b,N}$ and I.

Extraterrestrial solar radiation

$$I = I_0\left[1 + 0.034 \cos\left(\frac{360n}{365.25}\right)^°\right]$$

$$= 1353 \text{ W/m}^2\left[1 + 0.034 \cos\left(\frac{360 \times 32}{365.25}\right)^°\right]$$

$$= 1392 \text{ W/m}^2$$

(Find k from Table 5.3 and assume C_n is 1)

$$I_{b,N} = C_n I e^{-k/\sin\alpha}$$

$$= 1392 e^{-0.144/\sin(41.7°)}$$

$$= 1121 \text{ W/m}^2 \text{ or } 355 \text{ Btu/h ft}^2$$

Beam radiation on the collector (Equation 5.45)

$$I_{b,c} = I_{b,N} \cos i$$

$$= (1121) \cos(23.4°)$$

$$= 1029 \text{ W/m}^2 \text{ or } 326 \text{ Btu/h ft}^2$$

Sky diffuse radiation on the collector (Equation 5.47)

$$I_{d,c} = C I_{b,N} \cos^2\left(\frac{\beta}{2}\right)$$

$$= (0.060)(1121)\cos^2\left(\frac{30}{2}\right)$$

(Find C from Table 5.3.)

$$63 \text{ W/m}^2 \text{ or } 20 \text{ Btu/h ft}^2$$

Ground-reflected radiation on the collector (Equation 5.49)

$$I_{r,c} = \rho I_{b,N}(\sin\alpha + C)\sin^2\left(\frac{\beta}{2}\right)$$

Assume

$$\rho = 0.2$$

$$I_{r,c} = (0.2)(1121)(\sin 41.7° + 0.060) \sin^2(15°)$$

$$= 11 \text{ W/m}^2 \text{ or } 3.5 \text{ Btu/h ft}^2$$

Total insolation on the collector

$$I_c = 1029 + 63 + 11 = 1103 \text{ W/m}^2 \text{ or } 350 \text{ Btu/h ft}^2$$

A convenient website that will provide the same information obtained in the preceding example can be found online at http://www.builditsolar.com/Tools/RadOnCol/radoncol. htm. The online calculator model requires merely inputting the latitude, collector tilt, collector azimuth, altitude of location, the month, and the day. With this information, the program provides the direct, diffuse, and total radiation perpendicular to the sun in the third, fourth, and fifth columns. The sixth column provides the total radiation on a horizontal surface. The seventh column gives the incident angle of the sun on the collector. The 9th, 10th, and 11th columns give the direct, diffuse, and total radiation falling on the collector surface. From this program, which was updated in 2009, the radiation falling on the collector at noon is 1037 W/m^2, which is about 4% less than the value obtained by the empirical procedure outlined in the book. The printout from the model for the day is shown in Table 5.4 with the radiation values in Btu/h-ft^2.

TABLE 5.4

Solar Radiation Output for Example 5.6 Using the Online Calculator

Solar radiation on collector for day of 2/1

Collector area	1 (ft^2)
Collector azimuth	0 (degree) measured from south + is to east
Collect tilt	40 (degree) measured from horizontal
Latitude	30 (degree)
Altitude	0 (ft) above sea level
Sun rise	6.7 (h) sunrise in solar time

			Sun					Collector		
			Dir.			Horz.	Incid.			
Time (h)	Az.	Elev.	Normal	Diffuse	Total	Tot.	Ang.	Direct.	Diffuse	Total
4	88.9	−34.1	0	0	0	0	90	0	0	0
5	82	−21.2	0	0	0	0	90	0	0	0
6	75.1	−8.4	0	0	0	0	87	0	0	0
7	67.7	3.9	47	3	50	6	73	14	2	16
8	59.2	15.5	228	13	241	74	59	119	12	131
9	48.8	26	280	16	297	139	44	201	15	216
10	35.6	34.8	302	18	320	190	30	262	16	278
11	19.1	40.8	312	18	330	222	16	300	16	316
12	0	42.9	315	18	333	233	7	312	16	329
13	−19.1	40.8	312	18	330	222	16	300	16	316
14	−35.6	34.8	302	18	320	190	30	262	16	278
15	−48.8	26	280	16	297	139	44	201	15	216
16	−59.2	15.5	228	13	241	74	59	119	12	131
17	−67.7	3.9	47	3	50	6	73	14	2	16
18	−75.1	−8.4	0	0	0	0	87	0	0	0
19	−82	−12.2	0	0	0	0	90	0	0	0
20	−88.9	−34.1	0	0	0	0	90	0	0	0
Day total			2653	156	2809	1496		2102	138	2240

Angles in degrees.
Radiations in Btu/h.
Day total in Btu/day.
1 Btu/h ft^2 = 3.152 W/m^2.

5.5.4 Monthly Solar Radiation Estimation Models

One of the earliest methods of estimating solar radiation on a horizontal surface was proposed by the pioneer spectroscopist Angström. It was a simple linear model relating average horizontal radiation to clear-day radiation and to the sunshine level, that is, percent of possible hours of sunshine. Since the definition of a clear day is somewhat nebulous, Page [11] refined the method and based it on extraterrestrial radiation instead of the ill-defined clear day:

$$\bar{H}_h = \bar{H}_{o,h}\left(a + b\frac{\bar{n}}{\bar{N}}\right)$$

$$= \bar{H}_{o,h}\left(a + b\frac{\overline{PS}}{100}\right) \tag{5.50}$$

where
$\bar{H}_h$ and $\bar{H}_{o,h}$ are the horizontal terrestrial and horizontal extraterrestrial radiation levels averaged for a month listed in Table 5.5
$\overline{PS}$ is the monthly averaged percent of possible sunshine (i.e., hours of sunshine/maximum possible duration of sunshine × 100)
a and b are constants for a given site
$\bar{n}$ and $\bar{N}$ are the monthly average numbers of hours of bright sunshine and day length, respectively

The ratio $\bar{n}/\bar{N}$ is also equivalent to the monthly average percent sunshine $(\overline{PS})$. $\bar{H}_{o,h}$ can be calculated by finding $H_{o,h}$ from Equation 5.51, using Equations 5.37 and 5.46 and averaging $I_{o,h}$ for the number of days in each month:

$$H_{o,h} = \int_{t_{sr}}^{t_{ss}} I \sin \alpha \, dt \tag{5.51}$$

Some typical values of a and b are given in Table 5.6 [12]. Additional values for worldwide locations from Refs. [35] and [37] can be found in Table W.5.2 on the website, http://www.crcpress.com/product/isbn/9781466556966

Example 5.7

Using the predictive method of Angström–Page, estimate the monthly horizontal solar radiation for Miami, Florida, located at a latitude of 25°N.

Solution
Using the climate data given, the expected monthly average horizontal radiation for Miami is calculated in Table 5.5 using $a = 0.42$ and $b = 0.22$ from Table 5.6.

TABLE 5.5

$\bar{H}_h$ $\bar{H}_{o,h}$ for Angstrom—Page Method (See Equations 5.50 and 5.51)

Month	$\overline{PS}/100$	$\bar{H}_{o,h}$[a] Wh/m²·Day	kJ/m²·Day	$\bar{H}_h$ kJ/m²·Day	Btu/ft²·Day
January	0.66	6,656	23,962	13,543	1193
February	0.68	7,769	27,968	15,931	1403
March	0.74	9,153	32,951	19,204	1691
April	0.76	10,312	37,123	21,799	1919
May	0.72	10,936	39,370	22,771	2005
June	0.68	11,119	40,028	22,800	2008
July	0.72	10,988	39,557	22,880	2015
August	0.71	10,484	37,742	21,747	1915
September	0.7	9,494	34,178	19,618	1728
October	0.7	8,129	29,264	16,798	1479
November	0.67	6,871	24,736	14,035	1236
December	0.63	6,284	22,622	12,637	1113

[a] Monthly averaged, daily extraterrestrial radiation.

TABLE 5.6

Coefficients a and b in the Angström–Page Regression Equation

Location	Climate[a]	Sunshine Hours in Percentage of Possible Range	Average	a	b
Albuquerque, New Mexico	BS-BW	68–85	78	0.41	0.37
Atlanta, Georgia	Cf	455–71	59	0.38	0.26
Blue Hill, Massachusetts	Df	42–60	52	0.22	0.50
Brownsville, Texas	BS	47–80	62	0.35	0.31
Buenos Aires, Argentina	Cf	47–68	59	0.26	0.50
Charleston, South Carolina	Cf	60–75	67	0.48	0.09
Dairen, Manchuria	Dw	55–81	67	0.36	0.23
EI Paso, Texas	BW	78–88	84	0.54	0.23
Ely, Nevada	BW	61–89	77	0.54	0.20
Hamburg, Germany	Cf	11–49	36	0.22	0.18
Honolulu, Hawaii	Af	57–77	65	0.14	0.57
Madison, Wisconsin	Df	40–72	58	0.30	0.73
Malange, Angola	Aw-BS	41–84	58	0.34	0.34
Miami, Florida	Aw	56–71	65	0.42	0.22
Nice, France	Cs	49–76	61	00.17	0.63
Poona, India (monsoon)	Am	25–49	37	0.30	0.51
(Dry)		65–89	81	0.41	0.34
Stanleyville, Congo	Af	34–56	48	0.28	0.39
Tamanrasset, Algeria	BW	76–88	83	0.30	0.43

Source: Löf, G.O.G. et al., World distribution of solar energy, University of Wisconsin, Madison, WI, Engineering Experiment Station Report 21, 1996.

[a] Af, tropical forest climate, constantly moist, rainfall all through the year; Am, tropical forest climate, monsoon rain, short dry season, but total rainfall sufficient to support rain forest; Aw, tropical forest climate, dry season in winter; BS, steppe or semiarid climate; BW, desert or arid climate; Cf, mesothermal forest climate, constantly moist, rainfall all through the year; Cs, mesothermal forest climate, dry season in winter; Df, microthermal snow forest climate, constantly moist, rainfall all through the year; Dw, microthermal snow forest climate, dry season in winter.

*5.6 Models Based on Long-Term Measured Horizontal Solar Radiation

Long-term measured solar radiation data are usually available as monthly averaged total solar radiation per day on horizontal surfaces. In order to use these data for tilted surfaces, the total solar radiation on a horizontal surface must first be broken down into beam and diffuse components. A number of researchers have proposed models to do that, prominent among them being Liu and Jordan, Collares-Pereira, and Rabl, and Erbs, Duffe, and Klein.

5.6.1 Monthly Solar Radiation on Tilted Surfaces

In a series of papers, Liu and Jordan (LJ) [13–17] have developed an essential simplification in the basically complex computational method required to calculate long-term radiation on tilted surfaces. This is called the LJ method. The fundamental problem in such calculations is the decomposition of long-term measured total horizontal radiation into its beam and diffuse components.

If the decomposition can be computed, the trigonometric analysis presented earlier can be used to calculate incident radiation on any surface in a straightforward manner. LJ correlated the diffuse-to-total radiation ratio $(\bar{D}_h/\bar{H}_h)$ with the *monthly clearness index* $\bar{K}_T$, which is defined as

$$\bar{K}_T = \frac{\bar{H}_h}{\bar{H}_{o,h}} \tag{5.52}$$

where
$\bar{H}_h$ is the monthly averaged terrestrial radiation per day on a horizontal surface
$\bar{H}_{o,h}$ is the corresponding extraterrestrial radiation, which can be calculated from Equation 5.51 by averaging each daily total for a month

The original LJ method was based upon the extraterrestrial radiation at midmonth, which is not truly an average.

The LJ correlation predicts the monthly diffuse $(\bar{D}_h)$ to monthly total $\bar{H}_h$ ratio. It can be expressed by the empirical equation

$$\frac{\bar{D}_h}{\bar{H}_h} = 1.390 - 4\,027\bar{K}_T + 5.531\bar{K}_T^2 - 3.108\bar{K}_T^3 \tag{5.53}$$

Note that the LJ correlation is based upon a solar constant value of 1394 W/m² (442 Btu/h·ft²), which was obtained from terrestrial observations, whereas the newer value, based on satellite data, is 1377 W/m² (437 Btu/h·ft²). The values of $\bar{K}_T$ must be based on this earlier value of the solar constant to use the LJ method. Collares-Pereira and Rabl [18] conducted a study and concluded that although LJ's approach is valid, their correlations would predict significantly smaller diffuse radiation components. They also concluded that Liu and Jordan were able to correlate their model with the measured data because they used the measured data that were not corrected for the shade ring (see solar radiation measurements).

Collares-Pereira and Rabl (C-P&R) also introduced the sunset hour angle h_{ss} in their correlation to account for the seasonal variation in the diffuse component. The C-P&R correlation is

$$\frac{\bar{D}_h}{\bar{H}_h} = 0.775 + 0.347\left(h_{ss} - \frac{\pi}{2}\right) - \left[0.505 + 0.02619\left(h_{ss} - \frac{\pi}{2}\right)\right]\cos(2\bar{K}_T - 1.8) \tag{5.54}$$

where h_{ss} is the sunset hour angle in radians. The C-P&R correlation agrees well with the correlations for India [19], Israel [20], and Canada [21] and is, therefore, preferred to Equation 5.53.

The monthly average beam component $\bar{B}_h$ on a horizontal surface can be readily calculated by simple subtraction since $\bar{D}_h$ is known:

$$\bar{B}_h = \bar{H}_h - \bar{D}_h \tag{5.55}$$

It will be recalled on an instantaneous basis from Equations 5.43 and 5.45 and Figure 5.19 that

$$I_{b,N} = \frac{I_{b,h}}{\sin\alpha} \tag{5.56}$$

$$I_{b,c} = I_{b,N}\cos i \tag{5.45}$$

where $I_{b,h}$ is the instantaneous horizontal beam radiation. Solving for $I_{b,c}$, the beam radiation on a surface,

$$I_{b,c} = I_{b,h}\left(\frac{\cos i}{\sin\alpha}\right) \tag{5.57}$$

The ratio in parentheses is usually called the beam radiation *tilt factor* R_b. It is a purely geometric quantity that converts instantaneous horizontal beam radiation to beam radiation intercepted by a tilted surface.

Equation 5.57 cannot be used directly for the long-term beam radiation $\bar{B}_h$. To be strictly correct, the instantaneous tilt factor R_b should be integrated over a month with the beam component $I_{b,h}$ used as a weighting factor to calculate the beam tilt factor. However, the LJ method is used precisely when such short-term data as $I_{b,h}$ are not available The LJ recommendation for the monthly mean tilt factor $\bar{R}_b$ is simply to calculate the monthly average of cos i and divide it by the same average of sin α. In equation form for south-facing surfaces, this operation yields

$$\bar{R}_b = \frac{\cos(L-\beta)\cos\delta_s\sinh_{sr} + h_{sr}\sin(L-\beta)\sin\delta_s}{\cos L\cos\delta_s\sinh_{sr}(\alpha=0) + h_{sr}(\alpha=0)\sin L\sin\delta_s} \tag{5.58}$$

where the sunrise hour angle h_{sr} ($\alpha = 0$) in radians is given by Equation 5.30 and h_{sr} is the min [$|h_s$ ($\alpha = 0$)$|$, $|h_s$ ($i = 90°$)$|$], respectively, and are evaluated at midmonth. Non-south-facing surfaces require numerical integration or iterative methods to determine $\bar{R}_b$. The long-term beam radiation on a tilted surface $\bar{B}_c$ is then

$$\bar{B}_c = \bar{R}_b\bar{B}_h \tag{5.59}$$

which is the long-term analog of Equation 5.45. Values of $\bar{R}_b$ are tabulated in Table 5.7.

TABLE 5.7

Solar Collector Tilt Factor $(\bar{R}_b)$

Month	L = 20° β = 20°	β = 40°	L = 30° β = 30°	β = 50°	L = 40° β = 40°	β = 60°	L = 50° β = 50°	β = 70°
January	1.36	1.52	1.68	1.88	2.28	2.56	3.56	3.94
February	1.22	1.28	1.44	1.52	1.80	1.90	2.49	2.62
March	1.08	1.02	1.20	1.15	1.36	1.32	1.65	1.62
April	1.00	0.83	1.00	0.84	1.05	0.90	1.16	1.00
May	0.92	0.70	0.87	0.66	0.88	0.66	0.90	0.64
June	0.87	0.63	0.81	0.58	0.79	0.60	0.80	0.56
July	0.89	0.66	0.83	0.62	0.82	0.64	0.84	0.62
August	0.95	0.78	0.93	0.76	0.96	0.78	1.02	0.83
September	1.04	0.95	1.11	1.00	1.24,	1.12	1.44	1.32
October	1.17	1.20	1.36	1.36	1.62	1.64	2.10	2.14
November	1.30	1.44	1.60	1.76	2.08	2.24	3.16	3.32
December	1.39	1.60	1.76	1.99	2.48	2.80	4.04	4.52

Source: Kreider, J.F. and Kreith, F., *Solar Heating and Cooling*, revised 1st edn., Hemisphere Publishing Corp., Washington, DC, 1977.

Notes: The solar collector tilt factor is the ratio of monthly beam insolation on a tilted surface to monthly beam insolation on a horizontal surface. Here β = collector tilt angle and L = collector latitude.

Diffuse radiation intercepted by a tilted surface differs from that on a horizontal surface, because a tilted surface does not view the entire sky dome, which is the source of diffuse radiation. If the sky is assumed to be an isotropic source of diffuse radiation, the instantaneous and long-term tilt factors for diffuse radiation, R_d and $\bar{R}_d$ respectively, are equal and are simply the radiation view factor from the plane to the visible portion of a hemisphere. In equation form,

$$R_d = \bar{R}_d = \cos^2 \frac{\beta}{2} = \frac{(1+\cos\beta)}{2} \tag{5.60}$$

In some cases where solar collectors are mounted near the ground, some beam and diffuse radiation reflected from the ground can be intercepted by the collector surface. The tilt factor $\bar{R}_r$ for reflected total radiation $(\bar{D}_h + \bar{B}_h)$ is then calculated to be

$$\bar{R}_r = \frac{\bar{R}}{\bar{D}_h + \bar{B}_h} = \rho \sin^2 \frac{\beta}{2} = \frac{\rho(1-\cos\beta)}{2} \tag{5.61}$$

in which ρ is the diffuse reflectance of the surface south of the collector assumed uniform and of infinite extent. A list of reflectances is provided in Table 5.8. For snow, $\rho \cong 0.75$; for grass and concrete, $\rho \cong 0.2$. The total long-term radiation intercepted by a surface $\bar{H}_c$ is then the total of beam, diffuse, and diffusely reflected components:

$$\bar{H}_c = \bar{R}_b\bar{B}_h + \bar{R}_d\bar{D}_h + \bar{R}_r(\bar{D}_h + \bar{B}_h) \tag{5.62}$$

TABLE 5.8

Reflectance Values for Characteristic Surfaces (Integrated over Solar Spectrum and Angle of Incidence)

Surface	Average Reflectivity
Snow (freshly fallen or with ice film)	0.75
Water surfaces (relatively large incidence angles)	0.07
Soils (clay, loam, etc.)	0.14
Earth roads	0.04
Coniferous forest (winter)	0.07
Forests in autumn, ripe field crops, plants	0.26
Weathered blacktop	0.10
Weathered concrete	0.22
Dead leaves	0.30
Dry grass	0.20
Green grass	0.26
Bituminous and gravel roof	0.13
Crushed rock surface	0.20
Building surfaces, dark (red brick, dark paints, etc.)	0.27
Building surfaces, light (light brick, light paints, etc.)	0.60

Sources: Hunn, B.D. and Calafell, D.O., *Solar Energy*, 19, 87, 1977; see also List, R.J., *Smithsonian Meterological Tables*, 6th edn., Smithsonian Institution Press, Washington, DC, 1949, pp. 442–443.

Using Equations 5.60 and 5.61, we have

$$\bar{H}_c = \bar{R}_b\bar{B}_h + \bar{D}_h \cos^2 \frac{\beta}{2} + (\bar{D}_h + \bar{B}_h)\rho \sin^2 \frac{\beta}{2} \tag{5.63}$$

in which $\bar{R}_b$ is calculated from Equation 5.58.

Example 5.8

Using the average horizontal monthly radiation, find the monthly average insolation per day on a south-facing solar collector tilted at an angle of 25° in Miami, Florida, for January and compare it with the value from NREL (http://rredc.nrel.gov/solar/old_data/nsrdb/redbook/atlas/). Assume a ground reflectance of 0.2.

Solution

The following solution is for the month of January. Values for other months can be found by following the same method.

From Table 5.5,

$$\overline{H_h} = 13,543 \frac{kJ}{m^2 \cdot day} = 3.66 \frac{kWh}{m^2 \cdot day}$$

$$\overline{H_{o,h}} = 23,962 \frac{kJ}{m^2 \cdot day} = 6.47 \frac{kWh}{m^2 \cdot day}$$

Therefore,

$$\overline{K}_T = \frac{\overline{H}_h}{\overline{H}_{o,h}} = 0.565$$

δ_s and h_{sr} can be found for the middle of the month (January 16):

$$\delta_s = 23.45° * \sin\left[\frac{360 \times (284 + 16)}{365}\right]$$

$$= -21.1°$$

$$h_{sr} (\alpha = 0) = -\cos^{-1}(\tan L \tan \delta)$$

$$= -79.6° = -1.389 \text{ rad}$$

$$h_{ss} = 1.389 \text{ rad}$$

Using Equation 5.54,

$$\frac{\overline{D}_h}{\overline{H}_h} = 0.775 + 0.347 \times (1.389 - 1.5708) - [0.505 + 0.0261 \times (1.389 - 1.5708)]$$

$$\times \cos(2 \times 0.678 - 1.8)$$

$$= 0.212$$

Therefore,

$$\overline{D}_h = 0.212 \times 13543 = 2867 \frac{\text{KJ}}{\text{m}^2 \cdot \text{day}}$$

and

$$\overline{B}_h = \overline{H}_h - \overline{D}_h = 10,676 \frac{\text{KJ}}{\text{m}^2 \cdot \text{day}}$$

Insolation on a tilted surface can be found from Equation 5.62. We need to find $\overline{R}_b$ from Equation 5.58.

Therefore,

$$\overline{R}_b = \frac{\cos(0°) \times \cos(-21.1°) \times \sin(-79.6°) - 1.389 \times \sin(0°) \times \sin(-21.1°)}{\cos(25°) \times \cos(-21.1°) \times \sin(-79.6°) - 1.389 \times \sin(25°) \times \sin(-21.1°)} = 1.47$$

$$\overline{R}_d = \cos^2\left(\frac{25}{2}\right) = 0.9553$$

$$\overline{R}_r = \rho \times \sin^2\left(\frac{\beta}{2}\right) \quad (\text{Assume } \rho = 0.2)$$

$$= 0.2 \times \sin^2(12.5°)$$

$$= 0.009$$

Therefore, from Equation 5.63,

$$\overline{H}_c = \left(\overline{R_b} \times \overline{B_h}\right) + \left(\overline{R_d} \times \overline{D_h}\right) + \left(\overline{R_r} \times \overline{H_h}\right)$$

$$= (1.47 \times 10,676) + (0.953 \times 2,867) + (0.009 \times 13,543)$$

$$= 18,548 \, \frac{\text{kJ}}{\text{m}^2 \cdot \text{day}} \times 2.7 \cdot 10^{-4} \, \frac{\text{kWh}}{\text{kJ}} \cong 5 \, \frac{\text{kWh}}{\text{m}^2 \cdot \text{day}}$$

According to the NREL map found in http://rredc.nrel.gov/solar/old_data/nsrdb/redbook/atlas/, the average solar radiation for the month of January with a tilt of +15° is between 4 and 5 kWh/m² · day to the readable accuracy. Using an average of 4.5 kWh/m² · day, this is equivalent to 16,830 kJ/m² · day. A correction in the difference in tilt angle would increase this value somewhat. The Angstrom–Page empirical method from Problem 5.8 gives a value of 5 kWh/m² · day or about 9% larger than the TMY3 result. This indicates the approximate accuracy that can be expected from different approaches of using empirical correlations. Note also that the TMY3 data are collected at Miami Airport, which is located at some distance from the center of the city.

5.6.2 Circumsolar or Anisotropic Diffuse Solar Radiation

The models described in the earlier sections assume that the sky diffuse radiation is isotropic. However, this assumption is not true because of circumsolar radiation (brightening around the solar disk). Although the assumption of isotropic diffuse solar radiation does not introduce errors in the diffuse values on horizontal surfaces, it can result in errors of 10%–40% in the diffuse values on tilted surfaces. A number of researchers have studied the anisotropy of the diffuse solar radiation because of circumsolar radiation. Temps and Coulson [22] introduced an anisotropic diffuse radiation algorithm for tilted surfaces for clear-sky conditions. Klucher [23] refined the Temps and Coulson algorithm by adding a cloudiness function to it:

$$R_d = \frac{1}{2}(1 + \cos\beta) M_1 M_2 \tag{5.64}$$

where

$$M_1 = 1 + F \sin^3\left(\frac{\beta}{2}\right) \tag{5.65}$$

$$M_2 = 1 + F \cos^2 i \sin^3(z) \tag{5.66}$$

and

$$F = 1 - \left(\frac{D_h}{H_h}\right)^2 \tag{5.67}$$

Examining F, we find that under overcast skies ($D_h = H_h$), R_d in Equation 5.64 reduces to the isotropic term of LJ. The Klucher algorithm reduces the error in diffuse radiation to about 5%.

In summary, monthly averaged, daily solar radiation on a surface is calculated by first decomposing total horizontal radiation into its beam and diffuse components using Equation 5.53 or 5.54. Various tilt factors are then used to convert these horizontal components to components on the surface of interest.

5.6.3 Daily Solar Radiation on Tilted Surfaces

Prediction of daily horizontal total solar radiation for sites where solar data are not measured can be done using the Angström–Page model. Instead of monthly values, however, daily values are used for percent sunshine PS and extraterrestrial radiation I_{day}. The results of using this simple model would be expected to show more scatter than monthly values, however.

All U.S. National Weather Service (NWS) stations with solar capability report daily horizontal total (beam and diffuse) radiation. Liu and Jordan have extended their monthly method described earlier to apply to daily data. The equation, analogous to Equation 5.53, used to calculate the daily diffuse component $\bar{I}_{d,h}$ is [24]

$$\frac{\bar{I}_{d,h}}{\bar{I}_h} = 1.0045 + 0.04349 K_T - 3.5227 K_T^2 + 2.6313 K_T^3, \quad \text{and} \quad K_T \leq 0.75 \tag{5.68}$$

where K_T is the daily clearness index analogous to the monthly $\bar{K}_T$. (In this section, overbars indicate daily radiation totals.) For values of $K_T > 0.75$, the diffuse-to-total ratio is constant at a value of 0.166. K_T is given by

$$K_T = \frac{\bar{I}_h}{\bar{I}_{o,h}} \tag{5.69}$$

The daily extraterrestrial total radiation $\bar{I}_{o,h}$ is calculated from Equation 5.51. Note that Equation 5.68 is based on the early solar constant value of 1394 W/m² (442 Btu/h ft²).

The daily horizontal beam component $\bar{I}_{b,h}$ is given by the following simple subtraction:

$$\bar{I}_{b,h} = \bar{I}_h - \bar{I}_{d,h} \tag{5.70}$$

The beam, diffuse, and reflected components of radiation can each be multiplied by their tilt factors R_b, R_d, and R_r to calculate the total radiation on a tilted surface

$$\bar{I}_c = R_b \bar{I}_{b,h} + R_d \bar{I}_{d,h} + R_r (\bar{I}_{b,h} + \bar{I}_{d,h}) \tag{5.71}$$

in which

$$R_b = \frac{\cos(L-\beta)\cos\delta_s \sinh_{sr} + h_{sr}\sin(L-\beta)\sin\delta_s}{\cos L \cos\delta_s \sinh_{sr}(\alpha=0) + h_{sr}(\alpha=0)\sin L \sin\delta_s} \tag{5.72}$$

$$R_d = \cos^2\frac{\beta}{2} \tag{5.73}$$

and

$$R_r = \rho\sin^2\frac{\beta}{2} \tag{5.74}$$

by analogy with the previous monthly analysis.

If daily solar data are available, they can be used for design, the same as monthly data. Daily calculations are necessary when finer timescale performance is required. In addition, daily data can be decomposed into hourly data, which are useful for calculations made with large computerized solar system simulation models.

5.6.4 Hourly Solar Radiation on Tilted Surfaces

Hourly solar radiation can be predicted in several ways. Correlations between hourly total and hourly diffuse (or beam) radiation or meteorological parameters such as cloud cover or air mass may be used. Alternatively, a method proposed by Liu and Jordan based on the disaggregation of daily data into hourly data could be used [13–17]. Even if hourly NWS data are available, it is necessary to decompose these total values into beam and diffuse components depending upon the response of the solar conversion device to be used by these two fundamentally different radiation types.

Randall and Leonard [25] have correlated historical data from the NWS stations at Blue Hill, Massachusetts, and Albuquerque, New Mexico, to predict *hourly beam radiation* I_b and *hourly total horizontal radiation* I_h. This method can be used to decompose NWS data into its beam and diffuse components.

The hourly beam radiation was found to be fairly well correlated by hourly *percent of possible insolation* k_t, defined as

$$k_t = \frac{I_h}{I_{o,h}} \tag{5.75}$$

in which $I_{o,h}$ is the hourly horizontal extraterrestrial radiation, which can be evaluated from Equation 5.51 using 1 h integration periods. Carrying out the integration over a 1 h period yields

$$I_{o,h} = I_o\left(1 + 0.034 \cos\frac{360n}{365}\right)(0.9972 \cos L \cos \delta_s \cosh_s + \sin L \sin \delta_s) \tag{5.76}$$

where the solar hour angle h_s is evaluated at the center of the hour of interest.

The direct normal-beam correlation based on k_t is [17]

$$I_{b,N} = -520 + 1800 k_t \ (W/m^2) \quad 0.85 > k_t \geq 0.30 \tag{5.77}$$

$$I_{b,N} = 0 \quad k_t < 0.30 \tag{5.78}$$

This fairly simple correlation gives more accurate $I_{b,N}$ values than the more cumbersome LJ procedure, at least for Blue Hill and Albuquerque. Vant-Hull and Easton [26] have also devised an accurate predictive method for beam radiation.

Randall and Leonard [25] have made a correlation of *total* horizontal hourly radiation I_h (W/m²) on the basis of *opaque cloud cover* CC and *air mass* m using data for Riverside, Los Angeles, and Santa Monica, California. Cloud cover is defined as CC = 1.0 for fully overcast and CC = 0.0 for clear skies. A polynomial fit was used:

$$I_h = \frac{I_{o,h}}{100}(83.02 - 3.847m - 4.407\text{CC} + 1.1013\text{CC}^2 - 0.1109\text{CC}^3) \tag{5.79}$$

The average predictive error for this correlation was ±2.3% of the NWS data; the correlation coefficient of I_h with CC is 0.76 for the data used. Equation 5.82 was used to predict I_h for Inyokern, California, a site not used in the original correlation. Predictions of solar radiation for Inyokern were within 3.2% of NWS. The diffuse radiation can be calculated from Equations 5.77 through 5.79:

$$I_{d,h} = I_h - I_b \sin \alpha \tag{5.80}$$

5.7 TMY Data to Determine Solar Radiation

Sandia National Laboratory prepared a data set summarizing long-term weather and solar data in 1978. This was updated by NREL and released for use by building designers and solar thermal modelers as the Typical Meteorological Year version 3 (TMY3) data set in 2007. A user's manual for the data set is available from NREL's website [27]. The following example demonstrates the use of the TMY3 data to obtain the insolation on a solar collector, which was calculated previously from empirical correlations based on limited measurements in Example 5.8.

Example 5.9

Using typical meteorological data (TMY3), find the total insolation and beam insolation on a south-facing solar collector tilted at an angle of 25° in Miami, Florida, for January 1. Also find the average total insolation per day for the entire month of January. Assume a ground reflectance of $\rho = 0.2$. Use site information for the Miami International Airport. Compare the results from TMY3 with those from the Armstrong–Page method used in Example 5.8.

Solution

TMY3 data can be accessed via the National Renewable Energy Laboratory's (NREL) website on the National Solar Radiation Data Base page at http://rredc.nrel.gov/solar/old_data/nsrdb/1991-2005/tmy3/. Under "The Data," select "In alphabetical order by state and city." This will link you to a list of all sites for which TMY3 data have been collected. Scroll down to Florida and select site number 722020 for the Miami International Airport. The TMY3 data can be downloaded as a.CSV file format that can be read by most spreadsheet programs.

We will only need three separate columns from the TMY3 table: DNI, DHI, and GHI. DNI is *direct normal irradiance* and is the amount of beam radiation received on a surface normal to the sun. DHI is *diffuse horizontal irradiance* and is the amount of sky diffuse radiation received on a horizontal surface. GHI is *global horizontal irradiance* and is the total amount of radiation received on a horizontal surface. DNI, DHI, and GHI correspond to $I_{b,N}$, $I_{d,h}$, and I_h respectively as used in the text (see Table 5.9).

To find the total insolation for January 1, you must sum the hourly values of I_c. Using Equation 5.44, I_c is the sum of $I_{b,c}$, $I_{d,c}$, and $I_{r,c}$ at each hour during the day. Table 5.9 shows how this can be approached using a spreadsheet. Summing the I_c column, the total insolation for January 1 is 4544 Wh/m² · day (16,800 kJ/m² · day). Summing the $I_{b,c}$ column, the total beam insolation is 2899 Wh/m² · day. Repeating this problem for the entire month of January, the total monthly insolation is 133,522 Wh/m², with a daily average of 4,307 Wh/m² · day. The Armstrong–Page method gave 5000 Wh/m² · day and the NREL map gave a value of 4500 Wh/m² · day.

TABLE 5.9

Calculation of Solar Insolation for January 1 in Miami, Florida, Using TMY3 Data

Time	DNI	DHI	GHI	h_s	α	a_s	i	$I_{b,c}$	$I_{d,c}$	$I_{r,c}$	I_c
h	W/m²	W/m²	W/m²	°	°	°	°	W/m²	W/m²	W/m²	W/m²
1	0	0	0	−165.00	−76.1	−81.7	149.9	0.0	0.0	0.0	0.0
2	0	0	0	−150.00	−62.6	−89.4	143.4	0.0	0.0	0.0	0.0
3	0	0	0	−135.00	−49.1	−84.0	131.0	0.0	0.0	0.0	0.0
4	0	0	0	−120.00	−35.8	−79.2	117.7	0.0	0.0	0.0	0.0
5	0	0	0	−105.00	−22.6	−74.4	104.1	0.0	0.0	0.0	0.0
6	0	0	0	−90.00	−9.8	−69.1	90.3	0.0	0.0	0.0	0.0
7	0	0	0	−75.00	2.5	−62.9	76.5	0.0	0.0	0.0	0.0
8	262	18	39	−60.00	14.1	−55.3	62.9	119.2	17.2	0.4	136.7
9	694	38	218	−45.00	24.6	−45.7	49.8	447.9	36.2	2.0	486.2
10	768	65	394	−30.00	33.2	−33.4	37.7	608.0	62.0	3.7	673.6
11	747	125	540	−15.00	39.1	−17.9	27.9	660.1	119.1	5.1	784.3
12	230	265	411	0.00	41.2	0.0	23.8	210.4	252.6	3.9	466.9
13	324	290	503	15.00	39.1	17.9	27.9	286.3	276.4	4.7	567.4
14	517	192	514	30.00	33.2	33.4	37.7	409.3	183.0	4.8	597.1
15	107	339	396	45.00	24.6	45.7	49.8	69.1	323.1	3.7	395.9
16	117	265	313	60.00	14.1	55.3	62.9	53.2	252.6	2.9	308.7
17	152	81	116	75.00	2.5	62.9	76.5	35.4	77.2	1.1	113.7
18	22	14	16	90.00	−9.8	69.1	90.3	−0.1	13.3	0.1	13.4
19	0	0	0	105.00	−22.6	74.4	104.1	0.0	0.0	0.0	0.0
20	0	0	0	120.00	−35.8	79.2	117.7	0.0	0.0	0.0	0.0
21	0	0	0	135.00	−49.1	84.0	131.0	0.0	0.0	0.0	0.0
22	0	0	0	150.00	−62.6	89.4	143.4	0.0	0.0	0.0	0.0
23	0	0	0	165.00	−76.1	81.7	149.9	0.0	0.0	0.0	0.0

Source: Wilcox, S. and Marion, W., *Users Manual for TMY3 Data Sets*, Technical Report NREL/TP-581-43156, National Renewable Energy Laboratory, Golden, CO, May 2008.

See Figure 5.19.

α, solar altitude (Equation 5.30).

a_s, solar azimuth (Equation 5.31).

i, incidence angle (Equation 5.46).

β, tilt angle of a surface from horizontal (Equation 5.16).

*5.8 Measurement of Solar Radiation

Solar radiation measurements of importance to most engineering applications, especially thermal applications, include total (integrated over all wavelengths) direct or beam and sky diffuse values of solar radiation on instantaneous, hourly, daily, and monthly bases. Some applications such as photovoltaics, photochemical, and daylighting require knowledge of spectral (wavelength specific) or band (over a wavelength range, e.g., ultraviolet, visible, infrared) values of solar radiation. This section describes some of the instrumentation used to measure solar radiation and sunshine, and some sources of long-term measured data for different parts of the world. Also described briefly in this section is the method of satellite-based measurements.

5.8.1 Instruments for Measuring Solar Radiation and Sunshine

There are two basic types of instruments used to measure solar radiation, *pyranometer* and *pyrheliometer*. A pyranometer has a hemispherical view of the surroundings and therefore is used to measure total, direct, and diffuse, solar radiation on a surface. A pyrheliometer, on the other hand, has a restricted view (about 5°) and is, therefore, often used to measure the direct or beam solar radiation by pointing it toward the sun. Pyranometers are also used to measure the sky diffuse radiation by using a shadow band to block the direct sun view. A detailed discussion of the instrumentation and calibration standards is given by Iqbal [28] and Zerlaut [29].

A pyranometer consists of a flat sensor/detector (described later) with an unobstructed hemispherical view, which allows it to convert and correlate the total radiation incident on the sensor to a measurable signal. The pyranometers using thermal detectors for measurements can exhibit serious errors at tilt angles from the horizontal due to free convection. These errors are minimized by enclosing the detector in double hemispherical high-transmission glass domes. The second dome minimizes the error due to infrared radiative exchange between the sensor and the sky. A desiccator is usually provided to eliminate the effect due to condensation on the sensor or the dome. Figure 5.20 shows pictures of typical commercially available precision pyranometers.

A pyranometer can be used to measure the sky diffuse radiation by fitting a shade ring to it, as shown in Figure 5.21, in order to block the beam radiation throughout the day. The position of the shade ring is adjusted periodically as the declination changes. Since the shade ring obstructs some diffuse radiation from the pyranometer, correction factors must be applied.

Geometric correction factors (GCFs) that account for the part of the sky obstructed by the shade ring can be easily calculated. However, a GCF assumes isotropic sky, which results in errors because of the circumsolar anisotropy. Eppley Corp. recommends additional correction factors to account for anisotropy as +7% for clear sky, +4% for partly cloudy condition, and +3% for cloudy sky. Mujahid and Turner [31] determined that these correction factors gave less than 3% errors on partly cloudy days but gave errors of −11% for clear-sky conditions and +6% on overcast days. They suggested correction factors due to anisotropy as tabulated in Table 5.10, which reduce the errors to less than ±3%. It must be remembered that these correction factors are in addition to the GCFs. Recently, a sun occulting disk has been employed for shading the direct sun.

Beam or direct solar radiation is usually measured with an instrument called a pyrheliometer. Basically a pyrheliometer places the detector at the base of a long tube. This geometry restricts the sky view of the detector to a small angle of about 5°. When the tube points toward the sun, the detector measures the beam solar radiation and a small part

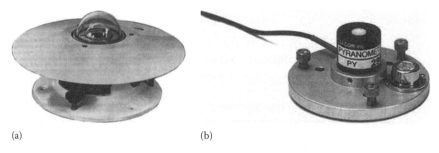

(a)   (b)

FIGURE 5.20
Typical commercially available pyranometers with (a) thermal detector and (b) photovoltaic detector.

FIGURE 5.21

A pyranometer with a shade ring to measure sky diffuse radiation.

TABLE 5.10

Shading Band Correction Factors due to Anisotropy

Solar Altitude Angle	k_T									
	0.0	0.1	0.2	0.3	0.4	0.5	0.6	0.7	0.8	0.9
<20°	0.0	0.0	0.0	0.0	0.015	0.06	0.14	0.23	0.24	0.24
20°–40°	0.0	0.0	0.0	0.0	0.006	0.05	0.125	0.205	0.225	0.225
40°–60°	0.0	0.0	0.0	0.0	0.003	0.045	0.115	0.175	0.205	0.205
60°+	0.0	0.0	0.0	0.0	0.0	0.035	0.09	0.135	0.17	0.17

Source: Gautier, C. et al., *J. Appl. Meteorol.*, 19(8), 1005, 1980.

of the diffuse solar radiation within the view angle. Figure 5.22 shows the geometry of a pyrheliometer sky occulting tube.

In this figure, the opening half angle

$$\theta_o = \tan^{-1}\frac{R}{L} \tag{5.81}$$

The slope angle

$$\theta_p = \tan^{-1}\left[\frac{(R-r)}{L}\right] \tag{5.82}$$

The limit half angle

$$\theta = \tan^{-1}\left[\frac{(R+r)}{L}\right] \tag{5.83}$$

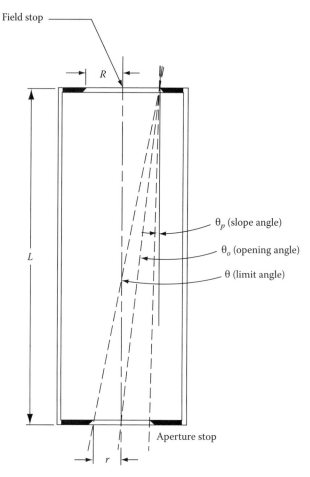

FIGURE 5.22
Geometry of a pyrheliometer sky occulting tube.

The field of view is $2\theta_o$. The World Meteorological Organization (WMO) recommends the opening half angle θ_o to be 2.5° [29] and the slope angle θ_p to be 1°.

Continuous tracking of the sun is required for the accuracy of the measurements. This is obtained by employing a tracking mechanism with two motors, one for altitude and the other for azimuthal tracking. Another problem is that the view angle of a pyrheliometer is significantly greater than the angle subtended by the solar disk (about 0.5°). Therefore, the measurements using a pyrheliometer include the beam and the circumsolar radiation. These measurements may present a problem in using the data for central receiver systems that use only direct beam radiation. However, this is not a significant problem for parabolic trough concentrators that in most cases have field of view on the order of 5°.

5.8.2 Detectors for Solar Radiation Instrumentation

Solar radiation detectors are of four basic types [28,29]: thermomechanical, calorimetric, thermoelectric, and photoelectric. Of these, thermoelectric and photoelectric are the most common detectors in use today.

A *thermoelectric detector* uses a thermopile that consists of a series of thermocouple junctions. The thermopile generates a voltage proportional to the temperature difference between the hot and cold junctions, which, in turn, is proportional to the incident solar radiation.

Photovoltaic detectors normally use silicon solar cells measuring the short circuit current. Such detectors have the advantage of being simple in construction. Because heat transfer is not a consideration, they do not require clear domes or other convection suppressing devices. They are also insensitive to tilt as the output is not affected by natural convection. One of the principal problems with photovoltaic detectors is their spectral selectivity. Radiation with wavelengths greater than the band gap of the photovoltaic detector cannot be measured. Silicon has a band gap of 1.07 eV corresponding to a wavelength of 1.1 μm. A significant portion of the infrared part of solar radiation has wavelengths greater than 1.1 μm. Therefore, photovoltaic detectors are insensitive to changes in the infrared part of solar radiation.

5.8.3 Measurement of Sunshine Duration

The time duration of bright sunshine data is available at many more locations in the world than the solar radiation. That is why a number of researchers have used these data to estimate the available solar radiation. Two instruments are widely used to measure the sunshine duration. The device used by the U.S. NWS is called a *sunshine switch*. It is composed of two photovoltaic cells—one shaded, the other not. During daylight, a potential difference is created between the two cells, which in turn operates the recorder. The intensity level required to activate the device is that just sufficient to cast a shadow. The other device commonly used to measure the sunshine duration is called the *Campbell–Stokes sunshine recorder*. It uses a solid clear glass sphere as a lens to concentrate the solar beam on the opposite side of the sphere. A strip of standard treated paper marked with time graduations is mounted on the opposite side of the sphere where the solar beam is concentrated. Whenever the solar radiation is above a threshold, the concentrated beam burns the paper. The length of the burned part of the strip gives the duration of bright sunshine. The problems associated with the Campbell–Stokes sunshine recorder include the uncertainties of the interpretation of burned portions of the paper, especially on partly cloudy days, and the dependence on the ambient humidity.

5.8.4 Measurement of Spectral Solar Radiation

Spectral solar radiation measurements are made with spectroradiometers. Full spectrum scanning is difficult, requires constant attention during operation, and is therefore expensive. Zerlaut [29] has described a number of solar spectroradiometers. These instruments consist basically of a monochromator, a detector–chopper assembly, an integrating sphere, and a signal conditioning/computer package. They have the capability of measuring solar radiation in the wavelength spectrum of 280–2500 nm.

5.8.5 Solar Radiation Data and Websites

Measured solar radiation data are available at a number of locations throughout the world. Data for many other locations have been estimated based on measurements at similar climatic locations. Some of the available data from various locations in the world are presented in Tables W.5.3 and W.5.4 in the website, http://www.crcpress.com/product/isbn/9781466556966. Table W.5.4 on the website also provides tables of modeled clear-sky data for various latitudes.

Solar radiation data for United States are available from the National Climatic Data Center (NCDC, http://www.ncdc.noaa.gov/oa/climate/climateinventories.html) of the National Oceanic and Atmospheric Administration (NOAA), and the NREL (http://www.nrel.gov/rredc/solar_resource.html). NOAA also provides data on the percent of possible sunshine for locations throughout the United States (http://www.ncdc.noaa.gov/oa/climate/online/ccd/pctpos.txt). In the mid-1970s, NOAA compiled a data base of measured hourly global horizontal solar radiation for 28 locations for the period 1952–1975 (called SOLMET) and of data for 222 additional sites (called ERSATZ) estimated from SOLMET data and some climatic parameters such as sunshine duration and cloudiness. NOAA also has two data sets of particular interest to engineers and designers: the typical meteorological year (TMY) and the Weather Year for Energy Calculations (WYEC) data sets. TMY data set represents typical values from 1952 to 1975 for hourly distribution of direct beam and global horizontal solar radiation. WYEC data set contains monthly values of temperature, direct beam, and diffuse solar radiation and estimates of *illuminance* (for daylighting applications). Illuminance is solar radiation in the visible range to which the human eye responds. Recently, NREL compiled a National Solar Radiation Data Base (NSRDB, http://rredc.nrel.gov/solar/old_data/nsrdb/) for 239 stations in the United States [32,33]. NSRDB is a collection of hourly values of global horizontal, direct normal, and diffuse solar radiation based on measured and estimated values for a period of 1961–1990. Since long-term measurements were available for only about 50 stations, measured data make up only about 7% of the total data in the NSRDB. A TMY data set from NSRDB is available as TMY2. NREL also provides maps with averaged daily solar radiation per month (http://rredc.nrel.gov/solar/old_data/nsrdb/redbook/atlas/).

The data for other locations in the world is available from national government agencies of most countries of the world. Worldwide solar radiation data are also available from the World Radiation Data Center (WRDC) in St. Petersburg, Russia, based on worldwide measurements made through local weather service operations [34]. WRDC, operating under the auspices of the WMO, has been archiving data from over 500 stations and operates a worldwide website in collaboration with NREL with an address of http://wrdc.mgo.nrel.gov. An International Solar Radiation Data Base was also developed by the University of Lowell [35].

Problems

5.1 Calculate the declination, the zenith angle, and the azimuth angle of the sun for New York City (latitude 40.77°N) on October 1 at 2:00 p.m. solar time.

5.2 A solar energy system in Gainesville, Florida, requires two rows of collectors facing south and tilted at a fixed 30° angle. Find the minimum normalized distance at which the second row should be placed behind the first row for no shading at noon at winter solstice. What percentage of the second row is shaded on the same day at 9:00 a.m. solar time?

5.3 Find the sunrise and sunset times for a location of your choice on September 1.

5.4 Construct a table of hourly sun angles for the 15th day of each month for a location of your choice. Also show the sunrise and sunset times for those days.

5.5 Referring to Figure 5.19, prove Equation 5.46 for the angle of incidence. (*Hint*: Use direction cosines of the sun-ray vector and a vector normal to the tilted surface to find the angle between them. Dot product of two unit vectors gives the cosine of the angle between them.)

5.6 Determine the following for a south-facing surface at 30° slope in Gainesville, Florida (latitude = 29.68°N, longitude = 82.27°W) on September 21 at noon solar time:

 a. Zenith angle

 b. Angle of incidence

 c. Beam radiation

 d. Diffuse radiation

 e. Reflected radiation

 f. Total radiation

 g. Local time

5.7 Show that the hourly averaged, extraterrestrial radiation for a given hour is the same, to within 1%, as instantaneous radiation at the hour's midpoint. This is equivalent to deriving Equation 5.76.

5.8 Prepare shadow maps for point P on the sun-path diagrams for 35°N and 40°N for the following three geometries shown in a, b, and c. Determine the hours of shading that occur each month.

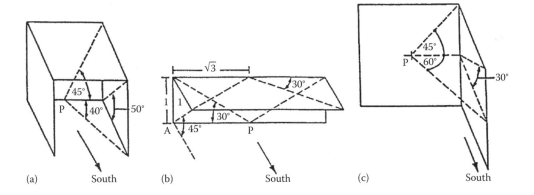

(a)   South   (b)   South   (c)   South

5.9 Repeat Problem 5.8c if the surface containing point P faces due west instead of due south for a 40°N location.

5.10 Calculate the incidence angle at noon and 9 a.m. on a fixed flat-plate collector located at 40°N latitude and tilted 70° up from the horizontal. Find i for June 21 and December 21.

5.11 a. If the surface in Problem 5.10 faces S 45° E, what are the incidence angles?

 b. If the collector in Problem 5.10 has a cylindrical surface, what are the incidence angles on June 21 and December 21?

5.12 Using a one-term Fourier cosine series, develop an empirical equation for solar declination as a function of day number counted from January 1 (see Table W.5.1 and Figure 5.7).

5.13 Derive an equation for the lines of constant declination in a sun-path diagram, for example, Figure 5.10d. Check your equation by plotting a few declination lines on a piece of polar coordinate graph paper.

5.14 Derive Equation 5.33 relating profile angle γ to azimuth angle α and altitude angle α.

5.15 Based on Equation 5.46, what value of β would result in the annual minimum value of the incidence angle i? Note that this tilt angle would result in maximum collection of beam radiation on a fixed, flat, south-facing surface. *Hint:* Use a double integration procedure.

5.16 At what time does the sun set in Calcutta ($23°N$) on May 1 and December 1?

5.17 What is the true solar time in Sheridan, Wyoming ($107°W$), at 10:00 a.m. Mountain Daylight Time on June 10? What is the true solar time at 10:00 a.m. Mountain Standard Time on January 10?

5.18 Using the Angström–Page method, calculate the average horizontal insolation in Hamburg in May and in October with $\overline{PS}$ 40% and 60%, respectively.

5.19 Equation 5.53 is based on an early solar constant value of 1394 W/m^2. Derive a modified form of Equation 5.53 based on the presently accepted value for the solar constant of 1353 W/m^2.

5.20 Predict the hourly beam and diffuse radiation on a horizontal surface for Denver ($40°N$) on September 9 at 9:30 a.m. on a clear day.

5.21 Derive an expression for the minimum allowable distance between east–west rows of solar collectors that will assure no shading of one row by the row immediately to the south. Use the law of sines and express the result in terms of the collector tilt and face length and the controlling value of the solar profile angle.

5.22 Using the $\overline{H}_h$ data calculated in Example 5.7 in place of the long-term measured data for the North Central Sahara Desert at latitude $25°N$, find the monthly averaged insolation per day on a south-facing solar collector tilted at an angle of $25°$ from the horizontal.

References

1. Howell, J.R., Siegel, R., and Menguc, M.P. (2010) *Thermal Radiation Heat Transfer*, McGraw-Hill Book Co., New York.
2. Kreith, F. (1962) *Radiation Heat Transfer for Spacecraft and Solar Power Plant Design*, International Textbook Co., Scranton, PA.
3. Kreith, F., Mangelik, R.M., and Bohn, S. (2011) *Principles of Heat Transfer*, 7th edn., Cengale Publishing Co., St. Paul, MN.
4. Sparrow, E M. and Cess, R.D. (1978) *Radiation Heat Transfer*, Wadsworth Publ. Co., Belmont, CA.
5. Threlkeld, J.L. and Jordan, R.C. (1958) Direct radiation available on clear days, *Transactions of ASHRAE*, 64, 45.
6. Frohlich, C. et al. (1973) The third international comparison of pyrheliometers and a comparison of radiometric scales. *Solar Energy*, 14, 157–166.
7. Quinlan, F.T. (ed.) (1977) *SOLMET Vol. 1: Hourly Solar Radiation Surface Meteorological Observations*, NOAA, Asheville, NC.
8. Quinlan, F.T. (ed.) (1979) *SOLMET Vol. 2: Hourly Solar Radiation Surface Meteorological Observations*, NOAA, Asheville, NC.

9. Kimura, K. and Stephenson, D.G. (1969) Solar radiation on cloudy days, *Transactions of ASHRAE*, 75, 227–234.
10. Thekaekara, M.P. (1973) Solar energy outside the Earth's atmosphere, *Sol. Energy*, 14, 109–127.
11. Page, J.K. (1966) The estimation of monthly mean values of daily total short-wave radiation on vertical and inclined surfaces from sunshine records for latitudes 40°N–40°S, *Proceeding of United Nations Conference on New Sources Energy*, Vol. 4, p. 378.
12. Löf, G.O.G. et al. (1966) World distribution of solar energy, University of Wisconsin, Madison, WI, Eng Expt. Station Rept. 21.
13. Liu, B.Y.H. and Jordan, R.C. (1961) Daily insolation on surfaces titled toward the equator, *Transactions of ASHRAE*, 67, 526–541.
14. Liu, B.Y.H. and Jordan, R.C. (1961) Daily insolation on surface titled toward the equator, *Transactions of ASHRAE*, 3(10), 53–59.
15. Liu, B.Y.H. and Jordan, R.C. (1967) Availability of solar energy for flat-plate solar heat collectors, in: *Low Temperature Engineering of Solar Energy*, Chap. 1, ASHRAE, New York; see also 1977 revision.
16. Liu, B.Y.H. and Jordan, R.C. (1963) A rational procedure for predicting the long-term average performance of flat-plate solar energy collectors, *Solar Energy*, 7, 53–74.
17. Liu, B.Y.H. and Jordan, R.C. (1960) The interrelationship and characteristic distribution of direct, diffuse and total solar radiation, *Solar Energy*, 4, 1–19. See also Liu, B.Y.H. (1960) Characteristics of solar radiation and the performance of flat plate solar energy collectors, PhD dissertation, University of Minnesota, Minneapolis, MN.
18. Collares-Pereira, M. and Rabl, A. (1979) The average distribution of solar radiation-correlation between diffuse and hemispherical, *Solar Energy*, 22, 155–166.
19. Choudhury, N.K.O. (1963) Solar radiation at New Delhi, *Solar Energy*, 7, 44.
20. Stanhill, G. (1966) Diffuse sky and cloud radiation in Israel, *Solar Energy*, 10, 66.
21. Ruth, D.W. and Chant, R.E. (1976) The relationship of diffuse radiation to total radiation in Canada, *Solar Energy*, 18, 153.
22. Temps, R.C. and Coulson, K.L. (1977) Solar radiation incident upon slopes of different orientations, *Solar Energy*, 19(2), 179–184.
23. Klucher, T.M. (1979) Evaluation of models to predict insolation on tilted surfaces, *Solar Energy*, 23(2), 111–114.
24. Budde, W. (1983) *Physical Detectors of Optical Radiation*, Vol. 4 of *Optical Radiation Measurements*, Academic Press, New York.
25. Randall, C.M. and Leonard, S.L. (1974) Reference insolation data base: A case history, with recommendations, in: *Rept. Recommendations Solar Energy Data Workshop*, pp. 93–103.
26. Vant-Hull, L. and Easton, C.R. (1975) Solar thermal power systems based on optical transmission, NTIS Rept. PB253167.
27. Wilcox, S. and Marion, W. (April/May 2008) *Users Manual for TMY3 Data Sets*, Technical Report NREL/TP-581-43156, National Renewable Energy Laboratory, Golden, CO.
28. Iqbal, M. (1983) *An Introduction to Solar Radiation*, Academic Press, New York.
29. Zerlaut, G. (1989) Solar radiation instrumentation, in: Roland L.H. (ed.), *Solar Resources*, MIT Press, Cambridge, MA.
30. Gautier, C., Diak, G., and Masse, S. (1980) A simple physical model to estimate incident solar radiation at the surface from GOES satellite data, *Journal of Applied Meteorology*, 19(8), 1005.
31. Mujahid, A. and Turner, W.D. (1979) Diffuse sky measurement and model, ASME Pap. No. 79-WA/Sol-5.
32. Maxwell, E.L. (1998) METSTAT—The solar radiation model used in the production of the national solar radiation data base (NSRDB), *Solar Energy*, 62, 263–279.
33. NSRDB. (1992) *Volume 1: Users Manual: National Solar Radiation Data Base (1961–1990), Version 1.0*, National Renewable Energy Laboratory, Golden, CO.
34. Voeikov Main Geophysical Observatory. (1999) Worldwide daily solar radiation. http://www.mgo.rssi.ru (accessed on April 18, 2013).

35. University of Lowell Photovoltaic Program. (1990) *International Solar Irradiation Database*, Version 1.0, University of Lowell Research Foundation, Lowell, MA.
36. Whillier, A. (1965) Solar radiation graphs, *Solar Energy*, 9, 165–166.
37. Kreider, J.F. and Kreith, F. (1982) *Solar Heating and Cooling*, Hemisphere Publishing Corp., Washington, DC.
38. Akinoglu, B.G. (1991) A review of sunshine-based models used to estimate monthly average global solar radiation, *Renewable Energy*, 1(3), 479–499.
39. Jain, S. and Jain, P.G. (1985) A comparison of the angstrom type correlations and the estimation of monthly average daily global irradiation, International Centre for Theoretical Physics, Internal Report, No. IC85-269.
40. Zabara, K. (1986) Estimation of the global solar radiation in Greece, *Solar Wind Technology*, 3, 267.
41. Fritz, S. (1949) Solar radiation in the United States, *Heater Venting*, 16, 61–64.
42. Garg, H.P. and Garg, S.N. (1985) Correlation of the monthly average daily global, diffuse and beam radiation with bright sunshine hours, *Energy Conversion and Management*, 25(4), 409–417.
43. Cano, D. et al. (1986) Method for the determination of the global solar radiation from meteorological satellite data, *Solar Energy*, 37(1), 31–39.
44. Dedieu, G., Deschamps, P.Y., and Kerr, Y.H. (1987) Satellite estimation of solar irradiance at the surface of the earth and of surface albedo using a physical model applied to METEOSAT data, *Journal of Climate and Applied Meteorology*, 26, 79–87.
45. Massaquoi, J.G.M. (1988) Global solar radiation in Sierre Leone (West Africa), *Solar Wind Technology*, 5, 281.
46. Jain, P.C. (1986) Irradiation estimation for Italian localities, *Solar Wind Technology*, 3, 323.
47. Khogali, A. (1983) Star radiation over Sudan: Comparison of measured and predicted data, *Solar Energy*, 31, 45.
48. Gopinathan, K. (1988) A general formula for computing the coefficients of the correlation connecting global solar radiation to sunshine duration, *Solar Energy*, 41, 499.
49. Hunn, B.D. and Calafell, D.O. (1977) Determination of average ground reflectivity for solar collectors, *Solar Energy*, 19, 87.
50. List, R.J. (1949) *Smithsonian Meteorological Tables*, 6th edn., Smithsonian Institution Press, Washington, DC, pp. 442–443.
51. Iqbal, M. (1979) Correlation of average diffuse and beam radiation with hours of bright sunshine, *Solar Energy*, 23, 169.
52. Islam, M.R. (1994) Evolution of methods for solar radiation mapping using satellite data. *RERIC International Energy Journal*, 16(2), 103–113.

6

Photovoltaics*

We do not inherit the earth from our ancestors, we borrow it from our children.

—**Chief Seattle**

Photovoltaic conversion is the direct conversion of sunlight into electricity with no intervening heat engine. Photovoltaic devices are solid-state devices; therefore, they are rugged and simple in design and require very little maintenance. Perhaps, the biggest advantage of solar photovoltaic devices is that they can be constructed as stand-alone systems to give outputs from microwatts to megawatts. That is why they have been used as the power sources for calculators, watches, water pumping, remote buildings, communications, satellites and space vehicles, and even megawatt-scale power plants. Photovoltaic panels can be made to form components of building skin, such as roof shingles and wall panels. With such a vast array of applications, the demand for photovoltaics is increasing every year. The market for photovoltaic solar cells is growing exponentially, increasing from 227 MW in 2000 to 3,800 MW in 2008 and to 20,500 MW in 2010 [1] and is expected to continue increasing exponentially. The global market created $82 billion in revenue in 2010 [1]. An example of a photovoltaic residential rooftop system is shown in Figure 6.1.

In the early days of solar cells in the 1960s and 1970s, more energy was required to produce a cell than it could ever deliver during its lifetime. Since then, dramatic improvements have taken place in the efficiencies and manufacturing methods. The energy payback periods have been reduced to about 2.5–5 years, depending on the location of use [2], while panel lifetimes were increased to over 25 years. The costs of photovoltaic panels have come down to less than $5 per peak watt over the last two decades and are targeted to reduce to around $1.00 per peak watt soon.

Historically, the photoelectric effect was first noted by Becquerel in 1839 when light was incident on an electrode in an electrolyte solution [3]. Adams and Day first observed the effect in solids in 1877 while working with selenium. Early work was done with selenium and copper oxide by pioneers such as Schottky, Lange, and Grandahl. In 1954, researchers at RCA and Bell Laboratories reported achieving efficiencies of about 6% by using devices made of p and n types of semiconductors. The space race between the United States and the Soviet Union resulted in dramatic improvements in photovoltaic devices. Reference [4] gives a review of the early developments in photovoltaic conversion.

6.1 Semiconductors

A basic understanding of the atomic structure is quite helpful in understanding the behavior of semiconductors and their use as the photovoltaic energy conversion devices. Any fundamental book on physics or chemistry generally gives adequate background

---

* Sections in this chapter marked with an asterisk may be omitted in an introductory course.

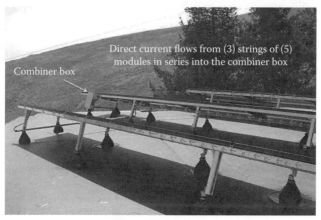

Combiner box
600 V Fuses protect each of the (3) strings of (5) modules
after the fuses the (3) circuits are combined in parallel
The combined circuit is then brought to the DC disconnect in the garage

FIGURE 6.1
(See color insert.) Kreith residence, 3 kW Sanyo high-efficiency photovoltaic system. (Courtesy of Chris Klinga, Lighthouse Solar, Boulder, CO. With permission.)

for basic understanding. Reference [4] presents an in-depth treatment of a number of topics in semiconductor physics.

For any atom, the electrons arrange themselves in orbitals around the nucleus so as to result in the minimum amount of energy. Table 6.1 shows the distribution of the electrons in various shells and subshells in light elements. In elements that have electrons in multiple shells, the innermost electrons have the minimum energy and, therefore, require the maximum amount of externally imparted energy to overcome the attraction of the nucleus and become free. Electrons in the outermost band of subshells are the only ones that participate in the interaction of an atom with its neighboring atoms. If these electrons are very loosely attached to the atom, they may attach themselves with a neighboring atom to give that atom a negative charge, leaving the original atom as a positively charged ion. The positively and negatively charged ions become attached by the force of attraction of the charges, thus forming *ionic bonds*. If the electrons in the outermost band do not fill the band completely but are not loosely attached either, they arrange themselves so that neighboring atoms can share them to make the outermost bands full. The bonds thus formed between the neighboring atoms are called *covalent bonds*.

Since electrons in the outermost band of an atom determine how an atom will react or join with a neighboring atom, the outermost band is called the *valence band*. Some electrons in the valence band may be so energetic that they jump into a still higher band and are so far removed from the nucleus that a small amount of impressed force would cause them to move away from the atom. Such electrons are responsible for the conduction of heat and electricity, and this remote band is called a *conduction band*. The difference in the energy of an electron in the valence band and the innermost subshell of the conduction band is called the *band gap* or the forbidden gap.

Materials whose valence bands are full have very high band gaps (>3 eV). Such materials are called *insulators*. Materials, on the other hand, that have relatively empty valence bands and may have some electrons in the conduction band are good *conductors*. Metals fall in this category. Materials with valence bands partly filled have intermediate band gaps (≤3 eV). Such materials are called *semiconductors* (Figure 6.2). Pure semiconductors are called *intrinsic semiconductors*, while semiconductors doped with very small amounts of impurities are called *extrinsic semiconductors*. If the dopant material has more electrons in the valence band than the semiconductor, the doped material is called an *n*-type of semiconductor. Such a material seems to have excess electrons available for conduction even though the material is electronically neutral. For example, silicon has four electrons in the valence band. Atoms of pure silicon arrange themselves in such a way that, to form a stable structure, each atom shares two electrons with each neighboring atom with covalent bands. If phosphorous, which has five valence electrons (one more than Si), is introduced as an impurity in silicon, the doped material seems to have excess electrons even though it is electrically neutral. Such a doped material is called *n*-type silicon. If on the other hand, silicon is doped with boron, which has three valence electrons (one less than Si), there seems to be a positive hole (missing electrons) in the structure, even though the doped material is electrically neutral. Such material is called *p*-type silicon. Thus, *n*- and *p*-type semiconductors make it easier for the electrons and holes, respectively, to move in the semiconductors.

6.1.1 *p–n* Junction

As explained earlier, an *n*-type material has some impurity atoms with more electrons than the rest of the semiconductor atoms. If those excess electrons are removed, the impurity atoms will fit more uniformly in the structure formed by the main semiconductor

TABLE 6.1

Electronic Structure of Atoms

Principal Quantum Number n			1	2		3			4	5
Azimuthal Quantum Number l			0	0	1	0	1	2	0	1
Letter Designation of State			1s	2s	2p	3s	3p	3d	4s	4p
Z	Symbol	Element	V_i (V)							
1	H	Hydrogen	13.60	1						
2	He	Helium	24.58	2						
3	Li	Lithium	5.39		1					
4	Be	Beryllium	9.32		2					
5	B	Boron	8.30		2	1				
6	C	Carbon	11.26	Helium core	2	2				
7	N	Nitrogen	14.54		2	3				
8	O	Oxygen	13.61		2	4				
9	F	Fluorine	17.42		2	5				
10	Ne	Neon	21.56		2	6				
11	Na	Sodium	5.14				1			
12	Mg	Magnesium	7.64				2			
13	Al	Aluminum	5.98				2	1		
14	Si	Silicon	8.15	Neon core			2	2		
15	P	Phosphorus	10.55				2	3		
16	S	Sulfur	10.36				2	4		
17	Cl	Chlorine	13.01				2	5		
18	A	Argon	15.76				2	6		
19	K	Potassium	4.34							1
20	Ca	Calcium	6.11							2
21	Sc	Scandium	6.56						1	2
22	Ti	Titanium	6.83						2	2
23	V	Vanadium	6.74						3	2
24	Cr	Chromium	6.76						5	1
25	Mn	Manganese	7.43						5	2
26	Fe	Iron	7.90						6	2
27	Co	Cobalt	7.86						7	2
28	Ni	Nickel	7.63	Argon core					8	2
29	Cu	Copper	7.72						10	1
30	Zu	Zinc	9.39						10	2
31	Ga	Gallium	6.00						10	2	1
32	Ge	Germanium	7.88						10	2	2
33	As	Arsenic	9.81						10	2	3
34	Se	Selenium	9.75						10	2	4
35	Br	Bromine	11.84						10	2	5
36	Kr	Krypton	14.00						10	2	6

Source: Moore, C.E., *Atomic Energy Levels*, Vol. 2, National Bureau of Standards Circular 467, U.S. Government Printing Office, Washington, DC, 1952. With permission.

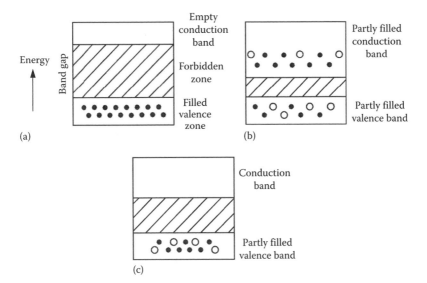

FIGURE 6.2
Electrical conduction is described in terms of allowed and forbidden energy bands. Band gap for insulators (a) is the highest, followed by metals (b) and semiconductors (c), respectively.

atoms; however, the atoms will be left with positive charges. On the other hand, a *p*-type material has some impurity atoms with fewer electrons than the rest of the semiconductor atoms. Therefore, these atoms seem to have holes that could accommodate excess electrons even though the atoms are electrically neutral (Figure 6.3). If additional electrons could be brought to fill the holes, the impurity atoms would fit more uniformly in the structure formed by the main semiconductor atoms; however, the atoms will be negatively charged.

The earlier scenario occurs at the junction when a *p*- and an *n*-type of material are joined together, as shown in Figure 6.4. As soon as the two materials are joined, "excess" electrons from the *n* layer jump to fill the "holes" in the *p* layer. Therefore, close to the junction, the material has positive charges on the *n*-side and negative charges on the *p*-side. The negative charges on the *p*-side restrict the movement of additional electrons from the *n*-side to the *p*-side, while the movement of additional electrons from the *p*-side to the *n*-side is made easier because of the positive charges at the junction on the *n*-side. This restriction makes the *p*–*n* junction behave like a diode. This diode character of a *p*–*n* junction is made use of in solar photovoltaic cells, as explained in the following text.

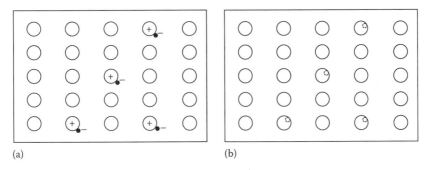

FIGURE 6.3
Representation of *n*- and *p*-type semiconductors: (a) *n*-type showing "excess" electrons as dots; (b) *p*-type showing "excess" positive holes as o.

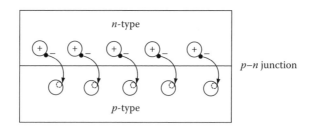

FIGURE 6.4
"Excess" electrons from *n* material jump to fill "excess" holes on the *p*-side of a *p–n* junction, leaving the *n*-side of the junction positively charged and the *p*-side negatively charged.

6.1.2 Photovoltaic Effect

When a photon of light is absorbed by a valence electron of an atom, the energy of the electron is increased by the amount of energy of the photon. If the energy of the photon is equal to or more than the band gap of these semiconductors, the electron with excess energy will jump into the conduction band where it can move freely. If, however, the photon energy is less than the band gap, the electron will not have sufficient energy to jump into the conduction band. In this case, the excess energy of electrons is converted to excess kinetic energy of electrons, which manifests in increased temperature. If the absorbed photon had more energy than the band gap, the excess energy over the band gap simply increases the kinetic energy of the electron. It must be noted that a photon can free up only one electron even if the photon energy is a lot higher than the band gap. This fact is a big reason for the low conversion efficiency of photovoltaic devices. The key to using the photovoltaic effect for generating useful power is to channel the free electrons through an external resistance before they recombine with the holes. This is achieved with the help of the *p–n* junction.

Figure 6.5 shows a schematic of a photovoltaic device. As free electrons are generated in the *n* layer by the action of photons, they can either pass through an external circuit, recombine with positive holes in the lateral direction, or move toward the *p* layer. The negative charges in the *p* layer at the *p–n* junction restrict their movement in that direction. If the *n* layer is made extremely thin, the movement of the electrons and, therefore, the probability of recombination within the *n* layer are greatly reduced unless the

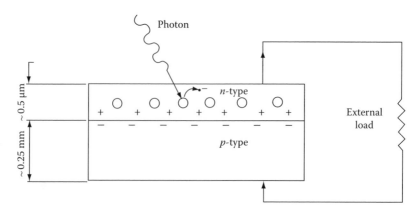

FIGURE 6.5
Schematic of a photovoltaic device.

external circuit is open. If the external circuit is open, the electrons generated by the action of photons eventually recombine with the holes, resulting in an increase in the temperature of the device.

In a typical crystalline silicon cell, the n layer is about 0.5 μm thick and the p layer is about 0.25 mm thick. As explained in Chapter 5, energy contained in a single photon E_p is given by

$$E_p = h\nu \tag{6.1}$$

where
  h is Planck's constant (6.625×10^{-34} J-s)
  ν is the frequency, which is related to the wavelength λ and the speed of light c by

$$\nu = \frac{c}{\lambda}$$

Therefore,

$$E_p = \frac{hc}{\lambda}. \tag{6.2}$$

For silicon, which has a band gap of 1.11 eV, the following example shows that photons of solar radiation of wavelength 1.12 μm or less are useful in creating electron–hole pairs. This spectrum represents a major part of the solar radiation. Table 6.2 lists some candidate semiconductor materials for photovoltaic cells along with their band gaps.

Example 6.1

Calculate the wavelength of light capable of forming an electron–hole pair in silicon.

Solution
The band gap energy of silicon is 1.11 eV. From Equation 6.2, we can write

$$\lambda = \frac{hc}{E_p}.$$

For $c = 3 \times 10^8$ m/s, $h = 6.625 \times 10^{-34}$ J-s, and 1 eV = 1.6×10^{-19} J, the earlier equation gives the required wavelength as

$$\lambda = \frac{(6.625 \times 10^{-34}\ \text{J-s})(3 \times 10^8\ \text{m/s})}{(1.11)(1.6 \times 10^{-19}\ \text{J})} = 1.12\ \mu\text{m}$$

Example 6.2

A monochromatic red laser beam emitting 1×10^{-3} W at a wavelength of 638 nm is incident on a silicon solar cell. Find

  a. The number of photons per second incident on the cell
  b. The maximum possible efficiency of conversion of this laser beam to electricity

TABLE 6.2

Energy Gap for Some Candidate
Materials for Photovoltaic Cells

Material	Band Gap (eV)
Si	1.11
SiC	2.60
$CdAs_2$	1.00
CdTe	1.44
CdSe	1.74
CdS	2.42
$CdSnO_4$	2.90
GaAs	1.40
GaP	2.24
Cu_2S	1.80
CuO	2.00
Cu_2Se	1.40
$CuInS_2$	1.01
$CuInTe_2$	0.90
InP	1.27
In_2Te_3	1.20
In_2O_3	2.80
Zn_3P_2	1.60
ZnTe	2.20
ZnSe	2.60
AlP	2.43
AlSb	1.63
As_2Se_3	1.60
Sb_2Se_3	1.20
Ge	0.67
Se	1.60

Source: Garg, H.P., *Advances in Solar Energy Technology,* Vol. 3, D. Reidel Publishing Company Dordrecht, the Netherlands, 1987. With permission.

Solution

a. The intensity of light in the laser beam (I_p) is equal to the energy of all the photons in it. If the number of photons is N_{ph}, then

$$I_p = N_{ph} \cdot E_p$$

$$1 \times 10^{-3} \text{ W} = N_{ph} \cdot E_p \tag{6.3}$$

$$E_p = \frac{hc}{\lambda}$$

$$= \frac{(6.625 \times 10^{-34} \text{ J-s}) \cdot 3 \times 10^8 \text{ m/s}}{638 \times 10^{-9} \text{ m}}$$

$$= 3.12 \times 10^{-19} \text{ J/photon}$$

$$\therefore N_{ph} = \frac{1 \times 10^{-3} \text{ J/s}}{3.12 \times 10^{-19} \text{ J/photon}} = 3.21 \times 10^{15} \text{ photons/s}$$

b. Assuming that each photon is able to generate an electron, a total number of N_{ph} electrons will be generated. Therefore, the electrical output will be equal to N_{ph} (n). Therefore, the maximum possible efficiency is

$$\eta_{max} = \frac{(N_{ph}) \cdot (\text{BG})}{(N_{ph}) \cdot E_p} = \frac{\text{BG}}{E_p}$$

$$= \frac{1.11 \times 1.6 \times 10^{-19} \text{ J}}{3.12 \times 10^{-19} \text{ J}} = 0.569\% \quad \text{or} \quad 56.9\% \tag{6.4}$$

From the earlier examples, it is clear that for a silicon solar cell, none of the photons of the sunlight over 1.12 μm wavelength will produce any electricity. However, photons of sunlight at a wavelength of 1.12 μm may be converted to electricity at a maximum efficiency of 100%, while photons at lower wavelengths will be converted at lower efficiencies. The overall maximum efficiency of a cell can be found by integrating the efficiency at each wavelength over the entire solar spectrum:

$$\eta = \frac{\int \eta_\lambda l_\lambda d\lambda}{\int l_\lambda d\lambda} \tag{6.5}$$

In addition, other factors such as probability of electron–hole recombination reduce the theoretical maximum achievable efficiency of a solar cell. Figure 6.6 shows a comparison of the maximum energy conversion of cells using different materials as well as the optimum band gap for terrestrial solar cells to be about the order of 1.5 eV.

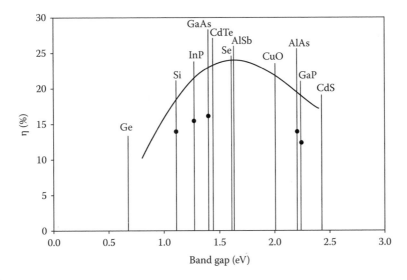

FIGURE 6.6
The maximum solar energy conversion efficiency as a function of the energy gap of the semiconductor. The curve has been calculated for an ideal junction outside the atmosphere.

6.2 Analysis of Photovoltaic Cells

This section presents an electrical analysis of photovoltaic cells, which will be useful in the design of photovoltaic devices for various applications. The physics leading to the expressions for the number density of electrons and holes in n and p materials at a temperature T will not be presented here. For such details, the reader is referred to books such as Refs. [3,4]. It would suffice to point out here that at the p–n junction, a current is generated called the *junction current*. The junction current J_j is the net current due to the J_o from the p-side to the n-side (called the *dark current* or the *reverse saturation current*) and a *light-induced recombination current* J_r from the n-side to the p-side. Based on the temperature T, a certain number of electrons in the p material exist in the conduction band. These electrons can easily move to the n-side to fill the holes created at the p–n junction, generating a current J_o. Normally, the electrons occupying the conduction band due to the temperature in the n material do not have enough potential energy to cross the p–n junction to the p-side. However, if a forward-bias voltage V is applied, which in a photovoltaic cell is due to the action of the photons of light, some of the electrons thus generated have enough energy to cross over and recombine with the holes in the p region. This gives rise to a light-induced recombination current J_r, which is proportional to J_o and is given by

$$J_r = J_o \exp\left(\frac{e_o V}{kT}\right)$$

(6.6)

where
e_o is the charge of an electron = 1.602×10^{-19} C or J/V
k is Boltzmann's constant = 1.381×10^{-23} J/K

The junction current J_j is the net current due to J_r and J_o.

$$J_j = J_r - J_o = J_o\left[\exp\left(\frac{e_o V}{kT}\right) - 1\right]$$

(6.7)

Referring to Figure 6.7, it is clear that the current generated in the cell has two parallel paths: one through the junction and the other through the external resistance, R_L. Figure 6.7 shows an equivalent circuit of a photovoltaic cell. It must be pointed out here that the current generated in a photovoltaic cell, including the junction current, is proportional to the area of the cell. Therefore, it is appropriate to analyze in terms of the current density J (current per unit area) instead of the current I. The relationship between the two is

$$I = J \cdot A$$

(6.8)

Referring to Figure 6.7, we can write

$$J_L = J_s - J_j = J_s - J_o\left[\exp\left(\frac{e_o V}{kT}\right) - 1\right]$$

(6.9)

where J_s is the short circuit current.
For short circuit, $V = 0$ and $J_L = J_s$.
For open circuit, $J_L = 0$ and $V = V_{oc}'$ which gives

$$0 = J_s - J_o\left[\exp\left(\frac{e_o V_{oc}}{kT}\right) - 1\right]$$

(6.10)

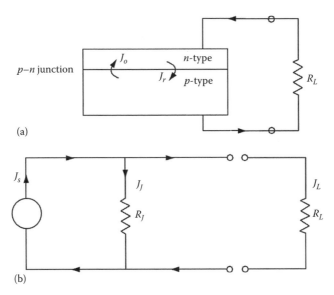

FIGURE 6.7
Equivalent circuit of a photovoltaic cell. (a) *p–n* junction. (b) Electrical circuit.

or

$$V_{oc} = \frac{kT}{e_o} \ln\left(\frac{J_s}{J_o} + 1\right)$$

Figure 6.8 shows a typical performance curve (*I–V*) of a solar cell. The power output is the product of the load current and voltage and is a function of the load resistance

$$P_L = AJ_L V = I_L V$$

$$= I_L^2 R_L \tag{6.11}$$

where *A* is the area of the cell.

The power output exhibits a maximum. To find the condition for the maximum power output (P_{max}), differentiate *P* with respect to *V* and equate it to zero:

$$\exp\left(\frac{e_o V_m}{kT}\right)\left(1 + \frac{e_o V_m}{kT}\right) = 1 + \frac{J_s}{J_o} \tag{6.12}$$

where V_m stands for voltage at maximum power. The current at maximum power condition $J_{L,m}$ and the maximum power P_{max} can be found from Equations 6.9 and 6.11, respectively.

$$J_{L,m} = J_s - J_o\left[\exp\left(\frac{e_o V_m}{kT}\right) - 1\right] \tag{6.13}$$

Combining Equation 6.12 and 6.13, $J_{L,m}$ is found to be

$$J_{L,m} = \frac{e_o V_m / kT}{1 + (e_o V_m / kT)}(J_s + J_o) \tag{6.14}$$

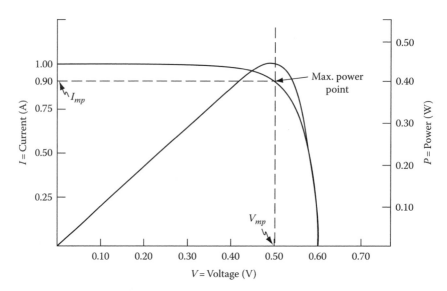

FIGURE 6.8
Typical current, voltage, and power characteristics of a solar cell.

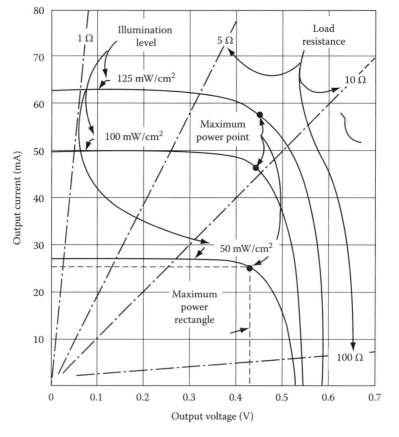

FIGURE 6.9
Typical current–voltage characteristics of a silicon cell showing the effects of illumination level and load resistance.

$$P_{max} = \frac{e_o V_m^2 / kT}{1 + (e_o V_m / kT)} (J_s + J_o) \cdot A \tag{6.15}$$

Figure 6.9 shows the effect of illumination intensity and the load resistance on the performance of a silicon cell. Temperature also affects the performance in such a way that the voltage and thus the power output decrease with increasing temperature.

Example 6.3

The dark current density for a silicon solar cell at 40°C is 1.8×10^{-8} A/m² and the short circuit current density is 200 A/m² when exposed to solar radiation of 900 W/m². Calculate

a. Open circuit voltage
b. Voltage at maximum power
c. Current density at maximum power
d. Maximum power
e. Maximum efficiency
f. The cell area required for an output of 25 W

Solution
Given

$$J_o = 1.8 \times 10^{-8} \text{A/m}^2$$

$$J_s = 200 \text{ A/m}^2$$

and

$$T = 40°C = 313 \text{ K}$$

a. Using Equation 6.10,

$$V_{oc} = \frac{kT}{e_o} \ln\left(\frac{J_s}{J_o} + 1\right)$$

Since (e_o/kT) will be needed for other parts of the problem solution, it will be evaluated separately:

$$\frac{e_o}{kT} = \frac{1.602 \times 10^{-19} \text{ J/V}}{(1.381 \times 10^{-23} \text{ J/K})(313 \text{ K})} = 37.06 \text{ V}^{-1}$$

Therefore,

$$V_{oc} = \frac{1}{37.60} \ln\left(\frac{200}{1.8 \times 10^{-8}} + 1\right) = 0.624 \text{ V}$$

b. Voltage at maximum power condition can be found from Equation 6.12 by an iterative or trial-and-error solution:

$$\exp(37.06 V_m)(1 + 37.06 V_m) = 1 + \frac{200}{1.8 \times 10^{-8}}$$

or $V_m = 0.542 \text{ V}$

c. Current density at maximum power can be found from Equation 6.14:

$$J_{L,m} = \frac{e_o V_m / kT}{1 + (e_o V_m / kT)} (J_s + J_o)$$

$$= \frac{(37.06) \cdot (0.542)}{1 + (37.06) \cdot (0.542)} (200 + 1.8 \times 10^{-8}) \, \text{A/m}^2$$

$$= 190.5 \, \text{A/m}^2$$

d. From Equation 6.15, the maximum power is

$$P_{max} = V_m \cdot J_m \cdot A$$

$$\frac{P_{max}}{A} = (0.542 \, \text{V}) \cdot (190.5 \, \text{A/m}^2)$$

$$= 103.25 \, \text{W/m}^2$$

e. The maximum efficiency is

$$\eta_{max} = \frac{103.25 \, \text{W/m}^2}{900 \, \text{W/m}^2} = 11.5\%$$

f. Cell area required

$$A = \frac{P_{out}}{P_{max}/A} = \frac{25 \, \text{W}}{103.25 \, \text{W/m}^2}$$

$$= 24.2 \, \text{cm}^2$$

6.2.1 Efficiency of Solar Cells

Theoretical limitation on the efficiency of a single-layer solar cell can be calculated using Equation 6.5. These efficiency limitation and the practical efficiencies of some of the cells are shown in Figure 6.6. Some of the reasons for the actual efficiency being lower than the theoretical limitation are

1. Reflection of light from the surface of the cell. This can be minimized by antireflection (AR) coating. For example, AR coating can reduce the reflection from a Si cell to 3% from 30% from an untreated cell.

2. Shading of the cell due to current collecting electrical contacts. This can be minimized by reducing the area of the contacts and/or making them transparent; however, both of these methods will increase the resistance of the cell to current flow.

3. Internal electrical resistance of the cell.

4. Recombination of electrons and holes before they can contribute to the current. This effect can be reduced in polycrystalline and amorphous cells by using hydrogen alloys.

6.2.2 Multijunction Solar Cells

The limits imposed on single-layer solar cells due to band gap can be partially overcome by using multiple layers of solar cells stacked on top of each other, with each layer having a band gap higher than the layer below it. For example (Figure 6.10), if the top layer is made from a cell of material A (band gap corresponding to λ_A), solar radiation with wavelengths less than λ_A would be absorbed to give an output equal to the hatched area A. The solar radiation with wavelength greater than λ_A would pass through A and be converted by the bottom layer cell B (band gap corresponding to λ_B) to give an output equal to the hatched area B. The total output and therefore the efficiency of this tandem cell would be higher than the output and the efficiency of each single cell individually. The efficiency would increase with the number of layers. For this concept to work, each layer must be as thin as possible, which puts a very difficult if not an insurmountable constraint on crystalline and polycrystalline cells to be made multijunction. As a result, this concept is being developed mainly for other materials, including thin-film amorphous solar cells.

6.2.3 Design of a Photovoltaic System

Solar cells may be connected in series, parallel, or both to obtain the required voltage and current. When similar cells or devices are connected in series, the output voltages and current are as shown in Figure 6.11. A parallel connection results in the addition of currents, as shown in Figure 6.12. If the cells or devices 1 and 2 have dissimilar characteristics, the output characteristics will be as shown in Figure 6.13. Cells are connected to form modules, modules are connected to form panels, and panels are connected to form arrays. Principles shown in Figures 6.11 and 6.12 apply to all of these connections.

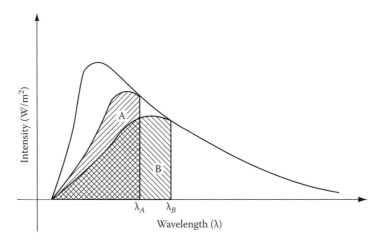

FIGURE 6.10
Energy conversion from a two-layered stacked cell. (From Goswami, Y. et al., *Principles of Solar Engineering*, 2nd edn., Taylor & Francis, Philadelphia, PA, 2000. With permission.)

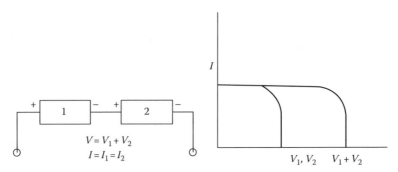

FIGURE 6.11
Characteristics of two similar cells connected in series.

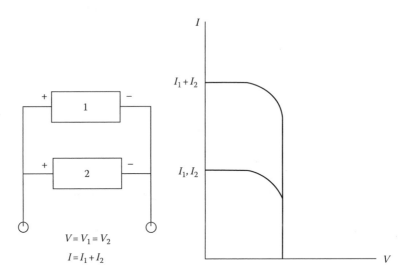

FIGURE 6.12
Characteristics of two similar cells connected in parallel.

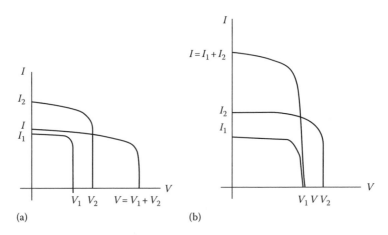

FIGURE 6.13
Characteristics of two dissimilar cells connected in (a) series and (b) parallel.

Example 6.4

An application requires 300 W at 14 V. Design a PV panel using solar cells from Example 6.3 each with an area of 25 cm².

Solution

Assuming that the cell will be operated at maximum power conditions, the voltage and current from each cell are

$$V_m = 0.542 \text{ V}, \quad I_m = \left(190.5 \ \frac{A}{m^2}\right)(0.0576 \text{ m}^2) = 10.97 \text{ A}$$

$$\text{Power/cell} = 0.542 \text{ V} \times 0.1143 \text{ A} = 5.95 \text{ W/cell} \approx 6 \text{ W/cell}$$

$$\text{Number of cell required} = \frac{300 \text{ W}}{6 \text{ W/cell}} = 50 \text{ cell}$$

$$\text{Number of cells in series} = \frac{\text{System voltage}}{\text{Voltage/cell}} = \frac{14 \text{ V}}{0.542 \text{ V}} \approx 26$$

$$\text{Number of rows of 26 cells connected in parallel} = \frac{50}{26} = 1.92 \approx 2$$

Since the number of rows must be a whole number, we may increase the number to 2 rows, which will give 312 W output (Figure 6.14).

A blocking diode is used in series with a module or an array to prevent the current from flowing backward, for example, from the battery to the cells under dark conditions. A bypass diode is used in parallel with a module in an array to bypass the module if it is shaded. A photovoltaic system may be connected to a DC or an AC load, as shown in Figure 6.15.

The nameplate rating for a PV system is in terms of the DC output. When the load requires an AC current, an inverter is required between the DC output and the AC load as shown in Figure 6.15. The inverter is a source of energy loss that appreciably decreases the net output of a PV system. There are a number of different inverters available, and the selection of the appropriate inverter for a given application depends on the requirements of the load, particularly the waveform, and on whether the system is grid-connected or stand-alone.

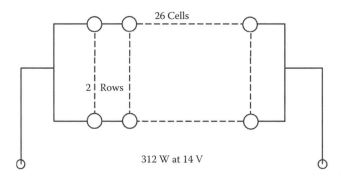

FIGURE 6.14
Connection of cells in rows and columns for Problem 6.4.

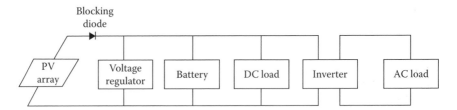

FIGURE 6.15
Schematic of a PV system.

TABLE 6.3

Inverter Characteristics

Parameter	Square Wave	Modified Sine Wave	Pulse Width Modulated	Sine Wave[a]
Output power range (W)	Up to 1,000,000	Up to 5000	Up to 20,000	Up to 500,000
Surge capacity (multiple of rated output power)	Up to 20 times	Up to 4 times	Up to 2.5 times	Up to 4 times
Typical efficiency over output range	70%–98%	>90%	>90%	>90%
Harmonic distortion	Up to 40%	>5%	<5%	<5%

Sources: Messenger, R.A. and Ventre, J. (2010) *Photovoltaic Systems Engineering*, 3rd edn., CRC Press, Boca Raton, FL, 2010; From data in Sandia National Laboratories, *Maintenance and Operation of Stand-Alone Photovoltaic Systems*, Albuquerque, NM, 1991; http://www.satcon.com/(Information on Satcon PV Inverters. With permission.)

[a] Multilevel H-bridge or similar technology to yield utility-grade sine wave output.

Table 6.3 summarizes currently available types of inverters. The performance of inverters is usually stated in terms of the rated power output, the surge capacity, the efficiency, and the harmonic distortion. Some loads have significant starting currents, and it is therefore important to provide adequate surge current capacity in the inverter to avoid overheating.

In selecting the appropriate inverter, some general observations are helpful. The square-wave inverter is the least expensive and is relatively efficient. It also has the best surge capacity, but suffers from harmonic distortion. The modified sine converter is fairly efficient but more complicated and expensive. The pulse-width modulated inverter is the most expensive, has the highest efficiency, and minimal distortion. The pure sine inverter has the least distortion and has the highest efficiency. Specifications of the losses incurred by an inverter for an AC load are included in the PV watts calculation illustrated later in the chapter. Detailed information about inverter construction and selection can be found in Ref. [6].

*6.3 Manufacture of Solar Cells and Panels

Manufacture of crystalline silicon solar cells is an outgrowth of the manufacturing methods used for microprocessors. A major difference is that silicon used in microprocessors is ultra pure, which is not needed for photovoltaic cells. Therefore, a major source of feedstock for silicon solar cells has been the waste material from the microelectronics industry. Solar cells are also manufactured as polycrystalline and thin films. Some of the common methods of manufacture of silicon solar cells are provided in the following text.

6.3.1 Single Crystal and Polycrystalline Cells

Single-crystal silicon cells are produced by a series of processes: (1) growing crystalline ingots of *p*-silicon, (2) slicing wafers from the ingots, (3) polishing and cleaning the surface, (4) doping with *n* material to form the *p–n* junction, (5) deposition of electrical contacts, (6) application of AR coating, and (7) encapsulation. Figure 6.16 illustrates the process.

The *Czochralski* method (Figure 6.17a) is the most common method of growing single-crystal ingots. A seed crystal is dipped in molten silicon doped with a *p*-material (boron) and drawn upward under tightly controlled conditions of linear and rotational speed, and temperature. This process produces cylindrical ingots of typically 10 cm diameter, although ingots of 20 cm diameter and more than 1 m long can be produced for other applications. An alternative method is called the *float zone* method (Figure 6.17b). In this method, a polycrystalline ingot is placed on top of a seed crystal and the interface is melted by a heating coil around it. The ingot is moved linearly and rotationally, under controlled conditions. This process has the potential to reduce the cell cost.

Polycrystalline ingots are produced by casting silicon in a mold of preferred shape (rectangular), as shown in Figure 6.18. Molten silicon is cooled slowly in a mold along one direction in order to orient the crystal structures and grain boundaries in a preferred direction. In order to achieve efficiencies of greater than 10%, grain sizes greater than 0.5 mm are needed, and the grain boundaries must be oriented perpendicular to the wafer. Ingots as large as 400 × 40 × 40 can be produced by this method.

Ingots are sliced into wafers by internal diameter saws or multiwire saws impregnated with diamond abrasive particles. Both of these methods result in high wastage of valuable crystalline silicon.

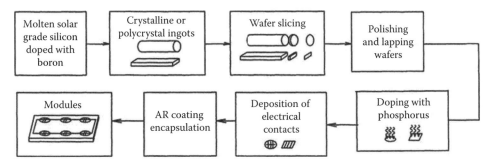

FIGURE 6.16
Series of processes for the manufacture of crystalline L polycrystalline cells.

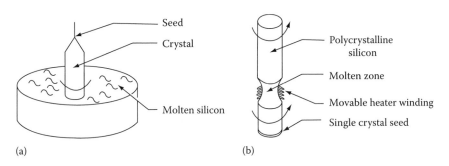

(a)   (b)

FIGURE 6.17
Crystalline silicon ingot production methods: (a) Czochralski method and (b) float zone method.

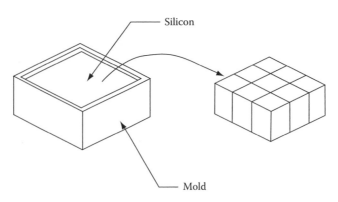

FIGURE 6.18
Polycrystalline ingot production.

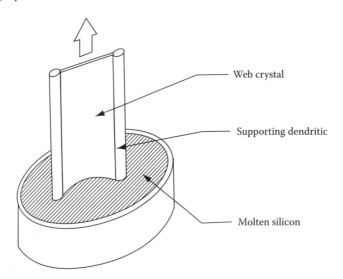

FIGURE 6.19
Thin-film production by dendritic web growth.

Alternative methods that reduce wastage are those that grow polycrystalline *thin films*. Some of the thin-film production methods include dendritic web growth (Figure 6.19), edge-defined film-fed growth (EFG) (Figure 6.20), ribbon against drop method, supported web method, and ramp-assisted foil casting technique (RAFT) (Figure 6.21).

A *p–n* junction is formed in the cell by diffusing a small amount of *n* material (phosphorous) in the top layer of a *p*-silicon wafer. The most common method is diffusion of phosphorous in the vapor phase. In this case, the backside of the wafer must be covered to prevent the diffusion of vapors from that side. An alternate method is to deposit a solid layer of the dopant material on the top surfaces followed by high-temperature (800°–900°) diffusion.

Electrical contacts are attached to the top surface of the cell in a grid pattern to cover no more than 10% of the cell surface, and a solid metallic sheet is attached to the back surface. The front grid pattern is made by either vacuum metal vapor deposition through a mask or by screen printing. Figure 6.22 shows how cells are connected to form modules.

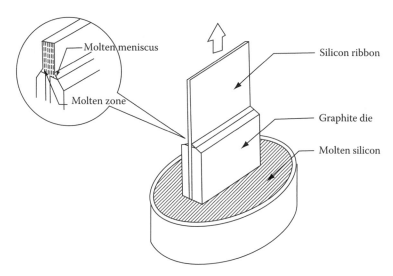

FIGURE 6.20
Thin-film production by edge-defined film-fed growth (EFG).

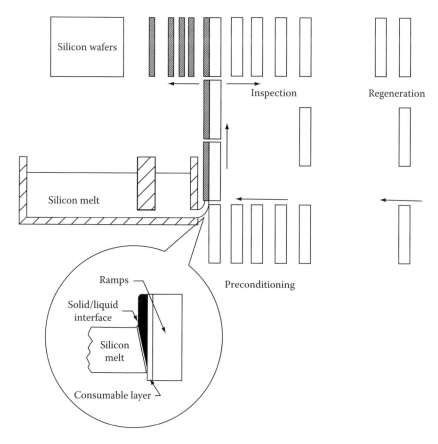

FIGURE 6.21
Schematic of RAFT processing.

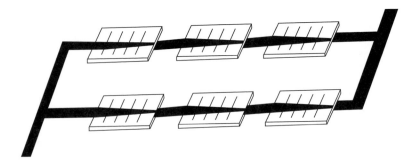

FIGURE 6.22
Assembly of solar cells to form a module.

AR coatings of materials such as silicon dioxide (SiO_2), titanium dioxide (TiO_2), and tantalum pentoxide (Ta_2O_5) are deposited on the cell surface to reduce refection from more than 30% for untreated Si to less than 3%. AR coatings are deposited by vacuum vapor deposition, sputtering, or chemical spraying. Finally, the cells are encapsulated in a transparent material to protect them from the environment. Encapsulants usually consist of a layer of either polyvinyl butyral or ethylene vinyl acetate and a top layer of low iron glass.

6.3.2 Amorphous Silicon and Multijunction Thin-Film Fabrication

Amorphous silicon (*a*-Si) cells are made as thin films of *a*-Si:H alloy doped with phosphorous and boron to make *n* and *p* layers, respectively. The atomic structure of an *a*-Si cell does not have any preferred orientation. The cells are manufactured by depositing a thin layer of *a*-Si on a substrate (glass, metal, or plastic) from glow discharge, sputtering, or chemical vapor deposition (CVD) methods. The most common method is by an RF glow discharge decomposition of silane (SiH_4) on a substrate heated to a temperature of 200°C–300°C. To produce *p*-silicon, diborane (B_2H_6) vapor is introduced with the silane vapor. Similarly, phosphine (PH_3) is used to produce *n*-silicon. The cell consists of an *n*-layer, an intermediate undoped *a*-Si layer, and a *p*-layer on a substrate. The cell thickness is about 1 μm. The manufacturing process can be automated to produce rolls

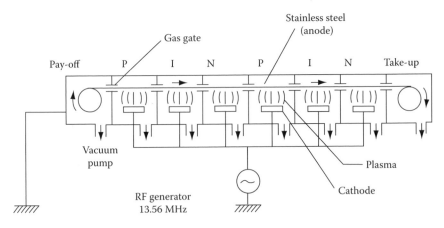

FIGURE 6.23
A schematic diagram of a roll-to-roll plasma CVD machine.

of solar cells from rolls of substrate. Figure 6.23 shows an example of roll-to-roll *a*-Si cell manufacturing equipment using a plasma CVD method. This machine can be used to make multifunction or tandem cells by introducing the appropriate materials at different points in the machine.

6.4 Design for Remote Photovoltaic Applications

Photovoltaic power may be ideal for a remote application requiring a few watts to hundreds of kilowatts of electrical power. Even where a conventional electrical grid is available, for some applications, where uninterruptible or emergency standby power is necessary, photovoltaic power would be appropriate. Some examples of remote PV applications include water pumping for potable water supply and irrigation, power for remote houses, street lighting, battery charging, telephone and radio communication relay stations, and weather stations. Examples of some other applications include electrical utility switching stations, peak electrical utility power where environmental quality is a concern, data acquisition systems, and speciality applications such as ventilation fans and vaccine refrigeration.

The design of a PV system is based on some basic considerations for the application:

1. Which is more important, the daily energy output or the power (average or peak)?
2. Is a backup energy source needed and/or available?
3. Is energy storage important? What type—battery, pumped water, etc.?
4. Is the power needed as AC or DC? What voltage?

There are three basic steps in the design of a PV system:

1. Estimation of load and load profile
2. Estimation of available solar radiation
3. Design of PV system, including area of PV panels, selection of other components, and electrical system schematic

Each of these steps will be explained in the following examples. These examples are based on Refs. [2,7].

6.4.1 Estimation of Loads and Load Profiles

Precise estimation of loads and their timings (load profile) are important for PV systems since the system is sized as the minimum required to satisfy the demand over a day. For example, if power is needed for five different appliances requiring 200, 300, 500, 1000, and 1500 W, respectively, so that only one appliance is on at any one time and each appliance is on for an average of 1 h a day, the PV system would be sized based on 1500 W peak power and 3500 Wh of daily energy requirement. The multiple loads on a PV system are intentionally staggered to use the smallest possible system, since the capital costs of a PV system are the most important as opposed to the energy costs in a conventional fuel-based system.

Example 6.5: Daily Load Calculations

How much energy per day is used by a remote weather station given the following load characteristics?

Load	Load Power (W)	Run Time (h/Day)
Charge controller	2.0	8
Data gathering	4.0	3
Modem (standby)	1.5	22.5
Modem (send/receive)	30.0	1.5

Solution

Daily energy $= (2.0 \text{ W})(8 \text{ h}) + (4.0 \text{ W})(3 \text{ h}) + (1.5 \text{ W})(22.5 \text{ h}) + (30.0 \text{ W})(1.5 \text{ h}) = 106.75 \text{ Wh}$

Daily energy use is about 107 Wh/day.

Example 6.6: Load Calculations

An owner of a remote cabin wants to install a PV power system. The loads in the home are described as follows. Assume that all lights and electronics are powered by AC. Find the daily and weekly peak and average energy use estimates. The system used is a 24 V DC system with an inverter.

Lights	4.23 W compact fluorescent bulbs	On at night for 5 h
Lights	6.13 W compact fluorescent bulbs	2 h each (daytime)
Stereo	110 W (amplifier), 15 W (other)	On for 8 h per week
Water pump	55 W (3.75 A start current)	Runs for 2 per day
Computer	250 W (monitor included)	On for 1½ h daily (weekend nights only)
Bathroom fan	40 W (3.5 A start current)	On for 1 h per day
Microwave	550 W (AC)–1000 W surge	On for 30 min per day

Solution
Loads need to be broken down according to (1) run time, (2) peak power, (3) night or day use, and (4) AC or DC loads. The load profile is as follows:

Load Name Description	Power (W)		Run Time (h)		Energy (Wh)	
	Average	Peak	Day	Week	Day	Week
Lights (AC)	(4) (23)	(4) (23)	5.0	35	460	3220
Lights (AC)	(6) (13)	(6) (13)	2.0	14	156	1092
Stereo (AC)	(1) (125)	(1) (125)	—	8	—	1000
Pump (DC)	(1) (55)	(3.75 A) (24 V)	2.0	14	110	770
Computer (AC)	(1) (250)	(1) (250)	1.5	3	—	750
Fan (DC)	(1) (40)	(3.5 A) (24 V)	1.0	7	40	280
Microwave (AC)	(1) (550)	(1) (1000)	0.5	3.5	275	1925

Average DC load: $[770 + 280]/7 = 150$ Wh/day.
Average AC load: $[3220 + 1092 + 1000 + 750 + 1925]/7 = 1141$ Wh/day.
Peak DC load: max $[\{(3.5)(24) + 55\} :: \{(3.75)(24) + 40\}] = 139$ W.
Peak AC load: $(1000) + \max [(4)(23) :: (6)(13)] + 250 + 125 = 1467$ W.

It can be assumed that the pump and fan will not start precisely at the same instant and that the night- and daylighting loads will not be on simultaneously.

6.4.2 Estimation of Available Solar Radiation

Methods of estimation of available solar radiation are described in Chapter 5. If long-term measured solar radiation values are available at a location, Equations 5.50 through 5.61 can be used to estimate the monthly average solar radiation per day. Otherwise, clear day data can be used along with the percent sunshine data to calculate the radiation. In addition, appropriate websites such as http://www.builditsolar.com/Tools/RadOnCol/radoncol.htm or http://rredc. nrel.gov/solar/old_data/nsrdb/1991-2005/ are useful resources for accurate approximations. For designing a PV system, a decision is made whether the PV panel will be operated as tracking the sun or will be fixed at a certain tilt and azimuth angle. For fixed panels, a tilt angle of latitude +15° works best for winter and latitude –15° for summer. To keep the panel fixed year round, an angle equal to the latitude provides the maximum yearly energy (see Figure 6.24).

6.4.3 PV System Sizing

If meeting the load at all times is not critical, PV systems are usually sized based on the average values of energy and power needed, available solar radiation, and component efficiencies. This is known as the *heuristic approach*. It is important to note that a system designed by this approach will not give the best design but may provide a good start for a detailed design. A detailed design accounts for the changes in the efficiencies of the components depending on the load and the solar radiation availability and whether the system is operating in a PV-to-load, PV-to-storage, or storage-to-load mode.

> **Example 6.7: Heuristic Approach to PV-System Sizing**
>
> A PV system using 50 W, 12 V panels with Trojan T-105 6 V, 125 Ah batteries is needed to power a home in Farmington, New Mexico, with a daily load of 1700 Wh. System voltage is 24 V. Assuming an average of 5 daylight hours in the winter, specify the collector and storage values for the system using the heuristic approach.
>
> **Solution**
> Load = 1700 Wh/day
> Daylight hours = 5 h/day
> Average panel output = 50 W
>
> $$\text{Number of panels} = \frac{1700 \text{ Wh/day}}{(5 \text{ h/day})(50 \text{ W/panel})} = 6.8, \text{ round off to 7 panels}$$

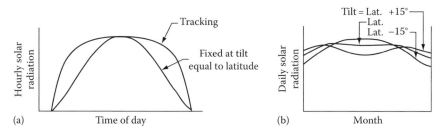

FIGURE 6.24
Solar radiation on panels at different tilt angles: (a) Hourly and (b) monthly.

Since the system voltage is 24 V, but each panel produces only 12 V, an even number of panels will be needed. Therefore, the number of panels = 8.

Farmington, New Mexico, is a very sunny location, so 3 days of storage are sufficient. Assuming a battery efficiency of 75% and a maximum depth of discharge of 70%,

$$\text{Storage} = (1700)(3)/(0.75 \times 0.7) = 9714 \text{ Wh}$$

$$\text{Number of batteries} = (9714 \text{ Wh})/(125 \text{ Ah} \times 6 \text{ V}) = 13 \text{ (rounded off to the next}$$
$$\text{whole number)}$$

Since the system voltage is 24 V, and each battery provides 6 V, the number of batteries is increased to 16. In a detailed design, the efficiencies of battery storage, inverter, and the balance of system (BOS) must be accounted for. The following example shows how these efficiencies increase the energy requirements of the PV panel.

Example 6.8: System Operating Efficiency

Using the cabin electrical system from Example 6.6, calculate the overall system efficiency for each operating mode possible for the system. Estimate the amount of energy required per day for the system. When load timing (day or night), assume half of the load runs during the day and half runs at night. The inverter used has a component efficiency of 91%, the battery efficiency is 76%, and the distribution system efficiency is 96%.

Solution
From the example, the loads are

Average DC load: 150 Wh/day
Average AC load: 1141 Wh/day

The various system efficiencies are

PV to load (DC):   0.96 (day, DC)
Battery to load (DC):   (0.76) (0.96) = 0.73 (night, DC)
PV to load (AC):   (0.96) (0.91) = 0.874 (day, AC)
Battery to load (AC):   (0.76) (0.91) (0.96) = 0.664 (night, AC)

Expected day and night loads are

Day (DC):   (0.5) (110) + (0.5) (40) = 75 Wh/day
Night (DC):   (0.5) (110) + (0.5) (40) = 75 Wh/day
Day (AC):   (156) + (0.5) (1000 + 750)/7 + (0.5) (275) = 418.5 Wh/day
Night (AC):   (460) + (0.5) (1000 + 750)/7 + (0.5) (275) = 722.5 Wh/day

Without considering system efficiency, the daily energy requirement is

$$E_{day} = (150) + (1141) = 1291 \text{ Wh/day}$$

The expected daily energy requirement is

$$E_{day} = \frac{75}{0.96} + \frac{75}{0.73} + \frac{418.5}{0.874} + \frac{722.5}{0.664}$$

$$E_{day} = 1725 \text{ Wh/day}$$

The actual energy requirement is 34% higher than that obtained in the approximate calculation in the previous example.

6.4.4 Water Pumping Applications

Water pumping for drinking water or irrigation at remote locations is an important application of PV. For a simple schematic shown in Figure 6.25, the power needed to pump water at a volumetric rate $\dot{V}$ is given by $\rho \dot{V} g H / \eta_p$ where ρ is the density of water, g is the acceleration due to gravity, H is the head loss the pump must over come, and η_p is the pump efficiency. The static head H_s is (A + B). In the case that the water level is drawn down, the static head would be (A + B + C). The pump must work against the total head H, which includes the dynamic head:

$$H = H_f + \frac{v^2}{2g}$$

where
   H_f is the frictional head loss in the pipe and the bends
   v is the velocity of the water at the pipe outlet

The pump efficiency η_p is a function of the load (head and flow rate) and is available as a characteristic curve from the manufacturer. There are two basic types of pumps: centrifugal and positive displacement. These pumps can be driven by AC or DC motors. DC motors are preferable for the PV applications, because they can be directly coupled to the PV array output. Centrifugal pumps with submersible motors are the optimum for PV applications because of their efficiency, reliability, and economy. However, for deep wells, Jack pumps may be necessary. Jack pumps are the piston type of positive displacement pumps that move chunks of water with each stroke. They require very large currents; therefore, they are connected through batteries.

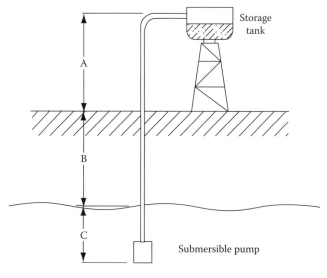

FIGURE 6.25
Water pumping using a submersible pump.

6.5 Thin-Film PV Technology

During the past few years, enormous advances have been made in the development of thin-film PV technologies. Although the market share worldwide for thin-film PV in 2010 was less than 11%, the market share of thin-film PV in the United States has grown rapidly and was reported at more than 40% of all photovoltaics in 2010 and expected to rise significantly [8,9]. As thin-film PV is developing, a number of applications are being pursued commercially, including building integrated photovoltaics, rooftop applications, and utility-scale applications.

Thin-film PV cells require multiple layers of precisely deposited thin films. An important criterion for the production of thin-film PV in a continuous process, such as a roller coater, is the percentage of the output that has an acceptable efficiency. For all of the thin-film technologies, it is possible to produce a so-called champion cell under laboratory conditions designed to achieve the highest cell efficiency possible (see Figure 6.26). These champion cells are indicative of what may be possible with a given technology, but as produced by laboratories such as National Renewable Energy Laboratory (NREL), they are not cost effective. However, they represent a goal that mass production facilities aim to achieve. The three primary thin-film PV technologies currently on the market are based on amorphous silicon (*a*-Si), cadmium telluride (CdTe), and copper indium gallium diselenide (CIGS). Each of these technologies has advantages and disadvantages, and the differences between these technologies are described briefly in the following text.

Amorphous silicon PV technology utilizes a widely available material, but the cells have a relatively low efficiency. An amorphous silicon cell prepared within a laboratory environment at the NREL achieved an efficiency of 12.3%. However, commercial production efficiencies for amorphous silicon typically only range from 6% to 8%. Furthermore, the capital equipment costs for manufacturing thin-film amorphous silicon are generally higher than that for other thin-film technologies.

Cadmium telluride (CdTe) thin-film PV modules have made impressive advances. As commercial-scale production has been approached, the manufacturing costs have dropped dramatically from $2.94/W in 2004 to about $1.25/W in 2009 with manufacturing target costs expected to be less than $1/W as a result of improvements in productivity, module efficiency, and yield. If this goal is met, CdTe thin-film PV modules would be competitive price-wise with average grid electricity. Colorado State University recently announced a spin-off plan for a manufacturing plant entitled, "Abound Solar," which can manufacture 3 million thin-film photovoltaic solar panels a year. The technology is based on CdTe, which is commonly found in copper/zinc mines. The cheap material coupled with an efficient process allows it to convert sheets of glass into solar panels at a cost of about $1/W at peak. One company actually claimed that their manufacturing technology can produce CdTe thin-film PV at $0.75/W, but this may not include installation and ancillary costs of a complete system [10].

CdTe champion cells produced by NREL have achieved efficiencies as high as 17.3% [10]. Commercially the production efficiencies for CdTe PV modules range from 9% to 11%. CdTe thin films have the advantages of low material costs and handling requirements associated with the manufacturing process. However, current production processes are limited to glass substrates, which are more limited in their applications than flexible films that can be used in buildings/integrated photovoltaic products, and can often be used to replace conventional building materials in some parts of a building.

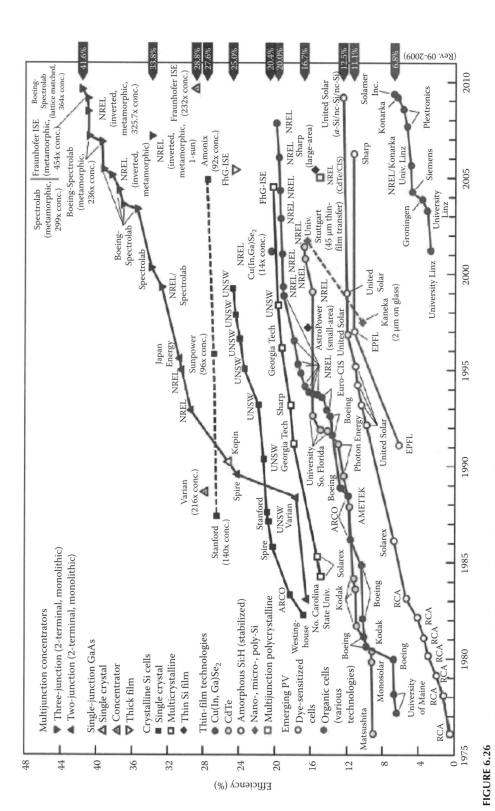

FIGURE 6.26

(See color insert.) Best research-cell efficiencies. (From Jayarama Reddy, P., Cell efficiencies, in *Science and Technology of Photovoltaics*, 2nd edn., CRC Press, Boca Raton, FL; Kurtz, S., Opportunities and challenges for development of a mature concentrating photovoltaic power industry, NREL, NREL/TP-520-43208, revised November 2009. With permission.)

An analysis of energy return on energy invested (EROI) of thin-film CdTe photovoltaic modules has recently been conducted under the order of the European research project, PVACCEPT [11]. In addition to the EROI, the study also investigated the environmental impact of CdTe thin-film modules. The following assumptions were made for the analysis:

- The expected lifetime of modules was assumed to be 20 years.
- The average insolation was assumed to be at 1700 kWh/(m²*/annum).
- A 20% efficiency loss was assumed with respect to nominal values for the modules.
- All waste materials generated in the production were assumed to be recycled and/or safely disposed of.
- Module decommissioning was not included because of a lack of data.

Using an efficiency of 8%, which is guaranteed at current by manufacturers, 10% for CIGS and 14% for poly-Si, which are typical values for the current state of actual production, the payback period for all three types of thin-film modules in years is shown in Figure 6.27.

The shaded area is the payback period for the module, and the white area accounts for the BOSs. The chart shows that the payback for the CdTe module itself is about 0.9 years, whereas the payback for the entire system is about 1.9 years. This compares to a payback period for the CIS system of 2.5 years and 4 years for the poly-Si system. Assuming a lifetime of 20 years, these payback periods result in EROIs of 10.5 for CdTe, 8 for CIGS, and 4 for poly-Si systems. The CdTe modules also have a very favorable environmental impact, generating only about 32 g of CO_2 equivalent per kWh of electricity produced.

CIGS thin-film photovoltaic products are available worldwide. The technology, however, is still in the development stage, and as many as 10 different deposition methods for growing the thin CIGS absorber layers are being pursued. *CIGS PV modules* have the best efficiency

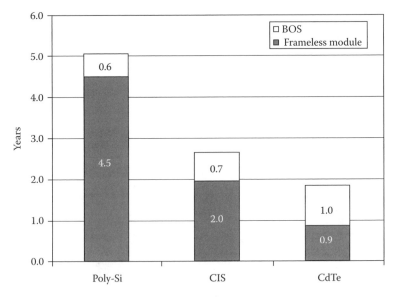

FIGURE 6.27
Payback times for three types of PV with frameless module and balance of systems. (From Raugei, M. et al., Energy and life cycle assessment of thin-film CdTe photovoltaic modules, accessed at http://www.abound. com/pdf/NREL_PV_Embodied_Energy.pdf. With permission.)

characteristics of the three technologies. An NREL champion cell achieved an efficiency of 20%. CIGS technology utilizes less raw material than the other two technologies and, therefore, achieves lower film thicknesses. Moreover, CIGS films have highlighted absorption capabilities. This technology appears to combine the highest solar conversion radiation efficiency with low manufacturing costs and the potential of being mass-produced with flexible substrates (see Figure 6.28).

The production of CIGS solar cells requires deposition of molybdenum, copper, indium, gallium, selenide, cadmium sulfide, and a transparent conductive oxide (TCO) onto a substrate. These manufacturing techniques are very demanding in accuracy, and a variety of different processes are under development. All current CIGS processes use sputtering during the molybdenum and TCO deposition steps. But the other CIGS compounds can be deposited with different technologies, including thermal evaporation, electrochemical plating, nanoparticle printing, or sputtering. The details of these technologies are under active development, and their eventual success will determine the cell efficiencies and cost achievable. A brief description of each of the four major deposition technologies is given as follows.

Sputtering: Sputtering is used extensively in the semiconductor industry to deposit thin films of various materials. The technology is a physical vapor deposition (PVD) method that deposits thin films by ejecting (called "sputtering") material from a "target" that is a source that then deposits onto a substrate such as a silicon wafer. Sputtering is used for high-volume and high-production manufacturing and has been proven effective in industrial-scale processes of coating glass and plasma display production. In this process, ions or reactive gases dislodge atoms from a solid plate of elements, which are deposited on an adjacent substrate. Targets in the PV industry are made up of materials such as CIGS and are attached to a reusable black back-plate installed within the PVD system.

Thermal evaporation: This is the most commonly used CIGS deposition technique. For this process, elements are heated in a vacuum and the resulting condensing vapor transfers these elements onto a substrate. Materials for this process are easy to obtain and use, but the process is often inefficient as far as material utilization is concerned, and it is difficult to control for the uniformity of deposition on large areas. Hence, the percent of spoilage can be large, although recent developments may improve this. The thermal evaporation process is also costly.

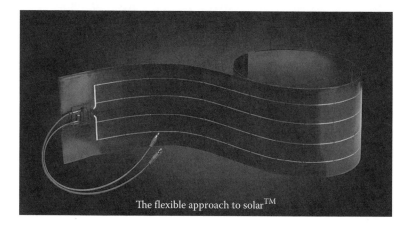
The flexible approach to solar™

FIGURE 6.28
(See color insert.) Thin-film flexible photovoltaic CIGS module. (From SoloPower, Inc., San Jose, CA. With permission.)

Electrochemical plating: In this process, a base metal is covered with a thin film of CIGS by submersing the substrate in a chemical solution through which an electric current flows. Although electrochemical plating has a high material utilization, the process may be slow and often results in nonuniform coverage. Hence, costly disposal of waste by-products from the process may be necessary.

Nanoparticle printing: This process is not yet in commercial production and poses considerable challenges for high-production operation. Ink-based technologies that can be deposited by low-cost inkjet or screen printing approaches have the potential for low cost with adequate resolution for PV applications. Using nanoparticulate precursors as the base for the ink may permit a degree of control in the composition process needed to meet electrical and stability requirements. Ink-depositing processes "print" CIGS semiconductor ink on a conductive foil substrate. The inks currently used for the CIGS absorber layer consist of oxides or selenide nanoparticles of the metals copper, indium, and gallium dispersed as a colloidal suspension in a solvent. The nanoparticles are intended to form a solid layer of CIGS. The majority of flexible substrates in use today are metal foils, but using polymer fibers such as polyimide that allow for high-speed roll-to-roll processing is gaining in acceptance.

6.6 Multilayer PV Technology

Advances in photovoltaic energy conversion have recently been summarized by Dan E. Arvizu, the director of the NREL [12]. As noted previously, the efficiency of a single-gap solar cell is limited as a result of its inability to efficiently convert the full range of energy of the photons emanating from the sun. Photons with energy below the band gap of the cell material either pass through the cell or are converted to heat within the material. Photons with energy above the band gap are also lost since only the energy necessary to generate the hole–electron pair is utilized while the remainder merely generates heat. The limitations of a single band gap solar cell can be overcome, however, by utilizing multiple junctions with several band gaps that can utilize a much broader range of photon energies from the sun. In the construction of multilayer cells, the layers are optically in series with the highest band gap material at the top. The first junction receives all of the spectrum where photons above the band gap of the first junction are absorbed. Photons below the band gap of the first layer pass through it to the second layer and are absorbed there. This is repeated for the two layers, and the third layer then absorbs as much of the remaining photons as possible. As many as five layers seem feasible.

   Multijunction photovoltaic cells use several layers of epitaxy deposited films. Epitaxy is the method of depositing a monocrystalline film on a monocrystalline substrate with the deposited film called the epitaxial layer. Triple-junction high-efficiency solar cells for terrestrial photovoltaic applications have been commercialized by a number of companies, including EMCORE. The cells are currently much more expensive than single band gap photovoltaic cells, but they can be combined with various types of concentrating devices that can capture solar radiation over a large area and then redirect it onto a small, high-efficiency solar cell. Figure 6.29 shows such a concentrating scheme with solar energy passing through a Fresnel lens that directs the radiation to a small, three-layer cell. A typical example of a three-layer cell is EMCORE T1000 that has achieved an efficiency better than 39% at a solar radiation concentration ratio of 503.

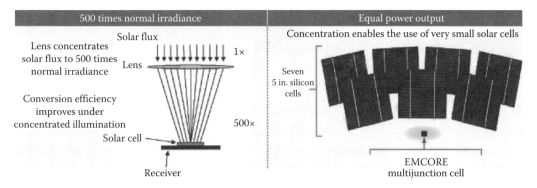

FIGURE 6.29
Schematic of multijunction solar cell with Fresnel concentrating lens arrangement. (From EMCORE, Albuquerque, NM, http://www.emcore.com/solar_photovoltaics/terrestrial_concentrator_photovoltaic_arrays. With permission.)

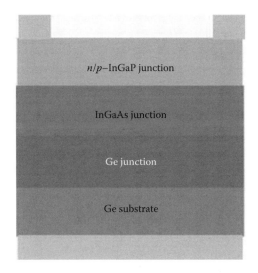

FIGURE 6.30
Physical cell structure. (From EMCORE, Albuquerque, NM, http://www.emcore.com/solar_photovoltaics/terrestrial_concentrator_photovoltaic_arrays. With permission.)

The physical cell structure is shown in Figure 6.30. It consists of an outer layer of antireflective coating that provides a low reflectance over wavelength in the range from 0.3 to 1.8 μm. The next three layers are an indium gallium phosphorous junction, followed by an indium gallium arsenide layer, and a Ge junction on a Ge substrate. The quantum efficiency of this type of three-layer cell is shown in Figure 6.31 as a function of wavelength, and the overall efficiency is shown as a function of concentration in Figure 6.32. It can be seen from the data in Figure 6.32 that the efficiency increases from 31.4% at 1× concentration to 39% at 503× concentration, where the peak occurs, and then decreases to about 33% at 1150× concentration. Typical current vs. voltage curves (*I–V* curves) at varying concentration levels and 25°C are shown in Figure 6.33.

There are a number of methods available for concentrating sunlight. Figure 6.34 shows a photo and a schematic of a PV concentration system using two reflections that direct the solar beam into a refractive rod that guides the radiation toward a triple-layer high-efficiency PV cell. This arrangement can achieve a concentration of 650 with a dual-axis

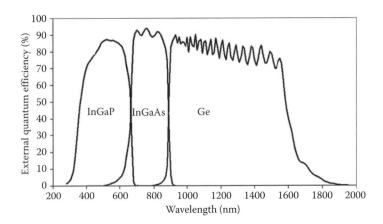

FIGURE 6.31
Quantum efficiency as a function of wavelength. (From EMCORE, Albuquerque, NM, http://www.emcore.com/solar_photovoltaics/terrestrial_concentrator_photovoltaic_arrays. With permission.)

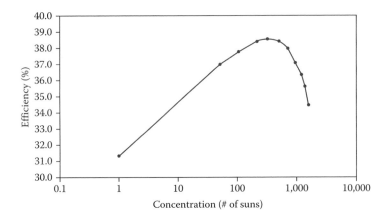

FIGURE 6.32
Overall cell efficiency as a function of concentration. (From EMCORE, Albuquerque, NM, http://www.emcore.com/solar_photovoltaics/terrestrial_concentrator_photovoltaic_arrays. With permission.)

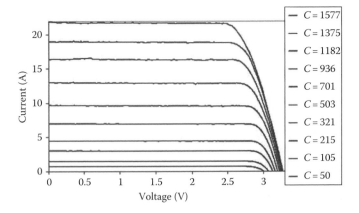

FIGURE 6.33
Typical *I–V* curves at varying concentration levels, 25°C. (From EMCORE, Albuquerque, NM, http://www.emcore.com/solar_photovoltaics/terrestrial_concentrator_photovoltaic_arrays. With permission.)

(a)

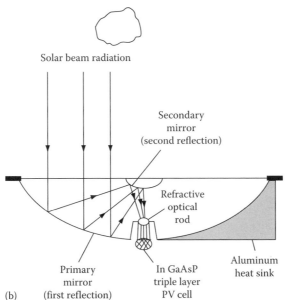

Solar beam radiation

Secondary
mirror
(second reflection)

Refractive
optical
rod

Aluminum
heat sink

Primary
mirror
(first reflection)

In GaAsP
triple layer
PV cell

(b)

FIGURE 6.34
Dual-refection-single-refraction PV concentrator system. (a) Photograph of SolFocus module. (b) Schematic of optical arrangement. (From SolFocus, Inc., Mountain View, CA. With permission.)

tracker according to Ref. [13]. Another arrangement (Figure 6.35a) with only one reflection and a secondary refractive (Figure 6.35b) lens to guide the radiation toward the PV cell is shown in the color photo with a novel dual-axis tracker developed at the University of Colorado (Figure 6.35c). This arrangement may be able to achieve a concentration as high as 1000.

The optimum concentration ratio is determined by economic, rather than optical, considerations. For a cell such as EMCORE T1000, the external dimensions are 12.58 mm × 12.58 mm while the total active area is only 108 mm^2 with a total cell thickness is 0.16 mm. The conversion efficiency of the design developed at the University of Colorado can be improved by a

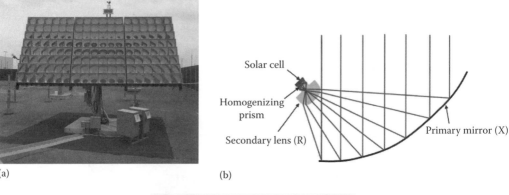

(a)   (b)

(c)

FIGURE 6.35
(See color insert.) Dual-axis rooftop tracker for Boeing high-concentration PV system developed at the University of Colorado, Boulder, CO. (a) Solar array mounted on conventional pedestal. (Courtesy of Adam Plesniak, Guy Martins, John Hall, Andy Messina and other members of the Boeing CPV Engineering Team.) (b) Ray trace pattern of the off-axis SMS3D nonimaging optical system in the Boeing CPV design. (Courtesy of Pablo Benitez, Professor, Universidad Polytechnic de Madrid and Light Prescription Innovators, LLC, Altadena, CA.) (c) Advanced dual-axis tracker with four of the six string Boeing high-concentration modules installed. Tracker was developed by Senior Design Teams at the University of Colorado, Boulder, CO, Dr. Frank Kreith, faculty advisor. (Photo courtesy of Kane Chinnel and Nate Bailey. With permission.)

finned heat pipe dissipation technique with the heat pipe attached to the bottom of the Ge substrate to avoid shading from the sun. Cooling of such a device is important because the efficiency decreases at about 0.06% per °C increase in operating temperature.

Improvements in multilayer photovoltaic cells are continuing, and in April 2008, the NREL announced that a world record in solar cell efficiency of 40.8% had been achieved at a concentration ratio of 326 suns with an inverted, metaphoric, triple-junction solar cell. This new design used a composition of a gallium indium phosphide and gallium indium arsenide to split the solar spectrum into three equal parts that are absorbed by each of the cells' three junctions for a higher efficiency. The construction was accomplished by growing the solar cell on a gallium arsenide wafer, flipping it over, and then removing the wafer [14]. The NREL record, however, was soon surpassed by another type of triple cell developed at the Fraunhofer Institute for Solar Energy Systems, which announced in January 2009 that an efficiency of 41.1% had been obtained at a concentration ratio of 545. Details of all of these triple cells, their manufacturing methodology, and costs are still privileged information.

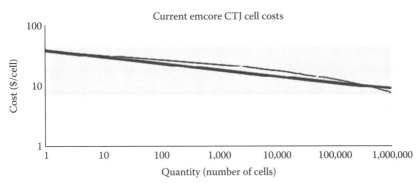

FIGURE 6.36
Multilayer cell costs as a function of production quantity. (From EMCORE representative, Personal communication, Albuquerque, NM. With permission.)

The cost of three-layer cells is still high and justifies the use of highly accurate CPC design with concentration ratios as high as 1000. But as shown in Figure 6.36, preliminary estimates of cell cost as a function of production by EMCORE indicate that cell cost is likely to decrease and concentrations beyond 500 suns may become counterproductive. Furthermore, since cell efficiency decreases with increase in cell temperature by 0.06% per °C, cooling may be important. Since higher concentration will also result in higher heat flux on the cell, cooling becomes more difficult for CPC high-concentration designs.

6.7 PVWatts for PV Performance Estimates

The NREL PVWatts calculator determines the energy production and economies of grid-connected photovoltaic energy systems throughout the world. It is publicly available at http://www.nrel.gov/rredc/pvwatts/. The calculator works by creating hour-by-hour performance simulations that provide estimated monthly and annual energy production in kilowatts. Users can select a location and choose either to use default values based on estimated averages from previous projects or they can input their own system parameters for size, electricity costs, array type, tilt angle, and azimuth angle. The program uses meteorological weather and solar data described previously in Chapter 5. From these data, the calculator determines the irradiation on the PV array and the average cell temperature for each hour of the year. The DC energy output for each hour is calculated from the PV system DC rating and the incident solar radiation, and then corrected for the PV cell temperatures. The AC energy for each hour is then calculated by multiplying the DC energy output by the overall DC to AC derate factor, including the inverter efficiency as a function of load. Hourly values of the AC energy are then summed to calculate monthly and annual AC energy production.

PVWatts calculators are available in two versions. Version 1 is the so-called Site Specific Data Calculator that allows the user to select a location from a map. Version 2, the Grid Data Calculator, allows a user to select any location in the United States from a list. The following example demonstrates the use of PVWatts in calculating the performance of a residential PV system. The same approach can be used to estimate the output of utility-scale PV applications.

Example 6.9

Using the PVWatts online software, estimate the incident solar radiation and AC energy output for a 6 kW solar PV panel in Phoenix, Arizona, during the month of October. Note that the 6 kW electric rating is for the DC output. Assume a DC-to-AC derate factor of 0.77 and an array tilt of 20°. What tilt angle will achieve the maximum annual energy output for this location?

Solution

To find solar information for Phoenix, choose the link labeled "Site Specific Data Calculator" on the PVWatts main page and then choose the link labeled "PVWatts Version 1 Calculator." From the PVWatts Version 1 web page, select Arizona from the map of the United States, then select Phoenix. Running the calculator for this location, the average incident solar radiation is 6.25 kWh/m²/day, and the AC energy output at a tilt angle of 20° is 787 kWh during the month of October. Using a "guess-and-check" method, a maximum annual energy output of 9705 kWh can be achieved with a tilt angle of 32°.

PVWatts can only be used to calculate the electrical power output of a system in a given location. To get information on the economics behind the system, another publicly available system called the System Advisory Model (SAM) has to be used. The following example shows how the economics of the same system in Example 6.9 can be analyzed.

Example 6.10

Estimate the levelized cost of electric power for the system specified in Example 6.9. Assume that the cost of the module is 2.00 $/W. The real discount rate is 6.00%, the inflation rate is 2.00%, and the lifespan of the system is 30 years.

Solution

To start a photovoltaic project in SAM, which is the program that will be used to solve this problem?

1. Start SAM
2. Under **Enter a new project name to begin,** type a name for your project. For example, "PV System"
3. Click Create New File
4. Under 1. Select a technology, click Photovoltaics
5. Click PVWatts System Model
6. Under 2. Select a financing option, click on Residential financing option
7. Click **OK**

Once the project is created, select the location where the PV system will be located under the "Climate" page. This is shown circled in the following figure.

Select Technology and Market...   [PVWatts, Residential]

System Summary

Climate

Location: PHOENIX, AZ
Lat: 33.4 Long: -112.0 Elev: 339.0 m

Utility Rate

Financing

Tax Credit Incentives

Payment Incentives

Annual Performance

Degradation: 0.5 % per year
Availability: 100 %

PV System Costs

Total: $ 22,856.00
Per Capacity: $ 5.71 per Wdc

PVWatts Solar Array

DC Rating: 4 kW
AC-DC Derate: 0.77

Choose Climate/Location

Filter locations by name:

SAM/AL Montgomery.tm2
SAM/AR Fort Smith.tm2
SAM/AR Little Rock.tm2
SAM/AUS_NT.Alice.Springs.Airport_RMY.epw
SAM/AZ Flagstaff.tm2
SAM/AZ Phoenix.tm2
SAM/AZ Prescott.tm2
SAM/AZ Tucson.tm2
SAM/CA Arcata.tm2
SAM/CA Bakersfield.tm2
SAM/CA Blythe 747188TY.epw

[Add/Remove...]
[Refresh list]
[Copy to project]
[Remove from project]
[Create TMY3 file]
[Location Lookup...]

Solar Advisor reads weather files in TMY2, TMY3, and EPW format. The default weather file library includes a complete set of TMY2 files for U.S.locations. To add files for other locations, use the web links below to find and download the files, and then click Add/Remove above to help SAM locate them on your computer.

Notes:

SAM looks for weather files in the specified folders. To change the search folders, click "Add/Remove". The prefix "SAM/" indicates a location from the standard SAM library, and those preceded by "USER/" are stored in your project file to facilitate sharing with other people.

Then, it is necessary to change the system parameters to match the system defined in the problem. To do this, we change the values circled in the following figure in the "PVWatts Solar Array" page.

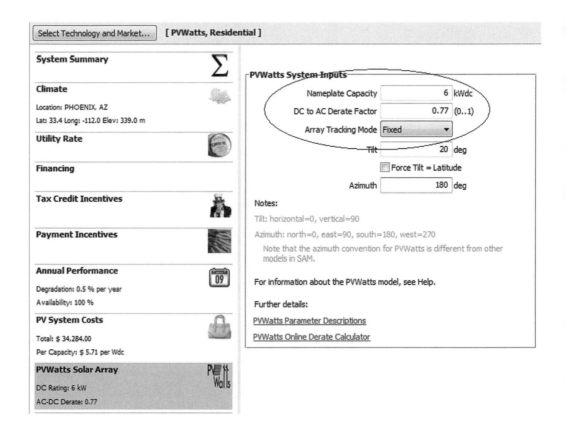

After the system is defined, the cost of the module needs to be changed accordingly. To do this, go to the "PV System Costs" page and change the circled value as follows.

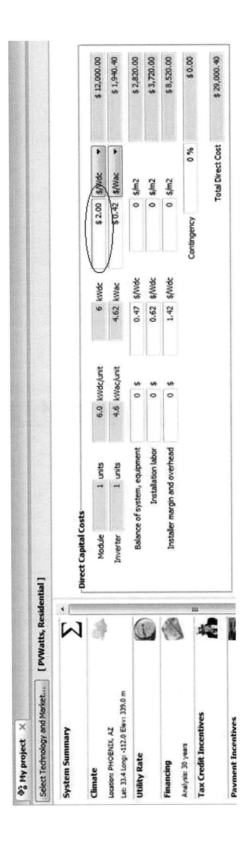

Similarly, for the financing options, go to the "Financing" page and change the circled values shown in the following figure.

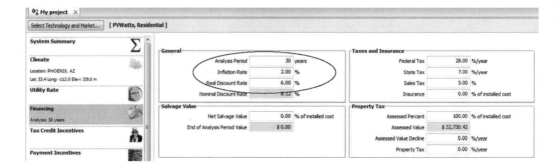

The parameters that were not mentioned in the problem can be left as default in SAM. Note that SAM does not allow you to change the total installed cost, but instead SAM allows you to change the price per Watt of the system. With all the parameters listed in the problem specified, click the green arrow on the bottom left corner of your screen and SAM will run the calculations. The following table should appear with various graphs.

Metric	Base
Net annual energy	9532 kWh
LCOE nominal	23.80 ¢/kWh
LCOE real	19.36 ¢/kWh
First year revenue without system	$0.00
First year revenue with system	$1143.85
First year net revenue	$1143.85
After-tax NPV	$−9472.51
Payback period	1e+099
DC-to-AC capacity factor	18.1 %
First year kWhac/kWdc	1589

NOTE: The "Net Annual Energy" output is the AC energy output of the system.

In the last row, the "First year kWhac/kWdc" should read 1.589.

The number of decimal figures in the program is unrealistic and should be at most three significant figures, but the accuracy is no better than 3% or 4%.

A payback period of "1e+099" indicates that the system will not pay for itself in the 30 year lifespan. However, if properly designed and maintained, the system is likely to have a life longer than 30 years. Moreover the cost of PV is coming down constantly.

To get the AC energy output for the month of October, click on the graph that is labeled "Monthly Output (Base Case)," which shows the AC energy output for each month. The graph is shown in the following figure.

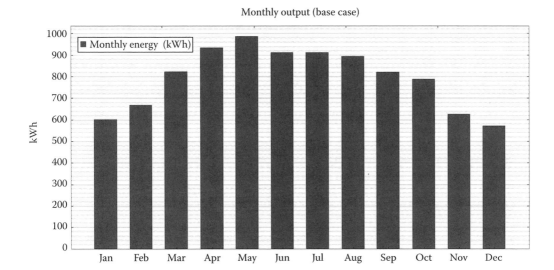

For the incident solar radiation, go to the "Climate" page and click on the button labeled "View Hourly Data." At the top of the page, click on the "Monthly" tab, and off to the right-hand side, choose to display the "Direct Normal Radiation" in W/m².

The program does not design to a specified voltage or amperage; in order to obtain the desired characteristics, it is necessary to put cells in series and/or parallel as discussed in Section 6.2.3.

Problems

6.1 Find the wavelength of radiation whose photons have energy equal to the band gap of GaAs.

6.2 What is the theoretical maximum efficiency of conversion if blue light of wavelength 0.45 μm is incident on a GaAs solar cell?

6.3 Find the theoretical maximum overall efficiency of GaAs solar cells in space.

6.4 The reverse saturation current I_o of a silicon cell at 40°C is 1.8×10^{-7} A. The short circuit current when exposed to sunlight is 5 A. From this information, compute

   a. Open circuit voltage.

   b. Maximum power output of the cell.

   c. The number of 4 × 4 cm cells needed to supply 100 W at 12 V. How must the cells be arranged?

6.5 At what efficiency is a photovoltaic array running if insolation on the collector is 650 W/m², the total collector area is 10 m², the voltage across the array is 50 V, and the current being delivered is 15 A?

6.6 If a PV array has a maximum power output of 10 W under an insolation level of 600 W/m², what must the insolation be to achieve a power output of 17 W? Would you expect the open circuit voltage to increase or decrease? Would you expect the short circuit current to increase or decrease?

6.7 A PV battery system has an end-to-end efficiency of 77%. The system is used to run an all-AC load that is run only at night. The charge controller efficiency is 96% and the inverter efficiency is 85%. How much energy will need to be gathered by the PV array if the load is 120 W running for 4 h per night?

6.8 If the average output of the PV system in Problem 6.7 is 200 W, the load is changed to run during the day, how much PV output energy is needed for the same load conditions? Assume that the battery bank is at 100% charge and that input efficiency is equal to output efficiency.

6.9 For the system in Problem 6.7, how many hours of sunlight are needed to ensure that the battery bank is at 100% charge at the end of the day assuming the same load?

Problems 6.10–6.14 The owner of a small cabin would like to convert her home to PV power. She has the following equipment and associated run times:

Household Equipment	Power (W)	Run Time (Day) (h)	Run Time (Night) (h)
Lighting (DC)	25	2	4
Stereo (AC)	40	3	2
Refrigerator (DC)	125	3[a]	3[a]
Water pump (DC)	400	1.5[b]	0.5[b]
Alarm clock (DC)	8	12	12
Computer (AC)	250	3	0
Printer (AC)	175	0.25	0
Outdoor safety lights (DC)	48	0[c]	8[c]
Answering machine (AC)	7	12	12
Coffee pot (AC)	1200	0.25	0

[a] The refrigerator is assumed to run 25% of the time.
[b] It can be assumed that the pump and fan will not start precisely at the same instant.
[c] It is assumed that the night- and daylighting loads will not be on simultaneously.

6.10 a. What is the homeowner's daily energy requirement as measured from the load?

  b. If she replaces her alarm clock with a wind-up clock, how much energy per day will she avoid using?

  c. What would you suggest she do to cut back her daily load?

6.11 How many 50 W panels will the owner require assuming battery storage is 75% efficient and all loads are DC (no inverter)?

  a. For a stationary system "seeing" 5 h of sunlight per day?

  b. For a tracking system "seeing" 8 h of sunlight per day?

6.12 For the loads listed:

  c. What size inverter (peak watts) should she purchase?

  d. If the inverter is 88% efficient, how much more daily energy is required from the PV array as compared to an all-DC system?

6.13 The homeowner decides to hire you to design a system for her. She has arranged with a local solar supplier for the following equipment. Specify the system and provide a line diagram.

PV panels	42 W, nominal 12 V
Batteries	125 Ah, 6 V, end-to-end efficiency = 72%
Charge controller	95% efficient, 12 V
Inverter	90% efficient, sizes of 500, 1000, 2000, and 4000 W available, 12 V input

6.14 Redesign the system in Problem 6.14 with the following equipment:

PV panels	51 W, nominal 12 V
Batteries	200 Ah, 12 V, end-to-end efficiency = 78%
Charge controller	97% efficient, 24 V
Inverter	91% efficient, sizes of 500, 1200, 2500, and 5000 W available, 24 V input

6.15 A flashing beacon is mounted on a navigation buoy in the shipping channel at a port at 30° N latitude. The load consists of a single lamp operating 1.0 s on and 3.6 s off during the hours of darkness. Hours of darkness vary from 9.8 h in July to 13.0 h in December. The lamp draws 2 A at 12 V when lighted. A flasher controls the lamp and draws 0.22 A when the lamp is on. There is a surge current of 0.39 A each time the flasher turns on. This current flows approximately 1/10th of the time the flasher is on. The design has 14 days of battery capacity. Provision has to be made to disconnect the load if the battery voltage drops below 11 V.

The available module has a rated voltage of 17.2 V at 25°C (15 V at 55°C) and 2.3 A at 1 kW/m². The available battery has a rated capacity of 105 Ah at 12 V. Assume that maximum depth of discharge is 0%. Design the PV system.

6.16 Design a PV system for the following application: A refrigerator/freezer unit for vaccine storage in a remote island of Roatan, Honduras (16°N latitude, 86°W longitude, temperature, range 15°C–30°C).

Two compressors—one each for refrigerator and freezer

Each compressor draws 5 A at 12 V

Compressors Remain on for	Summer (h/Day)	Winter (h/Day)
Refrigerator	9	5
Freezer	7	4

Design the PV system using the panels and the batteries described in Problem 6.15.

6.17 A homeowner in Santa Fe, New Mexico, is interested in having a solar PV system installed at his house. He would like the system to be able to cover all of his annual energy needs. The following is a table from his utility provider showing the monthly energy use in the household. His roof with solar access is facing due east at a 20° pitch with no shading. In addition, the local electric company is offering a rebate on residential solar installations of $2.00 per installed Watt, as well as a renewable energy credit rebate based on predicted system performance. The REC rebate is $2.50/W, which is then multiplied by the system performance. Since the installation is east facing, the predicted performance efficiency is 83% (with a system facing due south having a performance efficiency of 100%).

Consumption

	kWh
January	400
February	400
March	600
April	600
May	800
June	1000
July	1000
August	900
September	800
October	600
November	600
December	400

a. Design a system that will offset 100% of the home's consumption.

b. Assuming $0.10/W electricity expense with a 5% annual inflationary rate increase, and a $7.50/W installed cost, what will the payback be for this system after the rebate from the utility provider?

c. How do you wire the solar panel? Specifically, how many cells would you need and how would you wire them (series versus parallel) if the solar panel is to have a working voltage of 24.46 V and an amperage of 7.6 A? Assume the individual cells produce 98 V and 3.8 A.

Resources
The SMA string sizing program is found on the website—http://www.smaamerica.com/home
PV Watts is found on the NREL website.

6.18 A village in Antigua (17°N, 61°W, 15°C–30°C temperature, range), West Indies, requires 5000 gal of water per day for community water supply. Assuming a year-round average insolation of 8 kWh/day, design the system using the following components:

PV panels: Solarex panels 17.5 V and 3.6 A at 1000 W/m² and 25°C

Pump: Grundfos multistage pump input 105 V DC, 9 A, 30% efficiency

6.19 Prepare a preliminary design for a solar PV system to provide 1 kW of stand-alone 24 h/day power to a travel trailer. Make a schematic diagram of the overall system, a cost estimate, a cell array design, and a battery storage. Then discuss the assumptions you made for your design.

References

1. NPD Group Company. (2012) Global PV market, accessed at http://www.solarbuzz.com/facts-and-figures/market-facts/global-pv-market (accessed on April 13, 2013).

2. Nijs, J. et al. (1997) Energy payback time of crystalline silicon solar modules, in: K.W. Böer (ed.), *Advances in Solar Energy*, Vol. 11, Chap. 6, American Solar Energy Society, Boulder, CO, pp. 291–327.

3. Angrist, S.W. (1976) *Direct Energy Conversion*, 3rd edn., Allyn & Bacon, Boston, MA.
4. Bube, R.H. (1960) *Photoconductivity of Solids*, John Wiley & Sons, New York.
5. Garg, H.P. (1987) *Advances in Solar Energy Technology*, Vol. 3, D. Reidel Publishing Company, Dordrecht, the Netherlands.
6. Messenger, R.A. and Ventre, J. (2010) *Photovoltaic Systems Engineering*, 3rd edn., CRC Press, Boca Raton, FL.
7. Post, H.N. and Risser, V.V. (1995) *Stand-Alone Photovoltaic Systems—A Handbook of Recommended Design Practices*, Sandia National Laboratory Report SAND-87-7023, Sandia National Laboratories, Albuquerque, NM.
8. GreenTech Media. (2012) GTM research: CIGS will dominate thin-film PV in the future, accessed at http://www.photovoltaic-production.com/3271/gtm-research-cigs-will-dominate-thin-film-pv/ (accessed on April 13, 2013).
9. National Renewable Energy Laboratory. (2010) Technology and program market data— Solar energy technologie, accessed at http://www.nrel.gov/analysis/re_market_data_solar.html (accessed on April 13, 2013).
10. Andrew, B. (2011) First solar sets thin film CdTe solar cell efficiency world record, accessed at http://cleantechnica.com/2011/07/27/first-solar-sets-thin-film-cd-te-solar-cell-efficiency-world-record/ (accessed on April 13, 2013).
11. Rüb, C. (2005) PVACCEPT: Final report, accessed at http://www.pvaccept.de/ (accessed on April 13, 2013).
12. Arvizu, D.E. (2009) Let the solar revolution begin, *Solar Today*, March, 23(2), 24–27.
13. Masia, S. (2010) Concentrating photovoltaics, *Solar Today*, January/February, p. 42.
14. National Renewable Energy Laboratory. (2008) NREL solar cell sets world efficiency record at 40.8%, News Release NR-2708, August 13, 2008.
15. Goswami, Y. et al. (2000) *Principles of Solar Engineering*, 2nd edn., Taylor & Francis, Philadelphia, PA.
16. Moore, C.E. (1952) *Atomic Energy Levels*, Vol. 2, National Bureau of Standards Circular 467, U.S. Government Printing Office, Washington, DC, 1952.
17. Sandia National Laboratories (1991) *Maintenance and Operation of Stand-Alone Photovoltaic Systems*, Albuquerque, NM.

Suggested Readings

Crossley, P.A., Noel, G.T., and Wolf, M. (1968) Review and evaluation of past solar cell development efforts, RCA Astro-Electronics Division Report, AED R-3346, Contract NASW-1427, NASA, Washington, DC.

Florida Solar Energy Center. (1991) Photovoltaic system design, FSEC-GP-31-86, Florida Solar Energy Center, Cocoa Beach, FL.

Jayarama Reddy, P. (2009) Cell efficiencies, in: *Science and Technology of Photovoltaics*, 2nd edn., CRC Press, Boca Raton, FL.

Jayarama Reddy, P. (2010) *Science and Technology of Photovoltaics*, 2nd edn., CRC Press, 2010. This is a good basic textbook on the science and technology of solar photovoltaics.

Kurtz, S. (2009) Opportunities and challenges for development of a mature concentrating photovoltaic power industry, NREL, NREL/TP-520-43208, revised November 2009.

Lasnier, F. and Ang, T.G. (1990) *Photovoltaic Engineering Handbook*, A. Hilger Publishing, New York.

Markvart, T., ed. (2000) *Solar Electricity*, 2nd edn., John Wiley & Sons, Chichester, U.K.. This is a good general purpose, lower-level introduction.

Masters, G.M. (2001) *Renewable and Efficient Electric Power Systems*, John Wiley & Sons, Chichester, U.K. This is a comprehensive textbook for distributed electric power systems with a solutions manual for instructors.

Messenger, R.A. and Ventre, J. (2010) *Photovoltaic Systems Engineering*, 3rd edn., CRC Press, Boca Raton, FL. This is a useful book that focuses primarily on PV system analysis.

Nelson, J. (2003) Properties of semiconductor materials, in: *The Physics of Solar Cells*, Imperial College Press, London, U.K., 2003. This book emphasizes solar cell operation with an in-depth coverage of solar cells, including quantum effects.

Raugei, M., Bargigli, S., and Ulgaiati, S. (2007) Energy and life cycle assessment of the thin film CdTe photovoltaic modules, accessed at http://www.abound.com/pdf/NREL_PV_Embodied_Energy.pdf

7

Solar Heating and Cooling of Buildings*

I dreamed my genesis in sweat and death, fallen twice in the feeding sea, grown stale of Adam's brine until, vision of new man strength, I seek the sun.

—Dylan Thomas

Converting the sun's radiant energy to heat is a common and well-developed solar conversion technology. The temperature level and the amount of this converted energy are the key parameters that must be known to match a conversion scheme to a specified task effectively.

Heating and cooling of buildings offer great opportunities for energy conservation as well as utilization of solar energy. A first step in designing an energy-efficient building should be an assessment of the various options for conservation. These should include energy-efficient lighting, double- or triple-glazed windows, shading, increased insulation, and elimination of unnecessary air leakage to maintain interior comfort. These measures are usually the domain of the architect, but recently specialized energy conservation engineering for buildings has become a part of the architectural design.

The basic principle of solar thermal collection is that when solar radiation strikes a surface, part of it is absorbed, thereby increasing the temperature of the surface. The efficiency of that surface as a solar collector depends not only on the absorption efficiency, but also on how the thermal and reradiation losses to the surroundings are minimized and how the energy from the collector is removed for useful purposes. Various solar thermal collectors range from unglazed flat-plate-type solar collectors operating at about 5°C–10°C above the ambient, to central receiver concentrating collectors operating at above 1000°C. Table 7.1 lists various types of solar thermal collectors and their typical temperature and concentration ranges. Figures 7.1 and 7.2 show examples of some flat-plate collector systems.

This chapter and the next chapter analyze the thermal and optical performance of several solar thermal collectors. They range from nonconcentrating, flat-plate types to compound-curvature, continuously tracking types with concentration ratios up to 1000 or more. Applications of the energy converted by solar thermal collectors are described in this chapter for buildings and in Chapter 8 for power and process heat. The next section describes some fundamental radiative properties of materials, knowledge of which helps in the design of solar thermal collectors.

7.1 Radiative Properties and Characteristics of Materials

When radiation strikes a body, a part of it is reflected, a part is absorbed, and if the material is transparent, a part is transmitted, as shown in Figure 7.3. The fraction of the incident radiation reflected is defined as the reflectance ρ, the fraction absorbed as the absorptance

---

TABLE 7.1

Types of Solar Thermal Collectors and Their Typical
Temperature Range

Type of Collector	Concentration Ratio	Typical Working Temperature Range (°C)
Flat-plate collector	1	≤70
High efficiency flat-plate collector	1	60–120
Fixed concentrator	3–5	100–150
Parabolic trough collector	10–50	150–350
Parabolic dish collector	200–500	250–700
Central receiver	500–>3000	500–>1000

α, and the fraction transmitted as the transmittance τ. According to the first law of thermodynamics, these three components must add up to unity, or

$$\alpha + \tau + \rho = 1 \tag{7.1}$$

Opaque bodies do not transmit any radiation and $\tau = 0$.

The reflection of radiation can be *specular* or *diffuse*. When the angle of incidence is equal to the angle of refection, the refection is called specular; when the reflected radiation is uniformly distributed into all directions, it is called diffuse (see Figure 7.4). No real surface is either specular or diffuse, but a highly polished surface approaches specular refection, whereas a rough surface reflects diffusely.

Another important radiative property is called emittance, ε, which is the ratio of the radiative emissive power of a real surface to that of an ideal "black" surface.

All of the radiative properties of materials, α, τ, ρ, and ε can be functions of the wavelength and direction. In fact, such dependence is used in the design of solar energy devices and systems. For example, selective absorbers are used for solar collectors and passive heating systems, and glazing materials for daylighting and solar collectors.

The monochromatic directional emittance of a surface, $\varepsilon_\lambda(\theta, \phi)$, in a direction signified by an azimuth angle ϕ and a polar angle θ, is

$$\varepsilon_\lambda(\theta, \phi) = \frac{I_\lambda(\theta, \phi)}{I_{b\lambda}} \tag{7.2}$$

From the earlier equation, the total directional emittance $\varepsilon(\theta, \phi)$ over all the wavelengths or the monochromatic hemispherical emittance ε_λ can be obtained by the integration of $\varepsilon_\lambda(\theta, \phi)$ over all the wavelengths or the entire hemispherical space, respectively. The overall emittance, ε, is found by integrating the hemispherical emittance over all the wavelengths.

$$\varepsilon = \frac{1}{\sigma T^4} \int_0^\infty \varepsilon_\lambda E_{b\lambda} d\lambda \tag{7.3}$$

Observe that both ε_λ and ε are properties of the surface.

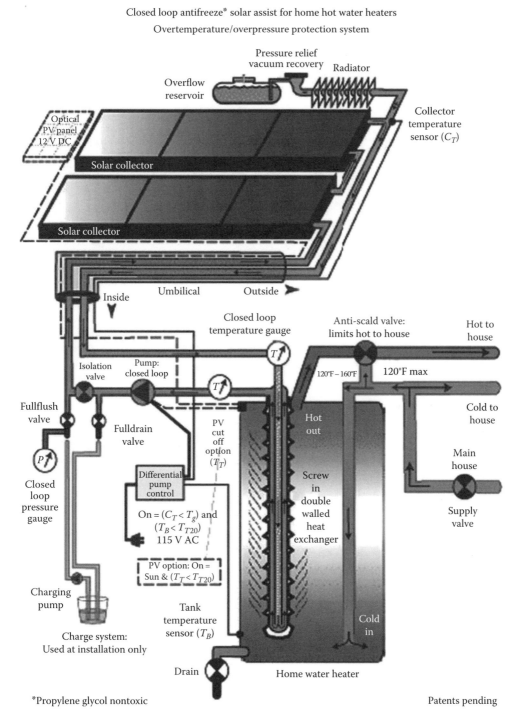

Closed loop antifreeze* solar assist for home hot water heaters
Overtemperature/overpressure protection system

*Propylene glycol nontoxic

Patents pending

FIGURE 7.1
(See color insert.) Schematic of an active flat-plate solar collector with storage tank and freeze protection.

SunMaxx M-2 flat-plate solar collector

Top header

Aluminum or steel or plastic housing — Fluid riser tubes

Foam or fiberglass insulation

Separator layer

Antireflection glass front cover

Selective solar absorber plate

Bottom header

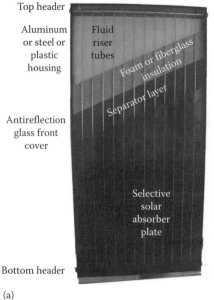
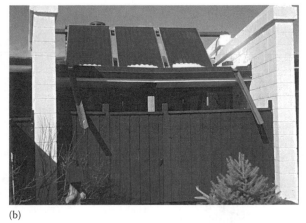

(a)　　　　　　　　　　　　　　(b)

FIGURE 7.2
(See color insert.) (a) Photo of an active flat-plate solar collector. (b) Photo of active flat-plate solar collector system on Kreith residence. (From Barry Butler, PhD, Butler Sun Solutions, Inc., Solana Beach, CA. With permission.)

Incoming radiation (=1)

ρ

α

τ

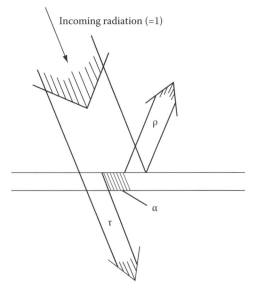

FIGURE 7.3
Schematic representation of transmittance τ, absorptance α, and reflectance ρ.

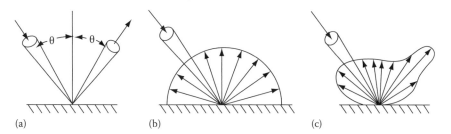

FIGURE 7.4
Reflections from (a) ideal specular, (b) ideal diffuse, and (c) real surfaces.

The next most important surface characteristic is the absorptance. We begin by defining the monochromatic directional absorptance as the fraction of the incident radiation at wavelength λ from the direction θ, ϕ that is absorbed, or

$$\alpha_\lambda(\theta,\phi) = \frac{I_{\lambda,a}(\theta,\phi)}{I_{\lambda,i}(\theta,\phi)} \tag{7.4}$$

where the subscript a and i denote absorbed and incident radiations, respectively. The monochromatic directional absorptance is also a property of the surface.

More important than $\alpha_\lambda(\theta, \phi)$ is the overall directional absorptance $\alpha(\theta, \phi)$, defined as the fraction of the total radiation from the direction θ, ϕ that is absorbed, or

$$\alpha(\theta,\phi) = \frac{\int_0^\infty \alpha_\lambda(\theta,\phi)I_{\lambda,i}(\theta,\phi)d\lambda}{\int_0^\infty I_{\lambda,i}(\theta,\phi)d\lambda} = \frac{1}{I_i(\theta,\phi)}\int_0^\infty \alpha_\lambda(\theta,\phi)I_{\lambda,i}(\theta,\phi)d\lambda \tag{7.5}$$

The overall absorptance is a function of the characteristics of the incident radiation and is, therefore, unlike the monochromatic absorptance, not a property of a surface alone. It is this characteristic that makes it possible to have selective surfaces that absorb the radiation from one source at a higher rate than from another. In other words, even though according to Kirchhoff's law the monochromatic emittance at λ must equal the monochromatic absorptance

$$\alpha_\lambda(\theta,\phi) = \epsilon_\lambda(\theta,\phi) \tag{7.6}$$

the overall emittance is not necessarily equal to the overall absorptance unless thermal equilibrium exists, and the incoming and outgoing radiations have the same spectral characteristics.

The effect of incidence angle on the absorptance is illustrated in Table 7.2, where the angular variation of the absorptance for a nonselective black surface, typical of those used on flat-plate collectors, is shown. The absorptance of this surface for diffuse radiation is approximately 0.90.

The third characteristic to be considered is the reflectance. Reflectance is particularly important for the design of focusing collectors. As mentioned previously, there are two limiting types of reflection: specular and diffuse. As illustrated in Figure 7.4a, when a ray of incident radiation at an angle θ is reflected at the same polar angle and the azimuthal angles differ by 180°, as for a perfect mirror, the reflection is said to be specular. The reflection is said to be diffuse if the incident radiation is scattered equally in all directions, as shown in Figure 7.4b. The reflection in Figure 7.4c is partially diffuse and partially specular.

TABLE 7.2

Angular Variation of Absorptance of Lampblack Paint

Incidence Angle i (°)	Absorptance $\alpha(i)$
0–30	0.96
30–40	0.95
40–50	0.93
50–60	0.91
60–70	0.88
70–80	0.81
80–90	0.66

Sources: Adapted from Löf, G.O.G. and Tybout, R.A., *Sol. Energy*, 16, 9, 1974; Löf, G.O.G. and Tybout, R.A., Model for optimizing solar heating design, ASME Paper 72-WA/SOL-8, 1972. With permission.

7.1.1 Selective Surfaces

Two types of special surfaces of great importance in solar collector systems are selective and reflecting surfaces. Selective surfaces combine a high absorptance for solar radiation with a low emittance for the temperature range in which the surface emits radiation. This combination of surface characteristics is possible because 98% of the energy in incoming solar radiation is contained within wavelengths below 3 μm, whereas 99% of the radiation emitted by black or gray surfaces at 400 K is at wavelengths longer than 3 μm. The dotted line in Figure 7.5 illustrates the spectral reflectance of an ideal, selective semi-gray surface having a uniform reflectance of 0.05 below 3 μm, but 0.95 above 3 μm. Real surfaces do not approach this performance. Table 7.3 lists properties of some selective coatings.

7.1.2 Reflecting Surfaces

Concentrating solar collectors require the use of reflecting surfaces with high specular reflectance in the solar spectrum or refracting devices with high transmittance in the

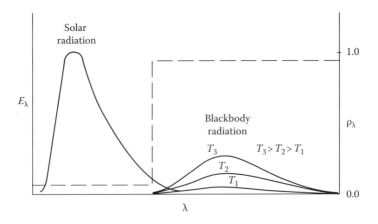

FIGURE 7.5

Illustrations of reflectance characteristics of an ideal selective surface. Shows radiation from ideal surfaces at different temperatures and solar radiation.

TABLE 7.3

Properties of Some Selective Plated Coating Systems

Coating[a]	Substrate	$\bar{\alpha}_s$	$\bar{\varepsilon}_i$	Breakdown Temperature (°F)	Humidity Degradation MIL STD 810B
				Durability	
Black nickel on nickel	Steel	0.95	0.07	>550	Variable
Black chrome on nickel	Steel	0.95	0.09	>800	No effect
Black chrome	Steel	0.91	0.07	>800	Completely rusted
	Copper	0.95	0.14	600	Little effect
	Galvanized steel	0.95	0.16	>800	Complete removal
Black copper	Copper	0.88	0.15	600	Complete removal
Iron oxide	Steel	0.85	0.08	800	Little effect
Manganese oxide	Aluminum	0.70	0.08		
Organic overcoat on iron oxide	Steel	0.90	0.16		Little effect
Organic overcoat on black chrome	Steel	0.94	0.20		Little effect

Source: U.S. Department of Commerce, Optical coatings for flat-plate solar collectors, NTIS No. PN-252–383, Honeywell, Inc., Morristown, NJ, 1975.

[a] Black nickel coating plated over a nickel–steel substrate has the best selective properties ($\bar{\alpha}_s$ = 0.95, $\bar{\varepsilon}_i$ = 0.07), but these degraded significantly during humidity tests. Black chrome plated on a nickel–steel substrate also had very good selective properties ($\bar{\alpha}_s$ = 0.95, $\bar{\varepsilon}_i$ = 0.09) and also showed high resistance to humidity.

solar spectrum. Reflecting surfaces are usually highly polished metals or metal coatings on suitable substrates. With opaque substrates, the reflective coatings must always be front-surfaced, for example, chrome plating on copper or polished aluminum. If a transparent substrate is used, however, the coating may be front- or back-surfaced. In any back-surfaced reflector, the radiation must pass through the substrate twice, and the transmittance of the material becomes very important.

Table 7.4 presents typical values for the normal specular reflectance of new surfaces for beam solar radiation.

TABLE 7.4

Specular Reflectance Values for Solar Reflector Materials

Material	ρ
Silver (unstable as front surface mirror)	0.94 ± 0.02
Gold	0.76 ± 0.03
Aluminized acrylic second surface	0.86
Anodized aluminum	0.82 ± 0.05
Various aluminum surfaces range	0.82–0.92
Copper	0.75
Back-silvered water-white plate glass	0.88
Aluminized type-C Mylar (from Mylar side)	0.76

7.1.3 Transparent Materials

The optical transmission behavior can be characterized by two wavelength-dependent physical properties—the index of refraction n and the extinction coefficient k. The index of refraction, which determines the speed of light in the material, also determines the amount of light reflected from a single surface, while the extinction coefficient determines the amount of light absorbed in a substance in a single pass of radiation.

Figure 7.6 defines the angles used in analyzing refection and transmission of light. The angle i is called the *angle of incidence*. It is also equal to the angle at which a beam is specularly reflected from the surface. Angle θ_r is the *angle of refraction*, which is defined, as shown in the figure. The incidence and refraction angles are related by Snell's law:

$$\frac{\sin(i)}{\sin(\theta_r)} = \frac{n_r'}{n_i'} = n_r \tag{7.7}$$

where
  n_i' and n_r' are the two refractive indices
  n_r is the index ratio for the two substances forming the interface

Typical values of refractive indices for various materials are shown in Table 7.5. For most materials of interest in solar applications, the values range from 1.3 to 1.6, a fairly narrow range.

By having a gradual change in the index of refraction, reflectance losses are reduced significantly. The reflectance of a glass–air interface common in solar collectors may be reduced by a factor of 4 by an etching process. If a glass is immersed in a silica supersaturated fluosilicic acid solution, the acid attacks the glass and leaves a porous silica surface layer. This layer has an index of refraction intermediate between glass and air. Figure 7.7 shows the spectral reflectance of a pane of glass before and after etching.

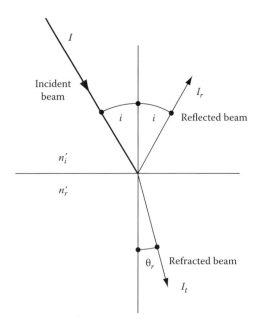

FIGURE 7.6
Diagram showing incident, reflected, and refracted beams of light and incidence and refraction angles for a transparent medium.

TABLE 7.5

Refractive Index for Various Substances in the Visible Range Based on Air

Material	Index of Refraction
Air	1.00
Clean polycarbonate	1.59
Diamond	2.42
Glass (solar collector type)	1.50–1.52
Plexiglass[a] (polymethyl methacrylate, PMMA)	1.49
Mylar[a] (polyethylene terephthalate, PET)	1.64
Quartz	1.55
Tedlar[a] (polyvinyl fluoride, PVF)	1.45
Tefon[a] (polyfluoroethylenepropylene, FEP)	1.34
Water–liquid	1.33
Water–solid	1.31

[a] Trademark of the DuPont Company, Wilmington, DE.

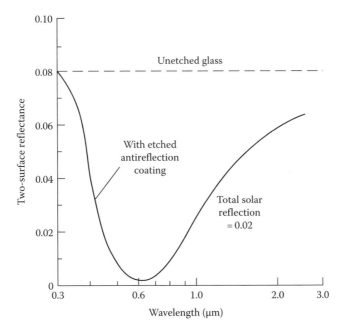

FIGURE 7.7
Refection spectra for a sample of glass before and after etching.

7.2 Flat-Plate Collectors

A simple flat-plate collector consists of an absorber surface (usually a dark thermally conducting surface); a trap for reradiation losses from the absorber surface (such as glass, which transmits shorter-wavelength solar radiation but blocks the longer-wavelength radiation from the absorber); a heat-transfer medium such as air, water, etc.; and some thermal insulation behind the absorber surface. Flat-plate collectors are used typically for temperature requirements up to 75°C although higher temperatures can be obtained from high-efficiency collectors. These collectors are of two basic types based on the heat-transfer fluid:

Liquid type, where heat-transfer fluid may be water, mixture of water and antifreeze oil, etc.

Air type, where heat-transfer medium is air, used mainly for drying and some space heating.

7.2.1 Liquid-Type Collectors

Figure 7.8 shows a typical liquid-type flat-plate collector. In general, it consists of the following:

1. *Glazing*: One or more covers of transparent materials like glass, plastics, etc. Glazing may be left out for some low-temperature applications.
2. *Absorber*: A plate with tubes or passages attached to it for the passage of a working fluid. The absorber plate is usually painted flat black or electroplated with a selective absorber.
3. *Headers or manifolds*: To facilitate the flow of heat-transfer fluid.
4. *Insulation*: To minimize heat loss from the back and the sides.
5. *Container*: Box or framewall casing.

7.2.2 Air-Type Collectors

Air types of collectors are most commonly used for agricultural drying and space-heating applications. Their basic advantages are low sensitivity to leakage and no requirement for an additional heat exchanger for drying and space-heating applications [20,21]. However, because of the low heat capacity of the air and the low convection heat-transfer coefficient between the absorber and the air, a large heat-transfer area and high flow rates are needed. Figure 7.9 shows some common configurations of air-heating collectors. Common absorber materials include corrugated aluminum or galvanized steel sheets, black metallic screens, or simply any black-painted surface. The most important

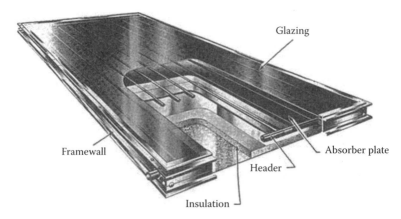

FIGURE 7.8
Typical liquid-type flat-plate collector. (From Morning Star Corporation, Orange Park, FL. With permission.)

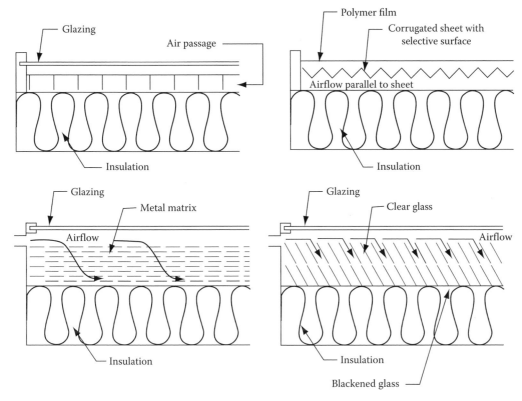

FIGURE 7.9
Some common configurations of air-heating collectors.

components, whose properties determine the efficiency of solar thermal collectors, are glazings and absorbers.

Unglazed, transpired solar air collectors offer a low-cost opportunity for some applications such as preheating of ventilation air and agricultural drying and curing [2]. Such collectors consist of perforated absorber sheets that are exposed to the sun and through which air is drawn. The perforated absorber sheets are attached to the vertical walls, which are exposed to the sun. Kutcher and Christensen [3] have given a detailed thermal analysis of unglazed transpired solar collectors.

7.2.3 Glazings

The purpose of a glazing or transparent cover is to transmit the shorter-wavelength solar radiation but block the longer-wavelength reradiation from the absorber plate, and to reduce the heat loss by convection from the top of the absorber plate. Consequently, an understanding of the process and laws that govern the transmission of radiation through a transparent medium is important. Section 7.1 describes in brief the transmission of radiation through materials.

Glass is the most common glazing material. Figure 7.10 shows transmittance of typical glass as a function of wavelength. Transparent plastics, such as polycarbonates and acrylics, are also used as glazings for flat-plate collectors. The main disadvantage of plastics is that their transmittance in the longer wavelength is also high; therefore, they are not as good a trap as glass. Other disadvantages include deterioration over a period of time due

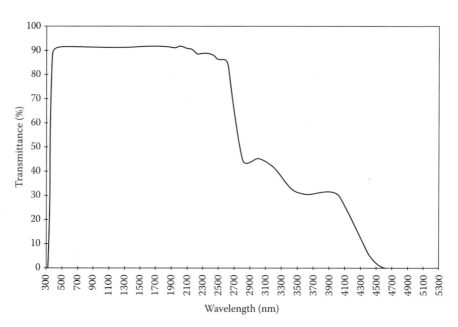

FIGURE 7.10
Spectral transmittance of 3 mm thick low-iron float glass.

to ultraviolet solar radiation. Their main advantage is resistance to breakage. Although glass can break easily, this disadvantage can be minimized by using tempered glass.

In order to minimize the upward heat loss from the collector, more than one transparent glazing may be used. However, with the increase in the number of cover plates, transmittance is decreased. Figure 7.11 shows the effect of the number of glass cover plates on transmittance.

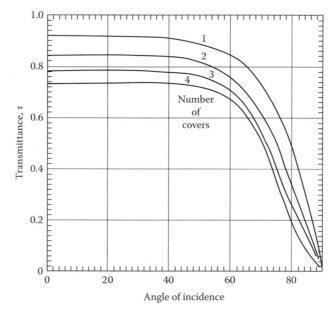

FIGURE 7.11
Transmittance of multiple glass covers vs. angle of incidence.

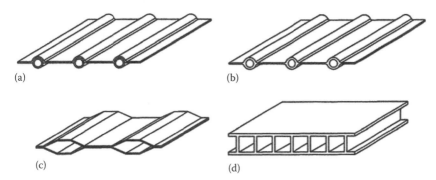

FIGURE 7.12
Common types of absorber plates: (a) Soldered, (b) extruded, (c) roll-bond, and (d) multichannel.

7.2.3.1 Absorbers

The purpose of the absorber is to absorb as much of the incident solar radiation as possible, reemit as little as possible, and allow efficient transfer of heat to a working fluid. The most common forms of absorber plates in use are shown in Figure 7.12. The materials used for absorber plates include copper, aluminum, stainless steel, galvanized steel, plastics, and rubbers. Copper seems to be the most common material used for absorber plates and tubes because of its high thermal conductivity and high corrosion resistance. However, copper is quite expensive. For low-temperature applications (up to about 50°C or 120°F), a plastic material called ethylene-propylene polymer (trade names EPDM, HCP, etc.) can be used to provide inexpensive absorber material. To compensate for the low thermal conductivity, a large surface area is provided for heat transfer.

In order to increase the absorption of solar radiation and to reduce the emission from the absorber, the metallic absorber surfaces are painted or coated with flat black paint or some selective coating. A selective coating has high absorptivity in the solar wavelength range (0.3–3.0 μm). Absorptivities and emissivities of some common selective surfaces are given in Table 7.3.

A simple and inexpensive collector consists of a black-painted corrugated metal absorber on which water flows down open, rather than enclosed in tubes. This type of collector is called a *trickle collector* and is usually built on site. Although such a collector is simple and inexpensive, it has the disadvantages of condensation on the glazing and a higher pumping power requirement.

7.2.4 Energy Balance for a Flat-Plate Collector

The thermal performance of any type of solar thermal collector can be evaluated by an energy balance that determines the portion of the incoming radiation delivered as useful energy to the working fluid. For a flat-plate collector of an area A_c, this energy balance on the absorber plate is

$$I_c A_c \tau_s \alpha_s = q_u + q_{loss} + \frac{de_c}{dt} \tag{7.8}$$

where
  I_c is the solar irradiation on a collector surface
  τ_s is the effective solar transmittance of the collector cover(s)
  α_s is the solar absorptance of the collector–absorber plate surface

q_u is the rate of heat transfer from the collector–absorber plate to the working fluid

q_{loss} is the rate of heat transfer (or heat loss) from the collector–absorber plate to the surroundings

de_c/dt is the rate of internal energy storage in the collector

The instantaneous efficiency of a collector η_c is simply the ratio of the useful energy delivered to the total incoming solar energy, or

$$\eta_c = \frac{q_u}{A_c I_c} \tag{7.9}$$

In practice, the efficiency must be measured over a finite time period. In a standard performance test, this period is on the order of 15 or 20 min, whereas for design, the performance over a day or over some longer period t is important. Then we have for the average efficiency

$$\eta_c = \frac{\displaystyle\int_0^1 q_u dt}{\displaystyle\int_0^1 A_c I_c dt} \tag{7.10}$$

where t is the time period over which the performance is averaged.

A detailed and precise analysis of the efficiency of a solar collector is complicated by the nonlinear behavior of radiation heat transfer. However, a simple linearized analysis is usually sufficiently accurate in practice. In addition, the simplified analytical procedure is very important because it illustrates the parameters of significance for a solar collector and how these parameters interact. For a proper analysis and interpretation of these test results, an understanding of the thermal analysis is imperative, although for design and economic evaluation, the results of standardized performance tests are generally used.

7.2.4.1 Collector Heat Loss Coefficient

In order to obtain an understanding of the parameters determining the thermal efficiency of a solar collector, it is important to develop the concept of *collector heat loss coefficient*. Once the collector heat loss coefficient U_c is known, and when the collector plate is at an average temperature T_c, the collector heat loss can be written in the simple form

$$q_{loss} = U_c A_c (T_c - T_a) \tag{7.11}$$

The simplicity of this relation is somewhat misleading because the collector heat loss coefficient cannot be specified without a detailed analysis of all the heat losses. Figure 7.13 shows a schematic diagram of a single-glazed collector, while Figure 7.14a shows the thermal circuit with all the elements that must be analyzed before they can be combined into a single conductance element shown in Figure 7.14b. The following analysis shows an example of how this combination is accomplished.

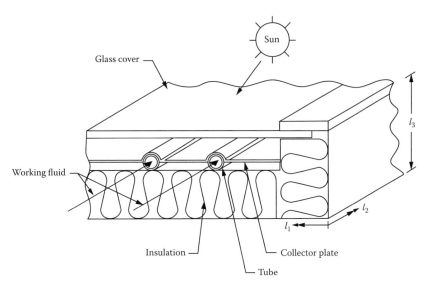

FIGURE 7.13
Schematic diagram of solar collector.

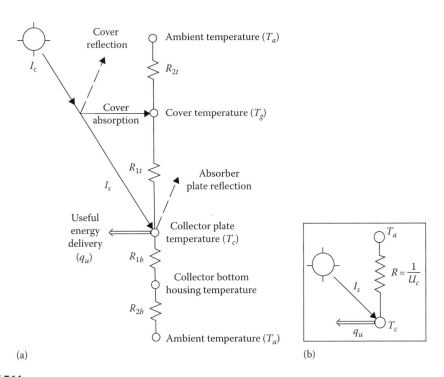

FIGURE 7.14
Thermal circuits for a flat-plate collector shown in Figure 7.13: (a) detailed circuit; (b) approximate, equivalent circuit to (a). In both circuits, the absorber energy is equal to $\alpha_s I_s$, where $I_s = \tau_s I_c$. Collector assumed to be at uniform temperature T_c.

In order to construct a model suitable for a thermal analysis of a flat-plate collector, the following simplifying assumptions will be made:

1. The collector is thermally in steady state.
2. The temperature drop between the top and bottom of the absorber plate is negligible.
3. Heat flow is one-dimensional through the cover as well as through the back insulation.
4. The headers connecting the tubes cover only a small area of the collector and provide uniform flow to the tubes.
5. The sky can be treated as though it were a black-body source for infrared radiation at an equivalent sky temperature.
6. The irradiation on the collector plate is uniform.

For a quantitative analysis, let the plate temperature be T_c and assume solar energy is absorbed at the rate $I_s\alpha_s$. Part of this energy is then transferred as heat to the working fluid, and if the collector is in the steady state, the other part is lost as heat to the ambient air if $T_c > T_a$. Some of the heat loss occurs through the bottom of the collector. It passes first through the back to the environment. Since the collector is in steady state, according to Equation 7.8,

$$q_u = I_c A_c \tau_s \alpha_s - q_{loss} \tag{7.12}$$

where q_{loss} can be determined using the equivalent thermal circuit, as shown in Figure 7.14:

$$q_{loss} = U_c A_c (T_c - T_a) = \frac{A_c (T_c - T_a)}{R} \tag{7.13}$$

There are three parallel paths to heat loss from the hot collector absorber plate at T_c to the ambient at T_a: top, bottom, and edges. Because the edge losses are quite small compared to the top and bottom losses, they are quite often neglected. However, they can be estimated easily if the insulation around the edges is of the same thickness as the back. The edge loss can be accounted for by simply adding the areas of the back (A_c) and the edges (A_e) for back heat loss. Therefore, the overall heat loss coefficient is

$$U_c A_c = \frac{A_c}{R_{1t} + R_{2t}} + \frac{A_c + A_e}{R_{1b} + R_{2b}} \tag{7.14}$$

The thermal resistances can be found easily from the definition. For example,

$$R_{1b} = \frac{l_i}{k_i} \quad \text{and} \quad R_{2b} = \frac{1}{h_{c,bottom}} \tag{7.15}$$

where
   k_i and l_i are, respectively, the thermal conductivity and thickness of the insulation
   $h_{c,bottom}$ is the convective heat-transfer coefficient between the collector and the air below the collector

In a well-insulated collector, R_{2b} is much smaller than R_{1b} and usually neglected. Referring to Figure 7.13,

$$A_e = 2(l_1 + l_2)l_3 \tag{7.16}$$

Since the heat loss from the top is by convection and radiation, it is more complicated than the bottom heat loss. Convection and radiation provide two parallel paths for heat loss from the absorber plate at T_c to the glass cover at T_g, and from the glass cover to the ambient. That is, the series resistance of R_{1t} and R_{2t} consists of
   Therefore,

$$\frac{1}{R_{1t}} = \frac{1}{R_{r,1}} + \frac{1}{R_{c,1}} = h_{r,1} + h_{c,1} \tag{7.17}$$

and

$$\frac{1}{R_{1t}} = \frac{1}{R_{r,\infty}} + \frac{1}{R_{c,\infty}} = h_{r,\infty} + h_{c,\infty} \tag{7.18}$$

Since thermal radiative heat transfer is proportional to the fourth power of the temperature, R_r and h_r are found as follows:
   Radiative heat transfer from the plate to the glass cover

$$q_{rc\to R} = \sigma A_c \frac{\left(T_c^4 - T_g^4\right)}{(1/\varepsilon_{p,i} + 1/\varepsilon_{s,i} - 1)} = h_{r,1} A_c (T_c - T_s) \tag{7.19}$$

where
   $\varepsilon_{p,i}$ is the infrared emittance of the plate
   $\varepsilon_{g,i}$ is the infrared emittance of the glass cover

Therefore,

$$h_{r,1} = \frac{\sigma(T_c + T_g)\left(T_c^2 + T_g^2\right)}{(1/\varepsilon_{p,i} + 1/\varepsilon_{g,i} - 1)} \tag{7.20}$$

Similarly, from the radiative heat transfer between the glass plate (at T_g) and the sky (at T_{sky}), we can find that

$$q_{rg\circledR sky} = \varepsilon_{g,i}\sigma A_c\left(T_g^4 - T_{sky}^4\right) = h_{r,\infty} A_c (T_g - T_a) \tag{7.21}$$

or

$$h_{r,\infty} = \varepsilon_{g,i}\sigma \frac{\left(T_g^4 - T_{sky}^4\right)}{(T_g - T_a)} \tag{7.22}$$

Evaluation of the collector heat loss coefficient defined by Equation 7.14 requires iterative solution of Equations 7.19 and 7.21, because the unit radiation coefficients are functions of the cover and plate temperatures, which are not known a priori. A simplified procedure for calculating U_c for collectors with all covers of the same material, which is often sufficiently accurate and more convenient to use, has been suggested by Hottel and Woertz [4] and Klein [5]. It is also suitable for the application to collectors with selective surfaces. For this approach, the collector top loss in watts is written in the form [6]

$$q_{top\,loss} = \frac{(T_c - T_a)A_c}{N/(C/T_c)\left[(T_c - T_a)/(N+f)\right]^{0.33} + 1/h_{c,\infty}}$$

$$+ \frac{\sigma\left(T_c^4 - T_a^4\right)A_c}{1/\left[\varepsilon_{p,i} + 0.05N(1-\varepsilon_{p,i})\right] + (2N+f-1)/\varepsilon_{g,i} - N} \qquad (7.23)$$

where
$f = (1 - 0.04h_{c,\infty} + 0.005h_{c,\infty}^2)(1 + 0.091N)$
$C = 250[1 - 0.0044(\beta - 90)]$
$N =$ number of covers
$h_{c,\infty} = 5.7 + 3.8V$
$\varepsilon_{g,i} =$ infrared emittance of the covers
$V =$ wind speed in m/s

The values of $q_{top\,loss}$ calculated from Equation 7.23 agreed closely with the values obtained from Equation 7.22 for 972 different observations encompassing the following conditions:

$320 < T_c < 420$ K
$260 < T_a < 310$ K
$0.1 < \varepsilon_{p,i} < 0.95$
$0 \le V \le 10$ m/s
$1 \le N \le 3$
$0 \le \beta \le 90$

The standard deviation of the differences in $U_c = q_{top\,loss}/A_c(T_c - T_a)$ was 0.14 W/m²·K for these comparisons.

7.2.5 Thermal Analysis of Flat-Plate Collector–Absorber Plate

In order to determine the efficiency of a solar collector, the rate of heat transfer to the working fluid must be calculated. If transient effects are neglected [5,7,8], the rate of heat transfer to the fluid flowing through a collector depends on the temperature of the collector surface from which heat is transferred by convection to the fluid, the temperature of the fluid, and the heat-transfer coefficient between the collector and the fluid. To analyze the rate of heat transfer, consider first the condition at a cross section of the collector with flow ducts of rectangular cross sections, as shown in Figure 7.15. Solar radiant energy impinges on the upper face of the collector plate. A part of the total solar radiation falls on the upper surface of the flow channels, while another part is incident on the plates

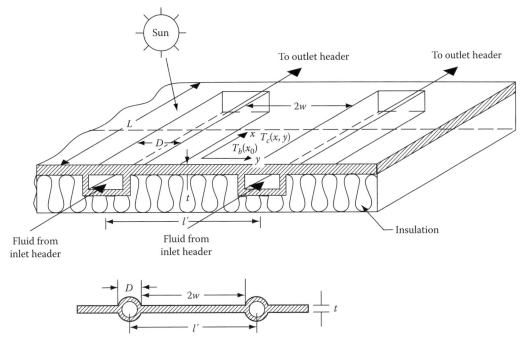

FIGURE 7.15
Sketch showing coordinates and dimensions for collector plate and fluid ducts.

connecting any two adjacent flow channels. The latter is conducted in a transverse direction toward the flow channels. The temperature is maximum at any midpoint between adjacent channels, and the collector plate acts as a fin attached to the walls of the flow channel. The thermal performance of a fin can be expressed in terms of its efficiency. The fin efficiency η_f is defined as the ratio of the rate of heat flow through the real fin to the rate of heat flow through a fin of infinite thermal conductivity, that is, a fin at a uniform temperature. We shall now derive a relation to evaluate this efficiency for a flat-plate solar collector.

If U_c is the overall heat loss coefficient from the collector-plate surface to the ambient air, the rate of heat loss from a given segment of the collector plate at x, y in Figure 7.15 is

$$q(x,y) = U_c \left[T_c(x,y) - T_a \right] dxdy, \tag{7.24}$$

where
  T_c is the local collector-plate temperature $(T_c > T_a)$
  T_a is the ambient air temperature
  U_c is the overall heat loss coefficient between the plate and the ambient air

U_c includes the effects of radiation and free convection between the plates, the radiative and convective transfer between the top of the cover and the environment, and conduction through the insulation. Its quantitative evaluation has been previously considered.

If conduction in the x direction is negligible, a heat balance at a given distance x_0 for a cross section of the flat-plate collector per unit length in the x direction can be written in the form

$$\alpha_s I_s dy - U_c (T_c - T_a) dy + \left(-kt \left. \frac{s T_c}{dy} \right|_{y,x_0} \right) - \left(-kt \left. \frac{d T_c}{dy} \right|_{y+dy,x_0} \right) = 0 \tag{7.25}$$

If the plate thickness t is uniform and the thermal conductivity of the plate is independent of temperature, the last term in Equation 7.25 is

$$\left. \frac{dT_c}{dy} \right|_{y+dy,x_0} = \left. \frac{dT_c}{dy} \right|_{y,x_0} + \left(\frac{d^2 T_c}{dy^2} \right)_{y,x_0} dy$$

and Equation 7.25 can be cast into the form of a second-order differential equation:

$$\frac{d^2 T}{dy^2} = \frac{U_c}{kt} \left[T_c - \left(T_a + \frac{\alpha_s I_s}{U_c} \right) \right] \tag{7.26}$$

The boundary conditions for the system described earlier at a fixed x_0 are

1. At the center between any two ducts, the heat flow is 0, or at $y = 0$, $dT_c = 0$.
2. At the duct, the plate temperature is $T_b(x_0)$, or at $y = w = (l' - D)/2$, $T_c = T_b(x_0)$ where $T_b(x_0)$ is the fin-base temperature.

If we let $m^2 = U_c/kt$ and $\phi = T_c - (T_a + \alpha_s I_s/U_c)$, Equation 7.26 becomes

$$\frac{d^2 \phi}{dy^2} = m^2 \phi \tag{7.27}$$

subject to the boundary conditions

$$\frac{d\phi}{dy} = 0 \quad \text{at} \quad y = 0$$

and

$$\phi = T_b(x_0) - \left(T_a + \frac{\alpha_s I_s}{U_c} \right) \quad \text{at} \quad y = w$$

The general solution of Equation 7.27 is

$$\phi = C_1 \sinh my + C_2 \cosh my \tag{7.28}$$

The constants C_1 and C_2 can be determined by substituting the two boundary conditions and solving the two resulting equations for C_1 and C_2. This gives

$$\frac{T_c - (T_a + \alpha_s I_s / U_c)}{T_b(x_0) - (T_a + \alpha_s I_s / U_c)} = \frac{\cosh my}{\cosh mw} \qquad (7.29)$$

From the preceding equation, the rate of heat transfer to the conduit from the portion of the plate between two conduits can be determined by evaluating the temperature gradient at the base of the fin, or

$$q_{fin} = -kt \left. \frac{dT_c}{dy} \right|_{y=w} = \frac{1}{m} \left\{ \alpha_s I_s - U_c \left[T_b(x_0) - T_a \right] \tanh mw \right\} \qquad (7.30)$$

Since the conduit is connected to fins on both sides, the total rate of heat transfer is

$$q_{total}(x_0) = 2w \left\{ \alpha_s I_s - U_c \left[T_b(x_0) - T_a \right] \right\} \frac{\tanh mw}{mw} \qquad (7.31)$$

If the entire fin were at the temperature $T_b(x_0)$, a situation corresponding physically to a plate of infinitely large thermal conductivity, the rate of heat transfer would be a maximum, $q_{total,max}$. As mentioned previously, the ratio of the rate of heat transfer with a real fin to the maximum rate obtainable is the fin efficiency η_f. With this definition, Equation 7.31 can be written in the from

$$q_{total}(x_0) = 2w\eta_f \left\{ \alpha_s I_s - U_c \left[T_b(x_0) - T_a \right] \right\} \qquad (7.32)$$

where $\eta_f \equiv \tanh mw / mw$.

The fin efficiency η_f is plotted as a function of the dimensionless parameter $w(U_c/kt)^{1/2}$ in Figure 7.16. When the fin efficiency approaches unity, the maximum portion of the radiant energy impinging on the fin becomes available for heating the fluid.

In addition to the heat transferred through the fin, the energy impinging on the portion of the plate above the flow passage is also useful. The rate of useful energy from region available to heat the working fluid is

$$q_{duct}(x_0) = D \left\{ a_s I_s - U_c \left[T_b(x_0) - T_a \right] \right\} \qquad (7.33)$$

Thus, the useful energy per unit length in the flow direction becomes

$$q_u(x_0) = (D + 2w\eta) \left\{ a_s I_s - U_c \left[T_b(x_0) - T_a \right] \right\} \qquad (7.34)$$

The energy $q_u(x_0)$ must be transferred as heat to the working fluid. If the thermal resistance of the metal wall of the flow duct is negligibly small and there is no contact resistance between the duct and the plate, the rate of heat transfer to the fluid is

$$q_u(x_0) = P\bar{h}_{c,i} \left[T_b(x_0) - T_f(x_0) \right] \qquad (7.35)$$

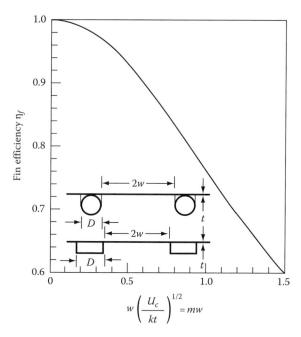

FIGURE 7.16
Fin efficiency for tube and sheet flat-plate solar collectors.

where P is the perimeter of the flow duct, which is $2(D + d)$ for a rectangular duct. Contact resistance may become important in poorly manufactured collectors in which the flow duct is clamped or glued to the collector plate. Collectors manufactured by such methods are usually not satisfactory.

7.2.6 Collector Efficiency Factor

To obtain a relation for the useful energy delivered by a collector in terms of known physical parameters, the fluid temperature, and the ambient temperature, the collector temperature must be eliminated from Equations 7.34 and 7.35. Solving for $T_b(x_0)$ in Equation 7.35 and substituting this relation in Equation 7.34 gives

$$q_u(x_0) = l'F'\left\{\alpha_s I_s - U_c\left[T_f(x_0) - T_a\right]\right\} \tag{7.36}$$

where
    F' is called the collector efficiency factor [9]
    $l' = (2w + D)$

F' is given by

$$F' = \frac{1/U_c}{l'[1/(U_c(D + 2w\eta_f)) + 1/(\bar{h}_{c,i}P)]} \tag{7.37a}$$

For a flow duct of circular cross section of diameter D, F' can be written as

$$F' = \frac{1/U_c}{l'\left[1/\left(U_c(D + 2w\eta_f)\right) + 1/(\bar{h}_{c,i}\pi D)\right]} \tag{7.37b}$$

TABLE 7.6

Typical Values for the Parameters That Determine the Collector
Efficiency Factor F' for a Flat-Plate Collector in Equation 7.37

		U_c
2 glass covers	4 W/m²·K	0.685 Btu/h·ft²°F
1 glass cover	8 W/m²·K	1.37 Btu/h·ft²°F
		kt
Copper plate, 1 mm thick	0.38 W/k	0.72 Btu/h°F
Steel plate, 1 mm thick	0.045 W/k	0.0866 Btu/h°F
		$h_{c,i}$
Water in laminar flow forced convection	300 W/m²·K	52 Btu/h·ft²°F
Water in turbulent flow forced convection	1500 W/m²·K	254 Btu/h·ft²°F
Air in turbulent forced convection	100 W/m²·K	17.6 Btu/h·ft²°F

Physically, the denominator in Equations 7.37 is the thermal resistance between the fluid
and the environment, whereas the numerator is the thermal resistance between the collector surface and the ambient air. The collector-plate efficiency factor F' depends on U_c, $\bar{h}_{c,i}$,
and η_f. It is only slightly dependent on temperature and can, for all practical purposes, be
treated as a design parameter. Typical values for the factors determining the value of F'
are given in Table 7.6.

The collector efficiency factor increases with increasing plate thickness and plate thermal conductivity, but it decreases with increasing distance between flow channels. Also
increasing the heat-transfer coefficient between the walls of the flow channel and the
working fluid increases F', but an increase in the overall conductance U_c will cause F' to
decrease.

7.2.7 Collector Heat-Removal Factor

Equation 7.36 yields the rate of heat transfer to the working fluid at a given point x along
the plate for specified collector and fluid temperatures. However, in a real collector, the
fluid temperature increases in the direction of flow as heat is transferred to it. An energy
balance for a section of flow duct dx can be written in the form

$$\dot{m}c_p(T_f\,|_{x+dx} - T_f\,|_x) = q_u(x)dx \tag{7.38}$$

Substituting Equation 7.36 for $q_u(x)$ and $T_f(x) + (dT_f(x)/dx)dx$ for $T_f|_{x+dx}$ in Equation 7.38 gives
the differential equation

$$\dot{m}c_p \frac{dT_f(x)}{dx} = l'F'\left\{\alpha_s I_s - U_c\left[T_f(x) - T_a\right]\right\} \tag{7.39}$$

Separating the variables gives, after some rearranging,

$$\frac{dT_f(x)}{T_f(x) - T_a - \alpha_s I_s/U_c} = \frac{l'F'U_c}{\dot{m}c_p}dx \tag{7.40}$$

Equation 7.40 can be integrated and solved for the outlet temperature of the fluid $T_{f,out}$ for a duct length L, and for the fluid inlet temperature $T_{f,in}$ if we assume that F' and U_c are constant, or

$$\frac{T_{f,out} - T_a - \alpha_s I_s / U_c}{T_{f,in} - T_a - \alpha_s I_s / U_c} = \exp\left(-\frac{U_c l' F' L}{\dot{m} c_p}\right) \tag{7.41}$$

To compare the performance of a real collector with the thermodynamic optimum, it is convenient to define the heat-removal factor F_R as the ratio between the actual rate of heat transfer to the working fluid and the rate of heat transfer at the minimum temperature difference between the absorber and the environment. The thermodynamic limit corresponds to the condition of the working fluid remaining at the inlet temperature throughout the collector. This can be approached when the fluid velocity is very high. From its definition, F_R can be expressed as

$$F_R = \frac{G c_P (T_{f,out} - T_{f,in})}{\alpha_s I_s - U_c (T_{f,in} - T_a)} \tag{7.42}$$

where G is the flow rate per unit surface area of collector $\dot{m}/A_c$. By regrouping the right-hand side of Equation 7.42 and combining with Equation 7.41, it can easily be verified that

$$F_R = \frac{G c_p}{U_c}\left[1 - \frac{\alpha_s I_s / U_c - (T_{f,out} - T_a)}{\alpha_s I_s / U_c - (T_{f,in} - T_a)}\right]$$

$$= \frac{G c_p}{U_c}\left[1 - \exp\left(-\frac{U_c F'}{G c_p}\right)\right] \tag{7.43}$$

It should be noted that F_R can be evaluated from design parameters. Inspection of the earlier relation shows that F_R increases with increasing flow rate and approaches as an upper limit F', the collector efficiency factor. Since the numerator of the right-hand side of Equation 7.42 is q_u, the rate of useful heat transfer can now be expressed in terms of the fluid inlet temperature, or

$$q_u = A_c F_R \left[\alpha_s I_s - U_c (T_{f,in} - T_a)\right] \tag{7.44}$$

If a glazing above the absorber plate has transmittance τ_s, then

$$q_u = A_c F_R \left[\tau_s \alpha_s I_c - U_c (T_{f,in} - T_a)\right] \tag{7.45}$$

and instantaneous efficiency η_c is

$$\eta_c = \frac{q_u}{I_c A_c} = F_R \left[\tau_s \alpha_s - \frac{U_c (T_{f,in} - T_a)}{I_c} \right] \tag{7.46}$$

Equation 7.46 is known as the *Hottel–Whillier–Bliss* equation. This is a convenient form for design, because the fluid inlet temperature to the collector is usually known or can be specified.

Example 7.1

Calculate the averaged hourly and daily efficiency of a water solar collector on January 15, in Boulder, CO. The collector is tilted at an angle of 60° and has an overall conductance of 8.0 W/m² K on the upper surface. It is made of copper tubes, with a 1 cm ID, 0.05 cm thick, which are connected by a 0.05 cm thick plate at a center-to-center distance of 15 cm. The heat-transfer coefficient for the water in the tubes is 1500 W/m² K, the cover transmittance is 0.9, and the solar absorptance of the copper surface is 0.9. The collector is 1 m wide and 2 m long, the water inlet temperature is 330 K, and the water flow rate is 0.02 kg/s. The horizontal insolation (total) I_h and the environmental temperature are tabulated later. Assume the diffuse radiation accounts for 25% of the total insolation.

Solution
The total radiation received by the collector is calculated from Equation 5.42, and neglecting the ground reflected radiation,

$$I_c = I_{d,c} + I_{b,c} = 0.25 I_h \cos^2\left(\frac{60}{2}\right) + (1 - 0.25) I_h \bar{R}_b$$

Time (h)	I_h(W/m²)	T_{amb} (K)
7–8	12	270
8–9	80	280
9–10	192	283
10–11	320	286
11–12	460	290
12–13	474	290
13–14	395	288
14–15	287	288
15–16	141	284
16–17	32	280

The tilt factor $\bar{R}_b$ is obtained from its definition in Chapter 5 (see Equation 5.58):

$$\bar{R}_b = \frac{\cos i}{\sin \alpha} = \frac{\sin(L - \beta)\sin \delta_s + \cos(L - \beta)\cos \delta_s \cos h_s}{\sin L \sin \delta_s + \cos L \cos \delta_s \cos h_s}$$

where $L = 40°$ $\delta_s = -21.1$ on January 15, and $\beta = 60°$. The hour angle h_s equals $15°$ for each hour away from noon.

The fin efficiency is obtained from Equation 7.32:

$$\eta_f = \frac{\tanh m(l' - D)/2}{m(l' - D)/2}$$

where

$$m = \left(\frac{U_c}{kt}\right)^{1/2} = \left(\frac{8}{390 \times 5 \times 10^{-4}}\right)^{1/2} = 6.4$$

and

$$\eta_f = \frac{\tanh 6.4(0.15 - 0.01)/2}{6.4(0.15 - 0.01)/2} = 0.938$$

The collector efficiency factor F' is, from Equation 7.37,

$$F' = \frac{1/U_c}{l'[1/(U_c(D + 2w\eta_f)) + 1/(h_{c,i}\pi D)]}$$

$$= \frac{1/8.0}{0.15[1/8.0(0.01 + 0.14 \times 0.938) + 1/1500\pi \times 0.01]} = 0.92$$

Then we obtain the heat-removal factor from Equation 7.43:

$$F_R = \frac{Gc_p}{U_c}\left[1 - \exp\left(-\frac{U_c F'}{Gc_p}\right)\right]$$

Time (h)	I_h (Wm²)	R_b	$I_{d,c}$ (W/m²)	$I_{b,c}$ (W/m²)	I_c (W/m²)	q_u (W)	T_{amb} (K)	η_c
7–8	12	10.9	1	98	99	0	270	0
8–9	80	3.22	5	193	198	0	280	0
9–10	192	2.44	12	351	363	0	283	0
10–11	320	2.18	20	523	543	148	286	0.137
11–12	460	2.08	29	718	747	482	290	0.322
12–13	474	2.08	30	739	169	512	290	0.333
13–14	395	2.18	25	646	671	351	288	0.261
14–15	287	2.49	18	525	543	175	288	0.162
15–16	141	3.22	9	341	350	0	284	0
16–17	32	10.9	2	261	263	0	280	0

$$F_R = \frac{0.001 \times 4184}{8.0}\left[1 - \exp\left(\frac{8.0 \times 0.922}{0.01 \times 4184}\right)\right] = 0.845$$

From Equation 7.45, the useful heat delivery rate is

$$q_u = 2 \times 0.845[I_c \times 0.81 - 8.0(T_{f,in} - T_{amb})]$$

The efficiency of the collector is $\eta_c = q_u / A_c I_c$ and the hourly averages are calculated in the earlier table.

Thus, $\sum I_c = 4546\,\text{W/m}^2$ and $\sum q_u = 1668\,\text{W}$. The daily average efficiency is obtained by summing the useful energy for those hours during which the collector delivers heat and dividing by the total insolation between sunrise and sunset. This yields

$$\bar{\eta}_{c,day} = \frac{\sum q_u}{\sum A_c I_c} = \frac{1668}{2 \times 4546} = 0.183 \quad \text{or} \quad 18.3\%$$

Example 7.2

Estimate the thermal energy output (which is equal to the electric energy saved) and the nominal levelized cost of saved energy (Equation 2.24) for a hot water solar collector system with two storage tanks near Boulder, CO, similar to the one in Example 7.1, but use the SAM program. Assume the cost of the collector is $200.00 per square meter, the pump and storage tank is $1200.00, and the installation and balance of system is $800.00 and $200.00 respectively. Except for the parameters specified, use the default values in the program. Assume that the average daily hot water usage is 200 kg/day and the collector area is 3 m² with a tilt angle of 30° and an azimuth of 0°. The parameters in the Hottel–Whillier–Bliss equation (Equation 7.45) are the same as in Problem 7.1: the FRta ($F_R\tau\alpha$) value is 0.845*0.9*0.9 = 0.684, the FRUL ($F_R U_c$) value is 0.845*8.0 = 6.76 W/m²*K, and the incidence angle modifier from Equation 7.66 is 0.15. The heat-exchanger efficiency is 0.5 and the auxiliary temperature in the electric water tank is set at 50°C. Use the default values for the economic parameters, but assume a real discount rate of 7% and an inflation rate of 2% with a lifetime of 30 years. To simplify the analysis, neglect any salvage value. A further discussion of the storage tanks in the system will be given in Section 10.7.6.

Solution

A solution for a very similar system is done using SAM, which can be accessed through https://sam.nrel.gov/. After downloading SAM from the website, to start a Solar Water Heating project

1. Start SAM
2. Under **Enter a new project name to begin,** type a name for your project. For example, "Solar Water Heater System"
3. Click **Create New File**
4. Under 1. **Select a technology,** click **Solar Water Heating**
5. Under 2. **Select a financing option,** click on **Residential** financing option
6. Click **OK**

The collector is located in Boulder, CO, so under the "Climate" tab, change the location to Boulder, CO. Under the "SWH System" tab, change the values that were quantified in the problem, which are circled in the following figures. To change the collector specifications, the "User Specified" button has to be selected.

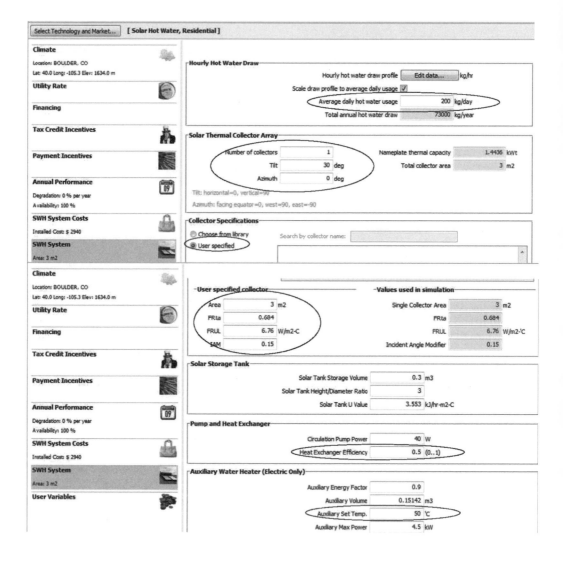

After defining the parameters of the solar water heater system, the financing options need to be adjusted to match the problem statement. Under the "Financing" tab and the "SWH System Costs" tab, the circled values in the following figures were changed to the numbers in the problem statement.

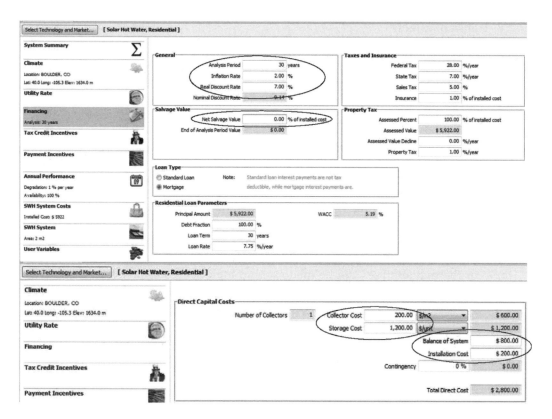

Click the green arrow button in the lower left-hand corner and the following table should show up.

Metric	Base
Annual energy saved (kWh)	1510
Aux with solar (kWh)	1556.4
Aux without solar (kWh)	3177.6
Solar fraction	0.565
LCOE nominal	12.86¢/kWh
LCOE real	10.51¢/kWh
First year revenue without system	$0.00
First year revenue with system	$181.23
First year net revenue	$181.23
After-tax NPV	$234.23
Payback period	16.1699 years
Capacity factor	11.9%
First year kWh/kW	1046

Note that the accuracy of solar performance predictions is generally no better than 10%, and the number of significant figures in the computer solution is not realistic. Note also that the levelized cost of the energy from this system is not likely to be economical without some tax incentives.

*7.2.8 Transient Effects

The preceding analysis assumed that steady-state conditions exist during the operation of the collector. Under actual operating conditions, the rate of insolation will vary, and the ambient temperature and the external wind conditions may change. To determine the effect of changes in these parameters on the performance of a collector, it is necessary to make a transient analysis that takes the thermal capacity of the collector into account.

As shown in [10], the effect of collector thermal capacitance is the sum of two contributions: the *collector–storage* effect, resulting from the heat required to bring the collector up to its final operating temperature, and the *transient* effect, resulting from fluctuations in the meteorological conditions. Both effects result in a net loss of energy delivered compared with the predictions from the zero capacity analysis. This loss is particularly important on a cold morning when all of the solar energy absorbed by the collector is used to heat the hardware and the working fluid, thus delaying the delivery of useful energy for some time after the sun has come up.

Transient thermal analyses can be made with a high degree of precision [11], but the analytical predictions are no more accurate than the weather data and the overall collector conductance. For most engineering applications, a simpler approach is therefore satisfactory [12]. For this approach, it will be assumed that the absorber plate, the ducts, the back insulation, and the working fluid are at the same temperature. If back losses are neglected, an energy balance on the collector plate and the working fluid for a single-glazed collector delivering no useful energy can be written in the form

$$(\overline{mc})_p \frac{d\overline{T}_p(t)}{dt} = A_c I_s \alpha + A_c U_p \left[\overline{T}_g(t) - \overline{T}_p(t) \right] \tag{7.47}$$

where
$(\overline{mc})_p$ is the sum of the thermal capacities of the plate, the fluid, and the insolation
I_s is the insolation on the absorber plate
U_p is the conductance between the absorber plate at $\overline{T}_p$ and its cover at $\overline{T}_g$

Similarly, a heat balance on the collector cover gives

$$(\overline{mc})_g \frac{d\overline{T}_g(t)}{dt} = A_c U_p \left[\overline{T}_p(t) - \overline{T}_g(t) \right] - A_c U_\infty \left[\overline{T}_g(t) - T_a \right] \tag{7.48}$$

where
$U_\infty = (h_{c,\infty} + h_{r,\infty})$ (see Equation 7.18)
$(\overline{mc})_g$ = thermal capacity of the cover plate

Equations 7.47 and 7.48 can be solved simultaneously, and the transient heat loss can then be determined by integrating the instantaneous loss over the time during which transient effects are pronounced. A considerable simplification in the solution is possible if one assumes that at any time, the collector heat loss and the cover heat loss are equal, as in a quasi-steady state, so that

$$U_\infty A_c \left[\overline{T}_g(t) - T_a \right] = U_c A_c \left[\overline{T}_p(t) - T_a \right] \tag{7.49}$$

Then, for a given air temperature, the differentiation of Equation 7.49 gives

$$\frac{d\bar{T}_g(t)}{dt} = \frac{U_e d\bar{T}_p(t)}{U_\infty dt} \tag{7.50}$$

Adding Equations 7.48 through 7.50 gives a single differential equation for the plate temperature:

$$\left[(\overline{mc})_p + \frac{U_c}{U_\infty}(mc)_R \right] \frac{d\bar{T}_p(t)}{dt} = \left[\alpha_c I_s - U_c \left(\bar{T}_p(t) - T_a \right) \right] A_c \tag{7.51}$$

Equation 7.51 can be solved directly for given values of I_s and T_a. The solution to Equation 7.51 then gives the plate temperature as a function of time, for an initial plate temperature $T_{p,0}$, in the form

$$\bar{T}_p(t) - T_a = \frac{\alpha_s I_s}{U_c} - \left[\frac{\alpha_s I_s}{U_c} - (T_{p,0} - T_a) \right] \exp\left[-\frac{U_c A_c t}{(\overline{mc})_p + (U_c/U_\infty)_c (mc)_g} \right] \tag{7.52}$$

Collectors with more than one cover can be treated similarly, as shown in [13].

For a transient analysis, the plate temperature $\bar{T}_p$ can be evaluated at the end of a specified time period if the initial value of $\bar{T}_p$ and the values of α_s, I_s, U_c, and T_a during the specified time are known. Repeated applications of Equation 7.52 provide an approximate method of evaluating the transient effects. An estimate of the net decrease in useful energy delivered can be obtained by multiplying the effective heat capacity of the collector, given by $(\overline{mc})_p + (U_c/U_\infty)(mc)_g$, by the temperature rise necessary to bring the collector to its operating temperature. Note that the parameter $[(\overline{mc})_p + (U_c/U_\infty)(mc)_g]/U_c A_c$ is the *time constant* of the collector [11,14], and small values of this parameter will reduce losses resulting from transient effects.

Example 7.3

Calculate the temperature rise between 8 and 10 a.m. of a 1×2 m single-glazed water collector with a 0.3 cm thick glass cover if the heat capacities of the plate, water, and back insulation are 5, 3, and 2 kJ/K, respectively. Assume that the unit surface conductance from the cover to ambient air is 18 W/m² · K, and the unit surface conductance between the collector and the ambient air is $U_c = 6$ W/m² · K. Assume that the collector is initially at the ambient temperature. The absorbed insolation $\alpha_s I_s$ during the first hour averages 90 W/m² and that between 9 and 10 a.m. is 180 W/m². The air temperature between 8 and 9 a.m. is 273 K and that between 9 and 10 a.m. is 278 K.

Solution

The thermal capacitance of the glass cover is $(mc)_g = (\rho V c_p)_g = (2500 \text{ kg/m}^3)$ (1 m × 2 m × 0.003 m) (1 kJ/kg·K) = 15 kJ/K. The combined collector, water, and insulation thermal capacity is equal to

$$(\overline{mc})_p + \frac{U_c}{U_\infty}(mc)_g = 5 + 3 + 2 + 0.3 + 15 = 15.5 \text{ kJ/K}$$

From Equation 7.52, the temperature rise of the collector is given by

$$T_p - T_a = \frac{\alpha_s I_s}{U_c} - \left[\frac{\alpha_s I_s}{U_c} - (T_{P,0} - T_a) \right] \exp\left[-\frac{U_c A_c t}{(\overline{mc})_p + (U_c/U_\infty)_c (mc)_g} \right]$$

At 8 a.m., $T_{p,0} = T_a$; therefore, the temperature rise, between 8 and 9 a.m., is

$$= \frac{90}{6}\left[1 - \exp\left(-\frac{2 \times 6 \times 3,600}{15,500} \right) \right] = 15 \times 0.944 = 14.2\,K$$

Thus, at 9 a.m., the collector temperature will be 287.2 K. Between 9 and 10 a.m., the collector temperature will rise as follows:

$$= 278 + \frac{180}{6} - (30 - 9.2)0.056 = 306.3\,K(91°F)$$

Thus, at 10 a.m., the collector temperature has achieved a value sufficient to deliver useful energy at a temperature level of 306 K.

7.3 Evacuated Tube Collectors

Two general methods exist for significantly improving the performance of solar collectors above the minimum flat-plate collector level. The first method increases solar flux incident on the receiver. It will be described in the next section on concentrators. The second method involves the reduction of parasitic heat loss from the receiver surface. This can be accomplished by placing a tubular receiver inside another transparent tube and evacuating the annular space between them to about 10–4 mmHg. This arrangement eliminates convection losses and is commonly referred to as an evacuated tube collector. Figure 7.17 is a schematic cross section of a single tube in an evacuated tube collector.

Evacuated tube devices have been proposed as efficient solar energy collectors since the early twentieth century. In 1909, Emmett [15] proposed several evacuated tube concepts for solar energy collection. Speyer [16] also proposed a tubular evacuated flat-plate design for high-temperature operation. With the recent advances in vacuum technology, evacuated tube collectors can be reliably mass-produced. Their high-temperature effectiveness is essential for the efficient operation of proposed solar air-conditioning systems and process heat systems.

Simple tubes with an evacuated annulus are suitable for various concentrating collectors such as parabolic troughs or compound parabolic concentrators to be discussed in Chapter 8. But for flat-plate stationary collectors, an arrangement with several tubes in parallel as shown in Figure 7.17 is required.

Since close packing of concentric-tube collector (CTC) tubes in an array can result in shading losses at any angle other than normal incidence, it is cost-effective to space the tubes apart and to use a back reflector in order to capture radiation passing between the tubes.

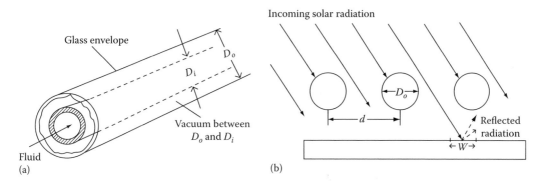

FIGURE 7.17
A schematic cross section of (a) single tube in an evacuated tube collector, and (b) a cross section of an assembly of tubes from a collector. (From Goswami, Y. et al., *Principles of Solar Engineering*, 2nd edn., Taylor & Francis, Philadelphia, PA, 2000. With permission.)

Figure 7.17 shows the geometry of part of a CTC array with tube spacing d; Figure 7.17 is a cutaway drawing of one tubular assembly. Beekley and Mather [17] have analyzed the CTC in detail and have shown that a tube spacing one envelope diameter apart (i.e., $d = 2D_o$) maximizes daily energy gain. The specular reflector improves performance by 10% or more.

CTC arrays can collect both direct and diffuse radiations. Each radiation component must be analyzed in turn. The optical efficiency η_o may be expressed as

$$\eta_o = \frac{\tau_e \alpha_r I_{eff}}{I_{b,c} + I_{d,c}} \tag{7.53}$$

where
I_{eff} is the effective solar radiation both directly intercepted and intercepted after refection from the back reflector
$I_{b,c}$ and $I_{d,c}$ are, respectively, the beam and diffuse radiation components intercepted per unit collector aperture area
τ_e and α_r are the glass envelope transmittance and receiver absorptance, respectively
The subscript e denotes envelope (tube) properties

The total effective insolation I_{eff} can be calculated by summing directly intercepted and reflected radiations, that is,

$$I_{eff} = I_b \left(\cos i + \cos i \frac{W}{D_o} \right) + I_{d,c} \left[\pi F_{TS} (1 + \rho \bar{F}) \right] \tag{7.54}$$

where
F_{TS} is the radiation shape factor from a tube to the sky dome
$\bar{F}$ is a measure of the shape factors of diffusely illuminated strips to collector tubes

For tubes spaced one diameter apart, $F_{TS} \cong 0.43$ and $\bar{F} \cong 0.34$. The optical efficiency η_o (Equation 7.53) is not, therefore, a simple collector property independent of operating conditions, but rather a function of time through the incidence angles.

*7.3.1 Thermal Analysis of an Evacuated Tube Collector

The heat loss, q_L, from an evacuated tubular collector occurs primarily by radiation from the absorber surface and can be expressed as

$$q_L = U_c(T_r - T_a) = \frac{T_r - T_a}{R_1 + R_2 + R_3} \tag{7.55}$$

where
   R_1 is the thermal radiative resistance between the surface of the tube and the inner surface of the glass tube
   R_2 is the thermal resistance of the glass tube
   R_3 is the thermal resistance between the outer surface of the glass tube and the environment by radiation and convection
   R_2 is negligible for commercial designs, and R_1 and R_3 are approximately given by

$$R_1 = \frac{(1/\varepsilon_r) + (1/\varepsilon_g) - 1}{\sigma(T_r + T_g)(T_r^2 + T_g^2)} \tag{7.56}$$

$$R_3 = \frac{1}{h_c + \sigma\varepsilon_g(T_g + T_a)(T_g^2 + T_a^2)} \tag{7.57}$$

where
   T_r is the receiver (absorber) temperature
   h_c is the external convection coefficient for the glass envelope

Tests have shown that the loss coefficient U_c for a single tube is between 0.5 and 1.0 W/m² °C, which is in agreement with the analysis.
   The CTC energy delivery rate q_u on an aperture area basis can be written as

$$q_u = \tau_e\alpha_r I_{eff} \frac{A_t}{A_c} - U_c(T_r - T_a)\frac{A_r}{A_c} \tag{7.58}$$

where
   A_t is the projected area of a tube (its diameter)
   A_r is the receiver or absorber area

The receiver-to-collector aperture area ratio is $\pi D_r/d$, where d is the center-to-center distance between the tubes. Therefore,

$$q_u = \frac{D_r}{d}\left[\tau_e\alpha_r I_{eff} - \pi U_c(T_r - T_a)\right] \tag{7.59}$$

For optimum performance, a specularly reflecting back surface and a tube spacing of one envelope diameter apart are recommended.

7.4 Experimental Testing of Collectors

The performance of solar thermal systems depends largely on the performance of solar collectors. Therefore, experimental measurement of thermal performance of solar collectors by standard methods is important and necessary. The experimentally determined performance data are needed for design purposes and for determining the commercial value of the collectors. The thermal performance of a solar collector is determined by establishing an efficiency curve from the measured instantaneous efficiencies for a combination of values of incident solar radiation, ambient temperature, and inlet fluid temperature. An instantaneous efficiency of a collector under steady-state conditions can be established by measuring the mass flow rate of the heat-transfer fluid, its temperature rise across the collector ($T_{f,out} - T_{f,in}$), and the incident solar radiation intensity (I_c) as

$$\eta_c = \frac{q_u}{A_c I_c} = \frac{\dot{m} C_P (T_{f,out} - T_{f,in})}{A_c I_c} \tag{7.60}$$

The efficiency, η_c, of a collector under steady state can also be written according to the Hottel–Whillier–Bliss equation (Equation 7.46) as

$$\eta_c = F_R \tau_s \alpha_s - F_R U_c \frac{(T_{f,in} - T_a)}{I_c} \tag{7.61}$$

where $\tau_s \alpha_s$ is the optical efficiency, η_{opt}.

Equation 7.61 suggests that for constant values of F_R and U_c, if η_c is plotted with respect to $(T_{f,in} - T_a)/I_c$, a linear curve will result, with a y intercept of $F_R \tau_s \alpha_s$ and a slope of $-F_R U_c$. Figure 7.18 shows a typical thermal performance curve for a flat-plate collector. Since τ_s and

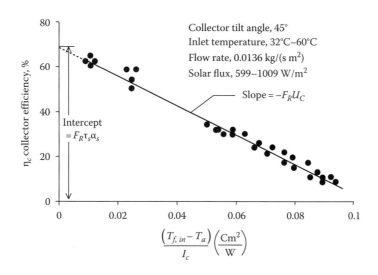

FIGURE 7.18
Experimental thermal efficiency curve for a double-glazed flat-plate liquid-type of solar collector. (From Goswami, Y. et al., *Principles of Solar Engineering*, 2nd edn., Taylor & Francis, Philadelphia, PA, 2000. With permission.)

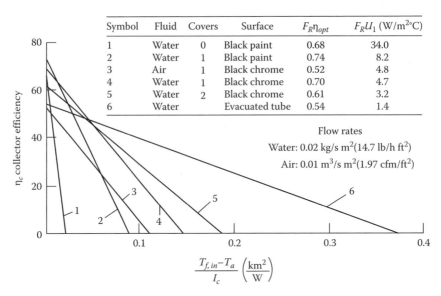

Symbol	Fluid	Covers	Surface	$F_R\eta_{opt}$	$F_R U_1$ (W/m²°C)
1	Water	0	Black paint	0.68	34.0
2	Water	1	Black paint	0.74	8.2
3	Air	1	Black chrome	0.52	4.8
4	Water	1	Black chrome	0.70	4.7
5	Water	2	Black chrome	0.61	3.2
6	Water		Evacuated tube	0.54	1.4

Flow rates

Water: 0.02 kg/s m²(14.7 lb/h ft²)

Air: 0.01 m³/s m²(1.97 cfm/ft²)

FIGURE 7.19
Typical performance curves for various flat-plate solar collectors.

α_s can be measured independently, a thermal performance curve of a flat-plate collector allows us to establish the value of F_R and U_c also. Figure 7.19 shows typical performance curves of various glazed and unglazed flat-plate solar collectors.

In Equation 7.61, the product $\tau_s\alpha_s$ will change with the angle of incidence. Since flat-plate collectors are normally fixed, the angle of incidence changes throughout the day. A relationship can be written between the actual or effective $(\tau_s\alpha_s)$ and $(\tau_s\alpha_s)_n$ for normal incidence as

$$(\tau_s\alpha_s) = (\tau_s\alpha_s)_n K_{\tau\alpha} \qquad (7.62)$$

where
$(\tau_s\alpha_s)_n$ is the value of the product for normal angle of incidence
$K_{\tau\alpha}$ is called the incidence angle modifier

Therefore, the thermal performance of a flat-plate collector may be written as

$$\eta_c = F_R\left[K_{\tau\alpha}(\tau_s\alpha_s)_n - U_c\frac{(T_{f,in} - T_a)}{I_c}\right] \qquad (7.63)$$

Thermal performance of collectors is usually found experimentally for normal angles of incidence in which case $K_{\tau\alpha}$ = 1.0. The incidence angle modifier is then measured separately. It has been established that $K_{\tau\alpha}$ is of the form

$$K_{\tau\alpha} = 1 - b\left(\frac{1}{\cos i} - 1\right) \qquad (7.64)$$

where
b is a constant
i is the angle of incidence

Figure 7.20 shows the effect of incidence angle on $K_{\tau\alpha}$ for flat-plate collectors.

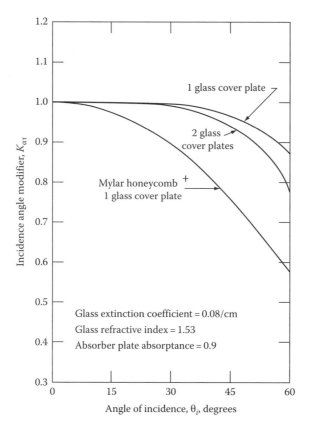

FIGURE 7.20
Incidence angle modifier for three flat-plate solar collectors. (From ASHRAE Standard 93-77, Methods of testing to determine the thermal performance of solar collectors, Atlanta, GA.)

7.4.1 Testing Standards for Solar Thermal Collectors

Standard testing procedures adopted by regulating agencies in various countries establish ways of comparing the thermal performance of various collectors under the same conditions. In the United States, the thermal performance standards established by the American Society of Heating, Refrigeration and Air-Conditioning Engineers (ASHRAE) have been accepted for thermal performance tests of solar thermal collectors. Similar standards have been used in a number of countries. The standards established by ASHRAE include

ASHRAE Standard 93-77, "Methods of testing to determine the thermal performance of solar collectors"

ASHRAE Standard 96-80, "Methods of testing to determine the thermal performance of unglazed solar collectors"

ASHRAE Standard 93-77 specifies the procedures for determining the time constant, thermal performance, and the incidence angle modifier of solar thermal collectors using a liquid or air as a working fluid. Figure 7.21 shows a schematic of a standard testing configuration for thermal performance testing. The tests are conducted under quasi-steady-state conditions.

ASHRAE 93-77 specifies that a curve for incidence angle modifier be established by determining the efficiencies of a collector for average angles of incidence of 0°, 30°, 45°, and

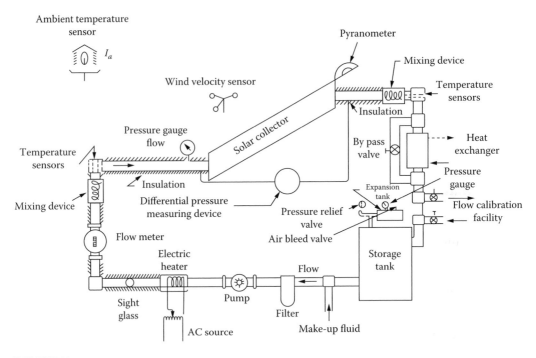

FIGURE 7.21
A testing configuration for a liquid-type solar collector. (From ASHRAE Standard 93-77, Methods of testing to determine the thermal performance of solar collectors, Atlanta, GA. With permission.)

60° while maintaining $T_{f,i}$ at ±1°C of the ambient temperature. Since $(T_{f,i} - T_a)$, 0, the incidence angle modifier according to Equation 7.63 is

$$K_{\tau\alpha} = \frac{\eta_c}{F_R(\tau_s\alpha_s)_n} \tag{7.65}$$

The denominator is the y intercept of the efficiency curve as shown in Figure 7.18.

In order to avoid transient effects, performance of a collector is measured and integrated over at least the time constant of the collector. The time constant is measured by operating the collector under steady or quasi-steady conditions with a solar flux of at least 790 W/m² and then abruptly cutting the incident flux to zero, while continuing to measure the fluid inlet $(T_{f,i})$ and exit $(T_{f,e})$ temperatures. The fluid inlet temperature is maintained at ±1°C of the ambient temperature. The time constant is then determined as the time required to achieve

$$\frac{T_{f,e} - T_{f,i}}{T_{f,e,initial} - T_{f,i}} = 0.30 \tag{7.66}$$

For determining the thermal performance of a collector, ASHRAE Standard 93-77 specifies test conditions to get at least four data points for the efficiency curve. The test conditions include

1. Near-normal incidence ($i \leq 5°$) angle tests close to solar noon time
2. At least four tests for each $T_{f,i}$, two before and two after solar noon
3. At least four different values of $T_{f,i}$ to obtain different values of $\Delta T/I_c$, preferably to obtain ΔT at 10%, 30%, 50%, and 70% of stagnation temperature rise under the given conditions of solar intensity and ambient conditions

*7.5 Calculations of Heating and Hot Water Loads in Buildings

Energy requirements for space heating or service water heating can be calculated from basic conservation of energy principles. For example, the heat required to maintain the interior of a building at a specific temperature is the total of all heat transmission losses from the structure and heat required to warm and humidify the air exchange with the environment by infiltration and ventilation.

Comfort in buildings has long been a subject of investigation by the ASHRAE. The ASHRAE has developed extensive heat load calculation procedures embodied in the *ASHRAE Handbook of Fundamentals* [18]. The most frequently used load calculation procedures will be summarized in this section; the reader is referred to the ASHRAE handbook for details.

Figure 7.22 shows the combinations of temperature and humidity that are required for human comfort. The shaded area is the standard U.S. comfort level for sedentary persons. Many European countries have human comfort levels from 3°C to 7°C below U.S. levels. If activity of a continuous nature is anticipated, the comfort zone lies to the left of the shaded area; if extra clothing is worn, the comfort zone is displaced similarly.

7.5.1 Calculation of Heat Loss

It is outside the scope of this book to describe the details of the heat load calculations for buildings. However, the method is described in brief in this section. For details, one should refer to the *ASHRAE Handbook of Fundamentals* [18] or some textbook on heating and air-conditioning. Table 7.7 lists the components of heat loss calculations of a building.

Complete tables of thermal properties of building materials are on the accompanying website, http://www.crcpress.com/product/isbn/9781466556966, and are numbered from Table W.7.1 to Table W.7.11. Transmission heat losses through attics, unheated basements, and the like are buffered by the thermal resistance of the unheated space. For example, the temperature of an unheated attic lies between that of the heated space and that of the environment. As a result, the ceiling of a room below an attic is exposed to a smaller temperature difference and consequent lower heat loss than the same ceiling without the attic would be. The effective conductance of thermal buffer spaces can easily be calculated by forming an energy balance on such spaces.

The following example is an illustration of the heat loss calculation method described in this section.

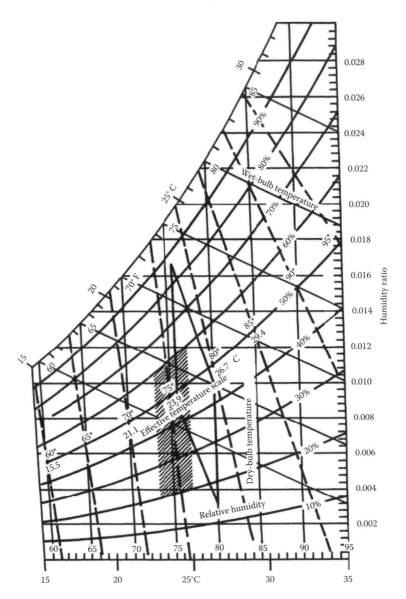

FIGURE 7.22
Heat loss calculations for buildings.

Example 7.4

Calculate the heat load on a house for which the wall area is 200 m², the floor area is 600 m², the roof area is 690 m², and the window area totals 100 m². Inside wall height is 3 m. The construction of the wall and the roof is shown in Figure 7.23.

Solution

The thermal resistance of the wall shown in Figure 7.23 can be found by the electrical resistance analogy as

$$R_{wa} = R_{outside\,air} + R_{wood\,siding} + R_{sheathing} + R_{comb} + R_{wall\,board} + R_{inside\,air}$$

TABLE 7.7

Heating Load Calculations for Buildings

Heating Load Component	Equations 7.67 through 7.70	Descriptions/References
Walls, roof, ceilings, glass	$q = U \cdot A(T_i - T_o)$ (7.67)	T_i, T_o are inside and outside air temperatures, respectively. U values of composite section are calculated from the thermal properties of components given in the tables on the accompanying website
Concrete floors on ground	$q_{fe} = F_e P_e (T_i - T_o)$ (7.68)	p_e is the perimeter of the slab. F_e values are given in the tables on the accompanying website
Infiltration and ventilation air	$q_{sensible} = Q \rho_a C p_a (T_i - T_o)$ (7.69) or $= 1200 * Q(T_i - T_o)$ W $q_{latent} = Q \rho_a h_{fg} \Delta W$ (7.70) or $= 2808^a Q \Delta W$ W	Q is the volume of airflow in m³/s. ρ_a and Cp_a are density and specific heat of air h_{fg} is the latent heat of water at room temperature. ΔW is humidity ratio difference between inside and outside air

a Assuming $\rho_a = 1.2$ kg/m³; $h_{fg} = 2340$ J/kg, $Cp_a = 1000$ J/kg°C.

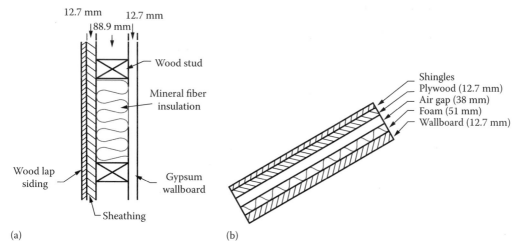

FIGURE 7.23
Cross sections of (a) the wall and (b) the roof for Example 7.4.

Combined thermal resistance for the studs and insulation (R_{comb}) is found as

$$\frac{1}{R_{comb}} = \left(\frac{A_{stud}}{R_{stud}} + \frac{A_{insulation}}{R_{insulation}} \right) \frac{1}{A_{stud} + A_{insulation}}$$

Assume that the studs occupy 15% of the wall area

$$\frac{1}{R_{comb}} = \frac{0.15}{0.77} + \frac{0.85}{1.94}$$

or

$$R_{comb} = 1.58 \, \text{m}^2 \text{°C/W}$$

Therefore, the wall thermal resistance, R_{wa}, can be found as

Element	Thermal Resistance (m²°C/W)
Outside air (6.7 m/s wind)	0.030
Wood bevel lap siding	0.14
12.7 mm sheathing	0.23
88.9 mm combined wood stud and mineral fiber insulation	1.58
1.7 mm gypsum wallboard	0.079
Inside air (still)	0.12
	$R_{wa} = 2.179$

Therefore,

$$U_{wa} = \frac{1}{R_{wa}} = \frac{1}{2.179} = 0.46\,\text{W/m}^2\text{°C.}$$

The heat loss through the windows depends on whether they are single- or double-glazed. In this example, single-glazed windows are installed, and a U factor equal to 4.7 W/m²°C is used. (If double-glazed windows were installed, the U factor would be 2.4 W/m²°C.)

The roof is constructed of 12.7 mm gypsum wall board, 51 mm foam insulation board, 38 mm still air, 12.7 mm plywood, and asphalt shingles (wooden beams and roofing paper are neglected for the simplified calculations here). Therefore,

$$U_{rf} = \frac{1}{\underset{\substack{outside\\air}}{0.030} + \underset{Shingels}{0.077} + \underset{Ply\,wood}{0.11} + \underset{Air\,gap}{0.17} + \underset{Foam}{2.53} + \underset{Wallboard}{0.079} + \underset{\substack{inside\\air}}{0.1}} = 0.32\,\text{W/m}^2\text{°C}$$

If the respective areas and U factors are known, the rate of heat loss per hour for the walls, windows, and roof can be calculated, assuming that floor heat loss is negligible:

$$\text{Walls:} \quad q_{wa} = (200\,\text{m}^2) \times 0.46\,\text{W/m}^2\text{°C} = 92\,\text{W/°C}$$

$$\text{Windows:} \quad q_{wi} = (100\,\text{m}^2) \times 0.47\,\text{W/m}^2\text{°C} = 470\,\text{W/°C}$$

$$\text{Roof:} \quad q_{rf} = (690\,\text{m}^2) \times 0.32\,\text{W/m}^2\text{°C} = 220\,\text{W/°C}$$

$$\text{Total} = 782\,\text{W/°C}$$

If double-glazed windows were used, the total heat loss would be reduced to 552 W/°C. As shown in Example 2.5, the installation of double-glazed windows is cost-effective and repays the energy invested in less than 2 years. Thus, this shows that before installing an active system to supply the heat necessary for maintaining adequate comfort level, energy conservation measures such as using double-glazed rather than single-glazed windows are the preferred option.

The infiltration and ventilation rate Q for this building is assumed to be 0.5 ACH (air changes per hour). The sensible and latent heat loads of the infiltration air may be calculated using the equations given in Table 7.7. Therefore,

$$Q = 0.5 \times (600\,\text{m}^2 \times 3\,\text{m}) = 900\,\text{m}^3/\text{h} = 0.25\,\text{m}^3/\text{s, volume}$$

$$q_{sensible} = 0.25\,\text{m}^3/\text{s} \times (1.2\,\text{kg/m}^3)(1000\,\text{J/kg°C}) = 300\,\text{W/°C.}$$

In residential buildings, humidification of the infiltration air is rarely done. Neglecting the latent heat, the total rate of heat loss q_{tot} is the sum of $q_{sensible}$ and q_{tr}:

$$q_{tot} = (782 + 300) = 1082 \text{ W/} ^\circ\text{C}$$

This calculation is simplified for purposes of illustration. Heat losses through the slab surface and edges have been neglected, for example.

More refined methods of calculating energy requirements on buildings do not use the steady-state assumption used earlier [19]. The thermal inertia of buildings may be expressly used as a load-leveling device. If so, the steady-state assumption is not met and the energy capacitance of the structure must be considered for accurate results. Many adobe structures in the U.S. Southwest are built intentionally to use daytime sun absorbed by 1 ft thick walls for nighttime heating, for example.

7.5.2 Internal Heat Sources in Buildings

Heat supplied to a building to offset energy losses is derived from both the heating system and internal heat sources. Table 7.8 lists the common sources of internal heat generation for residences. Commercial buildings such as hospitals, computer facilities, or supermarkets will have large internal gains specific to their function. Internal heat gains tend to offset heat losses from a building but will add to the cooling load of an air-conditioning system. The magnitude of the reduction in heating system operation will be described in the next section.

7.5.3 Degree-Day Method

The preceding analysis of heat loss from buildings expresses the loss on a per unit temperature difference basis (except for unexposed floor slabs). In order to calculate the peak load

TABLE 7.8

Some Common Internal Sensible Heat Gains That Tend to Offset the Heating Requirements of Buildings

Type	Magnitude (W or J/s)
Incandescent lights	Total W
Fluorescent lights	Total W
Electric motors	$746 \times (\text{hp/efficiency})$
Natural gas stove	$8.28 \times \text{m}^3/\text{h}$
Appliances	Total W
A dog	50–90
People	
Sitting	70
Walking	75
Dancing	90
Working hard	170

Source: ASHRAE, *Handbook of Fundamentals*, American Society of Heating, Refrigerating and Air-Conditioning Engineers, Atlanta, GA, 1997. With permission.

and total annual load for a building, appropriate design temperatures must be defined for each. The outdoor design temperature is usually defined statistically, such that the actual outdoor temperature will exceed the design temperature 97.5% or 99% of the time over a long period. The design temperature difference (ΔT) is then the interior building temperature minus the outdoor design temperature. The design ΔT is used for rating nonsolar heating systems, but is not useful for the selection of solar systems, since solar systems rarely provide 100% of the energy demand of a building at peak conditions.

A more useful index of heating energy demand is the total annual energy requirement for a building. This quantity is somewhat more difficult to calculate than the peak load. It requires knowledge of day-to-day variations in ambient temperature during the heating season and the corresponding building heat load for each day. Building heat loads vary with ambient temperatures, as shown in Figure 7.24. The environmental temperature T_{nl}, above which no heat need be supplied to the building, is a few degrees below the required interior temperature T_i because of internal heat-generation effects.

The no-load temperature at which internal source generation q_i just balances transmission and infiltration losses can be determined from the energy balance

$$q_i = \overline{UA}_B(T_i - T_{nl}) \tag{7.71}$$

where $\overline{UA}_B$ is the overall heat loss coefficient for the building (W/°C). Then

$$T_{nl} = T_i - \frac{q_i}{\overline{UA}_B} \tag{7.72}$$

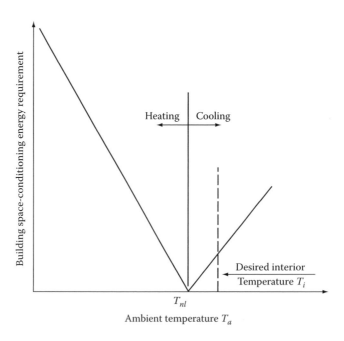

FIGURE 7.24
Building load profile versus ambient temperature showing no-load temperature T_{nl} and desired interior temperature T_i.

FIGURE 3.1
2.5 MW Liberty wind turbine (Clipper Windpower) near Medicine Bow, Wyoming. (Courtesy of DOE/NREL, Golden, CO. With permission.)

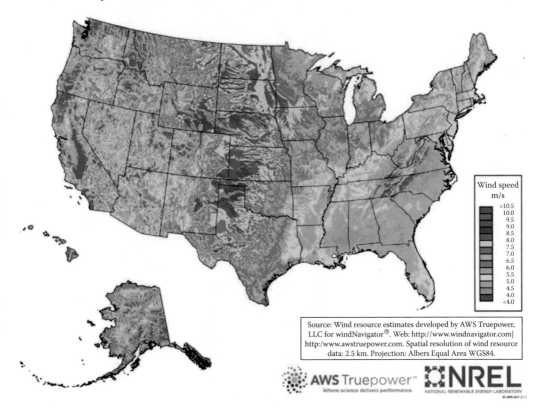

Wind speed
m/s
>10.5
10.0
9.5
9.0
8.5
8.0
7.5
7.0
6.5
6.0
5.5
5.0
4.5
4.0
<4.0

Source: Wind resource estimates developed by AWS Truepower, LLC for windNavigator®. Web: http://www.windnavigator.com| http:/www.awstruepower.com. Spatial resolution of wind resource data: 2.5 km. Projection: Albers Equal Area WGS84.

FIGURE 3.13
Wind resource map of the United States. This map shows the annual average wind power estimates at a height of 50 m. It is a combination of high-resolution and low-resolution datasets produced by NREL and other organization. The data were screened to eliminate areas unlikely to be developed onshore due to land-use or environmental issues. In many states, the wind resource on this map is visually enhanced to better show the distribution on ridge crests and other features. (Courtesy of U.S. Department of Energy, National Renewable Energy Laboratory, Washington, DC.)

Combiner box
600 V Fuses protect each of the (3) strings of (5) modules
after the fuses the (3) circuits are combined in parallel
The combined circuit is then brought to the DC disconnect in the garage

FIGURE 6.1
Kreith residence, 3 kW Sanyo high-efficiency photovoltaic system. (Courtesy of Chris Klinga, Lighthouse Solar, Boulder, CO. With permission.)

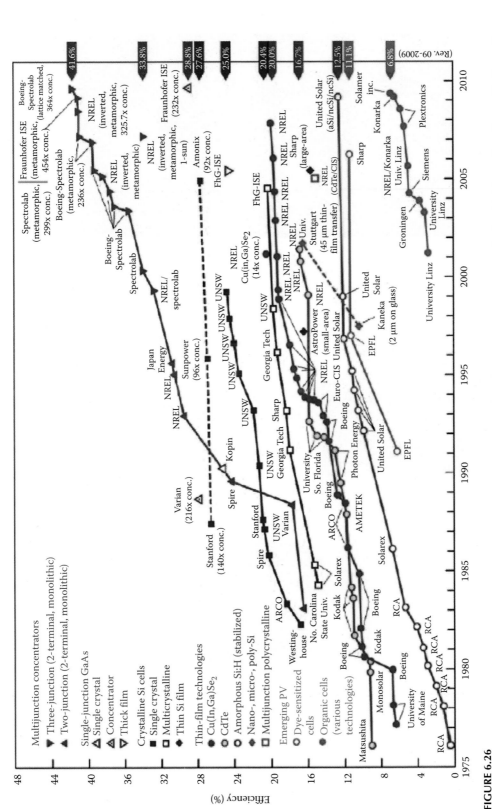

FIGURE 6.26
Best research-cell efficiencies. (From Jayarama Reddy, P., Cell efficiencies, in *Science and Technology of Photovoltaics*, 2nd edn, CRC Press, Boca Raton, FL; Kurtz, S., Opportunities and challenges for development of a mature concentrating photovoltaic power industry, NREL, NREL/TP-520-43208, revised November 2009. With permission.)

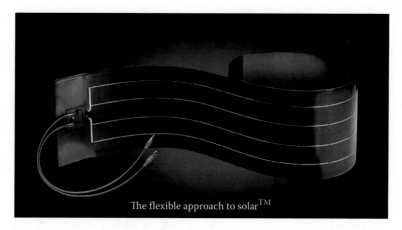

The flexible approach to solar^{TM}

FIGURE 6.28
Thin-film flexible photovoltaic CIGS module. (From SoloPower, Inc., San Jose, CA. With permission.)

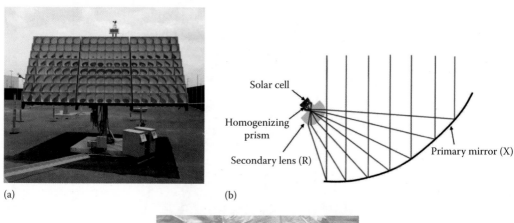

(a)

(b)

Solar cell

Homogenizing prism

Secondary lens (R)

Primary mirror (X)

(c)

FIGURE 6.35
Dual-axis rooftop tracker for Boeing high-concentration PV system developed at the University of Colorado, Boulder, CO. (a) Solar array mounted on conventional pedestal. (Courtesy of Adam Plesniak, Guy Martins, John Hall, Andy Messina and other members of the Boeing CPV Engineering Team.) (b) Ray trace pattern of the off-axis SMS3D nonimaging optical system in the Boeing CPV design. (Courtesy of Pablo Benitez, Professor, Universidad Polytechnic de Madrid and Light Prescription Innovators, LLC, Altadena, CA.) (c) Advanced dual-axis tracker with four of the six string Boeing high-concentration modules installed. Tracker was developed by Senior Design Teams at the University of Colorado, Boulder, CO, Dr. Frank Kreith, faculty advisor. (Photo courtesy of Kane Chinnel and Nate Bailey. With permission.)

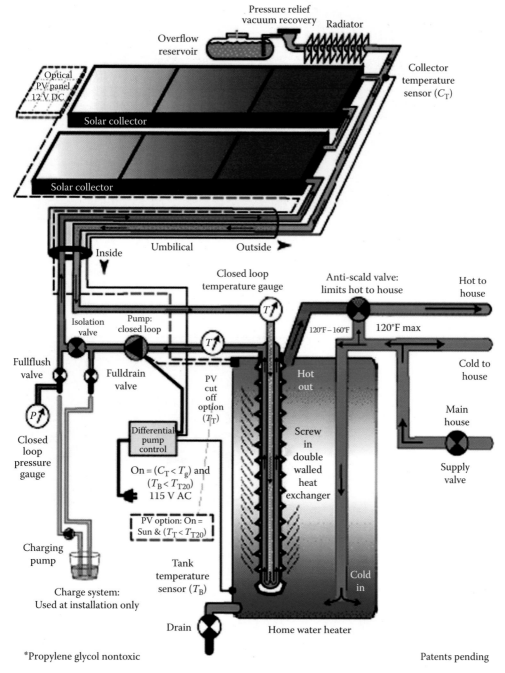

FIGURE 7.1
Schematic of an active flat-plate solar collector with storage tank and freeze protection.

SunMaxx M-2 flat-plate solar collector

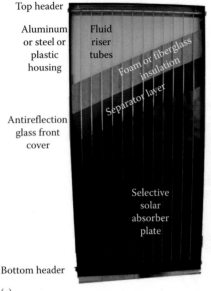

SunMaxx M-2 flat-plate solar collector

Top header

Aluminum or steel or plastic housing

Fluid riser tubes

Foam or fiberglass insulation

Separator layer

Antireflection glass front cover

Selective solar absorber plate

Bottom header

(a)

(b)

FIGURE 7.2
(a) Photo of an active flat-plate solar collector. (b) Photo of active flat-plate solar collector system on Kreith residence. (From Barry Butler, PhD, Butler Sun Solutions, Inc., Solana Beach, CA. With permission.)

Line focus systems

Parabolic trough

- Maximum operating temp. 393°C

- Turbine–steam rankine cycle

- Organic heat transfer fluid (HTF) in solar collector field

(a)

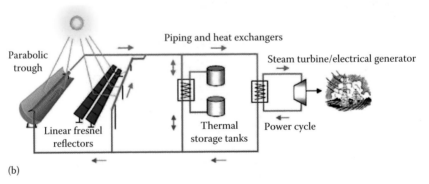

Piping and heat exchangers

Parabolic trough

Steam turbine/electrical generator

Linear fresnel reflectors

Thermal storage tanks

Power cycle

(b)

FIGURE 8.27
(a) Photograph of single axis-tracking parabolic trough collector field. (b) Schematic of parabolic trough line focus solar thermal power generation system. (Courtesy of Clifford Ho, Sandia National Laboratories.)

FIGURE 8.28
Sky Fuel solar sky trough near Golden, Colorado, viewed by Professor Frank Kreith (right) and National Institute of Standards and Technology researcher, Dr. Isaac Garaway (left). (Courtesy of CU Engineering, 2009, Casey A. Cass, photographer.)

Heliostats

FIGURE 8.31
Heliostats for power tower.

Point focus systems

Power tower
- Steam or molten nitrate salt HTF
- Max. operating temp. −560°C
- Steam rankine cycle

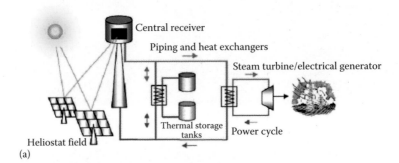

Central receiver

Piping and heat exchangers

Steam turbine/electrical generator

Thermal storage tanks

Power cycle

Heliostat field

(a)

(b)

FIGURE 8.32
(a) Schematic diagram of point focus solar thermal power tower system with a cavity receiver and (b) photograph of two Spanish Central Receiver Systems. (Courtesy of Clifford Ho, Sandia National Laboratories.)

(a)

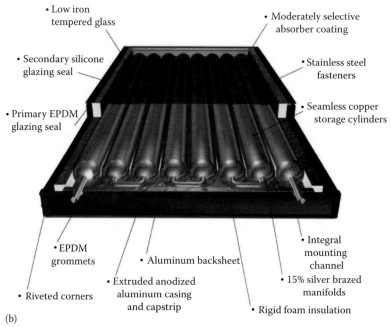

- Low iron
  tempered glass

- Moderately selective
  absorber coating

- Secondary silicone
  glazing seal

- Stainless steel
  fasteners

- Primary EPDM
  glazing seal

- Seamless copper
  storage cylinders

- EPDM
  grommets

- Aluminum backsheet

- Integral
  mounting
  channel

- Extruded anodized
  aluminum casing
  and capstrip

- 15% silver brazed
  manifolds

- Riveted corners

- Rigid foam insulation

(b)

FIGURE 9.1
CopperHeart passive solar collector with integrated storage. (Courtesy of SunEarth, Inc., Fontana, CA.)
(a) Photograph of installed module on residence in Arizona. (b) Schematic of CopperHeart integral collector
storage system.

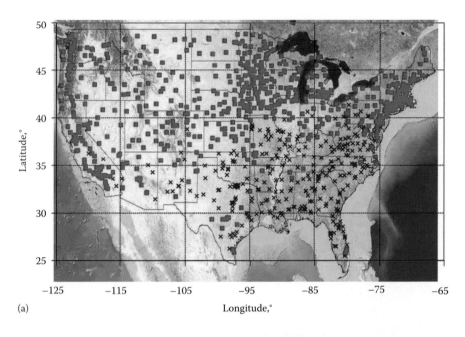

(a)

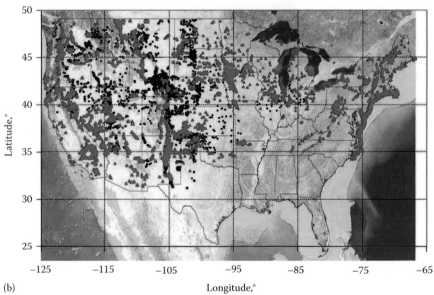

(b)

FIGURE 10.24
Optimum locations for (a) PV and (b) wind to provide 83% of electric load in the United States. (Courtesy of Victor Diakov, NREL.)

FIGURE 11.20
La Rance tidal power station. (From Khaligh, A. and Onar, O.C., *Energy Harvesting*, CRC Press, Boca Raton, FL, 2010, Figure 3.6.)

FIGURE 11.31
Pelamis WEC device at sea. (From Khaligh, A. and Onar, O.C., *Energy Harvesting*, CRC Press, Boca Raton, FL, 2010, Figure 4.72. With permission.)

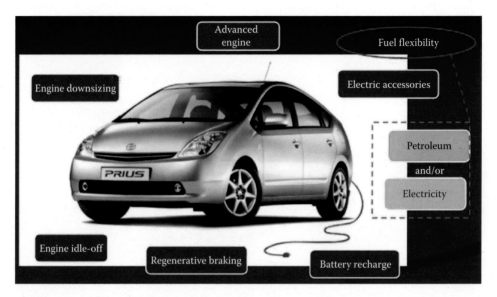

FIGURE 12.6
Plug-in hybrid electric vehicle. (Courtesy of National Renewable Energy Laboratory, Golden, CO.)

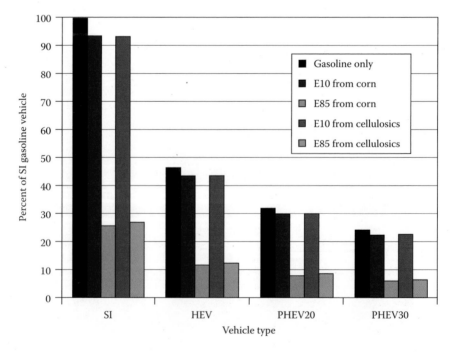

FIGURE 12.7
Petroleum requirement as a percentage of that for SI gasoline vehicle. (From Kreith, F. and West, R.E., *ASME J. Energy Resources Technol.*, 128(9), 236, September 2006. With permission.)

The total annual heat load on the building, Q_T, can be expressed as

$$Q_T = \int_{365\,days} \overline{UA}_B (T_{nl} - T_a)^+ dt \tag{7.73}$$

in which all arguments of the integral are functions of time. The superscript $^+$ indicates that only positive values are considered. In practice, it is difficult to evaluate this integral; therefore, three simplifying assumptions are made:

1. $\overline{UA}_B$ is independent of time.
2. T_{nl} is independent of time.
3. The integral can be expressed by the sum.

$$\overline{UA}_B \sum_{n=1}^{365} (T_{nl} - \overline{T}_a)_n^+ \tag{7.74}$$

where n is the day number, and the daily average temperature $\overline{T}_a$ can be approximated by $1/2(T_{a,max} + T_{a,min})$, in which $T_{a,max}$ and $T_{a,min}$ are the daily maximum and minimum temperatures, respectively.

The quantity $(T_{nl} - T_a)^+$ is called the *degree-day unit*. For example, if the average ambient temperature for a day is 5°C and the no-load temperature is 20°C, 15°C-days are said to exist for that day. However, if the ambient temperature is 20°C or higher, 0 degree-days exist, indicating 0 demand for heating that day. Degree-day totals for monthly $\Sigma_{month} (T_{nl} - T_a)^+$, and annual periods can be used directly in Equation 7.74 to calculate the monthly and annual heating energy requirements.

In the past, a single value of temperature has been used throughout the United States as a universal degree-day base, 65.0°F or 18.3°C. This practice is now outdated, since many homeowners and commercial building operators have lowered their thermostat settings in response to increased heating fuel costs, thereby lowering T_{nl}. Likewise, warehouses and factories operate well below the 19°C level. Therefore, a more generalized database of degree-days to several bases (values of T_{nl}) has been created by the U.S. National Weather Service (NWS, http://gis.ncdc.noaa.gov/map/viewer/#app=cdo&cfg=cdo&theme=normals&layers=01&node=gis&extent=–149.3:20.2:–60.1:69.6&custom=normals).

Example 7.5

A building located in Denver, CO, has a heat loss coefficient $\overline{UA}_B$ of 1000 kJ/h°C and internal heat sources of 4440 kJ/h. If the interior temperature is 20°C (68°F), what are the monthly and annual heating energy requirements? A gas furnace with 65% efficiency is used to heat the building.

Solution
In order to determine the monthly degree-day totals, the no-load temperature (degree-day basis) must be evaluated from Equation 7.72.

$$T_{nl} = 20 - \frac{4440}{1000} = 15.6°C \ (60°F)$$

TABLE 7.9

Monthly and Annual Energy Demands for Example 7.5

Month	°C-Days	Energy Demand[a] (GJ)
January	518	12.4
February	423	10.2
March	396	9.5
April	214	5.2
May	68	1.6
June	14	0.3
July	0	0
August	0	0
September	26	0.6
October	148	3.6
November	343	8.2
December	472	11.3
	2622	62.9

[a] Energy demand equals $\overline{UA_B} \times °\text{C-days} \times 24\,\text{h/day}$.

The monthly °C-days for Denver are taken from the U.S. NWS and given in Table 7.9. The energy demand is calculated as

$$\text{Energy demand} = \overline{UA_B} \times 24\frac{\text{h}}{\text{day}} \times °\text{C-days}$$

The monthly energy demand is given in Table 7.9.

If the annual energy demand of 62.9 GJ is delivered by a 65% efficient gas furnace, then the

$$\text{Average annual purchased energy} = \frac{62.9}{0.65}\text{GJ} = 96.8\,\text{GJ}$$

7.5.4 Service Hot Water Load Calculation

Service hot water loads can be calculated precisely with the knowledge of only a few variables. The data required for calculation of hot water demand are

Water source temperature (T_s)

Water delivery temperature (T_d)

Volumetric demand rate (Q)

The energy requirement for service water heating q_{hw} is given by

$$q_{hw}(t) = \rho_w Q(t) c_{pw} \left[T_d - T_s(t) \right] \tag{7.75}$$

where
ρ_w is the water density
c_{pw} is the specific heat of water

TABLE 7.10

Approximate Service Hot Water Demand Rates

	Demand per Person	
Usage Type	L/Day	gal/Day
Retail store	2.8	0.75
Elementary school	5.7	1.5
Multifamily residence	76.0	20.0
Single-family residence	76.0	20.0
Office building	11.0	3.0

The demand rate, $Q(t)$, varies in general with time of day and time of year; likewise, the source temperature varies seasonally. Source temperature data are not compiled in a single reference; local water authorities are the source of such temperature data.

Few generalized data exist with which to predict the demand rate Q. Table 7.10 indicates some typical usage rates for several common building types. Process water-heating rates are peculiar to each process and can be ascertained by reference to process specifications.

Example 7.6

Calculate the monthly energy required to heat water for a family of four in Nashville, TN. Monthly source temperatures for Nashville are shown in Table 7.11, and the water delivery temperature is 60°C (140°F).

Solution

For a family of four, the demand rate Q may be found using a demand recommended from Table 7.10:

$$Q = 4 \times 76 \text{ L/day} = 0.30 \text{ m}^3/\text{day}$$

The density of water can be taken as 1000 kg/m³ and the specific heat as 4.18 kJ/kg·°C.

Monthly demands are given by

$$q_m = (Q \times \text{days/month})(\rho_w c_{pw})[T_d - T_s(t)]$$
$$= (0.30 \times \text{days/month})(1000 \times 4.18)[60 - T_s(t)]$$

The monthly energy demands calculated from the earlier equation with these data are tabulated in Table 7.11.

TABLE 7.11

Water-Heating Energy Demands for Example 7.6

Month	Days/ Month	Demand (m³/Month)	Source Temperature°C	Energy Requirement (GJ/Month)
January	31	9.3	8	2.0
February	28	8.4	8	1.8
March	31	9.3	12	1.9
April	30	9.0	19	1.5
May	31	9.3	17	1.7
June	30	9.0	21	1.5
July	31	9.3	22	1.5
August	31	9.3	24	1.4
September	30	9.0	24	1.4
October	31	9.3	22	1.5
November	30	9.0	14	1.7
December	31	9.3	12	1.9

7.6 Solar Water-Heating Systems

Solar water-heating systems represent the most common application of solar energy at the present time. Small systems are used for domestic hot water applications while larger systems are used in industrial process heat applications. There are basically two types of water-heating systems: *natural circulation* or passive solar system (thermosyphon) and *forced circulation* or active solar system. The natural circulation systems are treated in Chapter 9. Natural circulation solar water heaters are simple in design and of low cost. Their application is usually limited to nonfreezing climates. The natural circulation systems are treated in Chapter 9. Forced-circulation water heaters are used in freezing climates and for commercial and industrial process heat.

7.6.1 Forced-Circulation Systems

If a thermosyphon system cannot be used for climatic, structural, or architectural reasons, a forced-circulation system is required. Figure 7.25 shows three configurations of forced-circulation systems: (1) open loop, (2) closed loop, and (3) closed loop with drainback. In an open-loop system (Figure 7.25a) the solar loop is at atmospheric pressure; therefore, the collectors are empty when they are not providing useful heat. A disadvantage of this system is the high pumping power required to pump the water to the collectors every time the collectors become hot. This disadvantage is overcome in the pressurized closed-loop system (Figure 7.25b) since the pump has to overcome only the resistance of the pipes. In this system, the solar loop remains filled with water under pressure.

In order to accommodate the thermal expansion of water from heating, a small (about 2 gal capacity) expansion tank and a pressure relief valve are provided in the solar loop. Because water always stays in the collectors of this system, antifreeze (propylene glycol or ethylene glycol) is required for locations where freezing conditions can occur.

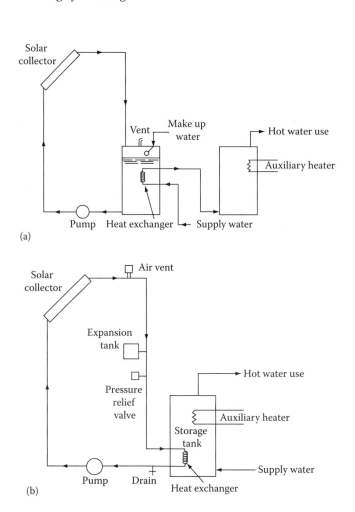

(a)

(b)

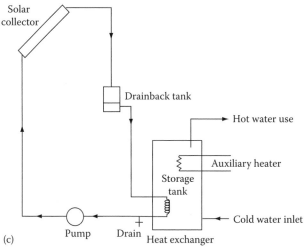

(c)

FIGURE 7.25
Typical configurations of solar water-heating systems: (a) open-loop system, (b) closed-loop system, and (c) closed-loop drainback system. (Adapted from Goswami, D.Y., *Alternative Energy in Agriculture*, vol. 1, CRC Press, Boca Raton, FL, 1986.)

During stagnation conditions (in summer), the temperature in the collector can become very high, causing the pressure in the loop to increase. This can cause leaks in the loop unless some fluid is allowed to escape through a pressure-release valve. Whether as a result of leaks or of draining, air enters the loop, causing the pumps to run dry. This disadvantage can be overcome in a closed-loop drainback system that is not pressurized (Figure 7.25c). In this system, when the pump shuts off, the water in the collectors drains back into a small holding tank while the air in the holding tank goes up to fill the collectors. The holding tank can be located where freezing does not occur, but still at a high level to reduce pumping power. In all three configurations, a differential controller measures the temperature differential between the solar collector and the storage, and turns the circulation pump on when the differential is more than a set limit (usually 5°C) and turns it off when the differential goes below a set limit (usually 2°C). Alternatively, a photovoltaic (PV) panel and a DC pump may be used. The PV panel will turn on the pump only when solar radiation is above a minimum level. Therefore, the differential controller and the temperature sensors may be eliminated.

For temperatures of up to about 100°C, required for many industrial process heat applications, forced-circulation flat-plate collector water-heating systems described earlier can be used. A schematic diagram for a complete liquid-based flat-plate solar heating system with antifreeze protection is shown in Figure 7.1. For higher temperatures, evacuated tube collectors or concentrating collectors must be used. Industrial process heat systems are described in more detail in Chapter 8.

7.7 Liquid-Based Solar Heating Systems for Buildings

Solar space-heating systems can be classified as active or passive depending on the method utilized for heat transfer. A system that uses pumps and/or blowers for fluid flow in order to transfer heat is called an active system. On the other hand, a system that utilizes natural phenomena for heat transfer is called a passive system. Examples of passive solar space-heating systems include direct gain, attached greenhouse, and storage wall (also called Trombe wall). Passive solar heating systems are described in Chapter 9. In this section, configurations, design methods, and control strategies for active solar heating systems are described.

7.7.1 Physical Configurations of Active Solar Heating Systems

Figure 7.26 is a schematic diagram of a typical space-heating system. The system consists of three fluid loops—collector, storage, and load. In addition, most space-heating systems are integrated with a domestic water-heating system to improve the year-long solar load factor.

Since space heating is a relatively low-temperature use of solar energy, a thermodynamic match of collector to task indicates that an efficient flat-plate collector or low-concentration solar collector is the thermal device of choice.

The collector fluid loop contains fluid manifolds, the collectors, the collector pump, and heat exchanger, an expansion tank, and other subsidiary components. A collector heat exchanger and antifreeze in the collector loop are normally used in all solar space-heating systems, since the existence of a significant heating demand implies the existence of some subfreezing weather.

The storage loop contains the storage tank and pump as well as the tube side of the collector heat exchanger. To capitalize on whatever stratification may exist in the storage tank,

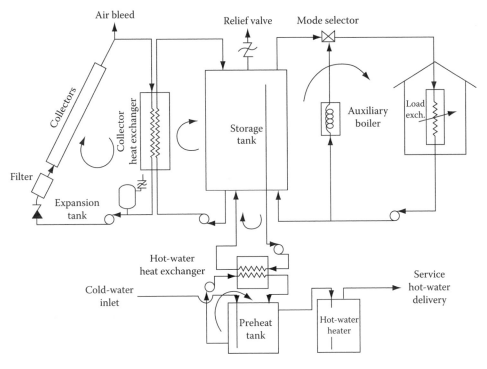

FIGURE 7.26
Typical solar thermal system for space heating and hot water heating showing fluid transport loops and pumps.

fluid entering the collector heat exchanger is generally removed from the bottom of storage. This strategy ensures that the lowest temperature fluid available in the collector loop is introduced at the collector inlet for high efficiency. The energy delivery-to-load loop contains the load device, baseboard heaters or fin-and-tube coils, and the backup system with a flow control (mode selector) valve.

7.7.2 Solar Collector Orientation

The best solar collector orientation is such that the average solar incidence angle is smallest during the heating season. For tracking collectors, this objective is automatically realized. For fixed collectors in the northern hemisphere, the best orientation is due south (due north in the southern hemisphere), tilted up from the horizon at an angle of about 15° greater than the local latitude.

Although due south is the optimum azimuthal orientation for collectors in the northern hemisphere, variations of 20° east or west have little effect on annual energy delivery [22]. Off-south orientations greater than 20° may be required in some cases because of obstacles in the path of the sun. These effects may be analyzed using sun-path diagrams and shadow-angle protractors as described in Chapter 5.

7.7.3 Fluid Flow Rates

For the maximum energy collection in a solar collector, it is necessary that it operates as closely as possible to the lowest available temperature, which is the collector inlet temperature. Very high fluid flow rates are needed to maintain a collector–absorber

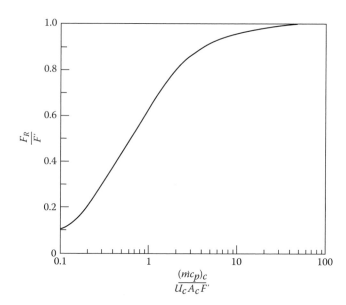

FIGURE 7.27
Effect of fluid flow rate on collector performance as measured by the heat-removal factor F_R; F' is the plate efficiency factor.

surface nearly isothermal at the inlet temperature. Although high flows maximize energy collection, practical and economic constraints put an upper limit on useful flow rates. Very high flows require large pumps and excessive power consumption and lead to fluid conduit erosion.

Figure 7.27 shows the effect of mass flow rate on annual energy delivery from a solar system. It is seen that the law of diminishing returns applies and that flows beyond about 50 kg/h m$_c^2$ ($\approx$ 10 lb/h ft$_c^2$) have little marginal benefit for collectors with loss coefficients on the order of 6 W/m$_c^2$°C ($\approx$ 1 Btu/h ft$_c^2$°F). In practice, liquid flows in the range of 50 – 75 kg/h m$_c^2$ (10 – 15 lb/h ft$_c^2$) of water equivalent are the best compromise among collector heat-loss coefficient, fluid pressure drop, and energy delivery. However, an infinitely large flow rate will deliver the most energy if pumping power is ignored for a nonstratified storage. If storage stratification is desired, lower flow rates must be used, since high flow destroys stratification. In freezing climates, an antifreeze working fluid is necessary for collectors. Attempts to drain collectors fully for freeze protection have not been successful.

*7.7.4 Unglazed Transpired Wall System for Air Preheating

Ventilation air preheating systems using wall-mounted unglazed transpired solar air collectors are the only active solar air-heating systems that have found market acceptance in commercial and industrial buildings [2]. Such systems preheat the ventilation air in a once-through mode without any storage. Figure 7.28 shows a transpired wall system in which the air is drawn through a perforated absorber plate by the building ventilation fan. Kutcher and Christensen [3] presented a thermal analysis of this system. From a heat balance on the transpired unglazed collector, the useful heat collected is

$$q_u = I_c A_c a_s - U_c A_c (T_{out} - T_a) \tag{7.78}$$

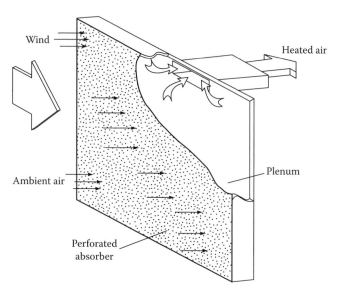

FIGURE 7.28
Unglazed transpired solar collector.

The overall heat loss coefficient U_c, which is due to radiative and convective losses, is given as

$$U_c = \frac{h_r}{\epsilon_{hx}} + h_c \tag{7.79}$$

where
   ϵ_{hx} is absorber heat-exchanger effectiveness
   h_r is a linearized radiative heat-transfer coefficient
   h_c is the convective heat loss coefficient

The heat-exchanger effectiveness for air flowing through the absorber plate is defined as

$$\epsilon_{hx} = \frac{T_{out} - T_\alpha}{T_c - T_a} \tag{7.80}$$

The forced convective heat loss coefficient due to a wind velocity of U_∞ is given as

$$h_c = 0.82 \frac{U_\infty v c_p}{L V_0} \tag{7.81}$$

where
   v is the kinematic viscosity of air in m²/s
   c_p the specific heat in J/Kg K
   V_0 the suction velocity in m/s
   L is the height of the collector in m

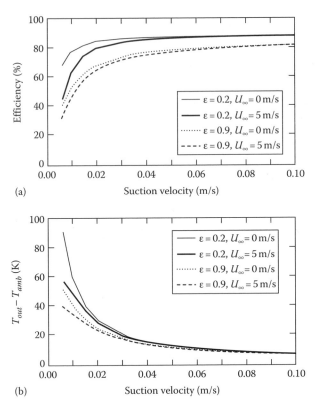

FIGURE 7.29
Predicted performance of unglazed transpired collector. (a) Efficiency vs. suction velocity and (b) temperature difference vs. suction velocity. (From Kutcher, C.F. and Christensen, C.B., Unglazed transpired solar collectors, In: Boer, K. (ed.), *Advances in Solar Energy*, vol. 7, pp. 283–307, 1992. With permission.)

Radiation heat loss occurs both to the sky and to the ground. Assuming the absorber is gray and diffuse with an emissivity ϵ_c, the radiative loss coefficient h_r is

$$h_r = \epsilon_c \sigma \frac{\left(T_c^4 - F_{cs}T_{sky}^4 - F_{cg}T_{gnd}^4\right)}{T_c - T_a} \tag{7.82}$$

where F_{cs} and F_{cg} are the view factors between the collector and the sky, and collector and the ground, respectively. For a vertical wall with infinite ground in front of it, both F_{cs} and F_{cg} will be 0.5 each. Using the earlier equations, Kutcher and Christensen [3] showed that the predicted performance matches the measured performances well. Figure 7.29 shows their predicted thermal performances.

*7.8 Methods of Modeling and Designing Solar Heating Systems

Several methods of modeling and design of solar space and water heating have been developed including *f*-chart, SLR, Utilizability, and TRNSYS. Klein and his coworkers [19,23–26] developed a method of simplified prediction of the performance of a solar heating system based on a large number of detailed simulations for various system configurations in various locations in the United States. The results from these simulations were then correlated with dimensionless parameters on charts that are general in form and usable anywhere. The charts are called f-charts, denoting a parameter f_s, the fraction of monthly load supplied by solar energy. The dimensionless groups used in the f-charts are derived from a nondimensionalization of the equations of governing energy flows. The f-chart method has been developed for standard solar heating and hot water system configurations. TRNSYS is a protected program from the University of Wisconsin and can be purchased from distributors.

The utilizability method is used to predict the long-term performance of a solar thermal system used for space heating, hot water, industrial process heat, or thermal power systems. It is based on finding the long-term utilizability, ϕ, of a thermal collector. ϕ is defined as the fraction of solar flux absorbed by a collector and delivered to the working fluid. The utilizability method is described in detail in Chapter 8 with applications in solar industrial process heat and solar power systems.

---

*7.9 Solar Cooling

The seasonal variation of solar energy is well suited to the space-cooling requirements of buildings, but this application of solar energy has so far not found much commercial success. Since the warmest seasons of the year correspond to periods of high insolation, solar energy is most available when comfort cooling is most needed. Moreover, the efficiency of solar collectors increases with increasing insolation and increasing environmental temperature. Consequently, in the summer, the amount of energy delivered per unit surface area of collector can be larger than that in winter.

There are several approaches that can be taken to solar space cooling and refrigeration. Because of the limited operating experience with solar-cooling systems, their design must be based on basic principles and experience with conventional cooling systems. The material presented in this chapter will therefore stress the fundamental principles of operation of refrigeration cycles and combine them with special features of the components in a solar system. This chapter presents the cooling requirements of buildings and the basics of active solar cooling techniques based on vapor-compression and vapor-absorption refrigeration cycles and desiccant humidification.

7.9.1 Cooling Requirements for Buildings

The cooling load of a building is the rate at which heat must be removed to maintain the air in a building at a given temperature and humidity. It is usually calculated on the basis of the peak load expected during the cooling season. For a given building, the cooling load depends primarily on

1. Inside and outside dry-bulb temperatures and relative humidities
2. Solar radiation heat load and wind speed
3. Infiltration and ventilation
4. Internal heat sources

A method of calculating the cooling load is presented in detail in [18].
  The steps in calculating the cooling load of a building are as follows:

1. Specify the building characteristics: wall area, type of construction, and surface characteristics; roof area, type of construction, and surface characteristics; window area, setback, and glass type; and building location and orientation.
2. Specify the outside and inside wet- and dry-bulb temperatures.
3. Specify the solar heat load and wind speed.
4. Calculate building cooling load resulting from the following: heat transfer through windows; heat transfer through walls; heat transfer through roof; sensible and latent heat gains resulting from infiltration and ventilation, sensible and latent heat gains (water vapor) from internal sources, such as people, lights, cooking, etc.

Equations 7.83 through 7.89 may be used to calculate the various cooling loads for a building. Cooling loads resulting from lights, building occupants, etc., may be estimated from [18]. For unshaded or partially shaded windows, the load is

$$\dot{Q}_{wi} = A_{wi}\left[F_{sh}\bar{\tau}_{b,wi}I_{h,b}\frac{\cos i}{\sin\alpha} + \bar{\tau}_{d,wi}I_{h,d} + \bar{\tau}_{r,wi}I_r + I_{wi}(T_{out}-T_{in}) \right] \tag{7.83}$$

For shaded windows, the load (neglecting sky diffuse and reflected radiations) is

$$\dot{Q}_{wi,sh} = A_{wi,sh}U_{wi}(T_{out}-T_{in}) \tag{7.84}$$

For unshaded walls, the load is

$$\dot{Q}_{wa} = A_{wa}\left[\bar{\alpha}_{s,wa}\left(\gamma_r + I_{h,d} + I_{h,b}\frac{\cos i}{\sin\alpha} \right) + U_{wa}(T_{out}-T_{in}) \right] \tag{7.85}$$

For shaded walls, the load (neglecting sky diffuse and reflected radiations) is

$$\dot{Q}_{wa,sh} = A_{wa,sh}\left[U_{wa}(T_{out}-T_{in}) \right] \tag{7.86}$$

For the roof, the load is

$$\dot{Q}_{rf} = A_{rf}\left[\bar{\alpha}_{s,rf}\left(I_{h,d} + I_{h,b}\frac{\cos i}{\sin\alpha} \right) + U_{rf}(T_{out}-T_{in}) \right] \tag{7.87}$$

Sensible-cooling load due to infiltration and ventilation is

$$\dot{Q}_i = \dot{m}_a(h_{out}-h_{in}) = \dot{m}_a Cp_a(T_{out}-T_{in}) \tag{7.88}$$

Latent load due to infiltration and ventilation is

$$\dot{Q}_w = \dot{m}_a(W_{out} - W_{in})\lambda_w \tag{7.89}$$

where
$\dot{Q}_{wi}$ is the heat flow through unshaded windows of area A_{wi}
$\dot{Q}_{wi,sh}$ is the heat flow through shaded windows of area $A_{wi,sh}$
$\dot{Q}_{wu}$ is the heat flow through unshaded walls of area A_{wa}
$\dot{Q}_{wu,sh}$ is the heat flow through shaded walls of area $A_{wa,sh}$
$\dot{Q}_{rf}$ is the heat flow through roof of area A_{rf}
$\dot{Q}_i$ is the heat load resulting from infiltration and ventilation
$\dot{Q}_w$ is the latent heat load
$I_{h,b}$ is the beam component of insolation on horizontal surface
$I_{h,d}$ is the diffuse component of insolation on horizontal surface
I_r is the ground-reflected component of insolation
W_{out}, W_{in} is the outside and inside humidity ratios
U_{wi}, U_{wa} are the overall heat-transfer coefficients for windows, walls, and roofs, including radiation
$\dot{m}_a$ is the net infiltration and ventilation mass flow rate of dry air
Cp_a is the specific heat of air (approximately 1.025 kJ/kg K for moist air)
T_{out} is the outside dry-bulb temperature
T_{in} is the indoor dry-bulb temperature
F_{sh} is the shading factor (1.0 = unshaded, 0.0 = fully shaded)
$\alpha_{s,wa}$ is the wall solar absorptance
$\alpha_{s,rf}$ is the roof solar absorptance
i is the solar incidence angle on walls, windows, and roof
h_{out}, h_{in} is the outside and inside air enthalpy
α is the solar altitude angle
λ_w is the latent heat of water vapor
$\overline{\tau}_{b,wi}$ is the window transmittance for beam (direct) insolation
$\overline{\tau}_{d,wi}$ is the window transmittance for diffuse insolation
$\overline{\tau}_{r,wi}$ is the window transmittance for ground-reflected insolation

ASHRAE handbooks recommend the use of the cooling load temperature difference method. For more details of the method, one should refer to [18].

Example 7.7

Determine the cooling load for a building in Phoenix, AZ, with the specifications tabulated in Table 7.12.

Solution
To determine the cooling load for the building just described, calculate the following factors in the order listed:

1. Incidence angle for the south wall i at solar noon can be written from Equations 5.28 and 5.44 as

$$\cos i = \cos\beta\cos(L - \delta_s) + \sin\beta\sin(L - \delta_s)$$

$$= 0.26 \tag{7.90}$$

TABLE 7.12

Specifications for Example 7.7

Factor	Description or Specification
Building characteristics	
Roof	
Type of roof	Flat, shaded
Area $A_{rf,sh}$, ft²	1700
Walls (painted white)	
Size, north and south, ft	8 × 60 (two)
Size, east and west, ft	8 × 40 (two)
Area A_{wa}, north and south walls, ft²	480 – A_{wi} = 480–40 = 440 (two)
Area A_{wa}, east and west walls, ft²	320 – A_{wi} = 320–40 = 280 (two)
Absorptance $\bar{\alpha}_{s,wa}$ of white paint	0.12
Windows	
Size, north and south, ft	4 × 5 (two)
Size, east and west, ft	4 × 5 (two)
Shading factor F_{sh}	0.20
Insolation transmittance	$\bar{\tau}_{b,wi} = 0.06$; $\bar{\tau}_{d,wi} = 0.81$; $\bar{\tau}_{r,wi} = 0.60$
Location and latitude	Phoenix, AZ; 33°N
Date	August 1
Time and local-solar-hour angle H_s	Noon; $H_s = 0$
Solar declination δ_s, deg	17°–55′
Wall surface tilt from horizontal β	90°
Temperature, outside and inside, °F	$T_{out} = 100$; $T_{in.} = 75$
Insolation I, Btu/h ft²	$I_{h,b} = 185$; $I_{h,d} = 80$; $I_r = 70$
U factor for walls, windows, and roof	$U_{wa} = 0.19$; $U_{wi.} = 1.09$; $U_{rf.} = 0.061$
Infiltration, lbm dry air/h	Neglect
Ventilation, lbm dry air/h	Neglect
Internal loads	Neglect
Latent heat load Q_w, %	30% of wall sensible heat load[a]

[a] Approximate rule of thumb for Phoenix.

2. Solar altitude a at solar noon (from Equation 5.28)

$$\sin\alpha = \sin\delta_s \sin L + \cos\delta_s \cos L \cos h_s = \cos(L - \delta_s) = \cos 15° = 0.966$$

3. South-facing window load (from Equation 7.83)

$$\dot{Q}_{wi} = 40\left\{\left(0.20 \times 0.6 \times 185 \frac{0.26}{0.966}\right) + (0.81 \times 80) + (0.60 \times 70)\right.$$

$$\left. + [1.09(100 - 75)]\right\} = 5600\,\text{Btu}/\text{h}$$

4. Shaded-window load (from Equation 7.84)

$$\dot{Q}_{wi,sh} = (3 \times 40)[1.09(100 - 75)] = 3270\,\text{Btu}/\text{h}$$

5. South-facing wall load (from Equation 7.85)

$$\dot{Q}_{wa} = (480 - 40)\left\{0.12\left[70 + 80\left(185\frac{0.26}{0.966}\right)\right] + 0.19(100 - 75)\right\}$$

$$= 12,610\,\text{Btu/h}$$

6. Shaded-wall load (from Equation 7.86)

$$\dot{Q}_{wa,sh} = [(480 + 320 + 320) - (3 \times 40)][0.19(100 - 75)] = 4750\,\text{Btu/h}$$

7. Roof load (from Equation 7.87)

$$\dot{Q}_{rf} = 1700[\bar{\alpha}_{s,rf} \times 0 + 0.067(100 - 75)] = 2600\,\text{Btu/h}$$

8. Latent heat load (30% of sensible wall load)

$$\dot{Q}_w = 0.3[(480 + 480 + 320 + 320) - (4 \times 40)][0.19(100 - 75)] = 2050\,\text{Btu/h}$$

9. Infiltration load

$$\dot{Q}_i = 0$$

10. Total cooling load for the building described in the example

$$\dot{Q}_{tot} = \dot{Q}_{wi} + \dot{Q}_{wi,sh} + \dot{Q}_{wa} + \dot{Q}_{wa,sh} + \dot{Q}_{rf} + \dot{Q}_w + \dot{Q}_i$$

$$\dot{Q}_{tot} = 30,880\,\text{Btu/h}$$

Looking at the various heat loads, it is apparent that the largest heat load comes from the south-facing walls. The heat load from these walls is three times larger than that from the shaded walls, and it would therefore behoove the architect or engineer to consider the options for reducing the heat load through the southern walls. Architectural options such as overhangs to shade these windows could be helpful. These options are discussed in Chapter 9.

7.9.2 Vapor-Compression Cycle

The two principal methods of lowering air temperature for comfort cooling are refrigeration with actual removal of thermal energy from the air or evaporation cooling of the air with adiabatic vaporization of moisture into it. Refrigeration can be used under any humidity condition of entering air, whereas evaporative cooling can be used only when the entering air has a comparatively low relative humidity.

The most widely used air-conditioning method employs a vapor-compression refrigeration cycle. Another method uses an absorption refrigeration cycle similar to that of the gas refrigerator. The vapor-compression refrigeration cycle requires energy input into the compressor, which may be provided as electricity from a PV system. Referring to Figure 7.30,

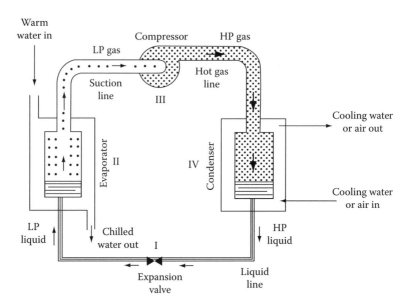

FIGURE 7.30
Schematic diagram illustrating the basic refrigeration vapor-compression cycle. (From Goswami, Y. et al., *Principles of Solar Engineering*, 2nd edn., Talyor & Francis, Philadelphia, 2000.)

the compressor raises the pressure of the refrigerant, which also increases its temperature. The compressed high-temperature refrigerant vapor then transfers thermal energy via a heat exchanger to the ambient environment in the condenser, where it condenses to a high-pressure liquid at a temperature close to, but above, the environmental temperature. The liquid refrigerant is then passed through the expansion valve where its pressure is reduced, resulting in a vapor–liquid mixture at a much lower temperature. The low-temperature refrigerant is then used to cool air or water in the evaporator where the liquid refrigerant evaporates by absorbing heat from the medium being cooled. The cycle is completed by the vapor returning to the compressor. If water is cooled by the evaporator, the device is usually called a chiller. The chilled water is then used to cool the air in the building.

The principle of operation of a vapor-compression refrigeration cycle can be illustrated conveniently with the aid of a pressure–enthalpy diagram, as shown in Figure 7.31. The ordinate is the pressure of the refrigerant in N/m^2 absolute, and the abscissa its enthalpy in kJ/kg. The Roman numerals in Figure 7.31 correspond to the physical locations in the schematic diagram of Figure 7.30.

Process I is a throttling process in which hot liquid refrigerant at the condensing pressure p_c passes through the expansion valve, where its pressure is reduced to the evaporator pressure p_e. This is an isenthalpic (constant enthalpy) process, in which the temperature of the refrigerant decreases. In this process, some vapor is produced, and the state of the mixture of liquid refrigerant and vapor entering the evaporator is shown by point A. Since the expansion process is isenthalpic, the following relation holds:

$$h_{ve}f + h_{le}(1-f) = h_{lc} \tag{7.91}$$

where
f is the fraction of mass in vapor state
subscripts v and l refer to vapor and liquid states, respectively
c and e refer to states corresponding to condenser and evaporator pressures, respectively

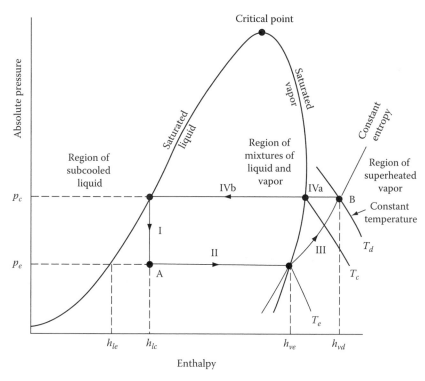

FIGURE 7.31

Simple refrigeration cycle on pressure–enthalpy diagram. (From Goswami, Y. et al., *Principles of Solar Engineering*, 2nd edn., Taylor & Francis, Philadelphia, 2000.)

and

$$f = \frac{h_{lc} - h_{le}}{h_{ve} - h_{le}} \tag{7.92}$$

Process II represents the vaporization of the remaining liquid. This is the process during which heat is removed from the chiller. Thus, the specific refrigeration effect per kilogram of refrigerant q_r is

$$q_r = h_{ve} - h_{lc} \text{ in kJ/kg (Btu/lb)} \tag{7.93}$$

In the United States, it is still common practice to measure refrigeration in terms of tons. One ton is the amount of cooling produced if 1 ton of ice is melted over a period of 24 h. Since 1 ton = 907.2 kg and the latent heat of fusion of water is 334.9 kJ/kg,

$$1\,\text{ton} = \frac{(907.2\,\text{kg}) \times (334.9\,\text{kJ/kg})}{(24\,\text{h}) \times (3,600\,\text{s/h})} = 3.516\,\text{kW} = 12,000\,\text{Btu/h} \tag{7.94}$$

If the desired rate of refrigeration requires a heat-transfer rate of $\dot{Q}_{r'}$, the rate of mass flow of refrigerant necessary $\dot{m}_r$ is

$$\dot{m}_r = \frac{\dot{Q}_{r'}}{(h_{ve} - h_{lc})} \tag{7.95}$$

Process III in Figure 7.31 represents the compression of refrigerant from pressure p_e to p_c. The process requires work input from an external source, which may be obtained from a solar-driven expander-turbine or a solar electrical system. In general, if the heated vapor leaving the compressor is at the condition represented by point B in Figure 7.31, the work of compression W_c is

$$W_c = \dot{m}_r(h_{vd} - h_{ve}) \tag{7.96}$$

In an idealized cycle analysis, the compression process is usually assumed to be isentropic.

Process IV represents the condensation of the refrigerant. Actually, sensible heat is first removed in the subprocess IVa as the vapor is cooled at a constant pressure from T_d to T_c, and latent heat is removed at the condensation temperature T_c, corresponding to the saturation pressure p_c in the condenser. The heat-transfer rate in the condenser $\dot{Q}_c$ is

$$\dot{Q}_c = \dot{m}_r(h_{vd} - h_{lc}) \tag{7.97}$$

This heat must be rejected into the environment, either to cooling water or to the atmosphere if no water is available.

The overall performance of a refrigeration machine is usually expressed as the ratio of the heat transferred in the evaporator $\dot{Q}_r$ to the shaft work supplied to the compressor. This ratio is called the *coefficient of performance* (COP), defined by

$$COP = \frac{\dot{Q}_r}{W_c} = \frac{h_{ve} - h_{lc}}{h_{vd} - h_{ve}} \tag{7.98}$$

The highest COP for any given evaporator and condenser temperatures would be obtained if the system were operating on a reversible Carnot cycle. Under these conditions [18],

$$COP(Carnot) = \frac{T_e}{T_d - T_e} \tag{7.99}$$

The earlier cycle has been idealized. In practice, the liquid entering the expansion valve is several degrees below the condensing temperature, while the vapor entering the compressor is several degrees above the evaporation temperature. In addition, pressure drops occur in the suction, discharge, and liquid pipelines, and the compression is not truly isentropic. Finally, the work required to drive the compressor is somewhat larger than W_c provided earlier, because of frictional losses. All of these factors reduce the COP below the maximum and must be taken into account in a realistic engineering design.

Example 7.8

Calculate the amount of shaft work to be supplied to a 1 ton (3.52 kW) refrigeration plant operation at evaporator and condenser temperatures of 273 and 309 K, respectively, using Refrigerant 134a (R-134a) as the working fluid. The properties of Refrigerant 134a are tabulated in Table 7.13. Also calculate the COP and the mass flow rate of the refrigerant based on the ideal cycle described earlier.

TABLE 7.13

Properties of Refrigerant 134a for Example 7.8

Temperature (K)	Absolute Pressure (kPa)	Vapor-Specific Volume (m³/kg)	Liquid Enthalpy (kJ/kg)	Vapor Enthalpy (kJ/kg)	Vapor Entropy (kJ/kg·K)
Saturated					
273	292.8	0.0689	50.02	247.2	0.919
309	911.7	0.0223	100.25	266.4	0.9053
Superheated					
308.5	900	0.0226	—	266.18	0.9054
313	900	0.0233	—	271.3	0.9217
312.4	1000	0.0202	—	268.0	0.9043
313	1000	0.0203	—	268.7	0.9066

Solution

From the property table, the enthalpies for process l are

Saturated vapor at 273 K h_{ve} = 247.2 kJ/kg
Saturated liquid at 309 K h_{lc} = 100.3 kJ/kg
Saturated liquid at 273 K h_k = 50.0 kJ/kg

Therefore, from Equation 7.92,

$$f = \frac{100.3 - 50.0}{247.2 - 50.0} = 0.255$$

The mass flow rate of refrigerant $\dot{m}_r$ is obtained from Equation 7.95 and the enthalpies given earlier, or

$$\dot{m}_r = \frac{3.52\,\text{kW}}{(247.2 - 100.3)\,\text{kJ/kg}} = 0.024\,\text{kg/s}$$

The specific shaft-work input required is

$$\frac{W_c}{\dot{m}_r} = h_{vd} - h_{ve}$$

The entropy s_e of the saturated vapor entering the compressor at 273 K and 292.8 kPa is 0.919 kJ/kg · K. From the property table, superheated vapor at a pressure of 911.7 kPa has an entropy of 0.919 kJ/kg · K at a temperature of 313 K with an enthalpy of 270.8 kJ/kg. Thus, the energy input to the working fluid by the compressor is

$$W_c = 0.024(270.8 - 247.2) = 0.566\,\text{kW}$$

Finally, the heat-transfer rate from the refrigerant to the sink, or cooling water in the condenser, is from Equation 7.97:

$$\dot{Q}_c = \dot{m}_r(h_{vd} - h_{lc}) = 0.024(270.8 - 100.3) = 4.09\,\text{kW}$$

The COP of the thermodynamic cycle is

$$COP = \frac{247.2 - 100.3}{270.8 - 247.2} = 6.2$$

whereas the Carnot COP is 273/36 or 7.6.

7.9.3 Absorption Air-Conditioning

In an absorption system, the refrigerant is evaporated or distilled from a less volatile liquid absorbent, the vapor is condensed in a water- or air-cooled condenser, and the resulting liquid is passed through a pressure-reducing valve to the cooling section of the unit. There it cools the water as it evaporates, and the resulting vapor flows into a vessel, where it is reabsorbed in the stripped absorbing liquid and pumped back to the heated generator. The heat required to evaporate the refrigerant in the generator can be supplied directly from solar energy, as shown in Figure 7.32.

Absorption air-conditioning is compatible with solar energy since a large fraction of the energy required is thermal energy at temperatures that flat-plate solar collectors such as previously described can provide. Figure 7.32 is a schematic of an absorption refrigeration system. Absorption refrigeration differs from vapor-compression air-conditioning only in the method of compressing the refrigerant. The compression process is shown to the left of the dashed line in Figure 7.33. The pressurization is accomplished by first dissolving the refrigerant in a liquid, called the absorbent, in the absorber section. This liquid or strong solution is then pumped to a high pressure with an ordinary liquid pump. The low boiling-point refrigerant is then driven from solution by the addition of heat in the generator. By this means, the refrigerant vapor entering the condenser is compressed without the large input of high-grade shaft work that the vapor-compression air-conditioning demands.

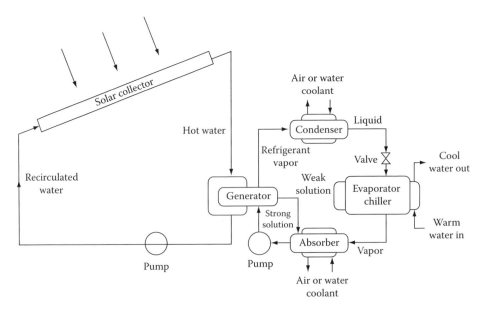

FIGURE 7.32
Schematic diagram of a solar-powered absorption refrigeration system.

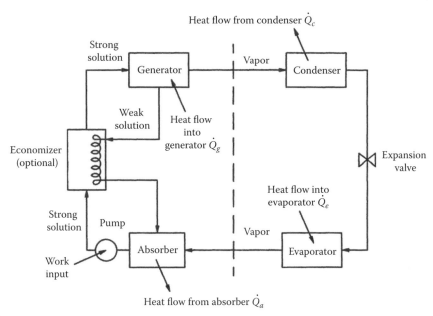

FIGURE 7.33
Diagram of heat and fluid flow of absorption air conditioner, with economizer.

The effective performance of an absorption cycle depends on the two materials that comprise the refrigerant–absorbent pair. Desirable characteristics for the refrigerant–absorbent pair follow:

1. The absence of a solid-phase absorbent.
2. A refrigerant more volatile than the absorbent so that separation from the absorbent occurs easily in the generator.
3. An absorbent that has a strong affinity for the refrigerant under conditions in which absorption takes place.
4. A high degree of stability for long-term operations.
5. Nontoxic and nonflammable fluids for residential applications. This requirement is less critical in industrial refrigeration.
6. A refrigerant that has a large latent heat so that the circulation rate can be kept low.
7. A low fluid viscosity that improves heat and mass transfer and reduces pumping power.
8. Fluids that must not cause long-term environmental effects.

Lithium bromide–water (LiBr–H$_2$O) and ammonia–water (NH$_3$–H$_2$O) are the two pairs that meet most of the requirements. In the LiBr–H$_2$O system, water is the refrigerant and LiBr is the absorber, while in the NH$_3$–H$_2$O system, ammonia is the refrigerant and water is the absorber. Because the LiBr–H$_2$O system has high volatility ratio, it can operate at lower pressures and, therefore, at the lower generator temperatures achievable by flat-plate collectors. A disadvantage of this system is that the pair tends to form solids. LiBr has a tendency to crystallize when air is cooled, and the system cannot be operated at or below the freezing point of water. Therefore, the LiBr–H$_2$O system is operated at evaporator temperatures of 5°C or higher. Using a mixture of LiBr with some other salt as the absorbent can overcome the crystallization problem. The NH_3–H_2O system has the advantage that it can be operated down to very low temperatures. However, for temperatures much below 0°C, water vapor must be removed from ammonia as much as possible to prevent ice crystals from forming. This requires a rectifying column after the boiler. Also ammonia is a safety Code Group B2 fluid (ASHRAE Standard 34-1992), which restricts its use indoors [18].

If the pump work is neglected, the COP of an absorption air conditioner can be calculated from Figure 7.33:

$$COP = \frac{\text{Cooling effect}}{\text{Heat input}} = \frac{\dot{Q}_e}{\dot{Q}_g} \tag{7.100}$$

The COP values for absorption air-conditioning range from 0.5 for a small single-stage unit to 0.85 for a double-stage steam-fired unit. These values are about 15% of the COP values that can be achieved by a vapor-compression air conditioner. It is difficult to compare the COP of an absorption air conditioner with that of a vapor-compression air conditioner directly because the efficiency of electric power generation or transmission is not included in the COP of the vapor-compression air-conditioning. The following example illustrates the thermodynamics of a LiBr–H_2O absorption refrigeration system.

Example 7.9

A LiBr–H_2O absorption refrigeration system such as that shown in Figure 7.34 is to be analyzed for the following requirements:

1. The machine is to provide 352 kW of refrigeration with an evaporator temperature of 5°C, an absorber outlet temperature of 32°C, and a condenser temperature of 43°C.
2. The approach at the low-temperature end of the liquid heat exchanger is to be 6°C.
3. The generator is heated by a flat-plate solar collector capable of providing a temperature level of 90°C.

Determine the COP, absorbent and refrigerant flow rates, and heat input.

Solution
For the analytical evaluation of the LiBr–H_2O cycle, the following simplifying assumptions are made:

1. At those points in the cycle for which temperatures are specified, the refrigerant and absorbent phases are in equilibrium.
2. With the exception of pressure reductions across the expansion device between points 2 and 3, and 8 and 9 in Figure 7.34, pressure reductions in the lines and heat exchangers are neglected.
3. Pressures at the evaporator and condenser are equal to the vapor pressure of the refrigerant, that is, water, as found in steam tables at http://www.ohio.edu/mechanical/thermo/property_tables/H2O/index.html.
4. Enthalpies for LiBr–H_2O mixtures are given in Figure 7.35.

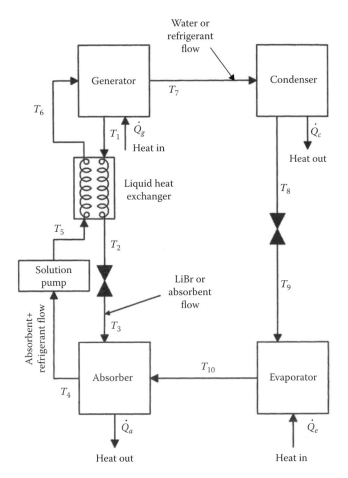

FIGURE 7.34
Li Br–H$_2$O absorption refrigeration cycle (see Table 7.14). *Note:* LP, low pressure; HP, high pressure. (From Klein, S.A. et al., *J. Eng. Power*, 96A, 109, 1974. With permission.)

As a first step in solving the problem, set up a table (Table 7.14) of properties, for example, given

Generator temperature = 90°C = $T_1 = T_7$
Evaporator temperature = 5°C = $T_9 = T_{10}$
Condenser temperature = 43°C = T_8
Absorber temperature = 32°C = T_4
Neglecting the pump work $T_5 \approx T_4 = 32°C$

Since the approach at the low-temperature end of the heat exchanger is 6°C,

$$T_2 = T_5 + 6°C = 38°C$$

and

$$T_3 \approx T_2 = 38°C$$

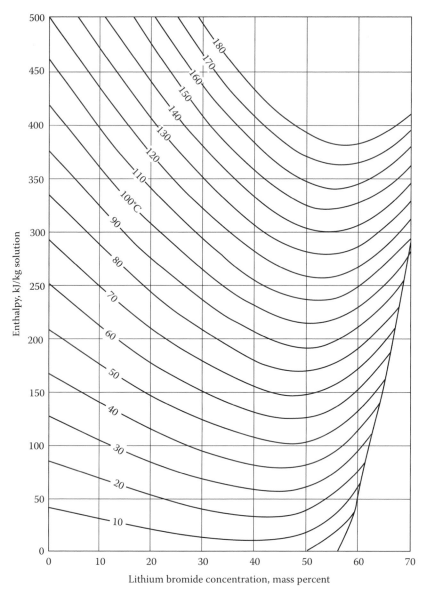

Equations   concentration range $40 < X < 70\%$ LiBr   Temperature range $15°C < t < 165°C$

$$h = \sum_0^4 A_a X^n + t \sum_0^4 B_n X^n + t^2 \sum_0^4 C_n X^n \text{ in kJ/kg, where } t = °C \text{ and } X = \% \text{ LiBr}$$

$A_0 = -2024.33$	$B_0 = 18.2829$	$C_0 = -3.7008214 \text{ E} - 2$
$A_1 = 163.309$	$B_1 = -1.1691757$	$C_1 = 2.8877666 \text{ E} - 3$
$A_2 = -4.88161$	$B_2 = 3.248041 \text{ E} - 2$	$C_2 = -8.1313015 \text{ E} - 5$
$A_3 = 6.302948 \text{ E} - 2$	$B_3 = -4.034184 \text{ E} - 4$	$C_3 = 9.9116628 \text{ E} - 7$
$A_4 = -2.913705 \text{ E} - 4$	$B_4 = 1.8520569 \text{ E} - 6$	$C_4 = -4.4441207 \text{ E} - 9$

FIGURE 7.35

Enthalpy–concentration diagram for LiBr–H_2O solutions. (From Goswami, Y. et al., *Principles of Solar Engineering*, 2nd edn., Taylor & Francis, Philadelphia, PA, 2000.)

TABLE 7.14

Thermodynamic Properties of Refrigerant and Absorbent for Figure 7.35

Condition No. in Figure 7.37	Temperature (°C)	Pressure (kPa)	LiBr Weight Fraction	Flow (kg/kg H₂O)	Enthalpy (kJ/kg)
1	90	8.65	0.605	11.2	215
2	38	8.65	0.605	11.2	110
3	38	0.872	0.605	11.2	110
4	32	0.872	0.53	12.2	70
5	32	0.872	0.53	12.2	70
6	74	8.65	0.53	12.2	162
7	90	8.65	0	1.0	2670
8	43	8.65	0	1.0	180
9	5	0.872	0	1.0	180
10	5	0.872	0	1.0	2510

Source: Kreider, J.F. and Kreith, F., *Solar Heating and Cooling*, revised 1st edn., Hemisphere Publishing Corporation, Washington, DC, 1977.

Since the fluid at conditions 7, 8, 9, and 10 is pure water, the properties can be found from the steam tables. Therefore,

$$P_7 = P_8 = \text{Saturation pressure of } H_2O \text{ at } 43°C = 8.65\,kPa$$

and

$$P_9 = P_{10} = \text{Saturation pressure of } H_2O \text{ at } 5°C = 0.872\,kPa$$

Therefore,

$$P_1 = P_2 = P_5 = P_6 = P_7 = 8.66\,kPa$$

and

$$P_3 = P_4 = P_{10} = 0.872\,kPa$$

Enthalpy

$$h_9 = h_g = 180\,kJ/kg \,(\text{saturated liquid at } 43°C)$$
$$h_{10} = 2510\,kJ/kg \,(\text{saturated vapor enthalpy at } 6°C)$$

and

$$h_7 = 2760\,kJ/kg \,(\text{superheated vapor at } 8.65\,kPa, 90°C)$$

For the LiBr–H₂O mixture, conditions 1 and 4 may be considered equilibrium saturation conditions, which may be found from Figures 7.35 and 7.36 as follows:
For

$$T_4 = 32°C \quad \text{and} \quad P_4 = 0.872\,kPa, \quad X_r = 0.53, \quad h_4 = 70\,kJ/kg\text{-sol.}$$

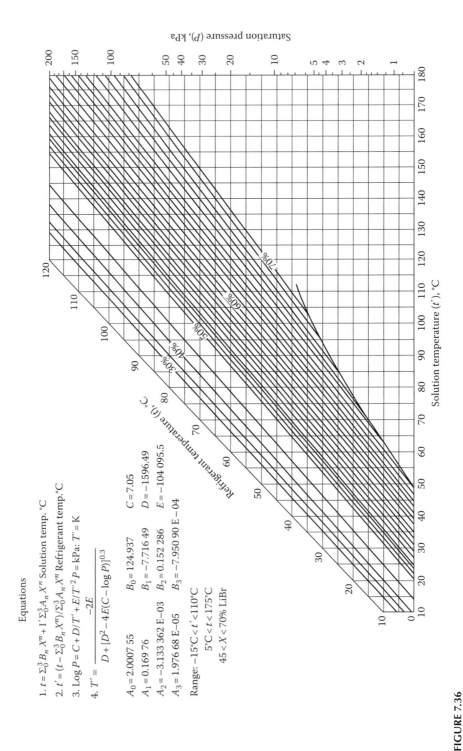

FIGURE 7.36

Equilibrium chart for LiBr–H₂O solutions. (From Goswami, Y. et al., *Principles of Solar Engineering*, 2nd edn., Taylor & Francis, Philadelphia, PA, 2000.)

Therefore,

$$h_5 = 70\,\text{kJ/kg-sol.}$$

and for

$$T_1 = 90°C \quad \text{and} \quad P_1 = 8.65\,\text{kPa}, \quad X_{ab} = 0.605, \quad h_1 = 215\,\text{kJ/kg-sol}$$

For

$$T_3 = 38°C, \quad X_3 = 0.605, \quad h_3 = 110\,\text{kJ/kg-sol}$$
$$h_2 = h_3 = 110\,\text{kJ/kg}$$

Relative flow rates for the absorbent (LiBr) and the refrigerant (H_2O) are obtained from material balances. A total material balance on the generator gives

$$\dot{m}_6 = \dot{m}_1 + \dot{m}_7$$

while a LiBr balance gives

$$\dot{m}_6 X_r = \dot{m}_1 X_{ab}$$

where
X_{ab} is the concentration of LiBr in absorbent of solution
X_r is the concentration of LiBr in refrigerant–absorbent of solution

Substituting ($\dot{m}_1 + \dot{m}_7$) for $\dot{m}_6$ gives

$$\dot{m}_1 X_5 + \dot{m}_7 X_5 = \dot{m}_1 X_{ab}$$

Since the fluid entering the condenser is pure refrigerant, that is, water, $\dot{m}_7$ is the same as the flow rate of the refrigerant $\dot{m}_r$:

$$\frac{\dot{m}_1}{\dot{m}_7} = \frac{X_s}{X_{ab} - X_s} = \frac{\dot{m}_{ab}}{\dot{m}_r}$$

where
$\dot{m}_{ab}$ is the flow rate of absorbent
$\dot{m}_r$ is the flow rate of refrigerant

Substituting for X_s and X_{ab} from the table gives the ratio of absorbent-to-refrigerant flow rate:

$$\frac{\dot{m}_{ab}}{\dot{m}_r} = \frac{0.53}{0.605 - 0.53} = 7.07$$

The ratio of the refrigerant–absorbent solution flow rate $\dot{m}_s$ to the refrigerant–solution flow rate $\dot{m}_r$ is

$$\frac{\dot{m}_s}{\dot{m}_r} = \frac{\dot{m}_{ab} + \dot{m}_r}{\dot{m}_r} = 7.07 + 1 = 8.07$$

Now Table 7.14 is complete except for T_6 and h_6, which may be found from an energy balance at the heat exchanger:

$$\dot{m}_s h_5 + \dot{m}_{ab} h_1 = \dot{m}_{ab} h_2 + \dot{m}_s h_6$$

Hence

$$h_6 = h_5 + \left[\frac{\dot{m}_{ab}}{\dot{m}_s}(h_1 - h_2)\right] = 70 + \frac{7.07}{8.07}[215 - 110] = 162\,\mathrm{kJ/kg\,of\,solution}$$

The temperature corresponding to this value of enthalpy and a LiBr mass fraction 0.53 is found from Figure 7.36 to be 74°C.

The flow rate of refrigerant required to produce the desired 352 kW of refrigeration is

$$\dot{Q}_e = \dot{m}_r (h_{10} - h_9)$$

where $\dot{Q}_e$ is the cooling effect produced by the refrigeration unit and

$$\dot{m}_r = \frac{352}{2510 - 180} = 0.15\,\mathrm{kg/s}$$

The flow rate of the absorbent is

$$\dot{m}_{ab} = \frac{\dot{m}_{ab}}{\dot{m}_r}\dot{m}_r = 7.07 \times 0.15 = 1.06\,\mathrm{kg/s}$$

while the flow rate of the solution is

$$\dot{m}_s = \dot{m}_{ab} + \dot{m}_r = 1.06 + 0.15 = 1.21\,\mathrm{kg/s}$$

The rate at which heat must be supplied to the generator $\dot{Q}_g$ is obtained from the heat balance

$$\dot{Q}_g = \dot{m}_r h_7 + \dot{m}_{ab} h_1 - \dot{m}_s h_6$$

$$= [(0.15 \times 2670) + (1.06 \times 215)] - (1.21 \times 1.62)$$

$$= 432\,\mathrm{kW}$$

This requirement, which determines the size of the solar collector, probably represents the maximum heat load that the collector unit must supply during the hottest part of the day.

The COP is

$$COP = \frac{\dot{Q}_e}{\dot{Q}_g} = \frac{352}{432} = 0.81$$

The rate of heat transfer in the other three heat-exchanger units—the liquid heat exchanger, the water condenser, and the absorber—is obtained from heat balances. For the liquid heat exchanger, this gives

$$\dot{Q}_{1-2} = \dot{m}_{ab}(h_1 - h_2) = 1.06[215 - 110] = 111 \text{kW}$$

where $\dot{Q}_{1-2}$ is the rate of heat transferred from the absorbent stream to the refrigerant–absorbent stream. For the water condenser, the rate of heat transfer $\dot{Q}_{7-8}$ rejected to the environment is

$$\dot{Q}_{7-8} = \dot{m}_r(h_7 - h_8) = 0.15(2670 - 180) = 374 \text{kW}$$

The rate of heat removal from the absorber can be calculated from an overall heat balance on this system:

$$\dot{Q}_a = \dot{Q}_{7-8} - \dot{Q}_g - \dot{Q}_e = 374 - 432 - 352 = -410 \text{kW}$$

Explicit procedures for the mechanical and thermal design as well as the sizing of the heat exchangers are presented in standard heat-transfer texts. In large commercial units, it may be possible to use higher concentrations of LiBr, operate at a higher absorber temperature, and thus save on heat-exchanger cost. In a solar-driven unit, this approach would require concentrator-type or high-efficiency flat-plate solar collectors.

Ammonia–water systems are quite similar to LiBr–H$_2$O systems except that a rectifier is needed after the boiler to condense water vapor. They are treated by Goswami et al. in Ref. [27].

*7.10 Solar Desiccant Dehumidification

In hot and humid regions of the world experiencing significant latent cooling demand, solar energy may be used for dehumidification using liquid or solid desiccants. Rangarajan et al. [28] compared a number of strategies for ventilation air-conditioning for Miami, Florida, and found that a conventional vapor-compression system could not even meet the increased ventilation requirements of ASHRAE Standard 62-1989. By pretreating the ventilation air with a desiccant system, proper indoor humidity conditions could be maintained and significant electrical energy could be saved. A number of researchers have shown that a combination of a solar desiccant and a vapor-compression system can save from 15% to 80% of the electrical energy requirements in commercial applications, such as supermarkets [29–37].

7.10.1 Solid Desiccant Cooling System

In a desiccant air-conditioning system, moisture is removed from the air by bringing it in contact with the desiccant and followed with sensible cooling of the air by a vapor-compression cooling system, vapor absorption cooling systems, or evaporative cooling system. The driving force for the process is the water vapor pressure. When the vapor pressure in air is higher than that on the desiccant surface, moisture is transferred from the air to the desiccant until equilibrium is reached (see Figure 7.37). In order to regenerate the desiccant for reuse, the desiccant is heated, which increases the water vapor pressure on its surface. If air with lower vapor pressure is brought in contact with this desiccant, the moisture passes from the desiccant to the air (Figure 7.37). Two types of desiccants are used: solids, such as silica gel and lithium chloride; or liquids, such as salt solutions and glycols.

The two solid desiccant materials that have been used in solar systems are silica gel and the molecular sieve, a selective absorber. Figure 7.38 shows the equilibrium absorption capacity of several substances. Note that the molecular sieve has the highest capacity up to 30% humidity, and silica gel is optimal between 30% and 75%—the typical humidity range for buildings.

Figure 7.39 is a schematic diagram of a desiccant cooling ventilation cycle (also known as Pennington cycle), which achieves both dehumidification and cooling. The desiccant bed is normally a rotary wheel of a honeycomb-type substrate impregnated with the desiccant. As the air passes through the rotating wheel, it is dehumidified while its temperature increases (processes 1 and 2) due to the latent heat of condensation. Simultaneously, a hot air stream passes through the opposite side of the rotating wheel, which removes moisture from the wheel. The hot and dry air at state 2 is cooled in a heat-exchanger wheel to condition 3 and further cooled by evaporative cooling to condition 4. Air at condition 3 may be further cooled by vapor-compression or vapor-absorption systems instead of evaporative cooling.

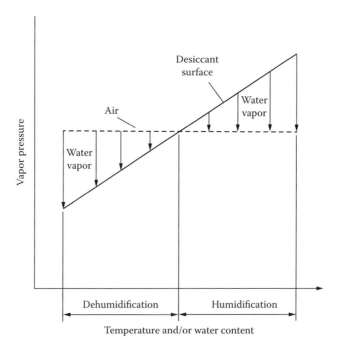

FIGURE 7.37
Vapor pressure versus temperature and water content for desiccant and air.

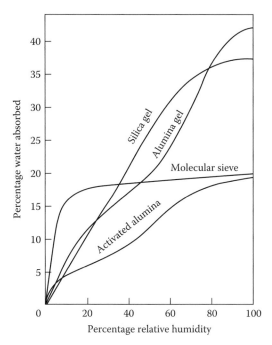

FIGURE 7.38
Equilibrium capacities of common water absorbents.

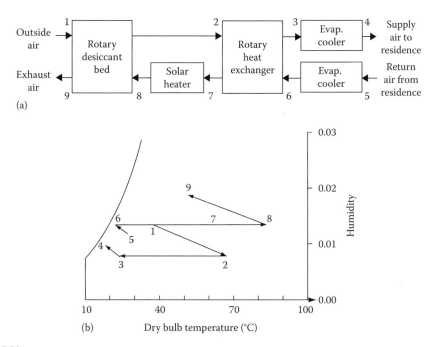

FIGURE 7.39
Schematic of a desiccant cooling ventilation cycle: (a) schematic of airflow and (b) process on a psychrometric chart.

The return air from the conditioned space is cooled by evaporative cooling (processes 5 and 6), which in turn cools the heat-exchanger wheel. This air is then heated to condition 7. Using solar heat, it is further heated to condition 8 before going through the desiccant wheel to regenerate the desiccant. A number of researchers have studied this cycle, or an innovative variation of it, and have found thermal COPs in the range of 0.5–2.58 [38].

7.10.2 Liquid Desiccant Cooling System

Liquid desiccants offer a number of advantages over solid desiccants. The ability to pump a liquid desiccant makes it possible to use solar energy for regeneration more efficiently. It also allows several small dehumidifiers to be connected to a single regeneration unit. Since a liquid desiccant does not require simultaneous regeneration, the liquid may be stored for later regeneration when solar heat is available. A major disadvantage is that the vapor pressure of the desiccant itself may be enough to cause some desiccant vapors to mix with the air. This disadvantage, however, may be overcome by proper choice of the desiccant material.

A schematic of a liquid desiccant system is shown in Figure 7.40. Air is brought in contact with concentrated desiccant in a countercurrent flow in a dehumidifier. The dehumidifier may be a spray column or packed bed. The packings provide a very large area for heat and mass transfer between the air and the desiccant. After dehumidification, the air is sensibly cooled before entering the conditioned space. The dilute desiccant exiting the dehumidifier is regenerated by heating and exposing it to a countercurrent flow of a moisture-scavenging air stream.

Liquid desiccants commonly used are aqueous solutions of LiBr, lithium chloride, calcium chloride, mixtures of these solutions, and triethylene glycol (TEG) [1]. Vapor pressures of these common desiccants are shown in Figure 7.41 as a function of concentration and temperature, based on a number of references [39–43]. Other physical properties important in the selection of desiccant materials are listed in Table 7.15. Although salt solutions and TEG have similar vapor pressures, salt solutions are corrosive and have higher surface tension. The disadvantage of TEG is that it requires higher pumping power because of its higher viscosity.

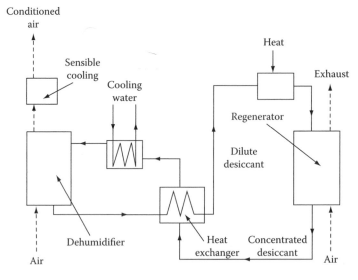

FIGURE 7.40
A conceptual liquid desiccant cooling system.

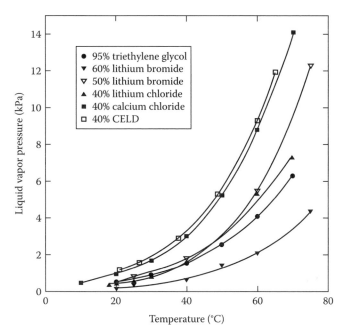

FIGURE 7.41
Vapor pressures of liquid desiccants.

TABLE 7.15

Physical Properties of Liquid Desiccants at 25°C

Desiccant	Density, ρ 10^{-3} (kg/nV)	Viscosity, $\mu \cdot 10^{-3}$ (N s/m²)	Surface Tension, γ 10^{-3} (N/m)	Specific Heat, c_p (kJ/Kg-°C)	Reference
95% by weight triethylene glycol	1.1	28	46	2.3	[22]
55% by weight LiBr	1.6	6	89	2.1	[43,44]
40% calcium chloride	1.4	7	93	2.5	[43,45,46]
40% by weight lithium chloride	1.2	9	96	2.5	[43]
40% by weight CELD	1.3	5	—	—	[47]

Source: Oberg, V. and Goswami, D.Y., *Sol. Eng.*, 128, 155, 1998.

Problems

7.1 Calculate the heat-removal factor for a collector having an overall heat loss coefficient of 6 W/m² K and constructed of aluminum fins and tubes. Tube-to-tube centered distance is 15 cm, fin thickness is 0.05 cm, tube diameter is 1.2 cm, and fluid tube heat-transfer coefficient is 1200 W/m² K. The cover transmittance to solar radiation is 0.9 and is independent of direction. The solar absorptance of the absorber plate is 0.9, the collector is 1 m wide and 3 m long, and the water flow rate is 0.02 kg/s. The water temperature is 330 K.

7.2 Calculate the efficiency of the collector described in Problem 7.1 on March 1 at a latitude of 40°N between 11 and 12 a.m. Assume that the total horizontal insolation is 450 W/m², the ambient temperature is 280 K, and the collector is facing south.

7.3 Calculate the plate temperature in Example 7.3 at 10 a.m., if the insolation during the first 3 h is 0, 150, and 270 W/m², respectively, and the air temperature is 285 K.

7.4 Calculate the overall heat-transfer coefficient, neglecting edge losses, for a collector with a double glass cover, with the following specifications:

Plate-to-cover spacing	3 cm
Plate emittance	0.9
Ambient temperature	275 K
Wind speed	3 m/s
Glass-to-glass spacing	3 cm
Glass emittance	5 cm
Back insulation thickness	5 cm
Back insulation thermal conductivity	0.04 W/m K
Mean plate temperature	340 K
Collector tilt	45°

7.5 The graph in Figure 7.18 gives the results of an ASHRAE standard performance test for a single-glazed flat-plate collector. If the transmittance for the glass is 0.90 and the absorptance of the surface of the collector plate is 0.92, determine

The collector heat-removal factor F_R

The overall heat loss conductance of the collector U_c in W/m² K

The rate at which the collector can deliver useful energy in W/m² K

When the insolation incident on the collector per unit area is 600 W/m² K, the ambient temperature is 5°C, and inlet water is at 15°C.

The maximum flow rate through the collector, with cold water entering at a temperature of 15°C, which will give an outlet temperature of at least 60°C if this collector is to be used to supply heat to a hot water tank.

7.6 Calculate the overall heat loss coefficient for a solar collector with a single glass cover having the following specifications:

Spacing between plate and glass cover	5 cm
Plate emittance	0.2
Plate absorptance at equilibrium temperature	0.2
Ambient air temperature	283 K
Wind speed	3 m/s
Back insulation thickness	3 cm
Conductivity of back insulation material	0.04 W/m K
Mean plate temperature	340 K
Collector tilt	45°

Note that the solution to this problem requires trial and error. Start by assuming a glass temperature and then determine whether the heat gained by the glass equals the heat loss from the glass at that temperature. If the heat gain is larger than the heat loss, repeat the calculations with a slightly higher temperature. After you have calculated the heat loss coefficient, compare your answer with the one obtained from Equation 7.23.

7.7 Standard tests on a commercially available flat-plate collector gave a thermal efficiency of

$$\eta = 0.7512 - \frac{0.138\left(T_{f,in} - T_a\right)}{I_c}$$

$$K_{\tau\alpha} = 1 - 0.15\left[\frac{1}{\cos(i)} - 1\right]$$

where $(T_{f,in} - T_a)/I_c$ is in K m²/W

Find the useful energy collected from this collector each hour and for the whole day in your city on September 15.

Assume that all the energy collected is transferred to water storage with no losses. Calculate the temperature of the storage for each hour of the day. Assume a reasonable ambient temperature profile for your city. Given

Collector area = 6 m²

Collector tilt = 30° (south facing in northern hemisphere, north facing in southern hemisphere)

Storage volume = 0.3 m³ (water)

Initial storage temperature = 30°C

7.8 The heat-removal factor F_R permits solar collector delivery to be written as a function of collector fluid *inlet* temperature T_f in Equation 7.46. Derive the expression for a factor analogous to F_R, relating collector energy delivery to fluid *outlet* temperature.

7.9 In nearly all practical situations, the argument of the exponential term in Equation 7.43 for F_R is quite small. Use this fact along with Taylor's series expansion to derive an alternate equation for F_R. Determine the range where the alternate equation and Equation 7.43 agree to within 1%.

7.10 What is the operating temperature for an evacuated tube collector operating at 50% efficiency if the insolation is 800 W/m²? Use data from Figure 7.19.

7.11 Derive an expression for the heat loss conductance U_c for a flat-plate collector in which convection and conductance are completely eliminated in the air layers by use of a hard vacuum.

7.12 Calculate the optical efficiency on November 13 of an evacuated tube collector array at noon and 2 p.m. if the direct normal insolation is 600 W/m² and the diffuse insolation is 100 W/m². The effective optical transfer function $\tau_e\alpha_r$ is 70j percent, and the tubes are spaced one diameter apart in front of a white painted surface with a reflectance ρ of 60%.

7.13 The no-load temperature of a building with internal heat sources is given by Equation 7.72. How would this equation be modified to account for heat losses through the surface of an unheated slab, the heat losses being independent of ambient temperature?

7.14 An unheated garage is placed on the north wall of a building to act as a thermal buffer zone. If the garage has roof area A_r, window area A_{wi}, door area A_d, and wall area A_{wa}, what is the effective U value for the north wall of the building if its area is A_n? The garage floor is well insulated and has negligible heat loss. Express the effective U value in terms of the U values and areas of the several garage surfaces.

7.15 What is the annual energy demand for a building in Denver, Colorado, if the peak heat load is 150,000 Btu/h based on a design temperature difference of 75°F? Internal heat sources are estimated to be 20,000 Btu/h, and the design building interior temperature is 70°F.

7.16 What is the January solar load fraction for a water-heating system in Washington, DC, using 100 m² of solar collector if the water demand is 4 m³/day at 65°C with a source temperature of 12°C? No heat exchanger is used, and the solar collector efficiency curve is given in Figure 7.19; the solar collector is tilted at an angle equal to the latitude.

7.17 Repeat Problem 7.16 for Albuquerque, NM, in July if the water source temperature is 17°C.

7.18 Explain how the *f*-chart (Figure 7.32) can be used *graphically* to determine the solar load fraction for a range of collector sizes once the solar and loss parameters have been evaluated for only one system size. *Hint*: Consider a straight line passing through the origin and the point (P_s, P_L).

7.19 In an attempt to reduce cost, a solar designer has proposed replacing the shell-and-tube heat exchanger in Figure 7.27 with a tube coil immersed in the storage tank. The shell-and-tube heat exchanger originally specified had a surface area of 10 m² and a U value of 2000 W/m² K to be used with a 100 m² solar collector. Using Equation 7.66 to estimate the U value of the submerged coil, how much length of 0.5 in. (1.27 cm) diameter copper pipe would be needed to achieve the same value of UA product as the shell-and-tube heat exchanger? What percentage of the storage tank volume would be consumed by this coil if 50 kg of water is used per square meter of collector? Use a storage water temperature of 60°C and a collector water outlet temperature of 70°C for the calculations.

7.20 If a solar system delivers 2500 MJ/m² ·year with a water flow rate of 30 kg/m$_c^2$ and a plate efficiency factor $F' = 0.93$, how much energy will it deliver if the flow rate is doubled? Neglect the effect of flow rate on F'. The collector has a heat loss conductance of 4 W/m²°C.

7.21 How large (MJ/h) should a heat-rejection system be if it must dump the entire heat production of a 1000 m² solar collector array in Denver, CO, on August 21 if the collector is at 100°C and the ambient temperature is 35°C? Use solar collector data in Figure 7.19 and NREL hourly solar radiation data that can be found on the Internet.

7.22 Use the *f*-chart to determine the amount of solar energy that can be delivered in Little Rock, AR, in January for the following solar and building conditions:

Building

    Load: 40 million Btu/month

    Latitude: 35°N

Solar system

    Collector tilt: 55°, facing south

    Area: 1000 ft²

Ambient temperature: 40.6°F

Collector efficiency curve: see Figure 7.19

No heat exchanger used

Nominal storage, flow rate, and load heat-exchanger values used

7.23 Estimate the electric power requirement, in kW, of a 1400 ft² floor area (three-bedroom home) with three occupants. Using your home power estimate, predict the power requirements for a city of 300,000 people. Use these results to estimate the area of silicon solar cells required to satisfy the community power requirements. Write a short narrative discussing your assumptions and analysis.

7.24 The *f*-chart is based on $F_R(\tau\alpha)$ and $F_R U_c$ values, which can be deduced from a plot of collector efficiency versus $(T_{fin} - T_a)/I$. If such a plot is not available but (a) a plot of efficiency versus $(\bar{T}_f - T_a)/I$ is available or (b) a plot of efficiency versus $(T_{fout} - T_a)/I$ is available, how can $F_R(\bar{\tau\alpha})$ and $F_R U_c$ be calculated from the slope and the intercept of these two curves? Express your results in terms of the slopes, intercepts, and fluid capacitance rate $\dot{m}c_p/A_c$.

7.25 The following schematic diagrams illustrate the operation of a solar-assisted heat-pump system and a solar system augmented with a heat pump. Discuss the advantages and disadvantages of each system with respect to different climatic conditions.

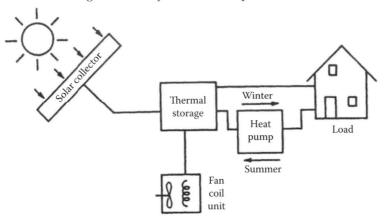

Solar-assisted heat-pump system

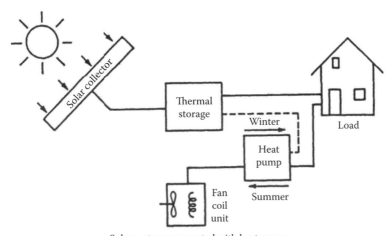

Solar system augmented with heat pump

7.26 The following table gives the characteristics of a building in Houston, TX. Determine the cooling load for July 30 at solar noon. Any information regarding the load not given may be assumed or neglected.

Factor	Description or Specification
Roof	
Type	Flat, shade
Area $_{A\,rf,sh}$ (m²)	250
U factor (W/m²K)	$U_{rf} = 0.35$
Walls	
Type	Vertical, painted white
Orientation, size (m × m)	North, south, 3 × 10; east, west, 3 × 25
U factor (W/m² K)	$U_{wa} = 1.08$
Windows	
Orientation, area (m²)	North, 8; south, 8; east, 20; west, 25
U factor (W/m² K)	$U_{wi} = 6.2$
Insolation transmittance	$\bar{\tau}_{b,wi} = 0.60;\ \bar{\tau}_{d,wi} = 0.80;\ \bar{\tau}_{r,wi} = 0.55$
Temperature	
Inside, outside (°C)	$T_{in} = 24;\ T_{out} = 37$
Insolation	
Beam, diffuse, reflected (W/m²)	$I_{h,b} = 580;\ I_{h,d} = 250;\ I_r = 200$

7.27 An air-conditioning system working in a vapor-compression cycle is used to manage the load for the building in Problem 7.26. If the high and low pressures in the cycle are 915 and 290 kPa, respectively, and the efficiency of the compressor is 90%, find the flow rate of Refrigerant 134a (R-134a) used for the equipment and the COP of the cycle.

7.28 Consider the absorption refrigeration cycle, shown in the following line diagram, which uses lithium bromide as carrier and water as refrigerant to provide 1 kW of cooling. By using steam tables at http://www.ohio.edu/mechanical/thermo/property_tables/H₂O/index.html and the chart giving the properties of lithium bromide and water, calculate first

a. Heat removed from the absorber

b. Heat removed from the condenser

c. Heat added to the evaporator

d. COP of the cycle

Then calculate, for a flat-plate collector with an F_R ($\tau\alpha$) intercept of 0.81 and an $F_R U_c$ of 3 W/m² K, the area required for operation in Arizona at noon in August for a 3-ton unit. Assume that the enthalpy of the water vapor leaving the condenser can be approximated by the equation

$$h_{vc} = 2463 + 1.9T_c \text{ kJ/kg}$$

and that the enthalpy of the liquid water is

$$h_{lc} = 4.2T_c \text{ kJ/kg}$$

where T_c is the temperature of the evaporator in °C. In the analysis, assume that evaporation occurs at 1°C and condensation at 32°C.

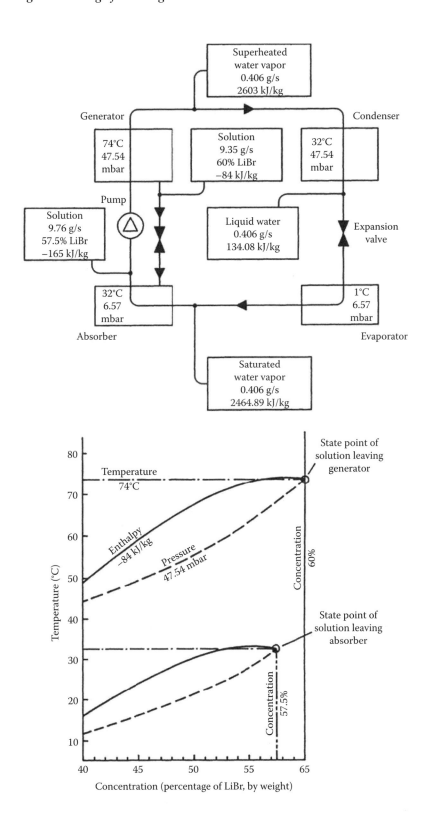

7.29 Make a preliminary design for a solar-driven Rankine refrigeration machine to provide the temperature environment required below the surface of a 20 × 40 m ice-skating rink that is to operate all year in the vicinity of Denver, CO. State all your assumptions.

7.30 The following table shows data from a dehumidification process using aqueous solution of lithium chloride (LiCl) in a packed tower. The desiccant leaving the dehumidifier is passed through another packed tower for regeneration. For the regeneration process, it is known that the rate of evaporation of water as a function of the concentration $X(kg_{LiCl}/kg_{sol})$ and temperature $T(°C)$ of the desiccant is given by the following equation:

$$m_{evap} = (a_0 + a_1T + a_2T^2) + (b_0 + b_1T + b_2T^2)X + (c_0 + c_1T + c_2T^2)X^2 (g/s)$$

where

$$a_0 = 285077, \quad b_0 = -1658652, \quad c_0 = 2412282$$

$$a_1 = -8992, \quad b_1 = 52326, \quad c_1 = -76112, \text{ and}$$

$$a_2 = 70.88, \quad b_2 = -412, \quad c_2 = 600$$

Find the temperature to which a flat-plate collector must raise the temperature of the desiccant for the regeneration process. Assume that the flow rate of liquid desiccant in both processes (dehumidification, regeneration) is the same.

	Air			Desiccant		
Variable	Inlet	Outlet	Inlet	Outlet	Variable	
Temperature (°C)	30.4	32.6	30	33.1	Temperature (°C)	
RH (%)	66.7	36.7	35	—	Concentration (%)	
Mass (kg/h)	260	—	850	—	Mass (kg/h)	

7.31 A survey of one million people near Miami, FL, shows that 80% of them use natural gas and 20% use electricity for their hot water needs in their homes. Estimate the saving in fossil fuel, surface area of the collectors, and annual reduction in CO_2 emissions if flat-plate solar water heaters similar to the one described in the text are used to help with hot water demands (for the one million people surveyed). Assume that Miami is at 24°N latitude. Suggestion... a spreadsheet will make this easier.

Assumptions

- Electric water-heating efficiency = 98%
- Natural gas water-heating efficiency = 75%

- Coal to electricity efficiency = 35%
- Average person uses 76 L of hot water per day (avg. temp demanded is 140°F)
- Ground refection is negligible
- Assume 11 h of daylight
- Assume mass flow rate of 0.01 kg/s
- I_h = average direct irradiation on a horizontal surface during daylight hours = 590 W/m²
- I_s = average direct irradiation on a 24° incline during daylight hours = 640 W/m²

Solar water heater design
Front view

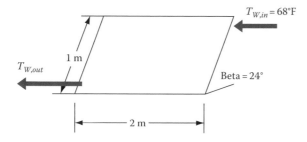

Cross-sectional view

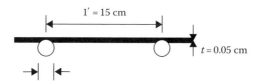

The following problem is designed for students interested in learning the fundamentals of a useful computer application entitled eQUEST. This program is widely used by architectural engineers.

7.32 A distant cousin Martha just passed away, peacefully in her sleep at the age of 105, and left you with one of her many real estate properties. The last time you had dinner with Martha, you described this great class you were taking on sustainable energy. She was thrilled to hear your passion and ideas about a sustainable energy future. Martha changed her will to give you an old building on the stipulation that you create an energy efficiency showcase restaurant/café/bar. For this homework, you are going to use eQUEST to determine the energy performance of the building as is (termed "baseline") and explore how specific building improvements will change that baseline energy use. See the following eQUEST Background for more information on this freeware. Please follow the given steps and then generate a report documenting the results. Your report should not be a bunch of disjointed bar charts but a few well-explained sample bar charts and a few tables that summarize the results. Follow the typical homework format.

eQUEST is a freeware created by the U.S. Department of Energy. You can go to www.doe2.om to download eQUEST and learn more about this freeware. eQUEST uses DOE-2.2 as a simulation engine and allows the user to model buildings either through the Schematic Design wizard (for preliminary analysis) or through the Design Development wizard (for more complex analysis). eQUEST also has a detailed editor to modify the input parameters (you need to be familiar with DOE-2 BDL to use the detailed editor). Parametric analysis can be performed by eQUEST using the Energy Efficiency Measure (EEM) wizard or a more advanced editor called Parametric Runs.

For this homework, you should download eQUEST version 3.6. You will be using the Schematic Design and EEM Wizards.

Step 1: Download eQUEST.

Step 2: Open eQUEST.
   eQUEST Startup Options. Select "Create a New Project via the Wizard"
   Which Wizard? Select "Schematic Design Wizard."

Step 3: Create baseline building using the Schematic Design Wizard
   eQUEST Schematic Design Wizard
   You will walk through 41 Wizard Screens—the building you have inherited is described by all the Wizard defaults except for data described later.

Wizard Screen 1 (General Information)
Project Name: Up to you—be creative
Building Type: *Select* "Restaurant, Bar/Lounge"
Building Location, Utilities, and Rates
Coverage: *Select* "All eQUEST Locations"
Region: *Select* "Colorado"
City: *Select* "Boulder" Areas and Floors
Building Area: *Enter* "3000"

Wizard Screen 2 (Building Footprint)
Footprint Shape: *Select* "'L' Shape"
Building Orientation
Plan North: *Select* "South" Floor Heights
Check Pitched Roof

Wizard Screen 3 (Building Envelope Constructions)
Add'l Insulation for Above Grade Walls: *Select* "R-11 batt"

Wizard Screen 6 (Exterior Doors)
Describe Up To 3 Door Types
Door Type South North West East

1. Glass 1010

2. Opaque 110.0 Door Dimensions and Construction

1. For Construction *select* "Double Clr/Tint(2006 Version)"

2. For Construction *select* "Wood, Solid core Flush, 1– 3/8 in."

Wizard Screen 7 (Exterior Windows)
Describe Up To 3 Window Types

1. Single Clr/Tint

2. Double Clr/Tint

Window Dimensions, Positions and Quantities South North West East
1.   20 20 20 20 2..0 0 0 0
    Click "Finish"

Step 4: Determine baseline building performance.

*   On the Actions navigation bar, *click* "Simulate Building Performance."
*   Once the simulation is complete, view the summary results/reports Pretty cool, huh?

Step 5: Use the EEM Wizard to explore changing the amount of window area and location of those windows. Try the following options:

1. Baseline
2. Option 1: S = 20%, N = 5%, W = 20%, E = 20%
3. Option 2: S = 50%, N = 20%, W = 20%, E = 20%
4. Option 3: S = 20%, N = 20%, W = 50%, E = 20%
5. Option 4: S = 50%, N = 20%, W = 50%, E = 20%

To use the EEM Wizard, return to the Building Description Mode. Click on EEM Wizard on the Actions navigation bar.
 Energy Efficiency Measure Creation Screen
 Measure Category: Select "Building Envelope"
 Measure Type: Select "Window Area"
 Apply Measure To: Select "baseline" (if asked)
 Click "OK"
 EEM Run Information Screen
 One Measure, "Window Area EEM," should appear in the Select Measure to View/ Edit window. Highlight that Measure and then change the name in the EEM Run Name to something like "window area option 1." *Click* on the "EEM Run Details ..." button and then edit the window area characteristics on that screen. You can use the "Create Run" button to add all the options—you will walk through the same process for each. *Click* "Finish" when you have created the options desired.
 On the Actions navigation bar, *click* "Simulate Building Performance."

Step 6: Use the EEM Wizard to determine the payback periods of the following measures:

1. Install better insulation in the walls; use R-19 at a cost of $0.25/ft².
2. Install better insulation in the ceiling; use R-30 at a cost of $0.35/ft².
3. Install high-efficiency furnace with a seasonal efficiency of AFUE = 0.90 with an incremental cost of $750.
4. 65°F when occupied and 55°F when unoccupied using a programmable thermostat at a cost of $100.
5. Install a daylighting photosensor at a cost of $100.
6. Implement all the measures listed earlier simultaneously.

For the economic analysis, use the simple payback period approach and use an electricity cost of $0.09/kWh and a natural gas cost of $1.25/therm.

References

1. Löf, G.O.G. and Tybout, R.A. (1972) Model for optimizing solar heating design, ASME Paper 72-WA/SOL-8.
2. Kutcher, C.F. (1996) Transpired solar collector systems: A major advance in solar heating, in: Paper presented at the *World Engineering Congress*, Atlanta, GA, November 6–8, 1996.
3. Kutcher, C.F. and Christensen, C.B. (1992) Unglazed transpired solar collectors, *Advances in Solar Energy*, vol. 7, pp. 283–307.
4. Hottel, H.C. and Woertz, B.B. (1942) Performance of flat-plate solar-heat collectors, *Trans. Am. Soc. Mech. Eng.* 64, 91.
5. Klein, S.A. (1975) Calculation of flat-plate collector loss coefficients, *Sol. Energy* 17, 79–80.
6. Agarwal, V.K. and Larson, D.C. (1981) Calculation of the top loss coefficient of a flat plate collector, *Sol. Energy* 27, 69–71.
7. Hottel, H.C. and Whillier, A. (1958) Evaluation of flat-plate collector performance, *Trans. Conf. Use Sol. Energy* 2(1), 74.
8. Vant-Hull, L.L. (1976) Development of the solar tower program in the United States. *Proc. SPIE Solar Energy Utilization Conf. II* 85, 104. See also Vant-Hull, L.L. 1976, 85, 111.
9. Bliss, R.W. (1959) The derivations of several plate efficiency factors useful in the design of flat-plate solar-heat collectors, *Sol. Energy* 3, 55.
10. Klein, S.A., Duffie, J.A., and Beckman, W.A. (1974) Transient considerations of flat-plate solar collectors, *J. Eng. Power* 96A, 109–114.
11. Kreith, F. and Bohn, M.S. (1997) *Principles of Heat Transfer*, 5th ed., PWS Publishers, Boston, MA.
12. Duffie, J.A. and Beckman, W.A. (1980) *Solar Energy Thermal Processes*, John Wiley & Sons, New York.
13. Collares-Pereira, M., Rabl, A., and Winston, R. (1977) Lens mirror combinations with maximal concentration, Enrico Fermi Institute Rept., No. EFI 77-20.
14. Duff, W.S. (1976) Optical and thermal performance of three line focus collectors, ASME Paper 76-WA/HT-15.
15. Emmett, W.L.R. (1911) Apparatus for utilizing solar heat, U.S. Patent 980,505.
16. Speyer, F. (1965) Solar energy collection with evacuated tubes, *J. Eng. Power* 87, 270.
17. Beekley, D.C. and Mather, G.R. (1975) *Analysis and Experimental Tests of a High Performance, Evacuated Tube Collector*, Owens-Illinois, Toledo, OH.
18. ASHRAE. (2009) *Handbook of Fundamentals*, American Society of Heating, Refrigerating and Air-Conditioning Engineers, Atlanta, GA.
19. Klein, S.A. et al. (1975) A method for simulation of solar processes and its application, *Sol. Energy* 17, 29–37.
20. Close, D.J. (1962) The performance of solar water heaters with natural circulation, *Sol. Energy* 6, 33.
21. Phillips, W.F. and Cook, R.D. (1975) Natural circulation from a flat plate collector to a hot liquid storage tank, ASME Paper 75-HT-53.
22. Kreider, J.F. and Kreith, F. (1977) *Solar Heating and Cooling*, revised 1st edn., Hemisphere Publishing Corporation, Washington, DC.
23. Klein, S.A. (1976) A design procedure for solar heating systems, PhD dissertation, University of Wisconsin, Madison, WI. For an approach similar to the *f*-chart for other solar-thermal systems operating above a minimum temperature above that for space-heating (~20°C), see Klein, S.A. and Beckman, W.A. (1977) A general design method for closed loop solar energy systems, in: *Proceedings of 1977 ISES Meeting*, Orlando, FL.
24. Beckman, W.A., Klein, S.A., and Duffie, J.A. (1977) *Solar Heating Design by the F-Chart Method*, John Wiley & Sons, New York.
25. Klein, S.A., Beckman, W.A., and Duffie, J.A. (1976) A design procedure for solar heating systems, *Sol. Energy* 18, 113.

26. Klein, S.A., Beckman, W.A., and Duffie, J.A. (1976) A design procedure for solar air heating systems, in: Paper presented at the *1976 ISES Conference*, American Section, Winnipeg, Manitoba, Canada, August 15–20.
27. Goswami, Y., Kreith, F., and Kreider, J. (2000) *Principles of Solar Engineering*, 2nd edn., Taylor & Francis, Philadelphia, PA.
28. Rangarajan, K., Shirley, D.B. III, and Raustad, R.A. (1989) Cost-effective HVAC technologies to meet ASHRAE Standard 62-1989 in hot and human climates, *ASHRAE Trans*. pt. 1, 166–182.
29. Meckler, H. (1994) Desiccant-assisted air conditioner improves IAQ and comfort, *Heat. Piping Air Cond*. 66(10), 75–84.
30. Meckler, M. (1995) Desiccant outdoor air preconditioners maximize heat recovery ventilation potentials, *ASHRAE Trans*. 101(2), 992–1000.
31. Meckler, M. (1988) Off-peak desiccant cooling and cogeneration combine to maximize gas utilization, *ASHRAE Trans*. 94(1), 575–596.
32. Meckler, M., Parent, Y.O., and Pesaran, A.A. (1993) Evaluation of dehumidifiers with polymeric desiccants, Gas Institute Report, Contract No. 5091-246-2247, Gas Research Institute, Chicago, IL.
33. Oberg, V. and Goswami, D.Y. (1998) Experimental study of heat and mass transfer in a packed bed liquid desiccant air dehumidifier, in: J.H. Morehouse and R.E. Hogan (eds.), *Solar Engineering*, ASME, New York, pp. 155–166.
34. Oberg, V. and Goswami, D.Y. (1998) A review of liquid desiccant cooling, *Adv. Sol. Energy* 12, ASES, 431–470.
35. Spears, J.W. and Judge, J. (1997) Gas-fired desiccant system for retail super center, *ASHRAE J*. 39, 65–69.
36. Thombloom, M. and Nimmo, B. (1994) Modification of the absorption cycle for low generator firing temperatures, in: *Joint Solar Engineering Conference*, ASME, New York.
37. Thombloom, M. and Nimmo, B. (1995) An economic analysis of a solar open cycle desiccant dehumidification system, in: *Solar Engineering—1995, Proceedings of the 13th Annual ASME Conference*, Maui, HI, vol. 1, pp. 705–709.
38. Pesaran, A.A., Penney, T.R., and Czanderna, A.W. (1992) Desiccant cooling: State-of-the art assessment, Report No. NREL/TP-254-4147, National Renewable Laboratory, NREL, Golden, Colorado.
39. Cyprus Foote Mineral Company. (1995) Technical data on lithium bromide and lithium chloride, Bulletins 145 and 151, Cyprus Foote Mineral Company, Kings Mountain, NC.
40. Dow Chemical Company. (1996) *Calcium Chloride Handbook*, Dow Chemical Company, Midland, MI.
41. Dow Chemical Company. (1992) *A Guide to Glycols*, Dow Chemical Company, Midland, MI.
42. Ertas, A., Anderson, E.E., and Kiris, I. (1992) Properties of a new liquid desiccant solution—Lithium chloride and calcium chloride mixture, *Sol. Energy* 49, 205–212.
43. Zaytsev, I.O. and Aseyev, G.G. (1992) *Properties of Aqueous Solutions of Electrolytes*, CRC Press, Boca Raton, FL.
44. Kettleborough, C.F. and Waugaman, D.G. (1995) An alternative desiccant cooling cycle, *J. Sol. Energy Eng*. 117, 251–255.
45. Close, D.J. (1965) Rock pile thermal storage for comfort air conditioning, *Mech. Chem. Eng. Trans. Inst. Eng. (Australia)* MC-1, 11.
46. Khan, A.Y. (1994) Sensitivity analysis and component modeling of a packed-type liquid desiccant system at partial load operating conditions, *Int. J. Energy Res*. 18, 643–655.
47. Löf, G.O.G. and Tybout, R.A. (1974) Design and cost of optimal systems for residential heating and cooling by solar energy, *Sol. Energy* 16, 9.

8

Solar Process Heat and Thermal Power*

> The illusion of unlimited powers, nourished by astonishing scientific and technological achievements, has produced the concurrent illusion of having solved the problem of production.
>
> —E.F. Schumacher

8.1 Historical Perspective

Attempts to harness the sun's energy for power production date back to at least 1774 [1], when the French chemist Lavoisier and the English scientist Joseph Priestley discovered oxygen and developed the theory of combustion by concentrating the rays of the sun on mercuric oxide in a test tube, collecting the gas produced with the aid of solar energy, and burning a candle in the gas. Also during the same year, an impressive picture of Lavoisier was published in which he stands on a platform near the focus of a large glass lens and is carrying out other experiments with focused sunlight (Figure 8.1).

A century later, in 1878, a small solar power plant was exhibited at the World Fair in Paris (Figure 8.2). To drive this solar steam engine, sunlight was concentrated from a parabolic reflector onto a steam boiler located at its focus; this produced the steam that operated a small reciprocating steam engine that ran a printing press. In 1901, a 10 hp solar steam engine was operated by A.G. Eneas in Pasadena, California [2]. It used a 700 ft² focusing collector in the shape of a truncated cone, as shown in Figure 8.3. Between 1907 and 1913, the American engineer F. Shuman developed solar-driven hydraulic pumps; in 1913, he built, jointly with C.V. Boys, a 50 hp solar engine for pumping irrigation water from the Nile near Cairo in Egypt (Figure 8.4). This device used long parabolic troughs that focused solar radiation onto a central pipe with a concentration ratio (CR) of 4.5:1.

With the increasing availability of low-cost oil and natural gas, interest in solar energy for power production waned. Except for C.G. Abbott, who exhibited in 1936 a 1/2 hp solar-powered engine at an *International Power Conference* in Washington, District of Columbia, and in 1938 in Florida, an improved, somewhat smaller version with a flash boiler, there was very little activity in the field of solar power between 1915 and 1950. Interest in solar power revived in 1949 when, at the centennial meeting of the American Association for the Advancement of Science in Washington, District of Columbia, one session was devoted to future energy sources. At that time, the potentials as well as the economic problems of solar energy utilization were clearly presented by Daniels [3]. Some important conferences that considered solar power generation were held by UNESCO in 1954, the Association for Applied Solar Energy in 1955, the U.S. National Academy of Sciences in 1961, and the United Nations in 1961. In addition, a research and development program supported by the National Aeronautics and Space

---

* Sections in this chapter marked with an asterisk may be omitted in an introductory course.

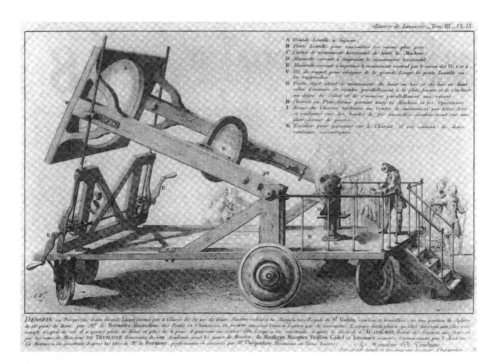

FIGURE 8.1
Solar furnace used by Lavoisier in 1774. (Photo Courtesy of Bibliotheque Nationale de Paris. Lavoisier, *Oeuvres*, Vol. 3. With permission.)

FIGURE 8.2
Parabolic collector powered a printing press at the 1878 Paris Exposition. (From Goswami, Y. et al., *Principles of Solar Engineering*, 2nd edn., Taylor & Francis, Philadelphia, PA, 2000. With permission.)

FIGURE 8.3
Irrigation pumps were run by a solar-powered steam engine in Arizona in the early 1900s. The system consisted of an inverted cone that focused rays of the sun on the boiler. (From Goswami, Y. et al., *Principles of Solar Engineering*, 2nd edn., Taylor & Francis, Philadelphia, PA, 2000. With permission.)

FIGURE 8.4
Solar irrigation pump (50 hp) operating in 1913 in Egypt. (From Goswami, Y. et al., *Principles of Solar Engineering*, 2nd edn., Taylor & Francis, Philadelphia, PA, 2000. With permission.)

Administration to build a solar electric power system capable of supplying electricity for the U.S. space program was undertaken in the 1960s. However, widespread interest developed only after research funds became available for the development of earth-bound solar electric power and process heat after the Arab oil embargo in 1973.

*8.2 Solar Industrial Process Heat

Industrial processes consumed over 30 quadrillion (10^{15}) Btus of energy in 2001 [4], and this amount is increasing. A study conducted for the Energy Research and Development Administration by the InterTechnology Corporation [5] indicated that solar energy has the potential of providing about 20% of this energy. The economic outlook for industrial solar heat appears to be extremely favorable because process heat solar collectors could be used throughout the year, and each system can be designed to fit the temperature level required for its specific applications, which is particularly important in the use of process heat. Table 8.1 shows the amount of heat used by selected industries in the United States.

TABLE 8.1

Summary of the U.S. Industrial Heat Usage by SIC Category for 1971 and 1994

	Quantities in 10^{12} kJ	
SIC Group	**1971**	**1994**
20 Food and kindred products	779	1,254
21 Tobacco products	14	W[a]
22 Textile mills	268	327
23 Apparel	22	W[a]
24 Lumber and wood products	188	518
25 Furniture	37	73
26 Paper and allied products	2,006	2,812
27 Printing and publishing	16	118
28 Chemicals	2,536	5,621
29 Petroleum products	2,576	6,688
30 Rubber	158	303
31 Leather	19	W[a]
32 Stone, clay, and glass	1,541	996
33 Primary metals	3,468	2,597
34 Fabricated metal products	295	387
35 Machinery	283	260
36 Electrical equipment	213	256
37 Transportation	310	383
38 Instruments	53	113
39 Miscellaneous	72	W[a]
Subtotal	14,854	22,854

Source: Intertechnology Corporation, Analysis of the economic potential of solar thermal energy to provide industrial process heat, ERDA Report No. C00/2829, 1997. With permission.

W[a], Withheld to avoid disclosing data for individual establishments.

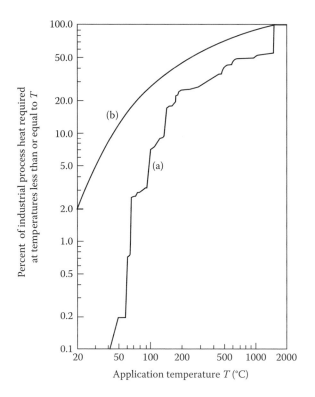

FIGURE 8.5
Distribution of the U.S. process heat use by required temperature level: (a) heat requirements; (b) IPH require-ments plus preheat from 15°C. (From Intertechnology Corporation, Analysis of the economic potential of solar thermal energy to provide industrial process heat, ERDA Report No. C00/2829, 1997. With permission.)

A majority of the heat is used in mining, food, textiles, lumber, paper, chemicals, petro-leum products, stone–clay–glass, and primary metals [6]. The breakdown of industrial energy usage is as follows [7]: process steam, 41%; direct process heat, 28%; shaft drive, 19%; feedstock, 9%; and others, 3%. In addition to the quantity of heat, quality (i.e., temper-ature) is also very important to match the proper solar collection system to the application. Figure 8.5 shows the cumulative process heat use by temperature requirement [7]. It is seen that about 25% of the heat is used at temperatures below 100°C, which may be provided by flat-plate collectors or compound parabolic concentrators (CPCs), about 50% of the heat is used at temperatures below 260°C, and 60% below 370°C. Therefore, a large percentage of all the process heat could be delivered by parabolic trough collectors (PTC).

The selection of the type of solar collectors depends on the process temperature require-ments. Table 8.2 gives the temperature requirements and the type of solar collectors suit-able for the process. The type of storage depends on the temperature requirement, the storage duration, the required energy density (space constraints), and the charging and discharging characteristics. These topics are discussed in detail in Chapter 10. The storage duration for solar industrial process heat (SIPH) systems is usually short since the solar systems are designed to displace part of the fossil fuel requirements. Land availability can be critical for SIPH for existing industries. However, in many cases roofs of industrial buildings can be utilized for this purpose.

Since SIPH systems use components and systems already described, this section will give some examples of SIPH systems. Most of the industrial process heat systems below

TABLE 8.2

Potential Applications for Various Solar IPH Systems

	Energy Form	Temperature (°C)	Simple Air Heaters	Flat Plates	Fixed Compound Surfaces	Single-Tracking Troughs	Central Receivers
Aluminum							
Bayer process digestion	Steam	216				X	
Automobile and truck manufacturing							
Heating solutions	Steam (water)	49–82	X	X			
Heating makeup air in paint booths	Air	21–29	X				
Drying and baking	Air	163–218			X	X	
Concrete block and brick							
Curing product	Steam	74–177			X	X	
Gypsum							
Calcining	Air	160			X	X	
Curing plasterboard	Steam (air)	299				X	X
Chemicals							
Borax, dissolving and thickening	Steam	82–99		X	X		
Borax, drying	Air	60–77	X	X			
Bromine, blowing brine/distillation	Steam	107			X		
Chlorine, brine heating	Steam (water)	66–93	X	X			
Chlorine, caustic evaporation	Steam	143–149			X	X	
Phosphoric acid, drying	Air	121			X		
Phosphoric acid, evaporation	Steam	160				X	
Potassium chloride, leaching	Steam	93		X	X		
Potassium chloride, drying	Air	121			X	X	
Sodium metal, salt purification	Steam	135			X	X	
Sodium metal, drying	Steam (air)	116			X	X	
Food							
Washing	Water	49–17	X	X	X		
Concentration	Steam (water)	38–43	X	X	X		
Cooking	Steam	121–188			X	X	
Drying	Steam (air)	121–232			X	X	

Process	Medium	Temperature					
Glass							
Washing and rinsing	Water	71–93	X	X			
Laminating	Air	100–177			X	X	
Drying glass fiber	Air	135–141		X	X	X	
Decorating	Air	21–93	X		X	X	
Lumber							
Kiln drying	Air	66–99	X	X	X	X	
Glue preparation/plywood	Steam	99–177			X	X	
Hot pressing/fiberboard	Steam	199			X		
Log conditioning	Water	82		X			
Mining (Frasch sulfur)							
Extraction	Pressurized Water	160–166			X	X	
Paper and pulp							
Kraft pulping	Steam	182–188				X	
Kraft liquor evaporation	Steam	138–143			X	X	
Kraft bleaching	Steam	138–143			X	X	
Papermaking (drying)	Steam	177				X	
Plastics							
Initiation	Steam	121–146			X	X	
Steam distillation	Steam	146			X	X	
Flash separation	Steam	216			X	X	
Extrusion	Steam	146			X	X	
Drying	Steam	188			X	X	
Blending	Steam	121			X	X	
Synthetic rubber							
Initiation	Steam (water)	121			X	X	
Monomer recovery	Steam	121			X	X	
Drying	Steam (air)	121			X	X	X
Steel							
Pickling	Steam	66–104		X	X	X	
Cleaning	Steam	82–93		X	X	X	

(continued)

TABLE 8.2 (continued)

Potential Applications for Various Solar IPH Systems

	Energy Form	Temperature (°C)	Simple Air Heaters	Flat Plates	Fixed Compound Surfaces	Single-Tracking Troughs	Central Receivers
Textiles							
Washing	Water	71–82	X	X			
Preparation	Steam	49–113	X	X	X		
Mercerizing	Steam	21–99	X	X	X		
Drying	Steam	60–135	X	X	X		
Finishing	Steam	60–149		X	X	X	

Source: Kreider, J.F., *Medium and High Temperature Solar Process*, Academic Press, New York, 1979. With permission.

Note: Mining, food, textiles, lumber, paper, chemicals, petroleum products, stone–clay–glass, and primary metals [6]. The breakdown of industrial energy usage is as follows [7]: process steam, 41%; direct process heat, 28%; shaft drive, 19%; feedstock, 9%; and other 3%. In addition to the quantity of heat, quality (i.e., temperature) is also very important to match the proper solar collection system to the application. Figure 8.5 shows the cumulative process heat use by temperature requirement [7]. It is seen that about 25% of the heat is used at temperatures below 100°C, which may be provided by flat-plate collectors or compound parabolic concentrators (CPCs), about 50% of the heat is used at temperatures below 260°C, and 60% below 370°C. Therefore, a large percent of all the process heat could be delivered by parabolic trough collectors (PTC8).

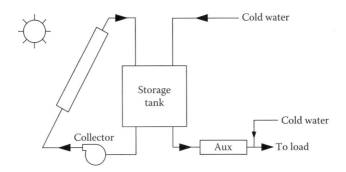

FIGURE 8.6
Schematic of a low-temperature solar industrial process heat system.

200°C require hot water, steam, or hot air. A typical low-temperature SIPH system is shown schematically in Figure 8.6. If the heat is needed in process air, water-to-air heat exchangers may be used or air-heating collectors and rock storage are used. Hot air is needed typically in agricultural drying, which may be provided by passive solar air heaters [8].

*8.2.1 SIPH System for Textile Industries

The textile industry is one of the 10 largest energy-consuming industries. Of all the energy used in the textile industry, 60%–65% is used in wet processing, including dyeing, finishing, drying, and curing. The energy for wet processing is used as hot water and steam. The textile industry in the United States uses approximately 500 billion liters of water per day, and approximately 25% of this water is used at an average temperature of 60°C. Tables 8.3 and 8.4 show typical calculations for determining energy consumption for jet dyeing [9] and tenter frame drying [10], respectively, for 100% textured polyester circular knit fabric [11]. In the analysis in Table 8.3, Wagner assumed 40% moisture content in the fabric. In carpet dyeing, moisture may be as much as 300% [12]. Drying involves the use of high-pressure (6 atm) steam

TABLE 8.3

Sample Calculations[a]: Heat for Pressure Jet Dyeing

Scour	Energy Consumption GJ/100 kg Fabric
Heat bath, 21°C–60°C	163
Heat cloth, 21°C–60°C	9
Raise bath to 129°C	293
Raise cloth to 129°C	14
Replace heat loss from radiation	
During cycle 21°C–129°C	23
During dyeing at 129°C	65
Scour at 60°C	172
Total	739

Sources: Wagner, R., *Chem. Colorists*, 9, 52, 1977; Goswami, B.C. and Langley, J., A review of the potential of solar energy in the textile industry, in *Progress in Solar Engineering*, Hemisphere Publishing Corporation, Washington, DC, 1987. With permission.

[a] Fabric load = 227 kg, bath ratio 10 of 1 cycle time = 2.75 h.

TABLE 8.4

Energy Consumption during Tenter Frame Drying

A. Steps	Energy Requirements (MJ/100 kg H$_2$O)
Evaporate water	
100 kg × 4.18 kJ/kg°C (100°C–21°C)	33
Latent heat of vaporization	226
Raise steam of 121°C	
100 kg × 1.9 kJ/kg°C (121°C–100°C)	4
Heat air (24 lb air/lb H$_2$O)	
2400 kg × 1.015 kJ/kg°C (121°C–21°C)	244
Heat fabric to 250	
100 kg × 100/40 × 2.08 × (121°C–21°C)	52
Dryer run at 121°C	559
Dryer run at 149°C	647 (+15%)

Effect of fabric moisture content on tenter frame energy demand

	Energy Requirements (MJ/100 kg Fabric)	
B. Steps	30%	80%
Evaporate water		
Raise temperature (21°C–100°C)	10.0	26.5
Latent heat of vaporization	67.7	180.5
Raise steam temperature to 121°C	1.4	3.5
Raise air to 121°C	73.3	195.2
Total	152.4	405.7

Sources: Hebrank, W.H., *Am. Dyestuff Rep.*, 63, 34, 1975; Goswami, B.C. and Langley, J., A review of the potential of solar energy in the textile industry, in *Progress in Solar Engineering*, Hemisphere Publishing Corporation, Washington, DC, 1987. With permission.

so that the condensed water may be used for other processes. There are no known examples of SIPH for textile drying, but there are a number of examples of SIPH for textile dyeing. One of those is for a dyeing operation at the Riegel Textile Corp. plant in LaFrance, South Carolina [11,13]. Figure 8.7 shows a schematic of the system.

The original system used evacuated tube collectors made by General Electric Corp. That system failed because of repeated tube breakage. Since the dyeing process used water at a temperature of 70°C, the evacuated tube collectors were replaced with flat-plate collectors. The flat-plate collectors at this plant have copper absorbers and tubes painted fat black, low iron textured and tempered glass cover, and bronze-enameled steel frames. The system has 621 m^2 of collector area.

The dyeing process at this plant is an atmospheric Dye Beck batch process at a maximum temperature of 90°C, which is also typical of the textile industry in the United States [11]. The batch dyeing process involves heating of approximately 4500 L of inlet water from 10°C–25°C to 90°C. The typical dye cycle is shown in Table 8.5. The refurbished SIPH system at Riegel Textile plant has been operating successfully. The system was simulated using TRNSYS computer model [15], and there was good agreement between the actual performance of the system and that predicted by TRNSYS. According to the measured performance of the system from August 14 to October 8, 1983, the system operated with an efficiency of 46% [11].

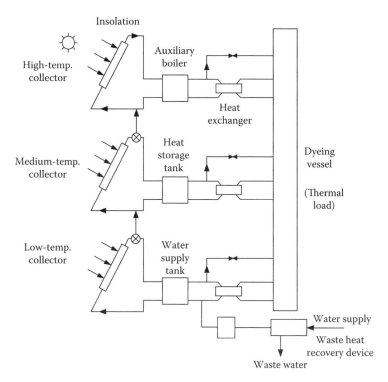

FIGURE 8.7
Schematic of solar energy system at Riegel Textile Corporation plant at LaFrance, South Carolina.

TABLE 8.5

Typical Dye Beck Heat Requirements

Dye Process Operation	Temperature of Dye Beck (°C)	Time Interval of Cycle (h)	Percent of Total Process Energy (%)
Heat the initial load	16–32	0.5	7
Dyeing preparation	32–88	0.5	22
Dye period	88	7–9	68
Cool and reheat for dye fixation	32–43	1.0	3

*8.2.2 SIPH System for Milk Processing

Food systems use approximately 17% of the U.S. energy [16], almost 50% of which is for food processing as hot water (<100°C) or hot air. Proctor and Morse [13] show that over 40% of the energy demand in the beverage industry in Australia was in the form of hot water between 60°C and 80°C. Considering the temperature and heat requirements of food processing, one would be tempted to conclude that SIPH would be ideal. However, many food processing requirements are seasonal, which may not be economical considering the present price of the conventional fuels unless the SIPH system can be used for a majority of the year. One application that is year-round is milk processing. Singh et al. [16] simulated an SIPH system for milk processing in the United States using TRNSYS. The unit operations (and their respective temperatures) for this plant compatible with solar thermal energy are

1. Boiler feed makeup water (100°C)
2. Pasteurizer makeup water (21°C)
3. Case washer and rinsing (49°C)
4. Cleanup (71°C)
5. High-temperature short-time cleanup (79°C)
6. Bottle water (93°C)

They estimated that for a plant producing 170,000 kg/week of milk and 98,000 kg/week of orange juice, a total of 621.3 GJ (or 80%) of energy demand was compatible with solar energy in summer and 724.9 GJ (or 93%) of energy demand was compatible in winter. Simulating a system similar to the one in Figure 8.6, they found that a 4000 m² collector area could provide about 30%–35% solar fraction for the milk-processing plant in Madison, Wisconsin, Fresno, California or Charleston, South Carolina [16].

8.3 Parabolic Concentrating Collectors

In order to provide thermal energy at temperatures above 100°C, it is necessary to use concentrating collectors. Concentration of solar radiation is achieved by reflecting or refracting the flux incident on an aperture area A_a onto a smaller receiver/absorber area A_r. An optical concentration ratio, CR_o, is defined as the ratio of the solar flux I_r on the receiver to the flux, I_a, on the aperture, or

$$CR_o = \frac{I_r}{I_a},$$
(8.1)

while a geometric CR is based on the areas shown in Figure 8.8, or

$$CR = \frac{A_a}{A_r}$$
(8.2)

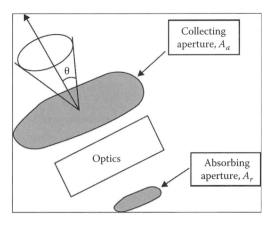

FIGURE 8.8
Definition of geometric concentration.

CR$_o$ gives a true CR because it accounts for the optical losses from the reflecting and refracting elements. However, since it has no relationship to the receiver area, it does not give an insight into the thermal losses that are proportional to the receiver area. In the analyses in this book, only geometric CR will be used.

Concentrators are inherently more efficient at a given temperature than are flat-plate collectors, since the area from which heat is lost is smaller than the aperture area. In the flat-plate device, both areas are equal in size. A simple energy balance illustrates this principle. The useful energy delivered by a collector q_u is given by

$$q_u = \eta_o I_c A_a - U_c(T_c - T_a)A_r \tag{8.3}$$

in which η_o is the optical efficiency, and other terms are as defined previously. The instantaneous collector efficiency is given by

$$\eta_c = \frac{q_n}{I_c A_a} \tag{8.4}$$

from which, using Equation 8.3,

$$\eta_c = \eta_o - \frac{U_c(T_c - T_a)}{I_c} \frac{1}{CR} \tag{8.5}$$

For the flat plate, $CR \cong 1$, and for concentrators $CR > 1$. As a result, the loss term (second term) in Equation 8.5 is smaller for a concentrator and the efficiency is higher. This analysis is necessarily simplified and does not reflect the reduction in optical efficiency that frequently, but not always, occurs because of the use of imperfect mirrors or lenses in concentrators. The evaluation of U_c in Equation 8.5 in closed form is quite difficult for high-temperature concentrators because radiation heat loss is usually quite important and introduces nonlinearities because radiation is proportional to T^4. One disadvantage of concentrators is that they can collect only a small fraction of the diffuse energy incident at their aperture. This property is an important criterion in defining the geographic limits to the successful use of concentrators and is described shortly.

From basic thermodynamics, it can be shown [14] that there exists a maximum geometric concentration allowed by physical conservation laws. This maximum concentration, C_{max}, is related to the angular field of view shown in Figure 8.8 [17] according to

$$C_{max} = \frac{n}{\sin \theta} \quad \text{for a two-dimensional concentrator such as a parabolic trough} \tag{8.6}$$

$$C_{max} = \frac{n^2}{\sin^2 \theta} \quad \text{for a three-dimensional concentrator such as a parabaloid dish} \tag{8.7}$$

where
  n is the index of refraction at the absorber surface
  θ is the half angle of acceptance seen from the collector aperture

A system that can obtain this concentration limit is referred to as "ideal." All conventional imaging systems fall short of this limit by a factor of at least 2–4.

The second law of thermodynamics prescribes not only the geometric limits of concentration as shown earlier, but also the operating temperature limits of a concentrator. The radiation emitted by the sun and absorbed by the receiver of a concentrator, q_{abs}, is

$$q_{abs} = \tau \alpha_s A_s F_{sa} \sigma T_s^4 \qquad (8.8)$$

where
T_s is the effective temperature of the sun
τ is the overall transmittance function for the concentrator, including the effects of any lenses, mirrors, or glass covers
A_s is the area of the sun
F_{sa} is the fraction of radiation emitted by A_s that reaches the area A_a

If the acceptance half angle θ_{max} is selected to just accept the sun's disk of angular measure θ_s ($\theta_s \sim \frac{1}{4}°$), we have, by reciprocity ($F_{as} A_a = F_{sa} A_s$), for an ideal concentrator with compound curvature,

$$q_{abs} = \tau \alpha_s A_s \sin^2 \theta_s \sigma T_s^4 \qquad (8.9)$$

If convection and conduction could be eliminated, all heat loss q_L is by radiation, and

$$q_L = \varepsilon_{ir} A_r \sigma T_r^4 \qquad (8.10)$$

where ε_{ir} is the infrared emittance of the receiver surface. Radiation inputs to the receiver from a glass cover or the environment can be ignored for this upper limit analysis.

An energy balance on the receiver is then

$$q_{abs} = q_L + \eta_c q_{abs} \qquad (8.11)$$

where η_c is the fraction of energy absorbed at the receiver that is delivered to the working fluid. Substituting for Equations 8.9 and 8.10 in Equation 8.11, we have

$$(1 - \eta_c)\tau \alpha_s A_a \sin^2 \theta_s \sigma T_s^4 = \varepsilon_{ir} A_r \sigma T_r^4 \qquad (8.12)$$

Because CR $= A_a/A_r$ and CR$_{max}$ $= 1/\sin^2 \theta_s$, the maximum temperature of the receiver is

$$T_r = T_s \left[(1 - \eta_c)\tau \frac{\alpha_s \text{CR}}{\varepsilon_{ir}\text{CR}_{max}} \right]^{1/4} \qquad (8.13)$$

8.3.1 Compound Parabolic Concentrators

Until recently, it was believed that no useful concentration could be achieved without tracking. However, the invention of Roland Winston [18] showed that the techniques of so-called nonimaging optics provide designs that can deliver moderate levels of concentration with completely stationary concentrators. Figure 8.9 shows a schematic cross section of the original compound parabolic concentrator (CPC) concept. The concentrator is

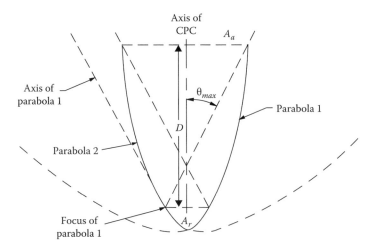

FIGURE 8.9
Schematic cross section of a CPC showing parabolic segments, aperture, and receiver. (From Goswami, Y. et al., *Principles of Solar Engineering*, 2nd edn., Taylor & Francis, Philadelphia, PA, 2000. With permission.)

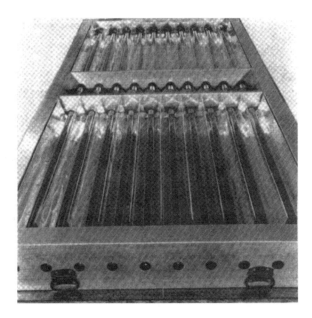

FIGURE 8.10
Commercial CPC trough collector module with evacuated tube receiver. (From Professor Roland Winston, University of California, Merced, CA. With permission.)

seen to be formed from two distinct parabolic segments, the foci of which are located at the opposing receiver surface end points. The axes of the parabolic segments are oriented away from the CPC axis by the acceptance angle θ_{max}. The slope of the reflector surfaces at the aperture is parallel to the CPC optical axis. Figure 8.10 is a photograph of a CPC collector. Different types of planar and tubular receivers have been proposed for CPCs. Of most interest in this book is the tubular-type receiver shown in Figure 8.11 since high-pressure heat-transfer fluid can flow through the tube.

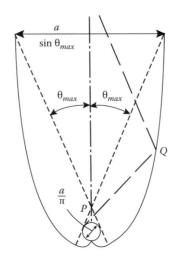

FIGURE 8.11
Cross section of CPC collector concept with tubular receiver.

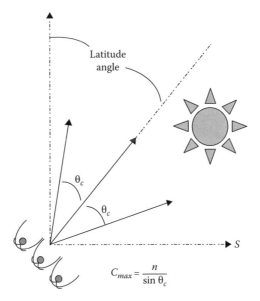

FIGURE 8.12
Schematic of an east–west aligned CPC.

The most common way to utilize a CPC collector is shown schematically in Figure 8.12 [17]. The long axis of the CPC trough is aligned in an east–west direction and that of the normal through the trough apertures is tilted downward from the zenith by an angle equal to the latitude. The angular acceptance is a wedge of half angle $\pm\theta_c$. Whenever the sun's path lies within this "orange slice," all of the direct solar radiation is collected, concentrated, and delivered to the absorber. In addition, some of the diffuse radiation entering through the aperture is also directed toward the absorber.

Advanced CPC collectors integrate the CPC optics into the design, as shown in Figure 8.13 [17]. The integrated CPC shown in Figure 8.13 has an acceptance angle of 70° and a CR of 1.6.

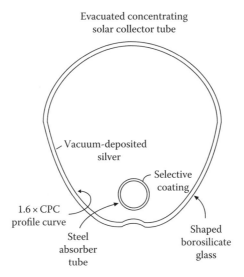

FIGURE 8.13
Schematic cross section of the advanced CPC integrated optics with evacuated tube receiver. (From O'Gallagher, J.J., *Nonimaging Optics in Solar Energy*, Morgan & Claypool Publishers, San Rafael, CA, 2008. With permission.)

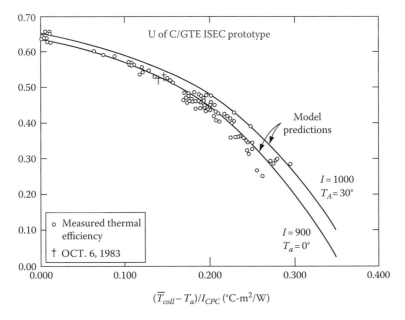

FIGURE 8.14
Comparison of the predicted and experimental performance of a nontracking CPC. (From O'Gallagher, J.J., *Nonimaging Optics in Solar Energy*, Morgan & Claypool Publishers, San Rafael, CA, 2008. With permission.)

A comparison of the predicted and experimental performance is shown in Figure 8.14. From these data, it appears that this type of nontracking CPC collector is a simple and effective method for delivering solar thermal energy in the temperature range between 100°C and 300°C. This type of collector has great potential in the field of SIPH as well as some solar cooling systems described in Chapter 7.

8.3.2 Single-Axis Tracking Parabolic Trough Collectors

Parabolic trough solar concentrators (PTSCs) have been found to be cost-effective solar conversion systems and are growing in importance. Figure 8.15a is a schematic diagram of this type of single-axis tracking collector, and Figure 8.15b is a photograph of a commercial PTSC.

PTSCs have been built in many sizes up to the 6 m width aperture system constructed by Sky Fuel in Colorado. PTCs have been used in one of the early solar hybrid power plants installed by LUZ in Barstow, California. This 355 MWe plant has operated successfully more than 40 years. At present, PTSCs are being installed as fuel-saving devices in existing natural gas power plants, as well as in new power plant construction. The following will illustrate the design, energy return on energy invested (EROI), and cost of energy delivered for PTSCs. Details on the mechanical aspects of design are presented in [19].

The geometric relations for a parabolic trough concentrator are shown schematically in Figure 8.16.

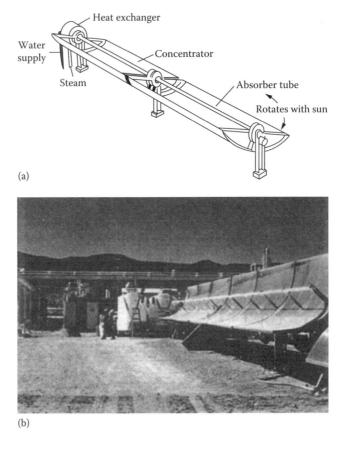

(a)

(b)

FIGURE 8.15
(a) Schematic diagram of single-axis tracking parabolic collector and (b) photograph of single-axis tracking parabolic collector.

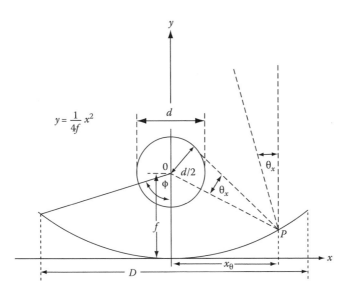

FIGURE 8.16
Geometric relations for a PTSC.

The surface of the reflector/concentrator has a parabolic cross section so that all rays of sunlight parallel to its axis are reflected from the surface to a focal line. The basic geometric relation that defines the parabolic shape for a point source is

$$y = \frac{x^2}{4f} \tag{8.14}$$

where f is the focal length of the concentrator. If D is the width of the concentrator aperture and d the absorber tube diameter, then the concentrator rim angle, ϕ, is

$$\tan\left(\frac{\phi}{2}\right) = \frac{D}{4f} \tag{8.15}$$

and the geometric CR, C, is

$$C = \frac{D}{\pi d} \tag{8.16}$$

In practice, parabolic troughs use concentrators with rim angles between 70° and 110°, as this range provides a balance between optical and structural requirements of the concentrator.

In addition to the optical error introduced by the finite specularity of the reflective surface, a number of other optical errors contribute to widening of the reflected beam at the receiver, such as concentrator slope error, tracking error, and receiver mislocation errors. Generally, these errors can be characterized by Gaussian distributions, so the total amount of beam spreading can be calculated as the root mean square (rms) value of all the optical errors.

The sun is not a point source, but the sun's finite size can be approximated with an rms width of $\sigma_{sun} = 2.8$ mrad (milli-radians). However, for line focus optics, the apparent width of the sun increases with incidence angle, and a good average of the all-day rms width is 5.0 mrad. The total beam spreading, σ_{total}, due to optical errors and the sun shape can be expressed as

$$\sigma_{total} = \left(\sigma_{optical}^2 + \sigma_{sun}^2 \right)^{1/2} \tag{8.17}$$

The optical error is given by

$$\sigma_{opt} = \left[4\sigma_{con}^2 + \sigma_{track}^2 + \lambda(\theta)(4\sigma_{con}^2 + \sigma_{spec}^2) + \sigma_{disp}^2 + \sigma_{spec}^2 \right]^{1/2} \tag{8.18}$$

where
σ_{con} is the rms angular deviation of concentrator from perfect parabola (slope error)
σ_{track} is the rms angular spread due to sun tracking error
σ_{spec} is the rms angular spread of reflected beam due to imperfect specularity of reflector
σ_{disp} is the equivalent rms angular spread, which accounts for the imperfect placement of the receiver
$\lambda(\theta)$ is the coefficient that accounts for the rim-angle-dependent contribution of longitudinal mirror errors to transverse beam spreading (*Note:* A value of 0.1 is recommended for $\lambda(\theta)$ for concentrators with rim angles between 80° and 100°.)

The first step in a PTSC design is to calculate the intercept factor, γ, defined as the fraction of the reflected beam radiation that is intercepted by the absorber tube. Bendt et al. [20] have shown that the intercept factor γ depends on the product ($\sigma_{total}C$) as well as the concentrator rim angle, ϕ, as shown in Figure 8.17.

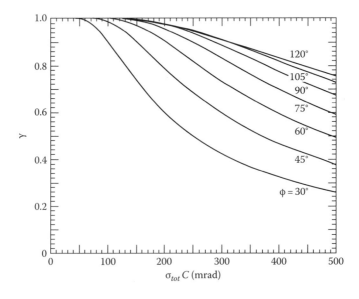

FIGURE 8.17
Intercept factor, γ, vs. product of concentration factor and optional error, $\sigma_{tot}C$, for different rim angles. (From Gee, R., Long term average performance benefits of parabolic trough improvements, SERI, Golden, CO, SERI/TR 632-439, March 1980. With permission.)

Once σ_{total} has been estimated, various CRs (and thus receiver diameters) can be evaluated for their resulting impact on the intercept factor. A reasonable value for the intercept factor from an engineering perspective is 95%, that is, 5% spill over losses.

The absorber/receiver for the parabolic trough system is usually a metal tube surrounded by a glass tube. The heat loss coefficient for this type of receiver/absorber is

$$U_L = \frac{\text{Rate of heat loss}}{(T_{absorber} - T_{amb})A_{absorber}} \tag{8.19}$$

Figure 8.18 shows experimentally measured absorber heat loss coefficients as a function of absorber temperature for various designs.

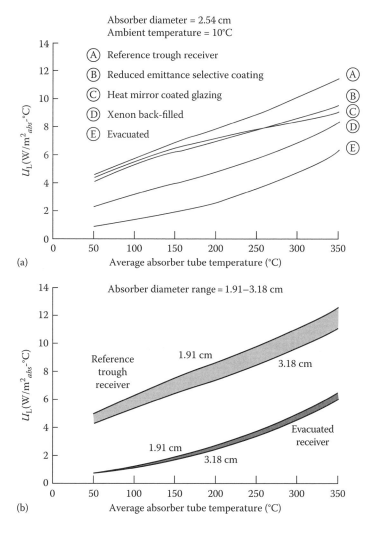

FIGURE 8.18
(a) Absorber heat loss coefficients vs. temperature and (b) heat loss coefficient variance with absorber diameter. (From Gee, R., Long term average performance benefits of parabolic trough improvements, SERI, Golden, CO, SERI/TR 632-439, March 1980. With permission.)

The reference trough receiver is the simplest construction with air in the annulus. The other curves show the improvement, that is the reduction in the heat loss coefficient, with other designs. The annulus between these two tubes can be evacuated (case E) to reduce heat losses to a minimum. This is called an evacuated tube parabolic collector.

To determine the collector efficiency equation, we can use the results of extensive tests conducted by Sandia National Laboratories. The results of these tests, as summarized by Gee and May [22], show that the collector efficiency, η_c, can be expressed by an equation of the form

$$\eta_c = \eta_{opt} - \frac{U_L(T_{absorber} - T_{ambient})}{I} \tag{8.20}$$

where
I is the insolation on the aperture
η_{opt} is the product of the reflectance of the concentrator reflector surface, ρ, the transmittance of the receiver glazing, τ, and the absorptance of the coating on the receiver tube, α

The heat loss coefficient from the absorber tube, U_L, can be related empirically to the absorber temperature, T_a, by the expression

$$U_L = A + B(T_a - T_\infty) \tag{8.21}$$

where
T_∞ is the ambient temperature
A and B are empirical coefficients to describe heat loss from the absorber

These coefficients are usually obtained from test data, as shown in Figure 8.18.

Example 8.1

Derive a relationship for the efficiency of a parabolic collector with a receiver/absorber tube corresponding to the reference case in Figure 8.18 as a function of insolation and temperature difference between the collector surface and ambient. The reflectance of the concentrator surface, ρ, equals 0.92, the transmittance of the glass tube surrounding the inner tube, τ, is 0.94, the absorptance of the coating on the receiver tub, α, e equals 0.95, and the interceptor factor, γ, equals 0.96. Assume that the average direct normal solar irradiance is 560 W/m², which is twice the NREL daily average spread over a 24 h day, $\sigma_{total} = 10.5$ mrad, and the axis of the collector has an east–west orientation.

Solution

The first step in the solution is to select a CR and rim angle for the collector that results in an intercept factor (that is, the portion of reflected sunlight that strikes the absorber tube) of at least 96%. For this selection, we use the results shown in Figure 8.17, which plots the intercept factor (γ) at a function of $\sigma_{tot} \times C$, where σ_{tot} is the total rms beam spread including the sun's angular width, which is taken as 0.6° or 10.5 mrad in accordance with the data. If we select a rim angle of 90° and an intercept factor of 0.96, the product $\sigma_{total} \times C$ is about 200. Hence, the CR equals

$$C = \frac{200}{10.5} = 19$$

Using the information in the problem statement, the optical efficiency equals

$$\eta_{opt} = \rho\tau\alpha\gamma = 0.92 \cdot 0.94 \cdot 0.95 \cdot 0.96 = 0.79$$

For the reference case shown in Figure 8.18, a linear curve fit for the relation between the heat loss coefficient, U_L, and the average absorber tube temperature yields the relationship

$$U_L = 3 + 0.025(T_{absorber} - T_{ambient}) = 3 + 0.025\Delta T$$

Hence, the efficiency for this collector, η_c, can be approximated by the equation

$$\eta_c = \eta_{opt} - \frac{(A + B\Delta T)\Delta T}{I} = 0.79 - \frac{(3 + 0.025\Delta T)\Delta T}{I}$$

where ΔT equals $(T_{absorber} - T_{ambient})$.

Note that the relation for the linear fit is based on a unit receiver area, whereas the efficiency equation is based on a unit aperture area. Since we need a relationship based on an aperture area, U_L needs to be divided by the CR that yields the following relation for the efficiency curve:

$$\eta_c = 0.79 - \frac{\left(\dfrac{3}{19} + \dfrac{0.025}{19}\Delta T\right)\Delta T}{I} = 0.79 - \frac{(0.16 + 0.0013\Delta T)\Delta T}{560}$$

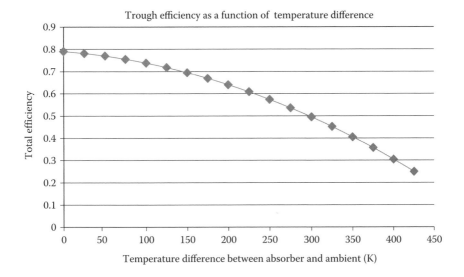

Trough efficiency as a function of temperature difference

Temperature difference between absorber and ambient (K)

The next example illustrates the economics and the EROI for a tracking parabolic trough solar thermal system.

Example 8.2

In 1986, a parabolic trough-type solar water heater was installed for a correctional facility in Boulder, Colorado. The system was designed to deliver 50°C water, and a performance model predicted that if energy for operating and maintenance (O&M) is subtracted from the gross output, a net of 1.1×10^6 kWh/year (3.666×10^9 Btu/year) will be delivered from the solar collector. The solar energy displaced natural gas at $5.00/GJ. The other pertinent information is listed as follows:

Total investment in the first quarter (1986 dollars)	$115,000
Value of fuel displaced in the first quarter (1986 dollars)	$19,250
Discount rate for present worth	10%

For this problem, omit effects of taxes, depreciation, and decommissioning to simplify analysis and emphasize the main points. Then calculate

   a. Simple rate of return
   b. Discounted payback time
   c. Discounted rate of return
   d. Energy payback time
   e. EROI using the cost breakdown in 1986 dollars in Table 8.7
   f. The levelized cost of energy

Solution

   a. The simple rate of return on investment approach does not take into account the time value of money. For this example, the simple rate of return is

$$\frac{\text{Fuel savings per year (1986\$)} \times 100}{\text{Initial investment (1986\$)}} = \frac{\$19,250}{\$115,000} \times 100 = 16.7\%/\text{year}$$

   b. The inverse of the fractional rate of return gives the simple economic payback time as 6 years.
   c. The present worth of the future savings is calculated by taking into account the time value of money and discounting future savings to dollars at the beginning of the project in 1986. Discount factors are shown in Table 8.6. For simplicity, annual discount factors are used, as though the fuel bill was paid only once at the end of each year. The calculations are shown in Table 8.6.

     When the cumulative present worth in 1986 dollars of the energy savings equals the initial investment of $115,000, then the original investment has been paid back in discounted dollars. From Table 8.6, this occurs at 9.6 years and that is the discounted payback time. The discounted return is the average annual present worth of cumulative savings over the payback time, that is, $115,000/9.6 = $11,980/year.
   d. The discounted rate of return is

$$\frac{\$11,980/\text{year}}{\$115,000} \times 100 = 10.4\%/\text{year}$$

   e. For an ER01 analysis of the parabolic trough solar water heating system, we use the estimated 1986 costs of materials and labors to contruct the system

TABLE 8.6

Present Worth Calculations for Example 8.2

Year	Discount Factor[a]	Present Worth of Fuel Savings in 1986 $	
		Annual	Cumulative
1	0.909	17,500	17,500
2	0.826	15,900	33,400
3	0.751	14,460	47,860
4	0.683	13,150	61,010
5	0.621	11,950	72,960
6	0.564	10,850	83,810
7	0.513	9,880	93,690
8	0.466	8,980	102,670
9	0.424	8,160	110,830
10	0.386	7,420	118,250

[a] A discount rate of 10% per year is used. The discount factor is given by $1/(1 + i)^N$, where i is the fractional discount rate, 0.10 per year, and N is the number of years.

are shown in the second column of Table 8.7. Annual O&M cost and energy input range from $0.13 to $1.34/ft^2/year (CGSSC, 1984). An average of 6.4% of the total capital investment has been used here. This value is conservative because it is considerably higher than the 3%/year often assumed for O&M. The energy requirement for O&M, estimated to be 234 million Btu/year, was subtracted from the estimated solar system output of 3900 million Btu/year to give the net energy output of 3666 million Btu/year. The energy embodied in all the materials and in the labor, per dollar of cost, is given in the third column. These values come from input/output energy analysis of the corresponding sector of the economy and are in 1977 dollars [8]. They must be converted to 1986 dollars by the inflation factor of 0.593. The 1986 product of cost and energy intensity gives the energy incorporated into the system with each material or labor item used and is shown in the fourth column. The sum of these individual items gives the total energy embodied in this system. Hence, the energy invested is equal to 5.43×10^9 Btu/year and the rate of return is thus

$$\frac{3.66 \times 10^9 \, \text{Btu}}{5.43 \times 10^9 \, \text{Btu}} \times 100 = 67.5\%/\text{year}$$

f. The levelized cost of energy delivered which corresponds to an energy payback time of 1.48 years when the net energy produced equals that invested. The EROI is the fractional rate of energy return times the project lifetime in years. Thus, for a 15 year life, the EROI is 10:1. The levelized cost of energy delivered by the system over 15 years can be obtained from Equation 2.24. The total life cycle cost presented in the following table is the sum of the initial investment and the total discounted O&M costs. The annual energy output is 1.1×10^6 kWh/year, and from Table 2.5, the capital recovery factor is 0.13147 for a 15 year life, $n = 15$, and a discount rate of 10%, $i = 10\%$.

Year	Investment	Discount Rate	O&M	
0	$115,000.00	0.000	$0	
1	$0.00	0.909	$6,800	
2	$0.00	0.826	$6,181	
3	$0.00	0.751	$5,619	
4	$0.00	0.683	$5,108	
5	$0.00	0.621	$4,644	
6	$0.00	0.564	$4,222	
7	$0.00	0.513	$3,838	
8	$0.00	0.467	$3,489	
9	$0.00	0.424	$3,172	
10	$0.00	0.386	$2,883	
11	$0.00	0.350	$2,621	
12	$0.00	0.319	$2,383	
13	$0.00	0.290	$2,166	
14	$0.00	0.263	$1,969	
15	$0.00	0.239	$1,790	TLCC
Total	$115,000.00		$56,893.47	$171,893

With these values, the levelized cost of energy, LCOE, is

$$\text{LCOE} = \left(\frac{\text{TLCC}}{Q}\right) \times \text{CRF} = \left(\frac{\$171,893}{1.1 \times 10^6 \text{ kWh}}\right) \times 0.13147 = \$0.02/\text{kWh}$$

The system on which this example is based [23] was built in 1986 and continued operation until 2006, when a road was built through the land on which the collector was located and the system had to be dismantled. Hence, the real life of the system was 20 years and could easily have been extended another 10 years. With a 30 year life, the EROI would have been 20:1. This shows that the claim often heard in some circles that solar energy systems take more energy to build than they deliver is patently false!

A significant advance in parabolic reflector technology has recently been reported by [24]. In this innovative parabolic trough technology, the traditional reflecting surface of silvered glass is replaced by a highly reflective and shatterproof silver polymer film, called ReflecTech® Mirror Film, which is laminated to a thin aluminum substrate. This film allows for larger and fewer panel segments than in previous trough designs and reduces the weight, the cost, as well as the amount of material required for the reflective surface. The technology is based upon a patent by NREL and ReflecTech [25]. The trough has undergone extensive tests at the NREL test site, and a preliminary estimate of its efficiency is shown in Figure 8.19.

*8.4 Long-Term Performance of SIPH Systems

In order to assess the economic viability of any solar process, its cumulative energy delivery over its economic life (in years or decades) must be known. It is very difficult to calculate this number accurately since (1) solar systems and their energy delivery are subject

TABLE 8.7

Parabolic Trough System Costs for Example 8.2

Item	Dollar Cost ($86)	Energy Intensity (Btu/$77)	Embodied Energy (10^6 Btu)
Collector materials			
Aluminum	16,250	158,201	1524
Steel	7,530	115,724	517
Borosilicate tubing	5,445	58,724	190
Adhesive	1,422	95,178	80
Wire rope	673	115,724	46
Fasteners	1,449	115,724	99
Solvent	335	171,572	34
Reflective film	18,103	99,470	1067
Nylon washers	135	32,765	3
Selective film	642	66,761	25
Electric motors	311	33,556	6
Pillow block bearings	2,320	42,638	59
Power jack	1,959	28,793	33
Shaft couplings	30	44,247	1
Silicone O-rings	270	53,271	9
Rebar	620	115,724	43
Bolts, nuts, washers	640	44,427	17
Concrete	1,310	216,631	168
Shipping	1,830	29,198	32
Fabrication	14,790	39,556	347
Installation	10,420	39,556	244
Collector total	86,484		4544
Balance of plant			
Steel piping	3,091	115,724	212
Pipe insulation	3,050	79,474	144
Flexhoses	2,610	27,116	42
Glycol	820	95,178	46
Pump and motor	532	27,731	9
Batteries	220	69,024	9
Expansion tank	110	42,621	3
Valves	1,580	36,093	34
Heat exchanger	1,280	42,621	32
Controls	2,300	23,412	32
Security fence	4,120	47,163	115
Plant installation	8,890	39,556	209
Plant total	28,603		887
Grand total	115,087		5431
Operation and maintenance (annual)	7,480	52,770	234

Source: Courtesy of Kenneth May, Private communication.

Notes: Dollar and energy costs for a 10,000 ft² parabolic trough system with north–south orientation and no thermal storage in Denver, Colorado. (A price deflator ratio of 0.593 was used to convert $86–$77.)

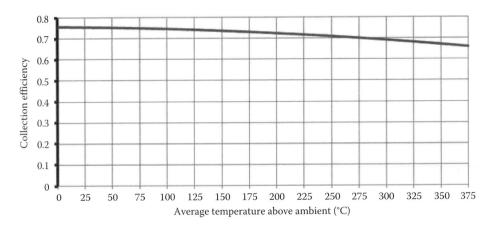

FIGURE 8.19
Projected efficiency of advanced parabolic trough collector. (From Farr, A. and Gee, R., The SkyTrough™ parabolic trough solar collector, in Paper *90090 ASME 3rd International Conference on Energy Sustainability*, San Francisco, CA, 2009. With permission; Courtesy of R. Gee.)

to the vagaries of local microclimate, which can change on a timescale on the order of hours and (2) future weather cannot be predicted at a high level of detail. The standard approach used to estimate future performance of a solar system is to use a typical year of past weather data and assume that it will represent the future on the average, to engineering accuracy. Two common methods used to predict long-term performance of SIPH systems are the utilizability and the TRNSYS methods [15,27]. These methods can also be used for the design of SIPH systems. The following sections describe both methods.

8.4.1 Utilizability Method

Utilizability, ø, has been used to describe the fraction of solar flux absorbed by a collector, which is delivered to the working fluid. On a monthly timescale

$$\bar{\phi} = \frac{\overline{Q_u}}{F\bar{\eta}_o}\bar{I}_c < 1.0 \tag{8.22}$$

where
  Overbars denote monthly means
  Q_u is the monthly averaged daily total useful energy delivery
  $\bar{\phi}$ is the fraction of the absorbed solar flux, which is delivered to the fluid in a collector operating at a fixed temperature T_c

The $\bar{\phi}$ concept does not apply to a system comprised of collectors, storage, and other components wherein the value of T_c varies continuously. The fixed-temperature mode will occur if the collector is a boiler, if very high flow rates are used, if the fluid flow rate is modulated in response to flux variations to maintain a uniform T_c value, or if the collector provides only a minor fraction of the thermal demand. However, if the flow is modulated, note that the value of F (i.e., F', F_R) may not remain constant to engineering accuracy.
  When T_c is not constant in time as in the case of a collector coupled to storage, the $\bar{\phi}$ concept cannot be applied directly. However, for most concentrators with CR > 10, the value

of U_c is small and the collector is relatively insensitive to a *small* change of operating temperatures. To check this assumption for a particular process, values of $\bar{\phi}$ at the extremes of the expected temperature excursion can be compared.

The value of $\bar{\phi}$ depends upon many system and climatic parameters. However, Collares-Pereira and Rabl [28] have shown that only three are of the first order—the clearness index $\bar{K}_T$ (see Chapter 5), the critical intensity ratio $\bar{X}$ (discussed in Section 8.4.2), and the ratio r_d/r_T (discussed in Section 8.4.3). The first is related to insolation statistics, the second to collector parameters and operating conditions, and the last to collector tracking and solar geometry.

8.4.2 Critical Solar Intensity Ratio X

The instantaneous efficiency equation for many solar collectors has been shown to be of the form

$$\eta_c = F\left(\eta_o - \frac{U_c \Delta T^+}{I_c}\right), \quad (\eta_c > 0) \tag{8.23}$$

where
ΔT^+ is the value of a collector to ambient temperature difference if positive η_c is zero otherwise
F is a heat-exchanger factor (F', F_R), the expression for which depends on the definition of ΔT^+

It is possible but not always economical to operate the solar collector system if $\eta_c > 0$. In practice, $\eta_c \geq \eta_{min} > 0$ is usually the system turn-on criterion since it is not worthwhile to operate collector loops for cases where η_c is very small.

Equation 8.23 can be used to determine the solar intensity level above which useful energy collection can take place. Solving Equation 8.23 for I_c, collector output begins when

$$I_c \geq \frac{U_c \Delta T^+}{\left(\eta_o - \eta_{min}/F\right)} \tag{8.24}$$

Since $\eta_{min} < 1$ and F is close to 1, for convenience, the second term in the denominator can be dropped and a dimensionless critical intensity ratio X defined as

$$X \equiv \frac{U_c \Delta T^+}{\eta_o I_c} \leq 1.0 \tag{8.25}$$

X is seen to be the ratio of collector heat loss to absorbed solar flux at $\eta_c = 0$, that is, at the no-net-energy-delivery condition. In many cases, the daily or monthly averaged daily critical intensity ratio $\bar{X}$ is of more interest and is defined as

$$\bar{X} \equiv \frac{U_c \overline{\Delta T^+} \Delta t_c}{\bar{\eta}_o \bar{I}} \tag{8.26}$$

where
$\bar{\eta}_o$ is the daily averaged optical efficiency
ΔT^+ is the daily mean temperature difference *during collection*

These can also be expressed as

$$\overline{\Delta T} = \frac{1}{\Delta t_c} \int_{t_o}^{t_o+\Delta t_c} (T_c - T)dt \tag{8.27}$$

$$\overline{\eta_o} = \frac{\displaystyle\int_{t_o}^{t_o\Delta t} \eta_o I_c dt}{\displaystyle\int_{t_o}^{t_o+\Delta t_c} I_c dt} \tag{8.28}$$

and

$$\overline{I_c} = \frac{1}{\Delta t_c} \int_{t_o}^{t_o=\Delta t_c} I_c dt \tag{8.29}$$

The collector cut-in time t_o and cutoff time $t_o + \Delta t_c$ are described shortly. The time $t = [0, 24]$ h and is related to the solar hour angle h_s by $t = (180 + h_s)/15$; Δt_c is the collection period in hours. In Equation 8.27, T_c can be collector surface, average fluid, inlet fluid, or outlet fluid temperature, depending upon the efficiency data basis.

8.4.3 Collection Period (Δt_c)

The collection period t_c can be dictated either by optical or thermal constraints. For example, with a fixed collector, the sun may pass beyond the acceptance limit or be blocked by another collector, and collection would then cease. Alternately, a high-efficiency solar-tracking concentrator operating at relatively low temperature might be able to collect from sunrise to sunset. A third scenario would be for a relatively low concentration device operating at high temperature to cease to have a positive efficiency during daylight at the time that heat losses are equal to absorbed flux. In this case, the cutoff time is dictated by thermal properties of the collector and the operating conditions.

Collares-Pereira and Rabl [28] have suggested a simple procedure to find the proper value of Δt_c. Useful collection Q_c is calculated using the optical time limit first, that is, $\Delta t_c = 2$ min $\{[h_{sr}(\alpha = 0), h_{sr}(i = 90)]/15\}$. Second, Q_u is calculated for a time period slightly shorter, say by one-half hour, than the optical limit. If this value of Q_u is larger than that for the first optically limited case, the collection period is shorter than the optical limit. The time period is then further reduced until the maximum Q_u is reached.

The earlier method assumes that collection time is symmetric about solar noon. This is almost never the case in practice since the heat collected for an hour or so in the morning is required to warm the fluid and other masses to operating temperature. A symmetric phenomenon does not occur in the afternoon. If the time constant of the thermal mass in the collector loop is known, the collection period may be assumed to begin at t_o, $\Delta t_c/2$ h (from the earlier symmetric calculation), before noon decreased by two or three time constants. Another asymmetry can occur if solar flux is obstructed during low sun angle periods in winter. It is suggested that r_T and r_d from Table 8.8 under asymmetric collection conditions be calculated from

$$r_T = \frac{\left[r_T\left(h_{s,stop}\right) + r_T\left(h_{s,start}\right)\right]}{2} \tag{8.30}$$

TABLE 8.8

Parameters r_T and r_d Used to Calculate Monthly Solar Flux Incident on Various Collector Types[a]

Collector Type	r_T^{b-d}	r_d^e
Fixed aperture concentrators that do not view the foreground	$\left[\dfrac{\cos(L-\beta)}{(d\cos L)}\right]\left\{\begin{array}{l}\left[\dfrac{-ah_{coll}\cos h_{sr}(i=90)+[a-b\cos h_{sr}(i=90)]\sin h_{coll}}{\cos L}\right]\\+\left(\dfrac{b}{2}\right)\sin h_{coll}\cos h_{coll}+h_{coll}\end{array}\right\}$	$\left(\dfrac{\sin h_{coll}}{d}\right)\dfrac{\{[\cos(L+\beta)/\cos L)-[1/CR)]\}+\left(\dfrac{h_{coll}}{d}\right)}{\{[\cos h_{sr}(\alpha=0)/CR)]-[\cos(L-\beta)/\cos L)/\cos h_{st}(i=90)]\}}$
East–west axis tracking[f]	$\dfrac{1}{d}\displaystyle\int_0^{h_{coll}}\left\{\left[\dfrac{(a+b\cos x)}{\cos L}\sqrt{\cos^2 x+\tan^2\delta_s}\right]\right\}dx$	$\dfrac{1}{d}\displaystyle\int_0^{h_{coll}}\left\{\left[\dfrac{(a+b\cos x)}{\cos L}\sqrt{\cos^2 x+\tan^2\delta_s}-\left[\dfrac{1}{(CR)}\right][\cos x-\cos h_{sr}(\alpha=0)]\right]\right\}dx$
Polar tracking	$(ah_{coll}+b\sin h_{coll})/(d\cos L)$	$(h_{coll}/d)\left\{\left(\dfrac{1}{\cos L}\right)+[\cos h_{sr}(\alpha=0)/(CR)]-\sin h_{coll}/[d(CR)]\right\}$
Two-axis tracking	$(ah_{coll}+b\sin h_{coll})/(d\cos\delta_s\cos L)$	$(h_{coll}/d)\{[1/\cos d_s\cos L]+[\cos h_{sr}(a=0)/(CR)]-\sin h_{coll}/[d(CR)]$

[a] The collection hour angle value h_{coll} not used as the argument of trigonometric functions is expressed in radians. Note that the total interval $2h_{coll}$ is assumed to be centered about solar noon.

[b] $a = 0.409 + 0.5016 \sin[h_{sr}(\alpha=0) -60°]$.

[c] $b = 0.6609 - 0.4767 \sin[h_{sr}(\alpha=0)-60°]$.

[d] $d = \sin h_{sr}(\alpha=0) - h_{sr}(\alpha=0)\cos h_{sr}(\alpha=0)$.

[e] CR is the collector concentration ratio.

[f] Elliptic integral tables to evaluate terms of the form of $\displaystyle\int_0^h\sqrt{\cos^2 x+\tan^2\delta_s}\,dx$ contained in r_T and r_d can be found in standard handbooks. Use the identity $\cos\delta_s = \sin(90° - \delta_s)$ and multiply the integral by $\cos\delta_s$, a constant. For computer implementation, a numerical method can be used. For hand calculations, use Weddle's rule or Cote's formula.

$$r_d = \frac{\left[r_d\left(h_{s,stop}\right) + r_d\left(h_{s,start}\right)\right]}{2} \tag{8.31}$$

where the collection starting and stopping hour angles account for transients, shading, etc., as described earlier:

$$h_{s,start} = 180 - 15t_o \tag{8.32}$$

$$h_{s,stop} = 180 - 15(t_o + \Delta t_c) \tag{8.33}$$

8.4.4 Empirical Expressions for Utilizability

Empirical expressions for $\bar{\phi}$ have been developed for several collector types [28].
For nontracking collectors,

$$\bar{\phi} = \exp\left\{-\bar{X} - \left(0.337 - 1.76\bar{K}_T + 0.55r_d/r_T\right)\bar{X}^2\right\} \tag{8.34}$$

$$\text{for } \bar{\phi} > 0.4, \quad \bar{K}_T = [0.3, 0.5], \quad \text{and} \quad \bar{X} = [0, 1.2]$$

and

$$\bar{\phi} = 1 - \bar{X} + \left(0.50 - 0.67\bar{K}_T + \left(0.25r_d/r_T\right)\right)\bar{X}^2 \tag{8.35}$$

$$\text{for } \bar{\phi} > 0.4, \quad \bar{K}_T = [0.5, 0.75], \quad \text{and} \quad \bar{X} = [0, 1.2]$$

The $\bar{\phi}$ expression for tracking collectors (CR > 10) is

$$\bar{\phi} = 1.0 - (0.049 + 1.44\bar{K}_T)\bar{X} + 0.341\bar{K}_T\bar{X}^2 \tag{8.36}$$

$$\text{for } \bar{\phi} > 0.4, \quad \bar{K}_T = [0, 0.75], \quad \text{and} \quad \bar{X} = [0, 1.2]$$

Also

$$\bar{\phi} = 1.0 - \bar{X} \tag{8.37}$$

for $\bar{\phi} > 0.4$, $\bar{K}_T > 0.75$ (very sunny climate), and $\bar{X} = [0, 1.0]$ for any collector type.

Equations 8.34 through 8.37 were developed using curve-fitting techniques emphasizing large $\bar{\phi}$ values since this is the region of interest for most practical designs. Hence, they should be considered accurate to ±5% only for $\bar{\phi} > 0.4$.

In order to use this method to find monthly average daily solar radiation on a tracking solar concentrator, the tilt factor must be continuously changed due to tracking. However, Collares-Pereira and Rabl [28] have given an expression for various modes of tracking that can be used to calculate the average insolation $\bar{H}_c$ given as follows:

$$\bar{H}_c = \bar{r}_T\bar{H}_h - \bar{r}_d\bar{D}_h \tag{8.38}$$

Equations for $\bar{r}_T$ and $\bar{r}_d$ are given in Table 8.8.

An example calculation to illustrate the use of the long-term method will be worked in stepwise fashion. The sequential steps are as follows:

1. Evaluate $\bar{K}_T$ from terrestrial $\bar{H}_h$ data and extraterrestrial $\bar{H}_{o,h}$ data.
2. Calculate r_d/r_T for the CR and tracking mode for the collector.
3. Calculate the critical intensity ratio $\bar{X}$ from Equation 8.26 using a long-term optical efficiency value $\bar{\eta}_o$ and monthly average collector plane insolation:

$$\bar{I}_c = \left(r_T - \frac{r_d \bar{D}_h}{\bar{H}_h} \right) \bar{H}_h \tag{8.39}$$

The collection time Δt_c may need to be determined in some cases for nontracking, low-concentration collectors by an iterative method, as described in the next section.

Example 8.3

Find the thermal energy delivery of a polar-mounted parabolic trough collector operated for 8 h per day ($\Delta t_c = 8$) during March in Albuquerque, New Mexico ($L = 35°$N). The collector has an optical efficiency $\bar{\eta}_o$ of 60%, a heat loss coefficient $U_c = 0.5$ W/m²°C, CR = 20, and heat removal factor $F_R = 0.95$. The collector is to be operated at 150°C. The ambient temperature is 10°C.

Solution

Following the earlier three-step procedure, the clearness index is calculated:

$$\bar{H}_{o,h} = 8.15 \frac{\text{kWh}}{\text{m}^2 \cdot \text{day}}$$

$$\bar{H} = 19.31 \frac{\text{MJ}}{\text{m}^2 \cdot \text{day}} = 5.36 \frac{\text{kWh}}{\text{m}^2 \cdot \text{day}}$$

$$\bar{K}_T = \frac{\bar{H}_h}{\bar{H}_{o,h}} = \frac{5.36}{8.15} = 0.66$$

The geometric factors r_d and r_T are calculated from expressions in Table 8.8:

$$r_T = \frac{(ah_{coll} + b\sin h_{coll})}{d\cos L}$$

$$r_d = \left(\frac{h_{coll}}{d}\right) \times \left(\frac{1}{\cos L} + \frac{\cos h_{sr}(\alpha = 0)}{\text{CR}}\right) - \left(\frac{\sin h_{coll}}{d \times \text{CR}}\right)$$

where

$$h_{coll} = \frac{\Delta t_c}{2} \times 15° = 60° = 1.047 \text{ rad}$$

If the collection period is centered about solar noon,

$$h_{sr}(\alpha = 0) = 90° = 1.571 \text{rad}$$

$$a = 0.409 + 0.5016 \sin 30° = 0.66$$

$$b = 0.6609 - 0.4767 \sin 30° = 0.42$$

$$d = \sin 90° - 1.571 \cos 90° = 1.0$$

in which case

$$r_T = \frac{(0.66 \times 1.047 + 0.42 \times \sin 60°)}{1.0 \cos 35°} = 1.29$$

$$r_d = \left(\frac{1.047}{1.0}\right) \times \left(\frac{1}{\cos 35°} + \frac{\cos 90°}{20}\right) - \left(\frac{\sin 60°}{1.0 \times 20}\right) = 1.23$$

Finally, the critical intensity ratio is

$$\bar{X} = \frac{U_c \overline{\Delta T^+} \Delta t_c}{\bar{\eta}_o \bar{I}_c}$$

and the collector plane insolation $\bar{I}_c$ from Equation 8.39 is

$$\bar{I}_c = (1.29 - 1.23 \times 0.27) \times 5.36 = 5.13 \frac{\text{kWh}}{\text{m}^2}$$

so

$$\bar{X} = \frac{0.5 \times (150 - 10) \times 8}{0.6 \times 5130} = 0.182$$

The utilizability $\bar{\phi}$ from Equation 8.36 is

$$\bar{\phi} = 1.0 - (0.049 + 1.44 \times 0.66)(0.182) + 0.314 \times 0.66 \times 0.182^2 = 0.83$$

Finally, the useful energy is

$$Q_u = F_R \bar{\eta}_o \bar{I}_c \bar{\phi} = 0.95 \times 0.6 \times 5.13 \times 0.83 = 2.43 \frac{\text{kWh}}{\text{m}^2 \cdot \text{day}}$$

for the month of March on the average.

Example 8.4

Calculations in Example 8.3 were based on $\Delta t_c = 8$ h. Repeat for 10 h to see the effect of collection time if a symmetric collection period about noon is used.

Solution

The values of rt and r_d for $h_{coll} = 75° = 1.31$ rad are

$$r_T = \frac{(0.66 \times 1.31 + 0.42 \times \sin 75°)}{1.0 \cos 35°} = 1.55$$

$$r_d = \left(\frac{1.31}{1.0}\right) \times \left(\frac{1}{\cos 35°} + \frac{\cos 90°}{20}\right) - \left(\frac{\sin 75°}{1.0 \times 20}\right) = 1.55$$

The collector plane insolation is then

$$\bar{I}_c = (1.55 - 1.55 \times 0.27) \times 5.36 = 6.06 \frac{\text{kWh}}{\text{m}^2}$$

and

$$\bar{X} = \frac{0.5 \times (150 - 10) \times 10}{0.6 \times 6065} = 0.192$$

Then $\bar{\phi}$ is 0.82 from Equation 8.36, and the useful energy delivery is 2.83 kWh/m² day. Hence, it is worthwhile operating the collector for at least 10 h. The calculation can be repeated by the reader for an asymmetric case 4 h before noon and 6 h after to determine the effect of warm-up.

8.4.5 Yearly Collector Performance

Procedures for determining the daily and monthly thermal collector output have been discussed in Section 8.4.4. However, in many cases, it is the yearly useful energy delivered by a thermal collector that is of interest for economic assessment and evaluation of sustainability.

The yearly total useful energy (Q_{CY}) delivered by a solar collector can be determined if the inlet temperature to the collector is approximately constant during the year. Using results of hour-by-hour simulations for 26 stations in the United States combined with typical mean year radiation data [29], Rabl [30] proposed empirical correlations to determine Q_{CY} for different types of solar collectors. Rabl found that, for a given collector type (fat plate, CPC one-axis tracking, two-axis tracking), it is only necessary to know the latitude, L, the critical radiation level, I_c, and the average clearness index, K, to estimate the yearly collectible solar energy with relative ease. The critical radiation level (or minimum threshold radiation) can be calculated from the Hottel–Whillier–Bliss equation by setting $q_C = 0$. This yields

$$I_C = \frac{U_L(T_{ci} - T_{amb})}{\eta_o} \tag{8.40}$$

where

T_{ci} is the inlet temperature to the collector
T_{amb} is the yearly average ambient temperature
U_L is the collector/receiver heat loss coefficient
η_o is the optical efficiency

For equator-facing fat-plate collectors with tilt equal to latitude, the average beam radiation at normal incidence, I_{bn}, is related to the average annual clearness index, $\bar{K}$, by the empirical correlation

$$I_{bn} = 137\bar{K} - 0.34 \tag{8.41}$$

where I_{bn} is in kW/m². For this type of fat-plate collector, Q_{CY} is given by the following empirical relation:

$$Q_{CY} = A_C F_R \eta_n$$

$$\times \left\{ \begin{array}{l} \left[5.215 + 6.973\tilde{I}_{bn}) + (-5.412 + 4.293\tilde{I}_{bn})L + (1.403 - 0.899\tilde{I}_{bn})L^2 \right] \\ + \left[(-18.596 - 5.931\tilde{I}_{bn}) + (15.498 + 18.845\tilde{I}_{bn})L + (-0.164 - 35.510\tilde{I}_{bn})L^2 \right] I_C \\ + \left[(14.601 - 3.570\tilde{I}_{bn}) + (-13.675 - 15.549\tilde{I}_{bn})L + (1.620 + 30.564\tilde{I}_{bn})L^2]I_C^2 \right] \end{array} \right\} \tag{8.42}$$

where Q_{CY} is in GJ/m² y, I_c and I_{bn} in kW/m², and the latitude, L, in radians.

Similar correlations for other collector types shown in Figure 8.20 are presented in Table 8.9.

It should be noted that, for a given location, Equation 8.42 reduces to a simple quadratic equation of the form

$$Q_{CY} = A_C F_R \eta_n \left(\tilde{a} + \tilde{b} I_C + \tilde{c} I_C \right) \tag{8.43}$$

The corresponding correlation of the yearly collectible energy for flat-plate collectors is given by Equation 8.43.

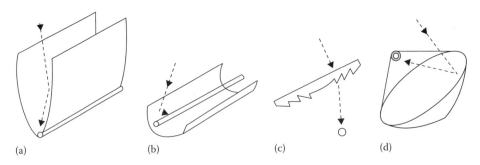

(a)　　　　　　　　　(b)　　　　　　　　　(c)　　　　　　　　　(d)

FIGURE 8.20
Commonly used concentrating collector designs: (a) compound parabolic collectors ($C = 11–12$); (b) parabolic trough ($C = 2–40$); (c) Fresnel lens ($C = 50–500$); and (d) paraboloids ($C = 100–10,000$) (Table 8.9).

TABLE 8.9

Correlations of the Yearly Collectible Energy for Different Collector Types

Collector Type	$\dfrac{Q_{CY}}{A_a F_R \eta_n}$
CPC (geometric concentration ratio $C = 1.5$, acceptance half angle $35°$, tilt equal to latitude, due equator-facing)	$[(1.738 + 11.758\tilde{I}_{bn}) + (1.990 - 8.875\tilde{I}_{bn})L + (-3.236 + 7.617\tilde{I}_{bn})L^2]$ $+ [(-13.240 - 14.688\tilde{I}_{bn}) + (3.979 + 43.653\tilde{I}_{bn})L + (7.345 - 52.556\tilde{I}_{bn})L^2]I_c$ $+ [(14.015 - 1.437\tilde{I}_{bn}) + (-11.884 - 25.852\tilde{I}_{bn})L + (-0.0079 + 39.538\tilde{I}_{bn})L^2]$
Collector with aperture tracking about horizontal east–west axis	$[-0.098 + 11.944\tilde{I}_{bn} -] + [-0.599 - 30.363\tilde{I}_{bn} +]I_c + [1.093 + 17.606\tilde{I}_{bn} - 17.290\tilde{I}_{bn}^{-2} - 17.290\tilde{I}_{bn}^{-2}]\tilde{I}_{bn}^2$
Collector with aperture tracking about horizontal north–south axis	$[(0.640 + 11.981\tilde{I}_{bn}) + (-2.365 + 7.979\tilde{I}_{bn})L + (2.380-)L^2]$ $+ [(-6.021 - 19.086\tilde{I}_{bn}) + (-4.592 + 20.298\tilde{I}_{bn})L + (10.570 - 22.978\tilde{I}_{bn})L^2]I_c$ $+ [(6.440 + 5.219\tilde{I}_{bn}) + (6.986 - 30.500\tilde{I}_{bn})L + (-14.095 + 40.089\tilde{I}_{bn})L^2]I_c^2$
Collector with aperture tracking about polar axis	$[-0.075 + 14.432\tilde{I}_{bn} - 0.592\tilde{I}_{bn}^2] + [-0.780 - 35.67\tilde{I}_{bn} +]I_c + [1.373 + 19.604\tilde{I}_{bn} - 23.965\tilde{I}_{bn}^2]I_c^2$
Two-axis tracker	$[-0.147 + 16.084\tilde{I}_{bn} - 0.7921\tilde{I}_{bn}^2] + [-0.886 - 37.659\tilde{I}_{bn} + 26.983\tilde{I}_{bn}^2]I_c$ $+ [1.700 + 18.883\tilde{I}_{bn} - 24.887\tilde{I}_{bn}^2]I_c^2$

Source: Gordon, J.M. and Rabal, A., *Appl. Opt.,* 31, 7332, 1992. With permission.

Notes: Q_{CY}, Yearly collectible energy (GJ/m²); F_R, collector heat removal factor; η_n, collector optical efficiency at normal solar incidence; A_a, collector net aperture area (m²); $\tilde{I}_{bn}$, annual average beam radiation at normal solar incidence (kW/m²); L, latitude (rad); I_c, critical radiation level defined by Equation 8.39 (kW/m²).

Example 8.5

Obtain a quadratic expression for the yearly total energy delivered by a solar collector located in Athens, Greece ($L = 38°$, which is equal to 0.66 rad), with an average clearness index, $\bar{K} = 0.626$.

Solution

From the data given, we obtain

$$\tilde{I}_{bn} = 1.37 \times 0.626 - 0.34 = 0.518 \text{ kW/m}^2$$

Using the earlier values and L and I_{bn}, Equation 8.43 reduces to

$$\frac{Q_{CY}}{A_C F_R \eta_n} = \frac{7.122 - 13.095 I_C + 6.017}{I_C^2}$$

where
  Q_{CY} is in GJ/(m² y)
  I_C is in kW/m²

Similar quadratic relations can be obtained for other collectors at a given location from the equations in Table 8.9. It should be noted that this procedure is empirical in nature, and its accuracy is limited to approximately ±10%. If higher accuracy is required, the reader is referred to [31] or to obtain daily or monthly average values for the collectible energy.

8.5 Thermal Power Fundamentals

Electricity is fast becoming the energy form of choice all over the world even, unfortunately, for space- and water-heating applications. There are two basic approaches to solar electric power generation. One is by photovoltaic process, a direct energy conversion. The other approach is to convert sunlight to heat and then heat to mechanical energy by a thermodynamic power cycle and, finally, convert the mechanical energy to electricity. This indirect approach, called solar thermal power, is based on well-established principles of fossil fuel thermal power. A vast majority of electricity in the world is produced by thermal power conversion. Most of the thermal power production in the world is based on the Rankine cycle and to a smaller extent on the Brayton cycle. Both of these are applicable to solar thermal power conversion, with Rankine cycle being the most popular. Stirling cycle has also shown great potential, and solar thermal power systems based on this cycle are under development.

8.5.1 Rankine Cycle

Most existing thermal power plants are based on the Rankine cycle. The basic ideal Rankine cycle is shown schematically in Figure 8.21, which also shows a temperature–entropy (T–S) diagram for steam as a working fluid. The ideal cycle consists of the following processes:

Process

  1–2 Saturated liquid from the condenser at state 1 is pumped to the boiler at state 2 isentropically.

  2–3 Liquid is heated in the boiler at constant pressure. The temperature of the liquid rises until it becomes a saturated liquid. Further addition of heat vaporizes the

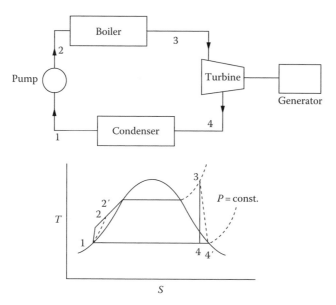

FIGURE 8.21
Basic Rankine power cycle.

liquid at constant temperature until all of the liquid turns into saturated vapor. Any additional heat superheats the working fluid to state 3.

3–4 Steam expands isentropically through a turbine to state 4.

4–1 Steam exiting the turbine is condensed at constant pressure until it returns to state 1 as saturated liquid.

In an actual Rankine cycle, the pumping and the turbine expansion processes are not ideal. The actual processes are 1–2′ and 3–4′ respectively. For the earlier cycle,

$$\text{Turbine efficiency } \eta_{turbine} = \frac{h_3 - h_{4'}}{h_3 - h_4} \tag{8.44}$$

$$\text{Pump efficiency } \eta_{pump} = \frac{h_1 - h_2}{h_1 - h_{2'}} \tag{8.45}$$

$$\text{Net work output} = (h_3 - h_{4'}) - (h_{2'} - h_1) \tag{8.46}$$

$$\text{Heat input} = h_3 - h_{2'} \tag{8.47}$$

$$\text{Pump work} = h_{2'} - h_1 = \frac{v(P_2 - P_1)}{\eta_{pump}} \tag{8.48}$$

$$\text{Cycle efficiency} = \frac{\text{Net work output}}{\text{Heat input}} = \frac{(h_3 - h_{4'}) - (h_{2'} - h_1)}{h_3 - h_{2'}} \tag{8.49}$$

where
  h represents enthalpy
  v is the specific volume at state 1

Example 8.6

In a simple steam Rankine cycle, shown in Figure 8.21, steam exits the boiler at 7.0 MPa and 540°C. The condenser operates at 10 kPa and rejects heat to the atmosphere at 40°C. Find the Rankine cycle efficiency, η_R, and compare it to the Carnot cycle efficiency. Both pump and turbine operate at 85% efficiencies.

Solution

The Rankine cycle efficiency is obtained as follows:

$$\eta_R = \frac{(h_3 - h_{4'}) - (h_{2'} - h_1)}{h_3 - h_{2'}}$$

$$\eta_{turbine} = \frac{h_3 - h_{4'}}{h_3 - h_4}$$

and

$$\eta_{pump} = \frac{h_2 - h_1}{h_{2'} - h_1} = \frac{v_1(P_2 - P_1)}{h_{2'} - h_1}$$

The enthalpies at the state points are found from standard steam tables.

$$h_1 = h_f(10 \text{ kpa}) = 191.8 \text{ kJ/kg}$$

Pump work from 1 to 2'

$$_1W_{2'} = \frac{v_1(P_2 - P_1)}{\eta_{pump}} = h_{2'} - h_1$$

$$h_{2'} = h_1 + \frac{v_1(P_2 - P_1)}{\eta_{pump}}$$

$$= 191.8 \text{ kJ/kg} + \frac{(0.00101 \text{ m}^3/\text{kg})(7000 - 10) \text{ kpa}}{0.85}$$

$$= 200.1 \text{ kJ/kg}$$

$$h_3(540°C, 7 \text{ MPa}) = 3506.9 \text{ kJ/kg}$$

States 3 and 4 have the same entropy.
Therefore,

$$s_3 = 6.9193 \text{ kJ/kg K} = s_4$$

Saturated vapor entropy at state 4 (10 kpa),

$$s_{4g} = 8.1502 \text{ kJ/kg K}$$

Since s_4 is less than s_{4g}, 4 is a wet state

$$s_4 = s_t + X s_{fg} \quad \text{or} \quad s_g - M s_{fg}$$

where X is the vapor quality and M is the moisture.

$$M_4 = \frac{s_g - s_4}{s_{fg}} = \frac{8.152 - 6.9193}{7.5009} = 0.1641 \text{ or } 16.41\%$$

$$h_4 = h_g - M h_{fg} = 2584.7 - 0.1641(2392.8) = 2192 \text{ kJ/kg}$$

Therefore,

$$h_{4'} = h_3 - \eta_{turbine}(h_3 - h_4)$$

$$= 3507 - 0.85(3507 - 2192) = 2389 \text{ kJ/kg}$$

Actual moisture at the turbine exhaust is

$$h_{4'} = h_g - M h_{fg}$$

$$M_{4'} = \frac{h_g - h_{4'}}{h_{fg}} = \frac{2584.7 - 2389}{2391}$$

$$= 0.0817 \text{ or } 8.17\%$$

$$\text{Net work} = (h_3 - h_{4'}) - (h_2 - h_1) = 1109 \text{ kJ/kg}$$

$$\eta = \frac{\text{Net work}}{h_3 - h_{2'}} = \frac{1109}{3507 - 200} = 0.3354 \text{ or } 33.54\%$$

$$\eta_{Carnot} = \frac{813 \text{ K} - 313 \text{ K}}{813 \text{ K}} = 0.615 \text{ or } 61.5\%$$

Several improvements can be made to the basic Rankine cycle in order to increase the cycle efficiency. For example, the pressure in the boiler can be increased. However, that will result in increased moisture in the steam exiting the turbine. In order to avoid this problem, the steam is expanded to an intermediate pressure and then reheated in the boiler. The reheated steam is expanded in the turbine until the exhaust pressure is reached.

Another improvement to the basic Rankine cycle is the regenerative cycle in which expanded steam is extracted at various points in the turbine and mixed with the condensed water to preheat it in the feedwater heaters. These modifications to improve the performance of the basic Rankine cycle are treated in standard thermodynamics texts such as [32] or *Mechanical Engineering Handbooks*.

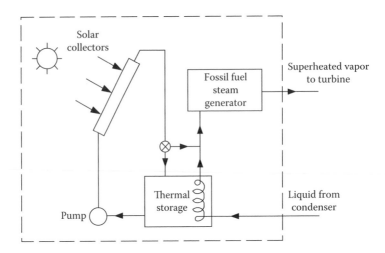

FIGURE 8.22
Schematic of a solar boiler.

8.5.2 Components of a Solar Rankine Power Plant

Major components of a Rankine power plant include boiler, turbine, condenser, pumps (condensate pump, feedwater booster, boiler feed pump), and heat exchangers (open heaters and closed heaters). All of the components of a solar thermal power plant are the same as those in a conventional thermal power plant except the boiler. The boiler in a solar thermal power plant includes a solar collection system, a storage system, an auxiliary fuel heater, and heat exchangers. Figure 8.22 shows a schematic representation of a solar boiler. The maximum temperature from the solar system depends on the type of solar collection system (PTCs, central receiver with heliostat field, parabolic dishes, etc.). If the temperature of the fluid from the solar system/storage is less than the required temperature for the turbine, an auxiliary fuel is used to boost the temperature. A fossil fuel or a biomass fuel may be used as the auxiliary fuel.

A condenser is a large heat exchanger that condenses the exhaust vapor from the turbine. Steam turbines employ surface-type condensers, mainly shell and tube heat exchangers operating under vacuum. The vacuum in the condenser reduces the exhaust pressure at the turbine blade exit to maximize the work in the turbine. The cooling water from either a large body of water such as a river or a lake, or from cooling tower, circulates through the condenser tubes. The cooling water is cooled in the cooling tower by evaporation. The airflow in the cooling tower is either natural draft (hyperbolic towers) or forced draft (see Figure 8.23). The condensate and feedwater pumps are motor-driven centrifugal pumps, while the boiler feed pumps may be motor- or turbine-driven centrifugal pumps.

Steam turbines and generators are described in detail in a number of thermodynamics books [33,34] and are not discussed here.

8.5.3 Choice of Working Fluid

Working fluid in a solar Rankine cycle is chosen based on the temperature from the solar collection system. The working fluid must be such that it optimizes the cycle efficiency based on the expected temperature from the source. Steam is the most common working fluid in a Rankine cycle. Its critical temperature and pressure are 374°C and 22.1 MPa.

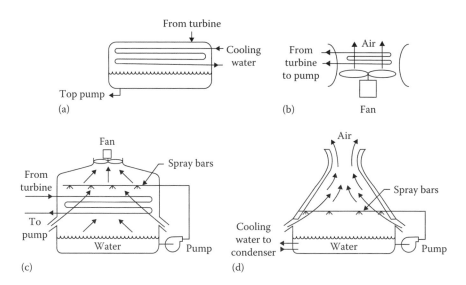

FIGURE 8.23
Types of condenser and/or heat rejection used in Rankine cycle solar power systems: (a) tube-and-shell condenser; (b) dry cooling tower; (c) wet cooling tower; and (d) natural-draft cooling tower. (Adapted from Vant-Hull, L.L., *Solar Today*, November/December issue, pp. 13–16, 1992. With permission.)

Therefore, it can be used for systems operating at fairly high temperatures. Systems employing parabolic trough, parabolic dish, or central receiver collection systems can use steam as a working fluid. Other major advantages of steam are that it is nontoxic, environmentally safe, and inexpensive. Its major disadvantage is its low molecular weight, which requires very high turbine speeds in order to get high turbine efficiencies.

Steam is a wetting fluid, that is, as it is expanded in a turbine, once it reaches saturation, any further expansion increases the moisture content. In other words, steam becomes wetter as it expands, as shown in Figure 8.24a. On the other hand, a fluid that has a *T–S* (temperature–entropy) diagram similar to that shown in Figure 8.24b is called a drying fluid. As seen from this figure, even though the working fluid passes through the two-phase region, it may exit the turbine as superheated. Normally, the turbine speed is so high that, in such a case, there is no condensation in the turbine. Examples of drying fluids include hydrocarbons (toluene, methanol, isobutane, pentane, hexane), chlorofluorocarbons (CFCs such as R-11, R-113), and

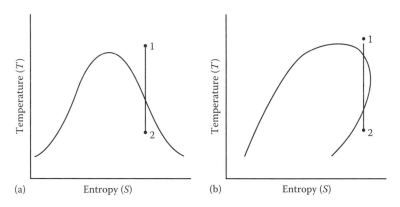

FIGURE 8.24
T–S characteristics of (a) wetting and (b) drying types of working fluids.

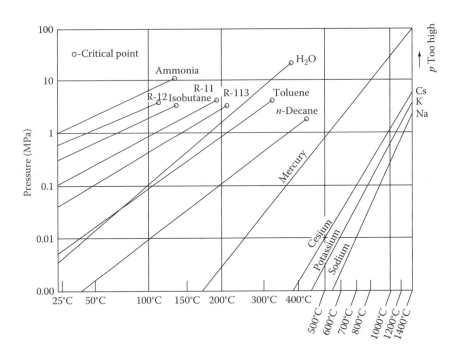

FIGURE 8.25
Saturation pressure–temperature relationships for potential Rankine cycle working fluids. (Adapted from Vant-Hull, L.L., *Solar Today*, November/December issue, pp. 13–16, 1992. With permission.)

ammonia. Since a drying fluid does not get wetter on expansion from a saturated vapor condition in an ideal or real process, it does not have to be superheated. Therefore, a Rankine cycle using a drying fluid may be more efficient than the cycle using a wetting fluid. In fact, a drying fluid may be heated above its critical point so that that upon expansion, it may pass through the two-phase dome. Because of the *T–S* characteristics, the fluid may pass through the two-phase region and still exit from the turbine as superheated. These characteristics can be used to increase the resource effectiveness of a cycle by as much as 8% [35].

Figure 8.25 and Table 8.10 give some characteristics of candidate working fluids for solar Rankine power cycles.

8.6 Solar Thermal Power Plants

There are several successful solar thermal power plants working on Rankine and Stirling cycles, using parabolic troughs, parabolic dishes, and central receiver towers. An example of each type is treated as follows [37].

8.6.1 Parabolic Trough–Based Power Plant

Luz Corporation developed components and commercialized PTC based solar thermal power by constructing a series of hybrid solar power plants from 1984 to 1991. Starting with their first 14 MW$_e$, solar electric generating station (SEGS I) in Southern California, Luz

TABLE 8.10

Physical and Thermodynamic Properties of Prime Candidate Rankine Cycle Working Fluids

Property	Water	Methanol	2-Methyl Pyridine, H_2O	Fluorinol 85	Toluene	R-113	Ammonia	Isobutane
Molecular weight	18	32	33	88	92	187	17	58
Boiling point (1 atm) (°C)	100	64	93	75	110	48	−330	−12
Liquid density (kg/m³)	999.5	749.6	934	1370	856.9	2565	682	594
Specific volume (saturated vapor at boiling point) (m³/kg)	1.69	0.80	0.87	0.31	0.34	0.14	1.124	0.35
Maximum stability temperature (°C)	—	175–230	370–400	290–330	400–425	175–230	300[a]	>200
Wetting–drying	W	W	W	W	D	D	D	D
Heat of vaporization at 1 atm (kJ/kg)	2256	1098	879	442	365	1370	1370	367
Isentropic enthalpy drop across turbine (kJ/kg)	348–1160	162–302	186–354	70–186	116–232	23–46	200–600	120–380

Source: Adapted from Kolb, G.J., *Sol. Energy*, 62(1), 51, 1998. With permission.

[a] Anhydrous ammonia in the presence of iron. Small trace of water increases this limit.

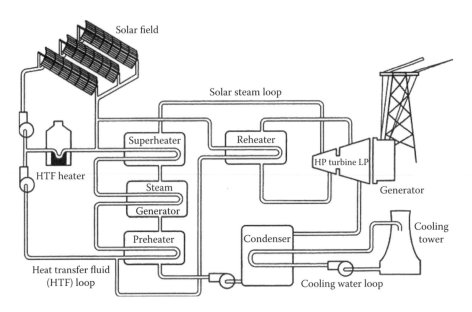

FIGURE 8.26
Flow of heat-transfer fluid through the SEGS VIII and IX plants. (Adapted from de Laquil, P. et al., *Solar Thermal Electric Technology in Renewable Energy*, Island Press, Washington, DC, 1993.)

added a series of SEGS power plants with total generating capacity of 354 MW$_e$. Figure 8.26 shows a schematic of the SEGS VIII and IX plants. Table 8.11 provides information about the solar field, power block, and other general information about SEGS I and SEGS IX plants. All of these plants use solar energy to heat the working fluid and natural gas as the auxiliary fuel to achieve the desired temperature. On average, 75% of the energy is obtained from the sun and 25% from natural gas. With power plant electrical conversion efficiencies on the order of 40% and the solar field efficiencies of 40%–50%, overall efficiencies for solar-to-electricity conversion on the order of 15% are being achieved in these plants. It is claimed that the cost of electricity from these plants has decreased from about 30 ¢/KWh to less than 10 ¢/KWh.

Recently, several solar trough power plants similar to the SEGS installation have been built and placed into operation. A notable example is Nevada Solar One. The plant is not 100% renewable, but has, in accordance with Nevada state law, a maximum of 2% fossil fuel backup from a small natural gas heater, which is also used to prevent the water from freezing and provide power on cloudy days or during peak demand periods.

Nevada Solar One went online on June 27, 2007, on the southeast fringe of Boulder City, Nevada. The plant has a nominal capacity of 64 MW and a maximum capacity of 75 MW. The project required an investment of $266 million, and electricity production is estimated to be 134 million kWh/year (http://www.acciona.us/Business-Divisions/Energy/Nevada-Solar-one). The time of construction for the plant was about 16 months, and it occupies approximately 400 acres (1.6 km²), while the solar collectors cover about 1.2 km². Nevada Solar One uses 760 parabolic troughs that concentrate solar radiation onto evacuated tube collectors that heat the heat-transfer fluid to 391°C (730°F). The heat-transfer fluid is a molten salt, which passes through a heat exchanger to produce steam that drives a conventional 70 MW reheat steam turbine.

TABLE 8.11

Important Characteristics of SEGS I to SEGS IX Plants

	Units	I	II	III	IV	V	VI	VII	VIII	IX
Power										
Turbine-generator output	Gross MW$_e$	14.7	33	33	33	33	33	33	88	88
Output to utility	Net MW$_e$	13.8	30	30	30	30	30	30	80	80
Turbine generator set										
Solar steam conditions										
Inlet pressure	Bar	35.3	27.2	43.5	43.5	43.5	100	100	100	100
Reheat pressure	Bar	0	0	0	0	0	17.2	17.2	17.2	17.2
Inlet temperature	°C	415	360	327	327	327	371	371	371	371
Reheat temperature	°C	N/A	N/A	N/A	N/A	N/A	371	371	371	371
Gas-mode steam conditions[a]										
Inlet pressure	Bar	0	105	105	105	105	100	100	100	100
Reheat pressure	Bar	0	0	0	0	0	17.2	17.2	17.2	17.2
Inlet temperature	°C	0	510	510	510	510	510	510	371	371
Reheat temperature	°C	N/A	N/A	N/A	N/A	N/A	371	371	371	371
Electrical conversion efficiency										
Solar mode[b]	Percent	31.5	29.4	30.6	30.6	30.6	37.5	37.5	37.6	37.6
Gas mode[c]	Percent	0	37.3	37.3	37.3	37.3	39.5	39.5	37.6	37.6
Solar field										
Solar collector assemblies										
LS 1 (128 m²)		560	536	0	0	0	0	0	0	0
LS 2 (235 m²)		0	0	0	0	32	0	184	852	888
LS 3 (545 m²)		0	0	0	0	32	0	184	852	888
Number of mirror segments		41,600	96,464	117,600	117,600	126,208	96,000	89,216	190,848	198,912
Field aperture area	m²	89,960	190,338	230,300	230,300	250,560	188,000	194,280	464,340	483,960
Field inlet temperature	°C	240	231	248	248	248	293	293	293	293
Field outlet temperature	°C	307	321	349	349	349	390	390	390	390
Annual thermal efficiency	Percent	35	43	43	43	43	42	43	53	50
Peak optical efficiency	Percent of peak	71	71	73	73	73	76	76	80	80

(continued)

TABLE 8.11 (continued)

Important Characteristics of SEGS I to SEGS IX Plants

	Units	I	II	III	IV	V	VI	VII	VIII	IX
System thermal losses	Percent of peak	71	71	73	73	73	76	76	80	80
System thermal losses		17	12	14	14	14	15	15	15	15
Heat-transfer fluid type										
Inventory		Esso 500	VP-1	VP-1	VP-1	VP-1	VP-1	VP-1	VP-1	VP-1
Thermal storage capacity	m³	3.213	416	403	403	461	416	416	1,289	1,289
	MWh$_t$	110	0	0	0	0	0	0	0	0
General										
Annual power outlet	Net MWh/year	30,100	80,500	91,311	91,311	99,182	90,850	92,646	252,842	256,125
Annual power use	10⁹ m³/year	4.76	9.46	9.63	9.63	10.53	8.1	8.1	24.8	25.2

[a] Gas superheating contributes 18% of turbine inlet energy.
[b] Generator gross electrical output divided by solar field thermal input.
[c] Generator gross electrical output divided by thermal input from gas-fired boiler or HTF-heater.

Example 8.7

The parabolic trough solar collector, whose efficiency was analyzed in Example 8.1, is to be used for electric power production. Assuming that the efficiency of the power generation unit equals one-half of the Carnot efficiency, determine

a. The temperature at which the solar trough power generation system achieves its highest efficiency
b. The levelized cost of energy from this trough generation system assuming a life of 25 years and an effective discount rate of 3%, and assuming that the total installed system cost is $600 per m² of collector aperture area. This includes the cost of the installed solar collector field, as well as the power block and balance of system (turbine, cooling tower, power transformers/interconnection, etc.).

Solution

According to thermodynamic principles, the conversion of thermal energy into electric energy increases in efficiency as the temperature of the thermal source increases. But Equation 8.20 shows that the efficiency of the collector decreases with increasing operating temperature. In order to obtain the optimum operating temperature, both of these relations need to be taken into account.

If we assume that the high-temperature source in the Carnot efficiency equals the operating temperature of the receiver/absorber, T_a, and the ambient temperature is 20°C, the electric power output, P_e, is equal to

$$P_e = I\eta_c \times 0.5\left(\frac{\Delta T_a}{T_\infty}\right) = I\left[0.79 - \frac{(0.16 + 0.0013\Delta T)\Delta T}{I}\right] 0.5\left(\frac{\Delta T_a}{T_\infty}\right)$$

P_e is shown in the following plot as a function of operating temperature. To get the annual energy collected into kWh/m²/year, it is necessary to multiply the third column by the number of daylight hours during the year, which is assumed to be 12 h/day.

Operating Temperature (°C)	Linearized U_L (W/m²-C)	Annual Coll. (W/m²)	Annual Coll. (kWh/m²/Year)	50% of Carnot	Annual Elec. (kWh/m²/Year)
100	0.26	421.35	1846.77	0.11	198.04
125	0.30	411.31	1802.79	0.13	237.81
150	0.33	399.64	1751.61	0.15	269.16
175	0.36	386.31	1693.22	0.17	292.91
200	0.39	371.35	1627.62	0.19	309.69
225	0.43	354.74	1554.81	0.21	320.02
250	0.46	336.48	1474.79	0.22	324.28
275	0.49	316.58	1387.56	0.23	322.84
300	0.53	295.03	1293.12	0.24	315.95
325	0.56	271.84	1191.48	0.26	303.85
350	0.59	247.01	1082.62	0.26	286.73
375	0.63	220.53	966.56	0.27	264.76
400	0.66	192.40	843.29	0.28	238.08

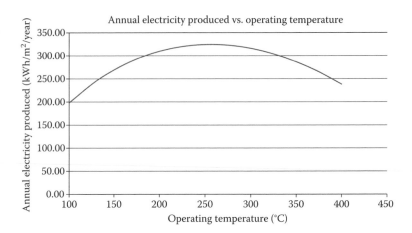

We observe that the peak annual electric generation occurs at about 250°C. The annual electricity generation at the optimum operating temperature, as shown in the last column, is approximately 324 kWh/m²/year.

To calculate the LCOE, we first assume that annual operation and maintenance costs are insignificant compared to the capital cost of the system. The levelized cost of energy delivered by the system over 25 years obtained from Equation 2.24 is

$$LCOE = \left(\frac{TLCC}{Q}\right) \times CRF = \left[\frac{600 \ \$/m^2}{324 \ kWh/m^2}\right] \times 0.064 = \$0.12/kWh$$

An advanced version of a solar trough and a line focus solar thermal power system are shown in Figures 8.27 and 8.28.

Line focus systems

Parabolic trough

- Maximum operating temp. 393°C
- Turbine–steam rankine cycle
- Organic heat transfer fluid (HTF) in solar collector field

(a)

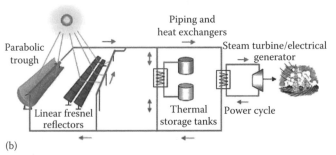

(b)

FIGURE 8.27
(See color insert.) (a) Photograph of single axis-tracking parabolic trough collector field. (b) Schematic of parabolic trough line focus solar thermal power generation system. (Courtesy of Clifford Ho, Sandia National Laboratories.)

FIGURE 8.28
(See color insert.) Sky Fuel solar sky trough near Golden, Colorado, viewed by Professor Frank Kreith (right) and National Institute of Standards and Technology researcher, Dr. Isaac Garaway (left). (Courtesy of CU Engineering, 2009, Casey A. Cass, photographer.)

The following example illustrates how the System Advisor Model (SAM) can be used to analyze the performance of a parabolic trough power system.

Example 8.8

The empirical parabolic trough model in SAM uses a set of equations based on empirical analysis of data collected from installed systems (the SEGS projects in the southwestern United States) to represent the performance of parabolic trough components. One of these systems is a SAM/CSP Trough Power Cycles/SEGS 30 MWe Turbine.

NOTE: Many of the input variables in the parabolic trough model are interrelated and should be changed together. For example, the storage capacity, which is expressed in hours of thermal storage, should not be changed without changing the tank heat loss value, which depends on the size of the storage system. Some of these relationships are described in this documentation, but not all. If you have questions about parabolic trough input variables, please contact SAM Support at sam.support@nrel.gov.

Evaluate the performance and cost of a 30 MWe solar trough power system in Daggett, California. The trough system is to be 200 m in length, 5 m in width, and has an aperture area of 940.6 m². For this problem, keep the default values for the mirror characteristics such as the dust on envelope and the cleanliness factor (see Equation 8.18). The heat collection element is the receiver tube shown in Figure 8.15. For this system, a 2008 Schott PTR70 Vacuum tube was selected. The heat loss coefficient in Equation 8.19 is provided by the manufacturer and given in SAM. This system has a power cycle similar to SEGS VI shown in Figure 8.26. The turbine specifications can be found in Table 8.11. The solar multiple, which is the solar field area divided by the exact area, which is the solar field area required to deliver sufficient solar energy to drive the power block at the turbine gross output level, is 1.7. For this design, no storage is provided. For the financial estimates, the real discount rate is 6%, the inflation rate is 2%, and the loan period is 30 years. Assume that the property tax is zero and use the default values for other financial parameters.

Note that the turbine performance parameters are measured values provided by the manufacturer, and heat loss coefficients for piping between the collector and the power block are either calculated or measured. For the values that are not specified in the problem such as heat-transfer fluid properties, use the default values given by SAM.

Solution
To evaluate the performance of the specified plant,

1. Start SAM
2. Under **Enter a new project name to begin**, type a name for your project. For example, "Parabolic Trough System."
3. Click Create New File
4. Under 1. Select a technology, click Concentrating Solar Power
5. Click Empirical Trough System
6. Under **2. Select a financing option**, click on **Commercial** financing option
7. Click **OK**

SAM creates a new. zsam file with a single case populated with default input values for a 100 MW parabolic trough system.

Start by specifying Daggett, California, as the location of the plant under the "Climate" tab. Then, the power block parameters have to be defined. Under the "Power Block" tab, the circled values in the following figure were changed based on the information given by the problem statement. From Table 8.11, the SEGS VI turbine has a gross output of 33 MWe and a net output of 30 MWe, which results in a 0.9 conversion factor. The power block for the system was chosen from the SAM library.

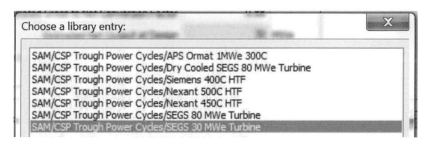

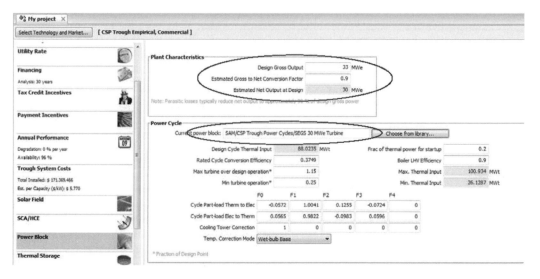

Then, the solar collector assembly parameters need to be adjusted according to the problem figures. The circled numbers in the following figure show the changes that were made under the "SCA/HCE" tab. The heat collection element was chosen from the SAM library. The parameters for this evacuated collector tube, as mentioned in the problem statement, have been provided by the manufacturer.

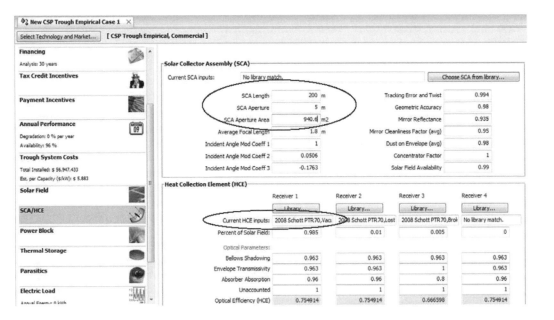

Then, some of the solar field parameters were changed according to the problem. The circled numbers in the following figure were adjusted under the "Solar Field" tab.

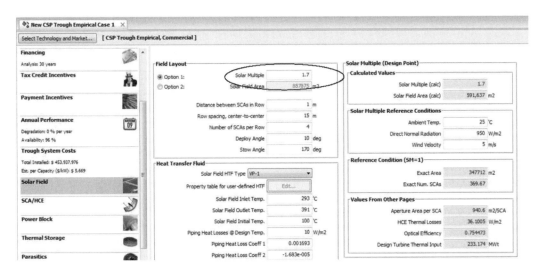

It was assumed that there was no thermal storage, thus the default value circled in the figure was changed to zero. This change reduces the cost, but restricts the system's availability.

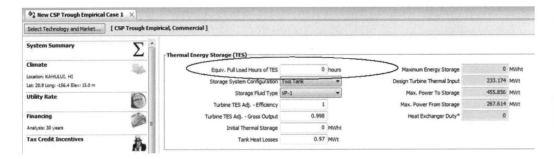

Under the "Financing" tab, the following circled values were changed according to the problem statement.

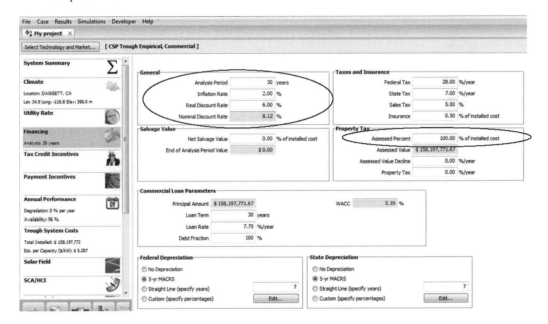

Once the green arrow is clicked, the following data table and figures appear.

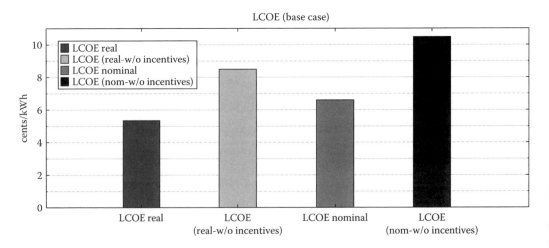

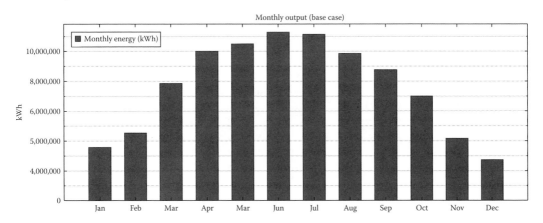

Metric	Base
Net Annual Energy	87,572,396 kWh
LCOE Nominal	6.53 ¢/kWh
LCOE Real	5.28 ¢/kWh
First Year Revenue without System	$0.00
First Year Revenue with System	$ 10,508,687.53
First Year Net Revenue	$ 10,508,687.53
After-tax NPV	$ 31,309,245.23
Payback Period	12.0331 years
Capacity Factor	33.4 %
Gross to Net Conv. Factor	0.89
Total Land Area	238.60 acres

To get the monthly thermal output of the system, click on the graph called "Monthly Output (Base Case)" on the bottom of screen.

For thermal applications such as industrial process heat (see Section 8.2), you have to go through the design as outlined earlier for an electrical system and then use the thermal output, which is an intermediate step in SAM. If the thermal output is specified, iteration may be necessary. SAM does not design a system for a specified thermal output directly. In the future, SAM may be expanded to apply for solar thermal processes.

8.6.2 Central Receiver Systems

In a power tower or central receiver system (CRS), the solar receiver is mounted on top of a tower, and solar radiation is concentrated on its surface by means of a large paraboloid consisting of a field of individual collector/reflector modules called heliostats, as shown in Figure 8.29.

FIGURE 8.29
Photograph of Solar One central receiver pilot plant in Barstow, California. (From Goswami, Y. et al., *Principles of Solar Engineering*, 2nd edn., Taylor & Francis, Philadelphia, PA, 2000. With permission.)

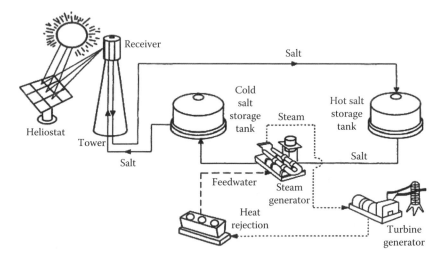

FIGURE 8.30
Schematic of solar central receiver plant configuration with thermal storage. (From Goswami, Y. et al., *Principles of Solar Engineering*, 2nd edn., Taylor & Francis, Philadelphia, PA, 2000. With permission.)

CRS has a great potential for future electricity production because it can achieve high temperatures in a working fluid, which leads to high thermodynamic efficiencies of power generation. Figure 8.30 is a schematic diagram of a CRS with thermal energy storage. Heliostats, shown in Figure 8.31, direct the sunlight to the receiver on top of the tower. A heat-transfer fluid is pumped through a riser to the receiver where it is heated and then via a downcomer to the thermal storage unit and/or a heat engine, which is usually located on the ground near the tower. Typical optical concentration factors range from 200 to 1000, and plant sizes from 10 to 200 MW$_e$ are typical for future systems. The heat flux impinging on the receiver

Heliostats

FIGURE 8.31
(See color insert.) Heliostats for power tower.

can be between 300 and 1000 kW/m^2 and allows the working fluid to achieve temperatures up to 1000°C. CRS can achieve high annual capacity factors, either by using thermal storage or by operating in a hybrid mode with a fossil fuel supplement [38]. With storage, CRS solar plants have been shown to operate more than 4500 h/year at nominal power [39].

The technology favored by industry in the United States is based on molten salt technology that utilizes large thermal storage capacity with molten nitrate salts as the working fluid. The decision to use molten salt was based on a 3 year long utility study designed to compare efficiencies and costs of various options [40]. The recommendations are based on experience with a 20 MW central receiver power plant, called Solar Two, which operated in California in 1999. Solar Two demonstrated advanced molten salt technology at a scale sufficient to allow commercialization of the technology. The plant operated successfully with solar energy collected efficiently over a broad range of operating conditions and showed that low-cost thermal energy storage systems can operate reliably and efficiently. The unique thermal storage capability allowed solar energy to be collected and stored under high insolation conditions so that electric power could also be generated at night or whenever required by the utility, even when the sun was not shining. Figure 8.30 is a schematic diagram of the molten salt CRS system, with cold and hot storage tanks, and Table 8.12 shows the characteristics of four sizes of molten salt CRS. Solar 100, shown in the fourth column, is based on a conceptual design of a 100 MW$_e$ commercialized CRS.

Two types of receiver configurations have been considered for a molten salt power plant. The cavity receiver consists of an insulated enclosure within which a heat-absorbing structure is located. Radiation enters through the aperture, is subjected to multiple reflection, and thus enhances the effective absorptivity of the cavity. Radiation losses can occur only through the aperture, and the absorbing area of the receiver can be enlarged without substantially increasing the radiation losses. However, convective losses increase as a result of a larger heated area, and cavity receivers tend to be large and heavy. Moreover, the aperture size restricts the field of view, and spillage can be considerable. In a plain or fat-plate

TABLE 8.12

Characteristics of Four Sizes of Molten Salt Power Towers

Parameter	Solar Two (Mature)	Solar Tres	Solar 50/Solar Cuatro	Solar 100
Plant rating	10	15	50	100
Location	Barstow, California	Córdoba, Spain	Southern Spain	Southwestern United States
Annual solar insolation, kWh/m²	2,700	2,067	2,067	2,700
Capacity factor, %	20	65	69	70
Field area, m²	81,400	263,000	971,000	1,466,000
Receiver thermal rating, MW	42	120	466	796
Thermal storage size, MWh	110	610	1,850	3,820
Steam generator rating, MW	35	37	130	254
Annual net energy production, MWh	16,600	75,500	302,000	613,000
Peak net efficiency	0.13	0.19	0.22	0.22
Annual net efficiency	0.08	0.14	0.15	0.16
Levelized energy cost, $/kWh	N/A	0.16	0.12	0.08

Source: Romero, M. et al., *J. Sol. Energy Eng.*, 124, 98, May 2002. With permission.

receiver, the aperture of a cavity is replaced by a fat heat-exchanging surface. The backside can be insulated, but overall losses are appreciable.

The external receiver consists of a cylinder with a tubular structure containing a working fluid passing through the pipes on the outside of a cylinder. It can receive solar radiation from all directions and is relatively simple and lightweight because no enclosure is required. The design can accommodate heat flux densities as high as 850 kW/m² with molten salt, and the tower needs to be only about 70% as tall as the tower for a cavity receiver, which reduces the cost appreciably. The disadvantage of an external receiver is that the reflected fraction of incident radiation is lost and the temperature of the molten salt heat-transfer fluid must be maintained above a certain limit to avoid freezing. This is accomplished, as shown in Figure 8.30, by having hot and cold storage tanks [41].

The sizing procedure of the receiver from optical and geometric consideration is presented in [42]. However, the details of the optical design for a CRS are beyond the scope of this book. But from the engineering perspective of sustainable energy design, it is possible to estimate the performance of a typical CRS thermodynamically, including the size of the receiver, the number of heliostats required, the height of the tower, and the ground area necessary for a given power output as well as the EROI for the system. In any high-temperature concentrating system, the receiver is the focal point of the design. Because radiation and convection losses are proportional to the area of the receiver, it should be as small as possible, which requires high concentration. But the receiver must also be large enough to intercept and absorb a large fraction of the solar radiation directed toward it by the heliostats and transfer the heat from the sun to a working fluid. If the designer specifies a more perfect optical system to reduce the receiver size, this will increase the cost of the heliostats and may complicate the thermal design.

The efficiency of the receiver η_{th} equals [1 − (thermal loss/solar input)], and it is given by the following equation:

$$\eta_{th} = 1 - \frac{A_{receiver}\left[\varepsilon\sigma(T^4 - T_a^4) + c(T - T_a)^{1.25}\right]}{G_b A_{collector}\eta_{col}\eta_{int}\alpha} \tag{8.50}$$

or

$$\eta_{th} = 1 - \frac{\sigma\left(T^4 - T_a^4\right) + (c/\varepsilon)(T - T_a)^{1.25}}{G_b C \eta_{col} \eta_{int}} \tag{8.51}$$

where

$$C = \frac{\alpha A_{collector}}{\varepsilon A_{receiver}} = C_g \frac{\alpha}{\varepsilon}$$

T is the receiver operating temperature (K)
T_a is the ambient temperature (K)
ε is the emissivity of the receiver at the temperature T
σ is the Stefan–Boltzmann constant, 5.67×10^{-8} W/m^2 K^4
c is a temperature-independent convective loss coefficient, typically between 1 and 25 W/m^2 K$^{1.25}$
G_b is the solar beam radiation, W/m^2

Also α is the absorptivity for direct beam solar radiation (G_b) and η_{col} is the collector efficiency, which includes reflectivity (typically 0.85–0.95), cosine of angle of incidence effects (typically 0.7–0.9), and any factor to account for shading of mirror surfaces or blocking of reflected sunlight (typically 0.9–1.0). It is conventional to use the actual mirrored area of the heliostats for $A_{collector}$. The fraction of reflected light that is intercepted by the receiver is η_{int}, representing scattering and spillage losses. We see that the geometric concentration, C, plays an important role in suppressing the effects of thermal losses.

Based on the experience gained from the operation of Solar Two, Vant-Hull [43] proposed the following procedure for estimating the size and the performance of a utility-scale 100 MW CRS. This approach circumvents the need for using advanced optical analysis in the calculations, but relies on some empirical constants based on the Solar Two system design parameters and performance. He proposed a nominal allowance for parasitic power of 10% including the cold salt water pumps (6.2 MW$_e$), hot salt pumps (0.7 MW$_e$), feedwater pumps (2.2 MW$_e$), heliostat field operation (0.8 MW$_e$), and some miscellaneous load estimated at 3.3 MW$_e$. The analysis also assumed that the ratio of thermal energy output from the receiver to the turbine inlet power rating should be 1.8 to allow filling the thermal storage when generating rated power on a design day. With a reheat turbine efficiency of 43.2%, the receiver design point thermal output power (DPTOP) necessary to power a 100 MW system is given by

$$\text{DPTOP} = \text{turbine output} + \text{parasitic energy} \times 1.8/\text{turbine efficiency}$$

$$= (100 + 10) \times 1.8 / 0.32 = 468 \text{ MW}_t \tag{8.52}$$

Some additional design parameters for this system are listed in Table 8.13.

Receiver thermal losses depend on the receiver area and the local surface temperature. A first estimate of the receiver area can be obtained from the allowable flux-density limit. Due to end effects, and the reduced allowable flux at the high-temperature outlet (on the south side of the receiver system), the average allowable flux was reduced from the specified 825 kW/m^2 to about 440 kW/m^2, resulting in a required receiver area, A_e, of $A_e = 468$ MW/0.44 MW/m$^2 = 1.064$ m^2.

TABLE 8.13

Solar Two and First Commercial Plant from Utility Study

	Solar Two	First Commercial Utility Study
Design point power (electric)	10 MW	100 MW
Scheduled acceptance test	June 1996	~2,000
Location	15 Km E of Barstow, California, 35°N, 117°W	Arizona, Nevada, or Southern California
Site elevation	593 m	400–800 m
Heliostat field	Glass/metal	Glass/metal, or membrane
Configuration	Radial stagger	Radial stager
Reflective area (supplement to Solar One field)	71,000 m² at 39 m² + 10,000 m² at 95 m²	884,000 m² at 150 m²
Constraint on field (system trade)	South field favored to reduce peak flux on north side of receiver	
Ground coverage	27% avg., 40%–18%	25% avg., 40%–16%
Collector area (land)	0.3 km², ~75 ac	3.54 km², ~875 ac
Receiver	External, salt in tube	External, salt in tube
Focal height above heliostats	7.04 m	185 m
Diameter	5.05 m	16.1 m
Length	6.21 m	21.0 m
Heat-absorbing area	98.6 m²	1,068 m²
Design point power (absorbed thermal)	42.2 MW	468 MW
Absorbed flux density, maximum	850[a] kW/m²	820[a] kW/m²
Average	430[a] kW/m²	440[a] kW/m²
Thermal storage unit	Hot and cold tank	Hot and cold tank
Medium	Draw salt	Draw salt
Rating (thermal)	3 h at 35 MW	6 h at 260 MW
Turbogenerator rating (nominal, thermal)	35 MW	260 MW
Turbogenerator rating (nominal, electric)	12.5 MW	110 Mw
Solar multiple (receiver/turbine)	1.2	1.8
Heat engine efficiency	35% rated	42.3% rated
Power generation efficiency[b]	5.8% annual avg.	14.6% annual avg.

Source: Arizona Public Service Company, Arizona public service utility solar central receiver study, Report No. DOE/AL/38741-1, Phoenix, AZ, November 1988. With permission.

[a] Solar Two final receiver design is more aggressive than the preliminary commercial design, which was completed several years earlier.

[b] Estimated as electrical output/(annual insolation · reflective area).

For the first commercial design, an external receiver of cylindrical shape with an aspect ratio of 1.3 was selected to allow the highly peaked flux distribution to be leveled by selection of a vertical distribution of aim points, resulting in a receiver radius of 16.1 m. To provide the required 1000°F steam temperature at the turbine inlet, the salt storage temperature was set at 1050°F, which must also be the receiver outlet temperature. The working fluid is draw salt that begins to freeze at 430°F. To prevent freezing, a feedwater preheat outlet temperature of 455°F and a cold salt storage tank temperature of 500°F were specified. This defined the receiver inlet temperature at 500°F. Allowing for a flux-dependent 100–200 K temperature rise through the receiver tube walls results in a receiver surface temperature ranging from 750 to 850 K. Using the average of those temperatures for the receiver surface temperature T, 300 K for the ambient (T_a), and 0.88 for ε (typical for rough oxidized steel), and an empirical relation of $c = 5$ W/m² K¹·²⁵, the thermal loss, T_L, from the receiver is

$$T_L = A_{receiver}\left[\varepsilon\sigma\left(T^4 - T_a^t\right) + c(T - T_a)^{1.25}\right]$$

$$T_L = 1,056\left[0.88\cdot 5.67\times 10^{-8}\cdot(800^4 - 300^4) + 5(800 - 300)^{1.25}\right] \qquad (8.53)$$

$$T_L = 1,056\left[20,000 + 11,600\right] = 31.6\text{ MW}$$

The resulting thermal efficiency is

$$\eta_{th} = 1 - \frac{TL}{\text{incident power}} = 1 - \frac{TL}{DPTOP + TL} \qquad (8.54)$$

$$\eta_{th} = 1 - \frac{31.6}{468 + 31.6} = 0.937 = 93.7\%$$

The required collector (heliostat) area at the design point is given by

$$\text{Absorbed solar power} = DPTOP + TL = GpA_c(CSB\ T\eta)\alpha \qquad (8.55)$$

Hence

$$A_c = \frac{(468 + 31.6)\cdot 10^6}{(950\cdot 0.92\cdot(0.83\cdot 0.92\cdot 0.99)\cdot 0.92)} = 884,000\text{ m}^2$$

Here
  G is the design point beam insolation (=950 W/m² reflector area)
  p is the average heliostat reflectivity
  CSB is the combined effect of cosine, shading, and blocking losses
  T is the transmission loss from the heliostat to the receiver
  η is the receiver interception factor
  α is the receiver absorptivity

Rightfully, the term in parenthesis should be computed for each heliostat and then averaged, but using field average values for each item is a reasonable approximation.

If the ground coverage, F, is about 25% of the heliostat/collector area, A_c, with a nominal 3% addition for roads and the power block, the land area for the field, A_f, is

$$A_f = (1.03)\frac{A_c}{F} = (1.03)\frac{884,000}{0.25} = 3.64\text{ km}^2 = 1.4\text{ mi}^2 = 900\text{ ac}$$

Although the field is biased toward the south to minimize north-to-south variations in the receiver flux, the south radius is still only about 70% of the north field radius. This is approximated by replacing π by 2.5.

$$A_f = \pi R^2 - \left(0.3\,R\cdot\frac{\pi R}{2}\right) \cong 2.5\,R^2 \qquad (8.56)$$

So

$$R = \left(\frac{A_f}{2.5}\right)^{0.5} = \left(\frac{3{,}640{,}000}{2.5}\right)^{0.5} = 1{,}210 \text{ m}$$

In this case, the range of receiver elevation angles from the edge of the field ($\theta_{rim} = 90° - \psi_{rim}$) is
In the north, $\theta_{rim} = \arctan(185/1210) = 8.7°$
Due south, $\theta_{rim} = \arctan(185/(0.7 \times 1210)) = 12.3°$
The optimum rim angle ψ_{rim} will depend on the relative cost of heliostats and will be decreased if an exceedingly high flux density is required. A higher tower allows a higher heliostat density and a smaller field with a higher concentration. Typical values range about ±2.5° from the preceding figures.

Central receiver tower-based solar thermal power plants have been actively pursued in the United States, Germany, Switzerland, Spain, France, Italy, Russia, and Japan since the concept was first proposed by Alvin Hildebrandt and Lorin Vant-Hull in the early 1970s [44]. Table 8.14 summarizes the central receiver plants built in these countries. Solar One is a 10 MW$_e$ power plant built in Southern California as a joint venture of the U.S. Department of Energy, Southern California Edison Company, and the State of California. Figure 8.29 shows a photograph of the Solar One plant. The solar field generated superheated steam at 510°C

TABLE 8.14

Central Receiver Pilot Plants Constructed in Several Countries

Name	Location	Size MW$_e$	Receiver Fluid	Start-Up	Sponsors
Eurelios	Adrano, Sicily	1	Water/steam	1981	European Community
SSPS/CRS	Tabernas, Spain	0.5	Sodium	1981	Austria, Belgium, Italy, Greece, Spain, Sweden, Germany, Switzerland, United States
Sunshine	Nio, Japan	1	Water/steam	1981	Japan
Solar One	Daggett, California	10	Water/steam	1982	U.S. DOE, SCE[a] CEC[a], LADWP[a]
Themis	Targasonne, France	2.5	Hitec salt	1982	France
CESA-1	Tabernas, Spain	1	Water/steam	1982	Spain
MSEE	Albuquerque, New Mexico	0.75	Nitrate salt	1984	U.S. DOE, EPRI[a], U.S. Industry and Utilities
C3C-5	Crimea, former Soviet Union	5	Water/steam	1985	Former Soviet Union
Solar Two	Daggett, California	10	Molten salt	1995	U.S. DOE, SCE, LADWP
Andasol I	Granada, Spain	50	Molten salt w/7.5 h storage in 2 tanks	2007	Solar Millennium, ACS Cobra
Andasol II	Granada, Spain	50	Molten salt w/7.5 h storage in 2 tanks	2008	Solar Millennium, ACS Cobra
Gemasolar	Seville, Spain	20	Molten salt w/17 h indirect storage	2011	Torresol Energy

Source: De Laquil, P. et al., Solar-thermal electric technology, in *Renewable Energy*, Island Press, Washington, DC, 1993.
[a] SCE, Southern California Edison Company; CEC, California Energy Commission; LADWP, Los Angeles Department of Water and Power; EPRI, Electric Power Research Institute.

and 10.3 MPa, which ran the turbine. The system had a crushed rock and sand storage system that was charged by steam through a heat-transfer oil. Solar One was operated successfully for 6 years and then shut down for a redesign. The redesigned plant is called Solar Two. A schematic of Solar Two is shown in Figure 8.30. Solar Two uses a nitrate salt as the working fluid in the solar loop and as the storage medium.

In April 2009, Abengoa Solar began commercial operation of a 20 MW solar power tower plant at the Solucar Platform, near Seville, Spain. PS20 features a higher-efficiency cavity receiver, improved control systems, and a better thermal energy storage system than the first tower, PS10. The second system brings the power capacity to 20 MW, double that of the original PS10 tower. A photograph of both of these is shown in Figure 8.32b. Additional plants are

Point focus systems

Power tower
- Steam or molten nitrate salt HTF
- Max. operating temp. −560°C
- Steam rankine cycle

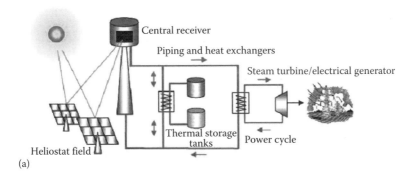

(a)

(b)

FIGURE 8.32
(See color insert.) (a) Schematic diagram of point focus solar thermal power tower system with a cavity receiver and (b) photograph of two Spanish Central Receiver Systems. (Courtesy of Clifford Ho, Sandia National Laboratories.)

being built and put into commercial operation in Spain. All of these later plants have several hours of storage in a two-tank system similar to that shown in Figure 8.32a.

Detailed EROI estimates for solar central receiver power plants have been made by Lorin Vant-Hull in a series of peer reviewed articles [37,45–47]. The EROI estimates were based on the detailed commercial central receiver plant design, which served as the reference for the 10 MW electric Solar One plant (incorporating 6 h of oil thermocline storage) and a similar analysis for the second-generation dispatchable plant incorporating two-tank molten salt storage (Solar Two). In both cases, the EROI was about 20 for electric power output to thermal (or fossil) energy input assuming a 30 year life and 6 h of storage. The heliostats represented 75% of the energy input. The discussion of the EROI included also consideration for recycling some of the components, and it was estimated that up to 41% of the initial capital energy may be recoverable.

Example 8.9

Use SAM to estimate the capacity factor, field area in km^2, annual thermal and electric power production in kWh and kWh$_e$ respectively, and the levelized electric power cost in \$/kWh for a 200 MW$_e$ molten draw salt power tower located near Daggett, CA. The unit is to have 6 h of storage capacity and an external receiver 19 m tall as shown in Figures 8.30 with a coating emittance on the receiver of 0.85. Use all the default values in SAM, but adjust the inflation to 2% and the real discount rate to 6%. SAM uses some concepts and definitions that are not in common use for solar thermal analysis. Their definitions can be found in the "Help" section of SAM, but to facilitate the inputs into the model, some of these definitions are given in the following table.

Term	Definition
Solar multiple	The solar multiple is the ratio of the receiver's design thermal output to the power block's design thermal input. In a system with a solar multiple larger than one, the excess energy goes to storage. A system with no storage should have solar multiple equal to one
Max receiver flux	The upper limit of solar radiation flux incident on the receiver surface that is allowed to be absorbed
Minimum turndown fraction	The minimum allowable fraction of the maximum salt flow rate to the receiver
Receiver design thermal power	The product of the solar multiple and the power design thermal power that can be found on the Power Cycle page
Receiver start-up delay time	The time in hours required to start the receiver. SAM calculates the start-up energy as the product of the available thermal energy, start-up delay time, and start-up delay energy fraction
Solar field land area (acres)	The actual aperture area converted from square meters to acres *Solar Field Area (acres) = Actual Aperture (m²) × Row Spacing (m)/ Maximum SCA Width (m) × 0.0002471 (acres/m²)*
Nonsolar field land area multiplier	Land area required for the system excluding the solar field land area, expressed as a fraction of the solar field aperture area. A value of one would result in a total land area equal to the total aperture area
Total land area (acres)	The land area required for the entire system including the solar field land area *Total land area (acres) = Solar field area (acres) × (1 + nonsolar field land area multiplier)*

*Note that SAM does not use the land area variables in any calculations. The shown values are only for your reference based on the default design for a 100 MW$_e$ system.

Solution

To evaluate the performance of a power tower power plant:

8. Start SAM
9. Under **Enter a new project name to begin**, type a name for your project. For example, "Power Tower System."
10. Click Create New File
11. Under 1. Select a technology, click Concentrating Solar Power
12. Click Power Tower System
13. Under 2. Select a financing option, click on Utility Independent Power Producer (IPP) financing option
14. Click **OK**

In the "Climate" page, change the location selection to "SAM/CA Daggett.tm2." This places the power tower in Daggett, CA.

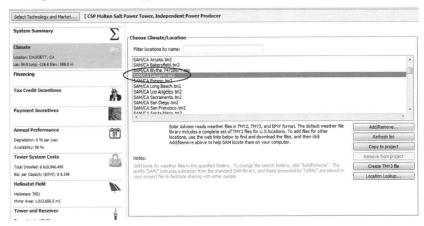

In the problem, it asks for a 200 MWe plant, so under the "Power Cycle" page, change the "Design Turbine Gross Output" to 200 MWe. This is the circled entry in the following figure.

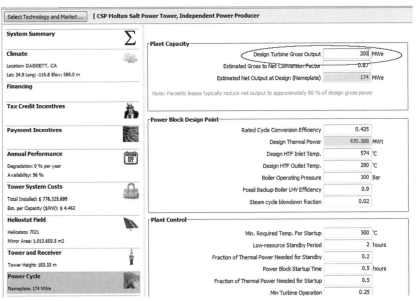

The receiver in Figure 8.30 is an external receiver with a coating emittance of 0.85 and a receiver height of 19 m, so this should be changed in SAM in the "Tower and Receiver" page. The values are circled in the following figure.

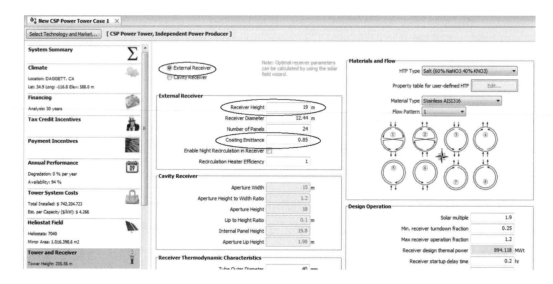

Figure 8.30 also showed a system with a two-tank storage system, and the problem stated that there was 6 h of storage. These properties can be changed under the "Thermal Storage" page and are circled as follows.

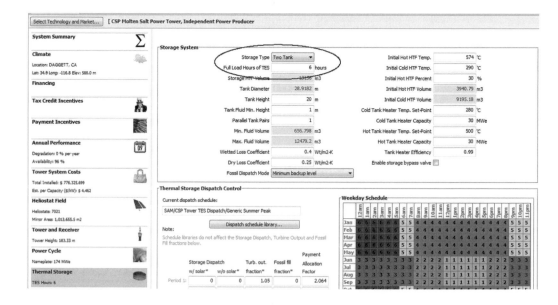

The problem also asks for an inflation rate of 2% and a real discount rate of 6%. These values can be changed in the "Financing" page and are circled in the following figure.

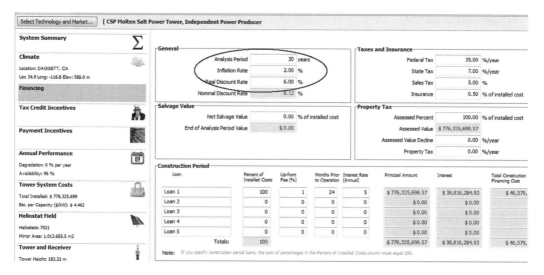

Once these changes are made, click the green arrow on the bottom left-hand corner, and the following summary characteristics of solar power tower should be shown. The following summary table shown contains numbers with three significant figures, which is the accuracy we should be analyzing these systems with.

Metric	Base
Net annual energy	327,000,000 kWh
PPA price	21.14 ¢/kWh
LCOE nominal	28.92 ¢/kWh
LCOE real	23.39 ¢/kWh
After-tax IRR	36.98%
Pre-tax min DSCR	1.4
After-tax NPV	$260,000,000.00
PPA price escalation	1.20%
Debt fraction	60.00%
Capacity factor	21.5%
Gross to net conv. factor	0.90
Annual water usage	1,190,000 m³
Total land area	1,280 acres

To find the annual thermal and electric power production of the system, select the graph called "Annual Energy Flow (Base Case)."

8.7 Parabolic Dish Systems and Stirling Engines

A parabolic dish–based solar thermal power plant was built in Shenandoah, Georgia. This plant was designed to operate at a maximum temperature of 382°C and to provide electricity (450 kW$_e$), air-conditioning, and process steam (at 173°C) for an industrial complex. The system consisted of 114 parabolic dish concentrators and was decommissioned in 1990.

Stirling engines have recently received increased attention because they can utilize the high temperature produced by concentrated solar energy, which can be achieved with parabolic dish concentrators. Stirling engines are a promising technology for solar applications because they have the potential of operating at a very high efficiency. In fact, the efficiency of the ideal Stirling cycle is the same as that of a Carnot cycle operating between the same temperatures. Since a Stirling engine is an external combustion engine, it can use any fuel or concentrated radiation. Thus, it could be operated at night with an auxiliary source such as natural gas or methane from a landfill. Stirling engines can operate at high temperatures, typically between 600°C and 800°C, resulting in a conversion efficiency of 30%–40%. Experience with Stirling engines that have recently been developed for cryogenic applications can be applied to solar-driven Stirling systems [50].

8.7.1 Thermodynamics of a Stirling Cycle

Figure 8.33 shows the thermodynamic process of an ideal Stirling cycle with a perfect gas as the working fluid on pressure–volume and temperature–entropy diagrams. The gas is compressed isothermally (at constant temperature) from state 1 to state 2 by means of heat rejection at the low temperature of the cycle, T_L. The gas is then heated at constant volume from state 2 to state 3, followed by an isothermal expansion from state 3 to state 4. During this expansion process, heat is added at the high temperature in the cycle, T_H. Finally, the gas is cooled at a constant volume from T_H to T_L during the process from state 4 to state 1. The cross-hatched areas in the temperature–entropy diagram during the constant volume process between states 2 and 3 represent the heat addition to the working gas while raising its temperature from T_L to T_H. Similarly, the cross-hatched area between states 4 and 1 represents the heat rejection as the gas is cooled from T_H to T_L. Note that the heat addition from state 2 to 3 is equal to the heat rejection from state 4 to 1 and that the processes occur between the same temperature limits. In the ideal cycle, the heat rejected between 4 and 1 is stored and transferred by perfect regeneration to the gas in process 2 to 3. Hence, the only external heat addition in the cycle occurs in the process between states 3 and 4 and is given by

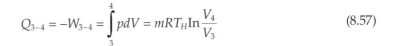

$$Q_{3-4} = -W_{3-4} = \int_3^4 pdV = mRT_H \ln\frac{V_4}{V_3} \qquad (8.57)$$

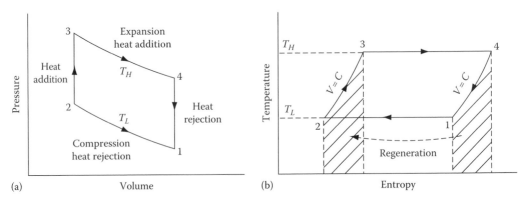

FIGURE 8.33
Thermodynamic diagrams of an ideal Stirling engine cycle. (a) Pressure–volume, (b) temperature–entropy.

The work input for compression from state 1 to 2 is

$$W_{1-2} = -\int_1^2 p\,dV = mRT_L \ln \frac{V_1}{V_2}$$

$$= -mRT_L \ln \frac{V_2}{V_1} \tag{8.58}$$

Noting that the ratio $V_2/V_1 = V_3/V_4$, and combining the earlier two equations, the net work output is

$$mR \ln \frac{V_3}{V_4}(T_H - T_L) \tag{8.59}$$

Therefore, the cycle efficiency

$$\eta = \frac{\text{Net work out}}{\text{Heat input}} = \frac{T_H - T_L}{T_L} \tag{8.60}$$

This efficiency, which is equal to the Carnot cycle efficiency, is based on the assumption that regeneration is perfect, which is not possible in practice. Therefore, the cycle efficiency would be lower than that indicated by the earlier equation. For a regeneration effectiveness e as defined later, the efficiency is given by [38]

$$\eta = \frac{T_H - T_L}{T_H + [(1-e)/(k-1)][T_H - T_L/\ln(V_1/V_2)]} \tag{8.61}$$

where
$e = T_R - T_L/T_H - T_L$ and T_R is the regenerator temperature
$k = c_p/c_v$ for the gas

For perfect regeneration ($e = 1$), the earlier expression reduces to the Carnot efficiency. It is also seen from the earlier equation that regeneration is not necessary for the cycle to work because even for $e = 0$, the cycle efficiency is not zero.

Example 8.10

A Stirling engine with air as the working fluid operates at a source temperature of 400°C and a sink temperature of 80°C. The compression ratio is 5.

Assuming perfect regeneration, determine the following:

1. Expansion work.
2. Heat input.
3. Compression work.
4. Efficiency of the machine.

If the regenerator temperature is 230°C, determine

5. The regenerator effectiveness.
6. Efficiency of the machine.
7. If the regeneration effectiveness is zero, what is the efficiency of the machine?

Solution

1. Expansion work per unit mass of the working fluid, assuming air as an ideal gas,

$$w_{34} = -\int_3^4 p\,dv = RT_H \ln\frac{v_4}{v_3} = (0.287)(400-273)\ln 5 = -310.9 \text{ kJ/kg}$$

Minus sign shows work output.

2. Heat input per unit mass of the working fluid

$$q_{34} = w_{34} = 310.9 \text{ kJ/kg}$$

3. Compression work per unit mass of the working fluid

$$w_{12} = -\int_1^2 P\,dv = RT_L \ln\frac{v_2}{v_1} = -(0.287)(80+273)\ln\left(\frac{1}{5}\right) = 163.1 \text{kJ/kg}$$

4. Efficiency of the machine

$$\eta = \frac{T_H - T_L}{T_H} = \frac{400-80}{(400-273)} = 0.475 = 47.5\%$$

5. The regenerator effectiveness

$$e = \frac{T_R - T_L}{T_H - T_L} = \frac{230-80}{400-80} = 0.469$$

6. Efficiency of the machine

$$\eta = \frac{T_H - T_L}{T_H + \dfrac{(1-e)(T_H - T_L)}{(k-1)\ln(v_1/v_2)}} = \frac{400-800}{(400+273)+\dfrac{(1-0)}{(1.4-1)}\dfrac{(400-80)}{\ln(5)}} = 0.341$$

$$\eta = 34.1\%$$

7. If the regeneration effectiveness is zero, the efficiency of the machine is

$$\eta = \frac{T_H - T_L}{T_H + \dfrac{(1-e)(T_H - T_L)}{(k-1)\ln(v_1/v_2)}} = \frac{400-80}{(400+273)+\dfrac{(1-0)}{(1.4-1)}\dfrac{(400-80)}{\ln(5)}} = 0.273$$

$$\eta = 27.3\%$$

A methodology for estimating the performance of a solar parabolic dish with a Stirling engine can be found in SAM.

8.7.2 Examples of Solar Stirling Power Systems

In order to understand how the Stirling cycle shown in Figure 8.33 may operate, in practice, the simple arrangement and sequence of processes shown in Figure 8.34 is helpful. In the proposed arrangement, two cylinders with pistons are connected via a porous media, which allows gas to pass through from one cylinder to the other. As the gas passes through the porous media, it exchanges heat with the media. The porous media, therefore, serves as the regenerator. In practice, this arrangement can be realized in three ways, as shown in Figure 8.35, by alpha, beta, and gamma types.

The choice of a working fluid for Stirling engine depends mainly on the thermal conductivity of the gas in order to achieve high heat-transfer rates. Air has traditionally been used as the working fluid. Helium has a higher ratio of specific heats (k), which lessens the impact of imperfect regeneration.

In the alpha configuration, there are two cylinders and pistons on either side of a regenerator. Heat is supplied to one cylinder, and cooling is provided to the other. The pistons move at

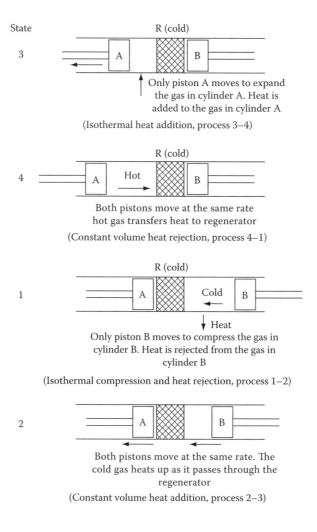

FIGURE 8.34
Stirling cycle states and processes with reference to Figure 8.33.

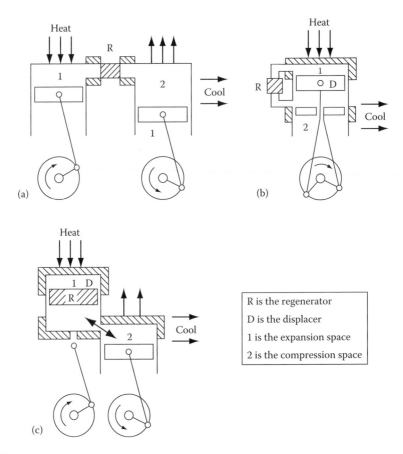

FIGURE 8.35
Three basic types of Stirling engine arrangements. (a) Alpha type, (b) beta type, and (c) gamma type.

the same speed to provide constant volume processes. When all the gas has moved to one cylinder, the piston of that cylinder moves with the other remaining fixed to provide expansion or compression. Compression is done in the cold cylinder and expansion in the hot cylinder. The Stirling Power Systems V-160 engine (Figure 8.36) is based on an alpha configuration.

The beta configuration has a power piston and a displacer piston, which divides the cylinder into hot and cold sections. The power piston is located on the cold side and compresses the gas when the gas is in the cold side and expands it when it is in the hot side. The original patent of Robert Stirling was based on beta configuration, as are free piston engines.

The gamma configuration also uses a displacer and a power piston. In this case, the displacer is also the regenerator, which moves gas between the hot and cold ends. In this configuration, the power piston is in a separate cylinder [49,50].

In a piston/displacer drive, the power and displacer pistons are designed to move according to a simple harmonic motion to approximate the Stirling cycle. This is done by a crankshaft or a bouncing spring/mass second-order mechanical system [51].

In a kinematic engine, the power piston is connected to the output shaft by a connecting rod crankshaft arrangement. Free piston arrangement is an innovative way to realize the Stirling cycle. In this arrangement, the power piston is not connected physically to an output shaft. The piston bounces between the working gas space and a spring (usually a gas spring). The displacer is also usually free to bounce. This configuration is called the

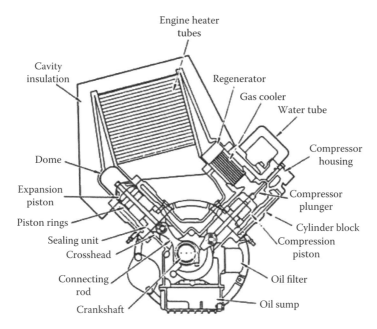

FIGURE 8.36
Stirling power systems/Solo Kleinmotoren V-160 alpha-configuration Stirling engine. (Adapted from Kistler, B.L., A user's manual for DELSOL3: A computer code for calculating the optical performance and optimal system design for solar thermal central receiver plants, Sandia National Labs, Albuquerque, NM, SAND86-8018, 1986.)

FIGURE 8.37
The McDonnell Douglas/United Stirling dish-Stirling 25 kWe module. (From Goswami, Y. et al., *Principles of Solar Engineering*, 2nd edn., Taylor & Francis, Philadelphia, PA, 2000. With permission.)

Beale free piston Stirling engine after its inventor, William Beale [51]. Since a free piston Stirling engine has only two moving parts, it offers the potential of simplicity, low cost, and reliability. Moreover, if the power piston is made magnetic, it can generate current in the stationary conducting coil around the engine as it moves. This is the principle of the free piston/linear alternator in which the output from the engine is electricity.

Stirling engines can provide very high efficiencies with high-concentration solar collectors. Since practical considerations limit the Stirling engines to relatively small sizes, a Stirling engine fixed at the focal point of a dual tracking parabolic dish provides an optimum match, as shown in Figure 8.37. Therefore, all of the commercial developments to date have been in parabolic dish–Stirling engine combination. The differences in the commercial systems have been in the construction of the dish and the type of Stirling engine. A thorough description of past and current dish–Stirling technologies is presented in Ref. [52].

Problems

8.1   Repeat Example 8.1 for a parabolic collector with an evacuated tube receiver (case E in Figure 8.18a).

8.2   Repeat Example 8.3, but for a collector located in Washington, DC.

8.3   Consider a Stirling cycle from Example 8.10 with imperfect regenerator (T_R = 230°C). Assume ideal gas.

   a.   Compute the efficiency from

$$\eta = \frac{Work_{net}}{Heat_{input}}$$

   where

$$Heat_{input} = q_{34} + mC_v(T_H - T_R)$$

   Verify that the efficiency you get is the same as the efficiency obtained by Equation 8.61.

   b.   Start from the earlier efficiency equation. Derive the efficiency in Equation 8.61.

8.4   A solar electric engine operating between 5°C and 95°C has an efficiency equal to one-half the Carnot efficiency. This engine is to drive a 4 kW pump. If the collector has an efficiency of 50%, calculate the area needed for operation in Egypt, where the insolation averages about 2800 kJ/m² h during the day. State any additional assumptions.

8.5   Prepare a thermal analysis matching a line focusing collector, a paraboloid dish collector, and a CPC collector with a CR of 3:1 to a suitable working fluid in Washington, DC, and Albuquerque, New Mexico. Comment on storage needs in both locations.

Low-temperature organic and steam efficiencies were computed with the following assumptions: expander efficiency, 80%; pump efficiency, 50%; mechanical efficiency, 95%; condensing temperature, 95°F; regeneration efficiency, 80%; high side pressure loss, 5%; low side pressure loss, 8%.

8.6 The following schematic diagram for a solar-driven irrigation pump was developed by Battelle Memorial Institute and uses tracking PTCs. Solar energy is used to heat the water in the collectors to 423 K, which then vaporizes an organic working fluid that powers the pump-turbine. Calculate the surface area needed to power a 50 hp pump capable of delivering up to 38,000 L/min of water at noon in Albuquerque, New Mexico.

8.7 Write a closed-form expression for the work output of a solar-powered heat engine if the energy delivery of the solar collector at high temperature is given approximately by the expression

$$q_u = (\tau\alpha)_{eff} I_c - \frac{\sigma \varepsilon \overline{T}_f^4}{CR}$$

(convection and conduction losses are neglected) where $\overline{T}_f$ is the average fluid temperature and CR is the CR. Two heat engines are to be evaluated:

1. Carnot cycle

$$\text{Cycle efficiency } \eta_c = 1 - \frac{T_\infty}{T_f},$$

2. Brayton cycle

$$\text{Cycle efficiency } \eta_B = 1 - \frac{C_B T_\infty}{T_f} C_B \geq 1,$$

where
$C_B = (r_p)^{(k-1)/k}$
r_p is the compressor pressure ratio
k is the specific heat ratio of the working fluid

Write an equation with $\overline{T}_f$ as the independent variable that, when solved, will specify the value of $\overline{T}_f$ to be used for maximum work output as a function of CR, surface emittance and $(\tau\alpha)_{eff}$ product, insolation level, and C_B. Optional: solve the equation derived previously for CR = 100, ε = 0.5, and $(\tau\alpha)_{eff}$ = 0.70, at an insolation level of 1 kW/m². What is the efficiency of a solar-powered Carnot cycle and a Brayton cycle for which C_B = 2?

8.8 Calculate the maximum power and LCOE for a parabolic collector similar to that in Example 8.1, but assume that an evacuated tube receiver is used (case E in Figure 8.18a).

8.9 Prepare a preliminary design for a CRS with a power capacity of 50 MW similar to Solar 2 for a location in Nevada.

8.10 It has been estimated that a 400 MW expansion of the existing SEGS solar plant in California would cost $2.4 billion. Estimate the expected capital costs of the generation facility in dollars/kW capacity and the minimum cost of electricity per kWh generated by the addition, if they are operated at full capacity for 6 h/day and have a lifetime of 30 years.

8.11 Using SAM, estimate the levelized cost for a 50 MW_e parabolic trough solar power plant in San Diego, CA, similar to the one shown in Figure 8.26. Also determine the thermal energy delivered to the power block during July and estimate the surface area required for the installation. Assume an inflation rate of 2% and look up the discount rate in the *Wall Street Journal* at the time of this assignment.

8.12 Using SAM, estimate the levelized cost for a central receiver solar power plant with 10 h of storage in Phoenix, AZ, similar to the one shown in Figure 8.30. Assume that the receiver is of external design and that heat rejection in the condenser is with a wet cooling tower configuration. Comment on the availability of cooling water and state all of your assumptions.

References

1. Trombe, F. (1956) High temperature furnaces, in: *Proceedings of the World Symposium on Applied Solar Energy*, Phoenix, AZ, 1955, Stanford Research Institute, Menlo Park, CA.
2. Daniels, F. (1964) *Direct Use of the Sun's Energy*, Yale University Press, New Haven, CT.
3. Daniels, F. (1949) Solar energy, *Science* 109, 51–57.
4. EIA. (2007) Industrial sector energy demand: Revisions for non-energy-intensive manufacturing. Accessed at http://wwweia.doe.gov/oiaf/aeo/otheranalysis/neim.html. (accessed on April 30, 2013).
5. Fraser, M.D. (1976) InterTechnology Corporation assesses industrial heat potential, *Solar Engineering* 1, pp. 11–12, September 1976.
6. Kreider, J.F. (1979) *Medium and High Temperature Solar Processes*, Academic Press, New York.
7. Intertechnology Corporation. (1977) Analysis of the economic potential of solar thermal energy to provide industrial process heat, ERDA Report No. COO/2829-1, 3 vols.
8. Hannon, B., Costanza, R., and Herendeen, R. (1986) Measures of energy cost and value in ecosystems, *Journal of Environmental Economics and Management* 13, 391–401.
9. Wagner, R. (1977) Energy conservation in dyeing and finishing, *Textile Chemists and Colorists* 9, 52.
10. Hebrank, W.H. (1975) Options for reducing fuel usage in textile finishing tenter dryers, *American Dyestuff Reporter* 63, 34
11. Goswami, B.C. and Langley, J. (1987) A review of the potential of solar energy in the textile industry, in: *Progress in Solar Engineering*, Hemisphere Publishing Corporation, Washington, DC.
12. Lowery, J.F. et al. (1977) Energy conservation in the textile industry, Technical Reports of Department of Energy Project No. Ey-76S-05-5099, School of Textile Engineering, Georgia Institute of Technology, Atlanta, GA, 1977, 1978, and 1979.
13. Proctor, D. and Morse, R.N. (1975) *Solar Energy for Australian Food Processing Industry*, CSIRO Solar Energy Studies, East Melbourne, VIC, Australia.
14. Goswami, D.Y., Kreith, F., and Kreider, J.F. (2000) *Principles of Solar Engineering*, 2nd edn., Taylor & Francis, Philadelphia, PA.

15. University of Wisconsin, TRNSYS: A transient simulation program, version 11.1 (2000), Engineering Experiment Station, Report 38, Solar Energy Laboratory, University of Wisconsin, Madison, WI.

16. Singh, R.K., Lund, D.B., and Buelow, F.H. (1986) Compatibility of solar energy with food processing energy demands, in: *Alternative Energy in Agriculture*, Vol. I, CRC Press, Boca Raton, FL.

17. O'Gallagher, J.J. (2008) *Nonimaging Optics in Solar Energy*, Morgan & Claypool Publishers, San Rafael, CA.

18. Winston, R. (1974) Principles of solar concentrators of a novel design, *Solar Energy* 16, 89.

19. Kreith, F. and West, R.E. (eds.) (1997) *CRC Handbook of Energy Efficiency*, CRC Press, Boca Raton, FL.

20. Bendt, P., Rabl, A., Gaul, H., and Reid, K. (1979) *Optical Analysis and Optimization of Line-Focus Solar Collectors*, SERI, Golden, CO, SERI/TR 34-092, September 1979.

21. Gee, R. (1980) Long term average performance benefits of parabolic trough improvements, SERI, Golden, CO, SERI/TR 632-439.

22. Gee, R. and May, K. (1997) Parabolic trough concentrating collectors component and system design, in: Kreith, F. and West, R.E. (eds.) *CRC Handbook of Energy Efficiency*, CRC Press, Boca Raton, FL.

23. Kreith, F.K., West, R.E., and Cleveland, C.J. (1987) Energy analysis for renewable energy technologies, *ASHRAE Transactions*, Vol. 93, Part 1, Atlanta, GA, ASHRE 1987.

24. DiGrazia, M., Gee, R., and Jorgensen, G. (2009) RefecTech mirror film: Attributes and durability for CSP applications, in: *Proceedings of the ASME 3rd International Conference on Energy Sustainability*, San Francisco, CA, July 19–23, 2009.

25. Jorgensen, G.J., Gee, R., and King, D.E. (2006) Durable corrosion and ultraviolet-resistant silver mirror, U.S. Patent No. 6989924. Issued January 24, 2006.

26. Farr, A. and Gee, R. (2009) The SkyTrough™ parabolic trough solar collector, in: Paper *90090 ASME 3rd International Conference on Energy Sustainability*, San Francisco, CA.

27. Reddy, T.A. (1987) *The Design and Sizing of Active Solar Thermal Systems*, Oxford University Press, Oxford, U.K.

28. Collares-Pereira, M. and Rabl, A. (1979) Simple procedure for predicting long term average performance of nonconcentrating and of concentrating solar collectors, *Solar Energy* 23, 235–253.

29. Hall, I., Prairie, R., Anderson, H., and Boes, E. (1978) *Generation of Typical Meteorological Years for 26 SOLMET Stations*, SAND78-1601, Sandia National Laboratories, Albuquerque, NM.

30. Rabl, A. (1981) *Active Solar Collectors and Their Applications*, University Press, Oxford, U.K.

31. Gordon, J.M. and Rabl, A. (1992) Nonimaging compound parabolic concentrator-type reflectors with variable extreme direction, *Applied Optics* 31, 7332–7338.

32. Cengel, Y.A. and Boles, M.A. (2006) *Thermodynamics: An Engineering Approach*, 6th edn., McGraw-Hill, New York.

33. Fitzgerald, A.E., Kingsley, C.F., and Kusko, A. (1971) *Electric Machinery*, 3rd edn., McGraw-Hill, New York.

34. Kreith, F. and Goswami, D.Y. (eds.) (1997) *The CRC Handbook of Mechanical Engineering*, CRC Press, Boca Raton, FL.

35. Goswami, D.Y., Hingorani, S., and Mines, G. (1991) A laser-based technique for particle sizing to study two-phase expansion in turbines, *Journal of Solar Energy Engineering, Transactions of the ASME* 113(3), 211–218.

36. Vant-Hull, L.L. (1992) Solar thermal central receivers, *Solar Today* Nov/Dec issue, 13–16.

37. Stine, W.B. and Harrigan, R.W. (1985) *Solar Energy Fundamentals and Design: With Computer Applications*, Wiley, New York.

38. Kolb, G.J. (1998) Economic evaluation of Solar One and hybrid power towers using molten salt technology, *Solar Energy* 62(1), 51–61.

39. Hillesland, T. and Weber, E.R. (1990) Utilities study of solar central receivers, in: Gupta, B.P. and Traugott, W.H. (eds.), *Proceedings of the 4th International Symposium Solar Thermal Technology: Research, Development and Applications*, Santa Fe, NM, Hemisphere Publishing Corp., New York, pp. 165–176.

40. Romero, M., Buck, R., and Pacheco, J.E. (2002) An update on solar central receiver systems, projects, and technologies, *Journal of Solar Energy Engineering* 124, 98–108.
41. Wagner, M.J., Klein, S.A., and Reindi, D.T. (2009) Simulation of utility-scale central receiver system power plants, in: *Proceedings of the 2009 ASME 3rd International Conference on Energy Sustainability*, San Francisco, CA, ES2009-90132.
42. Vant-Hull, L.L. (1991) Solar radiation conversion, in: Winter, C.J., Sizmann, R.L., and Vant-Hull, L.L. (eds.), *Solar Radiation Conversion*, Springer, Berlin, Germany, pp. 21–27.
43. Hildebrandt, A.F. and Vant-Hull, L.L. (1974) Tower-top focus solar energy collector, *Journal of Mechanical Engineering* 96(9), 23–27.
44. Vant-Hull, L.L. (1985) Solar thermal power generation, *Natural Resources Journal* 25, 1099–1111.
45. Vant-Hull, L.L. (1992–1993) Solar thermal electricity: An environmentally benign and viable alternative, *Perspectives in Energy* 2, 157–166.
46. Vant-Hull, L.L. (1992) Solar thermal receivers: Current status and future promise, *American Solar Energy Society* 7, 13–16.
47. Kistler, B.L. (1986) A user's manual for DELSOL3: A computer code for calculating the optical performance and optimal system design for solar thermal central receiver plants, Sandia National Labs, Albuquerque, NM, SAND86-8018.
48. De Laquil, P., Kearney, D., Geyer, M., and Diver, R. (1993) Solar-thermal electric technology, in: *Renewable Energy*, Island Press, Washington, DC.
49. Garaway, I. (2009) Personal communication.
50. Stine, W.B. (1998) Stirling engines, in: *The CRC Handbook of Mechanical Engineering*, CRC Press, Boca Raton, FL.
51. Mancini, T., Heller, P., Butler, B., Osborn, B., Schiel, W., Goldberg, V. et al. (2003) Dish-stirling systems: An overview of development and status, *JSEE* 125, 135–151.
52. Intertechnology Corporation. (1977) Analysis of the economic potential of solar thermal energy to provide industrial process heat, ERDA Report No. C00/2829.
53. Arizona Public Service Company. (1988) Arizona public service utility solar central receiver study, Report No.DOE/AL/38741-1, Phoenix, AZ, November 1988.

*9

Passive Solar Heating, Cooling, and Daylighting

For the well-being and health … the homesteads should be airy in summer and sunny in winter. A homestead promising these qualities will be longer than it is deep, and the main front would face south.

—Aristotle

9.1 Introduction

Passive systems are defined, quite generally, as systems in which the thermal energy flow is by natural means: by conduction, radiation, and natural convection. A *passive heating system* is one in which the sun's radiant energy is converted to heat upon absorption by the building. The absorbed heat can be transferred to thermal storage by natural means or used to directly heat the building. *Passive cooling systems* use natural energy flows to transfer heat to the environmental sinks: the ground, air, and sky.

If one of the major heat-transfer paths employs a pump or fan to force flow of a heat-transfer fluid, then the system is referred to as having an *active* component or subsystem. *Hybrid systems*—either for heating or cooling—are ones in which there are both passive and active energy flows. The use of the sun's radiant energy for the natural illumination of a building's interior spaces is called *daylighting*. Daylighting design approaches use both solar beam radiation (referred to as *sunlight*) and the diffuse radiation scattered by the atmosphere (referred to as *skylight*) as sources for interior lighting, with historical design emphasis on utilizing skylight.

9.1.1 Distinction between a Passive System and Energy Conservation

A distinction is made between energy conservation techniques and passive solar measures. *Energy conservation features* are designed to reduce the heating and cooling energies required to thermally condition a building: the use of insulation to reduce heating and cooling loads, and similarly, the use of window shading or window placement to reduce solar gains, reducing summer cooling loads. *Passive features* are designed to increase the use of solar energy to meet heating and lighting loads, plus the use of ambient "coolth" for cooling. For example, window placement to enhance solar gains to meet winter heating loads and/or to provide daylighting is passive solar use, and the use of a thermal chimney to draw air through the building to provide cooling is a passive cooling feature.

---

* Please note that this is an optional chapter and may be omitted in an introductory course.

9.2 Key Elements of Economic Consideration

The distinction between passive systems, active systems, or energy conservation is not critical for economic calculations, as they are the same in all cases: a trade-off between the life cycle cost of the energy saved (*performance*) and the life cycle cost of the initial investment, operating, and maintenance costs (*cost*).

9.2.1 Performance: Net Energy Savings

The key performance parameter to be determined is the net annual energy saved by the installation of the passive system. The basis for calculating the economics of any solar energy system is to compare it against a "normal" building; thus, the actual difference in the annual cost of fuel is the difference in auxiliary energy that would be used with and without solar energy. Thus, the energy saved rather than energy delivered, energy collected, useful energy, or some other energy measure must be determined.

9.2.2 Cost: Over and Above "Normal" Construction

The other significant part of the economic trade-off involves determining the difference between the cost of construction of the passive building and of the "normal" building against which it is to be compared. The convention, adopted from the economics used for active solar systems, is to define a "solar add-on cost." Again, this may be a difficult definition in the case of passive designs because the building can be significantly altered compared to typical construction, since in many cases, it is not just a one-to-one replacement of a wall by a different wall, but is more complex and involves assumptions and simulations concerning the "normal" building.

9.2.3 General System Application Status and Costs

Almost one-half million buildings in the United States were constructed or retrofitted with passive features in the 20 years after 1980. Passive heating applications are primarily in single-family dwellings and secondarily in small commercial buildings. Daylighting features, which reduce lighting loads and the associated cooling loads, are usually more appropriate for large office buildings.

A typical passive heating design in a favorable climate might supply up to one-third of a home's original load at a cost of $5–$10 per million Btu net energy saved. An appropriately designed daylighting system can supply lighting at a cost of 2.5–5¢ per kWh [1].

9.3 Solar Thermosyphon Water Heating

Solar hot water–heating systems are composed of a collector and a storage tank. A passive solar collector with integrated storage is shown in Figure 9.1. When the flow between the collector and the tank is by natural circulation, these passive solar hot water systems are referred to as thermosyphon systems. This ability of thermosyphon systems to heat water without an externally powered pump has spurred its use in both regions where power is unavailable and where power is very expensive and freezing is not a problem.

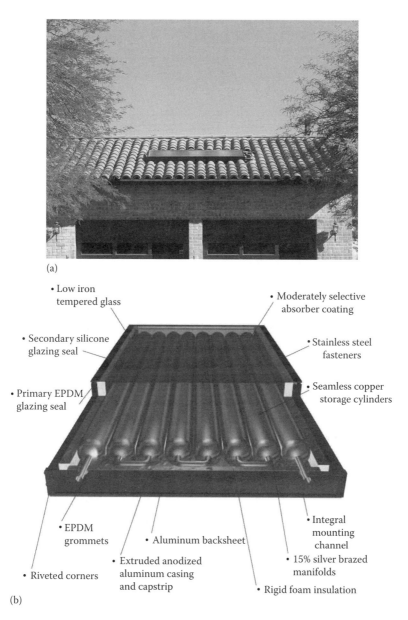

(a)

(b)

FIGURE 9.1
(See color insert.) CopperHeart passive solar collector with integrated storage. (Courtesy of SunEarth, Inc., Fontana, CA.) (a) Photograph of installed module on residence in Arizona. (b) Schematic of CopperHeart integral collector storage system.

9.3.1 Thermosyphon Concept

The natural tendency of a less dense fluid to rise above a more dense fluid can be used in a simple solar water heater to cause fluid motion through a collector. The density difference is created within the solar collector where heat is added to increase the temperature and decrease the density of the liquid. This collection concept is called a thermosyphon, and Figure 9.2 schematically illustrates the major components of such a system.

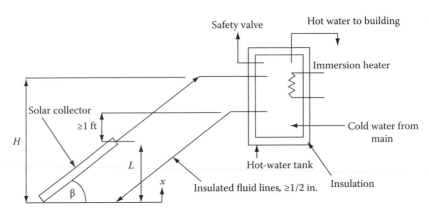

FIGURE 9.2
Schematic diagram of thermosyphon loop used in a natural circulation, service water-heating system. The flow pressure drop in the fluid loop must equal the buoyant force "pressure", $\left[\int_0^L g\rho(x)gdx + \rho_{stor}gL\right]$ where $\rho(x)$ is the local collector fluid density and ρ_{stor} is the tank fluid density, assumed uniform. (From Kreith, F. and West, R., *CRC Handbook of Energy Efficiency and Renewable Energy*, CRC Press, Boca Raton, FL, 2007. With permission.)

The flow pressure drop in the fluid loop (ΔP_{FLOW}) must equal the buoyant force "pressure difference" ($\Delta P_{BUOYANT}$) caused by the differing densities in the "hot" and "cold" legs of the fluid loop:

$$\Delta P_{FLOW} = \Delta P_{BUOYANT}$$

$$= \rho_{stor}gH - \left[\int_0^L \rho(x)gdx + \rho_{out}g(H-L)\right] \tag{9.1}$$

where
H is the height of the "legs" and L the height of the collector (see Figure 9.2)
$\rho(x)$ is the local collector fluid density
ρ_{stor} is the tank fluid density
ρ_{out} is the collector outlet fluid density (the latter two densities are assumed uniform)

The flow pressure term ΔP_{FLOW} is related to the flow loop system head loss, which is in turn directly connected to friction and fitting losses and the loop flow rate:

$$\Delta P_{FLOW} = \oint_{LOOP} \rho d(h_L) \tag{9.2}$$

where $h_L = KV^2$, with K being the sum of the component loss "velocity" factors (see any fluid mechanics text) and V the flow velocity.

9.3.2 Thermo-Fluid System Design Considerations

Since the driving force in a thermosyphon system is only a small density difference and not a pump, larger-than-normal plumbing fixtures must be used to reduce pipe friction losses. In general, one pipe size larger than would be used with a pump system is satisfactory.

Under no conditions should piping smaller than ½ in. (12 mm) national pipe thread (NPT) be used. Most commercial thermosyphons use 1 in. (25 mm) NPT pipe. The flow rate through a thermosyphon system is about 1 gal/ft² h (40 L/m² h) in bright sun, based on collector area.

Since the hot water system loads vary little during a year, the best angle to tilt the collector is that equal to the local latitude. The temperature difference between the collector inlet water and the collector outlet water is usually 15°F–20°F (8°C–11°C) during the middle of a sunny day [3]. After sunset, a thermosyphon system can reverse its flow direction and lose heat to the environment during the night. To avoid reverse flow, the top header of the absorber should be at least 1 ft (30 cm) below the cold leg fitting on the storage tank, as shown.

To provide heat during long cloudy periods, an electrical immersion heater can be used as a backup for the solar system. The immersion heater is located near the top of the tank to enhance stratification so that the heated fluid is at the required delivery temperature at the delivery point. Tank stratification is desirable in a thermosyphon to maintain flow rates as high as possible. Insulation must be applied over the entire tank surface to control heat loss. Figure 9.3 illustrates two common thermosyphon system designs.

Several features inherent in the thermosyphon design limit its utility. If it is to be operated in a freezing climate, a nonfreezing fluid must be used, which in turn requires a heat exchanger between the collector and the potable water storage. (If potable water is not required, the collector can be drained during cold periods instead.) Heat exchangers of either the shell-and-tube type or the immersion-coil type require higher flow rates for efficient operation than a thermosyphon can provide. Therefore, the thermosyphon is generally limited to nonfreezing climates. A further restriction on thermosyphon use is the requirement for an elevated tank. In many cases, structural or architectural constraints

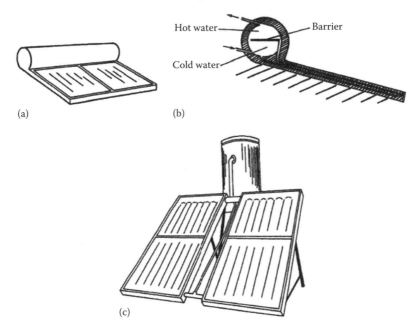

FIGURE 9.3
Passive solar water heaters: (a) compact model using combined collector and storage, (b) section view of the compact model, and (c) tank and collector assembly. (From Kreith, F. and West, R., *CRC Handbook of Energy Efficiency and Renewable Energy*, CRC Press, Boca Raton, FL, 2007. With permission.)

prohibit raised-tank locations. In residences, collectors are normally mounted on the roof, and tanks mounted above the high point of the collector can easily become the highest point in a building. Practical considerations often do not permit this application.

Example 9.1

Determine the "pressure difference" available for a thermosyphon system with 1 m high collector and 2 m high "legs." The water temperature input to the collector is 25°C, and the collector output temperature is 35°C. If the overall system loss "velocity" factor (K) is 15.6, estimate the system flow velocity.

Solution

Equation 9.1 is used to calculate the pressure difference, with the water densities being found from the temperatures (in steam tables):

$$\rho_{stor}(25°C) = 997.009 \text{ kg/m}^3; \quad \rho_{out}(35°) = 994.036 \text{ kg/m}^3;$$

$$\rho_{coll.ave.}(30°C) = 996.016 \text{ kg/m}^3 \, (Note: \text{ average collector temperature used in "integral")}$$

and with $H = 2$ and $L = $ m

$$\Delta P_{BUOYANT} = (997.009)9.81(2) - \left[(996.016)9.81(1) + (994.036)9.81(1)\right]$$

$$= 38.9 \text{ N/m}^2 \text{(Pa)}$$

The system flow velocity is estimated from the "system K" given, the pressure difference calculated earlier, taking the average density of the water around the loop (at 30°C), and substituting into Equation 9.2:

$$\Delta P_{BUOYANT} = (\rho_{loop.ave.})(h_L)_{loop} = (\rho_{loop.ave.})KV^2$$

$$V^2 = \frac{38.9}{(996.016)(15.6)}$$

$$V = 0.05 \text{ m/s}$$

9.4 Passive Solar Heating Design Fundamentals

Passive heating systems contain the five basic components of all solar systems, as described in Chapter 7. Typical passive realizations of these components are

1. Collector → windows, walls and floors
2. Storage → walls and floors, large interior masses (often these are integrated with the collector absorption function)
3. Distribution system → radiation, free convection, simple circulation fans

4. Controls → movable window insulation, vents both to other inside spaces or to ambient

5. Backup system → any nonsolar heating system

The design of passive systems requires the strategic placement of windows, storage masses, and the occupied spaces themselves. The fundamental principles of solar radiation geometry and availability are instrumental in the proper location and sizing of the system's "collectors" (windows). Storage devices are usually more massive than those used in active systems and are frequently an integral part of the collection and distribution system.

9.4.1 Types of Passive Heating Systems

A commonly used method of cataloging the various passive system concepts is to distinguish three general categories: direct, indirect, and isolated gain. Most of the physical configurations of passive heating systems are seen to fit within one of these three categories.

For direct gain (Figure 9.4), sunlight enters the heated space and is converted to heat at absorbing surfaces. This heat is then distributed throughout the space and to the various enclosing surfaces and room contents.

For indirect gain category systems, sunlight is absorbed and stored by a mass interposed between the glazing and the conditioned space. The conditioned space is partially enclosed and bounded by this thermal storage mass, so a natural thermal coupling is achieved. Examples of the indirect approach are the thermal storage wall, the thermal storage roof, and the northerly room of an attached sunspace.

In the thermal storage wall (Figure 9.5), sunlight penetrates the glazing and is absorbed and converted to heat at a wall surface interposed between the glazing and the heated space. The wall is usually masonry (Trombe wall) or containers filled with water (water wall), although it might contain phase-change material. The attached sunspace (Figure 9.6) is actually a two-zone combination of direct gain and thermal storage wall. Sunlight enters and heats a direct gain southerly "sunspace" and also heats a mass wall separating the northerly buffered space, which is heated indirectly. The "sunspace" is frequently used as

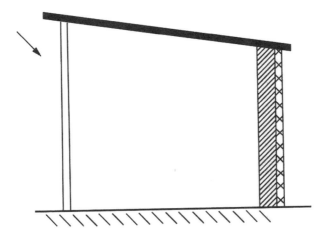

FIGURE 9.4
Direct gain. (From Kreith, F. and West, R., *CRC Handbook of Energy Efficiency and Renewable Energy*, CRC Press, Boca Raton, FL, 2007. With permission.)

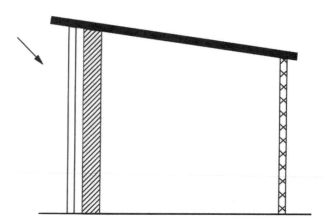

FIGURE 9.5
Thermal storage wall. (From Kreith, F. and West, R., *CRC Handbook of Energy Efficiency and Renewable Energy*, CRC Press, Boca Raton, FL, 2007. With permission.)

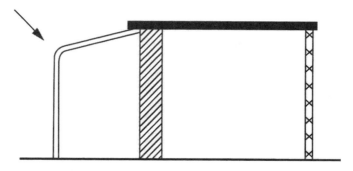

FIGURE 9.6
Attached sunspace. (From Kreith, F. and West, R., *CRC Handbook of Energy Efficiency and Renewable Energy*, CRC Press, Boca Raton, FL, 2007. With permission.)

a greenhouse, in which case the system is called an "attached greenhouse." The thermal storage roof (Figure 9.7) is similar to the thermal storage wall except that the interposed thermal storage mass is located on the building roof.

The isolated gain category concept is an indirect system except that there is a distinct thermal separation (by means of either insulation or physical separation) between the thermal storage and the heated space. The convective (thermosyphon) loop, as depicted in Figure 9.2, is in this category and, while often used to heat domestic water, is also used for building heating. It is most akin to conventional active systems in that there is a separate collector and separate thermal storage. The thermal storage wall, thermal storage roof, and attached sunspace approaches can also be made into isolated systems by insulating between the thermal storage and the heated space.

9.4.2 Fundamental Concepts for Passive Heating Design

Figure 9.8 is an equivalent thermal circuit for the building illustrated in Figure 9.5, the Trombe wall-type system. For the heat-transfer analysis of the building, three temperature nodes can be identified—room temperature, storage wall temperature, and the ambient temperature.

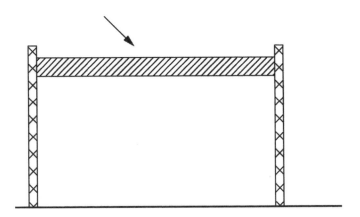

FIGURE 9.7
Thermal storage roof. (From Kreith, F. and West, R., *CRC Handbook of Energy Efficiency and Renewable Energy*, CRC Press, Boca Raton, FL, 2007. With permission.)

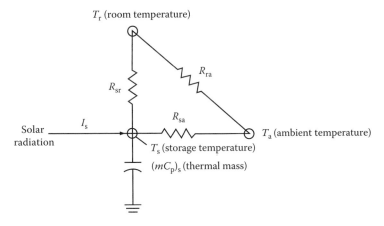

FIGURE 9.8
Equivalent thermal circuit for passively heated solar building in Figure 9.5. (From Kreith, F. and West, R., *CRC Handbook of Energy Efficiency and Renewable Energy*, CRC Press, Boca Raton, FL, 2007. With permission.)

The circuit responds to climatic variables represented by a current injection I_s (solar radiation) and by the ambient temperature T_a. The storage temperature T_s and room temperature T_r are determined by current flows in the equivalent circuit. By using seasonal and annual climatic data, the performance of a passive structure can be simulated, and the results of many such simulations correlated to give the design approaches are described in the following text.

9.5 Passive Design Approaches

Design of a passive heating system involves selection and sizing of the passive feature type(s), determination of thermal performance, and cost estimation. Ideally, a cost/performance optimization would be performed by the designer. Owner and architect ideas usually establish the passive feature type, with general size and cost estimation available. However, the thermal performance of a passive heating system has to be calculated.

There are several "levels" of methods that can be used to estimate the thermal performance of passive designs. First-level methods involve a rule of thumb and/or generalized calculation to get a starting estimate for size and/or annual performance. A second-level method involves climate, building, and passive system details, which allow annual performance determination, plus some sensitivity to passive system design changes. Third-level methods involve periodic calculations (hourly, monthly) of performance and permit more detailed variations of climatic, building, and passive solar system design parameters.

These three levels of design methods have a common basis in that they all are derived from correlations of a multitude of computer simulations of passive systems [4,5]. As a result, a similar set of defined terms is used in many passive design approaches:

- A_p, solar projected area, m² (ft²): The net south-facing passive solar glazing area projected onto a vertical plane.
- NLC, net building load coefficient, kJ/CDD (Btu/FDD): Net load of the nonsolar portion of the building per day per degree of indoor–outdoor temperature difference. The CDD and FDD terms refer to centigrade and Fahrenheit degree days, respectively.
- Q_{net}, net reference load, Wh (Btu): Heat loss from nonsolar portion of building is calculated by

$$Q_{net} = NLC \times (No. \ of \ degree \ days) \tag{9.3}$$

- LCR, load collector ratio, kJ/m² CDD (Btu/ft² FDD): Ratio of NLC to A_p,

$$LCR = \frac{NLC}{A_p} \tag{9.4}$$

- SSF, solar savings fraction, %: Percentage reduction in required auxiliary heating relative to net reference load,

$$SSF = 1 - \frac{Auxiliary \ heat \ required \ (Q_{aux})}{Net \ reference \ load \ (Q_{net})} \tag{9.5}$$

So using Equation 9.3,

$$Auxiliary \ heat \ required, \ Q_{aux} = (1 - SSF) \times NLC \times (No. \ of \ degree \ days) \tag{9.6}$$

The amount of auxiliary heat required is often a basis of comparison between possible solar designs as well as being the basis for determining building energy operating costs. Thus, many of the passive design methods are based on determining SSF, NLC, and the number of degree days in order to calculate the auxiliary heat required for a particular passive system by using Equation 9.6.

9.5.1 First Level: Generalized Methods

A first estimate or starting value is needed to begin the overall passive system design process. Generalized methods and rules of thumb have been developed to generate initial values for solar aperture size, storage size, solar savings fraction, auxiliary heat required, and other size and performance characteristics. The following rules of thumb are meant to be used with the defined terms presented earlier.

9.5.1.1 Load

A rule of thumb used in conventional building design is that a design heating load of 120–160 kJ/CDD per m^2 of floor area (6–8 Btu/FDD ft^2) is considered an energy-conservative design. Reducing these nonsolar values by 20% to solarize the proposed south-facing solar wall gives rule-of-thumb NLC values per unit of floor area:

$$\frac{\text{NLC}}{\text{Floor area}} = 100 - 130\,\text{kJ/CDD}\,\text{m}^2(4.8 - 6.4\,\text{Btu/FDD}\,\text{ft}^2) \tag{9.7}$$

9.5.1.2 Solar Savings Fraction

A method of getting starting-point values for the solar savings fraction is presented in Figure 9.9 [5]. The map values represent optimum SSF in % for a particular set of conservation and passive solar costs for different climates across the United States. With the Q_{net} generated from the earlier NLC rule of thumb and the SSF read from the map, the Q_{aux} can be determined.

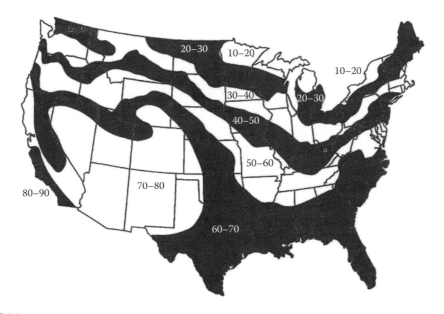

FIGURE 9.9
Starting-point values of SSF in percent. (From PSDH, *Passive Solar Design Handbook, Part One*: Total Environmental Action Inc., *Part Two*: Los Alamos Scientific Laboratory, *Part Three*: Los Alamos National Laboratory. Van Nostrand Reinhold, New York, 1984. With permission.)

9.5.1.3 LCR

The A_p can be determined using the NLC provided earlier, if the LCR is known. The rule of thumb associated with "good" values of LCR [5] differs depending on whether the design is for a "cold" or "warm" climate:

$$\text{"Good" LCR} = \begin{cases} \text{For cold climate: } 410\,\text{kJ/m}^2\,\text{CDD}\,(20\,\text{Btu/ft}^2\,\text{FDD}) \\ \text{For warm climate: } 610\,\text{kJ/m}^2\,\text{CDD}\,(30\,\text{Btu/ft}^2\,\text{FDD}) \end{cases} \qquad (9.8)$$

9.5.1.4 Storage

Rules of thumb for thermal mass storage relate storage material total heat capacity to the solar projected area [5]. The use of the storage mass is to provide for heating on cloudy days and to regulate sunny day room air temperature swing. When the thermal mass directly absorbs the solar radiation, each square meter of the projected glazing area requires enough mass to store 613 kJ/C. If the storage material is not in direct sunlight, but heated from room air only, then four times as much mass is needed. In a room with a direct sunlight-heated storage mass, the room air temperature swing will be approximately one-half the storage mass temperature swing. For room air–heated storage, the air temperature swing is twice that of the storage mass.

Example 9.2

A Denver, Colorado, building is to have a floor area of 195 m² (2100 ft²). Determine rule-of-thumb size and performance characteristics.

Solution
From Equation 9.5, the NLC is estimated as

$$\text{NLC} = (115\,\text{kJ/CDDm}^2) \times (195\,\text{m}^2)$$

$$= 22{,}400\,\text{kJ/CDD}(11{,}800\,\text{Btu/FDD})$$

Using the "cold" LCR value and Equation 9.4, the passive solar projected area is

$$A_p = \frac{\text{NLC}}{\text{LCR}} = \frac{(22{,}400\,\text{kJ/CDD})}{410\,\text{kJ/m}^2\text{CDD}}$$

$$= 54.7\,\text{m}^2(588\,\text{ft}^2)$$

Locating Denver on the map of Figure 9.9 gives an SSF in the 70%–80% range (use 75%). An annual °C-degree-day value can be found in city climate tables [5,6], and is 3491 CDD (6283 FDD) for Denver. Thus, the auxiliary heat required, Q_{aux}, is found using Equation 9.6:

$$Q_{aux} = (1 - .75)(22{,}400\,\text{kJ/CDD})(3{,}491\,\text{CDD})$$

$$= 19{,}600\,\text{MJ}\,(18.5 \times 10^6\,\text{Btu})\text{ annually}$$

The thermal storage can be sized using directly solar heated and/or room air heated mass by using the projected area. Assuming brick with a specific heat capacity of 840 J/kg°C, the storage mass is found by

$$A_p \times (613 \text{ kJ}/C) = m \times (840 \text{ J}/\text{kg} \,^\circ C)$$

$$m_d = 40,000 \text{ kg} \ (88,000 \text{ lbm}) \text{ Direct sun}$$

or

$$m_a = 160,000 \text{ kg} \ (351,000 \text{ lbm}) \text{ Air heated}$$

A more location-dependent set of rules of thumb is presented in [4]. The first rule of thumb relates solar projected area as a percentage of floor area to solar savings fraction, with and without night insulation of the solar glazing:

A solar projected area of (B1)% to (B2)% of the floor area can be expected to produce an SSF in (location) of (S1)% to (S2)%, or, if R9 night insulation is used, of (S3)% to (S4)%

where the values of B1, B2, S1, S2, S3, and S4 are found using Table 9.1 for a few typical locations. A complete list of these parameters is presented on the website, http://www.crcpress.com/product/isbn/9781466556966, in Table W.9.1. The thermal storage mass rule of thumb is again related to the solar projected area:

A thermal storage wall should have 14 kg × SSF (%) of water or 71 kg × SSF (%) of masonry for each square meter of solar projected area. For a direct gain space, the mass above should be used with a surface area of at least three times the solar projected area, and masonry no thicker than 10–15 cm. If the mass is located in back rooms, then four times the above mass is needed.

TABLE 9.1

Values to Be Used in the Glazing Area and SSF Relations Rules of Thumb

City	B1	B2	S1	S2	S3	S4
Birmingham, Alabama	0.09	0.18	22	37	34	58
Phoenix, Arizona	0.06	0.12	37	60	48	75
Denver, Colorado	0.12	0.23	27	43	47	74
Washington, District of Columbia	0.12	0.23	18	28	37	61
Miami, Florida	0.01	0.02	27	48	31	54
Des Moines, Iowa	0.21	0.43	19	25	58	75
Reno, Nevada	0.11	0.22	31	48	49	76
Forth Worth, Texas	0.09	0.17	26	44	38	64
Madison, Wisconsin	0.20	0.40	15	17	51	74
Cheyenne, Wyoming	0.11	0.21	25	38	47	74

Sources: PSDH, *Passive Solar Design Handbook. Volume One: Passive Solar Design Concepts*, DOE/CS-127/1, March 1980. Prepared by Total Environmental Action, Inc. (B. Anderson, C. Michal, P. Temple, and D. Lewis); *Volume Two: Passive Solar Design Analysis*, DOE/CS-0127/2, January 1980. Prepared by Los Alamos Scientific Laboratory (J.D. Balcomb, D. Barley, R McFarland, J. Perry, W. Wray and S. Noll). U.S. Department of Energy, Washington, DC, 1980. With permission.

Example 9.3

Determine size and performance passive solar characteristics with the location-dependent set of rules of thumb for the house of the previous example.

Solution
Using Table 9.1 with the 195 m² house in Denver yields

$$\text{Solar projected area} = 12\% - 23\% \text{ of floor area}$$

$$= 23.4 - 44.9\,\text{m}^2$$

$$\text{SSF (no night insulation)} = 27\% - 43\%$$

$$\text{SSF (R9 night insulation)} = 47\% - 74\%$$

Using the rule of thumb for the thermal storage mass,

$$m = 17\,\text{kg} \times 43\% \times 44.9\,\text{m}^2$$

$$= 33,000\,\text{kg} \ (72,000\,\text{lbm}) \quad \text{Thermal wall or direct gain}$$

Comparing the results of this example to those of the previous example, the two rules of thumb are seen to produce "roughly" similar answers. General system cost and performance information can be generated with results from rule-of-thumb calculations, but a more detailed level of information is needed to determine design-ready passive system type (direct gain, thermal wall, sunspace), size, performance, and costs.

9.5.2 Second Level: LCR Method

The LCR method is useful for making estimates of the annual performance of specific types of passive system(s) combinations. The LCR method was developed by calculating the annual SSF for 94 reference passive solar systems for 219 U.S. and Canadian locations over a range of LCR values. Table 9.2 includes the description of these 94 reference systems for use both with the LCR method and with the solar load ratio (SLR) method described later. Tables were constructed for each city with LCR versus SSF listed for each of the 94 reference passive systems. (*Note:* The SLR method was used to make the LCR calculations, and this SLR method is described in the next section as the third-level method.) While the complete LCR tables [5] include 219 locations, Table 9.3 only presents information for Medford, Oregon, to demonstrate the method. Table W.9.2 on the website includes also six "representative" cities (Albuquerque, Boston, Madison, Medford, Nashville, and Santa Maria). The LCR method consists of the following steps [5]:

1. Determine the building parameters:
   a. Building load coefficient, NLC
   b. Solar projected area, A_p
   c. LCR = NLC/A_p
2. Find the short designation of the reference system closest to the passive system design (Table 9.2).

TABLE 9.2

Designations and Characteristics for 94 Reference Systems

(a) Overall system characteristics

Masonry properties

Thermal conductivity (k)	
Sunspace floor	0.5 Btu/h/ft/°F
All other masonry	1.0 Btu/h/ft/°F
Density	(Q)150. lb/ft³
Specific heat (c)	0.2 Btu/lb/°F
Infrared emittance of normal surface	0.9
Infrared emittance of selective surface	0.1

Solar absorptances	
Waterwall	1.0
Masonry, Trombe wall	1.0
Direct gain and sunspace	0.8
Sunspace: water containers	0.9
Lightweight common wall	0.7
Other lightweight surfaces	0.3

Glazing properties

Transmission characteristics	Diffuse
Orientation	Due south
Index of refraction	1.526
Extinction of coefficient	0.5 in.$^{-1}$
Thickness of each pane	One-eighth inch
Gap between panes	One-half inch
Infrared emittance	0.9

Control range

Room temperature	65°F–75°F
Sunspace temperature	45°F–95°F
Internal heat generation	0

Thermocirculation vents (when used)

Vent area/projected area	
(sum of both upper and lower vents)	0.06
Height between vents	8 ft
Reverse flow	None

Nighttime insulation (when used)

Thermal resistance	R9
In place, solar time	5:30 PM to 7:30 AM

Solar radiation assumptions

Shading	None
Ground diffuse reflectance	0.3

(*continued*)

TABLE 9.2 (continued)

Designations and Characteristics for 94 Reference Systems

(b) Direct-gain (DG) system types

Designation	Thermal Storage Capacity[a] (Btu/ ft²/°F)	Mass Thickness[a] (in.)	Mass Area-to- Glazing Area Ratio	No. of Glazings	Nighttime Insulation
A1	30	2	6	2	No
A2	30	2	6	3	No
A3	30	2	6	2	Yes
B1	45	6	3	2	No
B2	45	6	3	3	No
B3	45	6	3	2	Yes
C1	60	4	6	2	No
C2	60	4	6	3	No
C3	60	4	6	2	Yes

(c) Vented trombe-wall (TW) system types

Designation	Thermal Storage Capacity[a] (Btu/ ft²/°F)	Wall Thickness[a] (in.)	ρck (Btu²/h/ ft⁴/°F²)	No. of Glazings	Wall Surface	Nighttime Insulation
A1	15	6	30	2	Normal	No
A2	22.5	9	30	2	Normal	No
A3	30	12	30	2	Normal	No
A4	45	18	30	2	Normal	No
B1	15	6	15	2	Normal	No
B2	22.5	9	15	2	Normal	No
B3	30	12	15	2	Normal	No
B4	45	18	15	2	Normal	No
C1	15	6	7.5	2	Normal	No
C2	22.5	9	7.5	2	Normal	No
C3	30	12	7.5	2	Normal	No
C4	45	18	7.5	2	Normal	No
D1	30	12	30	1	Normal	No
D2	30	12	30	3	Normal	No
D3	30	12	30	1	Normal	Yes
D4	30	12	30	2	Normal	Yes
D5	30	12	30	3	Normal	Yes
E1	30	12	30	1	Selective	No
E2	30	12	30	2	Selective	No
E3	30	12	30	1	Selective	Yes
E4	30	12	30	2	Selective	Yes

(d) Unvented trombe-wall (TW) system types

Designation	Thermal Storage Capacity[a] (Btu/ ft²/°F)	Wall Thickness[a] (in.)	Pck (Btu²/h/ ft⁴/°F²)	No. of Glazings	Wall Surface	Nighttime Insulation
F1	15	6	30	2	Normal	No
F2	22.5	9	30	2	Normal	No
F3	30	12	30	2	Normal	No

TABLE 9.2 (continued)

Designations and Characteristics for 94 Reference Systems

(d) Unvented trombe-wall (TW) system types

Designation	Thermal Storage Capacity[a] (Btu/ ft²/°F)	Wall Thickness[a] (in.)	Pck (Btu²/h/ ft⁴/°F²)	No. of Glazings	Wall Surface	Nighttime Insulation
F4	45	18	30	2	Normal	No
G1	15	6	15	2	Normal	No
G2	22.5	9	15	2	Normal	No
G3	30	12	15	2	Normal	No
G4	45	18	15	2	Normal	No
H1	15	6	7.5	2	Normal	No
H2	22.5	9	7.5	2	Normal	No
H3	30	12	7.5	2	Normal	No
H4	45	18	7.5	2	Normal	No
I1	30	12	30	1	Normal	No
I2	30	12	30	3	Normal	No
I3	30	12	30	1	Normal	Yes
I4	30	12	30	2	Normal	Yes
I5	30	12	30	3	Normal	Yes
J1	30	12	30	1	Selective	No
J2	30	12	30	2	Selective	No
J3	30	12	30	1	Selective	Yes
J4	30	12	30	2	Selective	Yes

(e) Waterwall (WW) system types

Designation	Thermal Storage Capacity[a] (Btu/ ft²/°F)	Wall Thickness[a] (in.)	No. of Glazings	Wall Surface	Nighttime Insulation
A1	15.6	3	2	Normal	No
A2	31.2	6	2	Normal	No
A3	46.8	9	2	Normal	No
A4	62.4	12	2	Normal	No
A5	93.6	18	2	Normal	No
A6	124.8	24	2	Normal	No
B1	46.8	9	1	Normal	No
B2	46.8	9	3	Normal	No
B3	46.8	9	1	Normal	No
B4	46.8	9	1	Normal	No
B5	46.8	9	2	Normal	No
C1	46.8	9	1	Selective	No
C2	46.8	9	2	Selective	No
C3	46.8	9	1	Selective	No
C4	46.8	9	2	Selective	No

(continued)

TABLE 9.2 (continued)

Designations and Characteristics for 94 Reference Systems

(f) Sunspace (SS) system types

Designation	Type	Tilt (°)	Common Wall	End Walls	Nighttime Insulation
A1	Attached	50	Masonry	Opaque	No
A2	Attached	50	Masonry	Opaque	Yes
A3	Attached	50	Masonry	Glazed	No
A4	Attached	50	Masonry	Glazed	Yes
A5	Attached	50	Insulated	Opaque	No
A6	Attached	50	Insulated	Opaque	Yes
A7	Attached	50	Insulated	Glazed	No
A8	Attached	50	Insulated	Glazed	Yes
B1	Attached	90/3	Masonry	Opaque	No
B2	Attached	90/3	Masonry	Opaque	Yes
B3	Attached	90/3	Masonry	Glazed	No
B4	Attached	90/3	Masonry	Glazed	Yes
B5	Attached	90/3	Insulated	Opaque	No
B6	Attached	90/3	Insulated	Opaque	Yes
B7	Attached	90/3	Insulated	Glazed	No
B8	Attached	90/3	Insulated	Glazed	Yes
C1	Semienclosed	90	Masonry	Common	No
C2	Semienclosed	90	Masonry	Common	Yes
C3	Semienclosed	90	Insulated	Common	No
C4	Semienclosed	90	Insulated	Common	Yes
D1	Semienclosed	50	Masonry	Common	No
D2	Semienclosed	50	Masonry	Common	Yes
D3	Semienclosed	50	Insulated	Common	No
D4	Semienclosed	50	Insulated	Common	Yes
E1	Semienclosed	90/3	Masonry	Common	No
E2	Semienclosed	90/3	Masonry	Common	Yes
E3	Semienclosed	90/3	Insulated	Common	No
E4	Semienclosed	90/3	Insulated	Common	Yes

Source: PSDH, *Passive Solar Design Handbook. Part One:* Total Environmental Action, Inc., *Part Two:* Los Alamos Scientific Laboratory, *Part Three:* Los Alamos National Laboratory. Van Nostrand Reinhold, New York, 1984. With permission.

[a] The thermal storage capacity is per unit of projected area, or, equivalently, the quantity $\rho c k$. The wall thickness is listed only as an appropriate guide by assuming $\rho c = 30$ Btu/ft^3/°F.

3. Enter the LCR Tables (Table 9.3):
   a. Find the city.
   b. Find the reference system listing.
   c. Determine annual SSF by interpolation using the LCR value from earlier text.
   d. Note the annual heating degree days (No. of degree of days).
4. Calculate the annual auxiliary heat required:

$$\text{Auxiliary heat required} = (1 - \text{SSF}) \times \text{NLC} \times (\text{No. of degree days})$$

TABLE 9.3

LCR Tables for Medford, Oregon

Medford, Oregon								4930	DD
WW A1	708	64	24	11	—	—	—	—	—
WW A2	212	73	38	22	13	7	3	—	—
WW A3	174	75	41	25	16	9	5	2	1
WW A4	158	74	43	27	17	11	6	3	2
WW A5	149	75	45	29	19	12	7	4	2
WW A6	144	75	46	30	20	13	8	4	2
WW B1	154	43	16	—	—	—	—	—	—
WW B2	162	80	48	31	21	14	9	6	3
WW B3	190	100	62	41	28	19	13	8	5
WW B4	171	99	65	45	32	23	16	11	7
WW B5	160	95	63	45	32	23	17	12	7
WW C1	205	108	67	45	31	21	15	10	6
WW C2	178	99	63	43	30	22	15	10	6
WW C3	189	117	80	57	42	31	23	16	10
WW C4	170	106	72	52	38	28	21	15	9
TW A1	607	63	25	12	5	—	—	—	—
TW A2	222	68	33	19	11	6	2	—	—
TW A3	175	67	36	21	13	8	4	2	—
TW A4	147	64	36	22	14	9	5	3	1
TW B1	327	61	27	14	7	3	—	—	—
TW B2	178	62	32	19	12	7	4	2	—
TW B3	154	60	33	20	12	8	4	2	1
TW B4	143	56	31	19	12	8	5	2	1
TW C1	212	56	27	15	9	5	2	—	—
TW C2	159	55	28	17	11	7	4	2	—
TW C3	154	52	27	16	10	6	4	2	1
TW C4	167	48	24	14	9	5	3	2	—
TW D1	112	34	14	—	—	—	—	—	—
TW D2	177	77	44	28	18	12	8	5	3
TW D3	180	85	50	32	21	14	9	6	3
TW D4	177	93	58	39	27	19	13	9	5
TW D5	168	92	58	40	28	20	14	10	6
TW E1	213	101	60	39	26	18	12	8	4
TW E2	194	98	59	39	27	19	13	9	5
TW E3	208	118	77	53	38	27	20	13	8
TW E4	186	108	71	49	36	26	19	13	8
TW F1	256	53	23	12	5	—	—	—	—
TW F2	153	56	29	17	10	5	2	—	—
TW F3	125	54	30	18	11	7	3	1	—
TW F4	102	48	28	18	11	7	4	2	1
TW G1	153	46	22	12	7	—	—	—	—
TW G2	109	46	25	15	9	5	3	1	—
TW G3	92	42	24	15	9	6	3	2	—

(continued)

TABLE 9.3 (continued)

LCR Tables for Medford, Oregon

Medford, Oregon								4930	DD
TW G4	74	35	20	13	8	5	3	2	—
TW H1	97	38	20	12	7	4	1	—	—
TW H2	75	34	19	12	7	5	3	1	—
TW H3	63	29	17	10	7	4	3	1	—
TW H4	49	23	13	8	5	3	2	1	—
TW I1	83	27	10	—	—	—	—	—	—
TW I2	133	64	38	24	16	11	7	4	2
TW I3	142	71	43	28	19	13	9	5	3
TW I4	146	80	51	35	25	17	12	8	5
TW I5	144	82	53s	37	26	19	13	9	6
TW J1	175	89	54	36	24	17	11	7	4
TW J2	158	85	53	36	25	18	12	8	5
TW J3	173	103	69	48	35	26	18	13	8
TW J4	160	96	64	45	33	24	17	12	8
DG A1	110	35	—	—	—	—	—	—	—
DG A2	142	58	32	18	9	—	—	—	—
DG A3	187	82	48	32	22	15	9	5	—
DG B1	110	40	15	—	—	—	—	—	—
DG B2	146	61	35	21	13	7	—	—	—
DG B3	193	84	51	34	24	17	12	7	3
DG C1	144	57	29	13	—	—	—	—	—
DG C2	177	75	44	28	19	12	6	—	—
DG C3	224	98	60	41	29	21	14	10	5
SS A1	415	110	51	28	16	9	4	2	—
SS A2	372	146	79	48	31	21	14	8	5
SS A3	397	96	42	21	10	—	—	—	—
SS A4	379	144	76	46	29	19	12	7	4
SS A5	732	111	45	23	12	5	—	—	—
SS A6	368	143	77	47	30	20	13	8	4
SS A7	846	90	33	14	—	—	—	—	—
SS A8	379	140	73	44	27	17	11	6	6
SS B1	274	81	38	21	12	6	3	—	—
SS B2	288	117	65	40	26	18	12	7	4
SS B3	249	71	33	17	8	—	—	—	—
SS B4	282	113	62	38	25	16	11	7	4
SS B5	368	72	30	15	7	—	—	—	—
SS B6	269	111	62	30	25	17	11	7	4
SS B7	323	58	23	10	—	—	—	—	—
SS B8	262	106	57	35	23	15	9	6	3
SS C1	153	62	33	19	11	5	—	—	—
SS C2	172	83	50	32	22	15	10	6	3
SS C3	166	51	24	13	7	3	—	—	—
SS C4	173	76	43	27	18	12	8	5	3
SS D1	367	129	65	37	22	13	7	3	1
SS D2	318	159	92	60	40	27	18	12	7

TABLE 9.3 (continued)

LCR Tables for Medford, Oregon

Medford, Oregon								4930	DD
SS D3	480	124	57	31	18	10	5	2	—
SS D4	328	153	89	57	38	26	17	11	6
SS E1	262	95	48	27	15	7	—	—	—
SS E2	257	124	73	47	31	21	14	9	5
SS E3	334	84	38	20	10	4	—	—	—
SS E4	269	118	67	42	27	18	12	7	4

Sources: PSDH, *Passive Solar Design Handbook. Volume One: Passive Solar Design Concepts*, DOE/CS-127/1, March 1980. Prepared by Total Environmental Action, Inc. (B. Anderson, C. Michal, P. Temple, and. D. Lewis); *Volume Two: Passive Solar Design Analysis*, DOE/CS-0127/2, January 1980. Prepared by Los Alamos Scientific Laboratory (J.D. Balcomb, D. Barley, R. McFarland, J. Perry, W. Wray and S. Noll). U.S. Department of Energy, Washington, DC, 1980. With permission.

If more than one reference solar system is being used, then find the "aperture area weighted" SSF for the combination. Determine each individual reference system SSF using the total aperture area LCR, then take the "area weighted" average of the individual SSFs.

The LCR method allows no variation from the 94 reference passive designs. To treat off-reference designs, sensitivity curves have been produced that illustrate the effect on SSF of changing one or two design variables. These curves are presented in Figure 9.10 for three "representative" cities, chosen for their wide geographical and climatological ranges. These SSF "sensitivity curves" are presented in Figure 9.10 for storage wall (a, b, c) and sunspace (d) design variations.

Example 9.4

The previously used 2,100 ft² building with NLC = 11,800 Btu/FDD is preliminarily designed to be located in Medford, Oregon, with 180 ft² of 12 in. thick vented Trombe wall and 130 ft² of direct gain, both systems having double glazing, nighttime insulation, and 30 Btu/ft² thermal storage capacity. Determine the annual auxiliary energy needed by this design.

Solution
Step 1 yields

$$NLC = 11,800 \, Btu/FDD$$

$$A_p = 180 + 130 = 20 \, ft^2$$

$$LCR = \frac{11,800}{320} = 36.8 \, Btu/FDD \, ft^2$$

Step 2 yields: From Table 9.2, the short designations for the appropriate systems are

TWD4 (Trombe wall)
DGA3 (direct gain)

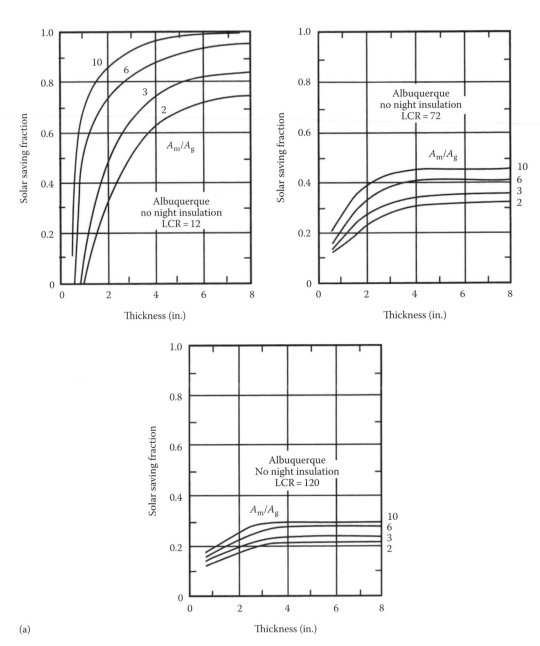

(a)

FIGURE 9.10
(a) Storage wall: Mass thickness sensitivity of SSF to off-reference conditions. (From PSDH, *Passive Solar Design Handbook.*, *Part One*: Total Environmental Action Inc., *Part Two*: Los Alamos Scientific Laboratory, *Part Three*: Los Alamos National Laboratory. Van Nostrand Reinhold, New York, 1984. With permission.)

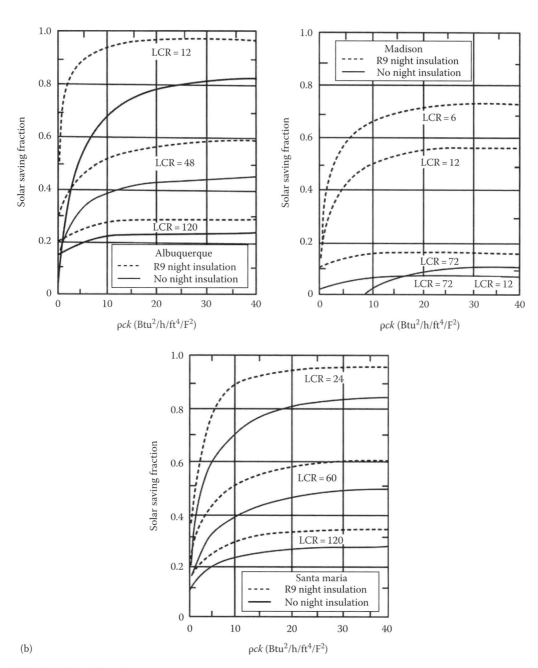

(b)

FIGURE 9.10 (continued)

(b) Storage wall: ρck product. (From PSDH, *Passive Solar Design Handbook.*, *Part One*: Total Environmental Action Inc., *Part Two*: Los Alamos Scientific Laboratory, *Part Three*: Los Alamos National Laboratory. Van Nostrand Reinhold, New York, 1984. With permission.)

(continued)

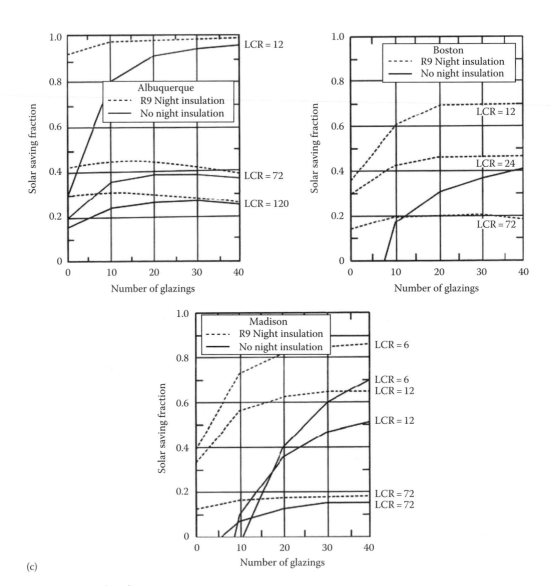

(c)

FIGURE 9.10 (continued)
(c) Storage wall: Number of glazings. (From PSDH, *Passive Solar Design Handbook.*, *Part One*: Total Environmental Action Inc., *Part Two*: Los Alamos Scientific Laboratory, *Part Three*: Los Alamos National Laboratory. Van Nostrand Reinhold, New York, 1984. With permission.)

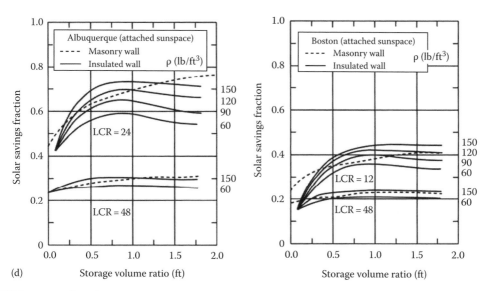

(d)

FIGURE 9.10 (continued)
(d) Sunspace: Storage volume to projected area ratio. (From PSDH, *Passive Solar Design Handbook.*, *Part One*: Total Environmental Action Inc., *Part Two*: Los Alamos Scientific Laboratory, *Part Three*: Los Alamos National Laboratory. Van Nostrand Reinhold, New York, 1984. With permission.)

Step 3 yields: From Table 9.3 for Medford, Oregon, with LCR = 36.8

TWD4: SSF(TW) = 0.42
DGA3: SSF(DG) = 0.37

Determine the "weighted area" average SSF:

$$SSF = \frac{180(0.42) + 130(0.37)}{320} = 0.39$$

Step 4 yields: Using Equation 9.6 and reading 4930 FDD from Table 9.3,
$Q_{aux} = (1 - 0.39) \times 11,800 \text{ Btu} \times 4,930 \text{ FDD} = 35.5 \times 10^6 \text{ Btu annually}$
Using the reference system characteristics yields the thermal storage size:

- Trombe wall ($\rho ck = 30$, concrete properties from Table 9.2c)

$$m(TW) = \text{density} \times \text{area} \times \text{thickness}$$

$$= 150 \text{lbm}/\text{ft}^3 \times 180 \text{ft}^2 \times 1 \text{ft}$$

$$= 27,000 \text{lbm}$$

- Direct gain ($\rho ck = 30$, concrete properties) using mass area to glazing area ratio of 6:

$$\text{mass area} = 6 \times 130 = 780 \text{ft}^2 \text{of 2 in. thick concrete}$$

$$m(DG) = 150 \text{lbm}/\text{ft}^3 \times 780 \text{ft}^2 \times 1/6 \text{ft}$$

$$= 19,500 \text{lbm}$$

Using the LCR method allows a basic design of passive system types for the 94 reference systems and the resulting annual performance. A bit more design variation can be obtained by using the sensitivity curves of Figure 9.10 to modify the SSF of a particular reference system. For instance, a direct gain system SSF of 0.37 would increase by approximately 0.03 if the mass area-to-glazing area ratio (assumed 6) were increased to 10 and would decrease by about 0.04 if the mass area-to-glazing area ratio were decreased to 3. This information provides a designer with quantitative information for making trade-offs.

9.5.3 Third Level: SLR Method

The SLR method calculates monthly performance, and the terms and values used are monthly based. The method allows the use of specific location weather data and the 94 reference design passive systems (Table 9.2). In addition, the sensitivity curves (Figure 9.10) can again be used to define performance outside the reference design systems. The result of the SLR method is the determination of the monthly heating auxiliary energy required, which is then summed to give the annual requirement for auxiliary heating energy. Generally, the SLR method gives annual values within ±3% of detailed simulation results, but the monthly values may vary more [5,7]. Thus, the monthly SLR method is more "accurate" than the rule-of-thumb methods, plus providing the designer with system performance on a month-by-month basis.

The SLR method uses equations and correlation parameters for each of the 94 reference systems combined with the insolation absorbed by the system, the monthly degree days, and the system's LCR to determine the monthly SSF. These correlation parameters are listed in Table 9.4 as A, B, C, D, R, G, H, and LCRs for each reference system [5]. The correlation equations are

$$SSF = 1 - K(1-F) \tag{9.9}$$

where

$$K = 1 + \frac{G}{LCR} \tag{9.10}$$

$$F = \begin{cases} AX, & \text{when } X < R \\ B - C \exp(-DX), & \text{when } X > R \end{cases} \tag{9.11}$$

$$X = \frac{S/DD - (LCR_S)H}{(LCR)K} \tag{9.12}$$

and X is called the generalized SLR. The term S is the monthly insolation absorbed by the system per unit of solar projected area. Monthly average daily insolation data on a vertical south-facing surface can be found and/or calculated using various sources [5,8], and the S term can be determined by multiplying by a transmission and an absorption factor and the number of days in the month. Absorption factors for all systems are close to 0.96 [5], whereas the transmission is approximately 0.9 for single glazing, 0.8 for double glazing, and 0.7 for triple glazing.

TABLE 9.4

SLR Correlation Parameters for the 94 Reference Systems

Type	A	B	C	D	R	G	H	LCR$_s$	STDV
WW A1	0.0000	1.0000	0.9172	0.4841	−9.0000	0.00	1.17	13.0	0.053
WW A2	0.0000	1.0000	0.9833	0.7603	−9.0000	0.00	0.92	13.0	0.046
WW A3	0.0000	1.0000	1.0171	0.8852	−9.0000	0.00	0.85	13.0	0.040
WW A4	0.0000	1.0000	1.0395	0.9569	−9.0000	0.00	0.81	13.0	0.037
WW A5	0.0000	1.0000	1.0604	1.0387	−9.0000	0.00	0.78	13.0	0.034
WW A6	0.0000	1.0000	1.0735	1.0827	−9.0000	0.00	0.76	13.0	0.033
WW B1	0.0000	1.0000	0.9754	0.5518	−9.0000	0.00	0.92	22.0	0.051
WW B2	0.0000	1.0000	1.0487	1.0851	−9.0000	0.00	0.78	9.2	0.036
WW B3	0.0000	1.0000	1.0673	1.0087	−9.0000	0.00	0.95	8.9	0.038
WW B4	0.0000	1.0000	1.1028	1.1811	−9.0000	0.00	0.74	5.8	0.034
WW B5	0.0000	1.0000	1.1146	1.2771	−9.0000	0.00	0.56	4.5	0.032
WW C1	0.0000	1.0000	1.0667	1.0437	−9.0000	0.00	0.62	12.0	0.038
WW C2	0.0000	1.0000	1.0846	1.1482	−9.0000	0.00	0.59	8.7	0.035
WW C3	0.0000	1.0000	1.1419	1.1756	−9.0000	0.00	0.28	5.5	0.033
WW C4	0.0000	1.0000	1.1401	1.2378	−9.0000	0.00	0.23	4.3	0.032
TW A1	0.0000	1.0000	0.9194	0.4601	−9.0000	0.00	1.11	13.0	0.048
TW A2	0.0000	1.0000	0.9680	0.6318	−9.0000	0.00	0.92	13.0	0.043
TW A3	0.0000	1.0000	0.9964	0.7123	−9.0000	0.00	0.85	13.0	0.038
TW A4	0.0000	1.0000	1.0190	0.7332	−9.0000	0.00	0.79	13.0	0.032
TW B1	0.0000	1.0000	0.9364	0.4777	−9.0000	0.00	1.01	13.0	0.045
TW B2	0.0000	1.0000	0.9821	0.6020	−9.0000	0.00	0.85	13.0	0.038
TW B3	0.0000	1.0000	0.9980	0.6191	−9.0000	0.00	0.80	13.0	0.033
TW B4	0.0000	1.0000	0.9981	0.5615	−9.0000	0.00	0.76	13.0	0.028
TW C1	0.0000	1.0000	0.9558	0.4709	−9.0000	0.00	0.89	13.0	0.039
TW C2	0.0000	1.0000	0.9788	0.4964	−9.0000	0.00	0.79	13.0	0.033
TW C3	0.0000	1.0000	0.9760	0.4519	−9.0000	0.00	0.76	13.0	0.029
TW C4	0.0000	1.0000	0.9588	0.3612	−9.0000	0.00	0.73	13.0	0.026
TW D1	0.0000	1.0000	0.9842	0.4418	−9.0000	0.00	0.89	22.0	0.040
TW D2	0.0000	1.0000	1.0150	0.8994	−9.0000	0.00	0.80	9.2	0.036
TW D3	0.0000	1.0000	1.0346	0.7810	−9.0000	0.00	1.08	8.9	0.036
TW D4	0.0000	1.0000	1.0606	0.9770	−9.0000	0.00	0.85	5.8	0.035
TW D5	0.0000	1.0000	1.0721	1.0718	−9.0000	0.00	0.61	4.5	0.033
TW E1	0.0000	1.0000	1.0345	0.8753	−9.0000	0.00	0.68	12.0	0.037
TW E2	0.0000	1.0000	1.0476	1.0050	−9.0000	0.00	0.66	8.7	0.035
TW E3	0.0000	1.0000	1.0919	1.0739	−9.0000	0.00	0.61	5.5	0.034
TW E4	0.0000	1.0000	1.0971	1.1429	−9.0000	0.00	0.47	4.3	0.033
TW F1	0.0000	1.0000	0.9430	0.4744	−9.0000	0.00	1.09	13.0	0.047
TW F2	0.0000	1.0000	0.9900	0.6053	−9.0000	0.00	0.93	13.0	0.041
TW F3	0.0000	1.0000	1.0189	0.6502	−9.0000	0.00	0.86	13.0	0.036
TW F4	0.0000	1.0000	1.0419	0.6258	−9.0000	0.00	0.80	13.0	0.032
TW G1	0.0000	1.0000	0.9693	0.4714	−9.0000				0.042
TW G2	0.0000	1.0000	1.0133	0.5462	−9.0000	0.00	0.88	13.0	0.035
TW G3	0.0000	1.0000	1.0325	0.5269	−9.0000	0.00	0.82	13.0	0.031

(continued)

TABLE 9.4 (continued)

SLR Correlation Parameters for the 94 Reference Systems

Type	A	B	C	D	R	G	H	LCR$_s$	STDV
TW G4	0.0000	1.0000	1.0401	0.4400	−9.0000	0.00	0.77	13.0	0.030
TW H1	0.0000	1.0000	1.0002	0.4356	−9.0000	0.00	0.93	13.0	0.034
TW H2	0.0000	1.0000	1.0280	0.4151	−9.0000	0.00	0.83	13.0	0.030
TW H3	0.0000	1.0000	1.0327	0.3522	−9.0000	0.00	0.78	13.0	0.029
TW H4	0.0000	1.0000	1.0287	0.2600	−9.0000	0.00	0.74	13.0	0.024
TW I1	0.0000	1.0000	0.9974	0.4036	−9.0000	0.00	0.91	22.0	0.038
TW I2	0.0000	1.0000	1.0386	0.8313	−9.0000	0.00	0.80	9.2	0.034
TW I3	0.0000	1.0000	1.0514	0.6886	−9.0000	0.00	1.01	8.9	0.034
TW I4	0.0000	1.0000	1.0781	0.8952	−9.0000	0.00	0.82	5.8	0.032
TW I5	0.0000	1.0000	1.0902	1.0284	−9.0000	0.00	0.65	4.5	0.032
TW J1	0.0000	1.0000	1.0537	0.8227	−9.0000	0.00	0.65	12.0	0.037
TW J2	0.0000	1.0000	1.0677	0.9312	−9.0000	0.00	0.62	8.7	0.035
TW J3	0.0000	1.0000	1.1153	0.9831	−9.0000	0.00	0.44	5.5	0.034
TW J4	0.0000	1.0000	1.1154	1.0607	−9.0000	0.00	0.38	4.3	0.033
DG A1	0.5650	1.0090	1.0440	0.7175	0.3931	9.36	0.00	0.0	0.046
DG A2	0.5906	1.0060	1.0650	0.8099	0.4681	5.28	0.00	0.00	0.039
DG A3	0.5442	0.9715	1.1300	0.9273	0.7068	2.64	0.00	0.00	0.036
DG B1	0.5739	0.9948	1.2510	1.0610	0.7905	9.60	0.00	0.00	0.042
DG B2	0.6180	1.0000	1.2760	1.1560	0.7528	5.52	0.00	0.00	0.035
DG B3	0.5601	0.9839	1.3520	1.1510	0.8879	2.38	0.00	0.00	0.032
DG C1	0.6344	0.9887	1.5270	1.4380	0.8632	9.60	0.00	0.00	0.039
DG C2	0.6763	0.9994	1.4000	1.3940	0.7604	5.28	0.00	0.00	0.033
DG C3	0.6182	0.9859	1.5660	1.4370	0.8990	2.40	0.00	0.00	0.031
SS A1	0.0000	1.0000	0.9587	0.4770	−9.0000	0.00	0.83	18.6	0.027
SS A2	0.0000	1.0000	0.9982	0.6614	−9.0000	0.00	0.77	10.4	0.026
SS A3	0.0000	1.0000	0.9552	0.4230	−9.0000	0.00	0.83	23.6	0.030
SS A4	0.0000	1.0000	0.9956	0.6277	−9.0000	0.00	0.80	12.4	0.026
SS A5	0.0000	1.0000	0.9300	0.4041	−9.0000	0.00	0.96	18.6	0.031
SS A6	0.0000	1.0000	0.9981	0.6660	−9.0000	0.00	0.86	10.4	0.028
SS A7	0.0000	1.0000	0.9219	0.3225	−9.0000	0.00	0.96	23.6	0.035
SS A8	0.0000	1.0000	0.9922	0.6173	−9.0000	0.00	0.90	12.4	0.028
SS B1	0.0000	1.0000	0.9683	0.4954	−9.0000	0.00	0.84	16.3	0.028
SS B2	0.0000	1.0000	1.0029	0.6802	−9.0000	0.00	0.74	8.5	0.026
SS B3	0.0000	1.0000	0.9689	0.4685	−9.0000	0.00	0.82	19.3	0.029
SS B4	0.0000	1.0000	1.0029	0.6641	−9.0000	0.00	0.76	9.7	0.026
SS B5	0.0000	1.0000	0.9408	0.3866	−9.0000	0.00	0.97	16.3	0.030
SS B6	0.0000	1.0000	1.0068	0.6778	−9.0000	0.00	0.84	8.5	0.028
SS B7	0.0000	1.0000	0.9395	0.3363	−9.0000				0.032
SS B8	0.0000	1.0000	1.0047	0.6469	−9.0000	0.00	0.87	9.7	0.027
SS C1	0.0000	1.0000	1.0087	0.7683	−9.0000	0.00	0.76	16.3	0.025
SS C2	0.0000	1.0000	1.0412	0.9281	−9.0000	0.00	0.78	10.0	0.027
SS C3	0.0000	1.0000	0.9699	0.5106	−9.0000	0.00	0.79	16.3	0.024
SS C4	0.0000	1.0000	1.0152	0.7523	−9.0000	0.00	0.81	10.0	0.025
SS D1	0.0000	1.0000	0.9889	0.6643	−9.0000	0.00	0.84	17.8	0.028

TABLE 9.4 (continued)

SLR Correlation Parameters for the 94 Reference Systems

Type	A	B	C	D	R	G	H	LCR$_s$	STDV
SS D2	0.0000	1.0000	1.0493	0.8753	−9.0000	0.00	0.70	9.9	0.028
SS D3	0.0000	1.0000	0.9570	0.5285	−9.0000	0.00	0.90	17.8	0.029
SS D4	0.0000	1.0000	1.0356	0.8142	−9.0000	0.00	0.73	9.9	0.028
SS E1	0.0000	1.0000	0.9968	0.7004	−9.0000	0.00	0.77	19.6	0.027
SS E2	0.0000	1.0000	1.0468	0.9054	−9.0000	0.00	0.76	10.8	0.027
SS E3	0.0000	1.0000	0.9565	0.4827	−9.0000	0.00	0.81	19.6	0.028
SS E4	0.0000	1.0000	1.0214	0.7694	−9.0000	0.00	0.79	10.8	0.027

Source: PSDH, *Passive Solar Design Handbook Part One*: Total Environmental Action, Inc., *Part Two*: Los Alamos Scientific Laboratory, *Part Three*: Los Alamos National Laboratory, Van Nostrand Reinhold, New York, 1984. With permission.

Example 9.5

For a vented, 180 ft², double-glazed with night insulation, 12 in. thick Trombe wall system (TWD4) in an NLC = 11,800 Btu/FDD house in Medford, Oregon, determine the auxiliary energy required in January.

Solution

Weather data for Medford, Oregon [5], yields for January (N = 31, days): daily vertical surface insolation = 565 Btu/ft² and 880 FDD, so S = (31) (565) (0.8) (0.96) = 13,452 Btu/ft²-month.

$$LCR = \frac{NLC}{A_p} = \frac{11,800}{180} = 65.6 \, Btu/FDD\,ft^2$$

From Table 9.4 at TWD4: A = 0, B = 1, C = 1.0606, D = 0.977, R = −9, G = 0, H = 0.85, LCRs = 5.8 Btu/FDD ft².
Substituting into Equation 9.10 gives

$$K = 1 + \frac{0}{65.6} = 1$$

Equation 9.12 gives

$$X = \frac{13,452/880 - (5.8 \times 0.85)}{65.6 \times 1} = 0.16$$

Equation 9.11 gives

$$F = 1 - 1.0606e^{-0.977 \times 0.16} = 0.09$$

and Equation 9.9 gives

$$SSF = 1 - 1(1 - 0.09) = 0.09$$

The January auxiliary energy required can be calculated using Equation 9.6:

$$Q_{aux}(\text{January}) = (1 - \text{SSF}) \times \text{NLC} \times (\text{No. of degree days})$$

$$= (1 - 0.09) \times 11{,}800 \times 880$$

$$= 9{,}450{,}000 \, \text{Btu}$$

As mentioned, the use of sensitivity curves [5] as in Figure 9.10 will allow SSF to be determined for many off-reference system design conditions involving storage mass, number of glazings, and other more esoteric parameters. Also the use of multiple passive system types within one building would be approached by calculating the SSF for each system type individually using a "combined area" LCR and then a weighted area (aperture) average SSF would be determined for the building.

9.6 Passive Space-Cooling Design Fundamentals

Passive cooling systems are designed to use natural means to transfer heat from buildings, including convection/ventilation, evaporation, radiation, and conduction. However, the most important element in both passive and conventional cooling design is to prevent heat from entering the building in the first place. Cooling conservation techniques involve building surface colors, insulation, special window glazings, overhangs and orientation, and numerous other architectural/engineering features.

9.6.1 Solar Control

Controlling the solar energy input to reduce the cooling load is usually considered a passive (versus conservation) design concern because solar input may be needed for other purposes, such as daylighting throughout the year and/or heating during the winter. Basic architectural solar control is normally "designed in" via the shading of the solar windows, where direct radiation is desired for winter heating and needs to be excluded during the cooling season.

The shading control of the windows can be of various types and "controllability," ranging from drapes and blinds, use of deciduous trees, to the commonly used overhangs and vertical louvers. A rule-of-thumb design for determining proper south-facing window overhang for both winter heating and summer shading is presented in Table 9.5. Technical details on calculating shading from various devices and orientations are found in [9,10].

9.6.2 Natural Convection/Ventilation

Air movement provides cooling comfort through convection and evaporation from human skin. ASHRAE Handbook [9] places the comfort limit at 79°F for an air velocity of 50 ft/min (fpm), 82°F for 160 fpm, and 85°F for 200 fpm. To determine whether or not comfort conditions can be obtained, a designer must calculate the volumetric flow rate, Q, which is

TABLE 9.5

South-Facing Window Overhang Rule of Thumb

$$\text{Length of the Overhang} = \frac{\text{Window Height}}{F}$$

(a) Overhang Factors

North Latitude	F^a	(b) Roof Overhang Geometry
28	5.6–11.1	
32	4.0–6.3	
36	3.0–4.5	
40	2.5–3.4	
44	2.0–2.7	
48	1.7–2.2	
52	1.5–1.8	
56	1.3–1.5	

Summer sun

Winter sun

Properly sized overhangs shade out hot summer sun but allow winter sun (which is lower in the sky) to penetrate windows

Source: Halacy, D.S., *Home Energy*, Rodale Press, Emmaus, PA, 1984. With permission.

[a] Select a factor according to your latitude. High values provide complete shading at noon on June 21: lower values, until August 1.

passing through the occupied space. Using the cross-sectional area, A_x, of the space and the room air velocity, V_a, required, the flow is determined by

$$Q = A_x V_a. \tag{9.13}$$

The proper placement of windows, "narrow" building shape, and open landscaping can enhance natural wind flow to provide ventilation. The air flow rate through open windows for wind-driven ventilation is given by [9]

$$Q = C_v V_w A_w \tag{9.14}$$

where
Q is the air flow rate, m³/s
A_w is the free area of inlet opening, m²
V_w is the wind velocity, m/s
C_v is the effectiveness of opening = 0.5–0.6 for wind perpendicular to opening and 0.25–0.35 for wind diagonal to opening

The stack effect can induce ventilation when warm air rises to the top of a structure and exhausts outside, while cooler outside air enters the structure to replace it. Figure 9.11 illustrates the solar chimney concept, which can easily be adapted to a thermal storage wall

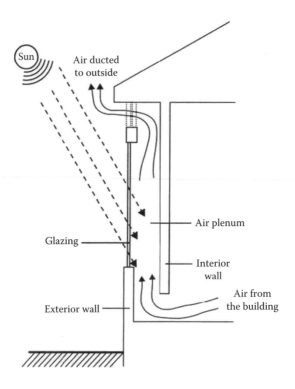

FIGURE 9.11

The stack-effect/solar chimney concept to induce convection/ventilation. (From PSDH. (1980) *Passive Solar Design Handbook*, Volume One: *Passive Solar Design Concepts*, DOE/CS-0127/1, March 1980. Prepared by Total Environmental Action, Inc. (B. Anderson, C. Michal, P. Temple, and D. Lewis); Volume Two: *Passive Solar Design Analysis*, DOE/CS-0127/2, January 1980. Prepared by Los Alamos Scientific Laboratory (J.D. Balcomb, D. Barley, R McFarland, J. Perry, W. Wray, and S. Noll). U.S. Department of Energy, Washington, DC. With permission.)

system. The greatest stack-effect flow rate is produced by maximizing the stack height and the air temperature in the stack, as given by

$$Q = 0.116 A_j \sqrt{h(T_s - T_o)} \qquad (9.15)$$

where
Q is the stack flow rate, m³/s
A_j is the area of inlets or outlets (whichever is smaller), m²
h is the inlet to outlet height, m
T_s is the average temperature in stack,°C
T_o is the outdoor air temperature,°C

If inlet or outlet area is twice the other, the flow rate will increase by 25%, and by 35% if the areas' ratio is 3:1 or larger.

Example 9.6

A two-story (5 m) solar chimney is being designed to produce a flow of 0.25 m³/s through a space. The preliminary design features include a 25 cm × 1.5 m inlet, a 50 cm × 1.5 m outlet, and an estimated 35°C average stack temperature on a sunny 30°C day. Can this design produce the desired flow?

Solution
Substituting the design data into Equation 9.15,

$$Q = 0.116(0.25 \times 1.5)[5(5)]^{1/2}$$

$$= 0.2 \text{ m}^3/\text{s}$$

Since the outlet area is twice the inlet area, the 25% flow increase can be used:

$$Q = 0.2 (1.25) = 0.25 \text{ m}^3/\text{s} \quad \text{(answer: Yes, the proper flow rate is obtained.)}$$

9.6.3 Evaporative Cooling

When air with less than 100% relative humidity moves over a water surface, the evaporation of water causes both the air and the water to cool. The lowest temperature that can be reached by this direct evaporative cooling effect is the wet-bulb temperature of the air, which is directly related to the relative humidity, with lower wet-bulb temperature associated with lower relative humidity. Thus, dry air (low relative humidity) has a low wet-bulb temperature and will undergo a large temperature drop with evaporative cooling, while humid air (high relative humidity) can only be slightly cooled evaporatively. The wet-bulb temperature for various relative humidity and air temperature conditions can be found via the "psychrometric chart" available in most thermodynamic texts. Normally, an evaporative cooling process cools the air only part of the way down to the wet-bulb temperature. To get the maximum temperature decrease, it is necessary to have a large water surface area in contact with the air for a long time, and interior ponds and fountain sprays are often used to provide this air–water contact area.

The use of water sprays and open ponds on roofs provides cooling primarily via evaporation. The hybrid system involving a fan and wetted mat, the "swamp cooler," is by far the most widely used evaporative cooling technology. Direct, indirect, and combined evaporative cooling system design features are described in [9,11].

9.6.4 Nocturnal and Radiative Cooling Systems

Another approach to passive convective/ventilative cooling involves using cooler night air to reduce the temperature of the building and/or a storage mass. Thus, the building/storage mass is prepared to accept part of the heat load during the hotter daytime. This type of convective system can also be combined with evaporative and radiative modes of heat transfer, utilizing air and/or water as the convective fluid. Work in Australia [3] investigated rock storage beds that were chilled using evaporatively cooled night air. Room air was then circulated through the bed during the day to provide space cooling. The use of encapsulated roof ponds as a thermal cooling mass has been tried by several investigators [12–14] and is often linked with nighttime radiative cooling.

All warm objects emit thermal infrared radiation; the hotter the body, the more energy it emits. A passive cooling scheme is to use the cooler night sky as a sink for thermal

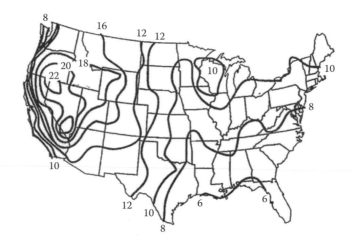

FIGURE 9.12
Average monthly sky temperature depression ($T_{air} - T_{sky}$) for July in°F. (From Martin, M. and Kreider, J.F. *Solar Energy*, 33, 321, 1984. With permission.)

radiation emitted by a warm storage mass, thus chilling the mass for cooling use the next day. The net radiative cooling rate, Q_r, for a horizontal unit surface [9] is

$$Q_r = \varepsilon\sigma\left(T_{body}^4 - T_{sky}^4\right) \tag{9.16}$$

where
Q_r is the net radiative cooling rate, W/m² (Btu/h ft²)
ε is the surface emissivity fraction (usually 0.9 for water)
$\sigma = 5.67 \times 10^{-8}$ W/m² K⁴ (1.714×20^{-9} Btu/h ft² R⁴)
T_{body} is the warm body temperature, Kelvin (Rankine)
T_{sky} is the effective sky temperature, Kelvin (Rankine)

The monthly average air–sky temperature difference has been determined [15], and Figure 9.12 presents these values for July (in°F) for the United States.

Example 9.7

Estimate the overnight cooling possible for a 10 m², 85°F water thermal storage roof during July in Los Angeles.

Solution
Assume the roof storage unit is black with $\varepsilon = 0.9$. From Figure 9.12, $T_{air} - T_{sky}$ is approximately 10°F for Los Angeles. From weather data for Los Angeles airport [5,9], the July average temperature is 69°F with a range of 15°F. Assuming night temperatures vary from the average (69°F) down to half the daily range (15.1/2), the average nighttime temperature is chosen as $69 - (1/2)(15/2) = 65$°F. So, $T_{sky} = 65 - 10 = 55$°F. From Equation 9.16,

$$Q_r = 0.9(1.714 \times 10^{-9})[(460 + 85)^4 - (460 + 55)^4]$$

$$= 27.6\,\text{Btu/h ft}^2$$

For a 10 h night and 10 m² (107.6 ft²) roof area,

$$\text{Total radiative cooling} = 27.6(10)(107.6)$$

$$= 29,700 \, \text{Btu}$$

NOTE: This does not include the convective cooling possible, which can be approximated (at its maximum rate) for still air [5] by

$$\text{Maximum total } Q_{conv} = hA(T_{roof} - T_{air})(\text{Time})$$

$$= 5(129)(85 - 55)(10)$$

$$= 161,000 \, \text{Btu}$$

This is the maximum since the 85°F storage temperature will drop as it cools—which is also the case for the radiative cooling calculation. However, convection is seen to usually be the more dominant mode of nighttime cooling.

9.6.5 Earth Contact Cooling (or Heating)

Earth contact cooling or heating is a passive summer cooling and winter heating technique that utilizes underground soil as the heat sink or source. By installing a pipe underground and passing air through the pipe, the air will be cooled or warmed depending on the season. A schematic of an open-loop system and a closed-loop air-conditioning system is presented in Figures 9.13 and 9.14, respectively [16].

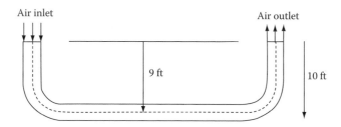

FIGURE 9.13
Open-loop underground air tunnel system. (From Goswami, D.Y. and Biseli, K.M. Use of underground air tunnels for heating and cooling agricultural and residential buildings. Report EES-78, Florida Energy Extension Service, University of Florida, Gainesville, FL, August 1994. With permission.)

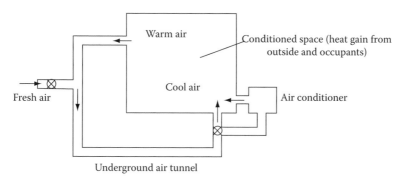

FIGURE 9.14
Schematic of closed-loop air-conditioning system using air tunnel. (From Goswami, D.Y. and Biseli, K.M. Use of underground air tunnels for heating and cooling agricultural and residential buildings. Report EES-78, Florida Energy Extension Service, University of Florida, Gainesville, FL, August 1994. With permission.)

The use of this technique can be traced back to 3000 BC when Iranian architects designed some buildings to be cooled by natural resources only. In the nineteenth century, Wilkinson [17] designed a barn for 148 cows where a 500 ft long underground passage was used for cooling during the summertime. Since that time, a number of experimental and analytical studies of this technique have continued to appear in the literature [18–20]. Goswami and Dhaliwal [21] have given a brief review of the literature as well as presented an analytical solution to the problem of transient heat transfer between the air and the surrounding soil as the air is made to pass through a pipe buried underground.

9.6.5.1 Heat-Transfer Analysis

The transient thermal analysis of the air and soil temperature fields [21] is conducted using finite elements with the convective heat transfer between the air and the pipe and using semi-infinite cylindrical conductive heat transfer to the soil from the pipe. It should be noted that the thermal resistance of the pipe (whether of metal, plastic, or ceramic) is negligible relative to the surrounding soil.

9.6.5.1.1 Air and Pipe Heat Transfer

The pipe is divided into a large number of elements and a psychrometric energy balance written for each, depending on whether the air leaves the element (a) unsaturated or (b) saturated.

a. If the air leaves an element as unsaturated, the energy balance on the element is

$$mC_p(T_1 - T_2) = hA_p(T_{air} - T_{pipe}) \tag{9.17}$$

T_{air} can be taken as $(T_1 + T_2)/2$

Substituting and simplifying, we get

$$T_2 = \frac{[(1 - U/2)T_1 + UT_{pipe}]}{(1 + U/2)} \tag{9.18}$$

where U is defined as

$$U = \frac{A_p h}{mC_p}$$

b. If the air leaving the element is saturated, the energy balance is

$$mC_p T_1 + m(W_1 - W_2)H_{fg} = mC_p T_2 + hA_p(T_{air} - T_{pipe}) \tag{9.19}$$

Simplifying, we get

$$T_2 = \frac{\left(1 - \dfrac{U}{2}T_1\right) + \dfrac{W_1 - W_2}{C_p}H_{fg} + UT_{pipe}}{(1 + U/2)} \tag{9.20}$$

The convective heat-transfer coefficient h in the preceding equations depends on Reynolds number, the shape, and roughness of the pipe. Using the exit temperature from the first element as the inlet temperature for the next element, the exit temperature for the element can be calculated in a similar way. Continuing this way from one element to the next, the temperature of air at the exit from the pipe can be calculated.

9.6.5.1.2 Soil Heat Transfer

The heat transfer from the pipe to the soil is analyzed by considering the heat flux at the internal radius of a semi-infinite cylinder formed by the soil around the pipe. For a small element, the problem can be formulated as

$$\frac{\partial^2 T(r,t)}{\partial r^2} + \frac{1}{r}\frac{\partial T(r,t)}{\partial r} = \frac{1}{\alpha}\frac{\partial T(r,t)}{\partial t} \tag{9.21}$$

with initial and boundary conditions as

$$T(r,0) = T_e$$

$$T(\infty,t) = T_e$$

$$-K\frac{\partial T}{\partial r}(r,t) = q''$$

where

T_e is the bulk earth temperature

q'' is also given by the amount of heat transferred to the pipe from the air by convection, that is, $q'' = h(T_{air} - T_{pipe})$

9.6.5.2 Soil Temperatures and Properties

Kusuda and Achenbach [22] and Labs [23] studied the earth temperatures in the United States. According to both of these studies, temperature swings in the soil during the year are dampened with depth below the ground. There is also a phase lag between the soil temperature and the ambient air temperature, and this phase lag increases with depth below the surface. For example, the soil temperature for light dry soil at a depth of about 10 ft (3.05 m) varies by approximately $\pm5°F$ (2.8°C) from the mean temperature (approximately equal to mean annual air temperature) and has a phase lag of approximately 75 days behind ambient air temperature [23].

The thermal properties of the soil are difficult to determine. The thermal conductivity and diffusivity both change with the moisture content of the soil itself, which is directly affected by the temperature of and heat flux from and to the buried pipe. Most researchers have found that using constant property values for soil taken from standard references gives reasonable predictive results [24].

9.6.5.3 Generalized Results from Experiments

Figure 9.15 presents data from [16] for an open system, 100 ft long, 12 in. diameter pipe, buried 9 ft deep. The figure shows the relationship between pipe inlet-to-outlet temperature

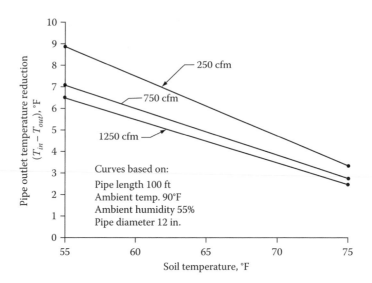

FIGURE 9.15
Air temperature drop through a 100 ft long, 12 in. diameter pipe buried 9 ft underground. (From Goswami, D.Y. and Biseli, K.M. Use of underground air tunnels for heating and cooling agricultural and residential buildings. Report EES-78, Florida Energy Extension Service, University of Florida, Gainesville, FL, August 1994. With permission.)

reduction $(T_{in} - T_{out})$ and the initial soil temperature with ambient air inlet conditions of 90°F and 55% relative humidity for various pipe flow rates.

Other relations from this same report that can be used with Figure 9.15 data include the following: (1) the effect of increasing pipe/tunnel length on increasing the inlet-to-outlet air temperature difference is fairly linear up to 250 ft and (2) the effect of decreasing pipe diameter on lowering the outlet air temperature is slight, and only marginally effective for pipes less than 12 in. diameter.

Example 9.8

Provide the necessary 12 in. diameter pipe length(s) that will deliver 1500 cfm of 75°F air if the ambient temperature is 85°F and the soil at 9 ft is 65°F.

Solution
From Figure 9.15, for 100 ft of pipe at 65°F soil temperature, the pipe temperature reduction is

$$T_{in} - T_{out} = 6° \text{ (at 250 cfm)}$$

$$= 5°F \text{ (at 750 cfm)}$$

$$= 4.5°F \text{ (at 1250 cfm)}$$

Since the "length versus temperature reduction" is linear (see the earlier text), the 10°F reduction required (85 down to 75) would be met by the 750 cfm case (5°F for 100 ft) if 200 ft of pipe is used. Then, two 12 in. diameter pipes would be required to meet the 1500 cfm requirement.

Answer
Two 12 in. diameter pipes, each 200 ft long. (*Note:* See what would be needed if the 250 cfm or the 1250 cfm cases had been chosen. Which of the three flow rate cases leads to the "cheapest" installation?)

9.7 Daylighting Design Fundamentals

Daylighting is the use of the sun's radiant energy to illuminate the interior spaces in a building. In the last century, electric lighting was considered an alternative technology to daylighting. Today the situation is reversed, primarily due to the economics of energy use and conservation. However, there are good physiological reasons for using daylight as an illuminant. The quality of daylight matches the human eye's response, thus permitting lower light levels for task comfort, better color rendering, and clearer object discrimination [25–27].

9.7.1 Lighting Terms and Units

Measurement of lighting level is based on the "standard candle," where the *lumen* (lm), the unit of *luminous flux* (ϕ), is defined as the rate of luminous energy passing through a 1 m^2 area located 1 m from the candle. Thus, a standard candle generates 4π lumens that radiate away in all directions. The *illuminance* (E) on a surface is defined as the luminous flux on the surface divided by the surface area, $E = \phi/A$. Illuminance is measured in either *lux* (lx), as lumens per square meter, or *footcandles* (fc), as lumens per square foot.

Determination of the daylighting available at a given location in a building space at a given time is important to evaluate the reduction possible in electric lighting and the associated impact on heating and cooling loads. Daylight provides about 110 lm/W of solar radiation, fluorescent lamps about 75 lm/W of electrical input, and incandescent lamps about 20 lm/W; thus daylighting generates only one half to one-fifth the heating that equivalent electric lighting does, significantly reducing the building cooling load.

9.7.2 Approach to Daylighting Design

Aperture controls such as blinds and drapes are used to moderate the amount of daylight entering the space, as are the architectural features of the building itself (glazing type, area, and orientation; overhangs and wing-walls; light shelves; etc.). Many passive and "active" reflective, concentrating, and diffusing devices are available to specifically gather and direct both the direct and diffuse components of daylight to areas within the space [28]. Electric lighting dimming controls are used to adjust the electric light level based on the quantity of the daylighting. With these two types of controls (aperture and lighting), the electric lighting and cooling energy use and demand, as well as cooling system sizing, can be reduced. However, the determination of the daylighting position and time illuminance value within the space is required before energy usage and demand reduction calculations can be made.

Daylighting design approaches use both solar beam radiation (referred to as *sunlight*) and the diffuse radiation scattered by the atmosphere (referred to as *skylight*) as sources for interior lighting, with historical design emphasis being on utilizing skylight. Daylighting is provided through a variety of glazing features, which can be grouped as *sidelighting* (light enters via the side of the space) and *toplighting* (light enters from the ceiling area). Figure 9.16 illustrates several architectural forms producing sidelighting and toplighting, with the dashed lines representing the illuminance

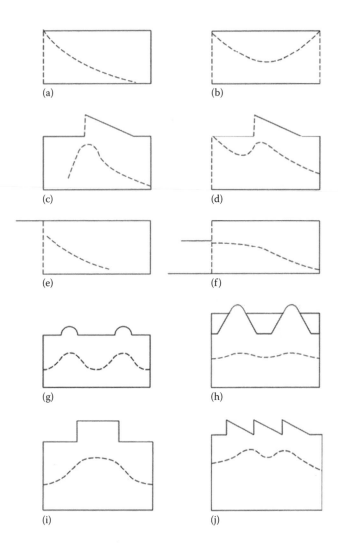

FIGURE 9.16
Examples of sidelighting and toplighting architectural features (dashed lines represent illuminance distributions). (a) Unilateral, (b) bilateral, (c) clerestory, (d) clerestory + unilateral, (e) overhang, (f) overhang + ground reflection, (g) skylight, (h) skylight + well, (i) roof monitor, and (j) sawtooth. (From Murdoch, J.B. *Illumination Engineering—From Edison's Lamp to the Laser*, Macmillan, New York, 1985. With permission.)

distribution within the space. The calculation of work-plane illuminance depends on whether sidelighting and/or toplighting features are used, and the combined illuminance values are additive.

9.7.3 Sun–Window Geometry

The solar illuminance on a vertical or horizontal window depends on the position of the sun relative to that window. In the method described here, the sun and sky illuminance values are determined using the sun's altitude angle (α) and the sun–window azimuth angle difference (Φ). These angles need to be determined for the particular time of day, day of year, and window placement under investigation.

9.7.3.1 Solar Altitude Angle (α)

The solar altitude angle is the angle swept out by a person's arm when pointing to the horizon directly below the sun and then raising the arm to point at the sun. The equation to calculate solar altitude, α, is

$$\sin \alpha = \cos L \cos \delta \cos H + \sin L \sin \delta \tag{9.22}$$

where
  L is the local latitude (degrees)
  δ is the earth–sun declination (degrees) given by $\delta = 23.45 \sin [360 (n - 81)/365]$
  n is the day number of the year
  H is the hour angle (degrees) given by

$$H = \frac{(12\,\text{noon} - \text{time})(\text{in min})}{4}; \quad (+\text{morning}, -\text{afternoon}) \tag{9.23}$$

9.7.3.2 Sun–Window Azimuth Angle Difference (Φ)

The difference between the sun's azimuth and the window's azimuth needs to be calculated for vertical window illuminance. The window's azimuth angle, γ_w, is determined by which way it faces, as measured from south (east of south is positive, westward is negative). The solar azimuth angle, γ_s, is calculated as

$$\sin \gamma_s = \frac{\cos \delta \sin H}{\cos \alpha} \tag{9.24}$$

The sun–window azimuth angle difference, Φ, is given by the absolute value of the difference between γ_s and γ_w:

$$\Phi = |\gamma_s - \gamma_w| \tag{9.25}$$

9.7.4 Daylighting Design Methods

To determine the annual lighting energy saved (ES_L), calculations using the lumen method described later should be performed on a monthly basis for both clear and overcast days for the space under investigation. Monthly weather data for the site would then be used to prorate clear and overcast lighting energy demands monthly. Subtracting the calculated daylighting illuminance from the design illuminance leaves the supplementary lighting needed, which determines the lighting energy required.

    The approach in the following method is to calculate the "sidelighting" and the "skylighting" of the space separately and then combine the results. This procedure has been computerized (Lumen II/Lumen Micro) and includes many details of controls, daylighting technologies, and weather. ASHRAE [9] lists many of the methods and simulation techniques currently used with daylighting and its associated energy effects.

9.7.4.1 Lumen Method of Sidelighting (Vertical Windows)

The lumen method of sidelighting calculates interior horizontal illuminance at three points, as shown in Figure 9.17, at the 30 in. (0.76 m) work plane on the room-and-window centerline. A vertical window is assumed to extend from 36 in. (0.91 m) above the floor to the ceiling. The method accounts for both direct and ground-reflected sunlight and sky-light, so both horizontal and vertical illuminances from sun and sky are needed. The steps in the lumen method of sidelighting are presented next.

As mentioned, the incident direct and ground-reflected window illuminances are nor-mally calculated for both a cloudy and a clear day for representative days during the year (various months), as well as for clear or cloudy times during a given day. Thus, the interior illumination due to sidelighting and skylighting can then be examined for effectiveness throughout the year.

Step 1: Incident direct sky and sun illuminances
The solar altitude and sun–window azimuth angle difference are calculated for the desired latitude, date, and time using Equations 9.22 and 9.25, respectively. Using these two angles, the total illuminance on the window (E_{sw}) can be determined by summing the direct sun illuminance (E_{uw}) and the direct sky illuminance (E_{kw}), each determined from the appropri-ate graph in Figure 9.18.

Step 2: Incident ground-reflected illuminance
The sun illuminance on the ground (E_{ug}), plus the overcast or clear sky illuminance (E_{kg}) on the ground, makes up the total horizontal illuminance on the ground surface (E_{sg}). A fraction of the ground surface illuminance is then considered diffusely reflected onto the vertical window surface (E_{gw}), where "gw" indicates from the ground to the window.

The horizontal ground illuminances can be determined using Figure 9.19, where the clear sky plus sun case and the overcast sky case are functions of solar altitude. The frac-tions of the ground illuminance diffusely reflected onto the window depend on the reflec-tivity (ρ) of the ground surface (see Table 9.6) and the window-to-ground surface geometry.

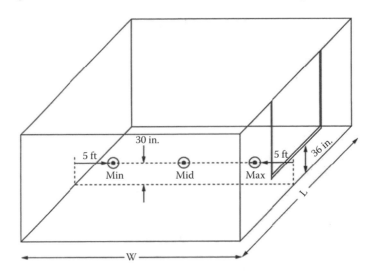

FIGURE 9.17
Location of illumination points within the room (along centerline of window) determined by lumen method of sidelighting. (From Kreith, F. and West, R., *CRC Handbook of Energy Efficiency and Renewable Energy*, CRC Press, Boca Raton, FL, 2007. With permission.)

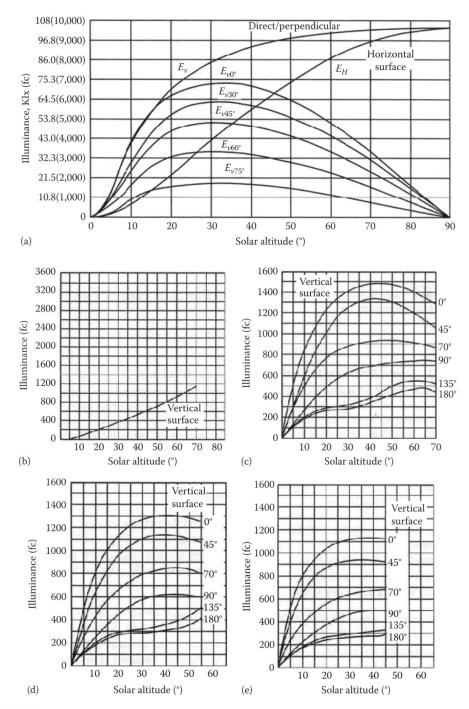

FIGURE 9.18

Vertical illuminance from (a) direct sunlight, (b) overcast skylight, (c) clear summer skylight (d) clear autumn skylight, and (e) clear winter skylight, for various sun–window azimuth angle differences. (Modified from Kaufman, J.E., ed., *The IES Lighting Handbook: Reference Volume*, IESNA, New York, 1984. With permission from the Illuminating Engineering Society of North America. With permission.)

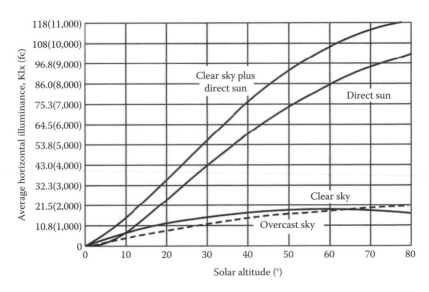

FIGURE 9.19
Horizontal illuminance for overcast sky, direct sun, and clear sky direct sun. (From Murdoch, J.B., *Illumination Engineering—From Edison's Lamp to the Laser*, Macmillan, New York, 1985. With permission.)

TABLE 9.6

Ground Reflectivities

Material	ρ (%)
Cement	27
Concrete	20–40
Asphalt	7–4
Earth	10
Grass	6–20
Vegetation	25
Snow	70
Red brick	30
Gravel	15
White paint	55–75

Source: Murdoch, J.B., *Illumination Engineering-From Edison's Lamp* to *the Laser*, Macmillan, New York, 1985. With permission.

If the ground surface is considered uniformly reflective from the window outward to the horizon, then the illuminance on the window from ground reflection is

$$E_{gw} = \frac{\rho E_{sg}}{2} \tag{9.26}$$

A more complicated ground-refection case is illustrated in Figure 9.20, with multiple "strips" of differently reflecting ground being handled using the angles to the window, where a strip's illuminance on a window is calculated:

$$E_{gw(strip)} = \frac{\rho_{strip} E_{sg}}{2} (\cos\theta_1 - \cos\theta_2) \tag{9.27}$$

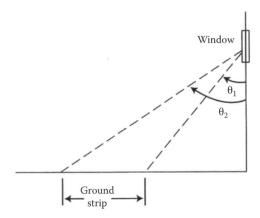

FIGURE 9.20
Geometry for ground "strips." (From Murdoch, J.B., *Illumination Engineering—From Edison's Lamp to the Laser*, Macmillan, New York, 1985. With permission.)

And the total reflected onto the window is the sum of the strip illuminances:

$$E_{gw} = \frac{\rho E_{sg}}{2} [\rho_1(\cos 0 - \cos \theta_1) + \rho_2(\cos \theta_1 - \cos \theta_2) + \cdots + \rho_n(\cos \theta_{n-1} - \cos 90)] \qquad (9.28)$$

Step 3: Luminous flux entering space
The direct sky–sun and ground-reflected luminous fluxes entering the building are attenuated by the transmissivity of the window. Table 9.7 presents the transmittance fraction (τ) of several window glasses. The fluxes entering the space are calculated from the total sun–sky and the ground-reflected illuminances by using the area of the glass, A_w:

$$\phi_{sw} = E_{sw}\tau A_w$$
$$\phi_{gw} = E_{gw}\tau A_w \qquad (9.29)$$

Step 4: Light loss factor
The light loss factor (K_m) accounts for the attenuation of luminous flux due to dirt on the window (WDD, window dirt depreciation) and on the room surfaces (RSDD, room surface dirt depreciation). The WDD depends on how often the window is cleaned, but a 6-month average for offices is 0.83 and for factories is 0.71 [29].

The RSDD is a more complex calculation involving time between cleanings, the direct–indirect flux distribution, and room proportions. However, for rooms cleaned regularly, RSDD is around 0.94, and for once-a-year-cleaned dirty rooms, the RSDD would be around 0.84.

The light loss factor is the product of the preceding two fractions:

$$K_m = (\text{WDD})(\text{RSDD}) \qquad (9.30)$$

TABLE 9.7

Glass Transmittances

Glass	Thickness (in.)	τ (%)
Clear	$\dfrac{1}{8}$	89
Clear	$\dfrac{3}{16}$	88
Clear	$\dfrac{1}{4}$	87
Clear	$\dfrac{5}{16}$	86
Gray	$\dfrac{1}{8}$	61
Gray	$\dfrac{3}{16}$	51
Gray	$\dfrac{1}{4}$	44
Gray	$\dfrac{5}{16}$	35
Bronze	$\dfrac{1}{8}$	68
Bronze	$\dfrac{3}{16}$	59
Bronze	$\dfrac{1}{4}$	52
Bronze	$\dfrac{5}{16}$	44
Thermopane	$\dfrac{1}{8}$	80
Thermopane	$\dfrac{3}{16}$	79
Thermopane	$\dfrac{1}{4}$	77

Source: Murdoch, J.B., *Illumination Engineering—From Edison's Lamp to the Laser*, Macmillan, New York, 1985. With permission.

Step 5: Work-plane illuminances
As discussed earlier, Figure 9.17 illustrates the location of the work-plane illuminances determined with this lumen method of sidelighting. The three illuminances (max, mid, min) are determined using two coefficients of utilization, the C factor and the K factor. The C factor depends on room length, width, and wall reflectance. The K factor depends on ceiling-floor height, room width, and wall reflectance. Table 9.8 presents C and K values for the three cases of incoming fluxes: sun plus clear sky, overcast sky, and ground-reflected. Assumed ceiling and floor reflectances are given for this case with no window controls (shades, blinds, overhangs, etc.). These further window control complexities can be found in IES [30], LOF [31], and others. A reflectance of 70% represents light-colored walls, with 30% representing darker walls.

TABLE 9.8

C and K Factors for No Window Controls for (a) Overcast Sky, (b) Clear Sky, and (c) Ground Illumination (Ceiling Reflectance, 80; Floor Reflectance, 20%)

C: Coefficient of Utilization

Room Length		Room Width (ft) 20 ft 70%	20 ft 30%	30 ft 70%	30 ft 30%	40 ft 70%	40 ft 30%
Wall Reflectance		70%	30%	70%	30%	70%	30%
(a) Illumination by overcast sky							
Max	20	0.0276	0.0251	0.0191	0.0173	0.0143	0.0137
	30	0.0272	0.0248	0.0188	0.0172	0.0137	0.0131
	40	0.269	0246	.0182	.0171	.0133	.0130
Mid	20	0.0159	0.0177	.0101	0.0087	0.0081	0.0071
	30	0.0058	0.0050	0.0054	0.0040	0.0034	0.0033
	40	0.0039	0.0027	0.0030	0.0023	0.0022	0.0019
Min	20	0.0087	0.0053	0.0033	0.0043	0.0050	0.0037
	30	0.0032	0.0019	0.0029	0.0017	0.0020	0.0014
	40	0.0019	0.0009	0.0016	0.0009	0.0012	0.0008
(b) Illumination by clear sky							
Max	20	0.0206	0.0173	0.0143	0.0123	0.0110	0.0098
	30	0.0203	0.0173	0.0137	0.0120	0.0098	0.0092
	40	0.0200	0.0168	0.0131	0.0119	0.0096	0.0091
Mid	20	0.0153	0.0104	0.0100	0.0079	0.0083	0.0067
	30	0.0082	0.0054	0.0062	0.0043	0.0046	0.0037
	40	0.0052	0.0032	0.0040	0.0028	0.0029	0.0023
Min	20	0.0106	0.060	0.0079	0.0049	0.0067	0.0043
	30	0.0054	0.0028	0.0047	0.0023	0.0032	0.0021
	40	0.0031	0.0014	0.0027	0.0013	0.0021	0.0012

K: Coefficient of Utilization

Ceiling Height		8 ft 70%	8 ft 30%	10 ft 70%	10 ft 30%	12 ft 70%	12 ft 30%	14 ft 70%	14 ft 30%
Wall Reflectance		70%	30%	70%	30%	70%	30%	70%	30%
Room Width (ft)									
Max	20	0.125	0.129	0.121	0.123	0.111	0.111	0.0991	0.0973
	30	0.122	0.131	0.122	0.121	0.111	0.111	0.0945	0.0973
	40	.145	.133	.131	.126	.111	.111	.0973	.0982
Mid	20	0.0908	0.0982	0.107	0.115	0.111	0.111	0.105	0.122
	30	0.156	0.102	0.0939	0.113	0.111	0.111	0.121	0.134
	40	0.106	0.0948	0.123	0.107	0.111	0.111	0.135	0.127
Min	20	0.0908	0.102	0.0951	0.114	0.111	0.111	0.118	0.134
	30	0.0924	0.119	0.101	0.114	0.111	0.111	0.125	0.126
	40	0.111	0.0926	0.125	0.109	0.111	0.111	0.133	0.130

(continued)

TABLE 9.8 (continued)

C and K Factors for No Window Controls for (a) Overcast Sky, (b) Clear Sky, and (c) Ground Illumination
(Ceiling Reflectance, 80; Floor Reflectance, 20%)

C: Coefficient of Utilization

Room Length	20 ft		30 ft		40 ft	
Wall Reflectance	70%	30%	70%	30%	70%	30%

(c) Ground illumination (ceiling reflectance, 80%; floor reflectance, 20%)

Room Width (ft)		70%	30%	70%	30%	70%	30%
Max	20	0.0147	0.0112	0.0102	0.0088	0.0081	0.0071
	30	0.0141	0.0012	0.0098	0.0088	0.0077	0.0070
	40	0.0137	0.0112	0.0093	0.0086	0.0072	0.0069
Mid	20	0.0128	0.0090	0.0094	0.0071	0.0073	0.0060
	30	0.0083	0.0057	0.0062	0.0048	0.0050	0.0041
	40	0.0055	0.0037	0.0044	0.0033	0.0042	0.0026
Min	20	0.0106	0.0071	0.0082	0.0054	0.0067	0.0044
	30	0.0051	0.0026	0.0041	0.0023	0.0033	0.0021
	40	0.0029	0.0018	0.0026	0.0012	0.0022	0.0011

K: Coefficient of Utilization

Ceiling Height	8 ft		10 ft		12 ft		14 ft	
Wall Reflectance	70%	30%	70%	30%	70%	30%	70%	30%

Room Width (ft)		70%	30%	70%	30%	70%	30%	70%	30%
Max	20	0.124	0.206	0.140	0.135	0.111	0.111	0.0909	0.0859
	30	0.182	0.188	0.140	0.143	0.111	0.111	0.0918	0.0878
	40	0.124	0.182	0.140	0.142	0.111	0.111	0.0936	0.0879
Mid	20	0.123	0.145	0.122	0.129	0.111	0.111	0.100	0.0945
	30	0.0966	0.104	0.107	0.112	0.111	0.111	0.110	0.105
	40	0.0790	0.0786	0.0999	0.106	0.111	0.111	0.118	0.118
Min	20	0.0994	0.108	0.110	0.114	0.111	0.111	0.107	0.104
	30	0.0816	0.0822	0.0984	0.105	0.111	0.111	0.121	0.116
	40	0.0700	0.0656	0.0946	0.0986	0.111	0.111	0.125	0.132

Source: IES, *Lighting Handbook, Applications Volume*, Illumination Engineering Society, New York, 1987. With permission.

The work-plane max, mid, and min illuminance are each calculated by adding the sun–sky and ground-reflected illuminances, which are given by

$$E_{sp} = \phi_{sw} C_s K_s K_m$$
$$E_{gp} = \phi_{gw} C_g K_g K_m \tag{9.31}$$

where the "sp" and "gp" refer to the sky-to-work-plane and ground-to-work-plane illuminances.

Example 9.9

Determine the clear sky illuminances for a 30-ft-long, 30-ft-wide, 10-ft-high room with a 20-ft-long window with a 3-ft sill. The window faces 10°E of South, the building is at 32° latitude, and it is January 15 at 2 PM. The ground cover outside is grass, the glass is 1/4 in. clear, and the walls are light-colored.

Solution
Following the steps in the "sidelighting" method,

Step 1: With $L = 32$, $n = 15$, $H = (12 - 14)60/4 = -30$,

$$\delta = 23.45 \sin[360(15 - 81)/365] = -21.3°$$

Then, Equation 9.22 yields $\alpha = 41.7°$, Equation 9.24 yields $\gamma_s = -38.7°$, and Equation 9.25 yields

$$\Phi = |-38.7 - (+10)| = 48.7°$$

From Figure 9.18 with $\alpha = 41.7°$ and $\Phi = 487°$,

a. For clear sky (winter, no sun): $E_{kw} = 875$ fc
b. For direct sun: $E_{uw} = 41.00$ fc
c. Total clear sky plus direct: $E_{sw} = 4975$ fc

NOTE: A high E_{uw} value probably indicates a glare situation!

Step 2: Horizontal illuminances from Figure 9.19: $E_{sg} = 7400$ fc
Then Equation 9.26 yields, with $p_{grass} = 0.06$, $E_{gw} = 222$ fc

Step 3: From Equation 9.29, with $\tau = 0.87$ and $A_w = 140$ ft^2,

$$\Phi_{sw} = 4{,}975(0.87)(140) = 605{,}955 \text{ lm}$$

$$\Phi_{gw} = 222(0.87)(140) = 27{,}040 \text{ lm}$$

Step 4: For a clean office room,

$$K_m = (0.83)(0.94) = 0.78$$

Step 5: From Table 9.8, for 30 ft width, 30 ft length, 10 ft ceiling, and wall reflectivity 70%,

a. Clear sky

$$C_s, \text{max} = 0.0137; \quad K_s, \text{max} = 0.125$$
$$C_s, \text{mid} = 0.0062; \quad K_s, \text{mid} = 0.110$$
$$C_s, \text{min} = 0.0047; \quad K_s, \text{min} = 0.107$$

b. Ground-reflected

$$C_g, \max = 0.0098; \quad K_g, \max = 0.140$$
$$C_g, \mathrm{mid} = 0.0062; \quad K_g, \mathrm{mid} = 0.107$$
$$C_g, \min = 0.0041; \quad K_g, \min = 0.0984$$

Then using Equation 9.31,

$$E_{sp}, \max = 605{,}955(0.0137)(0.125)(0.78) = 809 \text{ fc}$$
$$E_{sp}, \mathrm{mid} = 605{,}955(0.0062)(0.110)(0.78) = 322 \text{ fc}$$
$$E_{sp}, \min = 605{,}955(0.0047)(0.107)(0.78) = 238 \text{ fc}$$
$$E_{gp}, \max = 27{,}040(0.0098)(0.140)(0.78) = 29 \text{ fc}$$
$$E_{gp}, \mathrm{mid} = 27{,}040(0.0062)(0.107)(0.78) = 14 \text{ fc}$$
$$E_{gp}, \min = 27{,}040(0.0041)(0.0984)(0.78) = 9 \text{ fc}$$

Thus,

$$E_{\max} = 838 \text{ fc}$$
$$E_{\mathrm{mid}} = 336 \text{ fc}$$
$$E_{\min} = 247 \text{ fc}$$

9.7.4.2 Lumen Method of Skylighting

The lumen method of skylighting calculates the average illuminance at the interior work plane provided by horizontal skylights mounted on the roof. The procedure for skylighting is generally the same as that described earlier for sidelighting. As with windows, the illuminances from both overcast sky and clear sky plus sun cases are determined for specific days in different seasons and for different times of the day, and a judgment is then made as to the number and size of skylights and any controls needed.

The procedure is presented in four steps: (1) finding the horizontal illuminance on the outside of the skylight; (2) calculating the effective transmittance through the skylight and its well; (3) figuring the interior space light loss factor and the utilization coefficient; and finally, (4) calculating illuminance on the work plane.

Step 1: Horizontal sky and sun illuminances
The horizontal illuminance value for an overcast sky or a clear sky plus sun situation can be determined from Figure 9.19 knowing only the solar altitude.

Step 2: Net skylight transmittance
The transmittance of the skylight is determined by the transmittance of the skylight cover(s), the reflective efficiency of the skylight well, the net-to-gross skylight area, and the transmittance of any light-control devices (lenses, louvers, etc.).

TABLE 9.9

Flat-Plate Plastic Material Transmittance for Skylights

Type	Thickness (in.)	Transmittance (%)
Transparent	$\frac{1}{8} - \frac{3}{16}$	92
Dense translucent	$\frac{1}{8}$	32
Dense translucent	$\frac{3}{16}$	24
Medium translucent	$\frac{1}{8}$	56
Medium translucent	$\frac{3}{16}$	52
Light translucent	$\frac{1}{8}$	72
Light translucent	$\frac{3}{16}$	68

Source: Murdoch, J.B., *Illumination Engineering—From Edison's Lamp to the Laser*, Macmillan, New York, 1985. With permission.

The transmittance for several flat-sheet plastic materials used in skylight domes is presented in Table 9.9. To get the effective dome transmittance (T_D) from the fat-plate transmittance (T_F) value [32], use

$$T_D = 1.25T_F(1.18 - 0.416T_F) \tag{9.32}$$

If a double-domed skylight is used, then the single-dome transmittances are combined as follows [33]:

$$T_D = \frac{T_{D_1}T_{D_2}}{T_{D_1}T_{D_2} - T_{D_1}T_{D_2}} \tag{9.33}$$

If the diffuse and direct transmittances for solar radiation are available for the skylight glazing material, it is possible to follow this procedure and determine diffuse and direct dome transmittances separately. However, this difference is usually not a significant factor in the overall calculations.

The efficiency of the skylight well (N_w) is the fraction of the luminous flux from the dome that enters the room from the well. The well index (WI) is a geometric index (height, h; length, l; width, w) given by

$$WI = \frac{h\,(w + l)}{2wl} \tag{9.34}$$

and WI is used with the well–wall reflectance value in Figure 9.21 to determine well efficiency, N_w.

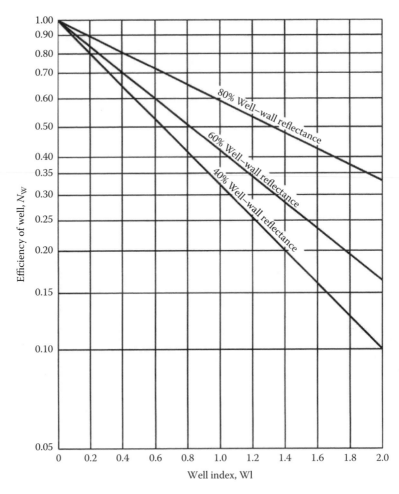

FIGURE 9.21
Efficiency of well versus well index. (Modified from Kaufman, J.E., ed., *The IES Lighting Handbook: Reference Volume*, IESNA, New York, 1984. With permission from the Illuminating Engineering Society of North America. With permission.)

With T_D and N_w determined, the net skylight transmittance for the skylight and well is given by

$$T_n = T_D N_w R_A T_C \tag{9.35}$$

where
   R_A is the ratio of net to gross skylight areas
   T_C is the transmittance of any light-controlling devices

Step 3: Light loss factor and utilization coefficient
The light loss factor (K_m) is again defined as the product of the RSDD and the skylight direct depreciation (SDD) fractions, similar to Equation 9.30. Following the reasoning for the sidelighting case, the RSDD value for clean rooms is around 0.94, and 0.84 for dirty rooms. Without specific data indicating otherwise, the SDD fraction is often taken as 0.75 for office buildings and 0.65 for industrial areas.

The fraction of the luminous flux on the skylight that reaches the work plane (K_u) is the product of the net transmittance (T_n) and the room coefficient of utilization (RCU). Dietz et al. [34] developed RCU equations for office and warehouse interiors with ceiling, wall, and floor reflectances of 75%, 50%, and 30%, and 50%, 30%, and 20%, respectively.

$$RCU = \frac{1}{1 + A(RCR)^B}, \quad \text{if RCH} < 8 \tag{9.36}$$

where
  $A = 0.0288$ and $B = 1.560$ (offices)
  $A = 0.0995$ and $B = 1.087$ (warehouses)

Room cavity ratio (RCR) is given by

$$RCR = \frac{5h_c(l + w)}{lw} \tag{9.37}$$

with h_c the ceiling height above the work plane and l and w being room length and width, respectively.

The RCU is then multiplied by the previously determined T_n to give the fraction of the external luminous flux passing through the skylight and incident on the workplace:

$$K_u = T_n(RCU) \tag{9.38}$$

Step 4: Work-plane illuminance
The illuminance at the work plane (E_{TWP}) is given by

$$E_{TWP} = E_H \left(\frac{A_T}{A_{wp}} \right) K_u K_m \tag{9.39}$$

where
  E_H is the horizontal overcast or clear sky plus sun illuminance from Step 1
  A_T is the total gross area of the skylights (number of skylights times skylight gross area)
  A_{WP} is the work-plane area (generally room length times width)

Note that in Equation 9.39, it is also possible to fix the E_{TWP} at some desired value and determine the skylight area required.

Rules of thumb for skylight placement for uniform illumination include 4%–8% of roof area and spacing less than 1.1/2 times ceiling-to-work-plane distance between skylights [29].

Example 9.10

Determine the work-plane "clear sky plus sun" illuminance for a 30' × 30' × 10' office with 75% ceiling, 50% wall, and 30% floor reflectance with four 4' × 4' double-domed skylights at 2 p.m. on January 15 at 32° latitude. The skylight well is 1' deep with 60% reflectance walls, and the outer and inner dome flat-plastic transmittances are 0.85 and 0.45, respectively. The net skylight area is 90%.

Solution

Follow the four steps in the lumen method for skylighting.

Step 1: Use Figure 9.19 with the solar altitude of 41.7° (calculated from Equation 9.26) for the clear sky plus sun curve to get horizontal illuminance:

$$E_H = 7400 \text{ fc}$$

Step 2: Use Equation 9.32 to get domed transmittances from the flat-plate plastic transmittances given:

$$T_{D_1} = 1.25(0.85)\,[1.18 - 0.416(0.85)] = 0.89$$

$$T_{D_2}(T_F = 0.45) = 0.56$$

and Equation 9.33 to get total dome transmittance from the individual dome transmittances:

$$T_D = \frac{(0.89)(0.56)}{(0.89) + (0.56) - (0.89)(0.56)} = 0.52$$

To get well efficiency, use WI = 0.25 from Equation 9.34 with 60% wall reflectance in Figure 9.21 to give $N_w = 0.80$. With $R_A = 0.90$, use Equation 9.35 to calculate net transmittance:

$$T_n = (0.52)(0.80)(0.90)(1.0) = 0.37$$

Step 3: The light loss factor is assumed to be from "typical" values in Equation 9.30: $K_m = (0.75)(0.94) = 0.70$. The room utilization coefficient is determined using Equations 9.36 and 9.37:

$$\text{RCR} = \frac{5(7.5)(30 + 30)}{(30)(30)} = 2.5$$

$$\text{RCU} = [1 + 0.0288(2.5)^{1.560}]^{-1} = 0.89$$

and Equation 9.37 yields $K_u = (0.37)(0.89) = 0.33$.

Step 4: The work-plane illuminance is calculated by substituting the earlier values into Equation 9.39:

$$E_{TWP} = 122 \text{ fc}$$

9.7.5 Daylighting Controls and Economics

The economic benefit of daylighting is directly tied to the reduction in lighting electrical energy operating costs. Also lower cooling system operating costs are possible due to the reduction in heating caused by the reduced electrical lighting load. The reduction in lighting and cooling system electrical power during peak demand periods could also beneficially affect demand charges.

The reduction of the design cooling load through the use of daylighting can also lead to the reduction of installed or first-cost cooling system dollars. Normally, economics dictate that an automatic lighting control system must take advantage of the reduced lighting/cooling effect, and the control system cost minus any cooling system cost savings should be expressed as a "net" first cost. A payback time for the lighting control system ("net" or not) can be calculated from the ratio of first costs to yearly operating savings. In some cases, these paybacks for daylighting controls have been found to be in the range of 1–5 years for office building spaces [35].

Controls, both aperture and lighting, directly affect the efficacy of the daylighting system. As shown in Figure 9.22, aperture controls can be architectural (overhangs, light shelves, etc.) and/or window shading devices (blinds, automated louvers, etc.). The aperture controls generally moderate the sunlight entering the space to maximize/minimize solar thermal gain, permit the proper amount of light for visibility, and prevent glare and beam radiation onto the workplace. Photosensor control of electric lighting allows the dimming (or shutting off) of the lights in proportion to the amount of available daylighting illuminance.

In most cases, increasing the solar gain for daylighting purposes, with daylighting controls, saves more in electrical lighting energy and the cooling energy associated with the lighting than is incurred with the added solar gain [35]. In determining the annual energy savings total from daylighting (ES_T), the annual lighting energy saved from daylighting (ES_L) is added to the reduction in cooling system energy (ΔES_C) and to the negative of the heating system energy increase (ΔES_H):

$$ES_T = ES_L + \Delta ES_C - \Delta ES_H \qquad (9.40)$$

A simple approach to estimating the heating and cooling energy changes associated with the lighting energy reduction is by using the fraction of the year associated with the

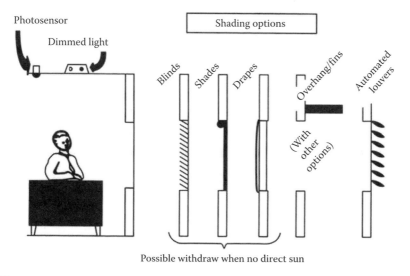

FIGURE 9.22
Daylighting system controls. (From Rundquist, R.A. *ASHRAE Journal*, 11, 30, 1991. With permission.)

cooling or heating season (f_C, f_H) and the seasonal coefficient of performance (COP) of the cooling or heating equipment. Thus, Equation 9.40 can be expressed as

$$ES_T = ES_L + \frac{f_C ES_L}{COP_C} - \frac{f_H ES_L}{COP_H}$$

$$ES_T = ES_L \left(1 + \frac{f_C}{COP_C} - \frac{f_H}{COP_H} \right)$$

(9.41)

It should be noted that the increased solar gain due to daylighting has not been included here but would reduce summer savings and increase winter savings. If it is assumed that the increased wintertime daylighting solar gain approximately offsets the reduced lighting heat gain, then the last term in Equation 9.41 becomes negligible.

To determine the annual lighting energy saved (ES_L), calculations using the lumen method described earlier should be performed on a monthly basis for both clear and overcast days for the space under investigation. Monthly weather data for the site would then be used to prorate clear and overcast lighting energy demands monthly. Subtracting the calculated (controlled) daylighting illuminance from the design illuminance leaves the supplementary lighting needed, which determines the lighting energy required.

This procedure has been computerized and includes many details of controls, daylighting methods, weather, and heating and cooling load calculations. ASHRAE [9] lists many of the methods and simulation techniques currently used with daylighting and its associated energy effects.

Example 9.11

A 30′ × 20′ space has a photosensor dimmer control with installed lighting density of 2.0 W/ft². The required workplace illuminance is 60 fc, and the available daylighting illuminance is calculated as 40 fc on the summer peak afternoon. Determine the effect on the cooling system (adapted from [35]).

Solution
The lighting power reduction is (2.0 W/ft²) (30 × 20) ft² × (40 fc/60 fc) = 800 W. The space-cooling load would also be reduced by this amount (assuming CLF = 1.0):

$$\frac{800\,W \times 3.413\,Btu\,h/W}{12,000\,Btu\,h/ton} = 0.23\,ton$$

Assuming 1.1/2 tons nominally installed for 600 ft² of space at \$2200/ton, the 0.23 ton reduction is "worth" 0.23 × \$2200/ton = \$506. The lighting controls cost about \$1/ft² of controlled area, so the net installed first cost is

$$Net\,first\,cost = \$600\,controls - \$500\,A/C\,savings = \$100$$

Assuming that the day-to-monthly-to-annual illuminance calculations gave a 30% reduction in annual lighting, the associated operating savings can be determined. Lighting energy savings are

$$ES_L = 0.30 \times 2.0\,W/ft^2 \times 600\,ft^2 \times 2500\,h/year = 900\,kWh$$

Using Equation 9.41 to also include cooling energy saved due to lighting reduction (with $COP_C = 2.5$, $f_C = 0.5$, and neglecting heating) gives

$$ES_T = 900\left(1 + \frac{0.5}{2.5} - 0\right) = 1080 \text{ kWh}$$

At $0.10 per kWh, the operating costs savings are $0.10/kWh × 1080 kWh = $108 per year. Thus, the simple payback is approximately 1 year (100/108) for the "net" situation, and a little over 5.1/2 years (600/108) against the "controls" cost alone. It should also be noted that the 800 W lighting electrical reduction at peak hours, with an associated cooling energy reduction of 800 W/2.5 COP = 320 W, provides a peak demand reduction for the space of 1.1 kW, which can be used as a "first-cost savings" to offset control system costs.

Problems

9.1 Explain how window placement in a building could be defined as (a) a passive solar feature, (b) an energy conservation technique, and (c) both of these.

9.2 Write an equation for calculating the cost of savings life cycle economics of a proposed passive solar system. Explain why it is important to be able to determine the auxiliary energy required for any given passive (or active) system design.

9.3 Referring to the thermal circuit diagram of Figure 9.7 for the thermal storage (Trombe) wall building, construct appropriate thermal circuits for (a) attached sunspace, (b) thermal storage, and (c) direct gain buildings.

9.4 Using rules of thumb for a 200 m² floor area Denver residence, determine (a) the auxiliary heating energy required, (b) the solar projected area, and (c) the concrete storage mass needed for a maximum 10°C daily temperature swing.

9.5 A 2000 ft² house in Boston is being designed with NLC—12,000 Btu/F-day and 150 ft² of direct gain. The direct gain system includes double glazing, nighttime insulation, and 30 Btu/ft² F thermal storage capacity. Using the LCR method, determine (a) the annual auxiliary heating energy needed by this design and (b) the storage mass and dimensions required.

9.6 Compare the annual SSF for 150 ft² of the following passive systems for the house in Problem 9.5: (a) direct gain (DGA3), (b) vented. Trombe wall (TWD4), (c) unvented Trombe wall (TW14), (d) waterwall (WWB4), and (e) sunspace (SSB4).

9.7 A design modification to the house in Problem 9.5 is desired. A 200 ft², vented, 12 in. thick Trombe wall is to be added to the direct gain system. Assuming the same types of glazing and storage as described earlier, determine (a) the annual heating auxiliary energy needed and (b) the Trombe wall mass.

9.8 Using the SLR method, calculate the auxiliary energy required in March for a 2000 ft², NLC 12,000 Btu/F-day house in Boston with a 150 ft², night-insulated double-glazed direct gain system with 6 in thick storage floors of 45 Btu/ft² F capacity.

9.9 Calculate the heating season auxiliary energy required for the Boston house in Problem 9.8.

9.10 Determine the length of the overhang needed to shade a south-facing 2 m high window in Dallas, TX (latitude 32°51′), to allow for both winter heating and summer shading.

9.11 A 10 mph wind is blowing directly into an open 3 ft × 5 ft window, which is mounted in a room's 8 ft high by 12 ft wide wall. If the wind's temperature is 80°F, are the room's occupants thermally comfortable?

9.12 Design a stack effect/solar chimney (vented Trombe wall) to produce an average velocity of 0.3 m/s within a 4 m wide by 5 m long by 3 m high room. Justify your assumptions.

9.13 Estimate the overnight radiant cooling possible from an open, 30°C, 8 m diameter water tank during July in Chicago. What would you expect for convective and evaporative cooling values?

9.14 For the buried pipe example (9.8) in the text, determine which of the three flow rate cases leads to the least expensive installation.

9.15 Using data from Figure 9.14, design a 9 ft deep ground-pipe system for Dallas in June to deliver 1000 cfm at 75°F when the outside air temperature is 90°F.

9.16 A 30 ft × 20 ft office space has a photosensor dimmer control working with installed lighting of 2 W/ft². The required workplace illuminance is 60 fc and the available daylighting is calculated as 40 fc on the summer peak afternoon. Determine the payback period for the dimmer control system assuming the following: 1–1/2 ton cooling installed for 600 ft² at $2200/ton, lighting control system cost at $1/ft², 30% reduction in annual lighting due to daylighting, $0.10/kW h electricity cost, and cooling for 6 months at a $COP_c = 2.5$.

9.17 Determine the illuminances (sun, sky, and ground-reflected) on a vertical, south-facing window at solar noon at 36°N latitude on June 21 and December 21 for (a) a clear day and (b) an overcast day.

9.18 Determine the sidelighting workplace illuminances for a 20 ft long, 15 ft wide (deep), 8 ft high light-colored room with a 15 ft long by 5 ft high window. Assume that the direct sun plus clear sky illuminance is 3000 fc and the ground-reflected illuminance is 200 fc.

9.19 Determine the clear sky day and the cloudy day work-plane illuminances for a 30 ft long, 30 ft wide, 10 ft high light-colored room. A 20 ft long by 7 ft high window with ¼-in clear glass faces 10°E of south, the building is at 32°N latitude, and it is January 15 at 2 p.m. solar. The ground outside is covered by dead grass.

9.20 Determine the clear day and cloudy day illuminances on a horizontal skylight at noon on June 21 and December 21 in (a) Miami, (b) Los Angeles, (c) Denver, (d) Boston, and (e) Seattle.

9.21 A 3 ft × 5 ft double-domed skylight has outer and inner fat-plate plastic transmittances of 0.8 and 0.7, respectively; a 2 ft deep well with 80% reflectance walls; and a 90% net skylight area. Calculate the net transmittance of the skylight.

9.22 Determine the number and roof placement of 10 ft × 4 ft skylights needed for a 50 ft × 50 ft × 10 ft high office when the horizontal illuminance is 6000 fc, the skylight has 45% net transmittance, and the required workplace illuminance is 100 fc.

9.23 What would be the procedure for producing uniform workplace illuminance when both sidelighting and skylighting are used simultaneously?

Defining Terms

Active system: A system employing a forced (pump or fan) convection heat-transfer fluid flow.

Daylighting: The use of the sun's radiant energy for illumination of a building's interior space.

Hybrid system: A system with parallel passive and active flow systems or one using forced convection flow to distribute from thermal storage.

Illuminance: The density of luminous flux incident on a unit surface illuminance is calculated by dividing the luminous flux (in lumens) by the surface area (m^2, ft^2). Units are lux (lx) (lumens/m^2) in SI and footcandles (fc) (lumens/ft^2) in English systems.

Luminous flux: The time rate of flow of luminous energy (lumens). A lumen (1 m) is the rate that luminous energy from a 1 candela (cd) intensity source is incident on a 1 m^2 surface 1 m from the source.

Passive cooling system: A system using natural energy flows to transfer heat to the environmental sinks (ground, air, and sky).

Passive heating system: A system in which the sun's radiant energy is converted to heat by absorption in the system, and the heat is distributed by naturally occurring processes.

Sidelighting: Daylighting by light entering through the wall/side of a space.

Skylight: The diffuse solar radiation from a clear or overcast sky, excluding the direct radiation from the sun.

Sunlight: The direct solar radiation from the sun.

Toplighting: Daylighting by light entering through the ceiling area of a space.

References

1. Larson, R., Vignola, F., and West, R. (eds.) (1992) *Economics of Solar Energy Technologies*, American Solar Energy Society (ASES), Boulder, CO, December 1992.
2. Kreith, F. and West, R. (2007) *CRC Handbook of Energy Efficiency and Renewable Energy*, CRC Press, Boca Raton, FL.
3. Close, D.J., Dunkle, R.V., and Robeson, K.A. (1968) Design and performance of a thermal storage air conditioning system, *Mechanical and Chemical Engineering Transactions*, Institute Eng Australia, MC4, 45.
4. PSDH (1980) *Passive Solar Design Handbook, Volume One: Passive Solar Design Concepts*, DOE/CS-0127/1, March 1980. Prepared by Total Environmental Action, Inc., B. Anderson, C. Michal, P. Temple, and D. Lewis; Volume Two: *Passive Solar Design Analysis*, DOE/CS-0127/2, January 1980. Prepared by Los Alamos Scientific Laboratory, J.D. Balcomb, D. Barley, R. McFarland, J. Perry, W. Wray, and S Noll. U.S. Department of Energy, Washington, DC.
5. PSDH (1984) *Passive Solar Design Handbook, Part One:* Total Environmental Action, Inc., *Part Two:* Los Alamos Scientific Laboratory, *Part Three:* Los Alamos National Laboratory. Van Nostrand Reinhold, New York.
6. NCDC (1992) National Climactic Data Center, *Climatography of the U.S. #81*, Federal Building, Asheville, NC.

7. Duffie, J.A. and Beckman, W.A. (1991) *Solar Engineering of Thermal Processes*, 2nd edn., Wiley, New York.
8. McQuiston, P.C. and Parker, J.D. (1994) *Heating, Ventilating, and Air Conditioning*, 4th edn. Wiley, New York.
9. ASHRAE Handbook. (1989, 1993, 1997). *Fundamentals*, I-P and S-I editions, American Society of Heating, Refrigerating and Air-Conditioning Engineers, Atlanta, GA.
10. Olgyay, A. and Olgyay, V. (1967) *Solar Control and Shading Devices*, Princeton University Press, Princeton, NJ.
11. ASHRAE Handbook. (1991, 1995) *Heating, Ventilating, and Air-Conditioning Applications*, I-P and S-I editions, American Society of Heating, Refrigerating and Air-Conditioning Engineers, Atlanta, GA.
12. Hay, H. and Yellott, J. (1969) Natural air conditioning with roof ponds and movable insulation, *ASHRAE Transactions* 75(1), 165–177.
13. Marlatt, W., Murray, C., and Squire, S. (1984) Roof Pond Systems Energy Technology Engineering Center. Rockwell International Report No. ETEC 6, April.
14. Givoni, B. (1994) *Passive and Low Energy Cooling of Buildings*, Van Nostrand Reinhold, New York.
15. Martin, M. and Berdahl, P. (1984) Characteristics of infrared sky radiation in the United States, *Solar Energy*, 33(314), 321–336.
16. Goswami, D.Y. and Biseli, K.M. (1994) *Use of Underground Air Tunnels for Heating and Cooling Agricultural and Residential Buildings*, Report EES-78, Florida Energy Extension Service, University of Florida, Gainesville, FL.
17. U.S.D.A. (1960) Power to produce, *1960 Yearbook of Agriculture*, U.S. Department of Agriculture, Washington, DC.
18. Krarti, M. and Kreider, J.F. 1996 Analytical model for heat transfer in an under-ground air tunnel, *Energy Conversion Management* 37(10), 1561–1574.
19. Hollmuller, P. and Lachal, B. (2001) Cooling and preheating with buried pipe systems: Monitoring, simulation and economic aspects, *Energy and Buildings* 33, 509–518.
20. De Paepe, M. and Janssens, A. (2003) Thermo-hydraulic design of earth-air heat exchanger, *Energy and Buildings* 35, 389–397.
21. Goswami, D.Y. and Dhaliwal, A.S. (1985) Heat transfer analysis in environmental control using and underground air tunnel, *Journal of the Solar Energy Engineering* 107(May), 141–145.
22. Kusuda, T. and Achenbach, P.R. (1965) Earth temperature and thermal diffusivity at selected stations in the United States, *ASHRAE Transactions*, 71(1), 965.
23. Labs, K. (1981) *Regional Analysis of Ground and above Ground Climate*, Report ORNAL/Sub-81/40451/1, Oak Ridge National Laboratory, Oak Ridge, TN.
24. Goswami, D.Y. and Ileslamlou, S. (1990) Performance analysis of a closed-loop climate control system using underground air tunnel, *Journal of the Solar Energy Engineering* 112(May), 76–81.
25. Robbins, C.L. (1986) *Daylighting—Design and Analysis*, Van Nostrand, New York.
26. McCluney, R. (1998) Advanced fenestration daylighting systems, in: *International Daylighting Conference'98*, Natural Resources Canada/CETC, Ottawa, Ontario, Canada.
27. Clay, R.A. (2001) Green is good for you, *Monitor on Psychology* 32(4), 40–42.
28. Kinney, L., McCluney, R., Cler, G., and Hutson, J. (2005) New designs in active daylighting: Good ideas whose time has (finally) come, in: *Proceedings of the 2005 Solar World Congress*, Orlando, FL.
29. Murdoch, J.B. (1985) *Illumination Engineering—From Edison's Lamp to the Laser*, Macmillan, New York.
30. IES. (1987) *Lighting Handbook, Applications Volume*, Illumination Engineering Society, New York.
31. LOF (1976) *How to Predict Interior Daylight Illumination*, Libbey-Owens-Ford Co., Toledo, OH.
32. AAMA. (1977) *Voluntary Standard Procedure for Calculating Skylight Annual Energy Balance*, Architectural Aluminum Manufacturers Association Publication 1602.1.1977, Chicago, IL.
33. Pierson, O. (1962) *Acrylics for the Architectural Control of Solar Energy*, Rohm and Haas, Philadelphia, PA.
34. Dietz, P., Murdoch, J., Pokski, J., and Boyle, J. (October 1981) A skylight energy balance analysis procedure, *Journal of the Illuminating Engineering Society* 11(1), 27–34.
35. Rundquist, R.A. (November 1991) Daylighting controls: Orphan of HVAC design, *ASHRAE Journal* 11, 30–34.

Suggested Readings

Anderson, E.E. (1983) *Fundamentals of Solar Energy Conversion*, Addison-Wesley, Reading, MA.
Anderson, B. and Wells, M. (1981) *Passive Solar Energy*, Brick House Publishing, Andover, MA.
Design Guidelines for the Passive Solar Home. (1980) Pacific Power and Light Pamphlet 6163 9/80 10 M ML.
Halacy, D.S., (1984) *Home Energy*, Rodale Press, Emmaus, PA.
Lebens, R.M. (1980) *Passive Solar Heating Design*, Applied Science Publishers Ltd., London, England.
Phillips, R.O. (1980) Making the best use of daylight in buildings, in: Cowen, H.J. (ed.), *Solar Energy Applications in the Design of Buildings*, Applied Science Publishers London, England, pp. 95–120.

Further Information

The most complete basic reference for passive system heating design is still the *Passive Solar Design Handbook*, all three parts. *Solar Today* magazine, published by the American Solar Energy Society, is the most available source for current practice designs and economics, as well as a source for passive system equipment suppliers.

The *ASHRAE Handbook of Fundamentals* is a general introduction to passive cooling techniques and calculations, with an emphasis on evaporative cooling. *Passive Solar Buildings* and *Passive Cooling*, both published by MIT Press, contain a large variety of techniques and details concerning passive system designs and economics. All the major building energy simulation codes (DOE-2, EnergyPlus, TRNSYS, TSB13, etc.) now include passive heating and cooling technologies.

The *Illumination Engineering Society's Lighting Handbook* presents the basis for and details of daylighting and artificial lighting design techniques. However, most texts on illumination present simplified format daylighting procedures. Currently used daylighting computer programs include various versions of *Lumen Micro, Lightscape,* and *Radiance*.

Passive Solar Design Strategies: Guidelines for Homebuilders (Passive Solar Industries Council, Washington, DC, 1989) presents a user-friendly approach to passive solar design.

10

Energy Storage*

There must surely come a time when heat and power will be stored in unlimited quantities…all gathered by natural forces.

—Thomas A. Edison, 1910

10.1 Overview of Storage Technology

In June 2008, the American Institute of Chemical Engineers published a white paper entitled, "Massive electricity storage" [1]. This paper states that electricity is generated and consumed instantaneously and that the electric power grid essentially has little or no storage capacity to smooth out variations in demand. However, coal-fired and nuclear power plants have turbine generators that can deliver continuous and dispatchable power required by consumers, and these power plants can meet shifting demands by utilizing peaking generators or so-called spinning reserves. On the other hand, renewable power generated by wind or solar radiation is intermittent, and massive electric storage (MES) is the critical technology needed if renewable power is to become a major sustainable source. For system stability and load leveling, large-scale stored power is needed, and without sufficient MES, renewable energy cannot serve as a base-load energy supplier.

In 2006, the U.S. grid had approximately 1000 GW of installed capacity and delivered 4254 TWh of electricity. Of that total, the energy sources were 49% coal, 22% natural gas (NG), 19% nuclear, 7% hydroelectric, and less than 2.5% came from nonhydro renewable sources. Based on these data, the white paper made an estimate of the upper bound of the U.S. MES size and economic cost for various levels of grid penetration by renewable power. These estimates from Ref. [1] are shown in Table 10.1.

An inspection of these figures clearly indicates that, in addition to the installation of renewable energy sources, the investment required for storage is enormous and has to be included in planning a sustainable energy future.

Energy storage will play a critical role in any future renewable energy system. There are two principal reasons why energy storage will be of increasing importance to renewable energy:

- Many of the most significant renewable energy sources are intermittent and generate power when the weather or the movement of the sun allows it, rather than when energy is needed.
- A ground transportation system using cars and buses requires energy to be carried by the vehicles. As gasoline becomes more expensive and less available, other forms of fuel and energy storage for cars and buses will be needed.

---

* Sections in this chapter marked with an asterisk may be omitted in an introductory course.

TABLE 10.1

Upper Bound Estimate of U.S. MES Size and Cost

Grid penetration by renewable power (%)	20	50	75
Firm renewable demand (GW)	200	500	750
Nameplate renewable installed capacity (GW)	570	1430	2150
Capital investment for installed capacity ($ billion)	860	2150	3220
MES power capacity (GW)	114	285	428
MES power storage capacity (GWh)	912	2280	3424
MES capital investment ($ billion)	342	855	1284

Energy can be stored in many forms: as mechanical energy in rotating, compressed, or elevated substances; as thermal or electrical energy that can be released as needed; or as electrical charges ready to travel from positive to negative poles as demanded. Storage technologies that can absorb and release electricity have great value because electricity can be converted easily into mechanical or thermal energy. Electricity is also the output of the most promising renewable energy technologies: wind turbines, solar thermal power plants, and photovoltaic cells. There are, however, some applications that can benefit from thermal or mechanical storage technology. For example, thermal energy storage (TES) is needed for the continuous operation of solar thermal power plants and overnight heat storage for heating or cooling of buildings.

The storage methods treated in this chapter can accept and deliver energy in three ways: electrical, mechanical, and thermal. When energy is stored electrically or mechanically, it can be converted to either of the other two forms quite efficiently. For example, electricity can drive a motor with only about 5% energy loss or can provide heat with a resistive heater with virtually no loss at all. The quality of TES depends mostly on its temperature. The efficiency of conversion of thermal energy into power is limited by thermodynamic considerations. The theoretical maximum quantity of useful work, W_{max} (mechanical energy), that can be extracted from a given quantity of heat, Q, is shown in Chapter 8 as

$$W_{max} = \frac{T_1 - T_2}{T_1} Q$$

where
  T_1 is the absolute temperature of the stored heat
  T_2 is the ambient absolute temperature

Any energy storage technology must be carefully chosen to accept and produce a form of energy consistent with both the energy source and the final application. Storage technologies that accept and/or produce thermal energy should only be used with heat energy sources or heat applications. Mechanical and electrical technologies are more versatile, but electrical storage technologies are more convenient than mechanical ones because electricity can be easily transmitted [2].

10.1.1 Applications

Table 10.2 presents an overview of the most important energy storage technologies. In this table, each technology is classified by its relevance to one of the following four principles.

TABLE 10.2

Overview of Energy Storage Methods and Their Applications

	Utility Shaping	Power Quality	Distributed Grid	Transportation
Direct electric				
Ultracapacitors		✓		✓
SMES		✓		
Electrochemical				
Batteries				
Lead–acid	✓	✓	✓	
Lithium ion	✓	✓	✓	✓
Nickel–cadmium	✓	✓		
Nickel–metal hydride				✓
Zebra				✓
Sodium–sulfur	✓	✓		✓
Flow batteries				
Vanadium redox	✓			
Polysulfide bromide	✓			
Zinc bromide	✓			
Electrolytic hydrogen				✓
Mechanical				
Pumped hydro	✓			
Compressed air	✓			
Flywheels		✓		✓
Direct thermal				
Sensible heat				
Liquids			✓	
Solids			✓	
Latent heat				
Phase change	✓		✓	
Hydration–dehydration	✓			
Chemical reaction	✓		✓	
Thermochemical				
Biomass solids	✓		✓	
Ethanol	✓			✓
Biodiesel				✓
Syngas	✓			✓

Utility shaping: For this application, large-capacity storage devices are required to respond to varying electricity demands. Examples would be a solar thermal power plant, which delivers energy generated during the day to a thermal storage device for use at night time or a wind turbine that compresses air during high winds to power a gas turbine when the wind stops.

Power quality: Storage technologies to smooth power delivery during short periods such as outages or switching events must be capable of providing large changes in power output over very short periods. Ultracapacitors and batteries are in this category.

Distributed grid technology: Storage devices for distributed grid technologies must be able to generate energy and storage at customer locations rather than at a central utility power station. Technologies in this category include batteries for photovoltaic power installed on the roof of a building. Other systems include solar thermal or geothermal power that can be implemented on a distributed scale. The required power generation is small compared to utility shaping. They require capacities in the 1–50 kWh range, while utility shaping requires capacities in the 1000 MWh range.

Transportation technologies: The devices that can be used for storage in automobiles and trucks include all types of batteries. Examples are hybrid electric vehicles such as the Prius, and plug-in electric vehicles such as the Volt. Large fleets of plug-in-hybrid-electric vehicles could also help in utility shaping by utilizing off-peak power to charge the batteries and feeding the grid system from the batteries during peak demand.

   The physical modes of energy storage may involve one or more mechanical, thermal, or electromagnetic forms. An energy storage technology may require both a storage reservoir and a converter and transmission system for moving the power to and from the reservoir to its destination and use. A pumped hydro system, for example, contains both elements using potential energy to store and a turbine to deliver the energy. A molten salt thermal system for a power tower, on the other hand, requires only one reservoir for storing and delivering energy.

10.1.2 Technology Characterization

Every energy storage technology can be characterized essentially by three important parameters: self-discharge time (SDT), storage size, and efficiency. Within a category, the final selection of a specific storage technology can be made by the consideration of cycle life, specific energy, specific power, energy density, and power density.

   SDT is the time required for a fully charged, self-standing storage device to reach a certain depth of discharge (DOD). DOD can be described as a percentage of the storage device's useful capacity. For example, 80% DOD means that 20% of the device's energy capacity is left. It should be noted that the relation between SDT and DOD is, in general, not linear and must therefore be measured. SDTs vary greatly from a few minutes to weeks, as, for example, in battery or thermal storage.

   Storage size describes the intrinsic scale of the technology. It is the most difficult to define of the three parameters. Some technologies have a fairly large storage size but cannot provide small-scale energy storage. Storage size is often called energy storage capacity and is measured in kWh or Btu.

   Efficiency is the ratio of the energy output from the device to the energy input. As for energy density and specific energy, the system boundaries must be carefully specified when determining efficiency. It is also important to note the form of energy required at the input and output interconnections and to include the entire system necessary to attach these interconnections. For example, if the system is used to store energy from a utility wind farm, then both the input and the output will be AC electricity. However, when comparing a battery with a fuel cell, it is necessary to include the efficiency of an DC-to-AC rectifier for the battery, whereas an AC-powered hydrogen power generation system is required for the fuel cell. DC-to-AC converters are associated with both. Efficiency is also related to the SDT. For example, if the discharge of a thermal storage system occurs much later than charging, the apparent efficiency will be lower because a significant amount of the thermal energy may be lost if the interim between charge and discharge is large.

Cycle life is the number of consecutive charge/discharge cycle that a storage system can undergo while maintaining the system's specifications. Cycle life specifications are made against a selected DOD, depending on the application of the storage device. For example, a battery used in a hybrid electric vehicle may consume only 10%–20% of the energy stored during most of the discharge cycle. The advantage of a thermal storage system is that it can undergo many charge/discharge cycles without significant waste.

Specific energy is a measure of how heavy the system is. It is measured in units of energy per mass, and here we will represent this quantity in terms of MJ/kg or kWh/kg. The higher the specific energy, the lighter the device. For transportation applications, a high specific energy is necessary, whereas for utility application, specific energy is relatively unimportant.

Energy density is how much space the system occupies. It is measured in units of energy per volume, MJ/L or kWh/m^3.

Storage energy modes depend on the end-use application, but the magnitude of the power load and the time scale are important. Figure 10.1 shows the energy storage technology characteristics of the various storage technologies to be discussed in this chapter according to size and discharge time. For example, to provide electric power for short power outages requiring a large power flux for a short period of time, such as seconds or minutes, capacitors and flywheels are well suited. On the other hand, when large amounts of electric or thermal energy are needed for periods of hours or longer, pumped hydro power, compressed air storage, or thermal storage are the appropriate technologies. Distributed grid, automotives, and building applications require a moderate discharge time and a moderate unit size. Figure 10.2 shows the fields of application in terms of discharge time and power rating.

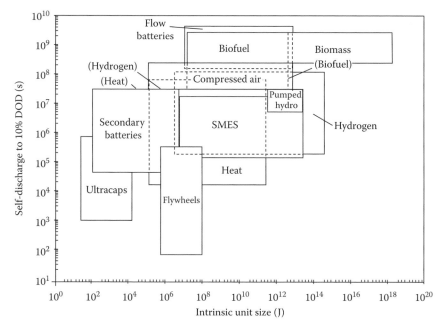

FIGURE 10.1
Self-discharge time to 10% DOD in seconds versus intrinsic unit size in Joules. Not all hidden lines are shown.

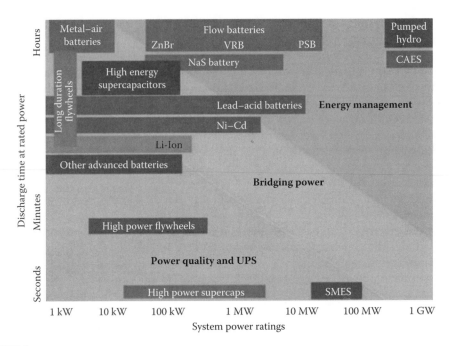

FIGURE 10.2

Discharge time vs. power rating for various energy storage methods. (From Technology comparisons | Ratings, Electricity Storage Association, www.esa.org, 2003. With permission.)

10.2 Mechanical Technologies

Two of the mechanical storage systems discussed later are at this time the only technologies available for large-scale and long-term utility size storage, as shown in Figure 10.2: pumped hydro and compressed air energy storage (CAES).

10.2.1 Pumped Hydroelectric Energy Storage

Pumped hydroelectric energy storage (PHES) is a mature technology that has been deployed for over a century. Examples as early as the 1890s exist in Italy and Switzerland. PHES uses electricity to pump water uphill into a reservoir where it can be stored. The energy can be recaptured when the water is released and runs back downhill through a turbine. These storage systems allow management of water as well as energy resources. PHES systems are very efficient, and with modern turbine technologies, they can approach 80%–85% round-trip efficiencies.

Figure 10.3 is a schematic diagram of a PHES installation at Raccoon Mountain, which is operated by the Tennessee Valley Authority. The Raccoon Mountain pumped storage plant is widely cited as an example of excellent engineering for PHES. The plant, which was completed in 1978, has a generating capacity of about 1,600 MW and can run for 22 h to supply about 35,000 MWh of electricity.

The power and capacity of a PHES system is a function of the hydraulic head, the water flow rate, and the efficiency of the pump and the turbine. The head is given by the upper elevation to which the water is pumped and the lower elevation at which the turbine is located. The flow rate can be regulated by valve action. The maximum flow that can be

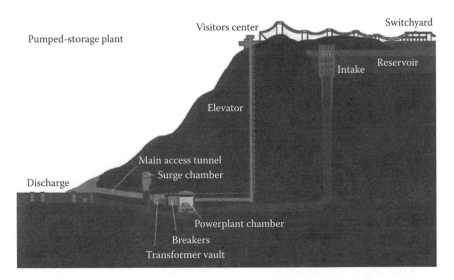

FIGURE 10.3
Schematic of the Raccoon Mountain pumped storage plant. (From Tennessee Valley Authority, The mountain-top marvel, www.tva.gov/heritage/mountaintop/index.htm, accessed October 12, 2010. With permission.)

obtained is limited by the reservoir size. The maximum flow rate is dictated by the reservoir size divided by the desired storage time to yield the available flow per unit time, minus a 15% reserve operating condition.

Given the hydraulic head, an upper bound on flow rate, and an efficiency of the plant, the power-generating capacity of a pumped hydroelectric installation is given by the following equation:

$$P = QH\rho g\eta \tag{10.1}$$

where
P is the generated output power in Watts (W)
H is the height differential in meters (m)
Q is the fluid flow rate in cubic meters per second (m³/s)
ρ is the fluid density in kilogram per cubic meter (kg/m³), about 1000 kg/m³ for water
g is the gravitational constant (9.81 m/s² on Earth)
η is the overall efficiency

Given the power capacity of the plant, the limiting value of the upper or lower reservoir dictates the energy capacity. Potential energy generation in kWh is calculated by power output times the operating time. Figure 10.4 shows the power as a function of flow rate and head in engineering units.

Example 10.1

Calculate the power generated by a PHES with a potential hydraulic height differential between reservoirs of 100 m and a volumetric flow rate of 1000 m³/s through a hydraulic turbine with an efficiency of 0.8.

Solution
We use Equation 10.1, $P = QH\rho g\eta$. By substituting the appropriate values to find that
$P = 1,000 \text{ m}^3/\text{s} \times 100 \text{ m} \times 1,000 \text{ kg/m}^3 \times 9.81 \text{ m/s}^2 \times 0.8 = 784,800,000 \text{ We} = 785 \text{ MWe}.$

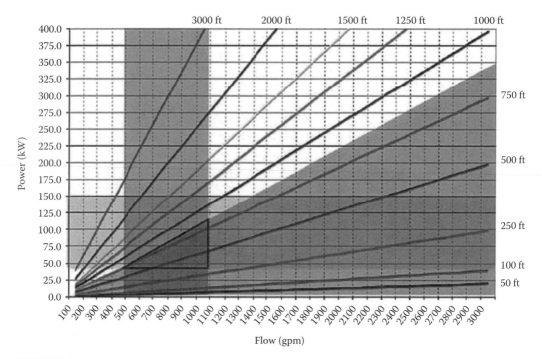

FIGURE 10.4

Pumped hydro turbine power output as a function of flow rate for various hydraulic heads, efficiency = 70%. (From Levine, J.G., Pumped hydroelectric energy storage and spatial diversity of wind resources as methods of improving utilization of renewable energy sources, Thesis, University of Colorado, Boulder, CO, 2007. With permission.)

10.2.1.1 Turbines

Hydro turbines have in the past been used mostly for hydropower generation from large dams. Hydropower produces about 20% of the world's electricity in 150 countries. The technology has evolved for more than 100 years, and the literature of hydroelectric power generation can be used as a guide for designing pumped storage facilities.

Conventional hydraulic turbines are either impulse or reaction machines (see Figure 10.5). Impulse turbines, often called Pelton turbines, convert the potential energy of a fluid into kinetic energy by expansion in a stationary nozzle to form a jet, which is then directed toward buckets attached to a rotating wheel to create shaft work that can be extracted. Reaction turbines are of the Francis or Kaplan type and utilize both hydraulic pressure and kinetic energy to create rotating shaft work. Each of the turbines has a specific operating range in terms of hydraulic head and power output. The choice of which turbine to use depends on the appropriate characteristics.

Pelton wheels were used in the early days of hydropower. They have a lower efficiency at low hydraulic head in comparison to reaction machine. Pelton wheels are still used, however, for high head resources (about 200 m) and small power outputs (less than 5 MW).

Francis turbines use a set of fixed veins that guide the fluid to the buckets that make up the turbine runner and are mounted on a central shaft. Francis turbines have a large operating range with heads from 40 to 500 m and unit sizes from 10 to 1000 MWe. Kaplan turbines are similar to Francis turbines, but the turbine blade angle can be adjusted to

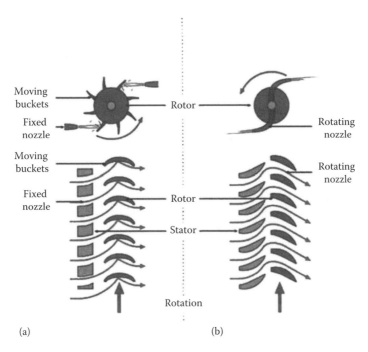

Moving buckets

Fixed nozzle

Moving buckets

Fixed nozzle

Rotor

Rotor

Stator

Rotation

Rotating nozzle

Rotating nozzle

(a)

(b)

FIGURE 10.5
Schematic diagram of (a) impulse and (b) reaction turbines.

improve performance under different flow conditions. They work well at low heads, less than 10 m to about 100 m, with power outputs of the order of 200 MWe.

Once built, hydro power installations have operated successfully for more than a century. It is therefore expected that also pumped storage facilities will have a long life. The main drawback of pumped storage is the limited availability of suitable sites. According to Schoenung et al. [4],

> There have been a number of studies that have indicated that the availability of storage can be a key element in the successful operation of a robust electric power system. Pumped hydro-storage systems are therefore in widespread use worldwide. In 1996, more than 300 systems with capacities ranging from about 20 MWe to more than 2000 MWe were operating.

10.2.2 Compressed Air Energy Storage

Compressed air energy storage (CAES) is a thermodynamic storage mechanism that resembles a gas turbine power plant. Figure 10.6 schematically shows side by side the operation of both a gas turbine and a CAES system.

The operation of a CAES can be demonstrated by comparing this with the gas turbine power plant. In a gas turbine power plant, atmospheric pressure air passes through a compressor, which is then delivered to a combustion chamber where fuel is injected and the high-pressure (hp) air is heated. The hp heated gas from the combustion chamber then passes through the blades of a turbine whose rotating shaft is connected to an electricity generator. The air from the turbine is exhausted at approximately atmospheric pressure. Approximately two-thirds of the energy generated by the turbine is used to drive the compressor, whereas the remaining one-third is converted to generate electric power.

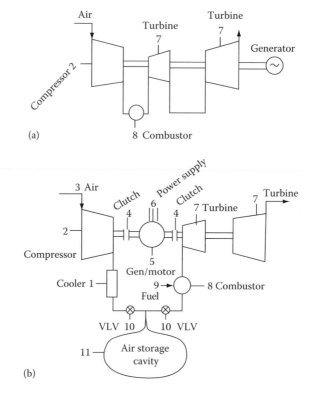

FIGURE 10.6
Comparison of (a) gas turbine and (b) compressed air storage system.

The components in a CAES are similar to those in a gas turbine, but clutches are added in order that the compressor and the turbine may be connected separately to the generator, which also has to work as a motor. In addition, provision is made to store the hp air from the compressor in a suitable cavity. The air from the cavity is heated in a combustor similarly to the gas turbine power plant, but the timing of discharging the stored hp air can be controlled. Since compression and expansion of the air can take place at different times, the entire output from the turbine is available to generate electricity as needed. In other words, compressed air can be placed in the cavity where it is stored until power is needed. This arrangement is of utmost utility for renewable energy systems such as wind turbines that do not operate continuously. But by using any excess power when available to compress and store energy, the compressed air in the cavity becomes available when winds die down but electricity is needed.

The technology of storing energy in the form of compressed air in underground cavities has been demonstrated to be economical and technically viable. Although the air stored at hp in the cavity could be used directly with a low-pressure (lp) turbine, this approach has a low overall thermodynamic efficiency, whereas the hybrid system as described previously has an excellent energy utilization factor.

The volume of the air storage is determined by the amount of energy to be stored. In a wind system, for example, the wind energy should be sufficient to supply electric power for an average period consistent with past variation in the available wind power. The rated capacity of the compressor depends on the required length, during which the reservoir is recharged, but the delivery and charging times are not the same. For example,

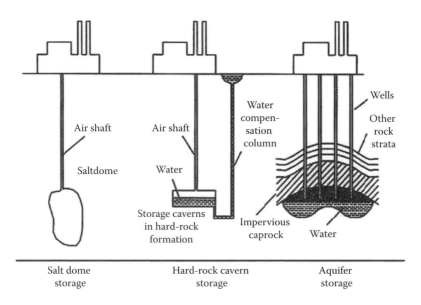

FIGURE 10.7
Methods for producing CAES cavity.

the volume of the reservoir may be sufficient to operate the turbine for 1 h at full load, while the compressor could be designed to recharge the reservoir in 4 h. In this arrangement, the compressor could be sized for one quarter of the air passing through the turbine, and a charging ratio of 1–4 would be obtained. The design of a CAES system offers considerable flexibility in size and location. Moreover, on a specific cost per rated power capacity, the main part of the equipment costs only a fraction of that for an equivalent gas turbine, since the electrical output of CAES can be several times larger than that of a simple gas turbine alone. There are three basic methods for producing a cavity suitable for CAES air storage, as shown in Figure 10.7.

The best developed of these is "solution mining" of cavities in salt formations. It involves dissolving salts with fresh water and removing the solution after it is saturated. The solution mining technology has been used to store gas and oil. Salt caverns are virtually leak tight, and suitable sites are widely available.

The second method is to actually produce a cavity in hard rock. This is an expensive operation, but its cost may be offset by using a water compensation leg to make it possible to operate at a constant pressure.

The third option can be used in aquifer regions when there is suitable domed caprock. It requires drilling into the aquifer to develop an air bubble to replace some of the water. This is a slow process, but after the air bubble is created, since air is less viscous than water, it is possible to achieve the required charge and discharge rates. Experience has shown that as many as 50 injection wells may be required to keep the total pressure difference between the charge and discharge cycles to between 10 and 20 bar. This may be the most expensive part of this low-cost aquifer air storage method.

10.2.2.1 Round-Trip Efficiency of CAES

CAES is a hybrid generation/storage system because it uses NG as fuel in the round-trip process. However, the NG heat input per unit power is much less than for a conventional gas turbine, and other heat sources, for example, syngas from biomass, could be used. CAES requires approximately 0.7–0.8 kWh of off-peak or excess electricity in addition to about 4,300 Btu (1.26 kWh) of NG heat input to produce 1 kWh of dispatchable electricity. This compares with a heat rate of 11,000 Btu per kWh for a conventional turbine powered by NG.

The round-trip efficiency for CAES, defined as [energy out/energy in], can be calculated in two different ways. The first method assumes that all energy inputs, electricity and NG, are of the same quality. In other words, 1 kWh of electricity output requires 0.75 kWh of electricity plus 4300 Btu (1.26 kWh) of NG. In this method, the total energy input is twice the energy output, for an efficiency of 50%. However, the exergy, or quality, of electric energy, is higher than that of NG because, if the NG were used to generate electricity, it would have a conversion efficiency of 30%–40%. Hence, by comparing electrical energy input to electrical energy output using a heat rate of 11,000 Btu per kWh, the gas turbine would produce 0.39 kWh of electricity. Adding that to the 0.75 kWh of off-peak electricity, the total theoretical electricity input is 1.14 kWh to produce 1 kWh of dispatchable electricity. Hence, the round-trip efficiency with this perspective, that is, the ratio of electricity in to electricity out, yields an efficiency of about 88%.

10.2.2.2 Comparison between CAES and PHES

Compressed air and pumped hydro are the only storage options for large-scale solar or wind power applications. Each has distinct advantages and disadvantages. PHES has very good round-trip efficiency, does not use any auxiliary fuel, and can be built for medium- and large-sized applications, but finding suitable sites and obtaining water rights may be difficult. In comparison, CAES has a much smaller visible footprint, does not require water rights or dams, is considerably cheaper, and can be built in more places. The major disadvantage of CAES is that it requires an NG booster, which leads to a lower round-trip efficiency than PHES unless one compares the efficiency as electricity input to electricity output.

Capital costs for pumped hydro and CAES are shown in Table 10.3. CAES has a low capital investment cost compared to other storage technologies. The capital cost for CAES is between $600 and $700 per kW, excluding the cavern. The PHES cost is somewhat higher due to the extra equipment and controls necessary, as well as the cost of a reservoir. Both methods are superior to battery storage.

The primary requirement for siting a CAES plant is a suitable geology with a cavern to hold air at hp. A secondary requirement for the site is vicinity to the generation source, which is likely to be a wind turbine. Typical cavern pressures for CAES range from 500 psi

TABLE 10.3

Capital Cost Comparison for Pumped Hydro and CAES

	Pumped Hydroelectric (1000 MW)	CAES (300 MW)	Sodium Sulfur Battery (10 MW)
Capital cost: Capacity ($/kW)	600	580	~1200
Capital cost: Energy ($/kWh)	37.5	1.75	~200
Hours of storage	10	40	6–9
Total capital cost ($/kW)	975	650	~3200

Source: Succar, S., Baseload power production from wind turbine arrays coupled to compressed air energy storage, PhD thesis, Princeton University, Princeton, NJ, 2008.

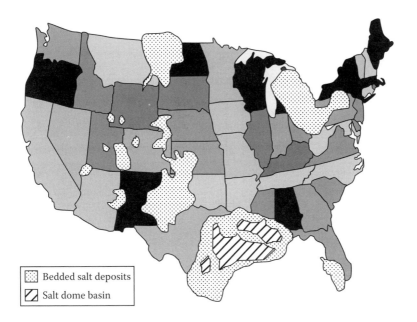

Bedded salt deposits

Salt dome basin

FIGURE 10.8
Map of potential salt cavities for CAES in the United States. (Adapted from Cohn, A. et al., Applications of air saturation to integrated coal gasification CAES power plants, ASME 91-JPGC-GT-2, 1999.)

when fully discharged to 1200 psi when fully charged. According to Ref. [4], suitable geological salt rock formations for building a CAES cavity can be found in 85% of the United States. Figure 10.8 shows a map of the United States with suitable sites for potential CAES locations. A simplified estimate of a CAES system in Colorado with a cavern size that allows holding enough air to produce 525 MWh of electricity from a turbine and an electricity cost of 8.8 cents per kWh showed a simple payback period (without considering the cost of money) of approximately 5 years [5].

10.2.2.3 CAES Volumetric Energy Density

One of the keys to assessing the geologic requirements for CAES is to determine how much electrical energy can be generated per unit volume of storage cavern capacity (E_{GEN}/V_S). The electrical output of the turbine (E_{GEN}) is given by

$$E_{GEN} = \eta_M \eta_G \int_0^t \dot{m}_T W_{CV,TOT} dt \qquad (10.2)$$

where
  the integral is the mechanical work generated by the expansion of air and fuel in the turbine
  $W_{CV,TOT}$ is the total mechanical work per unit mass generated in this process
  $\dot{m}_T$ is the air mass flow rate
  t is the time required to deplete a full storage reservoir at full output power
  η_M is the mechanical efficiency of the turbine
  η_G is the electric generator efficiency

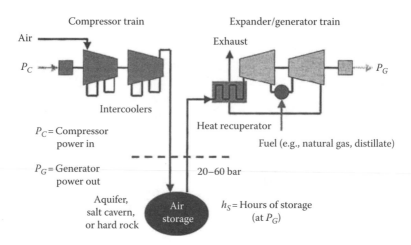

FIGURE 10.9
Typical CAES system.

All the CAES systems to date are based on two expansion stages, as shown in Figure 10.9. Hence, the work output can be expressed as the sum of the output from the two stages.

The first term in Equation 10.3 represents the work output from the hp turbine that expands the air from the hp turbine inlet pressure (p_1) to the lp turbine inlet pressure (p_2), while the second term is the expansion work of the lp turbine from p_2 to ambient pressure (p_a):

$$W_{CV,TOT} = w_{CV1} + w_{CV2} = -\int_{p_1}^{p_2} v\, dp - \int_{p_2}^{p_a} v\, dp \tag{10.3}$$

Consider first the work output from the first expansion stage. Assuming adiabatic compression and that the working fluid is an ideal gas with a constant specific heat (so that $pv^k = a$ constant, where $k = c_p/c_v$), the work per unit mass for both stages is

$$W_{CV,TOT} = c_p T_2 \left(\frac{c_{p1} T_1}{c_{p2} T_2} \left[1 - \left(\frac{p_2}{p_1}\right)^{\frac{k-1}{k}} \right] + \left[1 - \left(\frac{p_a}{p_2}\right)^{\frac{k-1}{k}} \right] \right) \tag{10.4}$$

The total mass flow through the turbine can be expressed as separate air ($\dot{m}_A$) and fuel ($\dot{m}_F$) input terms:

$$\dot{m}_T = \dot{m}_A + \dot{m}_F = \dot{m}_A \left(1 + \frac{\dot{m}_F}{\dot{m}_A} \right) \tag{10.5}$$

Finally, the electric power output per unity cavity volume, V_S, is

$$\frac{E_{GEN}}{V_S} = \frac{\alpha}{V_S} \int_0^t \dot{m}_A \left(\beta + 1 - \left(\frac{p_b}{p_2}\right)^{\frac{k-1}{k}} \right) dt \tag{10.6}$$

where

$$\alpha = \eta_M \eta_G c_{p2} T_2 \left(1 + \frac{\dot{m}_F}{\dot{m}_A} \right)$$

and

$$\beta = \frac{c_{p1} T_1}{c_{p2} T_2} \left[1 - \left(\frac{p_2}{p_1} \right)^{\frac{k-1}{k}} \right]$$

The existing CAES plants operate by throttling the reservoir pressure p_s to the hp turbine inlet pressure p_1. If the mass flow and expansion work output are constant, the integral representing the mechanical work in turbine expansion can be reduced to a simple time average, but the net air mass withdrawn from storage is a function of the storage pressure change over the range p_{S2} to p_{S1}:

$$\dot{m}_T = \frac{\Delta m_A}{t} \left(1 + \frac{\dot{m}_F}{\dot{m}_A} \right) \tag{10.7}$$

where

$$\Delta m_A = \frac{V_S p_{S2}}{R T_{S1}} - \frac{V_S p_{S2}}{R T_{S2}} \left(1 - \left[\frac{p_{S1}}{p_{S2}} \right]^{\frac{1}{k}} \right)$$

Substituting the previous into Equation 10.5 yields

$$\frac{E_{GEN}}{V_S} = \frac{\alpha M_W p_{S2}}{R T_{S2}} \left(\beta + 1 - \left(\frac{p_b}{p_2} \right)^{\frac{k-1}{k}} \right) \left(1 - \left[\frac{p_{S1}}{p_{S2}} \right]^{\frac{1}{k_S}} \right) \tag{10.8}$$

More detailed information can be found in Ref. [6].

10.2.2.4 Existing CAES Plants

Two CAES systems have previously been designed and operated for many years. One of these is located in Huntorf, Germany, and the other one is in McIntosh, Alabama. The experience gained from these existing CAES systems will be valuable in the design and operation of future compressed air storage.

10.2.2.4.1 Huntorf

The Huntorf CAES plant, completed in 1978 near Bremen, Germany, was the world's first CAES facility. The 290 MW plant was designed to provide start up during a complete grid outage for nuclear power plants near the North Sea and to provide inexpensive peak power. It has operated successfully for three decades primarily as a peak shaving unit and to supplement hydroelectric storage facilities on the system. Availability and starting reliability for this unit are reported as 90% and 99%, respectively. The plant can provide up to 3 h of storage and has recently been used to help balance the rapidly growing wind output from North Germany [7,8]. The underground portion of the plant consists of two salt caverns (310,000 m³ total) designed to operate between 48 and 66 bar. The compression and expansion sections draw 108 and 417 kg/s of air, respectively, and are each comprised of two stages. The first turbine stage expands air from 46 to 11 bar. Because gas turbine technology was not compatible with this pressure range, in 1979, a steam turbine was chosen for the hp expansion stage. Although the plant could have operated at a lower heat rate if equipped with heat recuperators to recover exhaust heat from the lp turbine for preheating the gas entering the hp turbine, this feature was omitted in order to minimize system start-up time.

10.2.2.4.2 McIntosh

The 110 MW McIntosh plant was built by the Alabama Electric Cooperative on the McIntosh salt dome in southwestern Alabama and has been in operation since 1991 [8]. It was designed for 26 h of generation at full power and uses a single salt cavern (560,000 m³) designed to operate between 45 and 74 bar.

The operational aspects of the plant (inlet temperatures, pressures, etc.) are similar to those of the Huntorf plant. The facility does, however, include a heat recuperator that reduces fuel consumption by approximately 22% at full load output and features a dual-fuel combustor capable of burning No. 2 fuel oil in addition to NG [9]. Over 10 years of operation, the plant achieved 92% average starting reliability as well as up to 99.5% average running reliability for the generation cycle and compression cycle, respectively [9].

Figure 10.10 is the schematic of a facility proposed for Iowa, which incorporates all the essential features of a CAES system.

10.2.3 Flywheels

Flywheels have been used for a long time for short-term energy storage. Modern flywheel energy storage systems consist of a massive rotating cylinder on a shaft that is supported on a stator by magnetically levitated bearings that reduce bearing wear and increase system life. By operating a flywheel system in a vacuum environment, drag can be drastically reduced. To complete the system, the flywheel is connected to a motor/generator that is mounted on the stator that, through power electronics, can interact with the utility grid.

The amount of energy stored in a rotating flywheel is given by

$$E = \frac{I\omega^2}{2} \tag{10.9}$$

where
I is the flywheel's moment of inertia in kg/m²
ω is its angular velocity in 1/s
E is in joules

In order to obtain a high specific energy for flywheels, the flywheel speed is increased, since the energy stored increases with the square of the velocity, whereas the specific energy only increases in direct proportion with its mass.

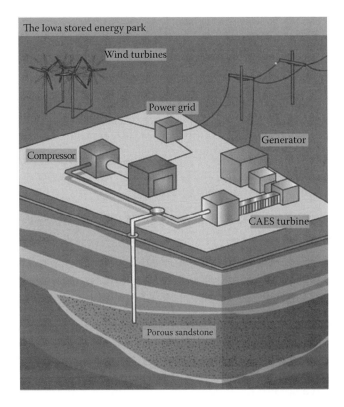

FIGURE 10.10
Schematic diagram of a CAES system proposed for Iowa. (From Harris, W. How the Iowa Stored Energy Park will work. HowStuffWorks.com 2008.)

Some of the advantages of flywheels are their low maintenance requirements, long cycle life (better than 10,000 cycles), and long lifetime (in excess of 20 years). High-speed and low-mass flywheels, which are the modern approach to flywheel storage, are made from composites such as carbon fiber. Flywheels can have an energy efficiency of better than 90%, but their energy storage capacity is limited to a short period. Their low energy density and specific energy limit them to voltage regulation and UPS capability.

10.3 Direct Electrical Technologies

10.3.1 Ultracapacitors

An ordinary capacitor stores energy in the electric field between two oppositely charged conductors with a dielectric between them. The dielectric prevents arcing between the plates and permits the plates to hold more charge, thus increasing the maximum potential energy storage. The ultracapacitor, also known as super capacitor, differs from the traditional capacitor in that it uses an extremely thin electrolyte instead of a dielectric. This electrolyte is only a few Angstrom thick, thereby making it possible to increase the energy density of the device. The electrolyte can be made either of an organic or an aqueous material. The aqueous design can operate over a larger temperature range but has a smaller energy density than the organic design. The electrodes are usually made of a porous carbon that increases their surface area as well as the energy density, compared to a traditional capacitor.

Ultracapacitors can effectively equalize voltage variation quickly, which makes them useful for power quality management and for regulating the voltage in automobiles during regular driving. Ultracapacitors can also work in tandem with batteries to relieve peak power needs. They also exhibit very high cycle life of greater than 500,000 cycles and a life span of more than 10 years. The main limitation of ultracapacitors is their ability to maintain charge voltage over any significant period, losing up to 10% of the charge per day.

10.3.2 Superconducting Magnetic Energy Storage

Superconducting magnetic energy storage (SMES) systems can store and discharge energy at high rates. They store energy in the magnetic field created by direct current in a coil of cryogenically cooled superconducting material. The advantage of a cryogenically cooled superconducting material over copper is that it reduces electrical resistance to nearly zero. SMES recharge quickly and can repeat the charge–discharge sequence thousands of times without degrading the magnet. They can also achieve full power rather quickly, within 100 ms. Theoretically, a coil of around 150–500 m radius would be able to support a load of 18,000 GJ at 1,000 MW, depending on the peak field and the ratio of the coil's height and diameter. Since no conversion of energy to other forms is involved (e.g., mechanical), the energy is stored directly and round-trip efficiency is expected to be very high. It is believed that mature, commercialized SMES can operate with round-trip efficiency as high as 97% and are a superb technology for providing reactive power on demand. The downside of SMES is the need to cool the coil to cryogenic temperatures, which requires a cryogenic refrigeration system and uses large amount of energy [10].

10.4 Fundamentals of Batteries and Fuel Cells

Both batteries and fuel cells date back to the early 1800s. Alexander Volta described an electrochemical cell in a letter to the British Royal Society in 1800. This invention and the development of the telegraph in the 1830s led to the deployment and manufacture of batteries, which provided the electric energy for the telegraph, as well as for other inventions. The fuel cell was first demonstrated by Sir William Grove in 1839, but it was not until recently that significant progress toward practical applications occurred.

The operation of batteries and fuel cells is based on similar electrochemical principles and processes. However, batteries are devices for the storage of energy directly, whereas fuel cells are continuous energy converters who by themselves have no inherent storage capability and need hydrogen or some other energy source to operate. The operation of a typical battery is shown schematically in Figure 10.11.

The battery has a fixed amount of chemicals that spontaneously produce a flow of electrons if a conducting path is connected to its terminals. Fuel cells require a continuous external supply of chemical reactants or hydrogen, and the products are eliminated continuously. The operation of the fuel cells is dependent on a supply of hydrogen, which can be stored externally. Hence, the combination of stored hydrogen and a fuel cell that produces electricity can be considered a storage system, which will be considered later on.

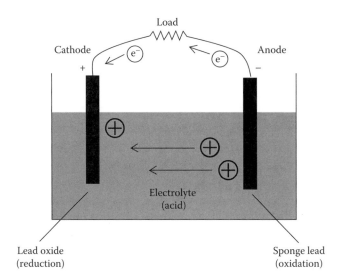

Load

Cathode

Anode

+

−

Electrolyte
(acid)

Lead oxide
(reduction)

Sponge lead
(oxidation)

FIGURE 10.11
Schematic diagram of a battery.

The reactions in both batteries and fuel cells are oxidation–reduction. Both contain electron pairs in contact with an electrolyte, which is the charge-carrying medium, as shown in Figure 10.11. The oxidation reaction takes places at the negative electrode called the *anode* whereas the reduction occurs at the positive electrode called the *cathode*. The anode delivers electrons to the external circuit, which flow from the negative to the positive electrode. However, the conventional current flows in the opposite direction from high to low potential. The electrolyte is usually a liquid solution through which positively charged ions pass from the anode to the cathode. At the cathode, electrons from the external load are neutralized by reacting with positive ions from the electrolyte.

Batteries designed for a single discharge cycle, such as flashlight batteries, are called primary cells, whereas those that can be recharged are called secondary cells. Fuel cells, on the other hand, utilize hydrogen to provide a continuous supply of direct current. There are many different types of batteries and fuel cells available, and we will consider the most important of these in the following sections.

10.4.1 Principles of Battery Operation

Batteries find a variety of applications today from tiny button cells in wrist watches to ignition batteries in automobiles and, finally, to large storage batteries in utility power supplies. At present, batteries are considered for storing energy from renewable sources such as solar and wind in order to provide a continuous supply of energy. Batteries are devices for storing an electric charge and automatically delivering an electric current flow on demand. A battery system consists of one or more cells connected in series to provide an open circuit voltage or electromagnetic force (EMF). The exact value of the EMF depends on the specific cell reactants used. The operation of the battery can be visualized by means of a simple conceptual model shown in Figure 10.12.

When current is drawn from a battery, the voltage drops below its EMF across the terminals, largely due to the internal voltage drop associated with the battery's internal

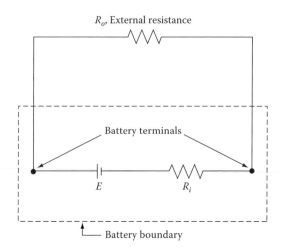

FIGURE 10.12
Simplified battery model.

resistance. As shown in Figure 10.12, the terminal voltage in this model is the difference between the EMF and the potential drop across the internal resistance, R_i:

$$V = E - IR_i \tag{10.10}$$

and the current, I, is

$$I = \frac{E}{R_i + R_o} \tag{10.11}$$

The following example illustrates the simplified model for a conventional lead–acid battery.

Example 10.2

Estimate the internal and external resistances for a lead–acid battery with an EMF of 12.7 V. When delivering a current of 50 A through an external resistance of 3 Ω, the voltage measured in operation is 11.1 V. Draw the voltage vs. current characteristics.

Solution
From Equations 10.10 and 10.11, we obtain the internal and external resistances as shown in the following:

$$R_i = \frac{E - V}{I} = \frac{(12.7 \text{ V} - 11.1 \text{ V})}{50 \text{ A}} = 0.032 \text{ Ω}$$

$$R_o = \frac{E}{I} - R_i = \frac{12.7}{50} - 0.032 = 0.222 \text{ Ω}$$

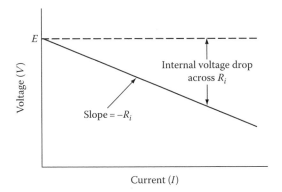

FIGURE 10.13
Current vs. voltage characteristics for simplified battery in problem solution.

The current flow through a 3 Ω external resistance is

$$I = \frac{E}{R_i + R_o} = \frac{12.7}{0.032 + 3} = 3.825 \text{ A}$$

and the voltage is

$$V = IR_o = (3.825 \text{ A})(3 \text{ Ω}) = 11.5 \text{ V}$$

The voltage vs. current characteristics for the simplified model are shown in Figure 10.13.

The linear model is useful, but represents real battery characteristics only qualitatively. Real batteries can provide current and power to a wide range of resistances with little voltage drop, but only for short periods of time. The linear model fails for longer periods of time and when large currents are drawn because it does not account for the finite amount of stored chemical energy and the rate of reaction limits of real batteries. In real batteries, the terminal voltage and power output drop drastically as the chemicals of the cell are consumed.

For utility-scale electric power storage and automotive propulsion, secondary cells offer great opportunities. In secondary cells, the reactions are approximately internally reversible, and with an external current source, the internal battery reactions may be reversed during a process called charging. For example, in a solar thermal power plant, the batteries can be charged during periods of sunshine and the energy stored in the battery can then be used for power applications when the sun goes down. In an electric car, the battery can be charged at night and used to power the vehicle in the day.

10.4.2 Cell Physics

When the current is small, the voltage is close to the cell EMF and the power output is small. To analyze the cell physics, assume that the chemical reaction in the battery consumes reactants at an electrode at the rate n_c mol per second. The electrons released in the reaction flow from the electrode through the external load at a rate proportional to the rate of reaction, jn_c, where j is the number of moles of electrons released per mole of reactant. Thus, jn_c is the rate of flow of electrons from the cell in moles of electrons per second.

Since there are 6.023×10^{23} electrons per gram-mole and each electron has a charge of 1.602×10^{-19} Coulomb (C), the product is $96,488 \times 10^{4}$ C/g-mol. The charge transported by a gram-mole of electrons is a constant named in honor of the pioneer of electrochemistry, Michael Faraday, Faraday's constant, F,

$$F = (6.023 \times 10^{23})(1.602 \times 10^{-19}) = 96,488 \times 10^{4} \, \text{C/g-mol}$$

With this model, the electric current from a cell, I, can be related to the reaction rate in the cell according to

$$I = jn_c F \tag{10.12}$$

and the instantaneous power, P, delivered by the cell is

$$P = jn_c FV \tag{10.13}$$

It can be seen that the cell electrode, the electrolyte material, and the nature and rate of chemical reaction control the maximum cell voltage, cell current, and maximum power output. It is also apparent that the amount of consumable reactants in the battery sets a limit to the battery capacity.

10.5 Rechargeable Batteries

Rechargeable storage batteries offer great opportunities for storing electric power, and many different types of batteries are under extensive research at the present time. Although the method of operation for all, except so-called flow batteries, is similar, many different materials and assemblies are being tried. Since at this time it is not clear which battery or groups of batteries will eventually achieve marketability, we will here simply discuss the method of operation of rechargeable batteries and then list the pros and cons of the various types with their potential applications. Figure 10.14 shows ranges of operation for different battery types as specific energy vs. specific power.

Lead–acid batteries have been the workhorse for electric storage for many decades, but they are too heavy and need to be recharged too often for many current applications, especially in the field of transportation. As a result, a worldwide effort is underway to improve and develop new batteries that can meet the demands of various industries including electric utilities and electric automobiles. The field of batteries is changing rapidly, and new batteries are constantly entering the market. At this time, it is impossible to pick winners and losers; therefore, the following merely outlines the potential applications as well as advantages and disadvantages of various typical batteries that are vying for a market share of the future. Only the lead–acid battery will be discussed in detail in order to demonstrate the basic method of operation of rechargeable batteries.

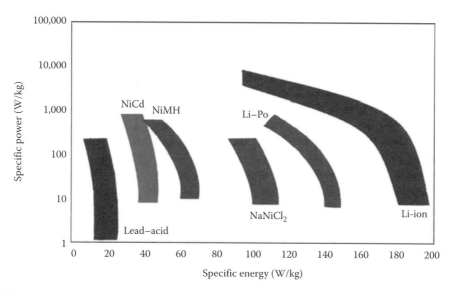

FIGURE 10.14
Approximate operating characteristics of different batteries in terms of specific energy and specific power. (Modified from de Guilbert, A. Batteries and supercapacitors cells for the fully electric vehicle. Saft Groupe SA, Brussels, 2009.)

10.5.1 Lead–Acid Batteries

In order to illustrate the operation of the charging cycle of a secondary battery, we will analyze the operation of a common lead–acid battery shown in Figure 10.15 in some detail.

For such a battery, each cell consists of a porous sponge lead anode and a lead-dioxide cathode, which is usually in the form of a plate. The two electrodes are separated by porous membranes in a sulfuric acid solution. The aqueous electrolyte contains positive

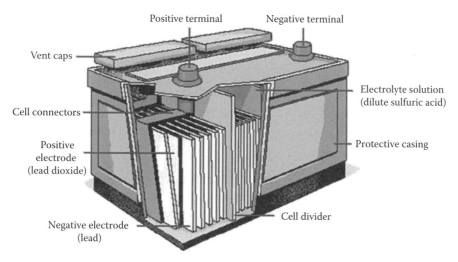

FIGURE 10.15
Schematic diagram of a conventional lead–acid battery.

hydrogen ions and negative sulfate ions, resulting from disassociation of the sulfuric acid in solution as shown by the following chemical equation:

$$2H_2SO_4 \rightarrow 4H^+ + 2SO_4^{2-} \tag{10.14}$$

As the lead is converted to lead sulfate at the anode, it releases two electrons to the external load as follows:

$$Pb(s) + SO_4^{2-}(aq) \rightarrow PbSO_4(s) + 2e^- \tag{10.15}$$

Hence, 2 mol of electrons are produced for each mole of sulfate ions and sponge lead reacted, or $j = 2$ for this battery.

In the reduction reaction at the cathode, lead dioxide, using hydrogen and sulfate ions from the electrolyte and electrons from the external circuit, is converted to lead sulfate, as shown by the following chemical equation:

$$PbO_2(s) + 4H^+(aq) + SO_4^{2-}(aq) + 2e^- \leftrightarrow PbSO_4(s) + 2H_2O(l) \tag{10.16}$$

We observe that the sulfate radical, that is, the sulfuric acid, is consumed at both electrodes and that hydrogen ions combine with oxygen from the cathode material to form water that remains in the cell. The water that is formed as the battery discharges makes the sulfuric acid electrolyte more dilute, as can be seen from the following net cell reaction. The net cell reaction, obtained as the sum of the electrode and electrolyte reactions, is

$$Pb(s) + PbO_2(s) + 2H_2SO_4(aq) \leftrightarrow 2PbSO_4(s) + 2H_2O(l) \tag{10.17}$$

In this and all other rechargeable battery chemistries, left to right indicates a battery discharging process and right to left indicates recharging. When any of the reactant is depleted, the battery ceases to function. However, the battery can be charged by applying an external power source. This reverses the electron flow, and the reactions given earlier are also reversed, as water is consumed to produce sulfate and hydrogen ions from the lead sulfate on the electrodes. Thus, the battery regains its charge and upon demand can again produce electric current.

There are three types of lead–acid batteries on the market: the wet cell, the sealed gel cell, and the sealed absorbed glass mat (AGM). Each of them has certain advantages and disadvantages. The wet cell has a liquid electrolyte, which must be replaced periodically to replenish hydrogen and oxygen escaping during the charge cycle. The sealed gel cell has a silica component added to the electrolyte to stiffen it. The AGM uses a fiberglass-like separator to hold the electrolyte in close proximity to the electrodes. This increases the efficiency. For both the gel and the AGM configuration, the risk of hydrogen explosion is reduced and corrosion is lessened. However, these two types require a slower charging rate. Both the gel cell and the AGM batteries are sealed and pressurized so the oxygen and hydrogen produced during charging are recombined into water.

The advantages of the lead–acid battery technology are low cost and high power density. However, its application for utility storage is very limited because of its short cycle life. A typical installation survives only a maximum of 1500 deep cycles. Therefore, lead–acid batteries have not found much use in commercial and large-scale energy management applications. The largest existing facility is a 140 GJ system in Chino, California, built in 1988 [10]. Because of the low specific energy at only 0.8 MJ/kg, lead–acid batteries are not a viable storage option for automobiles.

10.5.2 Nickel Metal (Ni–Cd and Ni–MH)

Description: The nickel–cadmium battery is the precursor to the nickel–metal hydride battery, but Ni–Cd batteries are losing popularity due to the toxicity of cadmium. Nickel–metal hydride (Ni–MH) batteries operate similar to other secondary batteries. The cathode in a Ni–MH battery uses a hydrogen-absorbing alloy whereas cadmium is used in a Ni–Cd battery. During cell discharge, hydrogen ions flow from the anode to the cathode through the separator.

Structure: The cathode of a Ni–MH battery is composed of nickel oxyhydroxide. The anode is generally an alloy of nickel, cobalt, manganese, and/or aluminum combined with a mixture of rare earth metals. The electrolyte is an alkaline solution, generally potassium hydroxide.
   During discharge, nickel oxyhydroxide is reduced to nickel hydroxide.

$$NiOOH + H_2O + e \rightarrow Ni(OH)_2 + OH^- \quad E'' = 0.52 \text{ V} \tag{10.18}$$

And the metal hydride, MH, is oxidized to the metal alloy, M.

$$MH + OH^- \rightarrow M + H_2O + e^- \quad E'' = 0.83 \text{ V} \tag{10.19}$$

The overall reaction on discharge is

$$MH + NiOOH \rightarrow M + Ni(OH)_2 \quad E'' = 1.35 \tag{10.20}$$

The process is reversed during charge.

Advantages of Ni–MH vs. Ni–Cd

- Higher energy density than standard Ni–Cd (30–80 Wh/kg).
- Periodic deep discharge cycling is required less often.
- Much more environmentally friendly than Ni–Cd. Ni–MH is easily recyclable.
- Long shelf life.

Disadvantages

- Service life limited with repeated deep cycling. More prevalent with high load currents.
- Low discharge current when compared to Li-ion batteries (0.2 C–0.5 C recommended).
- High self-discharge.
- Requires full discharge to prevent memory effect where the capacitance and the voltage drop until the battery is charged to its correct voltage and then discharged fully.

General uses

- Small electronics
- Hybrid electric and electric cars
- Wireless communications

10.5.3 Lithium Ion

Pure lithium batteries are primary batteries and cannot be recharged due to the instability of lithium. Li-ion batteries are secondary batteries and use lithium compounds instead of pure metallic lithium for their anode. Li-ion batteries, similar to other rechargeable batteries, have three primary functional components: the anode, the cathode, and the electrolyte. Lithium ions move from the anode to the cathode during discharging and the opposite way during charging. Most Li-ion batteries use a liquid organic electrolyte, but lithium polymer (Li–Po) batteries use a solid polymer composite that makes them easier to manufacture and more durable. Research on how to increase the power output of Li-ion batteries by decreasing the internal resistance of the battery is currently in progress. Li-ion batteries can be discharged at a continuous rate of 5–20 C with peak currents reaching 40 C. Li–Po batteries can achieve similar rates.

Structure: The cathode is normally made of a metal oxide such as lithium cobalt oxide ($LiCoO_2$) or lithium manganese oxide ($LiMn_2O_4$) attached to a current collector made of aluminum foil. The anode is generally made from layered graphitic carbon with a copper current collector. Recent research and development has shown that silicon nanowire anodes hold more energy and allow greater battery life (http://www.sciencedaily.com/releases/2007/12/071219103105.htm). The electrolyte is a lithium salt such as $LiPF_6$, $LiBF_4$, or $LiClO_4$. These salts are dissolved in an organic solvent, such as ether. In Li–Po batteries, the electrolyte is a composite of a gelled polymer and one of the lithium salts.

The chemical reactions for a typical Li-ion battery are shown as follows:

$$\text{Positive: } LiCoO_2 \leftrightarrow Li_{1-x}CoO_2 + xLi^+ + xe^- \tag{10.21}$$

$$\text{Negative: } xLi^+ + xe^- + 6C \leftrightarrow Li_xC_6 \tag{10.22}$$

$$\text{Overall: } LiCoO_2 + 6C \Leftrightarrow Li_{1-x}CoO_2 + Li_xC_6 \tag{10.23}$$

Note that the C in the earlier equations stands for a very carbon-rich negative material such as graphite. These equations are written in units of moles, using x as a coefficient.

Advantages of Li-ion batteries compared to Ni–MH

- Higher energy densities than other battery types (130–200 Wh/kg) compared to Ni–MH (30–80 Wh/kg)
- Generally much lighter
- Low maintenance (Ni–MH batteries have "memory" effect and have to be fully cycled occasionally)
- Low self-discharge rate—typically 5% per month

Disadvantages of Li-ion batteries

- Shelf life—the battery's capacity will slowly decay over time regardless of the number of charge/discharge cycles. Depending on conditions, the battery can lose anywhere from 6% to 40% of its capacity over 1 year. (If stored at 40% charge level and 25°C, the battery will lose 4% of its capacity after 1 year.)

- Requires a protective circuit to make sure it is not overcharged or over-discharged as this degrades the battery by causing deposits of material that can short the battery (over-discharge) or cause excessive heat (overcharge).

- If damaged, combustion of the electrolyte, anode, and cathode can occur due to the interaction of lithium and the water vapor in the air.

- Currently is expensive to manufacture although expected to become cheaper as the technology matures.

General uses

- Electric/hybrid cars (future potential)
- Cell phones and laptops

10.5.4 Flow Batteries

Description: Flow batteries are a novel type of secondary battery in which charged electrolytes are passed through a cell, producing electric charge (see Figure 10.16). The vanadium redox (VR) battery is the most common type of flow battery. The VR battery uses an electrolyte created from electrolytically dissolving vanadium pentoxide (V_2O_5) in sulfuric acid (H_2SO_4). The battery consists of two tanks of electrolyte and a membrane connected to a load, as shown in Figure 10.16.

Charging the electrolyte causes the cathode to produce VO^{2+} ions and the anode to produce V^{2+}. During discharge, the charged electrolyte is pumped across a membrane, creating VO_2^+ on the positive tank side and V^{3+} on the negative tank side. This produces an open

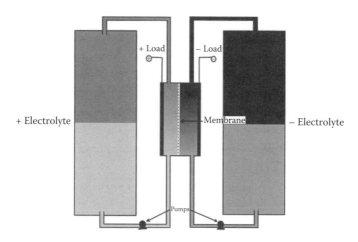

FIGURE 10.16
Schematic diagram of a typical flow battery.

circuit voltage. Since the chemical structure of the electrolyte does not undergo a change, this battery is very stable.

At the positive electrode

$$VO_2^+ + 2H^+ + e \underset{\text{charge}}{\overset{\text{discharge}}{\longleftrightarrow}} VO^{2+} + H_2O \quad E'' = 1.00 \text{ V} \tag{10.24}$$

At the negative electrode

$$V^{3+} + e \underset{\text{discharge}}{\overset{\text{charge}}{\longleftrightarrow}} V^{2+} \quad E'' = -0.26 \text{ V} \tag{10.25}$$

The potential generated by the cathode and the anode combine to create the overall potential for the battery.

Advantages

- Very environmentally friendly—the vanadium electrolyte does not break down and can theoretically last forever thus reducing the waste to near zero for the whole system. Vanadium is a nontoxic element so catastrophic failure of containment is not an issue.
- Capable of very large energy storage. Only limiting factor is how large the electrolyte tanks are. Can reach MWh range (http://www.vrb.unsw.edu.au/).
- Capable of very fast change in power.
- Can be fully charged/discharged without degradation.
- High cycle life (>10,000) after which only the membrane needs to be replaced.
- Very low self-discharge rate—electrolytes are stable at charged levels.
- Cost/kWh decreases as storage size increases.
- Possibility of instant recharge by replacing electrolyte.
- Already scaled to industrial size (http://www.vrb.unsw.edu.au/).
- High efficiency (between 80% and 90% in larger installations).

Disadvantages

- Energy density only 10–20 Wh/kg and 15–25 Wh/L. Second-generation V/Br batteries can achieve 35–50 Wh/kg with higher concentrations of vanadium in the solution.
- System is complex in larger storage sizes.

Uses

- Load leveling
- Off-peak production storage for renewable energy
- Possible electric car applications (fill up like gasoline, but no burning of fuel, simply recharged at the station)

10.6 Fuel Cells and Hydrogen

Electricity generation from hydrogen requires a fuel cell as an intermediate step, as discussed in Chapter 1. Details of the principles of operation of the fuel cell and the use of hydrogen as a storage medium are discussed in the following text [11].

10.6.1 Principles of Fuel Cell Operation

The operation of a fuel cell is shown schematically in Figure 10.17.

A fuel cell consists, like a battery, of two electrodes separated by an electrolyte, which transmits ions but is impervious to electrons. In a fuel cell, hydrogen, or another reducing agent, is supplied to the negative electrode, while oxygen (or air) goes to the positive electrode. A catalyst on the porous anode promotes dissociation of hydrogen molecules into hydrogen ions and electrons. The H^+ ions migrate through the electrolyte to the cathode where they react with electrons supplied through the external circuit and with oxygen to form water.

Although a fuel cell converts chemical energy into electricity directly, without an intermediate combustion cycle, it does not approach its theoretical efficiency of 100% in practice. Similar to batteries, fuel cells have an efficiency on the order of 40% in practice, but this efficiency is independent of the amount of power generated. Since the efficiency of an assembly of fuel cells is approximately equal to that of a single cell, there are no economies of scale. Therefore, fuel cells are promising for small, localized plants in the range of 1–100 kW capacity. Since fuel cells also reject heat, they can operate as a cogeneration system and could supply a building with both electricity and heat.

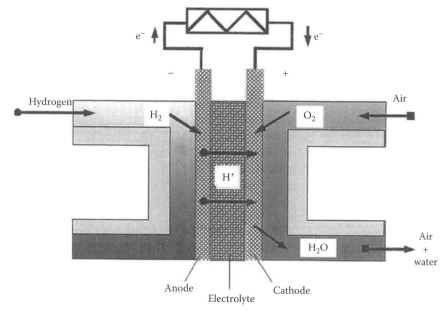

FIGURE 10.17
Schematic diagram of fuel cell operation.

10.6.2 Types of Fuel Cells

There are six principal types of fuel cells, as shown in Table 10.4.

- Proton exchange membrane fuel cell (PEMFC)
- Direct methanol fuel cell (DMFC)
- Molten carbonate fuel cell (MCFC)
- Alkaline fuel cell (AFC)
- Phosphoric acid fuel cell (PAFC)
- Solid oxide fuel cell (SOFC)

These types of fuel cells differ in operating temperatures, efficiency, and cost. Their operating characteristics and some potential applications are summarized in Table 10.4.

PEMFC cells operate at relatively low temperatures (about 80°C), have a high power density, and can change rapidly to meet varying power demand or a fast start-up. These features make PEMFC the favorite for automobile transportation. The membrane of a PEMFC is made of thin perfluorosulfonic acid sheets, which acts as an electrolyte and allows the passage of hydrogen ions only. The membrane is coated on both sides with highly dispersed metal alloy particles (currently mostly platinum) that act as catalysts.

DMFC cells also use a polymer membrane as an electrolyte. The most attractive feature of DMFCs is that they can use liquid methanol as a fuel. The anode of a DMFC can draw hydrogen from liquid methanol and thus eliminates the use of a fuel reformer.

MCFC cells use a molten carbonate salt as the electrolyte. They are flexible in their fuel supply and can use coal-derived gases, methane, or NG as fuel. The efficiency of these fuel cells is about 60%, but if operating as a cogeneration unit that uses the waste heat, the efficiency overall can be even higher.

PAFC cells use an anode and a cathode of finely dispersed platinum catalyst on carbon and a silicon carbide structure that contains phosphoric acid as electrolyte.

AFC cells are commercially available. They operate at low temperatures, below 100°C, and are used in submarines and for space power applications.

TABLE 10.4

Operating Characteristics of Fuel Cells

	PEMFC	DMFC	MCFC	AFC	PAFC	SOFC
Electrolyte	Ion exchange membrane	Polymer membrane	Immobilized liquid molten carbonate	Immobilized liquid phosphoric acid	Potassium hydroxide	Ceramic
Operating temperature (°C)	80	60–130	650	200	60–90	800–1000
Efficiency (%)	40–60	40	45–60	35–40	45–60	50–65
Typical electrical power	Up to 250 kW	<10 kW	>1 MW	>50 kW	Up to 20 kW	>200 kW
Possible applications	Vehicles	Portables	Power stations	Small power stations	Submarines, spacecraft	Power stations

SOFC cells work at high temperatures (between 800°C and 1000°C) and utilize a solid ceramic electrolyte such as zirconium oxide stabilized with yttrium oxide instead of a liquid. The efficiency of these cells can reach as high as 60%, and they have the potential of being used for large-scale energy storage or generating electricity industrially. They can also be used for cogeneration systems.

The major challenges for widespread application of fuel cells are their cost, durability, and their need for a continuous and an inexpensive supply of hydrogen.

10.6.3 Generation of Hydrogen

There are currently two different methods used for the commercial production of hydrogen. The most widely used is NG reforming. In this process, called steam methane reforming, NG is mixed with steam and the resulting mixture is introduced into a catalytic reforming reactor where it is passed through a nickel catalyst and converted at about 900°C into CO and H_2 according to the equation

$$CH_4 + H_2O \rightarrow 3H_2 + CO \quad \Delta H^0 = 206 \text{ kJ/mol} \tag{10.26}$$

The other method is electrolysis, where an electric current is passed through water to separate the hydrogen from the oxygen. Today most commercial water electrolyzers use an alkaline electrolyte system, as shown in Figure 10.18.

In such an electrolytic cell, the following simplified reaction takes place:

In the electrolyte

$$2H_2O \rightarrow 2H_2 + 2OH^- \tag{10.27}$$

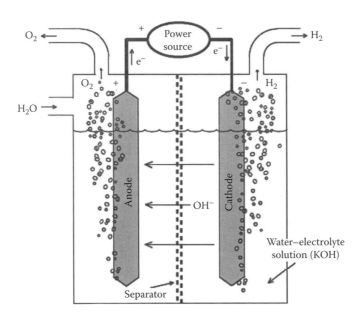

FIGURE 10.18
Schematic diagram of hydrogen generation by electrolysis.

At the cathode

$$2H^+ + 2e^- \rightarrow H_2 \tag{10.28}$$

At the anode

$$2OH^- \rightarrow \frac{1}{2}O_2 + H_2O + 2e^- \tag{10.29}$$

The overall reaction is

$$H_2O \rightarrow \frac{1}{2}O_2 + H_2 \quad \text{where } \Delta H_r = 286 \text{ kJ/mol} \tag{10.30}$$

According to Ref. [12], the estimated cost of hydrogen production by NG reforming or coal gasification is about \$1/kg, whereas the cost via a renewable technology, such as electrolysis with electricity from wind, is between \$6 and \$7/kg. The energy content of 1 kg of H_2 approximately equals the energy content of 1 gal of gasoline.

10.6.4 Storage and Transport

Once the hydrogen is generated, it can be stored and subsequently combusted to provide heat or work, or it can be used to generate electricity with a fuel cell. Gaseous hydrogen has a low specific density, and for storage purposes, it must be compressed or cryogenically liquefied. If hydrogen is compressed adiabatically to a pressure of 350 bar, a value often assumed suitable for automotive technologies, it consumes about 12% of its higher heating value (HHV). The loss can be reduced appreciably, however, if the compression approaches an isothermal process. If the hydrogen is stored as a liquid at cryogenic temperatures, the process takes about 40% of the HHV with current technology. Moreover, liquid storage is not feasible for automotive applications because mandatory boil-off from the storage container cannot be safely released in closed spaces such as garages or tunnels.

A recent technology proposed for storing hydrogen consists of bonding it into metal hydrides using an absorption process. The energy penalty of this type of storage, which requires only a pressure of about 30 bar, may be less. However, the density of the metal hydride can be between 20 and 100 times the density of the hydrogen stored [13]. Recently, carbon nanotubes have received attention as a potential hydrogen storage medium, but neither of the earlier processes has been fully developed and commercialized.

10.6.5 Thermodynamics and Economics

As an energy carrier, hydrogen should be compared with the only other widespread and viable alternative: electricity. Once generated and compressed, hydrogen could flow through large pipelines with a loss of approximately 0.8%/100 km [2,12]. According to Refs. [2,13], the energy loss experienced with high-voltage DC (long-distance) electric transmission lines is of the order of 0.6%/100 km. Hence, the losses incurred in the transmission of energy by hydrogen and electricity are of the same order of magnitude. However, the cost of building a transmission system for hydrogen has been estimated to be on the order of half a trillion dollars, and unless this up-front expenditure is made, transmission of

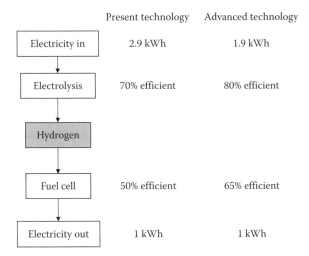

	Present technology	Advanced technology
Electricity in	2.9 kWh	1.9 kWh
Electrolysis	70% efficient	80% efficient
Hydrogen		
Fuel cell	50% efficient	65% efficient
Electricity out	1 kWh	1 kWh

FIGURE 10.19
Generating electricity with a fuel cell using hydrogen produced by electrolysis. (Adapted from Kreith, F. and West, R.E., *J. Energy Resourc. Technol.*, 126, 249, 2004.)

hydrogen would have to be accomplished by truck, which would incur enormous expenses and energy penalties. Moreover, as shown by Kreith and West [14], the extraction of electricity from hydrogen is associated with irreversible losses. Electrolysis of water to make hydrogen is commercially available at about 70% efficiency. The sequence of steps required of generating electricity for producing hydrogen and then using a fuel cell to generate electricity again is shown in Figure 10.19. Compressing hydrogen to a pressure of 350 bar for transporting it is estimated to be about 88% efficient. And, finally, conversion back into electricity with a fuel cell has an efficiency of approximately 60%, although the theoretical limit is somewhat higher. Taken together, generating power for electrolysis, compression, and generation of electricity in the fuel cell results in, at most, an overall efficiency of only about 50%, compared to the amount of electricity that could have been obtained from the original source. Hence, from a thermodynamic as well as from an economic perspective, the use of hydrogen as a storage and transmission system appears to be inferior to the use of electricity directly. It should also be noted that this loss of energy is the result of basic thermodynamic considerations and would occur irrespective of whether the original generation of electricity came from renewable sources, nuclear power, or fossil fuels. In addition, because two units of electricity are required to produce one unit of useful power at the end use, the amount of CO_2 generation/kW electricity would be double via the hydrogen path compared to any other electric power system.

10.7 Thermal Energy Storage

Solar thermal storage is a well-established technology, and the principles of design and operation in tanks or in the ground are extensively described in standard textbooks on heat transfer [15]. For solar thermal heat and power generation, thermal storage is important in applications such as storing heat overnight for buildings, seasonal heat storage for large-scale heating and/or cooling projects (called seasonal storage), and storage in solar thermal power plants such as solar troughs or central receivers. Solar storage can also be dispersed through the overall system as, for example, in the thermal mass of a building—an art that has been practiced for centuries. A particularly good example is the Mesa Verde buildings in Colorado, where heat is stored in rock walls, which form a part of the structures.

10.7.1 Sensible Heat

Thermal energy can be stored as sensible heat, as latent heat, or in reversible thermochemical equations. The sensible heat, ΔQ, gained or lost per unit mass by a material changing temperature from T_1 to T_2 is

$$\Delta Q = \int_{T_1}^{T_2} \rho c_p dT \tag{10.31}$$

Although both the density, ρ, and the specific heat, c_p, vary with temperature, the variations are usually small enough to use average values for engineering calculations. Equation 10.31 then simplifies so that $\Delta Q = c_p \Delta T$ per unit mass, or $\rho c_p \Delta T$ per unit volume. Commonly used figures of merits are $\$/c_p$ for relative media costs, ρc_p for volumetric heat capacity, using the lowest value of ρ at the highest expected operating temperature for relative container size.

Solid media: Table 10.5 shows the characteristics of some plentiful and economically competitive solid materials for sensible heat storage. Rocks, bricks, and cement are excellent candidates for overnight thermal storage in buildings.

TABLE 10.5

Physical Properties of Some Sensible Heat Storage Materials

Storage Medium	Temperature Range (°C)	Density (ρ) (kg/m³)	Specific Heat (c), (J/kg K)	Energy Density (ρc) (kWh/m³ K)	Thermal Conductivity (W/m K)
Water	0–100	1000	4190	1.16	0.63 at 38°C
Water (10 bar)	0–180	881	4190	1.03	—
50% ethylene glycol-50% water	0–100	1075	3480	0.98	—
Dowtherm A® (Dow Chemical, Co., Midland, Michigan)	12–260	867	2200	0.53	0.122 at 260°C
Therminol 66® (Monsanto Co., St. Louis, Missouri)	9–343	750	2100	0.44	0.106 at 343°C
Draw salt (50 NaNO₃–50 KNO₃)ᵃ	220–540	1733	1550	0.75	0.57
Molten salt (53 KNO₃/40 NaNO₃/7 NaNO₃)ᵃ	142–540	1680	1560	0.72	0.61
Liquid sodium	100–760	750	1260	0.26	67.5
Cast iron	m.p. (1150–1300)	7200	540	1.08	42.0
Taconite	—	3200	800	0.71	—
Aluminum	m. p. 660	2700	920	0.69	200
Fireclay	—	2100–2600	1000	0.65	1.0–1.5
Rock	—	1600	880	0.39	—

Note: m.p. = melting point.
ᵃ Composition in percent by weight.

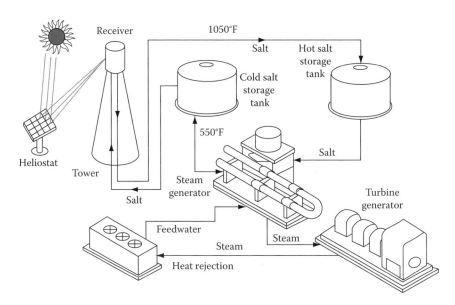

FIGURE 10.20
Two-tank storage for central receiver system.

Liquid media: Table 10.5 shows the characteristics of some useful liquid media for sensible heat storage. Water is clearly superior for temperatures below 100°C, whereas oils, molten salts, or liquid metals are used for solar thermal power applications such as parabolic trough or central receivers.

Dual media: Solids and liquid sensible heat storage materials can be combined in various ways. Rock beds and water tanks have been used in hybrid storage systems for space heating, whereas rocks and oils have been used in single vessels for solar thermal power applications to improve stratification. This approach could reduce the cost compared to relatively expensive dual-tank liquid storage systems where two vessels are necessary, as shown in Figure 10.20. However, experiences with single-tank storage to achieve stratification have not been reliable, and for solar thermal power applications, dual-tank storage is preferred.

10.7.2 Phase Change Heat Storage

Latent heat is the thermal energy released or absorbed by a material undergoing a phase transition, such as solid to liquid (melting) or liquid to gas (vaporization). Materials that have a high heat release or absorption during phase change are of interest for thermal storage. Latent heat storage material should have a large heat of transition, high density, appropriate transition temperature, low toxicity, and no gradation with multiple phase change. Materials attractive for storage are paraffin waxes and salt hydrates.

The advantage of using phase change material for storage is that a large amount of heat can be stored in small volumes, and the temperature of the material remains constant during phase change. The problem with phase change heat transfer is to find materials that have appropriate temperatures during phase change, and also the design of heat exchangers that transfer heat to or from the material during phase change is exceedingly difficult. Table 10.6 shows the material properties of several common TES materials that may potentially be useful for latent heat storage. The engineering design of latent heat storage systems is difficult and beyond the scope of this book.

TABLE 10.6

Physical Properties of Phase Change Storage Materials (PCMs)

Storage Medium	Melting Point (°C)	Latent Heat (kJ/kg)	Specific Heat (kJ/kg°C)		Density (kg/m³)		Energy Density (kWh/m³ K)	Thermal Conductivity (W/m K)
			Solid	Liquid	Solid	Liquid		
$LiClO_3 \cdot 3H_2O$	8.1	253	—	—	1720	1530	108	—
$Na_2SO_4 \cdot 10H_2O$ (Glauber's salt)	32.4	25.1	1.76	3.32	1460	1330	92.7	2.25
$Na_2S_2O_3 \cdot 5H_2O$	48	200	1.47	2.39	1730	1665	92.5	0.57
$NaCH_3COO \cdot 3H_2O$	58	180	1.90	2.50	1450	1280	64	0.5
$Ba(OH_3) \cdot 6H_2O$	90	163	1.56	3.68	1636	1550	70	0.611
$LiNO_3$	252	530	2.02	2.041	2310	1776	261	1.35
$LiCO_3/K_2CO_3$, (35:65)[a]	505	345	1.34	1.76	2265	1960	188	—
$LiCO_3/K_2CO/ Na_2CO_3$ (32:35:33)[a]	397	277	1.68	1.63	2300	2140	165	—
n-Tetradecane	5.5	228	—	—	825	771	48	0.150
n-Octadecane	28	244	2.16	—	814	774	52.5	0.150
HDPE (cross-linked)	126	180	2.88	2.51	960	90	45	0.361
Stearic acid	70	203	—	2.35	941	347	48	0.172ℓ

Note: ℓ = liquid.
[a] Composition in percent by weight.

10.7.3 Thermochemical Storage

Thermochemical energy is stored as the bond energy of a chemical compound. In thermochemical energy storage, atomic bonds are broken during a reversible chemical reaction and are catalyzed by an increase in temperature. After thermochemical separation occurs, the constituents are stored apart until the combination reaction is desired. Recombination of the bonds between atoms releases the stored thermochemical energy.

Reversible chemical reactions with reactants and products that can be easily stored as liquids and/or solids are of interest for thermochemical energy storage. Reactions that produce two distinct phases such as a solid and a gas are desirable since the separation of products to prevent back reaction is facilitated. Table 10.7 shows several common thermochemical storage reactions and their standard enthalpy change (ΔH^0) in kJ and their turning temperature (T') in K.

The turning temperature, T', in Table 10.7 is defined as the temperature at which the equilibrium constant is unity and is calculated using the ratio of the standard enthalpy change and the standard entropy change for the reaction. At this temperature, the reactants and products will be present in approximately equal quantities. When $T > T'$, the endothermic storage reaction is favored, meaning heat is necessary for and absorbed during the reaction. Conversely, for $T < T'$, the exothermic reaction dominates, meaning that heat is produced in the reaction.

The primary advantages of thermochemical storage include the high energy density and the long-term, low-temperature storage capability. However, the thermochemical process is very complex, and the thermochemical materials are often expensive and can be

TABLE 10.7

Thermochemical Storage Reactions

Reaction	ΔH^0 (kJ)	T' (K)
$NH_4F(s) \leftrightarrow NH_3(g) + HF(g)$	149.3	499
$Mg(OH)_2(s) \leftrightarrow MgO(s) + H_2O(g)$	81.1	531
$MgCO_3(s) \leftrightarrow MgO(s) + CO_2 (g)$	100.6	670
$NH_4HSO_4(l) \leftrightarrow NH_3(g) + H_2O(g) + SO_3(g)$	337.0	740
$Ca(OH)_2(s) \leftrightarrow CaO(s) + H_2O(g)$	109.3	752
$BaO_2(s) \leftrightarrow BaO(s) + \frac{1}{2}O_2(g)$	80.8	1000
$LiOH(1) \leftrightarrow \frac{1}{2}Li_2O(s) \frac{1}{2}H_2O(g)$	56.7	1000
$MgSO_4 \leftrightarrow MgO(s) + SO_3(g)$	287.6	1470

Source: Wyman, C. et al., *Sol. Energy*, 24(6), 517, 1980.

hazardous. There is considerable research activity in the field of thermochemical reactions for energy storage, but no large-scale commercial applications exist to date.

10.7.4 Applications

Many factors contribute to choosing an appropriate storage method and material for each TES application. Table 10.8 lists several TES options for solar thermal power and their appropriate storage materials. In the table, PCM stands for phase change material, HX means heat exchanger, and VHT means very high temperature.

10.7.5 Thermal Storage for Concentrating Collector Systems

The most important application of thermal storage is in high-temperature concentrator systems such as central receivers or parabolic troughs. Steam is generated in the solar receiver when the sun is up and either passes through the turbine or, when excess heat is available, is stored in a tank. The size of the tank will determine the number of hours that the system can operate when the sun is down. For a typical system of this type, steam is raised in the receiver at 40 bar and 250°C, and condensed at 0.6 bars and 5°C. After the steam is condensed, it is returned to the solar receiver as in any typical Rankine cycle. The drawback of this approach is that steam storage requires a large volume, which limits the length of time the turbine can operate when the sun is down.

A more significant application of thermal storage is the two-tank storage system shown in Figure 10.20 that was used in the Solar 2 central receiver. In this approach, a single heat-transfer fluid is used in the receiver and for storage. This approach eliminates the need for expensive heat exchangers and the degradation in the availability of the energy that occurs during heat transfer from one fluid to another. The fluid most commonly proposed for use in this kind of application is a molten salt, although other heat-transfer fluids such as oils or liquid metals are being considered.

When molten salts are used as the heat-transfer fluid, care must be taken to avoid freezing anywhere in the system. Molten salts freeze at a temperature between 120°C and 200°C (250°F–430°F). Thus, the Rankine cycle has to operate between a high temperature achievable in the receiver and a cold temperature that does not drop below the freezing point of the salt. The design of a typical two-tank direct system with approximate temperatures is shown in Figure 10.20. The temperature of the molten salt is raised in the receiver to about

TABLE 10.8

Options for TES in Solar Power Production

TES Options for Solar Thermal Power	Temperature (°C)	Storage Medium	Type
Small power plants and water pumps			
	100	Water in thermocline tank or two tanks	Sensible
Organic Rankine	300	Petroleum oil in the thermocline tank	Sensible
Steam Rankine with organic fluid receiver	375	Synthetic oil with trickle charge	Sensible
Dish-mounted engine generators (buffer storage only)			
Organic Rankine	400	Bulk PCM with indirect HX	Latent
Stirling and air Brayton	800	Bulk PCM with indirect HX	Latent
Advanced air Brayton	1370	Graphite	Sensible
		Graphite Encapsulated PCM	Latent
Larger power plants (typically 3–8 h storage)			
Steam Rankine with organic fluid receiver	300	Petroleum oil in thermocline tank or two tanks, evaporation only	Sensible
	300	Petroleum oil/rocks (dual medium) in thermocline tank	Sensible
		Petroleum oil in thermocline tanks, evaporation only	Sensible
		Petroleum oil/rocks (dual medium) in thermocline tank	Sensible
		Encapsulated PCM with evaporative HX	Latent
Steam Rankine with water–steam receiver		Bulk PCM with indirect HX	Latent
		Bulk PCM with direct HX	Latent
		Pressurized water above ground or underground	Latent
	540	Molten draw salt in thermocline tank or two tanks, superheat	Sensible
		Air/rocks	Sensible
		Bulk PCM with direct HX, evaporation stage	Latent
		Solid or liquid decomposition, evaporation stage	Thermochemical
Steam Rankine with molten draw salt receiver	540	Molten draw salt in thermocline tank or two tanks	Sensible
Steam Rankine with liquid metal receiver	540	Liquid sodium in one tank, mixed, buffer only	Sensible
		Liquid sodium in two tanks Air/rocks	Sensible
		Refractory or cast-iron	Sensible
Brayton with gas-cooled receiver	800	Bulk PCM with indirect HX Solid or liquid decomposition	Thermochemical
		VHT molten salt in two tanks	Sensible
Brayton with liquid-cooled receiver	800	VHT molten salt/refractory (dual medium) in thermocline tank	Sensible
	1100	Bulk glassy slag, liquid and solid bead storage, direct HX	Sensible and Latent

Note: TES, Thermal energy storage; PCM, phase change material; HX, heat exchange; VHT, very high temperature.

565°C and then passed to the hot salt storage tank. From this tank, salt is extracted to a heat exchanger, which raises the steam for a Rankine cycle power plant and then returned to the cold salt storage tank at about 290°C. The working fluid is then recirculated to the receiver in a virtually closed loop. A normal Rankine cycle is used for the generation of electricity. The length of time that the system can operate once the sun is down depends on the size of the storage tanks. The advantage of this approach is the simplicity of design, and the disadvantage is that a large amount of heat-transfer fluid is needed for the plant to operate.

In systems where the heat-transfer fluid is very expensive or cannot be used for some reason as the storage fluid, a two-tank indirect system is used. The two-tank indirect system functions the same way as the two-tank direct system, except that different fluids are used in the receiver heating loop and the storage. Therefore, the heat from the working fluid must be transferred to the storage fluid via a heat exchanger, in which the fluid enters at high temperature and leaves at low temperature. Storage fluid from the high temperature tank is then used to generate steam, as in the direct solar system.

10.7.6 Overnight Storage for Buildings and Domestic Hot Water

TES for domestic hot water or for overnight heating of buildings has been mentioned in Chapter 7. These applications require storing a liquid, almost always water, in a tank. A schematic diagram typical of such systems with electric backup is shown in Figure 10.21. Insulation of the storage tank is a simple heat-transfer problem treated in standard textbooks. The properties of insulation materials are given in Table 10.9.

The annual average insolation on a surface tilted at an angle equal to the latitude in the midlatitudes of the United States is about 23,000 kJ/m² per day (2,030 Btu/ft² per day). If the average collection efficiency is 30%, 6900 kJ/m² per day (610 Btu/ft² per day) is delivered by the collector system. The service hot water demand can be taken as 76 L/day (20 gal/day) per person. If the tap water has a temperature of 283 K (50°F), and the hot water 333 K (140°F), about 210 kJ/L (750 Btu/gal) is required for hot water heating. For a delivery of

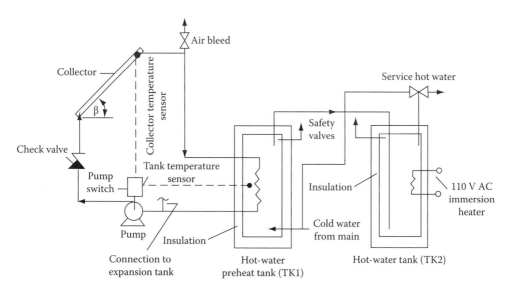

FIGURE 10.21
Schematic diagram of two-tank solar heating system with storage.

TABLE 10.9

Thermal Properties of Some Insulating and Building Materials

Material	Average Temperature (°F)	k [Btu/(h·ft·°F)]	c [Btu/(lb$_m$·°F)]	ρ (lb$_m$/ft³)	α (ft²/h)
Insulating materials					
Asbestos	32	0.087	0.25	36	~0.01
	392	0.12	—	36	~0.01
Cork	86	0.025	0.04	10	~0.006
Cotton fabric	200	0.046		—	—
Diatomaceous earth, powdered	100	0.030	0.21	14	~0.01
	300	0.036	—	—	—
Molded pipe covering	400	0.051	—	26	—
	1600	0.088	—	—	—
Glass wool					
fine	20	0.022	—	—	—
	100	0.031	—	1.5	—
	200	0.043	—	—	—
Packed	20	0.016	—	—	—
	100	0.022	—	—	—
Hair felt	100	0.027	—	8.2	—
Kaolin insulating brick	932	0.15	—	8.2	—
	2102	0.26	—	—	—
Kaolin insulating firebrick	392	0.05	—	19	—
	1400	0.11	—	—	—
Magnesia, 85%	32	0.032	—	17	—
	200	0.037	—	17	—
Rock wool	20	0.017	—	8	—
	200	0.030	—	—	—
Rubber	32	0.087	0.48	75	0.0024
Building materials					
Brick				—	—
Fire clay	392	0.58	0.20	144	0.02
	1832	0.95		—	—
Masonry	70	0.38	0.20	106	0.018
Zirconia	392	0.84	—	304	—
	1832	1.13	—	—	—
Chrome brick	392	0.82	—	246	—
	1832	0.96	—	—	—
Concrete					
Stone	~70	0.54	0.20	144	0.019
10% moisture	~70	0.70	—	140	~0.025
Glass, window	~70	~0.45	0.2	170	0.013
Limestone, dry	70	0.40	0.22	105	0.017

TABLE 10.9 (continued)

Thermal Properties of Some Insulating and Building Materials

Material	Average Temperature (°F)	k [Btu/(h·ft·°F)]	c [Btu/(lb$_m$·°F)]	ρ (lb$_m$/ft³)	α (ft²/h)
Sand					
Dry	68	0.20	—	95	—
10% water	68	0.60	—	100	—
Soil					
Dry	70	~0.20	0.44	—	~0.01
Wet	70	~0.15	—	—	~0.03
Wood					
Oak ⊥ to grain[a]	70	0.12	0.57	51	0.0041
‖ to grain	70	0.20	0.57	51	0.0069
Pine ⊥ to grain[a]	70	0.06	0.67	31	0.0029
‖ to grain	70	0.14	0.67	31	0.0067
Ice	32	1.28	0.46	57	0.048

Source: Kreith, F., *Principles of Heat Transfer*, Intext, New York, 1973.
[a] ⊥, perpendicular; ‖, parallel.

230 L/day (60 gal/day) for a family of three, 48.3 MJ/day (46,000 Btu/day) is needed. If the collector delivers 6900 kJ/m², a collector with a size of approximately 7 m² is needed. This collector is about 1/10th the size of a collector required for heating a typical building in the midlatitudes. A rule of thumb is that 1 m² of collector is required for each 30 L of hot water to be delivered. The storage tank should be large enough to hold about a 2 day supply of hot water, or about 500 L for a family of three.

Thermal storage tanks must be insulated to control heat loss. If a storage tank is located within a structure, any losses from the tank tend to offset the active heating demands of the building. However, such storage loss is uncontrolled and may cause overheating in the summer.

The amount of thermal storage used in a solar heating system is limited by the law of diminishing returns. Although larger storage results in larger annual energy delivery, the increase at the margin is small and hence not cost-effective. Seasonal storage is, therefore, usually uneconomic, although it can be realized in a technical sense. Experience has shown that liquid storage amounts of $50-75$ kg H_2O/m^2 ($10-15$ lb/ft$_c^2$) are the best compromise between storage tank cost and useful energy delivery.

Since solar energy heating systems operate at temperatures relatively close to the temperatures of the spaces to be heated, storage must be capable of delivering and receiving thermal energy at relatively small temperature differences. The designer must consider the magnitude of these driving forces in sizing heat exchangers, pumps, and air blowers. The designer must also consider the nonrecoverable heat losses from storage—even though storage temperatures are relatively low, surface areas of storage units may be large, and heat losses therefore appreciable.

Other mechanical components in solar heating systems include pumps, heat exchangers, air bleed valves, pressure release valves, and expansion tanks. Heat exchangers in solar systems are selected based on economic criteria. The best trade-off of energy delivery increase with increasing heat-exchanger size usually results from use of an exchanger with effectiveness in the range of 0.6–0.8. Counterflow heat exchangers are required for this level of effectiveness.

Achievement of the required effectiveness level may dictate fairly high flow rates in the storage tank side of the collector heat exchanger. Flows up to twice that in the collector side can improve exchanger performance significantly in many cases. Since the storage side loop is physically short and has a small pressure drop, increased flow in this loop increases pump energy requirements by a negligible amount. Typical solar heat-exchanger sizes range from 0.05 to 0.10 m^2 of heat-exchanger surface per square meter of net collector area.

In hydronic heating systems, it is essential that all air be pumped from the system. To facilitate this process, air bleed valves located at the high points in a system are used. These are opened during system fill and later if air should collect. Air bleeds are required at points of low velocity in piping systems where air may collect because the local fluid velocity is too low for entrainment.

Control strategies and hardware used in current solar system designs are quite simple and are similar in several respects to those used in conventional systems. The single fundamental difference lies in the requirement for differential temperature measurement instead of simple temperature sensing. In the space heating system shown in Figure 10.21, two temperature signals determine which of three modes is used. The signals used are the collector–storage differential and room temperature. The collector–storage difference is sensed by two thermistors or thermocouples, the difference being determined by a solid-state comparator, which is a part of the control device. Room temperature is sensed by a conventional dual-contact thermostat.

The control system operates as follows. If the first room thermostat contact closes, the mode selector valve and distribution pump are activated in an attempt to deliver the thermal demand from solar thermal storage. If room temperature continues to drop, indicating inadequate solar availability, the mode selector diverts flow through the backup system instead of the solar system, and the backup is activated until the load is satisfied.

The collector–storage control subsystem operates independently of the heating subsystem described earlier. If collector temperature, usually sensed by a thermistor thermally bonded to the absorber plate, exceeds the temperature in the bottom of the storage tank by 5°C–10°C (9°F–18°F), the collector pump and heat-exchanger pump (if present) are activated and continue to run until the collector and storage temperatures are within about 1°C–2°C (2°F–4°F) of each other. At this point, it is no longer worthwhile to attempt to collect energy, and the pumps are turned off. The collector–storage subsystem also has a high temperature cutout that turns the collector loop pump off when the storage temperature exceeds a set limit.

A heating load device transfers heat from the solar storage to the air in the space. Therefore, a liquid-to-air heat exchanger is sized based on the energy demand of a building. Several generic types of load devices are in common use:

1. Forced-air systems—tube-and-fn coil located in the main distribution duct of a building or zone of a building (see Figure 10.22).
2. Baseboard convection systems—tube-and-fn coils located near the floor on external walls. These operate by natural convection from the convectors to the room air.
3. Heated floors or ceilings—water coils. These transfer heat to large thermal masses that in turn radiate or convect into the space. This heating method is usually called radiant heating.

Each load device requires fluid at a different temperature in order to operate under design load conditions, as shown in Figure 10.23. Since baseboard heaters are small in

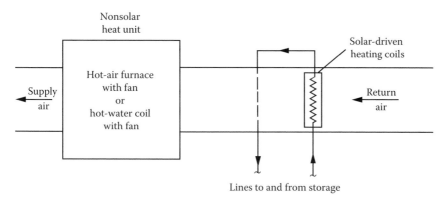

FIGURE 10.22
Forced-air heating system load device location upstream of nonsolar heat exchanger or furnace.

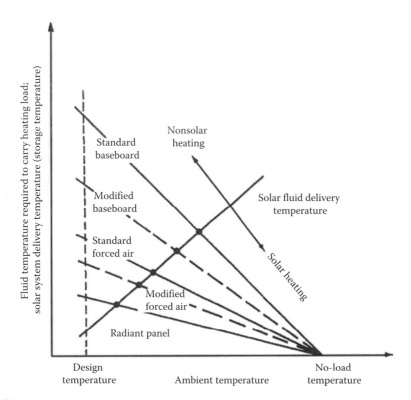

FIGURE 10.23
Heating load diagram for baseboard, forced-air, and radiant systems. Modified baseboard and forced-air systems are oversized in order to carry heating demands at lower temperature. Balance points are indicated by large dots at intersections.

heat-transfer area and rely on the relatively ineffectual mechanism of natural convection, they require the highest fluid temperature. Forced-air systems involve the more efficient forced-convection heat-transfer mode and, hence, are operable at lower fluid temperatures (see Figure 10.23). Radiant heating can use very large heat-transfer areas and is, therefore, operable at relatively low fluid temperatures.

In Figure 10.23, the intersection of the solar fluid temperature line and the load line for a specific configuration is called the *balance point*. At ambient temperatures below the balance point, solar energy cannot provide the entire demand, and some backup is required; above the balance point, solar capacity is sufficient to carry the entire load. Note that the load lines are specific to a given building. The solar fluid temperature line is not fixed for a building but depends on solar collector and storage size as well as local solar radiation levels. The line shown in Figure 10.23 is, therefore, an average line.

It is possible to modify load devices to lower the balance point, as shown in Figure 10.23. For example, a forced-air tube-and-fn exchanger can be enlarged by adding one or more additional rows of tubes. This increased heat-transfer area will permit the same energy delivery at a lower fluid temperature. The law of diminishing returns is evident as increasing effectiveness returns progressively less energy.

*10.8 Virtual Storage in the Electric Transmission Grid

Although it is generally recognized that solar and wind resources in the United States could provide the electric load requirements of the country, the variability of wind and solar irradiation is often cited by nuclear and fossil energy advocates as a major obstacle to employing these abundant energy resources. However, energy storage and load matching to offset the variability of wind and solar resources can overcome this obstacle. If there were an adequate electric distribution and transmission network, it would be possible to transmit excess energy generated by wind and/or solar in one part of the country to another part where the renewable resources cannot meet the load. This potential of using the grid for combined energy transmission/storage is called *virtual storage capacity*. It has been analyzed recently in a seminal paper entitled "Renewable Energy Load Matching for Continental US" by Short and Diakov [16]. The analysis is somewhat idealized in that it did not consider transmission constraints, economics, and siting issues, and selected sites only on the basis of their ability to contribute energy and electric capacity. It does, however, provide an indication of the potential of this virtual storage technology.

The authors of [16] used two different models; one was based on a quadratic objective function and the other was a linear program with the constraint that the load was met. For the latter approach, wind and solar sites were selected by the model to minimize the dispatchable generation (G_t) along with curtailment (C_t) of excess wind and solar generation and round-trip losses in a pumped hydro-storage system. These losses were taken as the difference between electric input to storage and the electricity removed from storage as determined by the roundtrip efficiency (eff). Curtailments of wind and solar output occur when the energy production exceeds the local demand at any given time, as, for example, by wind turbines late at night when winds are high and loads are low. The optimization criteria of the model asked that the sum of the power from conventional sources plus the curtailment from solar and wind plus the storage energy losses be minimized every hour. In equation form, the objective function

$$\sum_t \left[G_t + C_t + (1 - eff) * Sch_t \right] \tag{10.32}$$

is to be minimized and subject to the following five conditions:

$$l_t + C_t = \sum_i W_i * w_{it} + \sum_j P_j * p_{jt} + G_t + (Sdc_t - Sch_t) \quad \text{for all } t \qquad (10.33)$$

$$Sdc_t \le sCAP \quad \text{for all } t \qquad (10.34)$$

$$Sch_t \le sCAP \quad \text{for all } t \qquad (10.35)$$

$$0 \le E_0 + \sum_{t' < t} \left[eff * Sch_{t'} - Sdc_{t'} \right] \le sRES \quad \text{for all } t \qquad (10.36)$$

$$eff * \sum_t Sch_t = \sum_t Sdc_t \qquad (10.37)$$

where sCAP is the storage charge and discharge rate cap, sRES is the storage reservoir size, Sdc_t is the energy delivered from storage, Sch_t is the energy used to charge storage at time t, W_i is the action of the maximum potential built capacity at wind site i, w_{it} is the input generation that could be produced at hour t by wind resource at site i, P_j is the fraction of the maximum potential build capacity at the PV site j, and p_{jt} is the input generation that could be produced at hour t by the PV resource at site j.

Using wind and PV data from [17–19], the model assumed that any of these renewable resources could be utilized at full or fractional capacity and the model selected the optimal resource set to match the electric load as close as possible. But recognizing that wind and PV are variable and alone cannot meet all the load, the model asked what the maximum fraction was that wind and PV could meet with the support of some dispatchable generation from storage or gas turbines. The results of this analysis showed that, without any storage and no curtailment ($C_t = Sch_t = 0$ for all t), wind and PV can meet up to 50% of loads in the Western United States. For the no-storage case, it is beneficial to build more wind than PV because PV cannot meet any of the load after sundown. However, when storage is available and curtailments are allowed, the optimal mix has about 75% as much PV capacity as the wind, with the PV energy contributing 32% of the electricity produced from wind. Assuming that the current hydro pumped storage capacity can be doubled to 40 GW, as much as 83% of the load can be met with wind and PV sources according to the model. This compares to an energy availability factor for nuclear power plants of 81% in 2010 according to the IAEA [20]. At the same time, there are locations and times when PV and/or wind produce more energy than can be used or stored, and production must be curtailed. However, according to the model, less than 10% of the renewable energy potential through the year has to be curtailed, that is, dissipated because there is no suitable load, but that percentage can be reduced by increasing the pumped storage capacity or adding compressed air storage when NG or a biogas source is available. The big advantage of the PV/wind combination over any fossil or nuclear scenario is that the generation facilities require

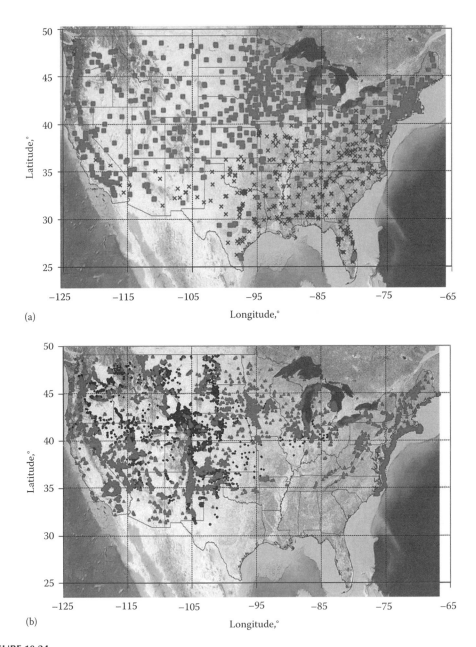

FIGURE 10.24
(See color insert.) Optimum locations for (a) PV and (b) wind to provide 83% of electric load in the United States. (Courtesy of Victor Diakov, NREL.)

virtually no water. This analysis shows that solar and wind can contribute close to a base-load generation electric system provided there is a robust transmission system where available. Figure 10.24 shows the renewable source locations for the high (83%) load matching scenario, and Figure 10.25 shows the locations of existing and potentially available pumped storage sites.

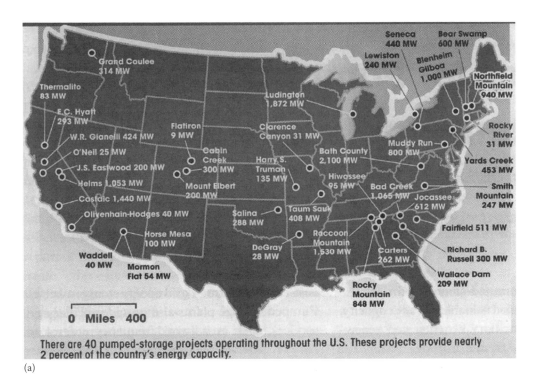

There are 40 pumped-storage projects operating throughout the U.S. These projects provide nearly 2 percent of the country's energy capacity.

(a)

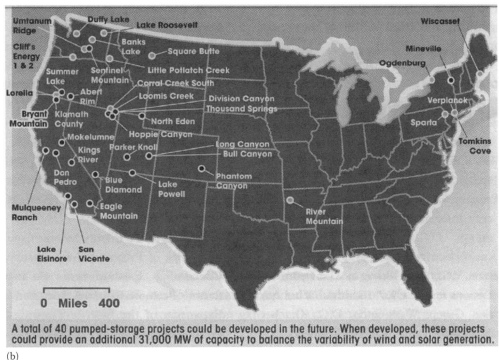

A total of 40 pumped-storage projects could be developed in the future. When developed, these projects could provide an additional 31,000 MW of capacity to balance the variability of wind and solar generation.

(b)

FIGURE 10.25
(a) Locations of existing pumped storage sites. (b) Locations of potentially available pumped storage sites. (Prepared by MWH for the USACE HDC, August 2009.)

In a follow-up paper, Short and Diakov showed that without energy storage, but with curtailment of the renewable sources incorporated, then 80% of the peak load in the Western United States could be met, and only 20% would have to be provided from storage and/or dispatchables such as NG turbines [21]. These results show that with a robust and efficient electric transmission system, renewable wind and PV could meet base-load requirements with relatively little additional dispatchable generation.

Problems (Some of These Problems are Open-Ended)

10.1 Design a compressed air energy storage system for a wind turbine. Assume that a cavity in an abandoned salt mine was available adjacent to the wind turbine. Use all the power generated above 5 kW to drive a compressor, which can compress the ambient air for the storage system.

10.2 A hydrogen/oxygen fuel cell stack produces 100 kW of DC power at an efficiency of 60% with water vapor as the product. Determine the hydrogen mass flow rate in g/s and the required cell voltage.

10.3 Derive an expression for the electric energy generation per unit of air volume stored in a cavity for a CAES system. Assume that there are two expansion stages and express the work output as the sum of the output from the two stages. The system operates with a constant cavern pressure, such as a hard rock cavern with a hydraulic compensation.

10.4 Supposing you own some land next to a 50 MW wind farm located on Colorado's Front Range. Your land has a fantastic cliff/ridgeline with an elevation difference of 250 m. Evaluate whether or not you can make money by building an energy storage system on your land that interfaces with the wind farm. Your plan is to capture excess energy from the wind farm or buy energy from the wind farm when electricity prices are low and sell that energy back to the grid when electricity prices are high (assume that the utility that owns the grid will grant you a fair contract). You find that a good power rating for your storage system is 50% of the rated wind farm output, or 25 MW. Also your storage system should be able to generate electricity at rated power for 8 h.

The first step in this problem is to evaluate the technical design of candidate energy storage systems.

a. Pumped hydroelectric. Determine the flow rate and reservoir size needed to accomplish the required power output and energy capacity. Pumped hydroelectric plants generally run at 80%–90% generating efficiency, depending on the size of the machinery. Suggest a reasonable surface area and depth for your two reservoirs.

b. Pulley and weight. You have a feeling that a simpler energy storage system is possible. What if you use a weight and a pulley that stores energy as the weight is lifted (using an electric motor) and energy is released when the weight falls (with an electric generator). You need to dig a shaft. Find the value of weight you would use, and the depth and diameter of the shaft. Is this viable?

c. Propose a new energy storage method. Be creative. Examples may be large springs, thermal storage, steam cycles, batteries, etc. Perform a simple analysis to size the system and say whether or not your creative idea is viable.

The second step is to evaluate the financial aspects of installing and operating your energy storage system to determine whether you could make money.

a. Pumped hydroelectric costs about $500 per kW to install your turbomachinery and penstocks, plus $2 per cubic yard to build reservoirs. Calculate the initial capital cost of your pumped hydroelectric system.

b. Suppose you can buy energy from the wind farm at $0.035/kWh between the hours of 10:00 PM and 8:00 AM to charge your storage and sell energy between 1:00 PM and 9:00 PM back to the grid at $0.1/kWh. Select a reasonable simple payback period that would motivate you to invest in energy storage. Calculate the maximum capital cost expenditure on your energy storage system that would allow this payback period.

c. Would you decide to build an energy storage system? Why or why not?

10.5 It has been proposed that the United States should strive toward a hydrogen economy. At present, the United States uses about 100 quads (10^{17} Btu) of energy per year. Suppose that 10% should be produced by 2015 with hydrogen fuel cells. The hydrogen is to be made by electrolysis of water in central stations located strategically. The electric power for electrolysis is to be supplied by nuclear power plants with a thermodynamic efficiency of 33% and a capacity factor of 0.90. Each plant should be of 500 MW capacity and operate on a simple Rankine cycle.

Estimate the following:

- The number of plants required.
- The amount of cooling water needed for the condensers (assuming evaporative cooling—2400 kJ cooling for each liter of cooling water).
- The amount of water needed for electrolysis assuming 1500 kJ is used by the fuel cell (i.e., only 750 kJ of energy is provided to the end user for each liter of water fed into the process).
- The total initial capital investment (if the cost of a nuclear plant is estimated at $6000 per kW and that of a fuel cell at $2100 per kW) [18].
- Comment on the viability of using hydrogen as an energy carrier on a global scale/niche application [18].

10.6 a. Design a compressed air energy storage system for a wind turbine you analyzed previously. Assume that a cavity in an abandoned salt mine is available adjacent to the wind turbine. Use all the power above 5 kW to drive a compressor, which can compress ambient air. Assume that NG is available for heating the air entering the compressor.

b. Repeat part 1, but assume that the wind turbine is located near a landfill where low Btu gas can be obtained for free. Compare this option with the design in part a, where the gas has to be purchased at $5 per 1000 standard ft³.

10.7 Derive an expression for the linear battery model power output as a function of the internal-to-external resistance ratio. Nondimensionalize the power output by dividing it by E^2/R_i. Tabulate the dimensional power function with a spreadsheet and then plot it. At what point is the maximum nondimensional power achieved?

10.8 Design a battery electric storage system for a power plant with 20 MW peak power delivery for a duration of 4 h and estimate the minimum number of batteries and the current during peak operation. Assume that the plant will use 600 A/h batteries, operating at 400 V DC.

10.9 A hydrogen/oxygen fuel cell operates with a voltage of 0.7 V with water vapor production. Calculate the kW/h of work per kg-mol of hydrogen and determine the cell efficiency.

10.10 You are asked to estimate the amount of heat that can be stored overnight in a solar-heated building, which has a storage space of 6 m^3. The solar system uses water with antifreeze in the collector, and there are two options for storing the heat overnight. One option is to use water with an allowable temperature change of 30°C. The other is to use a Glauber salt, which undergoes a phase change at approximately 30°C with a heat of fusion of 329 kJ/L. In both instances, a heat exchanger will be necessary in order to meet building code. Discuss the advantages and disadvantages of the two options.

10.11 A neighborhood fuel cell power plant is to be designed for an electric power output of 2000 kW with liquid–water product. Estimate the flow rates of hydrogen and oxygen during peak power production, assuming an 80% efficient power conditioner is used to convert DC to AC power and the fuel cell efficiency is 55%. What is the plant heat rate?

References

1. Lee, B. and Gushee, D. (June 2008) Massive electricity storage, White paper, AICHE, New York.
2. Hammerschlag, R., Pratt, R., and Schaber, C.P. (2004) Energy storage, in: F. Kreith and D.Y. Goswami (eds.), *Handbook of Energy Efficiency and Renewable Energy*, CRC Press, Boca Raton, FL, 2004.
3. Levine, J.G. (2007) Pumped hydroelectric energy storage and spatial diversity of wind resources as methods of improving utilization of renewable energy sources, Thesis, University of Colorado, Boulder, CO.
4. Cohn, A. et al. (1999) Applications of air saturation to integrated coal gasification CAES power plants, ASME 91-JPGC-GT-2.
5. Schoenung, S., Eyer, J.M., Iannucci, J.J., and Horgan, S.A. (1996) Energy storage for a competitive power market, *Annual Review of Energy and the Environment* 21, 347–370.
6. Succar, S. (September 2008) Baseload power production from wind turbine arrays coupled to compressed air energy storage, PhD thesis, Princeton University, Princeton, NJ.
7. Crotogino, F., Mohmeyer, K.-U., and Scharf, R. (2001) Huntorf CAES: More than 20 years of successful operation, in: *Solution Mining Research Institute Meeting*, Orlando, FL.
8. Van der Linden, S. (December 2006) Bulk energy storage potential in the USA, Current developments and future prospects, *Energy* 31, 3446–3457.
9. Davis, L. and Schainker, R. (2006) Compressed air energy storage (CAES): Alabama electric cooperative McIntosh plant—Overview and operational history, in: *Electricity Storage Association Meeting 2006: Energy Storage in Action*, Energy Storage Association, Knoxville, TN.
10. EPRI-DOE (2003) *Handbook of Energy Storage for Transmission and Distribution Applications*, EPRI/DOE, Palo Alto, CA/Washington, DC.

11. Gupta, R.G. (ed.) (2008) *Hydrogen Fuel—Production, Transport, and Storage*, CRC Press, Boca Raton, FL.
12. Bossel, U. et al. (2003) *The Future of Hydrogen Economy: Bright or Bleak*, Oberrohreorf, Switzerland.
13. Rabinwitz, M. (2000) Power systems of the future, Parts 1–3, *IEEE Power Engineering Review* 20(135), 5–16, 10–15, 21–24.
14. Kreith, F. and West, R.E. (2004) Fallacies of a hydrogen economy: A critical analysis of hydrogen production, *Journal of Energy Resources Technology* 126, 249.
15. Kreith, F. and Bohn, M.S. (2001) *Principles of Heat Transfer*, 7th edn., Brooks/Cole, Pacific Grove, CA.
16. Short, W. and Diakov, V. (November 2011) Renewable energy load matching for continental U.S., *Proceedings of the ASME 2011 International Mechanical Engineering Congress & Exposition*, Denver, CO.
17. Potter, C.W., Lew, D., McCaa, J., Cheng, S., Eichelberger, S., and Grimit, E. (2008) Creating the dataset for the western wind and solar integration study (U.S.A.). *Wind Engineering* 32(4), 325–338.
18. EnerNex Corp. (February 2011) Eastern wind integration and transmission study. Subcontract report NREL/SR-5500-47-86.
19. Wilcox, S., Anderberg, M., George, R., Marion, W., Myers, D., Renne, D., Lott, N. et al. (July 2007) Completing production of the updated National Solar Radiation Database for the United States. NREL Rep. CP-581-41511.
20. European Nuclear Society (May 2012) Nuclear power plants, world-wide, accessed at http://www.euronuclear.org/info/encyclopedia/n/nuclear-power-plant-world-wide.htm
21. Short, W. and Diakov, V. (2012) Matching Western U.S. Electric Consumption with Wind and Solar Resources, *Wind Energy*, doi: 10.1002/we.1513
22. Wyman, C., Castle, J., and Kreith, F. (1980) A review of collector and energy storage technology for intermediate temperature applications, *Solar Energy*, 24(6), 517–540.

Battery Resources

Battery technology is undergoing intensive research, and some resources to follow development are listed here.

Li-Ion
http://www.sciencedaily.com/releases/2007/12/071219103105.htm
http://www.a123systems.com/products
http://www.sciencedaily.com/releases/2008/11/081120103802.htm
http://electronics.howstuffworks.com/lithium-ion-battery.htm
http://e-articles.info/e/a/title/The-Lithium-Ion-Battery/
http://e-articles.info/e/a/title/Advantages-and-Limitations-of-the-Lithium-Polymer-Battery/
http://www.batteryuniversity.com/parttwo-34.htm
http://www.kokam.com/english/product/battery_main.html
http://www.gpbatteries.com/html/pdf/Li-ion_handbook.pdf

Ni–MH
http://e-articles.info/e/a/title/Advantages-and-disadvantages-of-the-Nickel-Metal-Hydride-(Ni-H)-Battery/
http://www.batteryuniversity.com/partone-4.htm

Vanadium Redox
http://www.vrb.unsw.edu.au/
http://thefraserdomain.typepad.com/energy/2006/01/vandium_refux_.html

General

http://www.batteryuniversity.com/partone-9.htm

Hydrogen

A useful reference regarding the potential of hydrogen is a report by the National Research Council. [Ref. National Research Council, *The Hydrogen Economy, Opportunities, Costs, Barriers, and R&D Needs*, U.S. Department of Energy Office of Basic Energy Sciences Committee on Alternatives and Strategies for Future Hydrogen Production and Use, National Academic Press, Washington, DC, Available at http://www.nap.edu/catalog/10922.html].

*11

Ocean Energy Conversion

> The offshore ocean area under U.S. jurisdiction is larger than our land mass, and teems with … energy sources.
>
> **—Tom Allen**

The oceans contain a great deal of energy, but the conversion of this energy into a useful form such as electricity requires advanced technology that is still under development. However, many demonstration prototypes have been designed and built. Valuable information is being obtained from the operation of these plants, and the results should lead to the development of large-scale units that are expected to become economically competitive in the future.

The key conversion technologies of ocean energy are

1. Ocean thermal energy conversion (OTEC)
2. Tidal energy
3. Wave energy

The basic elements of each of these three technologies will be presented in this chapter.

11.1 Ocean Thermal Energy Conversion

Oceans cover 71% of the earth's surface and receive the majority of the solar energy incident upon the earth. In semitropical and tropical oceans, the available temperature difference between the surface, which is warmed by the sun, and the lower depth (ΔT_0) is sufficient to operate a Rankine-type power plant using fluids such as ammonia. Since the temperature difference is available 24 h/day virtually throughout the year, OTEC has vast potential for implementation in many parts of the world. If the energy in a band ±10° latitude near the equator could be extracted with an energy efficiency of 0.5% of the incident solar energy, more than 7×10^5 MW power could be generated.

The main drawback of an OTEC system is that the overall thermal efficiency of the plant is low, and therefore large seawater pumps and large heat exchangers are required to produce adequate power. The maximum efficiency of any type of thermal power plant is the Carnot cycle efficiency, η_c, which for a typical OTEC plant is equal to

$$\eta_c = \frac{T_h - T_c}{T_h} = \frac{\Delta T_0}{T_h}$$

where

T_h is the hot surface water absolute temperature (about 298 K)
T_c is the absolute temperature of the cold water from below (about 278 K)

---

* Please note that this is an optional chapter and may be omitted in an introductory course.

Thus, the Carnot efficiency is only about 6.7%. This ideal efficiency does not allow for parasitic losses or the temperature differences between the seawater and the working fluid required for evaporation and condensation. A rough rule of thumb is that about 25% of ΔT_0 needs to be allowed for the evaporation and 25% for the condensing process, leaving only about half of the total temperature difference for the power turbine. For an 80% overall efficiency of the turbine generator, the gross power output efficiency would therefore only be about 3%. Parasitic load takes another 30% or so of the overall thermal efficiency, leaving a net power delivery efficiency of only about 2%–3%.

Figure 11.1 illustrates the four plant types that have been suggested: land-based (or shallow-water), shelf-mounted, floating, and moored plant types relative to the continent/ocean floor profile. The greatest number of potential sites exists for the floating plant with a suspended cold water pipe (CWP). Shelf-mounted or land-based plant types reduce electricity and water transportation, as well as the ocean dynamics/plant operation problems. Except for the CWP, the design and construction of floating and moored types are within current state-of-the-art technology. However, longer pipes are required, which increase costs and reduce the number of potential sites.

Disadvantages of the floating and moored plant types are related to electricity and water transportation and the effect of ocean dynamics on plant operation. For example, efficient transportation of electricity (by cable) and/or freshwater (by pipeline or barge) from a plant far offshore may be difficult. An example of the effect of ocean dynamics on plant design and operation is related to the evaporator. The evaporator produces steam as a result of the difference between the operating pressure and the saturation pressure of the warm seawater. Since this difference corresponds to only about 0.1 m of water, it is critical that the warm seawater supply pressure remains essentially constant, in spite of the wave action and platform motion, which may be difficult to achieve.

There are two basic technical approaches to producing power from the ocean thermal resources. Both are based on the Rankine thermodynamic cycle: one uses a closed system with a pressurized working fluid (typically ammonia or a fluorocarbon), and the second is an open system using the seawater itself as the working fluid. The concept of a closed-cycle

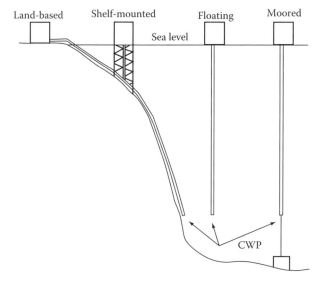

FIGURE 11.1
Basic OTEC plant configurations with cold water pipe (CWP).

OTEC (CC-OTEC) system was originally proposed by D'Arsonval [1] in 1881, and in 1930, his student, Claude [2], demonstrated the operation of an open-cycle OTEC (OC-OTEC) plant off the coast of Cuba.

11.1.1 Closed-Cycle Ocean Thermal Energy Conversion

Figure 11.2 is a schematic diagram of a closed-cycle ocean thermal conversion system. Warm surface water is used in the boiler to evaporate a working fluid; the sensible heat liberated by the seawater is the heat input, P_0, for the cycle

$$P_0 = \rho c Q \Delta T_0$$

where
  ρ is the density of seawater in kg/m³
  c is the specific heat of seawater in J/kg K
  Q is the flow rate in m³/s

The vapor produced in the boiler passes through a turbine to the condenser, and the expansion process generates power for electrical generation. The condenser uses cold water obtained at a depth of about 1000 m. Obviously, a long section of pipe and substantial pumping power are required for the condensation process.

The mechanical features of the closed cycle are very similar to a refrigeration cycle. Except for preventing biofueling of the heat exchanger and the CWP, the technology is currently available. It should also be noted that a CC-OTEC system can also be used to produce potable water [3].

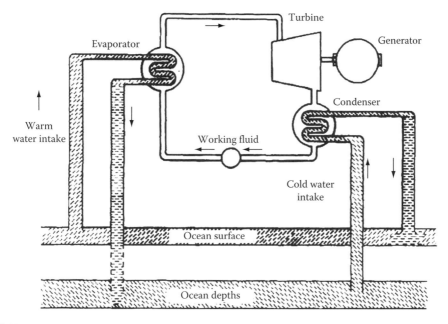

FIGURE 11.2
Schematic diagram of CC-OTEC.

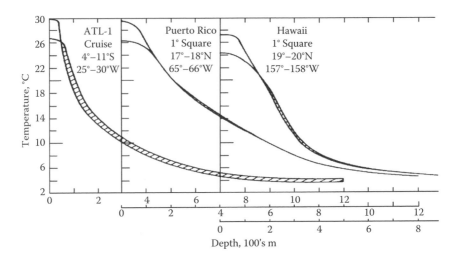

FIGURE 11.3
Temperature profiles vs. depth for Hawaii, Puerto Rico, and Recife, Brazil (ATL-1). (According to Dugger, G. et al., Ocean thermal energy conversion, in Kreider, J.F. and Kreith, F. (eds.), *Solar Energy Handbook*, McGraw-Hill, New York, Chapter 19. With permission.)

Figure 11.3 shows typical temperature vs. depth profiles near Hawaii, Puerto Rico, and Recife, Brazil (designated ATL-1), and Table 11.1 presents seasonal variations in ΔT_0 for the same three sites. The sites near Puerto Rico and Hawaii are of particular interest to the United States, because for these islands, all of the fuel needs to be imported. If movable or so-called grazing plants were to be installed, they would seek out the highest ΔT based on historical plots of monthly surface temperature contours, as well as daily information from satellites, which would indicate the optimum path for the platform on which the

TABLE 11.1

Monthly Values of ΔT_0 and Power P_{net} (Percentage of Design Value) for Selected Sites

	ATL-1		Puerto Rico		Hawaii	
Month	ΔT_0 (°C)	P_{net} (%)	ΔT_0 (°C)	P_{net} (%)	ΔT_0 (°C)	P_{net} (%)
January	23.5	96	21.3	88	20.6	90
February	24.3	104	20.9	84	20.1	85
March	25.2	115	20.9	84	20.0	84
April	25.2	115	21.2	87	20.6	90
May	24.2	103	22.2	99	21.3	99
June	24.4	105	22.6	104	21.3	99
July	23.6	97	23.0	108	22.4	112
August	24.3	104	23.0	108	22.7	116
September	22.4	84	23.6	116	22.7	116
October	23.3	94	23.9	119	22.5	114
November	22.6	86	23.3	112	21.5	101
December	23.3	94	22.3	100	20.7	92
Average	23.9	99.8	22.3	100.7	21.4	99.8
$P_{net\ avg\ rel}$	100		83.7		74.2	

For details on the power, see [3].

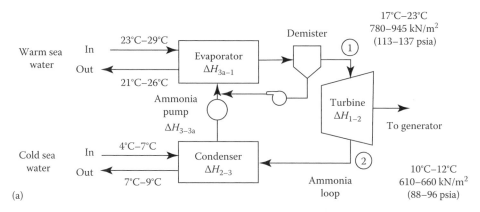

FIGURE 11.4
(a) Simplified diagram for typical CC-OTEC using ammonia as the working fluid and (b) a photo of the CC-OTEC operated in 1979 off Hawaii. (According to Dugger, G. et al., Ocean thermal energy conversion, in Kreider, J.F. and Kreith, F. (eds.), *Solar Energy Handbook*, McGraw-Hill, New York, Chapter 19. With permission.)

OTEC plant is based. A mini OTEC plant, as shown in Figure 11.4, operated off of Keahole Point, Hawaii, in 1977 and generated 50 kW (gross) and 10 kW (net) from a temperature difference of 21.2°C (38.2°F). The system contained all the essential components required to build a full-scale CC-OTEC plant. It used a titanium plate–type heat exchanger, and the plant was mounted on a barge moored by a single anchor, which incorporated a 24 in. OD polyethylene CWP as an integral member of the moor. A flexible rubber hose connected the top of the CWP to a moon pool in the center of the barge.

Example 11.1

Estimate the size of heat exchanger and the cold water flow rate required for a 10 MW CC-OTEC plant.

Assume that the cold water comes from a depth of 1000 m at a temperature T_{wc} of 5°C, while the hot water temperature, T_{wh}, is 25°C. Assume that the evaporator and the condenser operate at an effective temperature difference of ΔT of 4 K, with an overall heat transfer coefficient U of 6000 W/m² K. Use the Carnot efficiency for the power plant as an approximation.

Solution
The rate of heat transfer to the working fluid in the evaporator of area A is approximately

$$q_{in} = UA\Delta T$$

The rate of enthalpy increase of the water is

$$q_{in} = \dot{Q}\rho c(T_h - T_c) = \dot{Q}\rho c\left[(T_{wh} - \Delta T) - (T_{wc} + \Delta T)\right] = \dot{Q}\rho c(T_{wh} - T_{wc} - 2\Delta T)$$

where
ρ is the density (1.0 kg/m³)
c is the specific heat (4.2 × 10³ J/kg K)

The Carnot efficiency η_c of the energy conversion process is

$$\eta_c = \frac{T_h - T_c}{T_h} = \frac{T_{wh} - T_{wc} - 2\Delta T}{T_{wh} - \Delta T} = \frac{12}{294} = 0.041$$

The power output, P_0, is given by

$$P_0 = \frac{\dot{Q}\rho c(T_{wh} - T_{wc} - 2\Delta T)^2}{(T_{wh} - \Delta T)}$$

and the water flow rate by

$$\dot{Q} = \frac{P_0(T_{wh} - \Delta T)}{\rho c(T_{wh} - T_{wc} - 2\Delta T)^2} = \frac{P_0}{\rho c\eta_c}$$

$$= \frac{10 \times 10^3 \text{ J/s } 294 \text{ K}}{10^3 \text{ kg/m}^3 \times 4.2 \times 10^3 (\text{J/kg}) \times (12 \text{ K})^2}$$

$$= 4.86 \text{ m}^3/\text{s}$$

This is a large flow rate. If the CWP were 1 m in diameter, the water velocity, V, would be

$$V = \frac{4\dot{Q}}{\pi d^2} = \frac{4 \times 4.86}{\pi 1^2} = 6.2 \text{ m/s}$$

For a rate of heat transfer, q, the heat exchanger area is

$$A = \frac{q}{U\Delta T} = \frac{\dot{Q}\rho c(T_{wh} - T_{wc} - 2\Delta T)}{U\Delta T}$$

$$= \frac{P_0}{U\Delta T\eta_c}$$

$$= \frac{10 \times 10^6 \text{ W}}{6,000 \text{ (W/m}^2 \text{ K)}(4 \text{ K}) \times 0.041}$$

$$= 10,162 \text{ m}^2$$

This is a huge and expensive heat exchanger surface area.
The heat transfer and power generation processes are shown schematically in Figure 11.5.

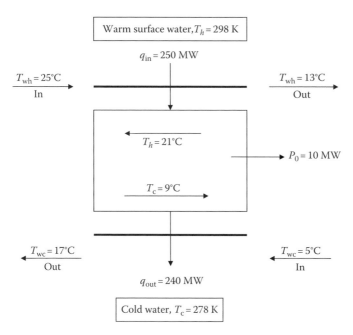

FIGURE 11.5
Schematic diagram for the solution of Example 11.1.

11.1.2 Open-Cycle Ocean Thermal Energy Conversion

OC-OTEC systems can generate electricity and/or freshwater using the temperature difference between surface and deep ocean seawater without the need for an expensive heat exchanger [4]. It is an attractive concept due to the vast amount of ocean energy available and the minimal ecological impact. OC-OTEC is ideally suited for locations that have a need for alternative electricity and freshwater, and are also close to an ocean thermal resource. Table 11.2 provides information about potentially suitable OC-OTEC sites in the United States.

The OC-OTEC concept is shown schematically in Figure 11.6. Warm (25°C typical) surface seawater is introduced into the evaporator, which operates at a pressure slightly below the saturation pressure corresponding to the warm seawater temperature. A small percentage

TABLE 11.2

Oceanographic Data for Potential OC-OTEC Sites in the United States

Site	Latitude (°N)	Longitude (°W)	Water Depth (m)	Distance from Shore (km)	Temperature Differential[a] (°C)
Puerto Rico	18.0	65.8	1000	3.1	22.3
Miami	25.4	79.9	740[b]	28.4	21.2
Key West	24.2	81.3	914[b]	60.0	21.5
Tampa	27.8	85.2	1000	281.6	20.9
New Orleans	28.8	88.8	1000	80.5	20.8
Island of Hawaii	19.9	156.1	1000	2.3	21.4

[a] Average.
[b] Maximum depth at location.

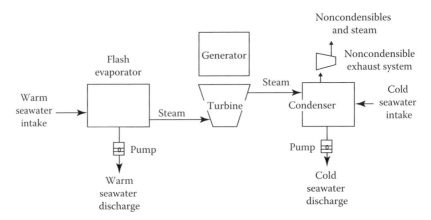

FIGURE 11.6
Conceptual schematic of OC-OTEC system.

(≈0.5%) of the seawater flashes to produce steam, which is expanded through a turbine connected to an electrical generator. Most of the steam is condensed in a direct-contact condenser supplied with cold (typically 5°C) seawater from depths of approximately 1000 m. A surface condenser is used only if freshwater production is desired, whereas a more efficient direct-contact condenser is used for electricity production. The remaining steam and noncondensibles (released from the seawater in the evaporator and direct-contact condenser and from possible leaks in the containment structure) are removed from the condenser by the noncondensible exhaust system. Variations of the basic cycle shown in Figure 11.6 include multiple stages of (evaporator, turbine, condenser) seawater pre-deaeration, and a combination of freshwater and direct-contact condenser with freshwater to seawater heat exchanger.

In an OC-OTEC system, seawater itself is used as the working fluid. A schematic diagram showing a plant cross section for a 100 MW electric open-cycle system proposed some time ago by Westinghouse is shown in Figure 11.7. The Westinghouse design proposed turbines with 140 ft tip diameter that were to operate at 225 rpm and deliver up to 100 MW of electric power under optimum conditions (Table 11.3). The hull in this design is located around the vertical axis turbine in an integrated design. An open channel evaporator is located axis-symmetrically around the turbine with warm seawater entering the evaporator through a peripheral submerged sluice. Vapor flashes from the warm water flow along a bent path toward the turbine blade, while the spent vapor leaving the turbine rotor is condensed in a surface condenser centrally located underneath the turbine. More recently, it has been proposed to utilize a direct-contact condenser, which increases the efficiency of the system [5]. The warm seawater leaving the evaporator and the cold seawater leaving the condenser are rejected into the ocean through a common pipe. Depending upon whether the condenser is a surface condenser or a direct-contact condenser, seawater distillation is possible with this approach, and Figure 11.7 illustrates schematically such a dual-purpose system. A detailed description of the OC-OTEC system using a falling jet evaporator and direct-contact condenser, a more advanced version of the Westinghouse concept, has been developed at the Solar Energy Research Institute and is described in [5] (Figure 11.8).

An OC-OTEC plant can be designed for electricity and/or freshwater production. The selection of the seawater flow rates and the temperature drops across various components of the system depends on the desired products. For example, a trade-off must be made

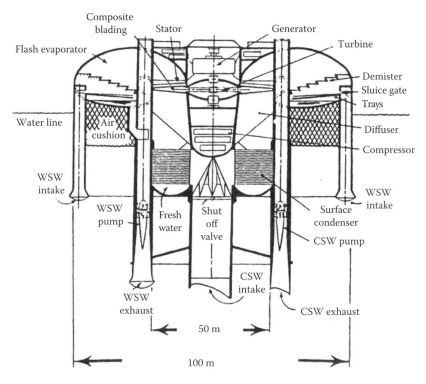

FIGURE 11.7
100 MW$_e$ OC-OTEC plant cross section (as proposed by Westinghouse). (From Rabas, T.J. et al., OTEC 100-MWe alternate power systems study, Report by Westinghouse Electric Corp. to U.S. DOE under contract EG/77/C/05/1473, March 1979. With permission.)

TABLE 11.3

Design and Performance Specifications for a 10 MWe, OC-OTEC Plant

Description	Specification
Warm water temperature	25°C
Warm water flow rate	62,100 kg/s
Warm water pipe diameter	9.27 m
Warm water pump power	1.03 MW
Evaporator size (platform area)	113 m²
Turbine (number and rotor diameter)	(3) 5.3 m
Gross power	15.77 MW
Cold water temperature	5°C
Cold water flow rate	35,590 kg/s
CWP diameter	4.7 m
Cold water pump power	2.27 MW
Condenser size (platform area)	158 m²
Purge system flow	1.30 kg/s at 1.14 kPa
Compressor power	0.76 MW
Net power produced	11.7 MW

Source: Rabas, T.J. et al., OTEC 100-MWe alternate power systems study, Report by Westinghouse Electric Corp. to U.S. DOE under contract EG/77/C/05/1473, March 1979.

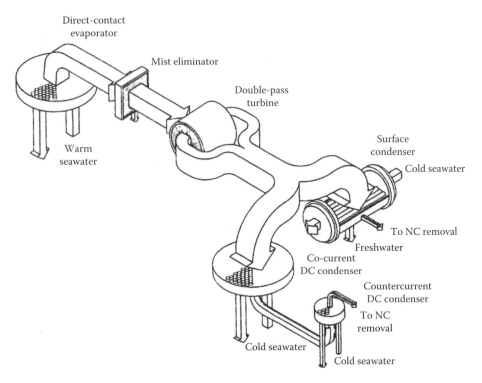

FIGURE 11.8
Schematic of OC-OTEC for power and freshwater production.

between the temperature drop across the evaporator (which affects the amount of steam generated) and the temperature drop across the turbines (which affects the amount of electricity produced). The optimum turbine temperature drop (based on plant economics) is approximately one half of the total available temperature drop for a single-stage electricity-producing plant, but only about one-fourth for a freshwater-producing plant [3].

11.1.2.1 Direct-Contact Evaporation and Condensation

The mechanism by which steam is generated in an OC-OTEC system is called direct-contact or "flash" evaporation. This process occurs when warm seawater is introduced into the evaporator at a pressure below the vapor pressure corresponding to the liquid inlet temperature, and steam is produced by the combined action of boiling and surface evaporation. Flash evaporation is usually quite violent as a result of the explosive growth of vapor bubbles from nucleation sites in the liquid. The growth of these bubbles shatters the liquid continuum and yields a wide range of droplet sizes. Because of the irregular geometry of the interface, it is practically impossible to define or measure the surface area from which evaporation takes place. Heat transfer in flash evaporation can therefore not be described in terms of a conventional heat transfer coefficient. To quantify the process and present experimental data, another parameter is used: "the effectiveness, ε." The effectiveness of flash evaporation is defined as the ratio of the temperature difference between the inlet and outlet liquid streams to the temperature difference between the inlet stream and the vapor temperature corresponding to the chamber saturation pressure. This definition is similar to the effectiveness of a conventional heat exchanger, since it is the ratio of the

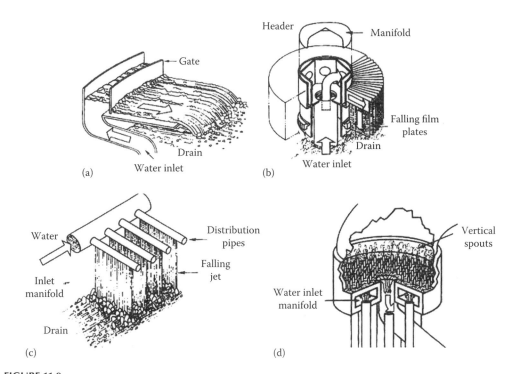

FIGURE 11.9
Contending direct-contact heat exchanger geometries for OC-OTEC: (a) open channel flow, (b) falling films, (c) falling jets, and (d) vertical spouts.

temperature difference actually achieved to the maximum temperature difference thermodynamically available.

Several evaporator options have been proposed and analyzed for OC-OTEC. They include open channel flow, falling films, falling jets, and spout evaporators. These are shown schematically in Figure 11.9. Based upon extensive experiments at the SERI test facility [6], the single-spout evaporator was found to be the most suitable configuration. The effectiveness measured in experiments with and without screens placed above the pipe outlet varied between 0.9 and 0.97, as shown in Figure 11.10, with a liquid-side pressure drop of 0.7 m (spout height + kinetic energy loss).

OTEC direct-contact condensation is more complicated than for conventional steam turbine applications, because ocean water contains a large amount of noncondensable gases that may come out of solution and the need to reduce parasitic power to a minimum. Based on extensive experiments, it was found that packed columns with structured packing are the best technology for OC-OTEC. The analysis for these configurations is presented in [4]. The most effective configuration was found to be an initial cocurrent condenser, followed by a countercurrent section in which noncondensable gases can be removed, as shown in Figure 11.8.

11.1.3 Comparison of Open- and Closed-Cycle OTEC Systems

For a given temperature potential, the OC-OTEC system is thermodynamically superior to the closed-cycle system because a larger portion of the temperature difference between the hot and the cold water is available to produce net power [5]. This is because of the direct-contact heat transfer mode of the evaporator and condenser, as used in the open-cycle

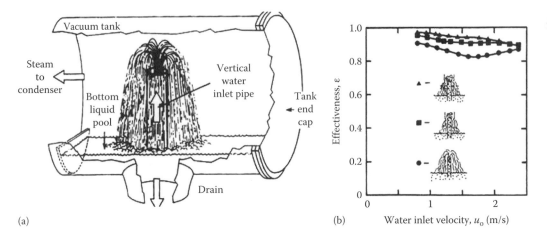

(a)   (b)   Water inlet velocity, u_o (m/s)

FIGURE 11.10
(a) Sketch of single-spout evaporator in SERI test chamber. (b) Effectiveness of spout evaporator with and without screens vs. water velocity. (From Bharathan, D. and Penney, T., Flash evaporation from falling turbulent jets, SERI-TP, 252-1853, Solar Energy Research Institute, Golden, CO, 1983; Bharathan, D. et al., Measured performance of direct contract jet condensers, SERT-TP, 252-1437, Solar Energy Research Institute, Golden, CO, 1982. With permission.)

system. The steam in the vacuum chamber can attain the temperature of the warm water discharged from the evaporator. At the same time, the exhaust steam from the turbine can be condensed at a temperature approaching the condenser cold water exit temperature.

In a closed-cycle system, the working fluid passing through the turbine is separated from the warm and cold waters by a solid wall. Hence, in both the evaporator and the condenser, a temperature drop is necessary first to transfer the heat from the water through the wall to produce the vapor in the boiler and then to condense the vapor of the working fluid at the outlet from the turbine by cold seawater. Consequently, the heat exchangers for CC-OTEC plants require very large surface areas to minimize the thermodynamically required temperature losses. These exchange surfaces are continuously exposed to corrosion and biofouling in the harsh ocean environment. On the other hand, the turbine in a CC-OTEC system is relatively small and is available as an "off-the-shelf" item. But because of the low pressure in which the turbine must operate in the open-cycle system, the turbine is very large, and so far, no full-scale turbines for plants 1 MW and greater have been designed and operated.

Any OTEC system faces the enormous challenge of producing net power with a temperature difference on the order of 20°C. Since the Carnot efficiency of such a plant is only on the order of 6% or 7%, any successful ocean energy conversion system must use as much of the temperature potential as possible and, at the same time, reduce parasitic losses to a minimum. The major parasitic losses in ocean thermal systems are the pumping power required to bring the cold water from a depth of about 1000 m to the surface, the power required to remove noncondensable gases in the open-cycle system, and the degradation of the available thermal energy by temperature drops due to fouling and corrosion across the heat transfer surfaces of a closed-cycle system.

11.1.4 Cold Water Pipe and Pumping Requirements

The cost of the CWP is the dominant feature in the overall cost for an OTEC system [7]. This pipe is subject to stresses that include drag from ocean current, oscillating forces due to vortex shedding, and possibly forces due to the drift and motion of the platform.

Moreover, the weight of the pipe is substantial, and it is not clear whether it should be made of a rigid or flexible material such as polythane. There may also be problems in assembling and positioning the pipe because of its enormous length. A premature failure of the CWP could be a financial blow to the entire OTEC system.

To transport the large quantities of cold and hot water necessary to operate an OTEC system requires a good deal of pumping power. This power will have to be supplied from the gross power output of the OTEC system and constitutes a loss in net energy. The frictional loss in pumping the water through the CWP can be reduced by increasing the pipe diameter, but this will increase the cost and weight of the pipe. It can also lead to instabilities that could prematurely destroy the cold water support. It should also be noted that the power output estimated in Example 11.2 is based on the efficiency of a Carnot engine, whereas a real turbine is less efficient and thereby reduces the net power output.

Example 11.2: Pumping Power Requirements

Estimate the power required to move water from a depth of 1000 m for a 10 MW OTEC system described in Example 11.1. Assume that the diameter of the pipe is 1 m.

Solution

The average water velocity is

$$\bar{u} = \frac{\dot{Q}}{A} = \frac{4.86 \,(\mathrm{m^3/s})}{\pi 0.5^2 \,\mathrm{m^2}} = 6.2 \,\mathrm{m/s}$$

Hence, the Reynolds number is

$$\mathrm{Re}_D = \frac{uD}{v} = \frac{6.2 \,(\mathrm{m/s}) \, 1\,\mathrm{m}}{1 \times 10^{-6} \,\mathrm{m^2/s}} = 6.1 \times 10^6$$

The flow is turbulent, and the friction factor can be estimated from the empirical relation given as follows:

$$f = \frac{0.184}{\mathrm{Re}_D^{0.2}} = \frac{0.184}{(6.1 \times 10^6)^{0.2}} = 0.0081$$

The head loss in the pipe equals

$$\Delta H = \frac{2fLu^2}{Dg} = \frac{2 \times 0.0081 \times 10^3 \,\mathrm{m} \times 6.2^2 \,\mathrm{m^2/s^2}}{1.0\,\mathrm{m} \times 9.8 \,\mathrm{m/s^2}} = 62.8\,\mathrm{m}$$

Hence, the pumping power required is

$$P_{pump} = \rho \dot{Q} g \Delta H = 1 \times 10^3 \,\mathrm{kg/m^3} \times 4.86 \,\mathrm{m^3/s} \times 9.8 \,\mathrm{m/s^2} \times 62.8\,\mathrm{m}$$

$$= 3 \times 10^6 \,\mathrm{kW} = 3\,\mathrm{MW}$$

The pumping power requirements could be reduced substantially by using a 2 m diameter pipe.

11.1.5 Economics

Cost estimates have been performed for both open- and closed-cycle systems [7]. By and large, they have indicated similar results. The main conclusion of one of the most detailed investigations is shown in Figures 11.11 and 11.12. Figure 11.11 shows the capital costs of OC-OTEC plants optimized for power production for a single-stage direct-contact condenser, a double-stage direct-contact condenser, and a single-stage surface condenser. For all three of these configurations, it can be seen that the capital cost decreases rapidly as net power output increases and levels off at net power above 20 MW electric. These results clearly indicate that in order for OTEC to be economically viable, extremely large plants would be necessary. For such large plants, the initial investment is large, and this has, so far, discouraged investment in this technology.

Figure 11.12 shows the fractional costs in percentage for the various parts of the OTEC system. These estimates indicate that the largest costs associated with an OTEC plant are the discharge and CWPs, with the CWP costing approximately 40% of the total at a net power output of 20 MW electric, while the discharge pipe would require about half that amount. These cost estimates have been made for several configurations. In the single-stage plants, all of the steam is produced in one flash evaporator, which expands through a single turbine, and is condensed afterward. The difference between the two single-stage plants analyzed is the type of condenser. One plant had a surface condenser allowing for the production of both electricity and freshwater, while the other plant had a direct-contact condenser and could only produce electricity. A direct-contact condenser is cheaper than an equivalent surface condenser, and there are trade-offs between freshwater production capabilities and plant capital costs depending upon the demand. Double-stage plants have potentially slightly higher efficiency and can provide greater flexibility of operation, but they are more complex and also more expensive.

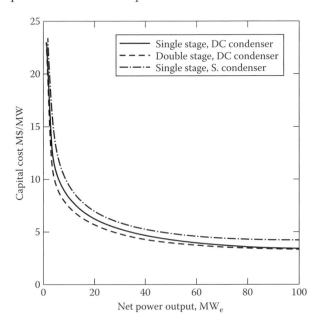

FIGURE 11.11

Capital cost estimates for three types of OC-OTEC plants vs. net power output optimized for power generation ($1983). (From Block, D.L. and Valenzuela, J.A., Thermoeconomic optimization of OC-OTEC electricity and water production plants, SERI/STR-251-2603, U.S. Department of Energy, Washington, DC, May 1985. With permission.)

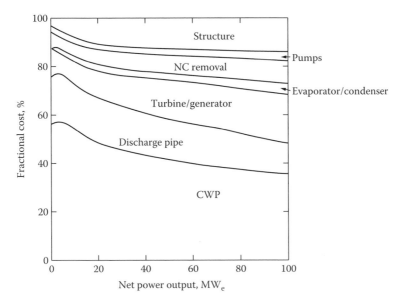

FIGURE 11.12
Fractional component cost of single-stage, direct-contact condenser plant optimized for power output. (From Block, D.L. and Valenzuela, J.A., Thermoeconomic optimization of OC-OTEC electricity and water production plants, SERI/STR-251-2603, U.S. Department of Energy, Washington, DC, May 1985. With permission.)

11.2 Tidal Energy

11.2.1 Introduction

Tidal energy has a long history. Small tidal mills were used for grinding corn in Europe in the Middle Ages, but more recently, using tidal energy on a large scale to generate electricity has been proposed for many parts of the world. Tidal energy can be harnessed by building large barrages—essentially long low dams—across the inlet of suitable estuaries. When the tides come in to the shore, they can be trapped by means of sluices in reservoirs called tidal basins behind these barrages. When the tide ebbs, the water behind the barrage can be let out via a turbine somewhat similar to a hydroelectric power plant. However, it is important to distinguish tidal power from hydropower. The former is the result of the interaction of the gravitational pull of the moon and the sun on the oceans, while the latter is derived from the hydrological climate cycle powered by solar energy. A simplified analysis of the earth–moon interaction that produced the tidal behavior is presented in [8].

Tidal energy technology relies on the twice-daily tides with the concomitant upstream flows and downstream ebbs in estuaries. To extract energy from the rise and fall of the tides, barrages are built across the inlet to the tidal basin, and turbines are located in water passages inside the barrages. The potential energy in the difference in water levels across the barrage due to the tidal motion is converted to kinetic energy as fast-moving water passes through the turbine. The rotational kinetic energy of the blades of the turbine can then be used to drive a generator to produce electricity. The average power output from a tidal system is roughly proportional to the square of the difference in water level between high and low tides. A tidal range of at least 5 m is considered the minimum for viable power generation. Sometimes tidal flows can be

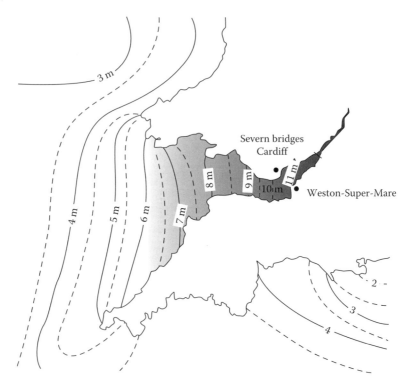

FIGURE 11.13
The effects of concentration of tidal flow in the Severn Estuary (tidal ranges in meters).

increased in height using the effect of concentration, as shown schematically [9] in the Severn Estuary near Cardiff in Wales (Figure 11.13).

Although many tidal systems have been proposed, the only major tidal power plant that has been constructed and operated is the La Rance Tidal Power Plant in France (Figure 11.14). The barrage for the plant was built in the 1960s near St. Maillot. A dam, 330 m long, was built in front of a 22 km² basin at a location where the tidal range was 8 m. The plant consists of 24 bulb-type turbine generators, 5.35 m in diameter, rated at 10 MW each. Power can be generated irrespective of whether the tide is going in or out. The plant has now been in operation for 40 years, during which time 20 billion kilowatt hours of electricity has been generated without major incident or mechanical breakdown.

11.2.2 Tidal Power

To estimate the amount of power output from a tidal barrage, assume that in front of the barrage is an infinite ocean, whereas behind it, a rectangular basin delineates the volume into which the water flows. The volume has a surface area A, and the range between high and low tide is H, as shown in Figure 11.15.

The incoming tide flows freely into the basin, but when the tide goes out, the water is held in the basin by closing the sluice gates at a level H above the level of the ocean. Given a rectangular basin, the center of gravity of the mass of water in the basin will be at a height $H/2$ above the sea level. The total volume in the basin will thus be AH and its mass will be ρAH. All of the water stored in the basin could now be made to flow through a turbine

(a)

(b)   (c)

FIGURE 11.14
A view of La Rance tidal scheme. (a) Aerial photo of La Rance Tidal Power Plant, (b) photo of La Rance showing dam, and (c) turbine wheel. (Courtesy of Renewable Energy UK, http://www.reuk.co.uk/La-Rance-Tidal-Power-Plant.htm)

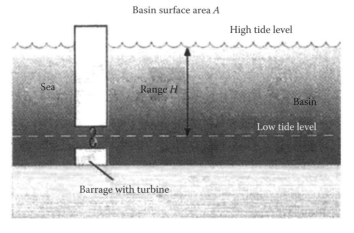

FIGURE 11.15
Nomenclature for tidal power analysis.

from the high to the low water level. The maximum potential energy E available per tide, assuming all the water passes through a height $H/2$, is therefore

$$E = \rho A H g \left(\frac{H}{2} \right) \tag{11.1}$$

where g is the gravitational attraction. The main periods of tides are diurnal at about 24 h and semidiurnal at about 12 h 25 min [9].

If the basin is allowed to fill again with the next incoming tide and the cycle repeats with a period t, the average potential energy, P, that could be extracted becomes

$$P = \frac{\rho A H^2 g}{2t} \tag{11.2}$$

For example, if $A = 10$ km², $H = 5$ m, $t = 12$ h 25 min, $P = 26.6$ MW.

However, the difference between high and low tidal levels, H, varies throughout the year from a maximum H_{max} in the spring, to a minimum, H_{min}, for the neap tides. As shown in [8], this variation in H can be accounted for by using a mean range for all the tides, $\bar{H}$. The mean power production per month, $\bar{P}$, then becomes

$$\bar{P} \sim \frac{\rho A g}{2g} \left(\frac{H_{max}^2 + H_{min}^2}{2} \right) \tag{11.3}$$

For example, if $H_{max} = 5$ m, $H_{min} = 2.5$ m, $A = 10$ km², $\rho = 1030$ kg/m³, and $t = 12$ h 25 min (4.47 × 10⁴ s), then $\bar{P} \cong 16.1$ MW.

Tidal power availability is dependent on local conditions, and there are many potential sites for tidal barrages around the world, as shown in Table 11.4, with an estimated total potential of the order of 300 TW hours per year. The total energy dissipated from tides globally is approximately 3000 GW, of which approximately a third is in accessible locations. The realistically recoverable resource has been placed at one-tenth of this amount, or 100 GW, due to practical limitations such as access to major tides [10].

There are essentially three ways to generate power from a tidal system [9]. In the most common method, called ebb generation, the incoming tide passes through the sluice gate in the barrage, and the water is trapped in the tidal basin by closing the sluices. The head of water in the basin can then be passed back through the turbine during the outgoing ebb tide, as shown in Figure 11.16. Alternatively, flood generation can use the incoming tide to generate electricity as it passes through the turbines in the barrage (see Figure 11.17). In either of these two cases, two bursts of power are produced during every 24.8 h period. This creates problems because the power production is not uniform. However, the electricity generation can be smoothed out by two-way operation, using energy both during the ebb and the flood, as shown in Figure 11.18. Reference [9] presents details of various schemes proposed for two-way operation.

The most popular scheme of power generation is the bulb system, as shown in Figure 11.19. In this system, the turbine generator is sealed in a bulb-shaped enclosure, which is mounted in the flow channel. Water flows around the large bulb and turns the turbine runner or blades, after passing through a distributor. The main disadvantage of this scheme is that to attain access for maintenance of the generator requires cutting off the

TABLE 11.4

Some World Locations for Potential Tidal Power Projects

Country	Mean Tidal Range (m)	Basin Area (km²)	Installed Capacity (MW)	Approx. Annual Output (TWh Per Year)	Annual Plant Load Factor (%)
Argentina					
San José	5.8	778	5,040	9.4	21
Golfo Nuevo	37	2,376	6,570	168	29
Rio Deseado	36	73	180	0.45	28
Santa Cruz	7.5	222	2,420	6.1	29
Rio Gallegos	7.5	177	1,900	4.8	29
Australia					
Secure Bay	7.0	140	1,480	2.9	22
Walcott Inlet	7.0	260	2,800	5.4	22
Canada					
Cobequid	12.4	240	5,338	14.0	30
Cumberland	10.9	90	1,400	3.4	28
Shepody	10.0	115	1,800	4.8	30
India					
Gulf of Kachchh (Kutch)	5.0	170	900	1.6	22
Gulf of Cambay (Khambat)	7.0	1,970	7,000	15.0	24
Korea (Rep)					
Garolim	4.7	100	400	0.836	24
Cheonsu	4.5	—	—	1.2	—
Mexico					
Rio Colorado	6–7	—	—	54	—
United States					
Passamaquoddy	5.5	—	—	—	—
Knik Arm	7.5	—	2,900	7.4	29
Turnagain Arm	7.5	—	6,500	16.6	29
Russian Federation					
Mezeh	6.7	2,640	15,000	45	34
Tugur[a]	6.8	1,080	7,800	16.2	24
Penzhinsk	11.4	20,530	87,400	190	25

Sources: Adapted from World Energy Council, http://worldenergy.org (accessed December 23, 2003); Elliot, D., Tidal power, in G. Boyle (ed.), *Renewable Energy: Power for a Sustainable Future*, 2nd edn., Oxford University Press, New York, 2004, Chapter 6. With permission.

[a] 7000 MW variant also studied.

flow of water. This could be avoided by the use of a rim generator or tubular turbine configurations described in [8].

11.2.2.1 Economics of Tidal Power

The cost of energy from a tidal system will depend on several factors such as location, capital cost, interest rate, turbine performance, length of time for amortization, and difference in

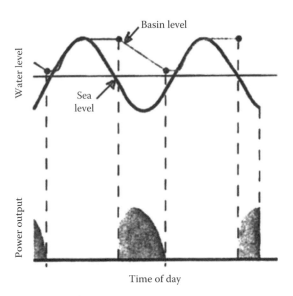

FIGURE 11.16
Schematic diagram of water levels and power outputs for an ebb generation scheme.

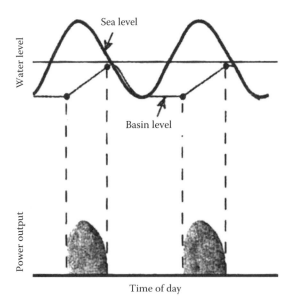

FIGURE 11.17
Schematic diagram of water levels and power outputs for a flood generation scheme.

water height between high and low tides. The last factor is particularly important because, as shown by Equation 11.2, the power output from a tidal system is a function of the square of the tidal ebb. If amortized over a life of 40 years, the cost of power from the La Rance tidal system will be one of the lowest of any renewable energy schemes (see Figure 11.20).

One of the most promising large-scale tidal systems, which has received considerable interest, is the so-called Severn barrage in the United Kingdom [11]. The proposed system would produce 8.6 GW, stretching 16 km across the Severn Estuary, and would generate 17 TWh of electric energy per year. The system is shown schematically in Figure 11.21.

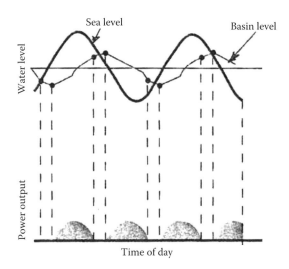

FIGURE 11.18
Schematic diagram of water levels and power outputs for a two-way generation scheme.

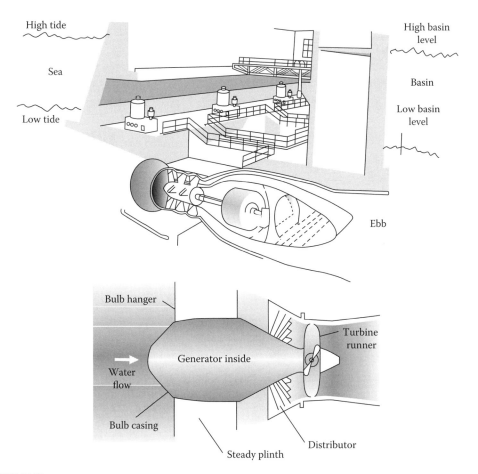

FIGURE 11.19
Conceptual layout of a tidal power generation scheme with details of a bulb turbine as used at La Rance.

FIGURE 11.20
(See color insert.) La Rance tidal power station. (From Khaligh, A. and Onar, O.C., *Energy Harvesting*, CRC Press, Boca Raton, FL, 2010, Figure 3.6.)

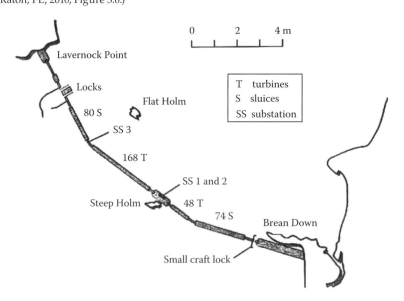

FIGURE 11.21
Layout of the proposed Severn barrage tidal system. (Adapted from Department of Energy, *The Severn Barrage Project: General Report*, Energy Paper 57, HMSO, London, U.K., 1989. With permission.)

Several other possible sites for large-scale projects exist all around the world, but information about their potential, economic as well as technical, is not generally available.

A detailed analysis of capital and running costs was made in 1989 for the Severn barrage tidal system in Great Britain. At that time, a detailed study, shown in Table 11.5, indicated that the total capital cost was estimated at £8.28 billion. This estimate was updated to £10–15 billion in 2002. It should be noted that these estimates do not include interest on borrowed capital accrued during construction. This is an important factor, because construction of a tidal power plant may take several years, and during this time, money is expended but no power is produced. It should further be noted that since the system would only operate during tidal cycles, the 8.6 GW turbine capacity of the Severn barrage would only be equivalent to about 2 GW of generating capacity of a conventional base-load power plant because the system functions only intermittently. The precise capacity credit that can be attributed to a tidal system, that is the value-as-replacement for a conventional plant, will depend, in practice, on the scale and the timing of the outputs of the plant. In other words, the barrage load factor, that is, the percentage of time that the plant can deliver power, was estimated to be about 23% for the Severn barrage. This is

TABLE 11.5

Severn Barrage "Reference Project" Capital and Annual Recurring Costs

	Cost/£ Million[a]
Pre-construction phase	
Feasibility and environmental studies, planning, and parliamentary costs	60
Design and engineering	130
Barrage construction	
Civil engineering works[b]	4900
Power generation works	2400
On-barrage transmission and control	380
Management, engineering, and supervision	300
Land and urban drainage sea defenses	30
Effluent discharge, port works	80
Barrage capital cost total	8280
Off-barrage transmission and grid reinforcement	
With all transmission line overhead	850
Extra cost for 10% of transmission lines underground	380
Annual costs	
Barrage operation and maintenance	40
Off-barrage costs	30
Annual costs total	70

Sources: Department of Energy, *The Severn Barrage Project: General Report*, Energy Paper 57, HMSO, London, U.K., 1989; Elliot, D., Tidal power, in *Renewable Energy*, 2nd edn., G. Boyle (ed.), Oxford University Press, New York, 2004, Chapter 6. With permission.

[a] The costs presented in Table 6.1, and in Box 6.6, are in 1988 money terms. In 1991, these estimates were updated, in line with inflation, with the total being put at over £10 billion. More recently, an STPG report, *The Severn Barrage—Definition Study for a New Appraisal of the Project*, commissioned by DTI to feed into the 2002 Energy Review [31], concluded that the barrage would cost between £10 and £15 billion at current prices.

[b] Excluding public road across the barrage (estimated to cost from £135 million to £207 million depending on links provided into road network).

about the same as the average output from an intermittent wind power plant, but does not approach that of a nuclear power plant, for which the load factor is about 90%, or of a combined cycle gas turbine with a load factor of 85% or more. A study conducted for the House of Commons Select Committee on Energy [11] indicated that the capital cost for the Severn Tidal Plant was about £10,000 million, the running cost per annum £86 million, and the electricity price about 5 pence/kWh, assuming a 20 year payback. These figures show that capital intensive projects such as tidal power plants are a good long-term investment, but may require support from government agencies because of their large initial capital requirements and relatively slow payback. However, like many other renewable energy systems, once the capital and interest costs have been paid off, tidal barrages can generate virtually free power without CO_2 generation for the rest of their life, which has been estimated to be at 100 years. Operating costs are relatively small, and since tidal barrages, like other renewable energy systems, require no fuel input, they are excellent prospects for future sustainable energy worldwide. References [12–17] provide additional perspectives for tidal power generation.

11.3 Ocean Wave Energy

Ocean waves are generated by wind blowing over water. Since wind is driven by solar energy, the energy in ocean waves is essentially a stored, moderately high-density form of solar energy. Typical average solar power levels of the order of 100 W m can be transformed into waves with power levels of over 100 kW/m of crest length. The size of the waves generated by wind depends upon the wind speed and its duration. The distance over which wind energy is transferred into the ocean to form waves is called the fetch. Greater amplitude waves contain more energy per meter of crest length than small waves. In practice, however, the power that can be extracted from the waves rather than their energy content is the useful parameter for engineers.

11.3.1 Deepwater Wave Power

A typical ocean wave can be approximated by a sinusoidal shape of height, a, as shown in Figure 11.22. Defining the average height above the water surface, a (m), the distance between peaks as the wavelength, λ (m), the speed at which the peaks move across the surface of the ocean as the wave velocity, v (m/s), and the time for successive peaks to pass a fixed point as period, P (s), and the wave frequency as $v = 1/P$, we can determine important

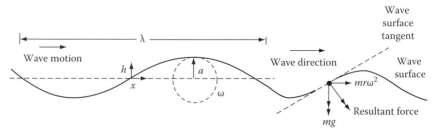

FIGURE 11.22
Characteristics of a sinusoidal wave.

wave characteristics. For example, using this type of simplified and idealized wave motion, Twidell and Weir [18] have proposed a procedure to estimate the power in deep ocean waves, based on theoretical foundations laid in 1977 by Coulson and Jeffrey [19]. The distinctive properties of deepwater waves are summarized as follows:

1. Surface waves consist of unbroken sine waves of irregular wavelength, phase, and direction.

2. The motion of any particle of water is circular. Whereas the surface forms of a wave show a definite progression, water particles have no progression.

3. Surface water remains on the surface.

4. The amplitudes of the water particle motions decrease exponentially with depth. At a depth of $\lambda/2\pi$ below the mean surface, the amplitude is reduced to $1/e$ of the surface amplitude.

5. The amplitude, a, of the surface wave is essentially independent of the wavelength, λ, velocity, c, or period T, of the wave and depends on the history of the wind above the surface.

Since most devices proposed for energy extraction are designed for deepwater waves, it may be assumed that the two dominant forces are the result of gravity and circular motion, as shown in Figure 11.23. An average wave for power generation has a wavelength ~100 m, an amplitude ~3 m, and behaves as a deepwater wave of depth greater than ~30 m with a circular motion of water particles as illustrated in Figure 11.23. The circles approximate the motion of a particle in a deepwater wave. This circular motion has an angular velocity, ω (rad/s), and an amplitude, a (m), that decreases exponentially with depth, D, and becomes negligible when $D > (\lambda/2)$. It should be noted that there is no net motion of water in deepwater waves, and an object suspended in water will move in circles, as shown in Figure 11.23.

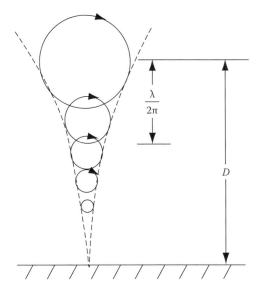

FIGURE 11.23
Motion of a particle in deepwater waves. The motion is circular with an angular speed ω (rad/s) and a radius a (m).

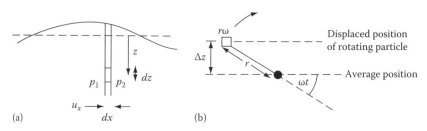

FIGURE 11.24
Local pressure fluctuation in a deepwater wave: (a) pressure in the wave and (b) local displacement of water particle.

To calculate the power that can be extracted from deepwater waves, we shall simplify the rigorous mathematical treatment. It shows that for a surface wave of amplitude a and wave number k, the radius r of particle motion below the surface is

$$r = ae^{kz} \tag{11.4}$$

where
$k = 2\pi/\lambda$
z is the distance below the surface, as shown in Figure 11.24

The vertical displacement from the average position as a function of time t is

$$\Delta z = r\sin\omega t = ae^{kz}\sin\omega t \tag{11.5}$$

and the horizontal velocity component, $u_{x'}$ is

$$u_x = r\omega\sin\omega t = \omega ae^{kz}\sin\omega t \tag{11.6}$$

Then, P^1, the power in the wave at x per unit width of wave-front is given by

$$P^1 = \int_{z=-\infty}^{z=0} (p_1 - p_2)u_x dz \tag{11.7}$$

where $(p_1 - p_2)$ is the pressure difference experienced by the element of height dz in the horizontal direction. This pressure difference is equal to the change in potential energy of particles rotating in the circular paths, or

$$p_1 - p_2 = pg\Delta z = pgae^{kz}\sin\omega t \tag{11.8}$$

Hence, substituting the earlier relation in Equation 11.7 yields

$$P^1 = \int_{z=-\infty}^{z=0} (pgae^{kz}\sin\omega t)(\omega ae^{kz}\sin\omega t)dz = pga^2\omega \int_{z=-\infty}^{z=0} e^{2kz}\sin^2\omega t\, dz \tag{11.9}$$

Since the time average over many periods of $\sin^2 \omega t$ can be shown to equal ½

$$P^1 = \left(\frac{\rho g a^2 \omega}{2} \right) \int\limits_{z=-\infty}^{z=0} e^{2kz} dz = \left(\frac{\rho g a^2 \omega}{2} \right)\left(\frac{1}{2k} \right) \qquad (11.10)$$

Introduce the definition of the phase velocity of a wave, c, given by

$$c = \left(\frac{\omega}{k} \right) = \left(\frac{gT}{2\pi} \right) = \left(\frac{g\lambda}{2\pi} \right) \qquad (11.11)$$

where T is the period of a sinusoidal wave given by

$$T = \frac{2\pi}{\omega} = 2\pi \left(\frac{\lambda}{2\pi g} \right)^{1/2} \qquad (11.12)$$

With these equations, the power in watts is

$$P' = \left(\frac{\rho g a^2 \omega}{2} \right)\left(\frac{1}{2} \right)\left(\frac{gT}{2\pi} \right) = \frac{\rho g a^2 T}{8\pi} \qquad (11.13)$$

or, as a function of the wavelength, it has the form

$$P' = \left(\frac{\rho g a^2}{8\pi} \right)\left(\frac{2\pi\lambda}{g} \right)^{1/2}$$

We can see that waves of great height and long wavelength are important parameters for effective wave power production.

11.3.2 Surface Wave Power

The waves in the ocean are irregular and are composed of many individual components. The combination of these waves is observed on the surface of the ocean, and the total power per meter of wavefront is the sum of the powers of all the components. It is impossible to measure the height and periods of each of these components, and an averaging process is therefore necessary to estimate the total power.

The variation in surface levels over a period of time can be obtained by means of wave-rider buoys. Although the average height of these waves is zero, a meaningful figure of the wave height is obtained by means of the significant wave height, H_s, which is defined as 4× the root-mean-square of the water elevation. In other words, the instantaneous elevations are squared, which makes the values positive, and then the mean over a number of waves is calculated. Finally, the significant wave height is 4× the square root of the mean. The significant wave height is approximately equal to the average of the highest one-third of the waves.

The second important parameter is the zero-up-crossing period, T_e. It is defined as the average time counted over 10 or more crossings between upward movements of the surface through the mean level. For a typical irregular sea, it can then be shown that the average total power per unit of wave crest, P, is given by

$$P = \frac{\left(H_s^2 T_e\right)}{2}$$

where P is in kW/m.

Figure 11.25a illustrates a typical wave record of the significant wave height and zero-crossing period as a function of time in seconds, whereas Figure 11.25b shows two wave records for the same location but on different days.

To obtain the wave power at any location, the ocean states at that location are measured over a whole year. Characterizing the states by the values of H_s and T_e, it is possible to obtain a statistical picture of the distribution of local wave conditions. This picture can be represented as a scatter diagram and gives the relative occurrences in parts per thousands of the contributions of H_s and T_e. An example of such a scatter diagram is shown in Figure 11.26. For this particular location at a zero-crossing period of 10 and a significant wave height of 6 m, the power density is 200 kW/m.

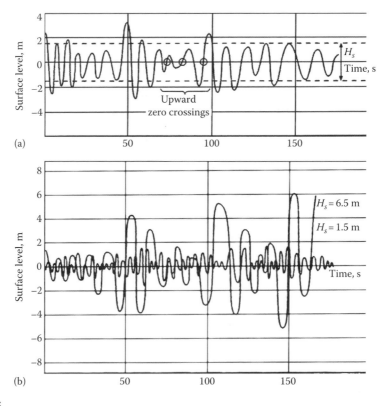

FIGURE 11.25
(a) A typical wave record. In this example, the significant wave height, $H_s = 3$ m. The successive upward movements of the surface are indicated with small circles. In this case, there are 15 crossings in 150 s, so $T_e = 10$ s. From this, P (kW/m) $= (3^2 \times 10)/2$ (kW/m) $= 45$ kW/m. (b) Two wave records are shown here for the same location but represent recordings taken on different days. (From Duckers, L., Wave energy, in Boyle, G. (ed.), *Renewable Energy*, 2nd edn., Oxford University Press, Oxford, U.K., 2004, Chapter 8. With permission.)

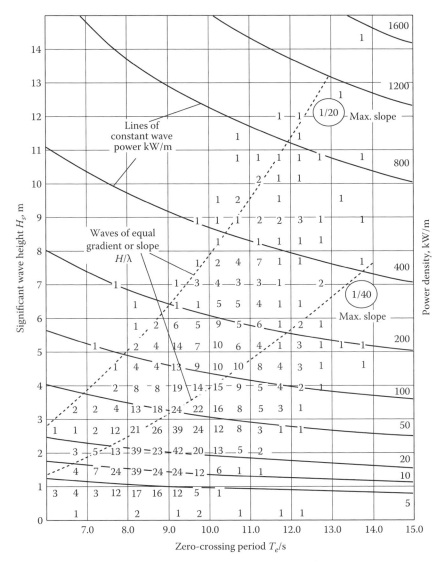

FIGURE 11.26

Scatter diagram of significant wave height (H_s) against zero-crossing period (T_e) for 58°N 19°W in the north Atlantic. The numbers on the graph denote the average number of occurrences of each H_s and T_e in each 1000 measurements made over 1 year. The most frequent occurrences are $H_s \sim 2$ m, $T_e \sim 9$ s. (From Boyle, G., ed., *Renewable Energy: Power for a Sustainable Future*, 2nd edn., Oxford University Press, New York, 2004. With permission.)

11.3.3 Wave Power Devices

Wave power generation has not yet reached the same state of development as wind power, but there have been many proposals for capturing energy from waves, and quite a few prototypes have been built, tested, and put into operation. So far, however, there have been no large-scale wave energy systems built and operated over any length of time. Consequently, it is not possible to decide which of the various devices have the greatest promise for eventual deployment. This section will therefore restrict itself to describing a few typical devices that have been built and operated.

11.3.3.1 Wave Capture Systems

Wave capture systems are relatively simple. When waves break over a sea wall or dam near the seashore, water is impounded at a height above the average sea level and can be captured in a reservoir. The water from the reservoir can then be returned to the ocean through a conventional low-head turbine attached to a generator, similar to a tidal system described in Section 11.2. The flow pattern of a wave capture system, however, is less regular than that of a tidal system.

A 350 kW wave capture system was built and operated successfully in Norway in 1985. A schematic diagram of the system, which was called Tapchan, is shown in Figure 11.27. In this system, waves were funneled through a tapered channel so that larger waves were able to flow over the walls immediately, while the smaller waves increased in height as they moved up in the channel.

A site for a successful wave capture system should have the following characteristics:

1. Deep water close to shore
2. A tidal range of less than 1 m
3. A convenient site for constructing the reservoir, preferably a natural gully in a rock formation near shore
4. Continuous waves with large amounts of energy

Although these types of wave capture systems are conceptually very simple, they suffer in general from a low head, which requires a large amount of water flow for a respectable energy output. On the other hand, the Tapchan concept has very few moving parts, its maintenance costs are low, and its reliability is high. Moreover, the storage reservoir helps to smooth the electrical output.

11.3.3.2 Oscillating Water Column

Oscillating water column (OWC) systems are simple and robust and have been deployed in a number of countries. They probably represent the most common form of wave energy

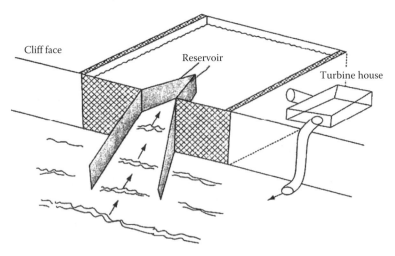

FIGURE 11.27
Schematic diagram of the Tapchan wave capture system in Norway.

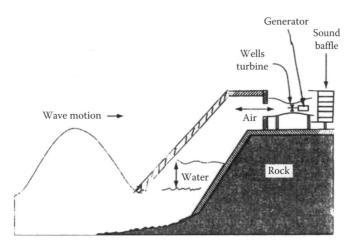

FIGURE 11.28
Schematic diagram of an onshore wave power system using an oscillating water column similar to Limpet.

conversion. A successful OWC system, called the Limpet, was constructed in the year 2000 at the Scottish island Islay. A schematic diagram of this onshore wave power system demonstrating its operation is shown in Figure 11.28. Air is compressed and decompressed by the OWC, and this causes air to be forced through a Wells turbine located above. The Wells turbine is capable of rotating in the same direction, regardless of the direction of the airflow. Thus, it generates power irrespective of the direction of movement of the water column. The water column rises with each incoming wave and compresses the air above, which is thus forced to flow through the turbine. As the water column goes down, air is drawn in the opposite direction through the Wells turbine, which generates electricity via a generator. The waves are captured by a reinforced concrete wall set into an excavated rock face on the shore. The existing Limpet system, shown in Figure 11.29, has a capacity of 500 kW and has been successful in generating electricity for the grid on the island for almost 10 years. The electricity is exported to the utility grid within the Scottish Administration's Renewable Obligation Program.

Power extraction by OWC can be enhanced by reducing the cross-sectional area of the channel approaching the turbine. This couples the slow motion of waves to the faster rotation of the turbine without mechanical gears. Another advantage is that the electrical generator is not in contact with saline water and, therefore, requires relatively little maintenance.

Although a majority of wave power devices are placed onshore, the Japanese have demonstrated a floating device called the Whale, placed off Gokashoa in 1998. The Whale has been tested as a full-scale prototype consisting of a massive structure 500 m in length, 1000 ton in weight, and rated at 110 kW. The wave conditions at Gokashoa Bay were found to be approximately 1 m in wave height, 5–8 s in frequency, and having a mean power density of 4 kW/m. The mean power output was 6–7 kW in an energy period of 6–7 s, yielding an overall efficiency of approximately 15% [21]. Other systems similar in concept to the Whale have been proposed and are described in [18].

11.3.4 Wave Profile Devices

These types of devices float on or near the surface of the ocean and respond to the shape of the waves, rather than the vertical up and down motion. The original idea was conceived

FIGURE 11.29
Photograph of Limpet. (From Boyle, G., ed., *Renewable Energy: Power for a Sustainable Future*, 2nd edn., Oxford University Press, New York, 2004. With permission.)

by Professor Steven Salter at Edinburgh University in the 1970s and is often called the "Salter's Duck." The Duck was designed to match the orbital motion of water particles. The concept is theoretically highly efficient but has not been fully developed.

The most advanced concept of a wave profile device is the Pelamis, or sea snake, shown schematically in Figure 11.30. It consists of a number of cylindrical sections hinged together as an attenuator device. The wave-induced motion of the cylinders is resisted at the joints by hydraulic rams that pump high-pressure oil through hydraulic motors via smoothing accumulators. The hydraulic motors then drive electric generators to produce electricity. Several devices can be connected in an array that itself is linked to shore through a seabed cable.

Each Pelamis is held in position by a mooring system, which restrains the device but at the same time allows the machine to swing head-on to oncoming prevailing waves as the snake spans successive wave crests. A 750 kW electric prototype 120 m long and 3.5 m in diameter was installed in 2005 off the shores of the main island of Orkney in Northern Scotland (see Figure 11.31).

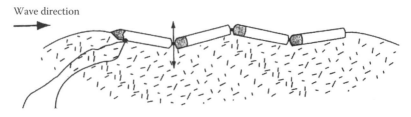

Wave direction

FIGURE 11.30
Schematic side view of a Pelamis wave power device.

FIGURE 11.31
(See color insert.) Pelamis WEC device at sea. (From Khaligh, A. and Onar, O.C., *Energy Harvesting*, CRC Press, Boca Raton, FL, 2010, Figure 4.72. With permission.)

Problems

11.1 Explain in your own words the manner in which the Pelamis wave power device operates. Note that the motion at the hinges produces hydraulic power that is fed into electric generators. A motion picture of the operation of the Pelamis can be viewed at http://www.youtube.com/watch?v=JYzocwUfpNg

11.2 Estimate the period and phase velocity of a deepwater wave of 100 m wavelength.

11.3 Using the phase velocity calculated in Problem 11.2, estimate the power in a deepwater wave of 1.5 m amplitude and 100 m wavelength.

11.4 Estimate the mean tidal power produced by a plant with an area of 10 km², a maximum range of 5 m, and a minimum range of 2.5 m (an average range of 3.7 m).

11.5 A 100 MW ocean thermal gradient power plant is to be designed for a location where the ocean surface temperature is 300 K (80°F) and water at a lower depth is available at 278 K (40°F). If the heat exchangers are sized to operate a power plant using R-22 as the working fluid between 294 and 284 K (70°F and 50°F), calculate the flow rates of R-22 and water required, assuming that the condenser heat exchangers have an overall conductance of 1000 W/m²·K (176 Btu/h·ft² °F) and an effectiveness of 100%. Also calculate the plant efficiency and the surface area of the condenser. The saturation pressure of R-22 at 294 K is 9.38×10^5 N·m² (136.1 psia), the enthalpy is 256.8 kJ/kg (110.4 Btu/lb$_m$), and the entropy is 0.900 kJ/kg·K (0.215 Btu/lb$_m$°R). The enthalpy of the saturated liquid at 284 K is 56.45 kJ/kg (24.27 Btu/lb) and the heat of vaporization is 197.0 kJ/kg (85.68 Btu/lb$_m$). The saturation pressure at 284 K is 6.81×10^5 N·m² (98.73 psia), while the entropy at 284 K of the saturated liquid is 0.217 kJ/kg K (0.0519 Btu/lb$_m$°R).

Assume efficiencies for the pump and the turbine are 80% and 90%, respectively. The specific volume of the saturated R-22 liquid at 284 K is 0.000799 m³/kg (0.0128 ft³/lb). A schematic diagram of the system is shown as follows:

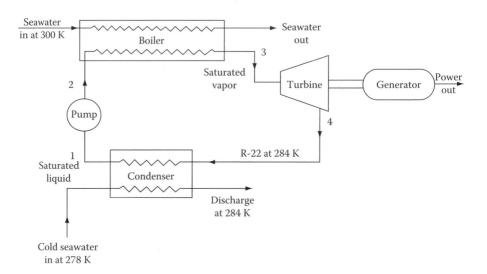

11.6 If the sea power plant described in Problem 11.5 is to deliver power at $8/10^6$ Btu, estimate the maximum permissible cost of the condenser and evaporator heat-exchanger surface in dollars per square foot, assuming a 20-year life, 10% discount rate, and 7% yearly fuel inflation.

The next two problems are open-ended with no solutions:

11.7 It has recently been proposed that OTEC could be used to produce ammonia, a chemical currently manufactured from natural gas at a total energy consumption of about 2.5% of the U.S. natural gas consumption. Since the manufacture of ammonia could be accomplished anywhere, the plant could be located in a favorable location, for example, off the coast of Brazil, where the difference in temperature between the upper layer of the ocean and the layer at the 600 m depth is approximately 20 K. Using an OTEC power plant of 500 MWe capacity, estimate the number of tons of ammonia that could be produced per year. For a price of $165/ton of ammonia, estimate the time required to regain the capital investment in the OTEC production plant and compare your estimate with the claim that the capital could be repaid in 2.5 years.

11.8 Aluminum sells today at approximately $1850/ton. The production of aluminum is one of the most energy-intensive processes in the metallurgical industry, and it has recently been proposed that an OTEC plant could be used for the production of aluminum. Metallurgical estimates indicate that approximately 18 kWh of energy is required in the refining process for each kilogram of aluminum produced. Assuming a plant load factor of 85%, calculate the size of an OTEC plant required to produce 250,000 ton/year of aluminum, if there is a temperature gradient similar to that in Problem 11.7. Then calculate the time required to repay an investment cost in the plant and compare your results with the estimate of 3 years obtained in the literature.

References

1. D'Arsonval, A. (1891) Utilisation des forces naturelles, Avenir de l'electricite, *Revue Scientifique*, 17, 370–372.
2. Claude, G. (1930) Power from tropical seas, *Mechanical Engineering*, 52, 1039–1044.
3. Dugger, G., Neff, F., and Snyder, J.E. (1981) Ocean thermal energy conversion, in: Kreider, J.F. and Kreith, F. (eds.), *Solar Energy Handbook*, McGraw-Hill, New York, Chapter 19.
4. Kreith, F. and Barathon, D. (1988) Heat transfer research for ocean thermal energy conversion, *Journal of Heat Transfer*, 110, 5–22, February.
5. Barathon, D., Kreith, F., and Owens, W.L. (1987) An overview of heat and mass transfer in open-cycle OTEC systems, in: Mari, Y. and Yang, W.J. (eds.), *Proceedings of the ASME/JSME Thermal Engineering Joint Conference*, Honolulu, HI, vol. 2, pp. 301–314.
6. Kreith, F. (1981) An overview of SERI solar thermal research facilities, *Heat Transfer Engineering*, 2, 3–4, January–June.
7. Block, D.L. and Valenzuela, J.A. (1985) Thermoeconomic optimization of OC-OTEC electricity and water production plants, SERI/STR-251-2603, U.S. Department of Energy, Washington, DC, May.
8. Twidell, J. and Weir, T. (2006) *Renewable Energy Resources*, 2nd edn., Taylor & Francis, Oxon, U.K., Chapter 13.
9. Elliott, D. (2004) Tidal power, in: Boyle, G. (ed.), *Renewable Energy: Power for a Sustainable Future*, 2nd edn., Oxford University Press, New York, Chapter 6.
10. Jackson, T. (1992) Renewable energy: Summary paper for the renewable energy series, *Energy Policy*, 20(9), 861–883.
11. Department of Energy. (1989) *The Severn Barrage Project: General Report*, Energy Paper 57, HMSO, cited in Ref. [9].
12. Baker, A.C. (1991) *Tidal Power*, Peter Peregrinus, London, U.K.
13. Cavanagh, J.E., Clarke, J.H., and Price, R. (1993) Ocean energy systems, in: Johansson, T.B., Kelly, H., Reddy, A.K.N., Williams, R.H., and Burnham, L. (eds.), *Renewable Energy—Sources for Fuels and Electricity*, Earthscan, London, U.K.
14. Charlier, R.C. (2003) Sustainable co-generation from the tides: A review, *Renewable and Sustainable Energy Reviews*, 7, 187–213.
15. Hubbert, M.K. (1971) The energy resources of the Earth, *Scientific American*, 225, 60–87, September.
16. Sorensen, B. (2000) *Renewable Energy*, 2nd edn., Academic Press, London, U.K.
17. Webb, D.J. (1982) Tides and tidal power, *Contemporary Physics*, 23, 419–442.
18. Twidell, J. and Weir, T. (2006) *Renewable Energy Resources*, 2nd edn., Taylor & Francis, Oxon, U.K., Chapter 12.
19. Coulson, C.A. and Jeffrey, A. (1977) *Waves*, Longman, London, U.K.
20. Duckers, L. (2004) Wave energy, in: Boyle, G. (ed.), *Renewable Energy*, 2nd edn., Oxford University Press, Oxford, U.K., Chapter 8.
21. Washyo, Y., Osawa, H., and Ogata, T. (2001) The open sea test of the offshore floating type wave power device "mighty whale": Characteristics of wave energy absorption and power generation, in: *Ocean 2001 Symposium*, Honolulu, HI, cited in Ref. [18].

Additional Readings

Bharathan, D. and Penney, T. (1983) Flash evaporation from falling turbulent jets, SERI/TP, 252-1853, Solar Energy Research Institute, Golden, CO.

Bharathan, D., Olson, D.A., Green, H.G., and Johnson, D.H. (1982) Measured performance of direct contact jet condensers, SERI-TP, 252-1437, Solar Energy Research Institute, Golden, CO.

Khaligh, A. and Onar, O.C. (2010) *Energy Harvesting*, CRC Press, Boca Raton, FL, Figures 3.6, 4.72, and 4.73.

Rabas, T.J., Wittig, J.M., and Finsterwalder, K. (1979) OTEC 100-MWe alternate power systems study, Report by Westinghouse. Electric Corp. to U.S. DOE under contract EG/77/C/05/1473, March.

Renewable Energy UK, http://www.reuk.co.uk/La-Rance-Tidal-Power-Plant.htm (accessed on April 18, 2013).

Severn Tidal Power Group (2002) *The Severn Barrage—Definition Study for a New Appraisal of the Project*, ETSU REPORT NO. T/09/00212/00/REP.

World Energy Council, http://worldenergy.org (accessed December 23, 2003).

12

Transportation

12.1 Introduction

A viable transportation system is a crucial part of a sustainable energy future. Transportation is a complex interdisciplinary topic, which really deserves a book unto itself. However, some of the main issues related to transportation have to do with fuels and energy storage, which are the topics covered in this book. This section does not purport to be exhaustive, but examines some of the key issues related to a viable transportation future.

Gasoline and diesel are not only very convenient fuels for ground transportation, but also have a high energy density that permits storage in a relatively small volume—an important asset for automobiles. For example, these liquid fuels have a volumetric specific energy content of about 10,000 kWh/m^3, compared to hydrogen compressed to 100 bar at about 300 kWh/m^3. But known petroleum resources worldwide are being consumed rapidly, and future availability of these resources is bound to decline. At present, more than 97% of the fuel used for ground transportation in the United States is petroleum-based, more than half of it is imported, and the percentage of imported petroleum is continuously increasing. Moreover, the increase in cost of gas and oil has become of growing concern to average citizens, and the emission from current transportation systems is a major component of CO_2 pollution that produces global warming.

There is worldwide agreement among oil experts that global oil production will reach a peak sometime between 2005 and 2030. The predictions for the date of peak world oil production according to various estimates [1] are demonstrated in Figure 12.1. The oil production is shown as a function of time for three total amounts of recoverable oil that span the entire range of assumptions by experts. Although new oil fields may be discovered and fracking may access additional oil in shale, the total recoverable amounts are somewhere between 3,000 and 4,000 billion barrels of oil (bbl). The obvious conclusion to be drawn from these predictions is that the production peak is imminent and the price of oil will continue to escalate as supplies decline. Thus, planning for a sustainable transportation system that does not depend entirely on petroleum resources is an imperative segment of a sustainable energy future.

12.2 Alternative Fuels

Alternative fuels available to supplement oil as well as their feedstock are shown in Table 12.1 [2]. Inspection of the table shows that biodiesel, electricity, ethanol, and hydrogen (via electricity) are the fuels potentially independent of a petroleum resource such as oil or natural gas. The potential of producing liquid fuel from biomass has been treated in Chapter 4. There is

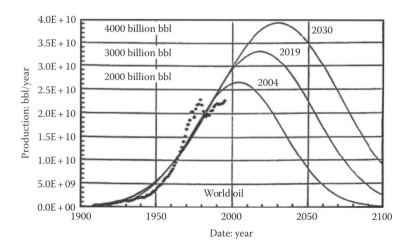

FIGURE 12.1
Oil production vs. time for various recoverable amounts of petroleum. (From Bartlett, A.A., *Math Goel.*, 32(1), 2000. With permission.)

TABLE 12.1

Feedstocks for Alternative Fuels

Fuel	Feedstock
Propane (LPG)	Natural gas (NG), petroleum
Compressed NG	NG
Hydrogen	NG or (water + electricity)
FT diesel	NG, coal, or algae
Methanol (M85)	NG or coal
Ethanol (E85)	Corn, sugarcane, or cellulosic biomass
Electricity	NG, coal, uranium, or renewables

yet no firm engineering picture of how biomass will provide a substantial part of the future transportation fuel, particularly because new conversion processes would be needed such as ethanol produced from cellulosic materials such as switchgrass or bio-waste, or diesel from algae. The reason for this is that ethanol produced by traditional methods from corn kernels has only an energy return on energy invested (EROI) on the order of 1.25, whereas the EROI for cellulosic ethanol is somewhere between 1 and 6 [3]. Diesel from algae may be even better. As shown in Figure 12.2, even cellulosic ethanol would not be able to replace oil, because growing it would require too large a percentage of all the arable lands in the United States, and the production of large amounts of ethanol would compete with production of food, which is of increasing importance for a socially sustainable energy system.

It has recently been proposed [4] to use biofuels derived from aquatic microbial oxygenic photoautotrophs (AMOPs), commonly known as algae. In this study, it was shown that AMOPs are inherently more efficient solar collectors, use less or no land, can be converted to liquid fuels using simpler technologies than cellulose, and offer secondary uses that fossil fuels do not provide. AMOPs have a 6- to 12-fold energy advantage over terrestrial plants because of their inherently higher solar energy conversion efficiency, which is claimed to be between 3% and 9%. Figure 12.2 compares the area needed for three different biomass sources. The data are for corn grain, switchgrass, mixed prairie grasses, and AMOPs. Each box superimposed on the map of the United States represents the area needed to produce

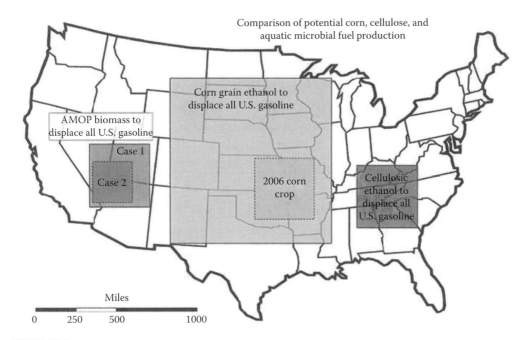

FIGURE 12.2
Relative land area requirement for various liquid fuel biosources. (From Dismukes, C. et al., *Curr. Opin. Biotechnol.*, 19, 235, June 2008. With permission.)

a sufficient amount of biomass to generate enough liquid fuel to displace all the gasoline used in the United States in the year 2007. The two boxes for AMOPs are for 30% and 70% conversion efficiency. The overall solar energy conversion to biofuels works out to about 0.05% for solar-to-ethanol from corn grain and roughly 0.5% from switchgrass-to-ethanol. Comparatively, this value is about 0.5%–1% for AMOPs-to-ethanol or biodiesel.

An even more favorable assessment for the potential of algae to produce biodiesel is presented in [5]. According to this study, microalgae may be a source of biodiesel that has the potential to displace fossil fuel. According to [5], microalgae grow extremely rapidly and are exceedingly rich in oil. Some microalgae double their biomass every day, and the oil content of microalgae can exceed 80% by weight. Table 12.2 shows a comparison of some sources of

TABLE 12.2

Comparison of Some Sources of Biodiesel

Crop	Oil Yield (L/ha)	Land Area Needed (Mha)[a]	Percent of Existing U.S. Cropping Area[a]
Corn	172	1540	846
Soybean	446	594	326
Canola	1,190	223	122
Oil palm	5,950	45	24
Microalgae[b]	136,900	2	1.1
Microalgae[c]	58,700	4.5	2.5

Source: Christi, Y., *Biotechnol. Adv.*, 25, 294, 2007. With permission.
[a] For meeting 50% of all transport fuel needs of the United States.
[b] 70% oil (by weight) in biomass.
[c] 30% oil (by weight) in biomass.

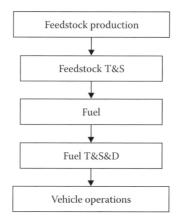

FIGURE 12.3
Steps in a well-to-wheel analysis for ground transportation vehicles.

biodiesel that could meet 50% of all of the transportation needs in the United States. According to estimates in [5], only a small percentage of U.S. cropping areas would be necessary to supply 50% of the entire transport fuel needs in the United States. However, no full-scale commercial algae biodiesel production facility has been built and operated for a sufficient time to make reliable predictions regarding the future of algae for a sustainable transportation system.

12.3 Well-to-Wheel Analysis

Rather than simply looking at the efficiency of an engine or a given fuel, a more comprehensive way to determine overall efficiency when evaluating the potential of any new fuel for ground transportation is to use what is called a *well-to-wheel analysis*. The approach to a well-to-wheel analysis is shown schematically in Figure 12.3 [6,16]. The well-to-wheel approach of a fuel cycle includes several sequential steps: feedstock production; feedstock transportation and storage; fuel production; transportation, storage, and distribution (T&S&D) of fuel; and finally vehicle operation. This approach is essential for a fair comparison of different options because each step entails losses. For example, whereas a fuel cell has a much higher efficiency than an internal combustion (IC) engine, it depends for its operation on a supply of hydrogen, which must be produced by several steps from other sources.

Moreover, there is no infrastructure for transporting and storing hydrogen, and this step in the overall well-to-wheel analysis has large losses and contributes to much larger energy requirements compared to a gasoline or electrically driven vehicle.

12.4 Mass Transportation

An opinion held widely among state governments and environmentalists is that mass transportation would greatly reduce the total energy consumption for the transportation sector. However, as shown in Table 12.3, based upon data collected by the U.S.

TABLE 12.3

Passenger Travel and Energy Use, 2002

	No. of Vehicles (Thousands)	Vehicle-Miles (Millions)	Passenger-Miles (Millions)	Load Factor (Persons/Vehicle)	Energy Intensities		Energy Use (Trillion Btu)
					(Btu per Vehicle-Mile)	(Btu per Passenger-Mile)	
Automobiles	135,920.7	1,658,640	2,604,065	1.57	5,623	3,581	9325.9
Personal trucks	65,268.2	698,324	1,201,117	1.72	6,978	4,057	4872.7
Motorcycles	5004.2	9,553	10,508	1.22	2,502	2,274	23.9
Demand response	34.7	803	853	1.1	14,449	13,642	11.6
Vanpool	6.0	77	483	6.3	8,568	1,362	0.7
Buses	a	a	a	a	a	a	191.6
Transit	76.8	2,425	22,029	9.1	37,492	4,127	90.0
Intercity[b]	a	a	a	a	a	a	29.2
School[b]	617.1	a	a	a	a	a	71.5
Air	a	a	a	a	a	a	2212.9
Certified route[c]	a	5,841	559,374	95.8	354,631	3,703	2071.4
General aviation	211.2	a	a	a	a	a	141.5
Recreation boats	12,409.7	a	a	a	a	a	187.2
Rail	18.2	1,345	29,913	22.2	74,944	3,370	100.8
Intercity[d]	0.4	379	5,314	14.0	67,810	4,830	25.7
Transit[e]	12.5	682	15,095	22.1	72,287	3,268	49.3
Commuter	5.3	284	9,504	33.5	90,845	2,714	25.8

Source: Davis, S. and Diegel, S., *Transportation Energy Data Book*, Oak Ridge National Laboratory / U.S. Department of Energy, Oak Ridge, TN, 2004. With permission.

a Data are not available.

b Energy use is estimated.

c Includes domestic scheduled service and half of international scheduled service. These energy intensities may be inflated because all energy use is attributed to passengers; cargo use not taken into account.

d Amtrak only.

e Light and heavy rail.

Transportation Department in this country, the energy intensity of intercity rail and transit buses, that is, the energy spent per passenger mile traveled, is virtually the same as the energy intensity of automobiles with at current occupancy rates. This is due to the urban sprawl in major cities that makes it difficult to reach outlying areas by a mass transport network. In other words, unless there are incentives for mass transport or disincentives to use the automobile, thereby achieving higher mass transport load factors—that is, more passengers per mile—on transit buses and light rail, the availability of mass transport systems will not materially change the overall energy use by the transportation sector. Light rail developments are capital intensive and must be carried out in a coordinated way with long term planning of urban forms that integrate high density and high intensity developments along the light rail corridors. Freight shipping and goods movements around cities uses nearly as much liquid fuel as personal transport. Moving freight by truck across long distances is at least four times more energy intensive than by rail. Many of the complex infrastructure and policy issues with shifting more freight to rail could have innovative engineering solutions [7].

12.5 Hybrid Electric Vehicles

Another obvious approach to ameliorating the expected increase in price and lack of availability of petroleum fuel is to increase the mileage of the vehicles. This can be achieved by improving the efficiency of the IC engine, as for instance by using advanced diesel engines that have a higher compression ratio than spark ignition (SI) engines, or by using hybrid electric vehicles (HEVs). Increasing the efficiencies of IC or diesel engines is a highly specialized topic and is not discussed here. However, HEVs offer a near-term option for utilizing improved battery technology, as discussed in Chapter 9.

An HEV is powered by the combination of a battery pack and electric motor—like that of an electric vehicle—and a power generation unit (PGU), which is normally an IC or diesel engine. Unlike electric vehicles, however, HEV batteries can be recharged by an onboard PGU, which can be fueled by existing fuel infrastructure.

HEVs can be configured in a parallel or a series design. The parallel design enables the HEV to be powered by both the PGU and the motor, either simultaneously or separately. The series design uses the PGU to generate electricity, which recharges the HEV battery pack and produces power via an electric motor. The key element of either design is that the battery pack, as well as the PGU, can be much smaller than those of a typical electric vehicle or a vehicle powered by an IC engine because the IC engine can be operated at its maximum efficiency nearly all the time.

Currently available HEVs, such as the Toyota Prius, use a parallel configuration, as shown in Figure 12.4. A parallel HEV has two propulsion paths: one from the PGU and one from the motor, while computer chips control the output of each. A parallel configuration HEV has a direct mechanical connection between the PGU and the wheels, as in a conventional vehicle, but also has an electric motor that can drive the wheels. For example, a parallel vehicle could use the electric motor for highway cruising and the power from IC engine for accelerating. The power produced by the PGU also drives the generator, which in turn can charge the battery as needed. The system to transfer electricity from the generator to the battery pack is exactly like that of an EV, with alternating current converted to DC by the inverter. HEV parallel designs also use a regenerative braking feature that converts energy stored in the inertia of the moving vehicle into electric power during deceleration (see Figure 12.5).

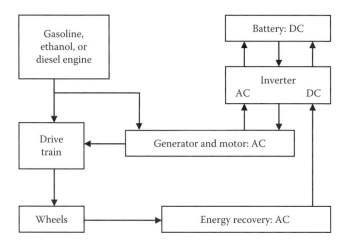

FIGURE 12.4
Schematic of a hybrid electric vehicle.

Some benefits of a parallel configuration vs. a series configuration include the following:

- The vehicle has more power because both the engine and the motor provide power simultaneously.
- Parallel HEVs do not need a separate generator.
- Power is directly coupled to the road, thus operating the vehicle more efficiently.

Energy and cost savings from an HEV depend on many factors, such as the overall car design, cost of fuel, and the cost and efficiency of the batteries. Preliminary estimates indicate that over 5–8 years, reduced size of the motor and savings in gasoline could pay for the additional cost of the batteries. Economically, an HEV would be beneficial when the price per gallon of gasoline or diesel exceeds $3.00/gal, but the life cycle of the batteries will also have to be considered. If batteries have a life cycle of 150,000 mi, as claimed by Toyota, they will need not to be replaced during the life of an average vehicle. On the other hand, if the battery life is considerably less and battery replacement is necessary during the expected 10 year life of the car, the operation cost over the life of an HEV would be considerably higher.

12.6 Plug-In Hybrid Electric Vehicles

Plug-in hybrid electric vehicles (PHEVs) have recently become available in the market in some locations. PHEVs have the potential of making the leap to the mainstream consumer market because they require neither a new technology nor a new distribution infrastructure. Like hybrids, which are already widely available, PHEVs have a battery and an IC engine for power, but the difference is that a PHEV has a larger battery capacity and a plug-in charger with which the battery can be recharged whenever the car is parked near a 210 V outlet. The so-called PHEV40 can travel the first 40 mi on grid-supplied electric power when fully charged. When that charge is depleted, the gas or diesel motor kicks

(a)

(b)

FIGURE 12.5
Fuel-efficient cars. (a) 2013 Toyota Plug-in Prius. The Prius is a "parallel" hybrid in which the electric motor and gasoline engine work together, outfitted with 4.4 kWh lithium-ion batteries, has according to EPA an expected range of 540 mi (870 km) and a maximum electric-only speed of 62 mph (100 km/h). The battery for the electric motor can be charged from an electric outlet in 180 min at 120 V or 90 min at 240 V. The EPA fuel economy rating is 95 mpg gasoline equivalent and a combined city/highway rating of 50 mpg, the same as the conventional Prius. (b) 2013 Ford Fusion. The Fusion can travel up to 47 mph (75 kph) on electric power alone. The motor is powered by a nickel–metal hydride battery (Ni–MH) pack, and the car's engine is a 2.5 L Atkinson Cycle I-4 gasoline engine. The Fusion gets 47 mpg in the city as well as highway. This makes it the most fuel efficient car of its class.

in and the vehicle operates like a conventional hybrid. Because most commuter trips are less than 40 mi, it is estimated that a PHEV could reduce gasoline usage by 50% or more for many U.S. drivers. Moreover, using electric energy is cheaper and cleaner than using gasoline in automobile-type ground transportation. Figure 12.6 shows a Prius with plug-in potential added. These plug-in additions are available commercially, but because they are not in mass production, their cost is high, and Toyota does not honor their warranty when plug-in features are added aftermarket.

An important feature of PHEVs is that their batteries can be charged at night when utilities have excess power available. Utilities have taken notice of the potential energy charging and storage capabilities of PHEVs because off-peak charging would help utilities to use low-cost base load generation more fully. Furthermore, more advanced

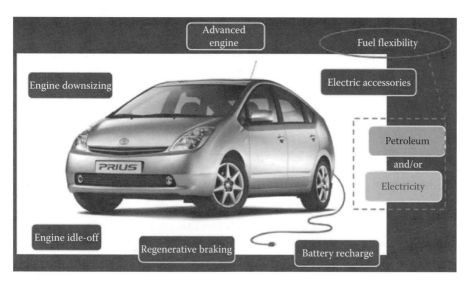

FIGURE 12.6
(See color insert.) Plug-in hybrid electric vehicle. (Courtesy of National Renewable Energy Laboratory, Golden, CO.)

vehicle-to-grid concepts would allow utilities to buy back energy from the batteries of vehicle owners during peak demand periods and thus make a fleet of PHEVs into a large, distributed storage-generation network. The arrangement would also enable renewable energy storage by charging PHEVs using solar- or wind-generated excess capacity. As mentioned in the previous section on batteries, lithium-ion and lithium–polymer batteries have the potential to store large charges in a lightweight package, which would make the HEVs even more attractive than the current technology that uses nickel–metal hydride (Ni–MH) batteries.

The efficiency of a PHEV depends on the number of miles the vehicle travels on liquid fuel and electricity, respectively, as well as on the efficiency of the prime movers according to the equation

$$\eta_{PHEV} = \frac{\text{Energy to wheels}}{\text{Energy from primary source}} = f_1\eta_1\eta_2 + f_2\eta_3\eta_4 \tag{12.1}$$

where
η_1 is the efficiency from the primary energy source to electricity
η_2 is the efficiency of transmitting energy to the wheels
f_1 is the fraction of energy supplied by electricity
f_2 is the fraction of energy supplied by fuel = $(1 - f_1)$
η_3 is the well-to-wheel efficiency
η_4 is the tank-to-wheel efficiency

PHEVs can be designed with different all-electric ranges. The distance, in miles, that a PHEV can travel on batteries alone is denoted by a number after PHEV. Thus, a PHEV20 can travel 20 mi on fully charged batteries without using the gasoline engine. According to a study by Electric Power Research Institute (EPRI) [8], on average, one-third of the annual mileage of a PHEV20 is supplied by electricity and two-third by gasoline. The percentage

TABLE 12.4

Net Value of Life Cycle Costs over 117,000 mi/10 Years
for Conventional Gasoline (IC), HEV, and PHEV20
Midsize Vehicles with Gasoline Costs at $1.75/gal

Vehicle Type	IC	HEV	PHEV20
Battery unit cost ($/kWh)		$385[a]	$316[a]
Incremental vehicle cost ($)		($547)	$224
Battery pack cost ($)	$60	$3,047	$3,893
Fuel costs ($)	$5,401	$3,725	$2,787
Maintenance costs ($)	$5,445	$4,733	$4,044
Battery salvage costs ($)		($54)	($43)
Total life cycle costs ($)	$10,906	$10,904	$10,905

Source: Extracted from EPRI, Advanced batteries for electric
drive vehicles: A technology and cost-effectiveness
assessment for battery electric vehicles, power assist
hybrid electric vehicles and plug-in hybrid electric
vehicles, EPRI Tech. Report 1009299, EPRI, Palo Alto,
CA, 2004. With permission.

[a] Battery module price at which life cycle parity with CV occurs.

depends, of course, on the vehicle design and the capacity of the batteries on the vehicle. A PHEV60 can travel 60 mi on batteries alone, and the percentage of electric miles will be greater as will be the battery capacity and weight. The tank-to-wheel (more appropriately battery-to-wheel) efficiency for a battery all-electric vehicle according to EPRI [8] is 0.82.

Given the potentials for plug-in hybrid vehicles, the EPRI [8] conducted a large-scale analysis of the cost, the battery requirements, and the economic competitiveness of plug-in vehicles today and within the near-term future. Table 12.4 presents the net present value of life cycle costs over 10 years for a midsized IC engine vehicle such as the Ford Focus [IC], HEVs such as the Prius [HEV], and a future PHEV20 plug-in electric vehicle. The battery module cost in dollars per kWh is the cost at which the total life cycle costs of all three vehicles would be the same. According to projections for the production of Ni–MH battery modules, a production volume of about 10,000 units per year would achieve the necessary cost reduction to make both an HEV and a PHEV20 economically competitive. The EPRI analysis was conducted in 2004 and is, therefore, extremely conservative because it assumed a gasoline cost of $1.75/gal. A reevaluation of the analysis based upon a gasoline cost of $2.50/gal showed that the permitted battery price at which the net present value of conventional IC vehicles and battery vehicles are equal for battery module costs $1,135 for an HEV and $1,648 for a PHEV20, respectively.

Table 12.5 shows the electric and plug-in hybrid vehicle battery requirements (module basis for the cost estimates in Table 12.4), and it is apparent that, even with currently available Ni–MH batteries, the cost of owning and operating HEVs and PHEVs is competitive with that of an average IC engine vehicle.

A cautionary note in the expectation of future ground transportation systems is the reduction in the rate of petroleum consumption that can be expected as HEVs and PHEVs are introduced into the fleet [9]. In these estimates, a rate of new vehicle sales of 7% of the fleet per year, a retirement rate of 5% per year resulting in a net increase in total vehicles of 2% per year, was assumed. This increase is in accordance with previous increase rates between 1966 and 2003. Based upon the existing mileage for IC engines, HEVs, and PHEVs, it was estimated that *even if all new cars were HEVs or PHEVs,* after 10 years, the annual

TABLE 12.5

NiMH Battery Cost Assumptions for Table 12.4

| Assumption | ARB 2000 Report for BEVs | | EPRI Assumptions |
	2003	Volume	
Module cost[a]	$300 per kWh	$235 per kWh	Varied[b]
Added cost for pack	$40 per kWh	$20 per kWh	$680 + $13 per kWh[c]
Multiplier for manufacturer and dealer mark-up	1.15	1.15	Varies[c]
Battery life assumptions	6 years	10 years	10 years

Note: BEV, Battery electric vehicles.

[a] Equivalent module costs for an HEV 0 battery is $480 for 2003 and $384 for volume. Equivalent module costs for a PHEV20 battery is $376 for 2003 and $301 for volume. HEV 0 and PHEV20 batteries have a higher power-to-energy ratio and are more costly.

[b] Battery module costs were varied in this analysis to determine the effect of battery module cost on life cycle cost.

[c] Manufacturer and dealer mark-up for HEV 0 battery modules estimated at $800, PHEV20 battery modules $850, pack hardware mark-up assumed to be 1.5. Method documented in 2001 EPRI HEV report.

gasoline savings as a percentage of the gasoline usage by an all-gasoline fleet in the same year would only be about 30% for the HEVs and 38% for the PHEVs. These relatively small reductions in the gasoline use are due to the fact that despite the introduction of more efficient vehicles, it takes time to replace the existing fleet of cars, and the positive effects will not be realized for some years.

12.7 Advanced Ground Transportation with Biomass Fuel

In the previous section, we have analyzed the potential of using batteries combined with traditional engines to reduce the petroleum consumption in the transportation system. However, the scenario used for this analysis can also be extended to determine the combination of PHEVs with biofuels, particularly ethanol made from corn or cellulosic biomass. No similar analysis for using diesel from algae is available at this time.

A scenario for a sustainable transportation system based on fuel from biomass has been presented in [10]. In this analysis, the following four vehicle types combined with various fuel options have been calculated. The preferred mixture in an economy based largely on ethanol (E85) would be 85% ethanol and 15% gasoline that could be used in Flex Fuel automobiles. Currently, the United States is using E10, a mixture of 90% gasoline and 10% ethanol, with ethanol produced from corn. The fuel types used in this analysis are gasoline only, E10 with ethanol made from either corn or cellulosic materials, and E85 with ethanol from either corn or cellulosic materials. The four vehicle combinations are a convention SI engine, an HEV similar to the Toyota Prius, a PHEV20, and a PHEV30. The analysis was based upon an average 2009 performance on the U.S. light vehicle fleet at approximately 20 mpg, for an HEV at 45 mpg, and for a PHEV20 at 65 mpg, according to [8]. For ethanol/gasoline-blended fuels, it was assumed that the gasoline and ethanol are utilized with the same efficiency. That is, the mileage per unit of fuel energy is the same for gasoline and ethanol.

Based on the earlier assumption, the following parameters were calculated:

1. The miles per gallon of fuel, including the gasoline used to make ethanols (mpg)
2. The petroleum required to drive a particular distance for a case vehicle as a percentage of the petroleum required to drive the same distance by a gasoline-fueled SI vehicle
3. The carbon dioxide emission rate for case vehicles as a percentage of that for SI gasoline-only

Based upon the earlier assumption, one can calculate the miles per gallon of fuel by the following equation:

Miles per gallon of fuel:

$$MF_{ij} = (mi/gal\ gas)$$

$$\times \left[(gal\ gas/gal\ fuel) + (1\text{-}gal\ gas/gal\ fuel) \times \frac{(energy, LHV/gal\ ethanol)}{(LHV/gal\ gasoline)} \right]$$

$$= MGO_i \times [FG_j + (1 - FG_j) \times (LHV\ ratio)] \tag{12.2}$$

where

MF_{ij} is the miles per gallon of fuel for vehicle type i and fuel type j
MGO_i is miles per gallon gasoline-only for vehicle i (see Table 12.4)
FG_j is volume fraction of gasoline in fuel type
$j(1 - FG_j)$ is the volume fraction of ethanol in fuel type j
the ratio of lower heating values (LHVs) is LHV ratio = (LHV/gal ethanol)/(LHV/gal gasoline) = 0.6625

The index i indicates the vehicle type as shown in Table 12.4, and the index j denotes the volume fraction of gasoline in an ethanol–gasoline blend as follows: for gasoline only, $FG_1 = 1$; for E10 (i.e., 10 vol% ethanol and 90 vol% gasoline), $FG_2 = 0.90$; and for E85 (85 vol% ethanol, 15 vol% gasoline), $FG_3 = 0.15$. Other ethanol concentrations could be used.

The miles per gallon of gasoline in the fuel, including the petroleum-based fuels used in the making of the ethanol in the fuel (by counting the energy of all petroleum-based fuels as gasoline), is given by

Miles per gallon of gasoline in fuel:

$$MG_{ijk} = \frac{MF_{ij}}{[FG_j + (1 - FG_j) \times (0.6625) \times (MJ\ gasoline\ used\ in\ making\ 1\ gal\ ethanol)/MJ/gal\ ethanol]}$$

$$= \frac{MF_{ij}}{FG_j + (1 - FG_j) \times (0.6625) \times Rk} \tag{12.3}$$

where Rk denotes the gallons of gasoline used to make 1 gal of ethanol. For corn-based ethanol, $k = 1$, $R_1 = 0.06$, while for cellulosic-based ethanol, $k = 2$, $R_2 = 0.08$, according to Ref. [11].

12.7.1 Petroleum Requirement

The petroleum required to produce an ethanol–gasoline blend, including the petroleum used to make the ethanol, is expressed as a percentage of the petroleum required for the same miles traveled by the same vehicle type using gasoline-only. For a gasoline-only-fueled vehicle of any type, this percentage is 100%.

A general equation for the percentage petroleum requirement is

$$\text{Petroleum requirement, } \% = 100 \times \frac{(MGO_i)}{(MG_{ijk})} \tag{12.4}$$

12.7.2 Carbon Dioxide Emissions

The CO_2 emissions, including those from making the ethanol and generating the electricity used from the grid by the vehicle, are expressed as a percentage of the emissions produced by the same type of vehicle fueled by gasoline-only traveling the same number of miles.

$$= 100 \times [FE_m \times (kWh/mi) \times (g\text{-carbon}/kWh)] + (1/MGO_i) \times (1 - FE_m)$$

$$\times [FG_j \times (g\text{-carbon}/MJ \text{ gasoline}) \times (MJ/gal \text{ gasoline})] + (1 - FG_j)$$

$$\times \frac{[(MJ/gal \text{ ethanol}) \times (g\text{-carbon}/MJ \text{ ethanol})]}{[(g\text{-carbon}/MJ \text{ gasoline used}) \times (MJ/gal \text{ gasoline}) / (MGO_1)]} \tag{12.5}$$

where FE_m is the fraction of the miles driven by electricity from the grid for a plug-in hybrid vehicle. For any non-PHEV, $m = 0$, and $FE_0 = 0$; for a PHEV20, $m = 1$, and $FE_1 = 0.327$ [12]; and for a PHEV30, $m = 2$, and $FE_2 = 0.50$ estimated by analogy with PHEV20. According to Ref. [12], the kWh/mi from the grid $= 0.2853$, the g-carbon emitted/kWh $= 157$ (146 average for all electricity generation [13] divided by 0.93, the average transmission efficiency), the g-carbon emitted/MJ gasoline $= 94$ [12], the MJ/gal gasoline $= 121$, the MJ/gal ethanol $= 80.2$, and CE_k the g-carbon emitted/MJ ethanol. For $k = 1$ (corn), $CE_1 = 87$, and for $k = 2$ (cellulosics), $CE_2 = 11$. So,

$$CO_2 \text{ production as } \% \text{ of that for a gasoline-only vehicle}$$

$$= \frac{100 \times \{FE_m \times (0.2853) \times (157) + (1/MF_{ij}) \times (1 - FE_m) \times [FG_j \times (94) \times 121 + (1 - FG_j) \times CE_k \times (80.2)]\}}{(94) \times (121)/21} \tag{12.6}$$

Example

Calculate the CO_2 production ratio relative to an IC vehicle for a PHEV30, E85 vehicle using ethanol from cellulosic material.

Solution
The relevant variables are

$$FE_m = 0.5$$

$$MF_{ij} = 85 \times (0.15 + 0.85 \times 0.6625) = 60.61$$

$$CE_k = 11$$

Using these parameters, the CO_2 production, as a % for that of gasoline-only vehicle is

$$= 100 \times \{(0.5 \times 0.2853 \times 157) + (1/60.61)$$

$$\times 0.5[0.15 \times 94 \times 121 + 0.85 \times 11 \times 80.2]\}/(94 \times 121/21)$$

$$= 4.13 + 2.60 + 1.14 = 7.87\%$$

Using the results of the earlier equation, the key conclusions of the analysis are shown in Figures 12.7 and 12.8. Figure 12.7 shows the gasoline requirement of any of the

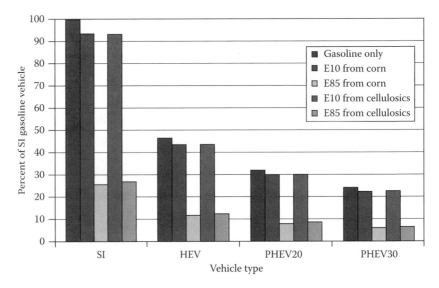

FIGURE 12.7
(See color insert.) Petroleum requirement as a percentage of that for SI gasoline vehicle. (From Kreith, F. and West, R.E., *ASME J. Energy Resources Technol.*, 128(9), 236, September 2006. With permission.)

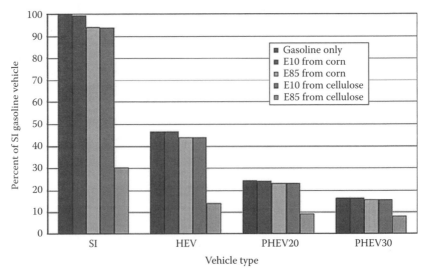

FIGURE 12.8
Carbon dioxide emissions as a percentage of emissions for SI gasoline vehicle.

20 cases considered as a percentage of that for an SI gasoline fleet. It can be seen that the gasoline consumption for a PHEV20 using E10 is about 30% of that for the gasoline fleet, whereas it decreases to about 8.8% for a fuel consisting of 85% ethanol from cellulosic with 15% from gasoline (E85 cellulosic). The potential reduction in CO_2 emission shown in Figure 12.8 is equally impressive. The CO_2 emission of a PHEV20 with E85 cellulosic is only about 10% of that for the SI fleet. Since, at present, emission from automobile engines is responsible for almost one-third of the total CO_2 emissions of the country, it is apparent that with the existing technology, substantial reduction in CO_2 generation could be achieved by using PHEVs, particularly if the fuel were to be manufactured from cellulosic material.

The final question to be asked in the utilization of PHEVs is whether or not the existing electricity system of the United States could handle the charging of the batteries in a PHEV-based ground transportation system. This question has recently been answered by an analysis in [14]. This analysis clearly showed that with a normal commuting scenario, which was based upon statistical information from a major city, if charging occurred during off-peak hours, no additional generational capacity or transmission requirements would be needed to charge a significant portion of the automotive fleet with PHEVs in the system, as illustrated in Figure 12.9. The analysis also showed that the off-peak charging scenario would also add to the profit of the electric utilities.

An unexpected result of the analysis was that, although the amount of CO_2 emitted by the electric utility will increase from PHEV charging, if this is compared to the corresponding reduction in tailpipe emission to assess the overall environmental impact, the generation/PHEV transportation system would substantially decrease CO_2 emission even if the current mix of coal, nuclear, and renewable generation facilities were unchanged. This result can be explained, however, because the average efficiency of the electric power system is on the order of 43%, whereas the average efficiency of IC engines is only on the order of 22%. Consequently, the net emission of CO_2 would be reduced by a hybrid system consisting of PHEVs and electric charging during off-peak hours. Moreover, the utility

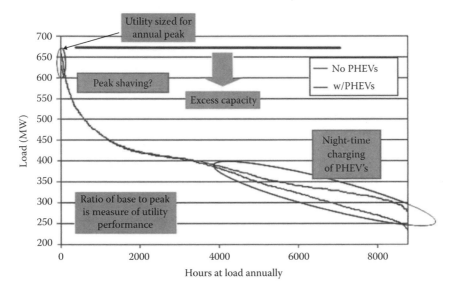

FIGURE 12.9
PHEV impact on utilities' load profile. (Courtesy of National Renewable Energy Laboratory, Golden, CO.)

generation profile would be evened out, and this would also contribute to reducing CO_2 emissions, as well as the cost of producing electricity.

In summary, in a PHEV transportation system, if the charging of batteries is limited to off-peak hours, this hybrid arrangement can

- Reduce the amount of petroleum consumed by the transportation sector
- Reduce the cost of driving
- Reduce CO_2 emission
- Improve the load profile of electric utilities

Although no quantitative analyses are as yet available, it is believed that the availability of electric storage in the batteries of a fleet of PHEVs could also be used to reduce the peak demand on electric utilities by utilizing the vehicle's batteries as a distributed storage system. Details of such an arrangement would have to be worked out by differential charges, incentives, and taxation arranged between utilities and owners of PHEVs.

12.8 Future All-Electric System

The next step in the development of a viable transportation system could be the all-electric car. All-electric vehicles were mandated in California as part of an effort to reduce air pollution about 25 years ago by the California Air Resources Board. Initially automakers embraced the idea, but it is likely that the acceptance by Detroit was the result of the mandate that required that at least 2% of all the cars sold in California by any one automaker had to be zero-emission vehicles. The only zero-emission vehicle available at the time was the electric car, and the mandate therefore required selling a certain number of all-electric vehicles. Battery technology at the time was nowhere near ready for commercialization, and in addition, there was no infrastructure available for charging vehicle batteries on the road. As a result, the mandate failed to achieve its objective, and as soon as the mandate was lifted, automakers ceased to make electric vehicles. In the meantime, however, battery technology has evolved to where one could potentially envision an all-electric ground transportation system. Some people propose that there should be enough charging stations built to make it possible to charge batteries anywhere in the country, whereas others propose that there should be, instead of gas stations, battery exchange stations that would simply replace batteries as they reach the end of their charge. At present, batteries take too long to be charged during a trip, and it may be necessary to combine the two ideas to evolve an all-battery transportation system sometime in the near future.

At this time, it is not possible to assess the future of electric vehicles. But at the start of 2010, the following information was collected from various sources on the Internet. It is subject to uncertainties, vagrancies, and changes, as the electric vehicle market develops. The student is encouraged to follow development on his or her own. The near-term availability of electric vehicles is difficult to predict. However, President Barack Obama announced in July 2009 a program with $2.4 billion in federal grants to develop next-generation vehicles and batteries in the United States. In Germany, one of the political party members proposed a €5,000 ($7,050) subsidy for people who buy electric cars. Japan's Nissan Motor Company unveiled, in July 2009, the LEAF, an electric car that is claimed to have a range of 100 mi (160 km) on a single battery charge. It has gone into mass production for a global market in 2012. Mitsubishi Motor Company launched, in June 2009, its own electric vehicle program, and the Chinese automaker, Dongfeng Motor Corporation, has announced that it will team up with a Dutch company to develop and produce electric vehicles. General Motors has announced its Chevrolet Volt, a PHEV that is claimed to get 200 mpg of gasoline. Finally, Daimler AG is working with the California-based electric car maker, Tesla Motors, Inc., on developing better batteries and electric drive systems for vehicles. The earlier summary is a snapshot at the time of preparing this text and is in a continuous state of flux. The reader is encouraged to follow development in the current literature included in Online Resources section.

12.9 Hydrogen for Transportation

Hydrogen was touted as a potential transportation fuel after former President George W. Bush said in his 2001 inaugural address that "a child born today will be driving a pollution-free vehicle … powered by hydrogen." This appeared to be welcome news. However, to analyze whether or not hydrogen is a sustainable technology for transportation, one must take into account all the steps necessary to make the hydrogen from a primary fuel source, get it into the fuel tank, and then power the wheels via a prime mover and a drive train. In other words, one must perform a well-to-wheel analysis, as shown in Figure 12.3. There is a loss in each step, and to obtain the overall efficiency, one must multiple the efficiencies of all the steps. Using natural gas as the primary energy source, a well-to-wheel analysis [15] showed that a hybrid SI car would have a wheel-to-energy efficiency of 32%; a hydrogen-powered fuel cell car with hydrogen made by steam reforming of natural gas would have a well-to-wheel efficiency of 22%; and a hydrogen fuel cell car with hydrogen made by electrolysis with electricity from a natural gas–combined cycle power plant with 55% efficiency would have a well-to-wheel efficiency of 12%, as shown in Table 12.6; and if the hydrogen were produced from photovoltaic cells, the well-to-wheel efficiency of the automobiles would be less than 5%. The estimates in Table 12.6 assume a fuel cell stack efficiency at a peak load of 44.5%, a part load efficiency factor of 1.1, and a transmission efficiency of 90%. For details, see [16].

TABLE 12.6

Well-to-Wheel Efficiency of Fuel-Cell Vehicle with Hydrogen Produced by Electrolysis

NG feedstock production efficiency	95%
Conversion efficiency (NG to electricity)	55%
Electrolysis efficiency (electricity to H_2)	63%
Storage and transmission	97%
Compression efficiency	87%
Overall efficiency of fuel production	28%
Total fuel cell well-to-wheel efficiency: $(0.28 \times 445 \times 1.1 \times 0.9)$	12%

In addition to the low overall efficiency of a hydrogen transportation system, as well as the high cost compared to other options, it should also be noted that before a hydrogen transportation system could be put into practice, an infrastructure for the distribution and storage of the hydrogen would have to be constructed. An extensive study of the comparative costs of fuel distribution systems conducted at Argonne National Laboratory in 2001 estimated that for a market penetration by the year 2030, 40% of hydrogen vehicles with a mileage of 2.5 times that of average conventional vehicles (i.e., about 55 mpg) would minimally cost $320 billion, but could be as high as $600 billion. Based upon these estimates, a national transportation plan based on hydrogen with any currently available technology would be wasteful and inefficient and should not be considered as a pathway to a sustainable energy future [17]. The U.S. Department of Energy, Secretary of Energy and Nobel Laureate, Dr. Steven Chu, concurred with this conclusion and, in 2009, cut off funds for the development related to hydrogen fuel cell vehicles in order to focus on other options [18].

12.10 Natural Gas as a Transitional Bridging Fuel

Despite its potential as a transportation fuel, except for large buses and trucks, natural gas has heretofore received relatively little attention from the U.S. automobile industry. The only major effort to use natural gas for transportation was the Freedom CAR initiative proposed by President George W. Bush in 2001. This program envisioned replacing gasoline with hydrogen, which at that time was largely produced from natural gas. More than a billion dollars was provided for R&D as well as generous tax incentives. But as shown in Table 12.6, the efficiency of a natural gas/hydrogen vehicle based on a well-to-wheel analysis is less than that of hybrid SI/natural gas vehicle, and the construction of a hydrogen distribution system would be extremely expensive. Thus the hydrogen via natural gas effort for transportation was a failure and was terminated in 2010. However, since 2009, new supplies of natural gas have become available in the United States, primarily as a result of fracking technology for natural gas extraction from oil shale deposits, have been developed. This development has heightened awareness of the potential of natural gas as a bridging fuel for transportation for an eventual zero-carbon future. Within this context, MIT has conducted an interdisciplinary investigation that addresses the question, "What is the role of natural gas in a carbon-constrained economy?" [19].

Natural gas is likely to find increased use in the transportation sector with compressed natural gas (CNG) playing an important role, particularly for high-mileage fleets. But the advantage of liquid fuel in transportation indicates that the chemical conversion of the gas into some form of liquid fuel may be a more desirable pathway for the future. It should be noted that a basic infrastructure for distributing natural gas exists, and if CNG were to be used to fuel automobiles, the only major addition at gas stations that have a natural gas outlet would be a compressor to increase the gas pressure above that in the natural gas automobile tank. However, the vast majority of natural gas supplies are delivered to markets by pipeline, and delivery costs typically represent a relatively large fraction of the total cost in the supply chain.

Natural gas vehicles have been in use for trucks and other large vehicles in many parts of the world for some time as shown in Table 12.7. But in the United States, Honda is at

TABLE 12.7

Number of NGV-Powered Vehicles in Top 10 NGV
Countries and the United States

Country	Number of Vehicles
1. Iran	2,859,386
2. Pakistan	2,850,500
3. Argentina	1,900,000
4. Brazil	1,694,278
5. India	1,100,000
6. China	1,000,000
7. Italy	779,090
8. Ukraine	390,000
9. Columbia	348,247
10. Thailand	300,581
17. United States	112,000

Source: NGV America, About NGVs https://www.ngvc.org/
about_ngv/, accessed May 5, 2013.

this time (2012) the only automobile company that sells a passenger-sized CNG powered car. The gasoline equivalent (GGE) of this car is 7.8, with natural gas stored at 3,600 psi. The car has a range of 218 miles on a full tank of CNG based upon EPA mileage ratings. A 7.8 GGE tank at 3,600 psi is equivalent to 30 gal in volume, which means that CNG needs a larger tank volume compared to a gasoline vehicle. At this point, it is not clear what the extra cost of a CNG or liquid natural gas (LNG) automobile would be. But the official list price for the Honda CX-CNG 2012 model was $25,490, with a fuel efficiency claimed to be 36 mpg, whereas the list price for the SI Honda Civic DX 2012 model was only $15,805 with an official performance at 39 mpg. This price difference suggests that the cost of new CNG models could be on the order of $10,000. In fact, the CNG 2012 model cost more than the Civic hybrid, which was listed at $24,050 with an official mileage of 44 mpg. Although there appears to be relatively little difference in the mpg, the cost of driving a CNG-powered car may be considerably less than that for an SI gasoline-engine car in the foreseeable future. Natural gas at the 2011 price was $4.30 per million Btu, while gasoline at $3.50 per gallon and 115,000 Btu per gallon equates to $30 per million Btu. Thus, even if natural gas prices should double, it is still likely to be considerably cheaper to drive a natural gas car than a gasoline SI version. However, the time required to recoup the extra initial investment and the availability of an adequate distribution infrastructure are uncertain.

As availability of natural gas wanes, a number of renewable sources for natural gas or biomethane are available. Biomethane can be produced from any organic matter. Nature produces it naturally in landfills, but it can also be produced, as discussed in Chapter 4, in anaerobic digesters or through pyrolysis from sewage, industrial, animal, or crop wastes or from specific energy crops. The biomethane from landfills used in NGVs reduces greenhouse gases by 90% according to the California Air Resources Board. After biomethane is produced, it can be injected into natural gas pipeline systems and sold to NGV station operators. A 5% or 10% blend of biomethane with natural gas would add to NGV's greenhouse gas potential. While many other alternative fuels are still in the R&D stage, NGVs are not in that category and are ready to go now.

Problems

12.1 On February 2006, President George W. Bush said, "All of a sudden, you know, we may be in the energy business by being able to grow grass on the ranch and have it harvested and converted into energy." Analyze and compare two of the following options:

1. Surveying the technical literature on the topic, we find an article by Schmer et al. [20]. According to this article, the net energy yield of switchgrass grown over 5 years on marginal crop land on farms in the mid-continental United States was 60 GJ/ha. This is equivalent to about 0.2 W/m^2, about 10 times the bioethanol from corn in the United States, and represents a potential source of ethanol sometime in the future.

2. Bioethanol can also be produced from sugarcane in climates such as in Brazil. For sugarcane, the production is 80 ton/ha.

3. Ethanol from corn using technology of 2009.

12.2 A BMW hydrogen car uses 2.5 kWh/km. Compare this performance with that of an average U.S. car that is able to get 22 mpg of gasoline.

12.3 From the information supplied in the following text, estimate the percentage of the total land area in the lower 48 states that would have to be used to replace all of the gasoline currently used for transportation with ethanol obtained from switchgrass. Do the same for the arable land in the lower 48 states.

Data:

- Estimated ethanol productivity = 1,000 gallons (gal) ethanol per hectare per year (based on switchgrass as the crop)—see, http://bioenergy.ornl.gov/paper/misc/switgrs.

- 1.51 gal of ethanol is required to obtain the same energy as 1 gal of gasoline, based on their respective LHVs.

- U.S. gasoline consumption in 2008 was 138 × 10^9 gal/year (information obtained from: http://eia.doe.gov/oil_gas/petroleum/inforglance/consumption/product_supplies/annual-thousand_barrels/finished_motor_gasoline).

- U.S. total land area (lower 48) = 1.92 × 10^9 ac (information obtained from: http://cia. gov/library/publication/the_world_fact_book).

- U.S. arable land area (suitable for crops) = 0.45 × 10^9 ac (information obtained from: http://cia.gov/library/publication/the_world_fact_book).

- 1 ha = 2.47 ac.

12.4 The combustion of ethanol goes according to the following equation:

$$C_2H_5OH + \underline{} O_2 \rightarrow \underline{} CO_2 + \underline{} H_2O \quad LHV = 21 \text{ MJ/L}$$

where LHV = lower heating value, the symbol "L" designates liter.

1. Calculate the amount of CO$_2$ produced in g per MJ of energy released (LHV). The density of ethanol at 16°C is 780 g/L.

   The growth of corn and conversion of the corn into ethanol produce 87 g CO$_2$/(MJ of ethanol) from the fossil fuels required. In addition, 35 g of CO$_2$/(MJ of ethanol) is produced during the fermentation of corn sugar into ethanol. However, each atom of carbon in the corn sugar, and thus each atom of carbon in the CO$_2$ released by fermentation of the sugar, originally came from an atom of carbon dioxide in the atmosphere. Therefore, the CO_2 from the fermentation is "carbon neutral," that is, the amount of CO_2 released by the fermentation is balanced by CO_2 removed from the atmosphere in making the corn sugar. [*Note:* The same is true for the fermentation of any biomass.] Likewise, each atom of C in the CO_2 produced by combustion of ethanol is also carbon neutral.

2. Calculate the total amount of CO_2 released to the atmosphere per MJ (LHV) of ethanol made from corn and combusted, and the net amount of CO_2 released per MJ (LHV) of ethanol made from corn and combusted. The net amount is the amount by which the production and combustion of ethanol increase the amount of CO_2 in the atmosphere.

   Gasoline is a mixture of many compounds of carbon and hydrogen, but may be represented by the formula C_8H_{16}, with an LHV of 32.8 MJ/L (L = liter).

3. Calculate the CO_2 emissions in g CO_2/MJ (LHV) with gasoline as fuel.

4. Estimate the total CO_2 emissions [g CO_2/MJ (LHV)] from "well-to-wheel" by including "well-to-tank" emissions with gasoline as a fuel. The well-to-tank energy efficiency for gasoline is approximately 90% (100 × (MJ energy to tank)/(MJ energy of petroleum "in the well")). Assume that the ratio of CO_2 emissions from tank-to-wheel is also 90% of the total CO_2 emissions.

12.5 A truck engine consumes diesel at a rate of 30 L/h and delivers 65 kW of power to the wheels. If the fuel has a heating value of 43.5 MJ/kg and a density of 800 kg/m³, determine

1. The thermal efficiency of the engine

2. The waste heat rejected by the engine

12.6 An automobile storage battery with an open circuit voltage of 12.8 V is rated at 260 A/h. If the internal resistance of the battery is 0.2 Ω, estimate the maximum duration of current flow and its value through an external resistance of 1.8 Ω.

12.7 Make a preliminary design for the power train of a hydrogen/oxygen fuel cell–powered automobile. Provide a system description with preliminary design estimates of the power and fuel supply arrangements.

12.8 Explain the difference in operation between HEVs with parallel and series designs.

Extra Credit Problems Supplied by Robert Kennedy, Member, ASME Energy Committee

12.9 Suppose you own and operate a good modern turbo-diesel automobile in Southern California. Unlike all your neighbors, your job is not so far away, plus you telecommute whenever you can, so you burn only 1 gal per day on average. Nevertheless, you are thinking of "going greener" by switching to an all-electric car, charging at night. In fact, "even greener," you are considering installing your own PV panels. You wonder when you should make these bets, if any.

According to the Bureau of Labor Statistics

1. The long-term trend (40 years, from 1973 to 2012) for price inflation for #2 diesel is 7.2% *per annum* (nominal dollars).

2. The long-term trend (40 years, from 1973 to 2012) for price inflation for grid electricity is 2.2% *per annum* (nominal dollars).

(For comparison, note that the prevailing Consumer Price Index is reckoned to be about 3%. YMMV.)

Assume an initial price of $4 per gal and an energy content of 135,000 Btu/gal (37 MJ/L).

Assume an initial fuel efficiency of 35 MPG (combined city + hwy). (*Note:* Per Kreith in this textbook, the "well-to-wheel" efficiency of a modern automobile with a turbo-diesel is about 33%, the best there is.)

Assume an initial residential off-peak rate of 10¢ per kWh, and peak daytime rate of 30¢ per kWh.

Assume 20% loss each way (in/out) for in-car battery storage, otherwise perfect drive train efficiency but zero regeneration.

Use NREL's solar map to get average full-sun-hours per day at your location, with flat plate tilted at latitude, southern azimuth, no seasonal bias. [*Hint:* it's 6.25 h/day.]

Assume an initial capital cost of $3.00 per DC-watt, 40 year economic life, with a capital cost reduction curve for installed PV capacity of 1% (relative, compounded) per month. Assume basic conversion efficiency of sunlight-to-DC-electric technology improves by +1% (absolute, i.e., 100 basis points) each year.

Assume no OpEx for the PV system.

Find in terms of cents per km.

1. Forecast the date at which specific motive energy from off-peak grid electricity will cross over to become nominally cheaper than that from a diesel engine.

2. Forecast the date at which specific motive energy from peak grid electricity will cross over to become nominally cheaper than that from a diesel engine.

3. Forecast the date at which specific motive energy from photovoltaic electricity will cross over to become nominally cheaper than that from a diesel engine.

Approach to solution

Step 1. Calculate the effective energy consumption for travel via present method, which is 2.53 MJ/km. Effective cost, present method and day, is 7.1¢ per km.

Step 2. Prepare a baseline series for the delivered cost of diesel-fueled travel, escalated over the next 50 years [nominal cents/km].

Step 3. Prepare a second series for the delivered cost of off-peak electric-fired travel, escalated over the next 50 years [nominal cents/km].

Step 4. See how many years from now when the terms in the first two series become equal. ANSWER (1).

Step 5. Modify the second series to a third series (simple) for the delivered cost of peak electric-fired travel, escalated over the next 50 years [nominal cents/km].

Step 6. See how many years from now when the terms in the first and third series become equal. ANSWER (2).

Step 7. Prepare a fourth series for the delivered cost of solar electric-fired travel, escalated over the next 50 years [nominal cents/km].

7A. Calculate the PV system capacity required now to provide 2.23 GJ of daily DC solar electricity input to the electric car [watts].

7B. Calculate the CapEx, then calculate the daily capital recovery charge to amortize CapEx over the 40 year life, using the methods presented in this chapter [cents/day].

7C. Divide by 56.3 km daily travel to get [cents/km].

7D. Apply the two improvement curves (conversion eff. and CapEx redux) to generate the fourth series over the next 50 years to forecast the trend in [nominal or "then-year" cents/km].

7E. See how many years from now when the terms in the first and fourth series become equal. ANSWER (3).

12.10 You are the president of the United States. You are thinking of implementing a revolutionary policy shift, affecting energy, foreign policy and trade, and national security to boot.

But Congress intends to impose a surtax on all motor fuel of fossil origin, collected at the pump as state and federal highway taxes are already, to fully recoup the presently unaccounted for externalities and opportunity costs of maintaining sufficient military presence in the Middle East to guarantee access to sources of crude oil there. Monies collected would defray the cost of repatriation, as well as be reinvested in domestic sustainable energy infrastructure/R&D.

Everything you need is in the *Statistical Abstract of the United States* (current edition online http://www.census.gov/compendia/statab/).

Find, in terms of dollars per gallon, what level of "war tax" at the pump would fully recoup the "annualized" present value of expected direct and indirect military and foreign policy "externalities" under business-as-usual?

(*Note:* "annualized present value" means to turn the present value into an annuity with infinite term.)

You're the boss, but state your assumptions.

Approach to solution

Comb the federal budget, especially the DoD, identifying those line items reasonably attributable to the United States's current overseas posture.

Example of direct externalities are overseas deployment cost, assumed US$1 million/soldier/year; equipment replacement; overseas MILCON (military construction).

Example of indirect externalities are VA cost for "wounded warriors" for the rest of their lives.

Extrapolate these annual costs over the next 30 years.

For a budget escalator, use the PPI, or whatever you can justify.

Discount these nominal cash flows and sum to one present value.

For inflation, use the CPI.

Convert this present value to an annuity, with infinite term.

For opportunity cost/cost of money, use long-term (30 year) bond interest rates.

Calculate the number of gallons of imported motor fuel of fossil origin burned per year in the United States. Assume Canada and Mexico are not "foreign."

Divide the annuity by the number of gallons.

References

1. Kreith, F. (1999) *Ground Transportation for the 21st Century*, National Conference of State Legislatures, Denver, CO and ASME Press, New York.
2. Kreith, F., West, R.E., and Isler, B. (2002) Legislative and technical perspectives for advanced ground transportation system, *Transportation Quarterly* 96(1), 51–73, Winter 2002.
3. Hammerschlag, R. (2006) Ethanol's energy return on investment: A survey of the literature 1990–present, *Environmental Science and Technology*, 40, 1744–1750.

4. Dismukes, C. et al. (2008) Aquatic phototrophs: Efficient alternatives to land-based crops for biofuels, *Current Opinions in Biotechnology* 19, 235–240, June 2008.
5. Christi, Y. (2007) Biodiesel from microalgae, *Biotechnology Advances* 25, 294–306.
6. Kreith, F. and West, R.E. (2003) Gauging efficiency: Well-to-wheel, *Mechanical Engineering Power*, pp. 20–23.
7. Davis, S. and Diegel, S. (2004) *Transportation Energy Data Book*, Oak Ridge National Laboratory/U.S. Department of Energy, Oak Ridge, TN.
8. EPRI (2004) Advanced batteries for electric drive vehicles: A technology and cost-effectiveness assessment for battery electric vehicles, power assist hybrid electric vehicles and plug-in hybrid electric vehicles, EPRI Tech Report 1009299, EPRI, Palo Alto, CA.
9. Kreith, F. and West, R.E. (2006) A vision for a secure transportation system without hydrogen or oil, *ASME Journal of Energy Resources Technology* 128(9), 236–243, September 2006.
10. Kreith, F. and West, R.E. (2008) A scenario for a secure transportation system based on fuel from biomass, *Journal of Solar Energy Engineering* 130, 1–6, May 2008.
11. Farrell, A.E., Plevin, R.J., Turner, B.T., Jones, A.D., O'Hare, M., and Kammen, D.M. (2006) Ethanol can contribute to energy and environmental goals, *Science* 311, 506–508.
12. EPRI (2004) Advanced batteries for electric drive vehicles, EPRI Tech. Report 1009299, EPRI, Palo Alto, CA.
13. EIA (2007) Annual Energy Outlook, www.eia.gov/fuelelectric/electricityinfocard2005
14. Himelich, J.B. and Kreith, F. (2008) Potential benefits of plug-in hybrid electric vehicles for consumers and electric power utilities, in: *Proceedings of ASME 2008 IMEC*, Boston, MA, October 31–November 6, 2008.
15. Kreith, F. and West, R.E. (2004) Fallacies of a hydrogen economy: A critical analysis of hydrogen production and utilization, *Journal of Energy Resources Technology* 126, 249–257.
16. Kreith, F., West, R.E., and Isler, B.E. (2002) Efficiency of advanced ground transportation technologies, *Journal of Energy Resources Technology* 24, 173–179.
17. Mince, M. (2001) Infrastructure requirement of advanced technology vehicles, in: *NCSL/TRB Transportation Technology and Policy Symposium*, Argonne National Laboratory, Argonne, IL.
18. M.L. Wald, *New York Times*, News Service, April 8, 2009.
19. Moniz, E. et al. (2011) The future of natural gas: An interdisciplinary MIT study, *MIT Energy Initiative*. Accessed at http://web.mit.edu/mitei/research/studies/natural-gas-2011.shtml (accessed on April 14, 2013).
20. Schmer, M.R. et al. (2008) Net energy of cellulosic ethanol from switchgrass, *PNAS* 105(2), 464–469, www.pnas.org/cgi/content/full/105/2/464
21. Bartlett, A.A. (2000) An analysis of U.S. and world oil production patterns using Hubbard-style curves, *Mathematical Geology* 32(1), 1–17.

Online Resources

http://www.nissanusa.com/leaf-electric-car/index
http://www.teslamotors.com/models
http://www.nrel.gov/sustainable_nrel/transportation.html
http://www.nrel.gov/learning/re_biofuels.html
http://www.nrel.gov/vehiclesandfuels/energystorage/batteries.html
http://cta.ornl.gov/vtmarketreport/index.shtml
http://cta.ornl.gov/data/download31.shtml (a large source of data that can be downloaded for free)

13

*Transition Engineering**

When one looks back over human existence, however, it is very evident that all culture has developed through an initial resistance against adaptation to the reality in which man finds himself.

Beatrice Hinkle

The purpose of this chapter is to present the fundamental ideas and methods of transition engineering. The majority of this book has provided technical information for renewable energy conversion. The first chapter gave an overview of the issues with the unsustainable exploitation of energy, water, and land, and the issues of population growth and environmental disruption. After studying the chapters of this book, you should have a good understanding that the availability of fossil fuels, water, land, and mineral resources will not continue growing into the future as they have grown in the past. You should also be able to quantitatively demonstrate how and why renewable energy cannot directly substitute for current patterns of fossil and nuclear fuel use. After studying this chapter, you will understand the complex and dynamic nature of energy and environment systems. You will be able to formulate engineering change projects and propose and evaluate innovate adaptive designs. You will learn several new strategic analysis methods for transition projects. You will also be able to use a holistic perspective without getting bogged down in conflicting ideologies. The relative roles of human behavior, policy, and economics should become clear, as will the critical role of the emergence of the field of transition engineering.

TRANSITION ENGINEERING

- *Issues and risks of unsustainability*: Assess the probability and impact of resource depletion or environmental impacts and quantify the risks to standard of living and essential activity systems.
- *Adaptation*: Assess adaptive capacity of existing energy activity systems and develop adaptive designs to accommodate the anticipated fossil energy and resource use reduction.
- *System redesign and change management*: Long-term concept design and innovation for prosperous systems that operate within resource and environmental constraints.

Every time a new set of problems or opportunities has become apparent, a new field of engineering has emerged. Air and waste management, reliability, and waste water engineering are surprisingly new fields, dealing primarily with issues of industrial development. Energy and power engineering, like transportation engineering, first emerged

---

* Sections in this chapter marked with an asterisk may be omitted in an introductory course.

to deliver reliable infrastructure at a pace and scale that could keep up with growth. After the 1970s' energy crisis, energy engineering expanded to include energy efficiency, energy management, and alternative energy technology. Transition engineering is the work of delivering change projects for existing systems that reduce energy consumption, resource use, waste production, and environmental impacts while maintaining access to essential activities, goods, and services. These kinds of short-term projects are necessary for prosperity as resource depletion effects become more pronounced. But transition engineering will also deliver strategic and innovative development of transformative infrastructure, products, and systems that will fit prosperous human activity systems into the resource and environmental constraints over the long term both locally and globally.

Key principles: *Transition engineering*

- Employ holistic or whole system perspectives
- Employ long-term perspectives
- Include complexity
- Upstream not tailpipe projects
- Use synergies of scale: time, place, relationships
- Change existing systems to reduce energy intensity into the future
- Change existing systems to reduce fossil fuel demand into the future
- Change existing systems to reduce resource and material intensity into the future
- Change existing systems to increase resilience
- Change existing systems to increase adaptive capacity
- Engage participation of stakeholders at all levels
- Active engagement, communication, and learning of all stakeholders
- Use strategic analysis and fact-based decision-making
- Develop capacity of stake holders and utilize knowledge and intelligence from all levels
- Use biophysical approach to economics (account for externalities and recognize physical limits)
- Gather and use scientific data and employ robust engineering methodologies
- Avoid social positionalism and polarization (I *don't believe* in global warming; I *believe* in the hydrogen economy; I am a Broncos fan—e.g., positions supported by belief rather than evidence)
- Acknowledge all of the problems and roadblocks, but do not give in to irony or cynicism

Transition engineering is emerging at a historical turning point, where we are beginning to recognize that the realities of the next 70 years will not be the same as the last 70 years. Engineering as a profession has always had a social responsibility to apply physical sciences, using accepted mathematical models of system behavior, to design and deliver systems that work. Transition engineering will make use of all of the successful engineering methodologies that have been previously developed. Transition engineers will be able to work in all fields, in much the same way safety engineers currently do. Most importantly, perhaps, transition engineering will use reliable science-based information about human needs, resource availability, and environmental impacts, and deliver adaptation of current systems for long-term global sustainability and prosperity, even if it is at the expense of short-term convenience or economic gain of some people.

13.1 Foundations of Transition Engineering

Human civilizations have always depended on ingenuity and adaptability for survival. Society has always depended on engineering for improving the standard of living and prosperity. Engineering is now a formal profession, but think about the long history of key innovations in energy like chimneys and furnaces that allowed buildings to be healthier and materials to be processed more efficiently. Food processing innovations like salting, drying, smoking, brining, and brewing were great advances that allowed people to safely store food from the harvest through the hard times of winter. Of course a lot of engineering advances were actually related to warfare, but on the whole, ingenuity and new technologies have increased the standard of living throughout history.

Standard of living refers to basic health, nutrition, education, and mortality. A prosperous society would have a high standard of living for a greater number of its members than an impoverished society. Lifestyle and quality of life are often confused with standard of living. However, lifestyle is related to factors of choice and freedom and only relevant for the people who already have a high standard of living. What if the successes of our industrial systems are actually a threat to our own well-being? This was definitely the case during the height of the industrial revolution. The factories of the 1800s provided new prosperity, but they were also very dangerous places to work, and the products and waste streams they produced were often deadly. The municipal and toxic waste streams and emissions from the new consumer economy of the 1950s and 1960s destroyed ecosystems in lakes and rivers and threatened many species with extinction. Engineering fields have emerged to deal with the risks associated with industrial success, and they solved problems and developed codes and standards that were later adopted as regulation. Transition engineering is emerging to deal with the *future* risks of both industrial and consumer activity.

PARAMETERS OF INDIVIDUAL AND SOCIAL WELL-BEING

- *Standard of living*: Clinical measures of the health, safety, security, and environmental carrying capacity. Rates of disease, infant mortality, abuse, violence. Nutrition, air and water quality, adequate housing, and sanitation. Access to medical care, legal protection, education. Biodiversity, habitat, and species population stability.
- *Quality of life*: Measures of prosperity and environmental quality. Public health, sports, recreation. Social assets, libraries, arts, leisure activities. Poverty rates, access to opportunities, level of education accessible to all economic sectors.
- *Lifestyle*: Economic measures of consumption. Energy and material intensity. Incomes, consumer spending, consumer goods, holiday travel, home size, rates of ownership.

The starting point for transition engineering is to accept the social responsibility of all engineering professionals to provide safety, security, and sustainability through research, testing, and expert consensus to develop standards. The next step is to understand the facts and the nature of the risks of unsustainable growth in the consumption of energy and other resources. Finally, we need to learn methods for complex problems that involve adaptation of established systems and entrenched ideas over a long-term planning horizon. The hardest part about getting started is that we have little or no experience with sustainability, and so far, all the historical growth in consumption beyond the planet's carrying capacity, expanding population, and increasing pressure on the environment has been good for the economy!

13.1.1 Social Mandate for Transition

Engineering ethics can be extended to individual and professional social responsibility for global sustainability. It is true that the profession has become commoditized and that most engineers work for employers who set the objectives for their work. But this section will show that transition engineering is founded on the same basic principle as safety engineering: prevention. Safety engineering was established in 1911 and has become so engrained that we may not even realize how revolutionary it is. Today, more than 30,000 American occupational safety, health, and environmental (SH&E) practitioners work in all workplaces and public areas to identify hazards, implement safety regulations, and continually advance the field. The SH&E professional is committed to protect people, property, and the environment through research, engineering, design, development and implementation of standards, continuous monitoring, management, and education. The first tenet of the profession is to be honest about hazards and possible remedies, and to bring issues to the attention of employers and the public. One hundred years after the emergence of SH&E engineering, it is hard to imagine how bad things were and to understand how far we have come. However, it is important for us to learn the history of safety engineering and other risk management professions, because of the amazing parallels with sustainability [1].

In the 5 years before 1911, 13,228 coal miners died on the job in America. The life expectancy of chemical, textile, and other factory workers was cut short due to exposure, respiratory damage, and accidents. There were no fire regulations, and no protection for workers from machinery, eye injury, or heat exposure. The workers wanted safer workplaces. The public was shocked and appalled at each new workplace disaster. Industry felt threatened by calls for change and widow's protests. Politicians were afraid to force changes when there were not many ideas about possible solutions, and there was no workable definition of safety. The public debate was becoming polarized between priorities of economics and humanity.

On March 25, 1911, 146 workers at the Triangle Shirtwaist Factory in New York City, most of them young girls, died over the course of the day when the factory they were working in caught fire [2]. There had been innumerable factory fires and loss of life prior to this tragedy. However, on this day, most of the girls died when they jumped 100 ft to the street below which was crowded with onlookers and photographers. Many of the girls jumped holding on to each other, with their hair and clothes ablaze. When the fire broke out, the workers found that most of the 27 water buckets, placed around the factory to douse the frequent ignition of fine cotton dust, were empty. Most of the factory doors were locked to keep workers from leaving without being checked for smuggling out bits of thread or ribbon, and the unlocked doors all opened inward, effectively being held shut by the press of girls trying to escape. The firefighters found that their ladders and water hoses could not reach the ninth floor windows of the factory, and there was no fire escape.

The infamous Triangle Shirtwaist Factory Fire tragedy galvanized worker's groups and led to political calls for improved worker safety. However, the real and lasting change occurred because the American Society of Safety Engineers was founded over the months following the fire by 62 industrial engineers who shared the belief that all of the deaths could have been prevented. They decided that it was the responsibility of the engineers who designed and built the factories to identify hazards and institute measures or make changes to prevent loss of life. The first standard developed by the organization was the rule that all doors must open outward and must be left unlocked. This first safety standard alone could have saved all of the lives lost in the Triangle Shirtwaist Factory fire.

The Great Depression and the First and Second World Wars disrupted the SH&E work, but the ASSE membership grew with new engineers from railroads, mining, chemical processing, and steel mills. After the wars, the ASSE started gaining support from industry, which hired members to conduct employee training and inspections. The ASSE also conducted research to develop new safety equipment like shatterproof glass for safety goggles, ventilators for chemical workers, and harnesses and hardhats for construction workers. Much of the research was funded by industry. In 1969, the nine-member Board of Certified Safety Professionals was formed with representation from all the fields of engineering to provide the certification and training of safety professionals. In 1970, the Occupational Safety and Health Act was passed, effectively establishing the standards developed by the SH&E profession as the standard for the whole country.

Over 5500 Americans died from work-related injuries in 2003, and in the decade ending in 2009, 354 miners died in accidents, all of which were deemed to be preventable. There will never be an "acceptable" number of workplace fatalities, but consider that a 96% reduction has been achieved over the past 100 years. The way that the extraordinary change has occurred is through continued, honest, and diligent work by SH&E engineers and professionals to identify hazards, assess risks, follow standards, train and inform workers, and make changes to reduce the risks. Engineers sometimes take safety shortcuts to save money, but in this country, we call that negligence. All of the engineering professions have a social mandate to follow existing safety regulations and to continue to work to prevent harm. What about the economics? Isn't it all about money? Don't engineers just have to do whatever their managers want? Responsible companies and communities are committed to safety, and the public expects and trusts that safety is a top priority and that profits can be made without sacrificing safety. In fact, a 2009 study by the Occupational Safety and Health Administration found that every $1 invested in safety returned $6 in benefits to society.

Figure 13.1 shows how safety, security, and sustainability are actually different scales of the overall continuum of survival. There are also different dimensions to survival. In the immediate time and space of the workplace, relationships between employees, managers, and application of safety rules are important for preventing accidents. In the longer time scale, security often depends on good relationships between trading partners and governments in different regions and negotiation of agreements. Peak oil, freight shipments, and power supply are security issues. On a very long, or continuous, time scale, sustainability has global and interconnected dimensions and depends on relationships between generations and species. Food, water, biodiversity, and soil condition over the long term are sustainability issues.

The key idea for transition engineering practice is that, like safety engineering, we are working to reduce and prevent failures of *unsustainability*. Sustainability advocates often

	Continuum of survival		
Dimensions	Safety	Security	Sustainability
Time	Daily–Annual	Lifetimes–Generations	Continuous
Space	Immediate–Local	Territorial–Regional	Global
Relationships	Individuals–Families	Organizations–Populations	Civilizations–Species

FIGURE 13.1
Safety, security, and sustainability are different scales in the continuum of survival.

make the mistake of pursuing sustainability as an absolute requirement. Green solutions have gained marketing success, but we cannot build a "sustainable" version of a car, house or office building, product, or an appliance because sustainability is a complex system condition over a long time frame. We can see why sustainability has been much more difficult to progress to the same practitioner level as safety engineering. New ideas for identifying and assessing long-term risks and for managing the relationships between species and generations need to be developed. The scientific community and the public are expressing growing alarm over the damage occurring to oceans, forests, wetlands, glaciers, etc., across the world. Just like 100 years ago, we know very well that there are problems, we know they are grave, and we know they are caused by industrial activities and the technology we produce. But just like 100 years ago, the most effective changes to reduce the threats to survival will not come from policy or business or even from public protest. Changes in industrial systems must be engineered.

The emerging multi-disciplinary field of transition engineering is needed to engineer the adaptive changes in complex energy–economic–social systems in order to reduce the risks to security and sustainability over the long term. Figure 13.2 shows existing engineering disciplines that work on different aspects of safety, security, and sustainability.

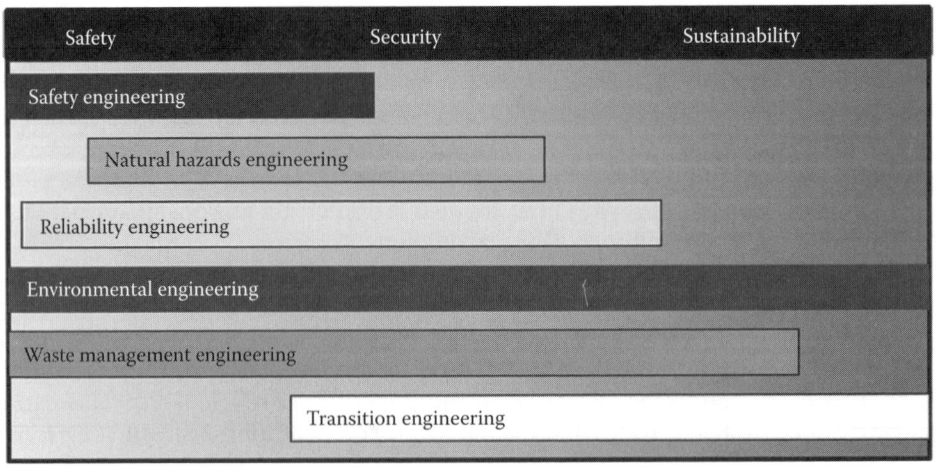

FIGURE 13.2
Interdisciplinary engineering disciplines in the context of the safety–security–sustainability continuum.

Each of the fields emerged to deal with problems and risks, most arising from industrial activities and man-made toxic waste and the hazards to public health. There is overlap between the existing fields and transition engineering. However, transition engineering deals more with adaptation of existing systems to manage energy and resource constraints. Transition engineering extends the other risk management fields into a discipline that delivers strategic analysis, adaptive design, change projects, and change management for long-term future prosperity.

The other interesting conclusion we can draw from understanding that safety, security, and sustainability are really different aspects of survival is that they all have the same definition. We often hear people say that we don't even know what sustainability means, or we don't have a definition for sustainability so we can't justify the costs of change. However, in the context of the underlying meaning of survival, we see that the definition actually depends on the negative. Survival can mean anything, except dead. We can't define what survival is, or what it would look like, but we can most certainly know when we do not have it. The workplace is safe, until there is an accident. Our supply of oil is secure until it is cut off. Ecosystems are sustainable unless we disrupt them. Our civilization is sustainable unless we collapse.

13.1.2 Identifying Issues

Unsustainability issues are clearly identified by scientific study. The World Watch Institute, the World Wildlife Fund, the United Nations Environment Programme, and numerous other organizations call for urgent action on climate change, toxic chemical exposure, air and water pollution, biodiversity and species loss, degradation of coral reefs, tundra, and rainforests. There continue to be warnings of degradation of natural resources and over-exploitation of soils, groundwater, wild fish stocks and native forests, grasslands, and wetlands. Engineers have been working on reducing air and water pollution, developing recycling and renewable energy technologies, and improving energy efficiency. However, engineers typically work on technology, manufacturing, materials, and infrastructure. We leave it to policy makers and the market to sort out complex issues. But new products like electronics are causing exposure to toxic chemicals, and new extractive techniques like tar sand processing are accelerating disastrous environmental damage. The history of safety engineering demonstrates that huge gains can be made economically and rapidly if we focus on identifying issues and work to prevent harm before it happens through changes in the existing system. We are doomed to fail if we only pursue "sustainable energy" as the growth in renewable energy and we don't focus on actively reducing the extraction and production of fossil fuels.

There are three primary unsustainable aspects of our current fossil-based energy system. The environmental damage done by extracting nonconventional fossil fuels is too great, the build-up of CO_2 in the atmosphere is too fast, and using all of the low-cost conventional fossil fuels now will leave future civilizations at a huge disadvantage. Building more wind turbines and solar panels and turning food crops into biofuel have not reduced the drive to extract fossil fuels. Today's buildings, factories, and transportation demand an unsustainable flow of fossil energy and resources. These end-use systems must change to use less fossil fuel.

The most immediate security issue for the United States is oil supply. There is no question that the U.S. transportation system has the highest energy intensity, and the economy has the highest oil dependency in the world. The only time that oil supply was reduced was the energy crisis caused by the OPEC oil embargo and that caused a serious economic

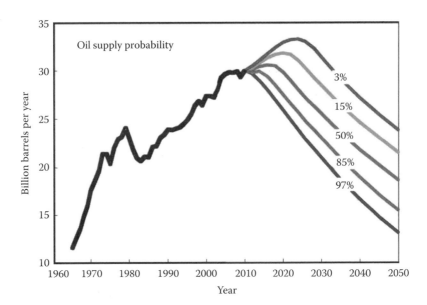

FIGURE 13.3
Probability of future oil supply generated from meta-analysis of oil expert analysis. (From Krumdieck, S.S. et al., *Transp. Res. A*, 44, 306, 2010.)

recession with high inflation. The last time that spare production capacity dipped below 1 million barrels per day (mbpd), there was a severe price spike, which contributed in a major way to the great recession of 2008–2009.

Prominent petroleum geologists like industry expert Dr. Colin Campbell [3], Princeton University Professor Kenneth Deffeyes [4], and Uppsala University Professor Kjell Aleklett [5] have provided analysis that showed conventional oil production peaking this decade. More reports and analysis have since been done by more experts using different assumptions and data giving a range of peak production dates and subsequent decline rates. Does disagreement among scientific experts mean uncertainty?

The suggested method for dealing with a multitude of experts is to include all of the information in a meta-analysis [6]. The method assumes that all of the experts have a reasonable accuracy if their assumptions and data are correct. A probability distribution function, like a Rayleigh distribution, is fit to all of the expert reports, and a probability space is generated. Figure 13.3 shows the cumulative probability of oil supply according to a recent survey of published expert analyses. The lowest curve is the oil supply that 97% of experts agree will be available. The top curve has the agreement of only 3% of experts. This type of analysis can be understood and used by decision-makers because they can decide their risk position and then use the associated future oil supply level over the planning horizon of interest.

Example 13.1: Future Road Planning

A small city has become a popular retirement destination and has been experiencing population growth of 4% per year. The new housing has mostly been on large wooded lots in the hills around the town. Traffic volume on the roads has increased by 25% between 1990 and 2010, resulting in heavy traffic and congestion. The mayor is pushing for a new road building investment to accommodate future growth to 2050, which is anticipated to continue at the same pace as the past resulting in a 50% increase in cars. This would mean expanding all of the current two-lane roads to four-lane roads at a cost of $1.8 billion over the next 10 years. One of the city councilors has read a book by K. Deffeyes and is worried about peak oil. She is an advocate of "smart growth." The idea of smart growth is to set boundaries on expansion of cities so that development of new homes and businesses must all be within the boundary. The councilor has hired your transition engineering consulting firm to provide an analysis to the council and recommendations.

Transition Engineering Solution
Looking at Figure 13.3, we can see that all of the published analyses of future oil supply agree that there is no possibility that the transport fuel supply could increase by 50% to 2050, thus 50% increase in travel demand using the same kind of transport modes as today is not likely. This would mean that the investment in road building in anticipation of growth in car traffic would be at risk if it were not needed. The current fleet efficiency is estimated at 21 mpg. If all vehicles were replaced over the next 20 years with vehicles that had an average of 42 mpg, then twice as many vehicles would use the same amount of fuel as today if the distances being driven did not get any greater—but the congestion on the roads would not increase. The combination of smart growth and doubling of fuel efficiency would mean that the fuel demand in 2050 would be the same as today. However, there is only a 3% probability that the amount of fuel available in 2050 would be 25% *less* than today. Nearly all oil experts agree that the oil supply will contract more than this. You ask what risk the council is willing to take with the investment in roads. The council responds that they are willing to accept a 50% risk. Again, from Figure 13.3, the 50% oil supply probability would represent a 40% reduction by 2050. We need to keep in mind that this kind of a decline in oil supply will probably not be evenly distributed across all end-use sectors. For example, will food production, construction, emergency response, and the military have the same proportional reduction as private vehicle transportation? Now we need to assess if the probability is greater than 50% that the number of vehicles will double and that 70% of all cars will be battery electric vehicles (BEVs). Long-range BEVs are more expensive than short range, so it is logical that a high proportion of cars could be electric only if the driving distances are small. You can present market research that shows a strong and growing demand for walkable neighborhoods and suggest that a low-risk investment would be in planning to redevelop the city for greater accessibility by walking and biking. The recommendations to the council would be that the smart growth plan is crucial and that the investment in increased road capacity has a higher risk than 50% of not being needed.

13.1.3 Depletion Issues

Consumption growth is not sustainable. Growth is always limited, temporary, and inevitably followed by decline. All resources are finite and limited. All growth requires resources. Thus, exponential growth in energy and resource use always has a peak and subsequent decline associated with one or more resource depletions. Equation 13.1 is a well-known Gaussian function, which is a good approximation of the annual production history, $P(t)$, of a finite resource of total initial quantity, R, which is consumed over a number of years, t, with the peak production occurring in year t_p when the inflection from growth to decline in the curve occurs.

$$P(t) = \frac{R}{\sigma\sqrt{2\pi}} \exp\left[\frac{-(t_p - t)}{2\sigma^2}\right] \tag{13.1}$$

where σ is a shape parameter of the Gaussian curve that adjusts the width for the total lifetime of the production. Equation 13.1 is also known as the production curve, the depletion curve, or the "Hubbert" curve because it was applied to U.S. oil reserves in the 1950s by petroleum geology expert, M. King Hubbert, who correctly predicted the 1970s' peak and decline in U.S. oil production [4]. As with the exponential growth function given in Chapter 1, Equation 1.1, the Hubbert curve can be used to gain quantitative understanding of resource and economic issues that usually become embroiled in politics and divisiveness.

Modeling of long-term effects of current decisions is a fundamental capability for transition engineering. Depletion curves can be constructed to understand the future availability of resources. However, the actual activity of extracting the resource and bringing it to the market is a function of economic factors and the regulatory environment. Consider the beginning of the extraction boom of some mineral like copper. The copper production initially grows, providing for demand growth, and the cost may even decline as production expands. The tailings from milling would be simply piled, and there would be little economic incentive for recovery and recycling. However, as the peak of the production curve is reached, the price climbs to the point where recycling becomes profitable, which begins to represent a new and significant production stream in the market. Also as new technologies are developed, the rare minerals in the tailings may become valuable and change the economics of copper mining altogether.

There can also be large-scale changes in the demand for a given mineral that reduce the production rate. For example, removal of lead from paint, gasoline, and now electronics together with regulations requiring recycling of batteries has accompanied reduced mining activities. Even for materials that cannot be recycled like fossil fuels, long before the last barrel of oil is extracted, the demand for the diminishing, and thus increasingly costly, resource will have also greatly diminished. We can't know for certain what the future economic, technology, and consumer demand situation will be, but we can use the depletion curve to understand that future growth in demand is not possible past the production peak.

Example 13.2: Exploitation of a Finite Resource: North Atlantic Cod Fish

For hundreds of years, the North Atlantic cod was a seemingly unlimited source of food and income [7]. The annual cod landings are estimated to have been below 0.2 ton per year, increasing slightly in the first half of the twentieth century. In the 1950s, new technology in the form of large factory trawlers and refrigeration equipment led to a dramatic increase in the tons of fish landed by the countries surrounding the fishery: the United States, Canada, Iceland, the United Kingdom, and Spain. New technology facilitated larger catches as well as access to much wider markets through an expanding transport network and refrigeration. The rapid growth of the industry meant booming growth of fishing industry towns, equipment suppliers, employment, and a rapid expansion in the fishing economies. The Canadian and U.S. fish management systems attempted to limit the catch to 20% of the resource per year, but the assessment of the resource became highly political, and it is now thought that the catch was actually more in the range of 60% of the resource. The science was too optimistic and was censored by the economic belief in unlimited resources and by politicians, whose main objectives were continued economic growth. The fish populations had fluctuated in the past, and there was a belief that the fish stocks would simply recover if catch was down in any given year. European trawlers were caught overfishing, and the new trawling technology actually ended up destroying the spawning grounds. A moratorium on cod fishing in 1993 has been followed by a permanent closure of the fisheries that were in fact destroyed and did not recover. The year that saw the greatest industry investment in bigger ships and new processing equipment to manage the massive growth in catch was 1968, the year that the fish landings peaked. There is no question that extracting all the cod from the North Atlantic made some people very rich and provided some jobs for a period of time. The political fight was focused on the figures produced by industry scientists regarding the total size of the resource in order to determine the size of the quotas. However, in retrospect, the only thing that could have saved the resource was regulation of the technology—basically limiting the size of the ships and forbidding the use of trawlers. How could that have been accomplished?

Transition Engineering Solution

Imagining that transition engineering was a mature field in 1949, what would the International Institution of Transition Engineers (IITE) have done to set standards for the fishing industry and issue permits for the manufacture of fishing vessels to ensure sustainable prosperity? In our revisionist history experiment, the IITE would have recognized the potential for a cod boom and called an extraction rush moratorium on manufacture of new fishing technology and new boat capacity until a production plan was established. The IITE would have gathered facts, species reproduction models, and data from scientists, marine biologists, and industry. The IITE working group on Fishing Technology would have developed a production standard for any new types of boats or equipment proposed by manufacturers based on balancing the catch capacity and externalities against the resource availability and ecosystem requirements. The special working group on cod would have learned about the spawning and lifecycle of the species and the ecosystem. They would have modeled the future fishery landings using Equation 13.1 as shown in Figure 13.4 using the potential catch rates facilitated by the new technologies. The model parameters are $t_p = 18$, $\sigma = 14$. The model shows that the peak value of the catch depends on what the assumed total resource value is, but the inevitable collapse of the resource does not. The destruction of the resource was ensured by deployment of technological capacity beyond the resource availability. The modeling

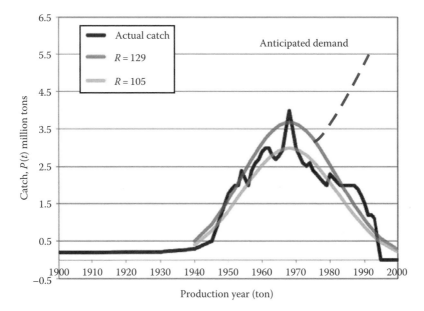

FIGURE 13.4
The actual annual North Atlantic Cod landings (Mt/year) and the modeled depletion curve for different values of total resource. Anticipated future demand drove massive investment in capacity just before the collapse of the fishery. (Data from the Food and Agriculture Organization of the United Nations, Rome, Italy, www.fao.org)

would have been used to set the standard for the maximum size of cod fishing ship that could be manufactured; it would have set a quota on the number of cod fishing ships that could be manufactured and placed in service. It would have also placed a total ban on the dangerous and destructive trawler technology. The cod fishery would have still been prosperous at any level (as it was during every year that it existed), and the policy makers would have adopted the IITE standards as regulations and could easily enforce the regulations through boat registrations. In our revisionist history solution, the cod landings may have climbed to 0.5 million tonnes per year, there would have been cod in the market today, and the IITE working group on fishing technology would have moved on to studying and regulating drift netting.

Clearly, alternative energy resources like wind and solar energy are not subjected to depletion effects. However, some of the critical materials, known as rare earth elements (REEs), are presenting challenges to expansion [8]. Alternative technologies like hybrid/electric vehicles require batteries and high-power magnets as do wind turbines. Solar photovoltaic (PV) panels and efficient lighting fixtures require dopants and phosphors. Table 13.1 gives some of the key REEs, their uses, and the price over the last few years. Rare earth mineral prices then collapsed by on average 70% in 2012 to about three times the 2009 price, due to a range of factors including release of privately held reserves. These elements are not rare in the sense of total quantity, but they are found in a limited number of deposits and there are challenges with extracting them from the mineral ore.

TABLE 13.1

Use of Rare Earth Elements in Alternative Energy Conversion, Air Pollution, Fuel Processing, and Energy-Efficient Technologies with the Price over the Past 3 Years Resulting from World Supply Shortage

Rare Earth Element	Uses in Energy-Related Technology	$/kg 2009	$/kg 2010	$/kg 2011
Lanthanum (La)	Lighter and longer-lasting La–NiMH batteries, catalysts for petroleum cracking, phosphors, H_2 storage	6.25	60	151.5
Cerium (Ce)	Efficient display screens, high-activity catalysts, lighter and stronger aluminum alloy, phosphors for fluorescent and halogen lamps	4.50	61	158
Neodymium (Nd)	High power, permanent NdFeB magnets, catalysts	14	87	318
Praseodymium (Pr)	Magnets for hybrid/electric vehicles, wind turbines	14	86.5	248.5
Samarium (Sm)	Magnets for hybrid/electric vehicles, wind turbines	3.40	14.40	106
Dysprosium (Dy)	Magnets for hybrid/electric vehicles, wind turbines	100	295	2510
Europium (Eu)	Magnets for hybrid/electric vehicles, wind turbines	450	630	5870
Terbium (Tb)	Magnets for hybrid/electric vehicles, wind turbines, phosphors for fluorescent and halogen lamps	350	605	4410

Very serious pollution issues have occurred in the places where REEs have been mined and milled in California and China. The eight elements listed account for 90% of all the REEs used. The elements are often found together in a mineral compound. Currently, nearly all of these elements are produced by China, which has set export quotas and applied new tariffs of 15%–25% resulting in shortages in 2011. Demand for REEs for magnets in motors was expected to increase by 10%–15% from 2010 to meet the growing demand for hybrid vehicles and wind turbines. A small amount of REE in iron increases the magnetic power, thus greatly reducing the size and weight of the electric motor. Most of these elements currently have no known substitute, alternative source, or potential for recycling. It may be that alternatives and new resources will enter the market to replace the Chinese supply. But it is important to note that even the supply of a material that is used in only trace amounts can represent an issue for energy technologies.

Oil supply is unambiguously a major issue. The historical approach to oil supply options is (1) drill deeper and farther to get more petroleum, (2) look to increase supply by developing alternatives and nonconventional hydrocarbons, or (3) develop alternative technology. It is clear that the producers of oil post-peak will see healthy profits, so option (1) has many proponents. Options (2) and (3) have great political and public appeal because they fit with our cultural heritage of technology development. However, increasing liquid fuel supplies through nonconventional sources is proving problematic. Enhanced oil recovery and nonconventional petroleum have lower energy return on energy invested (EROI) and higher environmental impacts. In Chapter 1, some of the issues around another alternative, biofuel expansion, were presented. Biofuels have a low EROI for ethanol to moderate EROI for biodiesel (EROI = 0.9–7.0). However, rapid development has led to food price rises and animal feed shortages. Increased water demand for ethanol processing is threatening crop water supplies. From an analytical engineering perspective, the quest for alternative transport fuel sources is not going to produce future growth in consumption and will actually not fill in the supply gap from declining conventional production.

The long-term prospects for continuation of the development patterns of the past are not looking positive if the only options available are extracting the remaining oil resources at a faster rate or finding alternatives. Of course there is also a segment of popular culture that responds to the peak oil issue with predictions of a total collapse of industrial society. Collapse is not an interesting engineering project, as engineers generally work to avert system failures, not just predict them. Ideally, we would like to manage resource exploitation to avoid the boom-and-bust pattern in order to gain much longer-term prosperity. However, many critical resources are already at or past the inflection point in the production curve. Transition engineering is emerging in the post-peak era to offer a fourth option for future development: (4) adaptation.

13.1.4 Price Issue

There is a long-standing belief that physical availability of a resource is not as important as the price of the resource. Classical economics began as the study of the growth of the wealth of nations and as the advocacy of policies to promote further growth. The study of the new economic system, capitalism that emerged to replace feudalism, observed that the free market of individuals acting in their own interest would establish a natural equilibrium between supply and demand. We should recall that before the mid-1700s, all economic decisions were either religious, based on tradition, or determined by aristocracy. For most of the historical period since the founding of the United States, there have been few physical constraints on growth. More recently, some constraints on resources have

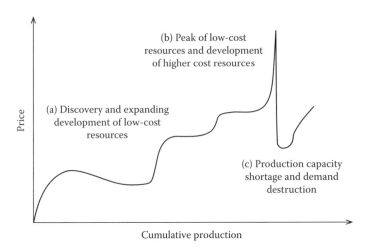

FIGURE 13.5
Commodity price vs. the cumulative production from natural resources over time.

brought in new models of how supply increases to meet demand, but economics still gives minimal consideration to actual physical limits, either voluntary or involuntary, and how policies could achieve prosperity within these limits.

Figure 13.5 shows the relationship between price and cumulative production of a commodity resource. Development stage (a) is the initial discovery and development of a resource, when there may be initial constraints in access to markets and extraction technologies. As markets are established, price stabilizes and production rate is high. Price may even trend downward with the expansion of low-cost resources. The cost–production relationship represents the "low hanging fruit" phase of development. Development stage (b) occurs when the conventional, lower-cost resources begin to decline. If the demand is inelastic to price, then a new higher price range for the commodity brings new, but more expensive production. This case describes what happened after the 1974 OPEC oil embargo, when the North Sea, Alaska, Mexico, and off-shore in America were brought into production, with the new threefold increase in the price range for petroleum. The post-peak development stage (c) is characterized by a series of jumps in price, which result in further cumulative production from increasingly costly resources. At some point, a shortfall in spare production capacity causes a price spike, resulting in demand destruction, which in turn causes a price collapse. The price collapse makes it uneconomical to resume production from the expensive resources that were developed during the previous price jumps. Then, the price climbs again as supplies remain tight. This situation is being played out in the oil and gas markets.

13.1.5 Diminishing Production and Declining EROI

The definition for EROI and explanation of the economic implications were given in Chapters 1 and 2 of this book. Diminishing EROI is an issue for fossil fuels and uranium, but also with the development of renewable energy because the energy return typically takes many years and there are limited sites for the most productive renewable resources, geothermal, hydro, and, to some degree, wind. Characterization of EROI is a growing area of study, like lifecycle cost assessment or carbon footprint, and more detailed analysis should become available in the future. However, we can understand the essence of the issue by the "low hanging fruit" analogy. The most profitable resources are always developed first. Even though there are significant resources beyond the easy-to-reach supply, more energy, equipment, and labor must be expended to bring the remaining resources to market. As we reach the "top of the tree," the economics may shift so that making better use of the remaining supplies is much more profitable than bringing new supplies to the market. Transition engineering will make use of EROI data to estimate the probability that a given new development will become available in the market. The most likely developments have both high EROI and cost savings. The next most likely developments have high EROI and low cost. The most unlikely developments have high EROI and high cost.

Declining EROI also means that future prosperity will not be possible without finding ways to reduce end-use consumption. When investments have a positive return, then the surplus can provide growth and prosperity. Growth means more buildings, more factories, more goods, and more energy-using appliances and vehicles that all support a higher energy-intensity lifestyle. Once more energy-consuming buildings and appliances are built, increased demand for energy is "locked in." If growth is focused on activities that consume energy and materials (and thus the embedded energy in extracting and processing resources), then there is a big difference between investments that are productive and those that are purely consumptive. For example, reclaiming and cleaning up a brown-field site provide improvement in quality of life, and building a vegetable hothouse produces food, but widening a freeway increases the capacity for more car travel. Some of the most important factors for prosperity do not necessarily require more energy or more material consumption: reduced infant mortality, improved public health, clean drinking water, good nutrition, more leisure time, higher education standard, and more access to arts, culture, recreation, social activities. Prosperity also requires that buildings, infrastructure, equipment, and cultural assets be maintained, that waste be processed, and that public and private places be kept safe and secure. We should note that warfare is a drain on both civilian growth and prosperity.

The cumulative investments in houses, buildings, roads, cars, trucks, air conditioners, and all of the other energy-consuming infrastructure over the past 70 years have resulted in an energy- and material-intensive lifestyle with very high energy demand. Over the next 70 years, EROI will decline [9,10], and fossil fuel extraction and production must be reduced to safeguard climate stability. There will be no net growth in energy supply from new production, yet maintenance and replacement of infrastructure and goods production systems require more energy.

The only reliable energy resource to secure essential goods and services is from reduced end-use consumption from nonproductive activities and goods. Figure 13.6 illustrates the energy balance for petroleum over the next 10 years as the EROI decreases from 25 to 15, and the production reduces by 25%. If end-use consumption is reduced by 30% from today as shown in Figure 13.6b, then growth will likely have slowed, but there will be sufficient supply available to maintain essential infrastructure and produce essential goods and services. This level of demand reduction could be achieved by energy efficiency improvements and consumption reduction in all sectors as highlighted in Chapter 1. Remember that consumption includes disposable materials and wasted food and water, which also have substantial embedded energy. Demand reduction can be achieved with lifestyle change and with adaptive designs that also improve quality of life. However, if there is no concerted effort to reduce demand, then Figure 13.6c shows the result, decline in prosperity. In this scenario, there is no growth, no energy available to maintain and replace aging infrastructure, and production would be impossible on an industrial scale. This illustrates the motivation for transition engineering projects, which focus on reducing end-use demand through efficiency, conservation, adaptation, redesign, and redevelopment.

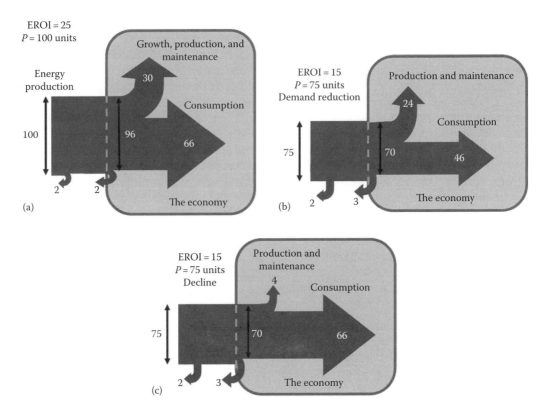

FIGURE 13.6
Illustration of petroleum production trends from the conditions of today (a) with aggregate EROI = 25 on the basis of 100 units of production, and two alternative development scenarios for 2022 when according to Figure 13.3, conventional production is reduced by 25% and EROI = 15. (b) Demand reduction scenario where nonessential and nonproductive consumption has been reduced in order to maintain prosperity. (c) Decline scenario where consumption has been maintained, but infrastructure and essential services are in decline.

13.1.6 Adaptive Capacity

Adaptive capacity is defined as energy and resource demand reduction without affecting the essential activities, goods, or services. Figure 13.7 illustrates the meaning of adaptive capacity and the different ways that demand declines with increasing pressure for change. Adaptive capacity of a system can be measured by various means including auditing, surveys, and modeling. A particular system has a threshold for demand reduction without impacting standard of living. For example, natural gas for heating may have adaptive capacity of 100% in San Diego without the risk of any freezing deaths, but in Minnesota, adaptive capacity of 20% might be the limit for current housing stock. Demand reduction below the adaptive capacity in Minnesota could be accomplished, but there would need to be redesign and redevelopment of the housing stock to improve energy efficiency. The most economically efficient projects involve adaptive designs that can both increase the adaptive capacity and develop the transition to lower demand in ways that produce multiple benefits including cost savings. In Minnesota, this might mean focusing on the oldest homes and bringing them up to the best standard.

Survival depends on adaptation. Species that do not adapt to changes in their environment must either move to an environment in which they can survive, or they will die out.

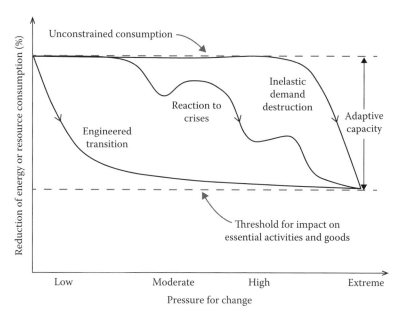

FIGURE 13.7
Adaptive capacity is defined as the energy or resource demand reduction in a system that can be realized without negatively impacting essential activities or goods. As supplies decline, systems will necessarily change in response to pressures like price or shortages. The objective of transition engineering is to achieve the demand reduction while realizing multiple benefits.

Adaptation is defined as changes in response to environmental change [11]. Successful adaptation is achieved by balancing benefits and risks. There is a risk to changing something that has previously worked well, as the result may not be better. But if the change produces clear benefits, then adaptation can occur quickly. In our economy, tight fuel, gas, and electricity supply causes price pressure. However, pressure that causes change can also come from policies, international events, taxes and regulations, as well as media coverage of scarcity or environmental issues.

Price elasticity is normally measured in percent change in demand for a given percent change in price. The way that households respond to fuel supply issues is complex and has been shown to depend more on perceptions than on actual price change. Firstly, the most dramatic and long-lasting fuel demand reductions in the United States occurred after physical fuel shortages and extensive media coverage of people waiting in line for hours to get a rationed quantity of fuel. This "panic" response appears to include reevaluation of travel modes, housing locations, and vehicle fuel efficiency [12]. A recent comprehensive review of data from 1968 to 2008 has identified key factors in personal transport activity and fuel consumption elasticity [13] as fuel tax, road tolls, parking fees, registration fees, distance-based insurance, or usage fees. The study showed that price increases that people considered to be durable had much higher effect on fuel demand reduction than market fluctuations.

The issue of peak oil may be having an influence on the short run fuel price elasticity. The fuel taxes in the United States have been reducing as a proportion of fuel price, but the elasticity has been increasing as oil supplies contract and as popular media coverage of the issue of peak oil has been increasing. Table 13.2 shows modeled elasticity in the United States for retail sales of transport fuel since the world oil production began to plateau in 2004 [14]. The increase in elasticity means that a price increase today causes a seven times greater reduction in fuel demand than before 2004.

TABLE 13.2

Short-Run Elasticity of Personal Transport Using Simple Model

Year	Average U.S. Fuel Price Regular Gasoline ($/gal)	High Weekly U.S. Fuel Price Regular Gasoline ($/gal)	Transport Fuel Elasticity (%Δgal per %Δ$)
2004	1.81	2.03	−0.04
2005	2.24	3.04	−0.08
2006	2.53	3.00	−0.12
2007	2.77	3.21	−0.16
2008	3.21	4.05	—
2009	2.32	2.66	—
2010	2.74	2.76	—
2011	3.48	3.91	−0.29

Sources: Komanoff, C., Gasoline price-elasticity spreadsheet, www.komanoff.net/oil_9_11/ Gasoline_Price_Elasticity.xls, 2008–2011; EIA, Gasoline and diesel report (1990–2012), http://www.eia.gov/petroleum/gasdiesel

Oil is used for manufacturing and as a feedstock for materials. There are currently virtually no substitutes for petroleum products that can be developed at a rate to replace the declining oil supply. Oil is also used in primary production for everything from fishing to farming and construction. There are currently no substitutes for diesel fuel that do not require diesel inputs and that can become available in the market at a rate equal to the supply decline rate. Thus, we can conclude that the largest demand reductions will be in personal transportation.

The adaptive capacity for personal transport is a function of geography, age, and availability of options that use less fuel [16]. The adaptive capacity for freight movement is a function of geography of primary production, market systems, logistics systems, and the availability of optional modes that use less fuel. The long-run adaptive capacity of the whole transport sector is critically dependent on network, infrastructure, and land-use decisions of the past. Figure 13.8 shows how the total energy consumption for a given city depends on many underlying and historical factors. It also illustrates an important fact for transition engineering work. While individual choices, lifestyle, and behavior are important factors in energy demand, travel choices can only be made in the context of the built environment, transport infrastructure, and vehicle technology available. The person km traveled or vehicle km traveled has a large influence on energy demand, as do the frequency of travel, the travel distances, and the modes of travel. However, all of these factors in turn are greatly affected by land-use patterns, the layout of the transportation networks of roads, trains, and walking and cycling paths, and the relationships between origins and destinations of the urban form.

13.1.7 Essentiality

Adaptation will have impacts that depend on how important the activity, services, or goods are to well-being. We need to recognize that not all consumption is equal. If we do not have the capacity to adapt, then some functions of the system will fail due to unsustainable processes. Loss of essential activities or goods is considered to have a high impact, while loss of an optional or discretionary activity or products would represent a low impact. Loss of an essential good would negatively impact standard of living. Loss of a necessary good would negatively impact quality of life. Loss of an optional good would alter lifestyle without causing any reduction in health or welfare. Essentiality can be measured for any type of end use and incorporated into the risk assessment analysis and adaptive design [6].

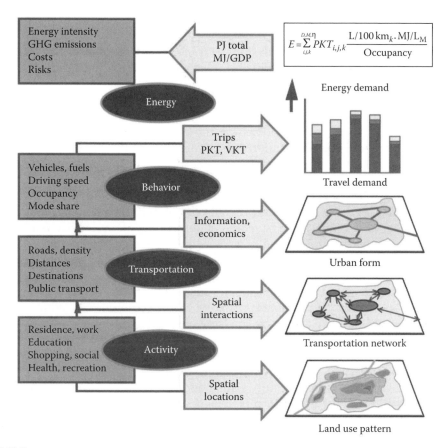

$$E = \sum_{i,j,k}^{D,M,\eta} PKT_{i,j,k} \frac{L/100\,km_k \cdot MJ/L_M}{Occupancy}$$

FIGURE 13.8
Energy demand for personal transportation depends on underlying factors that influence and limit choices and travel behavior.

CHARACTERIZATION OF ACTIVITIES, GOODS, AND SERVICES

- *Essential*: Loss or disruption would cause reduction in standard of living and reduction in well-being
- *Necessary*: Loss or disruption would cause decline in quality of life and possible reduction in well-being
- *Optional*: Loss or disruption would cause contraction in lifestyle and consumption but would not affect well-being

---

13.2 Anthropogenic System Dynamics

Thus far we have argued that there is a social mandate for the emergence of transition engineering to improve long-term sustainability in a way that is analogous to SH&E Engineering reducing workplace casualties. The case was made that safety, security, and sustainability are different scales of survival, not limited to any particular manifestation, but unambiguously defined by failure. We have seen that scientists identify the issues of unsustainability and that we can use meta-analysis of the range of scientific studies to provide a future probability space for the manifestation of the issue. We reviewed the Gaussian function as a model for consumption of finite resources. We then reviewed the issue of EROI and how depletion of higher EROI resources will cause the end of growth and decline of standard of living if we do not redesign and redevelop to reduce demand. We then looked at the adaptive capacity of end-use systems to reduced supply, and the implications of essentiality for the impact of supply risks. The purpose of this section is to provide another fundamental principle of transition engineering: engineered systems and technology determine the energy-consuming behavior. This may seem straightforward, but the conceptual framework presented is actually a significant departure from conventional thought about supply and demand and energy systems engineering. The model of anthropogenic system dynamics should help transition engineers understand complex systems in engineering terms.

What are the things that prevent our society from being sustainable? Your first answer might include politics, short-term thinking, economics, greed, or consumer behavior. These would be popular answers. Did you include in your list the purposeful decisions of the vast majority of people to not be sustainable? That would not seem logical. What civilization would purposefully choose collapse, by choosing unsustainable development activities? If you read media coverage of sustainability issues, you should get a pretty good sense that society does not want to experience failures of its energy, food, water, or ecosystems. In fact, this desire of the society to survive is what we have called the social mandate for sustainability. So what will change in the future so that our development proceeds on a long-term trajectory for sustainability? People in general believe that new technologies will be developed that will provide solutions. This expectation may be reasonable given the amazing advances of technology over the past 70 years. However, there is a real chance that we could be put at greater risk if we pursue technologies that are actually false hopes, have low return on investment, delay progress on effective changes, or cause more problems in the future.

13.2.1 Conventional Open-Flow Model of Energy Systems

Wind turbines, cars, and refrigerators are designed and manufactured like other products using the standard engineering approach. However, energy conversion devices are part of the whole energy system that extends from the natural resource to the end user. Figure 13.9 gives a schematic diagram of the conventional view of energy systems as

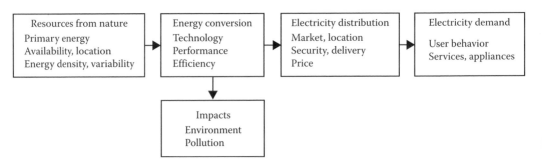

FIGURE 13.9
Open-flow model of the electricity generation system where the arrows represent energy and material flows.

open-flow systems. Energy is extracted, converted, transported, distributed, and used. Engineers know that the problem with an open-flow system like an irrigation system is that there is no feedback control. The only way to keep it working is to make sure that the inflow is always large enough that it is capable of meeting the outflow. In other words, the necessary condition for stable and secure operation of the open-flow energy system is sufficient spare supply capacity and unconstrained resources.

13.2.2 Feedback Control System Model

For most of the history of energy systems, the open-flow model was a reasonable framework. However, engineers should know very well the capabilities and advantages of feedback control in dynamic systems. Modeling, analysis, and control of dynamic engineering systems are accomplished through application of control system theory [17,18]. Control systems maintain the designed operation and system stability as long as the operational parameters of the system are not exceeded. How well a system responds to disturbances and to feedback parameters is a measure of the robustness and reliability of the system. The control system theory is a fundamental representation of dynamic system behavior, which can be applied, in principle, to mechanical, electrical, biological, and ecological systems.

Figure 13.10 shows the basic block diagram describing a feedback control system. There are various configurations for such a block diagram model, depending on the particular system of interest, but this representation includes most features according to the standards of the IEEE. With the exception of the external inputs/outputs, the arrows in the block diagram do not represent flows of materials, but are communication signals between dynamic elements. The control system is an integral part of a continuously operating dynamic system, for example, a cruise control system on a car traveling on a freeway.

The directive represents the motivating input to the system. It is independent of the actual performance, but expresses the required condition for system performance. The reference elements establish the values of the reference signal calibrated for a particular system in a particular situation. The reference signal is the particular form of the directive, which is directly useful to the system. The comparator performs the function of determining if the reference signal is equal to the feedback signal. If the signals are equivalent, then there is no change needed in the system in order to meet the operational directive.

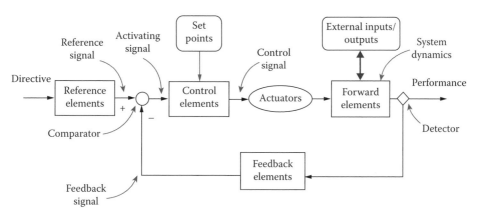

FIGURE 13.10
Feedback control system basic block diagram showing all of the major elements and signal paths.

Changes can occur for other reasons, but no changes are necessary. The activating signal to the control elements is used to determine operational changes that will push the system performance toward the reference signal values. The control elements generate the control signals according to the magnitude of the activating signal and according to pre-existing control design, strategies, and other set points. Control actuators cause physical changes, through existing connections and actuators. Forward elements represent the physical plant and the system dynamics that react to the actuators and affect the performance of the system. The performance is measured by detectors and represents the actual system behavior. Feedback elements translate measurements of the system performance into the feedback signal, which has the same calibration as the reference signal and thus can be used by the control elements to attenuate the system behavior.

In the example of an automobile cruise control system, the directive would be the desired speed set by the driver, the performance would be the actual speed, the set points would be relationships between controller signals and automobile dynamics, the actuator would be the fuel supply throttle and brake fluid, the external inputs would be fuel and air, and the forward elements would be the entire vehicle, including the engine thermodynamics, transmission, aerodynamics, and wheel tire friction with the road. This example illustrates how important the signal processors are in sustainable operation of a system. The driver sets the directive of a desired speed that is safe for the present driving conditions. That speed in the mind of the driver cannot be used by the microprocessor in the car unless it is first processed into an electronic reference signal. The measurement of vehicle speed is achieved through a transducer that produces an electronic signal that would not be recognized by the driver as speed unless that signal is processed through calibrated electronics, such as a Wheatstone bridge, and then sent to a speedometer and displayed as mph. Another important concept is that the controller acts in predictable ways in response to the control signal that is not the actual speed, but the difference between the reference and the feedback signals. The cruise control system for a given automobile was designed and calibrated, and the microprocessor controller was programmed according to the automobile specifications. It is vitally important to understand that the controller can only actuate the system according to the existing design and cannot operate the system in a way to change its design.

13.2.3 Feedback Control Model of Regional Energy Systems

The complex behavior of energy/economic/social/environment (EESE) systems can be understood by modeling the regional energy system as a feedback control system. The theoretical model shown in Figure 13.11 is a general description of an anthropogenic system irrespective of social structure, type of government, economic philosophy, or technological level. The model describes the structure and behavior of a continuous anthropogenic system at any particular point in time. Civilizations that have maintained a particular infrastructure, resource consumption, and activity level for several hundred years without unsustainable impact on the environment can be seen to have a robust, high-stability system, with all of the components and signal pathways working effectively. Societies that have collapsed due to environmental exhaustion can clearly be seen to have dysfunction in some part of the system. The model can also be used to understand an energy subsystem like transportation or residential heating.

The arrows represent information, connections, or actions, not energy or material flows, except for the system resource inputs and environmental impacts. We define the EESE system as any community of people, their relationships with each other through economic

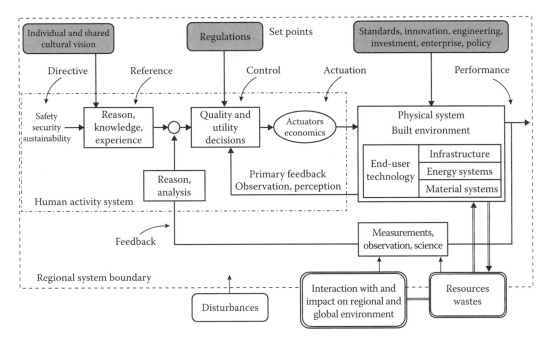

FIGURE 13.11
Conceptual framework of a regional energy/environment and anthropogenic system modeled as a feedback control system.

activities, the infrastructure that they use in these activities, including appliances, buildings, etc., within a given environment and resource setting. The model represents the dynamics involved in carrying out the society's activity systems at any point in time. Considering the example of the earlier automobile cruise control system, the same control theory model would apply to either a Ford or a BMW. However, the particular system model for each vehicle design would be unique.

The directive for society represents a shared cultural vision that transcends any details about the actual resources used, technology, or any particular activities. By logical deduction, if the regional energy system is safe, secure, and environmentally sustainable, then it can continue year after year. The reference signal represents sustainable resource consumption and environmental impacts of a particular community carrying out their normal activities using particular technologies. The higher-level directives are processed into the specific reference signal by means of the knowledge, reason, and education of the society. In preindustrial societies, sustainable, safe, and secure levels of consumption and impacts to support a certain level of activity were determined through experience and adherence to traditional practices. For example, the Anasazi civilization of the Southwestern United States maintained a civilization of dry-land farming of squash, maize, and beans, together with hunting and gathering of wild foods for nearly 700 years with almost no technological or resource consumption change [19]. Paleoanthropologists believe that the Anasazi, like most aboriginal people, had culturally integrated ideas concerning the use and allocation of natural resources. Their "way" of carrying out their activities was taught to new generations as part of a long tradition of "what works" in a particular environment.

There are usually several different feedback signals in control systems: primary and general. Primary feedback includes experience about price, convenience, functionality, and habits. The primary feedback is the main source of information for system control because it relates directly to particular activities. The general system-level feedback represents information about the aggregate performance and impact of activities on the environment and interactions with other regional energy systems. This information is usually not directly related to any particular actions by individuals. For example, the variation of river flows to match electricity demand from hydro generation has been shown to have an impact on riparian ecosystems. This is an aggregate effect that is caused by consumption behavior, but currently this information has no mechanism for changing individual behavior.

The comparator continuously evaluates feedback of actual measured consumption and impacts against the reference levels. Obviously, this type of analysis would require continuous monitoring of resource and environment systems in relation to human activities together with the knowledge of sustainable levels of consumption and impacts. It is simple to see how this would have worked for sustainable preindustrial societies as the impacts of people's activities on local resources would have been observable and understandable to people who relied on those resources for survival. In his book, *Collapse*, Jared Diamond sets out the theory that some societies choose to continue activity systems that lead to environmental collapse even though the problems were observable [20]. The regional energy system model can be seen to accurately represent this type of behavior. For example, Diamond describes the behavior of the Greenland Viking colonists who continued their shared cultural vision of behavior and resource utilization, even though it did not fit with the Greenland environment. He explains that in a land teeming with fish, the people starved to death rather than break with their traditions as cattlemen and hunters.

At this point, it may be evident that our modern society is in a similar quandary, where our shared cultural values and vision are not reconcilable with environmental sustainability. The public receive more and more feedback information about global climate change, for example, but that information does not have any existing control options and so the information cannot produce a signal to activate the controller to bring the system back into a safe, secure, and sustainable mode of operation. Clearly, there is a great opportunity for new technologies and systems in this area. However, if the built environment depends on unsustainable consumption, then the unsustainable condition will continue, even if the feedback signals are improved. This gets us to the main point of the anthropogenic system model of energy systems—the physical system must be redesigned and redeveloped for sustainable operation by the designers. Choice of the consumers or the regulators cannot make a system work in a sustainable manner when it was actually designed for unsustainable operation.

The controller for the EESE system is the aggregate decisions that individual people or organizations make on a continuous basis, given their existing knowledge and experience about what actions will provide the desired performance. An important input to these decisions is the primary feedback of their own experience and perception. For example, people have knowledge from their previous experiences of how to successfully carry out their activities. This body of experience and habit together with analytical capabilities provides essentially the same function as the electronic microprocessor in the cruise control system. The controller has preprogramming about the system signals and how to adjust fuel flow or apply braking in order to maintain the set speed. In the same way, individuals, and in the aggregate, the society, have a predetermined set of rather routine activities, and sufficient information and decision-making capabilities to manage variations in the patterns of daily life. Yet again, it is clear that very few of the individuals or organizations in a region will have the experience or the knowledge to redesign and redevelop the built environment to reduce unsustainable processes. We have seen recent developments like the Transition Town movement [21], where some people are breaking away from the established system and looking for eco-villages or self-sufficiency in response to information about the unsustainability of oil supply or CO_2 emissions.

A person's daily routine is essentially a set point. Our daily routines are formatted within the rules of governance and social behavior of our society, and utilize technologies and energy supplies that are already available. Except for extreme cases of totalitarian governments, the rules and regulations and acceptable social behaviors are, to some degree, expressions of the shared cultural values. This view of system control may represent a radical departure from the supply and demand economics view. The control theory indicates that people's rational behavior today is based mostly on the activities that were successful yesterday in meeting their needs. The primary factor in determining behavior is what people already know, not increasing their satisfaction through consumption. This type of control is true for any manifestation of built environment or technology level. In the control system model it is conceivable, indeed it is imperative that demand be controlled to match a safe, secure, and sustainable supply.

All of the activities in a community involve economic participation. Economic relationships between providers and consumers are the actuating elements of the system. Popular opinion might be that cost drives people's decisions about consumption. However, the control system model indicates that economic relationships are actuators that determine *how* people access the goods and services they decide to purchase to meet their needs and quality desires, not the *reason* they have desires or participate in activities. There may someday be a transition engineering innovation that conveys information about the resources being used to provide electricity, and people may develop a habitual pattern of minimizing end-use activities at peak times so that the system can maintain secure supply and reduce fossil-fueled generation. People might do this based primarily on the reference information that relates this cooperative behavior to the sustainable use of resources, and not based just on cost.

The physical system represents the generation technology, transmission circuits, appliances, and built environment that the community uses in the course of going about their normal activities. The physical system is basically the open-flow system shown in Figure 13.9. Some things like power plants and school buildings have collective uses, and others, like heat pumps and refrigerators, are used by households. If the built environment were designed to function within resource and environment constraints, then people's daily activities and decisions would be within the context of a potentially sustainable system, and their pursuit of individual desires within a free market would not necessarily threaten the environment [22,23].

We have already established that people learn what works to support their activities in a given built environment and then act rationally by using what they have learned. For example, consider a tourist from Dallas who visits Amsterdam. The tourist would quickly learn that driving a car in the way that is normal in Dallas is not a successful way to go out for meal compared to walking or taking the tram in Amsterdam. When the tourist rides the tram into the old city, then walks down the street to a café in Amsterdam, is he behaving to increase utility, or to reduce cost, or to reduce environmental impact? Or is the tourist behaving rationally given the physical system he is in because the tram and walking work better than driving? The point is that the built environment and the technology that already exist at a given time determine the behavior of the population.

The activities of people in a region require material inputs from the environment both from inside and from outside of the system boundary. The activities in the region may also produce products and wastes that move across the system boundary. Control system theory deals with these externalities as material inputs and outputs to the physical system. For the example of the cruise control system, the level of fuel in the tank is not a part of the speed control system. However, the constant speed can only be maintained as long as the fuel continues to flow to the engine.

Activities and technology can be changed in many ways, which we term disturbances, even including innovation and technology development. At any given time, the existing built environment and appliances are used as intended; people prepare their dinner, children read books, cafes serve coffee, factories make goods, etc. If a new regulation, a higher price, or a new behavioral pattern becomes part of the system, then it has essentially "disturbed" the original system and can represent adaptive changes. Changes in the physical system cause adaptation in the control and actuation. Thus, we conclude that the role of transition engineering is to provide the mechanism for adaptation of the different systems and subsystems, and the people will learn to use the new system at least as fast as a Texan learns to tram in Amsterdam.

*13.3 Risk Management

Sustainability of energy supply, natural resources, and the environment represent serious risk management issues. The nature and magnitude of the issues are unprecedented. The issues are complex, involving interrelated systems of government policy and regulation, technical innovation and professional standards, social behavior and lifestyle, economics, and affordability. We also must consider health, environment, equity, population, security, and safety. Energy engineers typically work on energy conversion technologies. However, it is imperative that energy engineers have knowledge and understanding about the whole system, both supply and demand, and including the expectations of society, the adaptive capacity, and the risks to essential services. Carrying capacity was discussed in Chapter 1. Clearly, there are risks associated with exceeding carrying capacity. Figure 13.12 illustrates how those risks escalate as growth pushes toward the carrying capacity for population, resources, and the environment. There are many historical examples of the problems encountered by societies that have exceeded carrying capacity in one or more respect.

Risk management is familiar in engineering practice. It involves the identification of issues, assessment of probability of occurrence, and quantification of impacts if the issue arises. Safety engineering involves identification of hazards (e.g., unsafe processes) and instituting changes to reduce the risks. Environmental engineering similarly works to identify, quantify, and reduce harmful emissions. However, greenhouse gas emissions and long-lived accumulating substances like mercury from burning coal are presenting new issues. The benefits of burning coal are immense, but the risks are on a global scale and long into the future. The immediate benefits of extracting and using energy are obvious. The risks of extracting and using energy are normally thought of in terms of the environmental impacts. However, security of energy supply and depletion of resources are also critical sustainability issues that have political, economic, and social costs that often do not become evident until much later if only economic observation is used.

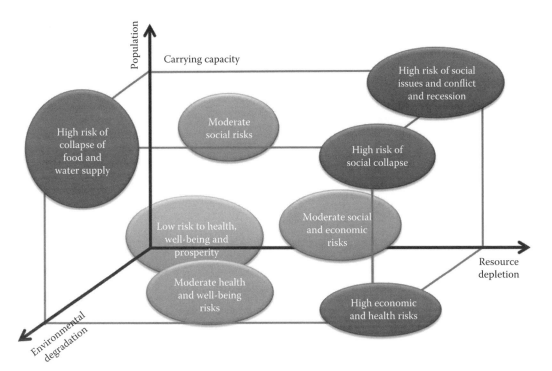

FIGURE 13.12
Growth that pushes population, environmental degradation, or resource depletion toward the physical limits poses social, economic, and well-being risks.

13.3.1 Supply–Demand Gap

The supply–demand gap is the most urgent issue that causes pressure for change in existing systems. The impact of the change depends on the adaptive capacity, readiness, resilience, and recovery of the system. Adaptive capacity depends on design, and readiness and resilience depend on forward planning and mitigation measures. Hirsch presented a "wedge" analysis for future oil supply to illustrate the effect of mitigation measures and the resulting supply shortfall [24].

The method is to start with the historical demand/supply of a critical resource and then extrapolate the demand growth into future years as shown in Figure 13.13. The actual availability is then estimated for future years using the depletion curve for finite resources or sustainable production estimates from scientific observation for renewable resources. The difference between the future supply and future demand is called the supply gap. The capability of different mitigation measures to "fill the gap" is estimated by assuming development, deployment, and uptake into the future. The remaining supply–demand gap is then clearly a pressure for change and will cause adaptation of extrapolated demand growth to the reduced available level. As mitigation measures take time to develop and implement, the timing of the gap analysis can be used to convey urgency to decision makers. The mitigating factors in the wedge analysis are portrayed as increased supply. The wedge analysis does recognize that there could be a peak and decline in the currently developed resource. However, by extrapolating a continued increasing demand, the wedge analysis misses the point of the depletion effect that growth is not a permanent condition, nor even an option once past the resource peak.

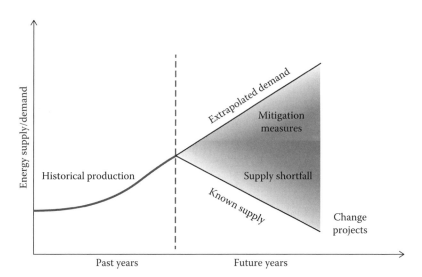

FIGURE 13.13
Wedge analysis of the future risk posed by the supply shortfall with extrapolated demand when possible mitigation measures are considered.

The supply–demand gap analysis conducted by Hirsch in 2005 concluded that there were no mitigation measures that could avert a shortfall in supply within the decade thus the result would be demand reduction. The mitigation measures examined included improved efficiency of vehicles, biofuels, and other synthetic transport fuels. If one looks at the extrapolated future demand from past IEA World Energy Outlook documents, it is clear that supply shortfall has not been mitigated and that demand has plateaued along with supply. The greatest risk of the supply shortfall is failure of systems due to energy shortage. There are also financial risks if infrastructure and business assets have been designed and built to accommodate the extrapolated demand growth. For example, a city may invest a lot of borrowed money to expand its airport in anticipation of the growth in visitor numbers in the future and the increased tax revenues they would generate. Investing in growth that does not occur represents a risk of debt accumulation.

13.3.2 Risks of Energy Myths

Myths are shared cultural stories, but they are also commonly held false beliefs. We have learned from the history of safety engineering and other risk management fields that being honest with stakeholders and the public is paramount. Perpetuating green energy or clean energy myths is not honest. Energy myths are hazards. The main risk is inaction on change and adaptation because the energy myth promises new energy sources or technology options that will not actually materialize. Energy myths can also take up research and development funding that could otherwise be used to work on transition change projects.

All myths and commonly held false beliefs have some anecdotal evidence or some believability. Myths are also very difficult to dispel once they become useful. Human society is susceptible to believing in myths, especially when the myth offers a solution to a complex problem. There are countless examples one can think of including all of the superstitions of the past and even historical medical cures. Imagine you were time-transported to 1590 and you got the flu. A doctor might be called to purge you and bleed you. You might want to decline this treatment because you know it is not really a cure. But your hosts would not be receptive to you questioning their only hope for medical treatment. They could all give you examples of when the cure worked. A recent example is the Y-2K myth that held that all of our modern systems were at risk because at the stroke of midnight on January 1, 2000, all of the computers in the world could fail. Computer experts around the world tried to explain that this would not happen, but at the time the myth could not be dispelled.

Consider that it is just as hard to dispel green energy myths that have been established in the popular culture. The hydrogen economy is an example of a green energy mythology. At a time when people were starting to become concerned about future energy supplies, the hydrogen economy mythology developed to provide a comfortable technology answer to an uncomfortable and uncertain future [25]. Hydrogen generation, storage, and fuel cell technology have taken up large amounts of research funding, even though relatively simple energy balance analysis shows conclusively that the very premise of the hydrogen economy is flawed. Hydrogen and fuel cells are known technologies, and alkaline fuel cells are used on the space shuttle. There are researchers working on hydrogen technologies. These truths are powerful reinforcement for the myth. Like other green energy myths, the hydrogen technology will always hold promise of overcoming our problems of dependence on foreign oil, will always be about 10 years in the future, and will always require more research to overcome the hurdles.

Important green energy supply mythologies include biofuels, wind, and solar PV. Of course, these technologies are currently viable, and they are being deployed and experiencing growth. The mythology of a transition to renewable energy is the promise that renewable energy can substitute for fossil fuels without massive demand reduction. Carbon capture and storage (CCS) is a high-risk environmental mitigation myth. The CCS mythology is that capturing CO_2 from combustion of coal or gas, compressing it, and sequestering it in geological formations deep underground will be a mitigating technology for the risks of climate change [26]. There are kernels of truth; CO_2 stripped from oil or gas is, in a few instances, compressed and injected into depleting oil fields to enhance the oil recovery. The amine temperature swing CO_2 absorption process is used on lean diesel combustion products to produce CO_2 for beverage carbonation. But there is no possibility that CCS will reduce the CO_2 emissions from the world's existing coal-fired power plants. The CCS myth is a risk because it delays work on changing demand to reduce coal combustion.

One of the most persistent myths is alternative personal vehicle platforms. The myth is that new vehicle platforms will develop so that the number of cars on the road and the amount of driving will continue to grow even as oil supplies decline. Again, there are some true pieces of the story—hybrid vehicles are available, although they make up less than 3% of new car sales and are not responsible for increasing vehicle numbers. American and other car companies are developing BEV platforms. Ethanol is available at many filling stations. However, the mythology is that electric cars will "replace" petroleum vehicles and cause a revolution to the economy as the power grid uses electric cars to store renewable energy [27]. This mythology is a problem because it encourages the continuation of the massive spending on expanding America's road and highway capacity to try to keep ahead of congestion. In fact congestion has not increased over the past decade and has been declining since 2008. The transportation systems of the country need to be adapting, along with the urban landscape to more walkable and lower traffic patterns. This will require a huge shift in thinking from designing and building the city infrastructure around personal vehicles powered by petroleum. It may be possible that electric vehicles have some role in the cities of the future, but considering the on-board energy density shown in Figure 13.14, it should be clear to any engineer that the average and maximum

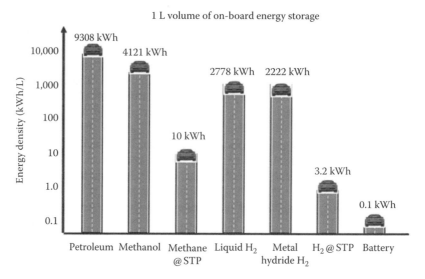

FIGURE 13.14
Comparison of on-board energy density for a range of personal vehicle platforms.

trip distance for electric vehicles must be greatly reduced from that designed for petroleum vehicles. This can only be done by large-scale change in land use and urban form.

The green energy mythologies are popular and become entrenched because they offer simple solution concepts to very complex and difficult problems. The transition engineer needs to have a solid factual understanding of all of the energy conversion technologies and all of the end-use systems. Many engineers are currently working on developing alternative energy and efficiency technologies. Ideally, the ideas of transition engineering will emerge to the point where the new profession becomes a way that all of the alternative technologies, end-use systems, behavior, policy, regulation, and economics can come together to begin developing adaptation projects based on hard realities and difficult choices concerning complex systems with a perspective including future prosperity. The methods for this kind of integrated approach are the subject of the remainder of this chapter.

*13.4 Framework of Change Projects

Engineering change projects are important for products, manufacturing, services, and infrastructure. Change projects can be motivated by risk or by opportunity. Change projects are always complex. Complex means there are many interrelated systems and subsystems, all with different dynamics. We must break down complex systems into component parts and subsystems, and we have to keep track of the interrelationships. Through continued research, we aim to understand, model, and ultimately predict the behavior of complex systems. A commercial or military aircraft is a complex system. Engineering change projects such as developing the next product version of software or the next version of a laptop computer are also known as transition engineering projects. Transition engineering change projects have primary objectives related to engineered systems or products, but another aspect of the project is the integration with the complicated human world.

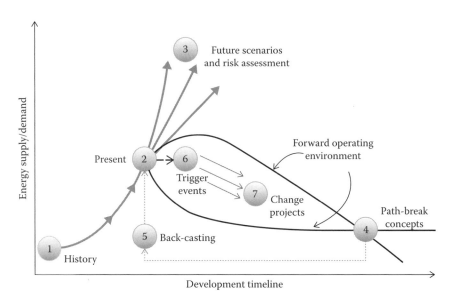

FIGURE 13.15
Diagram illustrating the transition engineering process for developing change projects in any given energy system.

The transition engineering approach to sustainable energy uses the product development, or change project approach illustrated in Figure 13.15. The seven steps are all important pieces of work that together lead to the best change projects. People typically work in isolation on different steps of the overall process, and they often have a perspective that their work alone is the answer to the question of sustainable development. For example, the study of peak oil is one part of this process as is the IEA's World Energy Outlook, where they develop different scenarios about the future. However, just like the old folk tale of the seven blind men and the elephant, different people working on different parts can be right about their own observations without having a clear vision of the whole system and how the parts fit together.

1. *History*: The first step is to gather all the facts and historical data. It is also important to work on understanding the system dynamics by observing past behavior. In energy systems, this means looking at both historical energy supply, end-use technology, and consumer behavior. Government agencies collect historical data, and an important transition engineering project is to look at the effects of different technologies or social developments on energy and resource demand.

2. *Present*: The second step is to take a stock take of the current position. What are the current investments, assets, and liabilities? What are the current capabilities and what do we not know? The stock take of current energy systems would include the age and condition of physical plant and supply chains, but also energy auditing and end-use behavior assessment. Adaptive capacity and resilience are also important to assess. Energy audits are an important tool for characterizing present energy demand, essential activities, and adaptive capacity.

3. *Future*: The third step is the use of scenarios to understand continuation and adjustment of current trends into the future. A large number of economic analysts are currently employed developing future scenarios of energy demand, many of which

are not well informed by technical reality or resource availability. An important aspect of the transition engineering framework is to characterize the forward operating environment which includes developing an energy or resource availability envelope as appropriate for the particular system. A finite resource is consumed at a certain rate each year. The consumption pattern is often depicted as a bell-shaped curve. However, it is possible to use less of the resource in the near term and leave some for later. The top line represents the aggressive production scenario where the resource is consumed at the fastest rate possible at all points in the development timeline, much the way bacteria populations flourish to consume a food source. The lower line represents the conservation-focused scenario where the resource consumption drops dramatically, but then continues on at a lower level for much longer than the production-focused scenario. Another part of the future scenario development is risk assessment. The risk to essential activities and goods is a result of the future expected and "locked-in" demand not being met by resource availability. Development scenarios within the forward operating environment resource availability envelope have managed risk of unsustainable operation. The adaptive capacity of the system and time for the adaptive capacity to be exceeded according to different scenarios are important in understanding possible trigger points.

4. *Path-break concepts*: The fourth step is the innovation process where path-break concepts are generated. As in product development, if all the conceivable modifications to the design do not meet the requirements, then a new perspective is taken. The path-break concepts are normally targeted to meet the critical requirements, but are not limited by any other current thinking, and so require highly creative and innovative thinking. Concept generation is not about modification of the existing system. The greatest challenge to the inventor is firstly to accept that there is no solution using any adaptation or modification of the existing system. The inventor needs to be able to mentally "let go" of the past, the present, and the possible scenarios forward. The inventor then needs to be able to accept that a solution exists and mentally set out to discover it, guided only by the system requirements, physical reality, the art of engineering, and unconstrained creativity. In energy systems, the critical requirement is that the energy system must be able to function using the amount of finite fuel available plus any renewable energy suited to the system loads. The other requirement for path-break concepts is that they must be technically feasible (e.g., no science fiction or green energy myths!).

5. *Back-casting*: The fifth step is back-casting of the innovative path-break concept systems to see how much they differ from the current systems and in what ways. Back-casting in product development is normally focused on assessing existing capabilities, supply chains, and manufacturing plant for retooling to produce the new product. Back-casting can also identify barriers and strategies for overcoming them. For energy systems, the back-casting exercise looks at the adaptive capacity of the end-use sectors and at the infrastructure and behavioral aspects that are fundamentally different in the path-break concept and how they could be modified from the present to develop in the direction of the path-break.

6. *Triggers*: The sixth step is the designation of the trigger points. Existing systems have great inertia in both investment and behavior. But they can also change very quickly through an effective change project or when a crisis occurs. The trigger event is the engineered change project initiation, or the collapse of one or more subsystems due to a crisis. An engineered change project trigger event involves

communication and gaining cooperation from all the stakeholders. The trigger event also represents the beginning of changes in relationships, the built environment, and technology. In product development, the trigger events could include buying a subsidiary company, retooling and opening of a new production line, hiring of new staff with new capabilities, or launching the new product in the market. In energy systems, the trigger events could be caused by an energy crisis as was seen in the 1980s with the shift away from diesel electricity generation or the large scale uptake of Japanese-manufactured compact cars. It may be that the Fukushima nuclear power plant disaster will become a trigger point for a shift from nuclear power. Transition engineering for energy systems seeks to use trigger events other than an energy or environmental crisis. Leadership can provide trigger events, and you can find examples on company websites, where a CEO or founder provides a directive for a company to undertake energy management projects. However, the challenge is to generate trigger points by communicating the nature of the energy and environment risks as well as the benefits of adaptive changes to a wide range of stakeholders.

7. *Change projects*: The final step is to develop and carry out change projects. Carrying out change projects may require several weeks or years, depending on the scale of the changes. In energy systems, there are many examples of demand side management projects that have helped to keep electricity demand within generation and distribution limits [28]. In the context of curtailing the growth of finite resource consumption, construction of new "green" buildings or new renewable energy power generation plants are not really considered change projects to improve global sustainability unless the whole project includes retiring of below-standard buildings or coal-fired power plants. It should be pointed out that collapse is not a prosperous change project.

PATH-BREAK ENERGY SYSTEM CHALLENGE

- *The 7th generation*: The design and planning of all developments and products explicitly, quantitatively, and legally include the interests of people in the 7th generation in the future
- *Consumption*: The energy system does not *consume* fossil fuels; it only *invests* them
- *Constraints*: Resource use is allocated according to what is available from renewable sources and according to essentiality as established by local democratic processes
- *Environment*: Emissions, land uses, agricultural practices, and resource use are regulated by local strong democratic processes, permitted according to environmental science and subject to monitoring
- *Long term*: Design and construction standards require passive design for 600 year useful life
- *Participation*: Behavior ensures maintenance of local environment, ecosystems, resources
- Waste is not acceptable
- Active mode accessibility is the core requirement of all urban design

13.5 Strategic Analysis of Complex Systems

The purpose of this section is to provide a method for the steps in the transition engineering process of generating path-break concepts, back-casting, identifying trigger events, and proposing transitional change projects. These steps present challenges for traditional engineering. This is because the methods to carry out this work are not mature or widely used in typical engineering practice. Concept generation is a normal part of product innovation and problem solving, but here we are talking about complex systems, not toasters. Back-casting is also known in product development and planning. Trigger or initiation planning is done in change management, but in transition engineering, the trigger can be a crisis as well as a planned project initiation.

A complex system has multiple dimensions and scales. Dynamics of the different parts of the system are interrelated. Climate, weather, and ecological systems are examples of complex systems. In the anthropogenic systems, there are different human perspectives and jargon used by the different participants. Getting data and modeling complex systems are not a mature field. Communication of technical information across sectors and dealing with nontechnical expectations are huge challenges being explored by transition engineering researchers. As with any engineering practice, development emerges with experience as more projects are carried out and more examples published. The experience of practitioners will in the future be organized into standards for development planning in different sectors.

The Strategic Analysis of Complex Systems approach (SACS) is an innovation that has been developed as a tool to manage concept generation, modeling, and back-casting [29]. The SACS method has been used for several large projects such as peak oil planning for the city of Dunedin, New Zealand [30]. The SACS method helps the engineering team organize the complex development issues into quantifiable units. It also has been found to be a very effective communication tool with all of the various stakeholders, including council workers, political leaders, businesses, and members of the public.

Before starting the strategic analysis process, as with all engineering analysis projects, we need to define a particular system, designate the system boundaries, and specify the requirements and goals of the path-break development in the context of the risks to the "business as usual" future scenarios. Thus, we need to have information about history, the present state, and implications of future trends in order to prepare the SACS exercise. We also need to define the time frame for the analysis, which usually involves setting a target date in the future. The SACS method is shown in Figure 13.16 and set out in the following steps.

13.5.1 Strategic Analysis of Complex System Method (Figure 13.16)

1. Define the system and the critical resource to be studied:
   a. The geographic area for study: the city, region, or other social structure
   b. The energy end-use system: the activities, goods, or services
   c. The infrastructure involved: buildings, roads, vehicles, appliances
   d. The resources involved: energy, water, land, materials
2. Define the time frame for analysis and set a target future date.
3. *Set resource constraint*: Use resource data, depletion curves, and information from natural resource sciences to model the resource availability envelope for the critical resource in question at the target future date.

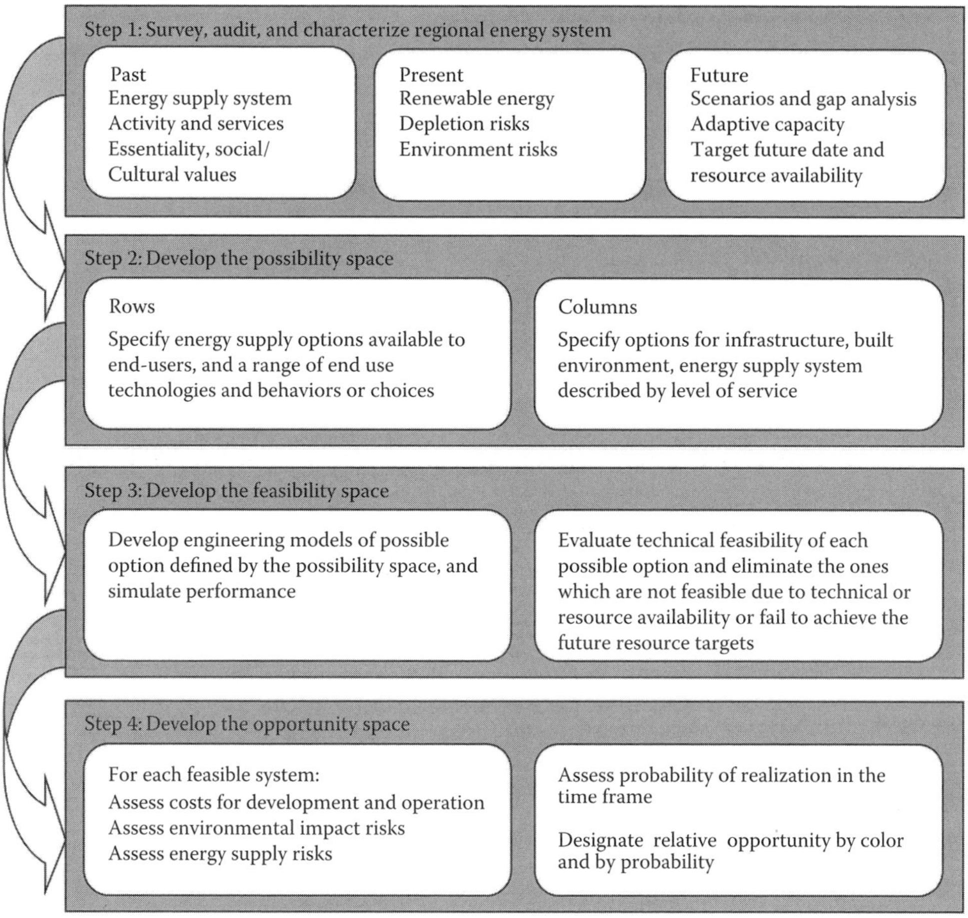

Step 1: Survey, audit, and characterize regional energy system

Past	Present	Future
Energy supply system	Renewable energy	Scenarios and gap analysis
Activity and services	Depletion risks	Adaptive capacity
Essentiality, social/	Environment risks	Target future date and
Cultural values		resource availability

Step 2: Develop the possibility space

Rows

Specify energy supply options available to end-users, and a range of end use technologies and behaviors or choices

Columns

Specify options for infrastructure, built environment, energy supply system described by level of service

Step 3: Develop the feasibility space

Develop engineering models of possible option defined by the possibility space, and simulate performance

Evaluate technical feasibility of each possible option and eliminate the ones which are not feasible due to technical or resource availability or fail to achieve the future resource targets

Step 4: Develop the opportunity space

For each feasible system:
Assess costs for development and operation
Assess environmental impact risks
Assess energy supply risks

Assess probability of realization in the time frame

Designate relative opportunity by color and by probability

FIGURE 13.16
SACS methodology for generating path-break concepts.

4. *Brainstorm development concepts*: Use participatory methods and surveys and include all ideas that are elucidated by stakeholders. Development concepts are in response to the issues of unsustainability.

5. *Generate the Possibility Space*: Arrange the concepts to form the *Possibility Space* by separating the concepts according to whether they involve the built environment or end-user choices. Make sure to characterize the options in terms of the system and the degree of the change over the time period of the analysis.

    a. Columns = infrastructure or built environment options

    b. Rows = end-use technologies, resources, or behavior changes

6. *Design and quantify option combinations*: Characterize the performance and energy use for each of the infrastructure development concepts with the particular technology or behavior adaptations. Use high-level modeling of steady-state behavior.

7. *Generate the feasibility space*: Assess the technical and resource feasibility of each opportunity combination to form the *feasibility space*. If the concept is not possible, then eliminate it from the possibility space by blacking out the space. Technical feasibility and

resource availability are the criteria for this step—at this point, do not consider cost or whether the necessary technology is available in the market at the current time.

8. *Assess development potential*: Assess the likelihood of each of the concepts being developed as modeled within the time frame of the analysis. This can include trigger analysis, supply chain, manufacturing, competition, government policy, replacement rate, etc. Rank each feasible possibility as

   a. Likely
   b. Possible
   c. Unlikely
   d. No possibility

9. *Cost estimates*: Assess the relative levelized cost of energy (LCOE) and EROI for each of the remaining feasible technologies in the context of the infrastructure concepts. Assess the relative costs of each of the infrastructure changes and the savings or costs from behavior changes.

10. *Assess the relative risks*: Environmental, social, operational, or other risks do not have to be quantitative. For example, the operational risk of a solar PV system is higher if the system has battery storage and an inverter than if the system can be used directly on site as DC.

11. *Concept evaluation*: Rank the feasible concepts and use the following colors to designate the overall risk level.

    Green: Low cost, high EROI (>15), low risk, renewable resources, recyclable, long life, enhances learning, improves community, improves health

    Yellow: Affordable, high EROI (>10), moderate risk, limited use of finite resources, reusable, no negative social impacts

    Orange: Expensive, modest EROI (<5), moderate risk, requires finite resources, negative social impacts, limited recyclability

    Red: High cost, low EROI (<2), high risk, requires finite resources, does not improve health, not fit with social values

12. *Generate opportunity space*: Use the resulting *opportunity space* to communicate which technologies, infrastructure investments, and behavior changes represent development opportunities to meet the transition development goals and how the infrastructure options interact or synergize with the technology and behavior options.

13.5.2 Possibility Space

The SACS method uses a matrix graphical format to organize the concepts and to define the systems for quantitative feasibility analysis. The matrix is also used to communicate the results to all stakeholders in a way that includes all of the ideas and concepts brought forward for consideration but makes it clear which options are possible, and which are actually good opportunities. Figure 13.17 gives the general layout for the *possibility space* generation. It is usually most helpful to reserve the first column for the current infrastructure and the first row for the current energy technologies and behaviors if there were no changes over the analysis period.

Different types of systems lend themselves to organization of the options in different ways. In general, the columns of the matrix are options that determine the demand,

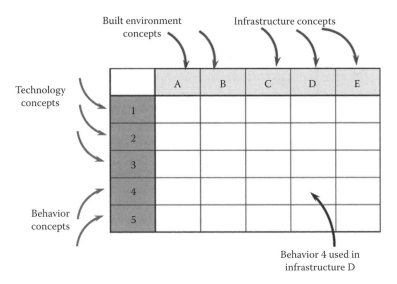

FIGURE 13.17
Matrix form of the possibility space generated by the SACS method.

and the rows are options that determine the supply or end use. We want to arrange the different options in ways that the options in columns represent the context that affects how well the options arranged in rows work. For example, in transportation systems, options in columns have been associated with the urban form, like electric tram system, cycle and walking infrastructure, and new walkable urban village developments, while options in rows have been associated with travel mode and travel choices, electric vehicles, and choices to live near destinations [31]. In the transport system, the combinations of options in each matrix cell were used to calculate fuel use. In a study of the food-carrying capacity on an island, the columns were options for energy intensity of the food supply and market system associated with the level of processing. For example, fresh produce from market gardens would have the lowest processing energy, and processed meals packaged in plastic trays and boxes and frozen for later reheating represent the highest level of processing energy. The rows were dietary options such as vegetarian, pescatarian, and the current beef and pork diet. The number of people who could be supported by the existing land and sea resources for each combination of options was calculated. In another example of electricity supply for residential consumers, we arranged different power supply system generation capacities in columns and different combinations of technologies for remote hybrid power systems in the rows [29].

13.5.3 Feasibility Space

The next step in the method is to evaluate the technical and resource feasibility of each option combination. The feasibility depends on how the calculated result fits with the constraints set for the analysis. At this point, cost is not considered. If the option combination is not feasible, then the cell is blacked out as shown in Figure 13.18. In the transportation project, the future target was set as a 50% petroleum demand reduction for residential personal transport for a 2050 planning exercise. The amount of gasoline savings was thus set, and the vehicle travel demand for each option combination was calculated to see if the target could be achieved. The option of having the 50% of transport

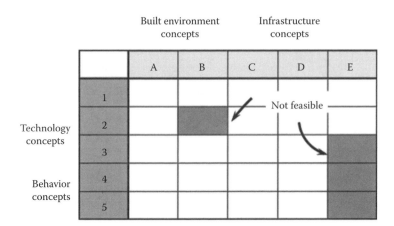

FIGURE 13.18
Feasibility space showing how options that do not meet the criteria set for the strategic analysis are deemed unfeasible and are not studied further.

fuel being supplied by ethanol was found to be unfeasible in the "business as usual" case where no other changes are made to the transport infrastructure because there was not enough land within several hundred miles to grow the crops for biofuel that would be needed. In the island food study, the current population of the island is one million. Thus, any diet and processing energy combination that could not support the population was marked as unfeasible. In the residential electric power analysis, unfeasible systems were those that could not provide the level of generation capacity using the resources and hybrid system. For example, the high generation capacity level could not be supplied by the all-solar PV system or by the all-wind system because of lack of suitable sites and resources.

13.5.4 Probability Space

The time frame set for the SACS project may present some challenges for achieving the different options. The next step in the process is to perform a qualitative assessment of the possibility that the system options could be developed and delivered into service within the designated time frame. Figure 13.19 shows the results of the time frame development analysis displayed for each of the technically feasible options. For example, apartment buildings can be planned, permitted, and built within 5 years. Thus, for the urban residential transportation project, the option to develop higher-density housing in the central business district over a 40 year time frame was deemed to be likely. On the other hand, an option for 50% of the car fleet to be BEVs was assessed as unlikely given that the average age of vehicles in the city was 15 years, the replacement rate for vehicles according to statistics was 2%, and the market penetration rate for BEVs was predicted by the AA to be less than 0.5% for the next decade. In the food-carrying capacity study, it was deemed unlikely that all of the land that grows crops for export would be changed to pasture for local beef consumption. The time frame for the residential electricity project was 60 years, so it was deemed possible that 30% of the homes older than 40 years could either have a major remodel or be replaced with much higher standard, passive designed residences.

		A	B	C	D	E
Technology concepts	1	Likely	No	Possible	No	No
	2	Possible		Possible	No	
	3	Possible	No	Possible	Unlikely	
Behavior concepts	4	Unlikely	No	Likely	Likely	
	5	Unlikely	No	Likely	No	

Built environment concepts: A, B, C; Infrastructure concepts: D, E

FIGURE 13.19
Time frame development probability evaluation of technically feasible options.

13.5.5 Opportunity Space

Costs and availability of each option are evaluated as well as any environmental or social risks. This next step in the analysis can be structured like any other concept evaluation. Different criteria and measures of merit can be established, and the relative score of each remaining option determined from the composite score. Often at the concept phase, costs are difficult to quantify, and EROI can vary greatly with small changes in assumptions about lifetime and utilization factor. We use relative EROI for current technologies if they are known. Figure 13.20 shows an example with the composite ranking for cost, EROI, environmental risk, social impacts, and further reliance on finite resources indicated by color. Red means high risk, and green means low risk. This color assignment is common in the risk analysis field.

In Figure 13.20, the change in built environment option C represents a good investment opportunity. All of the possible or likely technologies or behavior changes could achieve the

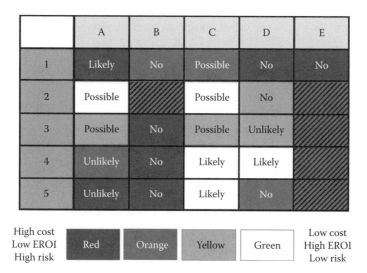

| High cost Low EROI High risk | Red | Orange | Yellow | Green | Low cost High EROI Low risk |

FIGURE 13.20
The opportunity space indicating that four of the options A-2, C-4, C-5, and D-4 are good opportunities.

target if development option C were pursued. Option D is only an opportunity if behavior change option 4 is achieved. Technology development option 1 is not an opportunity in nearly all infrastructure development scenarios. Infrastructure option B and built environment option E are not feasible and not viable options regardless of technology developments or behavior changes. An example of such a development would be exurban sprawl with low-density residential housing, car dependence, and long distances. On the other hand, option C could be redevelopment of gray-field areas in an urban area into walkable, high intensity urban activity hubs, and technology option 2 could be electric light rail or tram systems. It is well known that low-density sprawl cannot be served effectively by public transport.

In the transportation example, the costs of some of the options like electric bus and cycle paths were actually cost savings as the options reduced fuel use over 40 years and paid back the investment cost several times over. Other options, like developing urban village centers that concentrate destination activities in certain areas around transit hubs, had positive economic development and social benefits. Increased walking and cycling were one of the behavior options modeled, and this option provided reduced fuel and vehicle costs as well as health benefits. The food-carrying capacity project also had some options that had added health benefits.

13.6 Concluding Remarks

Transition engineering is focused on understanding issues for end-use systems and dealing with these issues through upstream redevelopment of the consumption systems to reduce demand to match the supply over a given future time frame. The past century has indeed demonstrated impressive technological advances and ever expanding ability to extract resources. However, the 10,000 years prior to the industrial revolution provide ample testimony to human ingenuity in being able to adapt to nearly every environment on the planet and to make use of local resources in ways that resulted in long-term security. In the distant past, adaptation was a matter of trial and error. Transition engineering takes on the processes of adaptation for our industrial society as engineering projects, bringing to bear all of the modern science, analytical and modeling tools, plus a revival of imagination, creativity, and innovation.

Problems

13.1 The economically destructive price spike of 2007–2008 occurred when spare production capacity fell below 1 mbpd, causing a run-up in oil price from $50 to $145 per barrel. An issue for world oil spare capacity is the internal consumption of OPEC producers. OPEC production has ranged between 30 and 33 mbpd since 2004. Internal use is currently about 25% of total OPEC production, but it is growing by 2% per year. The internal consumption growth is largely for nonelastic end uses such as electricity generation, water treatment, and new buildings. If the current world spare capacity is 2 mbpd and the OPEC total production does not increase beyond 33 mbpd, how many years will it take for the spare capacity to fall below 1 mbpd again?

13.2 We have a historical example of the kind of problem that the supply–demand gap can cause from the 1970s' oil shocks and the dynamics of the world oil supply over the past four decades. On October 5, 1973, the Yom Kippur War started when Syria and Egypt attacked Israel. The United States and most Western countries showed support for Israel. Several Arab nations and Iran imposed an embargo and curtailed their production by 5 mbpd. There had been no spare production in the United States, the largest oil-producing nation, since 1971, but other non-OPEC producers were able to increase production by 1 mbpd, resulting in a 7% decline in supply. By the end of 1974, the price of oil had risen from around $3.00 per barrel, where it had been since regulation began in 1910, to over $12 per barrel. Go to the website provided and read through the presentation by Dr. Robert Hirsch giving a review of the energy crisis (http://www.aspo2012.at/speakers/#hirsch).

(a) Ask your parents or anyone you know over 50 years of age what they remember about the energy crisis of the 1970s.

(b) Write a hypothetical proposal to your hometown city council for a transition engineering project that would assess the adaptive capacity for gas and diesel fuel reduction in the event of another energy crisis (limit 1 page).

(c) Give two things you would do to prepare for a fuel shortage and calculate how each would improve your adaptive capacity from today.

13.3 The International Energy Agency (IEA) was founded during the oil crisis in 1974 as an autonomous organization of 28 member countries. The stated aim of the IEA is to engender cooperation among the members, to provide data and statistics on oil supply, and to provide policy advice. The IEA staff analysts produce monthly reports and an annual review with forward projections using scenarios. The business-as-usual scenario has oil demand of 96 mbpd by 2035 including natural gas liquids, unconventional oil, and as yet undiscovered petroleum resources. The 450 Scenario, which has carbon reductions to limit climate change to 2°C warming, has demand declining to 78 mbpd by 2035. (IEA, 2011 *World Energy Outlook*, available at: www.worldenergyoutlook.org).

Consider the 55 mbpd or so of new crude oil fields yet to be found and developed that are needed to replace decline of known fields to 2035. Also consider that oil production has not increased since 2004 even though the price has gone from the $30 per barrel range to a new $100 per barrel range.

(a) Make a table with the first column being a set of end uses for petroleum organized by sector. In the next column, rank the different uses according to essentiality (Essential for well-being, Necessary for quality of life, Optional for lifestyle).

(b) In the next column, assign multipliers for how many times more expensive this petroleum end use could be and still be viable. For example, kidney dialysis ×100 and disposable cutlery ×2.

(c) In the next column, assign a percentage of the reduction in this end use that could be sustained and still maintain a high standard of living. For example, kidney dialysis −0% and disposable cutlery −100%.

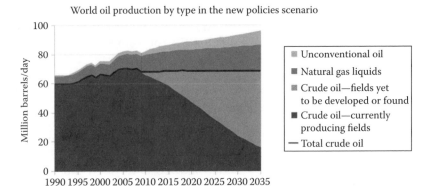

World oil production by type in the new policies scenario

Global oil production reaches 96 million barrels/day in 2035 on the back of rising output of
natural gas liquids and unconventional oil, as crude oil production plateaus

13.4 The Association for the Study of Peak Oil (ASPO) was founded by Professor Colin
Campbell and Jean Laherrère in 2001. Campbell and Laherrère had published a
ground-breaking article in 1998 in *Scientific American* called "The End of Cheap Oil"
based on sound data of petroleum geology and mathematical analysis. For nearly a
decade, the most important points of discussion among the relatively small group
of petroleum geologists and others seemed to be the year that the supply would
peak. There has been little or no political discussion, academic research, or industry
response to the issue. Visit the website for ASPO (http://www.peakoil.net). Read the
current front-page article and imagine you work for a major airline as a strategic
planner. Write a review in 500 words or less for an executive board at your airline
that summarizes the ASPO information, whether there is an issue that could affect
the airline, and what you would recommend as a next course of action if any.

13.5 Using the following energy flow diagram, compare the impact of different transition
engineering projects. Calculate the upstream effect on reduced demand for extraction of
petroleum, the capital cost, payback period, and the EROI for each option. Option 1 is a
substitution of end-use technology and transition to an alternative fuel stream. Option 2
is a demand management approach where the need to drive is reduced or the choice of
mode is changed to cycle or walk. Option 3 is an energy efficiency improvement approach.

U.S. statistics and data: 15,000 mi average travel per vehicle, average mileage
21 mpg, average cost of driving $0.60/mi for a petrol sedan.

Nissan Leaf MSRP $36,000 + home charging station $2,000, 80 kW motor.

Anderson Oil Boiler MSRP $2255, average household oil use 730 gal/year.

Embedded energy: electric car = 161 MBtu, bicycle = 3.5 MBtu, oil boiler = 6.1 MBtu.

Project (1): Invest in new electric vehicles and electric vehicle charging stations. The
development goal is to replace 1 Quad of transport fuel demand with electricity.

Project (2): Invest in travel demand reduction. This will require new innovations,
systems, urban developments ... that reduce miles traveled by 2% overall. You
can research and brainstorm possible measures and costs.

Project (3): Identify and replace 25% of the 8.1 million domestic oil boilers with the
lowest energy efficiency (pre-1970 AFUE = 60%) with new boiler with AFUE = 86%.

The following figure shows the petroleum energy flow diagram for the United States
in 2008 [quadrillion Btu].

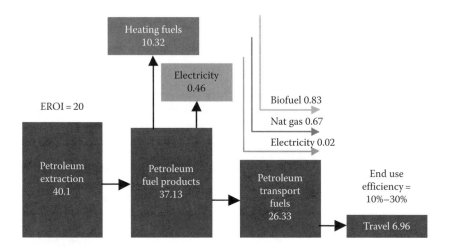

13.6 *Carry out a strategic analysis*: Consider the example of a city like Boulder, CO, which has a citizen initiative to transition to coal-free electricity. The citizens set the target date as 2020, which is a date set by NASA climate scientist, Dr. James Hansen for phase out of coal in order to reduce risks of climate disaster. The process of getting to the carbon-free electricity is not the question for the path-break strategic analysis. We want to know what would actually work to meet the target, then we will back-cast to look at how we might achieve it. The electricity supply is generated 66% from coal, 20% from natural gas, and 14% from renewable resources. The peak loads are on summer afternoons as temperatures can top 100°F. Assume that the city has 97,000 people (30,000 homes, 12,000 apartments). The housing stock has 30% electric heating, 75% electric cooking, 30% electric water heating, and 30% air conditioners, and 45% of electricity demand is residential. Residential average annual household demand is 9000 kWh, and average annual building electricity use for commercial buildings is 26 kWh/ft^2. There is an estimated 24 million square feet of nonresidential buildings in Boulder. Forty-five percent of homes were built before 1975, as are 40% of commercial and school buildings. Passive solar design is not evident in the architecture of the city either for winter heat or summer cooling. A meeting of concerned citizens was held, and the majority interest was for wind power with solar PV also in high favor. The local utility wants to repower a coal-fired power plant with natural gas (estimate that as 30% of the local coal-fired electricity). NREL suggests that commercial buildings with good energy management can achieve 17 kWh/ft^2 electricity use.

References

1. Krumdieck, S. (2011) The survival spectrum: The key to transition engineering of complex systems, *Proceedings of the ASME 2011*, November 11–17, 2011, Denver, CO, ICEME2011-65891.
2. ASSE, About the safety profession, http://www.asse.org/about/history.php (accessed April 2010).
3. Campbell, C. and J.H. Laherrare (1998) The end of cheap oil, *Scientific American*, March 1998.
4. Deffeyes, K. (2001) *Hubbert's Peak: The Impending World Oil Shortage*, Princeton University Press, Princeton, NJ.
5. Aleklett, K. and C.J. Colin (2003) The peak and decline of world oil and gas production, *Minerals & Energy*, 18, 5–20.

6. Krumdieck, S., S. Page, and A. Dantas (2010) Urban form and long term fuel supply decline: A method to investigate the peak oil risks to essential activities, *Transportation Research Part A* 44, 306–322.

7. Harris, M. (1999) *Lament for an Ocean—The Collapse of the Atlantic Cod Fishery—A True Crime Story*, M&S, Toronto, Ontario, Canada, 432pp.

8. Massari, S. and M. Ruberti (2012) Rare earth elements as critical raw materials: Focus on international markets and future strategies, *Resources Policy* 38(1), 36–43, http://dx.doi.org/10.1016/j.resourpol.2012.07.001

9. Dale, M., S. Krumdieck, and P. Bodger (2012) Global energy modeling—A biophysical approach (GEMBA) part 1: An overview of biophysical economics, *Ecological Economics* 73, 152–157.

10. Dale, M., S. Krumdieck, and P. Bodger (2012) Global energy modeling—A biophysical approach (GEMBA) part 2: Methodology and results. *Ecological Economics* 73, 158–167.

11. *Encarta Dictionary* online, www.dictionary.msn.com (accessed on October 2010).

12. Hughes, J., C. Knittel, and D. Sperling (2006) Evidence of a shift in the short-run price elasticity of gasoline demand, Working Paper No. 12530, National Bureau of Economic Research, Cambridge, MA, http://papers.nber.org/papers/W12530 (last accessed on October 30, 2012).

13. Li, S., J. Linn, and E. Muehlegger (2011) Gasoline taxes and consumer behavior, http://economics.stanford.edu/files/muehlegger3_15.pdf (last accessed on October 30, 2012).

14. Komanoff, C. (2008–2011) Gasoline price-elasticity spreadsheet, www.komanoff.net/oil_9_11/Gasoline_Price_Elasticity.xls (last accessed on October 30, 2012).

15. EIA, Gasoline and diesel report (1990–2012), http://www.eia.gov/petroleum/gasdiesel (last accessed on October 30, 2012).

16. Watcharasukarn, M., S. Krumdieck, and S. Page (2012) Virtual reality simulation game approach to investigate transport adaptive capacity for peak oil planning, *Transportation Research Part A*, 46, 348–367.

17. Palm, W.J. (2000) *Modeling, Analysis, and Control of Dynamic Systems*, John Wiley & Sons, Inc., New York.

18. D'Azzo, J.J. and C.H. Houpis (1981) *Linear Control System Analysis and Design, Conventional and Modern*, McGraw-Hill, Inc., New York.

19. Rohn, A.H. (1971) *Mug House Mesa Verde National Park*, Archaeological Research Series No 7-D, National Park Service, U.S. Department of the Interior, Washington, DC, pp. 255–265.

20. Diamond, J. (2005) *Collapse, How Societies Choose to Fail or Survive*, Penguin Group, Australia, Chapter 14.

21. Hopkins, R. (2008) *The Transition Handbook: From Oil Dependency to Local Resilience*, Chelsea Green Publishing, White River, VT.

22. Binswanger, H.C. (1998) Making sustainability work, *Ecological Economics* 27(1), 3–11.

23. Neumayer, E. (2000) Scarce or abundant? The economics of natural resource availability, *Journal of Economic Surveys* 14(3), 307–335.

24. The Hirsch Report (2005) *Peaking of World Oil Production: Impacts, Mitigation, & Risk Management*, NETL (http://www.netl.doe.gov/publications/others/pdf/oil_peaking_netl.pdf).

25. Rifkin, J. (2002) *The Hydrogen Economy*, Penguin Group, New York.

26. Rackley, S.A. (2010) *Carbon Capture and Storage*, Elsevier, Burlington, MA.

27. Billmaier, J. (2010) *JOLT! The Impending Dominance of the Electric Car and Why America Must Take Charge*, Advantage, Charleston, SC.

28. Kreith, F. and D.Y. Goswami (2007) *Energy Management and Conservation Handbook*, CRC Press, Taylor & Francis Group, Boca Raton, FL.

29. Krumdieck, S. and A. Hamm (2009) Strategic analysis methodology for energy systems with remote island case study, *Energy Policy*, 37(9), 3301–3313.

30. Krumdieck, S. (2010) Peak oil vulnerability assessment for Dunedin, http://www.dunedin.govt.nz/your-council/policies-plans-and-strategies/peak-oil-vulnerability-analysis-report (last accessed on October 30, 2012).

31. Krumdieck, S. (2011) Transition engineering of urban transportation for resilience to peak oil risks, *Proceedings of the ASME 2011*, ICEME2011-65836, November 11–17, 2011, Denver, CO.

Index